High Pressure in Semiconductor Physics I

SEMICONDUCTORS
AND SEMIMETALS
Volume 54

Semiconductors and Semimetals

A Treatise

Edited by R. K. Willardson
CONSULTING PHYSICIST
SPOKANE, WASHINGTON

Eicke R. Weber
DEPARTMENT OF MATERIALS
SCIENCE AND MINERAL
ENGINEERING
UNIVERSITY OF CALIFORNIA
AT BERKELEY

High Pressure in Semiconductor Physics I

SEMICONDUCTORS
AND SEMIMETALS

Volume 54

Volume Editors

TADEUSZ SUSKI

UNIPRESS
HIGH PRESSURE RESEARCH CENTER
POLISH ACADEMY OF SCIENCES
WARSAW, POLAND

WILLIAM PAUL

PHYSICS DEPARTMENT AND
DIVISION OF APPLIED SCIENCES
HARVARD UNIVERSITY
CAMBRIDGE, MASSACHUSETTS

ACADEMIC PRESS
San Diego   London   Boston   New York
Sydney   Tokyo   Toronto

This book is printed on acid-free paper

COPYRIGHT © 1998 BY ACADEMIC PRESS
ALL RIGHTS RESERVED.
NO PART OF THIS PUBLICATION MAY BE REPRODUCED OR TRANSMITTED IN ANY FORM OR BY ANY MEANS, ELECTRONIC OR MECHANICAL, INCLUDING PHOTOCOPY, RECORDING, OR ANY INFORMATION STORAGE AND RETRIEVAL SYSTEM, WITHOUT PERMISSION IN WRITING FROM THE PUBLISHER.

The appearance of the code at the bottom of the first page of a chapter in this book indicates the Publisher's consent that copies of the chapter may be made for personal or internal use of specific clients. This consent is given on the condition, however, that the copier pay the stated per-copy fee through the Copyright Clearance Center, Inc. (222 Rosewood Drive, Danvers, Massachusetts 01923), for copying beyond that permitted by Sections 107 or 108 of the U.S. Copyright Law. This consent does not extend to other kinds of copying, such as copying for general distribution, for advertising or promotional purposes, for creating new collective works, or for resale. Copy fees for pre-1998 chapters are as shown on the title pages; if no fee code appears on the title page, the copy fee is the same as for current chapters. 0080-8784/98 $25.00

ACADEMIC PRESS
525 B Street, Suite 1900, San Diego, CA 92101-4495, USA

http://www.apnet.com

ACADEMIC PRESS
24–28 Oval Road, London NW1 7DX, UK
http://www.hbuk.co.uk/ap/

International Standard Serial Number: 0080-8784
International Standard Book Number: 0-12-752162-3

Printed in the United States of America
98 99 00 01 02 BB 9 8 7 6 5 4 3 2 1

Contents

PREFACE xi
LIST OF CONTRIBUTORS xiii

Chapter 1 High Pressure in Semiconductor Physics: A Historical Overview
William Paul

- I. Introduction 2
- II. Experiments on Bulk Samples of the Zincblende Family to 1970 . . . 4
  - 1. Changes of Band Structure under Pressure: The Empirical Rule . . . 5
  - 2. Changes of Parameters Determining Single-band Transport under Pressure 8
  - 3. Applications of the Empirical Rule 11
- III. Theoretical Calculations of the Effect of Pressure on Bulk Band Structures 15
  - 1. Introduction 15
  - 2. Pseudopotential Calculations 15
  - 3. Calculations Using the Dielectric Theory of the Chemical Bond . . . 17
- IV. Current Studies of Bulk Semiconductors of the Zincblende Family . . . 18
  - 1. Introduction 18
  - 2. Deep Impurity Levels and Several Conduction Band Extrema . . . 19
  - 3. DX and EL2 20
  - 4. Other Deep Impurity Levels 24
- V. Quantum Wells, Heterostructures, and Superlattices 25
  - 1. Introduction 25
  - 2. Band Offsets 25
  - 3. Influence of the X-States on Phenomena in Heterostructures . . . 26
  - 4. Resonant Tunneling 28
  - 5. Heterostructures and DX Centers 28
- VI. Structural Phase Transitions 29
  - 1. Introduction 29
  - 2. The Family of Zincblende Structures at Atmospheric Pressure . . . 29
  - 3. S, Se, and Te 31
  - 4. Other Systems 31
  - 5. Lattice Dynamics: Inelastic Neutron Scattering 32
- VII. Other Materials 33
  - 1. Lead Chalcogenides 33
  - 2. Group 6 Elements 34

3. Amorphous Semiconductors	35
4. Nanocrystals and Porous Silicon	36
5. Semiconducting Nitrides	37
VIII. Other Matters	38
References	38

Chapter 2 Electronic Structure Calculations for Semiconductors under Pressure
N. E. Christensen

I. Introduction	49
II. Basic Theory	50
1. Zeroth Born–Oppenheimer Approximation	51
2. Density-Functional Theory	53
3. Self-Interaction Correction and Optimized Effective Potentials	57
4. Gradient Corrections	58
5. Excitation Energies: "GW"	60
6. "LDA Gap Error" and Gap Deformation Potentials	64
7. Ad Hoc LDA Gap Corrections	66
III. "ASA-LDA," "Frozen-Potential Method"	69
1. Atomic-Spheres Approximation	71
IV. Deformation Potentials	79
1. Shear Deformation Potentials in Cubic Compound Semiconductors	80
2. Hydrostatic Deformation Potentials	86
V. Pressure-Induced Structural Changes	98
1. B3 → B1 Pressure-Induced Transitions	99
2. B3 → Imma	105
3. Transitions in and from Wurtzite Structures	106
4. Pressure-Induced Transformations of Group IV Semiconductors	113
5. d-States	122
VI. Pressure Dependence of Phonon Frequencies	130
VIII. Concluding Remarks	134
References	136

Chapter 3 Structural Transitions in the Group IV, III-V and II-VI Semiconductors Under Pressure
R. J. Nelmes and M. I. McMahon

I. Introduction	146
II. Structures	150
1. Introduction	150
2. Diamond and Zincblende	152
3. Wurtzite and Lonsdaleite	154
4. β-tin and Diatomic β-tin	155
5. NaCl	157
6. NiAs	157
7. Imma, Imm2 and Immm	159

8. Simple Hexagonal 161
　　　9. Cinnabar. 163
　　10. Cmcm 164
　　11. C222₁ 168
　　12. CsCl and Body-Centered Cubic 169
　　13. ST12 169
　　14. BC8, SC16, and R8. 171
III. Individual Systems 175
　　　1. Introduction 175
　　　2. Silicon 177
　　　3. Germanium 181
　　　4. Boron-V Compounds 183
　　　5. Aluminum Nitride 183
　　　6. Aluminum Phosphide 184
　　　7. Aluminum Arsenide 186
　　　8. Aluminum Antimonide 187
　　　9. Gallium Nitride 189
　　10. Gallium Phosphide 190
　　11. Gallium Arsenide 191
　　12. Gallium Antimonide 198
　　13. Indium Nitride 200
　　14. Indium Phosphide 202
　　15. Indium Arsenide 203
　　16. Indium Antimonide 205
　　17. Zinc Oxide 211
　　18. Zinc Sulfide 212
　　19. Zinc Selenide 213
　　20. Zinc Telluride 215
　　21. Cadmium Oxide 218
　　22. Cadmium Sulphide 218
　　23. Cadmium Selenide 220
　　24. Cadmium Telluride. 222
　　25. Mercury Oxide 225
　　26. Mercury Sulphide 226
　　27. Mercury Selenide 228
　　28. Mercury Telluride 230
IV. Discussion 233
V. Concluding Remarks 244

Chapter 4 Optical Properties of Semiconductors Under Pressure

A. R. Goñi and K. Syassen

I. Introduction 248
II. Experimental Aspects 251
　　1. Diamond Anvil Cell 251
　　2. Pressure Medium 253
　　3. Pressure Measurement 254
　　4. Optical Spectroscopy at High Pressure 256

- III. Electronic Band Structure Under Pressure 259
  - 1. Linear Combinations of Atomic Orbitals and Tight-Binding Method 260
  - 2. Penn Model and the Dielectric Theory of the Covalent Bond. . . . 264
  - 3. the $\vec{k} \cdot \vec{p}$ Method 267
  - 4. Uniaxial Stress Effects 269
  - 5. Quantum Well Structures. 271
- IV. Refractive Index Dispersion 273
  - 1. Refractive Index and Optical Dispersion 274
  - 2. Volume Dependence of the Low-Frequency Dielectric Function . . . 277
  - 3. Method of Optical Interferences 279
  - 4. Experimental Results 282
- V. Optical Absorption Near the Fundamental Gap 285
  - 1. Absorption Spectral Functions 285
  - 2. Results and Discussion 289
- VI. Effect of Pressure on Exciton Absorption 297
  - 1. Energy Spectrum of Excitons 298
  - 2. Exciton Absorption 303
  - 3. Pressure Effects on Excitons 307
- VII. Photoluminescence Studies Under Pressure 321
  - 1. Optical Emission in Semiconductors 321
  - 2. Pressure Effects on PL Emission in Bulk Materials. 327
  - 3. Low-Dimensional Structures Under Pressure: PL Studies . . . 333
- VIII. Photomodulated Reflectance. 342
  - 1. Photomodulated Optical Response 342
  - 2. Results Bulk Materials 345
  - 3. Quantum Wells and Epilayers 346
- IX. Optical Properties of Electron Gases Under Pressure 351
  - 1. Elementary Excitations of the Electron Gas 352
  - 2. Inelastic Light Scattering by Elementary Excitations of the Electron Gas 358
  - 3. LO-phonon-plasmon Coupled Modes in Bulk Semiconductors . . . 363
  - 4. Elementary Excitations of the 2D Electron Gas Under Pressure . . . 370
  - 5. Electron-electron Interactions in Double-Layer 2D Electron Gases . . 377
- X. High Pressure Phases 385
  - 1. General Remarks 385
  - 2. Results for Polar Compounds 386
  - 3. Results for the Elemental Semiconductors 391
- A. Appendix: Linear Optical Response of Solids 392
  - A1. Electromagnetic Waves in a Medium 394
  - A2. Interband Absorption 397
  - A3. Reflectance and Transmittance 400
  - A4. Lorentz-Drude Oscillators 402
  - A5. Dispersion Relation and Sum Rules 405
  - A6. Analysis of High Pressure Reflectance Spectra 407
  - Acknowledgments. 409
  - References 410

Chapter 5   Defects

Chapter 5.1   Hydrostatic Pressure and Uniaxial Stress in Investigations of the EL2 Defect in GaAs

Pawel Trautman, Michal Baj, and Jacek M. Baranowski

I. Introduction	427
II. Basic Properties of the EL2 Defect	428
1. Occurrence of the EL2 Defect in GaAs	428
2. Optical Absorption Due to EL2	430
3. Metastable Properties of the EL2 Defect in GaAs	433
4. Discovery of the As_{Ga} Antisite by EPR Spectroscopy	435
III. Piezospectroscopic Investigations of No-Phonon Lines of EL2	436
IV. Studies of the EL2 Defect Under Hydrostatic Pressure	440
V. Microscopic Nature of the Metastable State of EL2	445
1. Theoretical Model of the Metastability of EL2	445
2. Determination of Symmetry of EL2 in the Metastable State	447
VI. Conclusions	453
References	453

Chapter 5.2   High-Pressure Study of DX Centers Using Capacitance Techniques

Ming-fu Li and Peter Y. Yu

I. Introduction	457
II. Techniques for Electrical Measurements on Samples Inside the DAC	459
1. Introducing Wires into the DAC	459
2. Performing Capacitance Measurements Inside the DAC	461
III. Introduction to Capacitance Transient Techniques	462
1. Capacitance Transients at Constant Temperature	463
2. Capacitance Transient when Scanning Temperature — Deep Level Transient Spectroscopy (DLTS)	465
3. Photo Capacitance Transient Measurements	466
IV. Experimental Studies of DX Centers	467
1. Introduction	467
2. Establishment of the DX Center as Due to Substitutional Donors	469
3. Models of the DX Center	474
V. Concluding Remarks	481
Acknowledgments	482
References	482

Chapter 5.3   Spatial Correlations of Impurity Charges in Doped Semiconductors

Tadeusz Suski

I. Introduction	485
II. Experimental Techniques	487

III. Donor-charges in HgSe:Fe 487
   IV. Donor States in GaAs and AlGaAs 490
   V. Experimental Studies of Spatial Correlations for DX-center Charges
       in Bulk Semiconductors 495
   VI. Spatial Correlations of Remote Impurity Charges 502
   VII. Remote-Charge Correlations and Quantum Transport of 2 DEG . . 505
   VIII. Concluding Remarks 508
        References 510

Chapter 6 Pressure Effects on the Electronic Properties of Diluted Magnetic Semiconductors

Noritaka Kuroda

   I. Introduction 513
   II. Structural Properties 515
       1. Under Ambient Conditions 515
       2. Elastic Properties 517
       3. Phase Transition 518
   III. Electronic Properties 523
       1. Energy Gap 523
       2. Exchange Interactions 524
       3. Intraion d-d Optical Transitions 540
   IV. Superstructures 543
       1. $ZnSe/Zn_{1-x}Mn_xSe$ 543
       2. $CdTeCd_{1-x}Mn_xTe$ 544
   V. Miscellaneous 545
       1. Raman Scattering 545
       2. Galvanomagnetic Effects in $Hg_{1-x}Fe_xSe$ 546
       Acknowledgments 546
       References 547

INDEX 551
CONTENTS OF VOLUMES IN THIS SERIES 561

Preface

In 1963 two review books were published which reported extensions to the monumental work contained in Bridgman's treatise, *The Physics of High Pressure* (1931, 1949). The first of these, *Solids under Pressure,* edited by Paul and Warschauer, included two chapters on semiconductors and the electronic properties of solids in addition to eleven chapters on atomic diffusion, equations of state, phase equilibria and phase transitions, magnetic properties, shock-wave techniques and geophysics. The second, *High-Pressure Physics and Chemistry,* in two volumes edited by Bradley, ranged more widely over topics in gases, liquids and solids, but also included seven chapters on solids, one of them on semiconductors. In the ensuing period the number of laboratories pursuing research in high pressures has much increased, in some part because of the easier availability of commercial piston-cylinder apparatus, such as that manufactured by the Unipress, in Warsaw, but most particularly because of the advent of the diamond anvil cell technique. This latter has permitted much higher hydrostatic pressures to be achieved in risk-free environments, and spectacular progress has been made in many areas of research. The work on solids has been reported at numerous conferences, sponsored, for example, by the AIRAPT (International Association for the Advancement of High Pressure Science and Technology), the EHPRG (European High Pressure Research Group), and by *ad hoc* groupings of research workers. There have been reviews and articles in journals such as *Review of Modern Physics, High-Pressure Research* and *High Pressures — High Temperatures,* and in serial books such as *Solid State Physics.* From the reports it is evident that the research on semiconductors has been one of the most enduring endeavors. Thus it is that in this volume we have gathered together a number of review articles on subjects of current interest. In doing so we have drawn heavily on the work reviewed in the biennial conferences titled *High Pressure Semiconductor Physics,* held since 1984, which in 1988 became satellite conferences to the International Conferences on the Physics of Semiconductors (ICPS). We are grateful to the organizers of these conferences for their diligence in establishing a historical record of this subject, which continues to contribute significantly to the understanding of semiconductors under ambient conditions. The book chapters indicate unambiguously the vitality of an ongoing investigation, and no doubt the details of the results reported in these chapters will need to be updated in a relatively short time. Nevertheless, we expect that these chapters of critical review, with their extensive bibliographies, will provide a source book for continued investigation for some time to come.

With minor exceptions, this book follows that of Paul and Warschauer in presenting only enough of the details of technique to make the subject matter understandable. The exceptions obviously lie in those areas where new strides have been made possible by discontinuous, nonincremental, advances in available techniques.

The editors wish to thank the contributors for their friendly collaboration: this is their book, and we have merely helped in assembling their work. The thanks of all of us are due to the staff of Academic Press for their patience and accommodation, and especially to Dr. Zvi Ruder.

This volume is the first of two volumes on the subject.

List of Contributors

Numbers in parenthesis indicate the pages on which the authors' contribution begins.

MICHAL BAJ (427), *Institute of Experimental Physics, Warsaw University, Warsaw, Poland*

JACEK M. BARANOWSKI (427), *Institute of Experimental Physics, Warsaw University, Warsaw, Poland*

N. E. CHRISTENSEN (49), *Institute of Physics and Astronomy, University of Aarhus, DK-8000 Aarhus, Denmark*

A. R. GOÑI (247), *Max-Planck-Institut für Festkörperforschung, Heisenbergstr.1, 70569 Stuttgart, Germany*

NORITAKA KURODA (513), *Institute for Materials Research, Tohoku University, Sendai, Japan*

MING-FU LI (457), *Department of Electrical Engineering, National University of Singapore, Singapore*

M. I. MCMAHON (145), *Department of Physics, The University of Liverpool, Liverpool, United Kingdom*

R. J. NELMES (145), *Department of Physics and Astronomy, The University of Edinburgh, Edinburgh, United Kingdom*

WILLIAM PAUL (1), *Harvard University, Department of Physics and Division of Applied Sciences, 229 Pierce Hall, Cambridge, MA 02138*

TADEUSZ SUSKI (485), *UNIPRESS, High Pressure Research Center, Polish Academy of Sciences, ul. Sokolowska 29, 01-142 Warsaw, Poland*

K. SYASSEN (247), *Max-Planck-Institut für Festkörperforschung, Heisenbergstr.1, 70569 Stuttgart, Germany*

PAWEL TRAUTMAN (427), *Institute of Experimental Physics, Warsaw University, Warsaw, Poland*

PETER Y. YU (457), *Department of Physics, University of California, Berkeley, CA 94720*

CHAPTER 1

High Pressure in Semiconductor Physics: A Historical Overview

William Paul

DEPARTMENT OF PHYSICS AND DIVISION OF ENGINEERING AND APPLIED SCIENCES
HARVARD UNIVERSITY
CAMBRIDGE, MA

I. INTRODUCTION	2
II. EXPERIMENTS ON BULK SAMPLES OF THE ZINCBLENDE FAMILY TO 1970	4
1. Changes of Band Structure under Pressure: The Empirical Rule	5
2. Changes of Parameters Determining Single-band Transport under Pressure	8
3. Applications of the Empirical Rule	11
III. THEORETICAL CALCULATIONS OF THE EFFECT OF PRESSURE ON BULK BAND STRUCTURES	15
1. Introduction	15
2. Pseudopotential Calculations	15
3. Calculations Using the Dielectric Theory of the Chemical Bond	17
IV. CURRENT STUDIES OF BULK SEMICONDUCTORS OF THE ZINCBLENDE FAMILY	18
1. Introduction	18
2. Deep Impurity Levels and Several Conduction Band Extrema	19
3. DX and EL2	20
4. Other Deep Impurity Levels	24
V. QUANTUM WELLS, HETEROSTRUCTURES, AND SUPERLATTICES	25
1. Introduction	25
2. Band Offsets	25
3. Influence of the X-States on Phenomena in Heterostructures	26
4. Resonant Tunneling	28
5. Heterostructures and DX Centers	28
VI. STRUCTURAL PHASE TRANSITIONS	29
1. Introduction	29
2. The Family of Zincblende Structures at Atmospheric Pressure	29
3. S, Se, and Te	31
4. Other Systems	31
5. Lattice Dynamics: Inelastic Neutron Scattering	32
VII. OTHER MATERIALS	33
1. Lead Chalcogenides	33
2. Group 6 Elements	34
3. Amorphous Semiconductors	35

4. Nanocrystals and Porous Silicon	36
5. Semiconducting Nitrides	37
VIII. OTHER MATTERS	38
References	38

I. Introduction

The wording of the title of this introductory chapter is deliberate. It will be less a survey of a separate world of experimentation than a description of the evolution of a discipline where the systematic exploitation of changes in properties with pressure has contributed vitally to the understanding of semiconductors *at atmospheric pressure*. That it can be so described is the result of a symbiotic development, on the one hand, of the theory of semiconductor energies and band structures, and on the other, of innovative experimentation on the structural, electrical, and optical properties, ingeniously replicated at high hydrostatic pressures. In no other field where high pressures are used has there developed such a high degree of mutually supportive theory and experiment.

This chapter will not review, for reasons of space, the very valuable field of semiconductors under nonhydrostatic stress. Carefully controlled and specified uniaxial stresses have been very useful in establishing the details of the band structure because they lift state degeneracies occurring in highly symmetrical cubic semiconductors. Thus, the twofold degeneracy (ignoring spin) at the maximum of the valence band of the zincblende family at the Γ-point may be lifted, and the properties of light and heavy holes differentiated. Similarly, the degeneracy at the L or X local extrema of the conduction band is lifted, permitting, *inter alia*, an identification of the location of the absolute extrema. Lifting of these degeneracies clarifies the fundamental properties and can improve the device capabilities. A discussion of these effects is given in Chapter 3 of Volume 55 by Anasstassakis and Cardona, in Chapter 4 by Pollak, and in Chapter 5 by Adams et al.[1]

There have been two major contributions to the development of apparatus in the post-Bridgman era which have led to considerable expansion of activity in the field. The first is the introduction of strong, flexible high-pressure tubing separating a massive pressure-generating system from the final experimental vessel, which may be at the exit of a spectrometer, between the poles of a magnet, or settled deep in the interior of a cryostat;

---

[1]Earlier reviews of the exploitable results of uniaxial stress are to be found in the writings of Keyes [1960], Fritzsche [1965], and Paul [1966].

the second is the advent of the compact diamond anvil cell which avoids the storage of large energies in compressed fluids. A discussion of the techniques of generation and measurement of high pressures using diamond anvil cells is given in Chapter 7 of Volume 55.[2] Generally speaking, using the former technique, all experiments that are feasible at atmospheric pressure can be repeated, at high or low temperatures, to at least 15 kbar. The diamond anvil cell has the advantage of providing reasonably easy access to much higher pressures, but its use is restricted mainly to optical experiments, albeit over a wide range of frequencies. The principal experiments rendered difficult are those requiring electrical contacts. Now, certain phenomena — not just the obvious ones of transitions of phase — are sufficiently nonlinear in pressure so as to require pressures above 15 kbar to make them observable at all. An example is the influence of higher conduction band extrema in the zincblende system. Here, the use of anvil cells is *almost* obligatory.[3] There is a need for more development of reliable techniques for electrical measurement in diamond or sapphire anvil cells at least in the lower pressure range between 20 and 50 kbar, where restrictions of cell design are less severe. See Erskine *et al.* [1987] and Li *et al.* [1987].

The first studies of (extrinsic) semiconductors under pressure were carried out by Bridgman as part of surveys involving many materials (black P [1921], AgS, Ge, PbTe [1935], Te [1938]) which were not followed up by analysis. The pressure dependence of the fundamental energy gap of Ge was determined by Miller and Taylor in 1949 from conductivity measurements, and by Hall, Bardeen, and Pearson in 1951 from analysis of changes in the current–voltage characteristics of carefully chosen p–n junctions. In 1951–53 Bridgman reported nonlinear behavior of the resistivity of n-type Ge in hydrostatic pressure experiments to 30 kbar and

[2]Earlier references to techniques for piston–cylinder apparati are to be found in Chapter 15 of *Solids under Pressure* [Paul and Warschauer, 1963] and in Martinez' review [1980]. A useful early review on the development of the diamond anvil cell has been given by Jayaraman [1983], and there are additions to technique in Martinez [1980]. It is also relevant to recognize that, prior to the advent of the diamond anvil cell, a huge amount of optical experimentation in large pressure apparati involving slightly nonhydrostatic stresses in NaCl was carried out by Drickamer and his collaborators [Babb and Robertson, 1963].

[3]Almost — but not totally. First, nonanvil apparatus, not immensely flexible, does exist for the pressure range above 15 kbar. Second, the judicious use of semiconductor alloying can shift the observation of phenomena requiring electrical contacts into the pressure range accessible to piston–cylinder apparati. Thus, the study of Ge–Si alloys shows that the higher lying (100) extrema in the Ge conduction band are shifted toward the lowest lying (111) extrema by about 0.1 eV for each 10% of Si added; this is the equivalent of a built-in pressure of roughly 15 kbar, and brings phenomena normally observable over above, say, 30 kbar, into observation in the sub-15-kbar range. Similar comment applies to other members of the zincblende family of semiconductors.

a maximum in resistivity in quasi-hydrostatic measurements near 50 kbar. These experiments were taken as the starting point for a systematic attack by Paul and collaborators [Paul and Warschauer, 1963] focused on the properties of Ge under pressure, the aim being to measure all accessible parameters by electric, magnetic, and optical experiments over a range of pressure to 30 kbar, and of temperature between 77 and 600 K, and to produce a self-consistent explanation of the properties in terms of the relevant parameters, including the alteration of the band structure. The family of semiconductors of which Ge is a representative member (group 4 elements, such as Si, group 3–5 and group 2–6 compounds such as GaAs and CdTe, and the related ternary and quaternary compounds of zincblende symmetry) has been by far the most investigated under pressure. At this date, the work of many groups has established the order of magnitude of the change with pressure of the many parameters affecting optical and transport properties. None of these changes were known in 1950, and it is a sobering thought that progress in fully understanding other systems — group 6 elements such as Se, group 4–6 compounds such as PbTe, organics, and transition metal oxides, will require similar ground work not yet completed.

In the next sections of this chapter I shall expand on these preliminary remarks, while keeping my account within bounds, and avoiding much detail, by giving liberal reference to reviews that fortunately are available. Sections II and III deal with the period before 1970, when the effect of pressure on the band structures of the zincblende family was established. They will be relatively detailed, while the modern developments which exploit the basic results will of necessity be treated more cursorily in subsequent sections.

II. Experiments on Bulk Samples of the Zincblende Family to 1970

Section II.1 discusses the changes with pressure of the band structure of intrinsic crystals, and the development of an Empirical Rule governing them. Section II.2 deals with the changes in the parameters determining extrinsic carrier transport. Section II.3 explains how the Empirical Rule can be used both to help determine the atmospheric pressure band structure and to elucidate phenomena which occur at atmospheric pressure. Consideration of the effect of pressure on impurity energy levels will be given in Section IV.

1. Changes of Band Structures under Pressure: The Empirical Rule

a. Transport Measurements

If we start from ideas underlying the tight-binding approximation, we expect that the effect of pressure will be to broaden allowed bands of energy and narrow forbidden ones. This is true in a gross sense, but the important band extrema may approach or separate for the small changes in lattice parameter available pressures produce. We illustrate this using the band structure of Ge shown in Fig. 1. It is sufficient, for the moment, to take note of the three conduction band minima at Γ, L, and Δ. We now know (and knew with less certainty in the early 1960s) that the band structures of all members of the zincblende family are perturbations on the scheme shown in Fig. 1. The minimum gap in Ge is between the $\Gamma_{25'}$, state in the valence band and the L_1 state in the conduction band. The focus of the earliest experiments in the 1950s was to determine the change

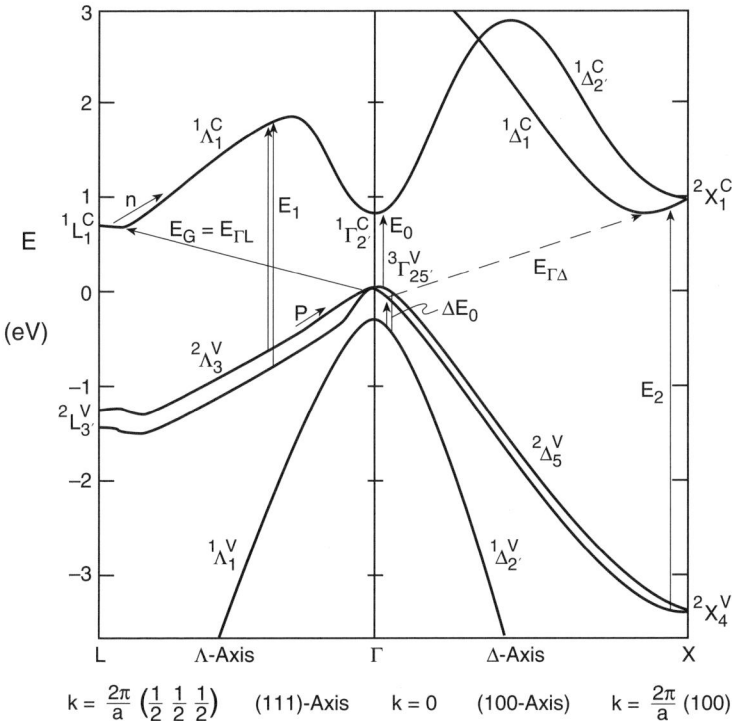

Fig. 1. Band structure of Ge.

with pressure of this gap (the state symmetry was not yet identified) from the change in the intrinsic conductivity.[4]

The intrinsic conductivity, σ_i, is given by

$$\sigma_i = en_i (\mu_n + \mu_p),$$

where $n_i = A(m^*, T) \exp(-E_g/2kT)$. Here μ_n and μ_p are the electron and hole mobilities, respectively, n_i is the intrinsic single-band carrier density, and A depends on an effective mass m^* which is a combination of electron and hole masses. The contributions of changes in μ_n and μ_p can be corrected for by measuring exclusively extrinsic conduction in n-type and p-type crystals of low impurity density, so that the scattering processes determining the mobilities are the same in both intrinsic and extrinsic samples. It is straightforward to correct for the presence of extrinsic carriers, but in any event, it is clear that the changes in σ_i are dominated by changes in E_g. These experiments give

$$\frac{dE_g}{dP} = 5 \times 10^{-6} \text{ eV/bar} \qquad \text{for Ge [Paul and Brooks, 1954]}$$

and

$$\frac{dE_g}{dP} = -1.5 \times 10^{-6} \text{ eV/bar} \qquad \text{for Si [Paul and Pearson, 1955]}.$$

The report that the gap changes in Ge and Si had opposite signs was the first indication in the literature that the band structures of Ge and Si were different; it was, even if unrecognized, a harbinger of the contributions to come from the pressure technique.

Guided by an accelerating decrease of the extrinsic conductivity above 15 kbar, Paul and Brooks [1954] interpreted their data on intrinsic and extrinsic samples in terms of the increasing occupation of a second set of conduction-band extrema where the carriers had low mobility. They found that the rate of increase of their effective energy gap slowed at pressures near 30 kbar, and this was soon identified as caused by the approach to the valence band of the second set of extrema at a rate essentially identical to the rate of decrease of the gap in Si. The later determination of the symmetry of the lowest extrema in Ge and Si, combined with studies of

---

[4]This is a simplified discussion. Readers interested in detailed arguments need to examine the original papers or the reviews by Paul and Brooks [1963], Paul and Warschauer [1963], or Paul [1966].

the variation of the smallest gap in Ge–Si alloys, soon revealed that the $\Gamma_{25'}$ to Δ_1 gap in Ge decreased with pressure at the same rate as the $\Gamma_{25'}$ to Δ_1 gap in Si. This was the first indication of an empirical rule governing the displacement under pressure of band features possessing the same symmetry.

b. Optical Measurements

Measurements of the change with pressure of the fundamental optical absorption edge followed. They had to be analyzed with care. If the *shape* of the curve of absorption coefficient versus photon energy changes with pressure, then isoabsorption measurements (change with pressure of the photon energy corresponding to an arbitrarily chosen fixed absorption coefficient) do not give the change in optical gap. If the energy gap is direct ($\Gamma_{25'}$ to $\Gamma_{2'}$, in Fig. 1), then the absorption edge shape will change somewhat because of changes in effective mass (here, we defer consideration of excitonic effects, an important caveat). If the energy gap is indirect ($\Gamma_{25'}$, to L_1 in Fig. 1), the shape of the edge can be very complex, since it is a superposition of processes involving (in principle) more than one type of intermediate state, and several types of phonon in absorption and emission. The principal intermediate state for the indirect transition in Ge is the $\Gamma_{2'}$ state, and the quantity ($E_{\Gamma_{2'}} - E_{L_1}$) appears in the denominator of the expression for the absorption. The upshot is that shape changes may occur. They are clearly seen in the data for Ge, which can however, be analyzed to determine the shift with pressure of both L_1 and $\Gamma_{2'}$, and to reconcile the optical and the transport data. Such analyses can be risky and uncertain, so the best approach, in optical measurements — an approach not available for the early studies of Ge and Si — is to measure the displacement under pressure of a sharp feature on the absorption edge, such as an excitonic absorption peak; the corrections necessary for the exciton energy are usually rather small. Prominent in the early measurements of absorption under pressure were the studies of Drickamer and his colleagues [1965], where the large pressure range used overwhelmed the corrections required for low-pressure measurements. These, of course, still require corrections in order to provide the most accurate pressure coefficients, and they face their own difficulties of nonlinear effects stemming either from changes in compressibility with pressure, or from nonlinearity in the dependence of deformation potential itself on volume. All that said, the observation by Drickamer that eventually, at sufficiently high pressure, the optical edge of all members of the Si family displaced toward the red provided very vital information regarding the band structure of the family members.

The changes in the optical gap were also determined from the displacement under pressure of the peak of the photoluminescence spectrum. Again, in principle, corrections are necessary dependent on the precise recombination process involved: it may be excitonic, it may involve optical phonon participation in absorption or emission, or it may involve shallow donor or acceptor states. In practice, each one of these corrections can usually be shown to be second-order compared to the changes in gap.

c. *The Empirical Rule*

From all of these measurements, first on Ge and Si, and subsequently on the 3–5 compounds, there emerged on empirical rule (Paul [1961]), at first applied only to the principal three minima in the conduction band at Γ, L, and X:

> Independent of the family member studied, the pressure coefficient of the direct energy gap at Γ falls in the range 10–15 meV/kbar, that of the indirect gap at L near 5 meV/kbar, and that of the indirect gap at X between -1 and -2 meV/kbar.

During the 1960s, there was much activity in correlating structure in the spectrum of reflectivity versus photon energy with critical points in the interband density of states (such as L_3 to L_1, see Fig. 1). The problem was to identify the observed peaks in reflectivity with specific interband transitions. A magnificent interplay of theory and experiment resulted: Simple band structure calculations were used to suggest candidates at specific energies, and a combination of experiments using magnetic fields or strain to induce fine structure was used to confirm the identification. Techniques to modulate the reflectivity, using electric fields, uniaxial stress, or temperature, added immensely to the accuracy of determination of the reflectivity features. In turn, the determination of the energy of identified interband separations led to refinement of the theoretical calculational technique (pseudopotential theory, see Section III). Repetition of the reflectivity or modulated reflectivity measurements under pressure established the interband deformation potentials for all the identified transitions [Zallen and Paul, 1967]. The results, not unexpected at this point from the developing theory of pseudopotentials, was an extension of the Empirical Rule to all conduction- and valence-band extrema separations.

2. CHANGES OF PARAMETERS DETERMINING SINGLE-BAND TRANSPORT UNDER PRESSURE

Thus far we have concentrated on determining the changes with pressure of the intrinsic band structure. In this section we shall discuss purely extrinsic transport under pressure under four heads: (1) the change in energy of shallow hydrogenic impurity levels, (2) the change in energy of deep levels, (3) the redistribution of a fixed number of carriers between inequivalent extrema with different pressure coefficients, and (4) the change of scattering parameters in a single band of constant carrier density.

a. Shallow Hydrogenic Impurity Levels

The exact expression governing the carrier density in a band supplied by a shallow impurity level depends on the density of dominant and compensating impurities. Where analytic expressions are appropriate, they contain the factor $\exp(-E_I/\alpha kT)$, where $\alpha = 1$ or 2 and E_I is the ionization energy of the shallow impurity. The ionization energy $E_I = E_H(m^*/K^2)$, where E_H is 13.6 eV, and m^* and K are an effective mass and dielectric constant, respectively. The resistivity of samples of Si doped with As, Al, and In was studied by Holland and Paul [1962]. The changes are very small and are of the same order of magnitude as changes in the mobility. The changes in the ionization energy are deduced to be consistent with measured changes in the dielectric constant and with estimates of the mass change from changes in the band structure. It is very doubtful that further work on the hydrogenic impurities would repay the effort.

b. Deep Impurity Levels

This subject is treated in greater depth in Section IV. Here we discuss some very early measurements which have not been analyzed theoretically up to the present time. Holland and Paul [1962] and Nathan and Paul [1962] investigated, as a function of pressure, the resistivity changes of Ge and Si doped with Au. The Au produces four energy levels in the gap of Ge, and two in the gap of Si; thus, the statistics of impurity level occupation are very complex. Nevertheless, the authors were able to find the pressure coefficients of the separation of the different levels from the relevant band edge. Two comments are pertinent. First, the pressure coefficients with respect to the valence band are much smaller than with respect to the conduction band; it is almost as if the level is tied to the valence band, even if its energy lies close to the conduction band. Second, Au in Si has an acceptor level almost exactly in mid-gap, so that appropriate counterdoping makes it behave as an acceptor or donor of electrons. Both of its ionization energies, with respect to either band edge, decrease with pressure, but again the coefficient with respect to the valence band edge is by far the smaller. The sum of the coefficients neatly fits the observed pressure dependence of the bandgap. These results have no obvious connection with those on "deep levels" to be discussed in Section IV. It is clear that this field is still wide open.

c. *Inequivalent Extrema with Different Pressure Coefficients*

We suppose that the total carrier density remains constant, but that carriers are shared differently between the inequivalent extrema as pressure changes their separation. The parameters determining the conductivity are evidently (1) the relative populations of the extrema, which depend on their separation in energy at atmospheric pressure, the ratio of their densities of states, and the pressure coefficient of their separation, and (2) the mobilities in the two sets of extrema which depend on the changes in intra- and inter-extrema scattering. The first of these scattering processes depends on the effective mass and whichever combination of scattering processes is operative, including scattering between *equivalent* extrema. The second depends on the density of states for scattering events which satisfy energy and momentum conservation; this scattering will be maximum when the extrema are at the same energy, and will give rise to a minimum in the conductivity.

It is evident that transport results may be fitted to give the unknowns of the initial energetic separation of the inequivalent extrema, their relative densities of states, and the pressure coefficient of their separation, as well as information about the inter-extrema scattering processes. This complex situation was first analyzed by Nathan *et al.* [1961] for *n*-type Ge. Similar analyses have been done for GaAs and GaSb, where a double transfer of electrons from Γ_1 to L_1 to X_1 occurs. A very extensive set of measurements on GaSb has been reported by Pitt and Lees [1970].

d. *Scattering in a Single Band*

Finally in this section, let us consider a single band with no intrinsic carriers and a constant extrinsic carrier density. The change in conductivity with pressure then depends on the change in the effective mass and the scattering processes. The best method to determine the change in mass appears to be through some type of magneto-optical experiment: Faraday rotation or Shubnikov–de Haas. Taking account of scattering processes is harder. The simplest imaginable scattering process is inside a single extremum (no *equivalent* extrema) by the longitudinal acoustic vibrations of long wavelength. For this case, Shockley and Bardeen [1950a,b] derived

$$\mu = \frac{2(2\pi)^{1/2}e\hbar^4 dc_s^2}{3(kT)^{3/2}m^{*5/2}E_1^2},$$

where d is the density, c_s an average sound velocity, m^* the isotropic effective mass, and E_1 the deformation potential, or the amount by which the energy of an electron at the band edge is changed by unit dilation of the crystal. If, fortuitously, the scattering of both electrons and holes were adequately described by this expression, then the sum or difference in deformation potentials deduced from the experimental mobilities should check with experimental determinations of the change with volume of the energy gap. I do not know of any such case.

There are, in fact, always other scattering processes present (e.g., by transverse acoustic modes, by optical modes, by impurities and defects, both charged and neutral) in unknown proportions, which render difficult a good understanding of the mobility at atmospheric pressure. Thus, even if one determined the changes with pressure of d, c_s, m^*, dielectric constant, other phonon energies, etc., any fit to the pressure dependence of conductivity will be uncertain. Obviously, scattering between *equivalent* extrema complicates the problem still more. Despite this bleak prognosis, there have been very noteworthy attempts to analyze measurements of conductivity, Hall effect, magnetoresistivity, and magnetoconductivity in terms of the changes in individual parameters. The early work has been reviewed by Paul and Brooks [1963]. Later work is found in articles by Pitt [1969], Pitt and Lees [1970], and Pitt [1977], and in several articles for a conference in Grenoble in 1969.

3. APPLICATIONS OF THE EMPIRICAL RULE

a. *Establishment of Band Structures*

The empirical rule has been used to help establish the band structures of bulk crystals, heterostructures, and superlattices. Since adequate reviews have already been given of its application to the determination of bulk band structures [Paul, 1966], only two brief illustrations will be given here.

The (100) minima. The experiments of Drickamer and collaborators (Drickamer [1965]) showed that the absorption edges of all semiconductors of the Si family shifted to lower photon energies at sufficiently high pressures. This negative pressure coefficient was argued to identify Δ_1 (or X_1) symmetry for the lowest conduction band extrema under these pressure conditions, so that the extrapolation of the high-pressure data gave their energetic positions at atmospheric pressure.

GaP. The band structure of GaP [Zallen and Paul, 1964] is illustrated in Fig. 2. It is constructed at least in part from high-pressure measurements.

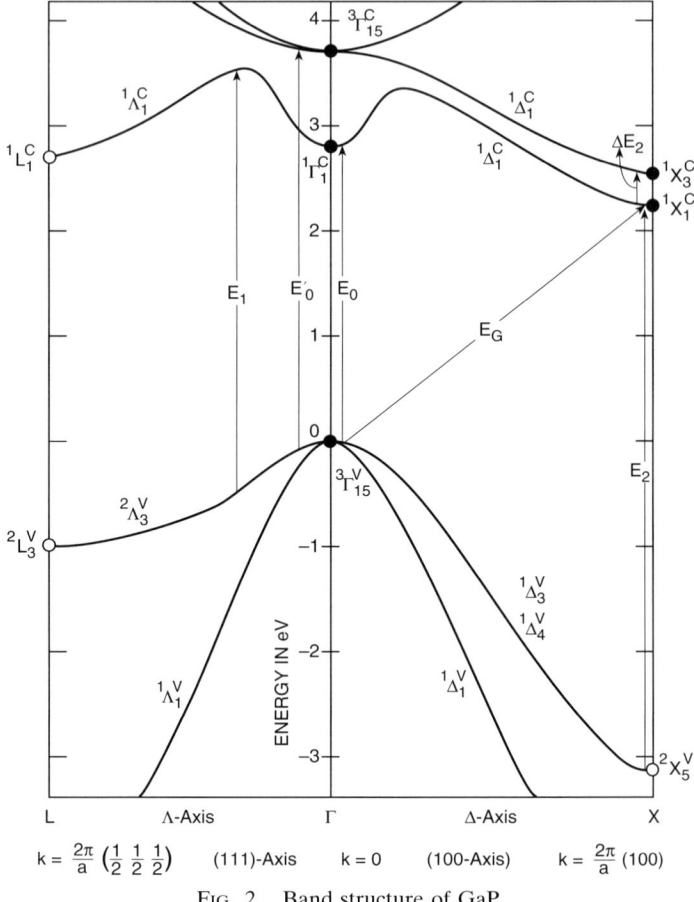

FIG. 2. Band structure of GaP.

First, the fundamental gap is between an X_1 conduction band state and a Γ_{15} valence band state. The first demonstration that this was appropriate came from the fact that the pressure coefficient of the smallest gap in this material was negative.

Second, the lowest conduction band state at the center of the zone is 2.8 eV above the VB maximum, or 0.6 eV above the state lying in the (100) direction. Again, the first quantitatively correct estimate of the energy of this state was made as a result of reflectivity measurements which showed structure at this energy that displaced to higher energies precisely at the rate expected from the empirical rule for the Γ_1 state.

Third, absorption measurements had suggested that the direct gap was about 2.6 eV, that is, that there was a separation of 0.4 eV between the Γ_1

state and that forming the lowest conduction band state. This deduction was made on the basis of the observation of a bump in the infrared absorption near 0.4 eV, which was attributed to the electron transfer from the lowest minimum into the Γ_1 state. Investigation of the displacement of this bump with pressure was decisive: if we accept the empirical rule that under pressure the Γ_1 and the X_1 states will separate at a net rate of about 14×10^{-6} eV/bar, then we can test the correctness of the assigned reason for the optical absorption by measuring the shift of the absorption bump under pressure. The result is that the observed displacement is more than an order of magnitude smaller than that predicted on the basis of the supposed identification. This measurement was used first to eliminate the original explanation of the bump, second to invalidate the identification of the energy of the Γ_1 minimum, and third to suggest that the transition was between the lowest X_1 minimum and another one at the same **k**-vector just directly above it (ΔE_2).

Other family members. There have been similar contributions to establishment of the band structures of other members of the zincblende family. Thus the accepted inverted band structure of α–Sn, with the $\Gamma_{25'}$, states lying higher in energy than the $\Gamma_{2'}$ states, was deduced to be the only structure capable of yielding the pressure dependences of the intrinsic and extrinsic conductivity and the Hall coefficient of α–Sn [Groves and Paul, 1963]. This inverted band structure also applies for HgSe, HgTe, and HgTe–CdTe alloys of high HgTe content, and the inverted structure was the inspiration for the now accepted inversion in band structure at the L-point between PbTe and SnTe.

The higher energy parts of the conduction band of GaAs and GaSb were also first suggested from the results of pressure measurements of the resistivity and/or Hall effect. These studies also established properties of the higher extrema not otherwise accessible, and established the basis for later understanding of hot electron effects in this family of semiconductors. The excellent work of Pitt and Lees [1970] is especially notable as an advanced analysis comparable to the multi-conduction-band extrema analysis carried out for Ge by Nathan *et al.* [1961] and Howard [1961].

Elucidation of phenomena. Knowledge of the shifts with pressure of the three dominant minima in the band structure of the zincblende family has led to clarification of phenomena occurring at atmospheric pressure, such as the tunnel diode effect and the properties of hot electrons. We use the latter as illustration. The first demonstration that high electric fields could transfer electrons into subsidiary conduction band minima in Ge was given by Koenig *et al.* [1960], who identified transfers into (100) minima through the change in high electric field current–voltage characteristics with hydrostatic pressure. A much more spectacular demonstration, however, came in the identification of the source of the transferred electron, or Gunn effect. This effect is due to the existence in some semiconductors of a range of electric fields in which the electron drift velocity decreases with increasing field (negative mobility). This leads to a situation where any space charge fluctuations grow in space and time. As a result, a uniform distribution of fields within the sample becomes unstable, and, as was shown by Gunn, high-field domains are created. The propagation of these domains at gigacycle frequency forms the basis for devices but here we concern ourselves with the conditions of band structure and scattering mechanisms responsible for the negative differential mobility. Although, in principle, several mechanisms are possible, Hutson *et al.* [1965] were able to identify that responsible in GaAs. They showed that the threshold electric field for the Gunn effect decreased with increasing pressure. This fit with a model that the field heats electrons in the Γ_1 minimum to energies higher than the subsidiary minima and causes transfer of the electrons to these minima. Since the electron mobility in the higher minima is lower than in Γ_1, the average mobility and drift velocity of electrons decrease with increasing applied field. The threshold field clearly depends on the separation in energy between the Γ_1 and the higher minima, and thus the reduced threshold is just what one would expect of either the L_1 or the X_1 minima. In fact, the quantitative data were interpreted to identify the X_1 minima as those operating at high pressures. The Gunn effect disappears at sufficiently high pressure, evidently when the carriers are evenly shared at zero field.

Consistent evidence, but with additional information, resulted from studies of the Gunn effect by Porowski *et al.* [1970] in n-InSb, where the band structure situation is quite different. The energy gap is smaller than the distance from the Γ_1 conduction band edge to the bottom of the next set of minima. As a result, an electron requires less energy to impact-ionize another electron from the valence band to the Γ_1 conduction band than to transfer into a higher-lying minimum. Such semiconductors were originally considered unsuitable for observation of the Gunn effect. However, the known displacement of the Γ, L, and X minima under pressure produces a favorable band structure, both by increasing the fundamental gap and by decreasing the energetic distance from Γ_1 to higher minima. The Gunn effect is observed. It is, however, found that the effect occurs right down to atmospheric pressure, where the gap is much smaller than the separation of the conduction bands. This implies that the cross-section for impact ionization is simply too small to prevent carrier heating into the higher minima at strong fields. It is clear that the use of pressure was invaluable in settling the basic property of transfer of electrons yielding the Gunn effect, but the explanation may be more complicated than that. In Section IV, we discuss the existence of states produced by certain donor impurities, which appear to be associated with specific minima, not necessarily the lowest, and to have energies at atmospheric pressure resonant with the conduction-band continuum. Such states could have the same pressure coefficients as the extrema they are linked to and provide the low (zero) mobility receptors of the transferred electrons.

III. Theoretical Calculations of the Effect of Pressure on Bulk Band Structures

1. INTRODUCTION

We restrict our discussion to a volume deformation potential, defined for a state $|i\rangle$ at wave vector $\mathbf{k}$ as $\partial_\epsilon^i(\mathbf{k})/(\partial \ln V)$, and assume it is the same for all states at the same symmetry point, that is, we ignore changes in effective mass. In principle the volume deformation potential might be found experimentally from measurements of carrier mobility if carrier scattering were restricted mainly to scattering by long-wavelength longitudinal acoustic vibrations, but this is not a reliable assumption, as was explained in Section II.2.d. The order of magnitude of a volume deformation potential may be crudely guessed from the size of shear deformation potentials, which *are* directly measurable, as about 10 eV. This implies that if 10 kbar of pressure changes the volume by 1%, then there will be a change in energy separation of the order of 0.1 eV, which is much larger than kT at room temperature. Thus, it can be argued that pressure may rearrange the order of local band extrema and cause large changes in transport. This is, in fact, the argument that pressure experiments are likely to produce large, measurable changes in properties.

2. PSEUDOPOTENTIAL CALCULATIONS

We refer here to the article by Martinez [1980], which is the major review of the field since the several reviews of the 1960s. Martinez assesses the results of various calculations based on the pseudopotential method of Phillips and Kleinman [1959]. This method of calculating band structures is based on the idea of a repulsive pseudopotential introduced to simulate the necessity of orthogonalizing the conduction band and valence band wave functions to core wave functions. When balanced against the attractive potential of the cores, the net effective potential is free-electron-like with a small perturbation having the symmetry of the lattice. As a result, the band structures of members of the same family are expected to be perturbed weakly among each other, and the deformation potentials of states of the same symmetry are expected to be about the same. This is essentially a statement of expectation of an empirical rule for deformation potentials. Quantitative calculations aim at estimating the deformation potential magnitudes and comparing them against experiment.

The earliest argument generally justifying an empirical rule for deformation potentials was given by Brooks (see Paul [1966]). The degree of agreement of calculated deformation potentials with experiment depends crucially on a good choice of the initial values for the pseudopotential form factors. Saravia and Brust [1969] were able to obtain fairly good agreement with experiment for Ge. Melz [1971] claimed similarly good agreement in determining the deformation potentials for Ge and Si, using the same model pseudopotential as Saravia and Brust, but a simpler calculational method with a smaller basis set than ordinarily used in calculating the band structure itself. Melz's results for Si appear to be comparable in agreeing with experiment to those of Herman *et al.* [1966], but his calculations for GaAs, and especially CdTe, do not agree well with experiment. He has pointed out that Zhang and Callaway [1969], using a method essentially identical to his for calculations on GaSb, found good agreement with experiment, and he has ascribed this difference in outcome to the superior pseudopotential form factors used by Zhang and Callaway. This suggests that the calculation of the deformation potentials per se is not the problem, and it also suggests that experimental deformation potential constants may be sensitive parameters to use in choosing the best pseudopotential form factors for application in the empirical pseudopotential method (EPM). This simply repeats the long-standing contention that *any* acceptable band structure calculational method should also deliver accurate pressure coefficients for band gaps.

Martinez concludes that the EPM as applied by Schluter *et al.* [1975] has given the best results to date. The EPM deals with the pseudopotential form factors as fitting parameters to reproduce established experimental data. The pseudopotential Hamitonian is written

$$\mathcal{H} = -(\hbar^2/2m)\nabla^2 + V(\mathbf{r}),$$

where the pseudopotential $V(\mathbf{r})$ is expanded in reciprocal lattice vectors $\mathbf{G}$ and can be expressed as the product of a structure factor $S(\mathbf{G})$ and a pseudopotential form factor $V_\mathbf{G}$. $V_\mathbf{G}$ depends only on $|\mathbf{G}|$. Then

$$V(\mathbf{r}) = \sum_{\alpha, |\mathbf{G}|, G_o} S_\alpha(\mathbf{G}) V G^\alpha,$$

where the summation runs over the different atoms α and is restricted to a few different values of $|\mathbf{G}|$, sufficient to reproduce the main features of the band structure determined by experiment. It is the judicious choice of the V_G which ultimately dictates the utility of the derived deformation potentials. Schluter *et al.* find, *inter alia*, that (1) the deformation potential is characteristic of the symmetry of the level in a given structure, (2) the deformation potentials are ordered in increasing magnitude for p-like and s-like states, so that the deformation potential for $\Gamma_{25'}$ (Γ_{15}) is smaller than that for L_1 and X_1, and those are smaller than for $\Gamma_{2'}$ (Γ_1), (3) the gap opened up by the antisymmetric part of the potential in compounds has a small pressure coefficient, and (4) the contribution of the spin–orbit interaction to splittings may be expected to vary as the inverse of the linear compressibility. Their first conclusion is a validation of the empirical rule; their second agrees with a conclusion of Brooks (see Paul [1966]); their third agrees with an experimental result found earlier by Zallen and Paul [1964] in a study of the infrared absorption of GaP; and the fourth appears to agree with a detailed study by Melz and Ortenburger [1971] on the effect of pressure on spin–orbit splittings in Ge, GaAs, and CdTe, using experimental electroreflectivity measurements and theoretical calculations by the relativistic orthogonalized plane-wave technique. These modulated reflectivity measurements yield more precise pressure coefficients for interband separations than ordinary absorption or reflection and so are able to determine pressure coefficients of the spin–orbit splitting 100 times smaller than for the direct transition in Ge. The theoretical and experimental results agree in indicating a smaller volume dependence of the splitting than expected if the electron density in the core increased in proportion to the decrease in crystal volume. This is an interesting result of basic significance which, however, demands a high level of *ab initio* calculation and the precise experimental coefficients best afforded by one of the modulation techniques.

3. CALCULATIONS USING THE DIELECTRIC THEORY OF THE CHEMICAL BOND

Martinez has also reviewed the calculations of Camphausen *et al.* [1971] on the pressure coefficients of five interband transitions and the refractive index in the zincblende family, using the dielectric theory of the chemical bond introduced by Phillips [1968a,b, 1969] and developed by van Vechten [1969a,b]. The reader is referred to the original papers, which demonstrate remarkable agreement between theory and experiment.

IV. Current Studies of Bulk Semiconductors of the Zincblende Family

1. INTRODUCTION

In this section we shall concentrate on investigating the properties of impurities not describable by a simple hydrogenic theory, a problem which has dominated studies on the bulk zincblende-type semiconductors under pressure since about 1970.

Practically all of the experiments carried out to characterize such impurities (see Baranowski and Grynberg [1992]) are easily repeatable at pressures up to about 15 kbar at high and low temperatures. At higher pressures, significant recent additions to the use of DAC apparatus include far-infrared absorption (Haller *et al.* [1996]) and far-infrared magneto-optical techniques (Dmochowski and Stradling [1993], Chen *et al.* [1996]).

Lannoo [1992] has reviewed the theoretical techniques applied to so-called "shallow" and "deep" impurities. For the shallow impurities, effective mass theory (EMT) envisages a slowly varying potential near an isolated donor or acceptor impurity, such as a group 4 atom substituting for a group 3 host in a 3–5 crystal. The solutions resemble those for an H atom, but are modified by an effective mass and the dielectric constant of the crystal. The application of this theory is simplest when the energy band has a single extremum, and it is best for the excited states, where the electron spends little time near the impurity. The first major correction to this hydrogenic picture is a central cell correction for the ground state — a so-called chemical shift. The EMT must be further modified when there are several equivalent extrema — conduction band minima, say — which augment the central cell correction and may split the ground state into a nondegenerate A_1 and a three-fold degenerate T_2 state, with the A_1 lower. Modifications to EMT theory are also necessary for the degenerate valence band maximum energy. Yet more modification occurs when there is appreciable electron–phonon interaction, which may produce severely localized states and a distortion in the impurity site. All of these EMT levels, with their ground state modifications, exist for each set of conduction band extrema, not only the lowest set. This leads to the possibility of discrete levels which are resonant with the lowest band continuum, levels estimated to be of the order of 10^{-6} to 10^{-5} eV wide and with a nanosecond lifetime.

The other extreme from EMT contemplates a very localized (one-cell) departure from a periodic potential and is treated by a local density (LD) or empirical tight-binding (TB) theory. This leads to energy levels normally deep in the bandgap, and a wave function involving **k**-vectors from all over the Brillouin zone. The intermediate case, which is that we have to deal with for the 3–5 and 2–6 semiconductors, may involve a shallow behavior–deep behavior instability, complicated by local distortions to minimize local free energy.

The theoretical understanding of deep impurities is still an open question. As soon as the nature of the central cell dominates the properties of the ground state, the solution for it is uncertain, a fact which is reflected in theories which reach quantitatively different conclusions. As we shall describe, pressure experiments have made, and continue to make, an important contribution to the elucidation of a problem of considerable fundamental interest and technological importance.

2. Deep Impurity Levels and Several Conduction Band Extrema

Based on an original suggestion of Bate [1962], Kosicki and Paul [1966] and Paul [1968] analyzed the optical and electrical properties of GaSb doped with S, Se, or Te in terms of impurity states of different energies whose wave functions were made up of band functions exclusively from any of the Γ_1, L_1, or X_1 extrema. No lattice distortion was envisaged in this treatment. A good fit of the pressure results was found when these impurity levels were given the same pressure coefficients with respect to the valence band as the Γ, L, or X extrema. This system was also studied extensively, and the preceding conclusions supported in general, if not always in detail, by Kosicki *et al.* [1968], Pitt [1969], Vul *et al.* [1970a], Hoo and Becker [1976] and Dmowski *et al.* [1977b]. Supporting evidence for these states linked to higher-lying extrema was adduced from experiments on GaAs by Sladek [1964] and Hutson *et al.* [1967], on GaAs–GaP alloys by Holonyak *et al.* [1966] and Craford *et al.* [1968], on InSb by Porowski *et al.* [1970], and on CdTe by Foyt *et al.* [1966]. Theory suggesting or supporting their existence was published by Kaplan [1963], Peterson [1964], Bassani *et al.* [1969], and Altarelli and Iadonisi [1971]. The three states could co-exist, hybridize to a certain extent, and be resonant in energy with the lowest conduction band. This early description may still be correct in its entirety for some impurities, although additional experiment and theory have by now suggested possible modifications, connected with the possibility of local distortions. What was not in doubt was the profound influence of some new type of impurity state, unrecognized previously, on photoluminescence and transport phenomena.

During the 1970s, there was increasing study of the results of impurity–lattice coupling, such as differences in the thermal and optical ionization energies, caused by a distortion of the lattice and a change in the lattice energy near the impurity, which was dependent on the occupation of the impurity state. Although this effect is small for the "normal" hydrogenic type of impurity (say, As in Si), it can be very large for other atoms, so that at low temperatures the local configuration cannot reach thermal equilibrium and metastable states are observed. Persistent photoconductivity, the aforementioned differences in thermal and optical ionization energies, and abrupt changes in properties at certain critical temperatures result. Pressure data must be analyzed taking into account such effects (Vul *et al.* [1970b], Iseler *et al.* [1972], Porowski *et al.* [1974], Burkey *et al.* [1976], and Dmowski *et al.* [1977a]). Historically, this extension and elaboration on the interpretation of the earliest work preceded the outburst of new activity that resulted from the proposal of a new entity (the DX state) by Lang and co-workers [1977, 1979, 1986], but it is now clear that the same continuing problem has been involved throughout.

3. DX AND EL2

Studies on the DX and EL2 centers have received the greatest attention in the past two decades. The DX center is ubiquitous where donor impurities in 3–5 and 2–6 compounds are required, and the EL2 center is responsible for the semi-insulating character of GaAs substrates used in many applications. They are extensively discussed in the theoretical article by Lannoo, as well as in reviews (for example) by Mooney [1990], Yu and Li [1990], Baj and Suski [1990], Dmochowski and Stradling [1993], Suski [1994], and Malloy and Khachaturyan [1993]. They are the subject matter of chapters in this book by Li and Yu and by Trautman *et al.* The important role that pressure has played over time has also been emphasized in the work of Porowski and Trzeciakowski [1985] and Stradling [1985].

The nomenclature DX arises from the study of a deep trap investigated by Lang and co-workers using deep-level transient spectroscopy (DLTS) applied to n-type AlGaAs alloys [1977]. The photoionization threshold of this center was about 10 times larger than its thermal ionization energy, and it possessed a very small capture cross-section which led to persistent photoconductivity. This suggested a model of the center as a very localized one with large lattice relaxation, consisting of a complex of a donor atom D and an unknown defect X, defects such as vacancies being considered very probable in the alloy structure. The center was not found in pure GaAs at atmospheric pressure; however, Mizuta *et al.* [1985] found that pressures exceeding 24 kbar applied to n-type GaAs:Si produced a DLTS signal which they attributed to the same DX center investigated by Lang and his co-workers. This suggested that a defect was not involved but that the Si donor might be undergoing a (shallow to deep) transformation coincidentally with, and perhaps as a result of, the known changes with pressure of the conduction band structure. It is now clear that the DX center, associated with all the group 4 and group 6 donors in GaAs, and with large lattice relaxation, is a possible metastable state resonant with the conduction band at atmospheric pressure. The effects on the band structure of alloying Al to the GaAs or of increasing the pressure are very similar: the energies of the Γ_1 electrons are raised with respect to those of the electrons at L_1 and X_1, 1 kbar of pressure being roughly equivalent to a 1% addition of Al. Apparently, consistent with this equivalence, sufficient Al addition or pressure forces the resonant state into the energy gap where it becomes the stable state of the donor.

A closer examination of the theory of the impurity state intermediate between EMT and very localized TB is required. EMT theory informs us that donors such as Si_{Ga} and Te_{As} produced by Si and Te in GaAs will produce sets of hydrogen-like states associated with the conduction band minima at Γ, L, and X. If lattice distortion leads to a wave function localized to only a few near neighbors, TB theory in the defect molecular model gives a nondegenerate A_1 antibonding state and a higher energy threefold degenerate T_2 antibonding state. First-principles LD calculations by Chadi and Chang [1988, 1989], for the unrelaxed condition of the donor, find the $A_1(ab)$ state resonant with the Γ band of GaAs, a result confirmed in the work of Yamaguchi et al. [1990], who found $A_1(ab)$ at $E_c + 0.17$ eV and the $T_2(ab)$ state 0.8 eV higher. Yamaguchi et al. also determined the shifts of the $A_1(ab)$ state with increased pressure and Al content in the GaAs: both changes bring the L_1 and X_1 states closer to the Γ_1, with the result that the $A_1(ab)$ state enters the forbidden gap for unalloyed GaAs near 25 kbar, or for 30% Al content at atmospheric pressure. Yamaguchi et al. also concluded that the distorted DX state suggested by Chadi and Chang (see next paragraph) lay at a higher energy than their calculation for $A_1(ab)$.

Chadi and Chang adopt a different model. The relaxation involved in the Yamaguchi calculation is small (SLR), but Chadi and Chang propose a massive local distortion (LLR) in which, for the case of a Si donor occupying a Ga site, the Si shifts about 1 Å in a (111) direction, becomes three-fold coordinated with As atoms, has two filled lone-pair states, and leaves a broken bond on an erstwhile As neighbor. They calculate that this model, which involves a negative correlation energy for the two lone-pair electrons, is a lower energy state than the $A_1(ab)$ state with small lattice relaxation of Yamaguchi. It is clear that it provides an explanation for the experimental observations of an emission barrier, a Franck–Condon shift, persistent photoconductivity, absence of electron spin resonance signal, and fine structure of the ground state in alloys because of the several possibilities for the number of *nnn* Al to a Si donor.

Indeed, the attributes of the Chadi-Chang model fit the experimental data so well that it has been widely accepted. The situation, however, may be more complicated. There are no obvious flaws in either of the calculations, yet their results are substantially different, and we are reminded that the LD theory does rather poorly in calculating excited states and energy bandgaps. Thus, theory is an uncertain guide, and both possibilities being admitted, experiment must decide. Experiment, however, does not, at present, make a clear choice: there is evidence in the literature for most of the states stemming from the EMT and TB theories. We summarize this in Sections IV.3.a–IV.3.e.

a. The EMT States

There is abundant evidence of hydrogen-like states associated with each inequivalent set of extrema in the conduction band of the 3–5 compounds and their alloys. The energies of each set of these states depend on the particular donor atom as well as the details of the band structure; it is, for example, different for S in GaAs and GaSb than it is for Te in these compounds. The ground state for any one of these sets may be in the energy gap, or it may be resonant with the conduction band, depending on the details of the band structure. When the energies of any one of these sets of impurity states is well separated from the others, its pressure coefficient is close to that of the extrema causing it; when the energies of the different sets are close, there are likely to be nonlinear shifts with pressure. See, for example, Paul [1968], Porowski *et al.* [1970], or Dmochowski *et al.* [1996].

b. The DX States

The LLR model of Chadi and collaborators successfully explains many observations. There is no apparent reason why it should not occur for appropriate pressure in all the 3–5 compounds and alloys. The ground-state energy E_{DX} depends on many parameters: (1) on the nature of the donor and the host crystal, so that it is very different, for example, in GaAs doped with Si, Ge, or Sn (Wisniewski *et al.* [1993]) or in GaAs and GaAlAs doped with the same donor; (2) on the surroundings of the donor, out to an undetermined small radius encompassing near neighbors, and hence its dependence on the different numbers of Al neighbors to a donor atom in GaAlAs alloys (Contreras *et al.* [1993]); and (3) on the separation of the atoms, which can be changed by pressure. The energy of the ground state can be plotted as a function of pressure (or x in $Ga_{1-x}Al_xAs$) for any donor and compound, along with the changes in the intrinsic band structure.

Although the changes in E_{DX} correlate in a general way with the changes in band structure — for example, the approach of the L_1 and X_1 extrema to the Γ_1 extremum with increase of pressure or Al content correlates with a shift of E_{DX} in the same direction with respect to Γ_1 — there is no established association of that energy with any individual extremum of the band structure. That is, the E_{DX} does not have the same pressure coefficient as any particular extremum. The E_{DX} may lie in the Γ-continuum or the energy gap, depending on all of the parameters cited; for example, it is in the continuum for GaAs: Si, Ge, Sn, but in the gap for $Ga_{0.7}Al_{0.3}As$: Si, both at atmospheric pressure. The DX state has a potential barrier for both electron emission and electron capture, and this barrier is also species-specific: estimates have been made of 0.33, 0.29, and 0.07 eV for DX centers produced by Si, Ge, and Sn in GaAs, respectively.

c. *The A_1 (ab) States*

The $A_1(ab)$ SLR ground state energy similarly depends on all the parameters already cited for the DX state. Its energy depends on the local surroundings, and they depend on the pressure and any alloying. The pressure and alloying also determine the band structure. Thus, its energy can be correlated with the band structure, but as with the DX, the energy is not correlated with any one feature of the band structure. The $A_1(ab)$ state differs from the DX in that the lattice distortion is small and there is no emission barrier into degenerate states of the band structure.

d. *Co-existence of EMT, DX, and $A_1(ab)$ States*

There is much evidence on the co-existence of at least two of the three types of impurity state. A usual experiment consists of applying pressure sufficient to make the DX state the lowest in energy, so that it captures the conduction band electrons; then lowering the temperature and reducing the pressure to ambient so that the DX state has its energy in the conduction band continuum, but the electrons are locked in place by an emission barrier; and then releasing the electrons in a controlled way by photoionization that inserts them into the ground (1s) EMT state. Under different conditions of pressure and alloying, the EMT state reached may be found to be linked to any of the conduction band extrema. See, for example, Goutiers *et al.* [1993].

The DX and A_1 states may also co-exist: an example is GaAs, as shown by Wisniewski *et al.* [1993], Baj *et al.* [1993], and Wasik *et al.* [1996]. At ambient pressure, the ground state energy levels are 110 and 80 meV, respectively, above the bottom of the conduction band. Their pressure coefficients are established as -9.0 meV/kbar and -7.5 meV/kbar with respect to Γ_1; these coefficients resemble those for Γ_1–X_1 (12–13 meV/kbar) and Γ_1–L_1, (7 meV/kbar), although, as we have already intimated, there is no special significance to be attached to that. By manipulating the pressure and temperature it has been demonstrated that the two types of localized states co-exist, one resembling the singly occupied $A_1(ab)$ state and the other the doubly occupied DX$^-$ state.

Finally, there is clear evidence that the $A_1(ab)$ and EMT states may co-exist from magneto-optical and photoluminescence experiments in the far infrared on GaAs doped with Si, Ge, Sn, S, and Se, as discussed by Wasilewski and Stradling [1986], Dmochowski and Stradling [1993], and Dmochowski et al. [1996].

e. Additional Comments

For reasons of space, we have not discussed at length the pressure experiments on the EL2 center (see Lannoo [1992]) or on 2–6 compounds (see Weinstein et al. [1996]). Nor have we discussed the fact that consideration of the DX and EL2 centers necessarily enters into analysis of many quantum-well and superlattice structures which require the materials of this section. In fact there has been a report of yet another negatively charged donor state in modulation-doped GaAs-based quantum wells, which is shallow and undistorted and co-exists with the DX$^-$ state (Huant et al., 1996).

The experimental possibilities may be greatly helped in the future by the availability of strong synchrotron radiation in the far infrared, which would be ideal for experiments in DACs. Another type of experiment would be very useful but difficult: electron-spin resonance (or other techniques) which would endeavor to establish more precisely the entity which is created and reported upon as a result of our usual doping techniques.

4. OTHER DEEP IMPURITY LEVELS

While by far the most activity on deep levels under pressure has been on the complex states provided by group 4 and group 6 donors in 3–5 semiconductors, and by group 3 and group 7 donors in 2–6 semiconductors, there was early activity on impurities such as Au in Si and Ge, as was mentioned in Section 2.b. Activity in this area has been desultory, but one should note measurements on transition metals in Si (Pfeiffer et al. [1993]), on rare earth complexes in InP and GaAs (Takarabe [1996]), on Mg in GaN (Teisseyre *et al.* [1996]), and on Si in GaN (Wetzel *et al.* [1996]).

V. Quantum Wells, Heterostructures, and Superlattices

1. INTRODUCTION

Research on semiconductor heterostructures has been reviewed in many books and journal articles since the original proposal by Esaki and Tsu [1970]. See for example, Bastard [1988], Bastard *et al.* [1991], and articles by Brillson, Voisin and Bastard, McCombe and Petrou, and Meyer *et al.* in Volumes 1, 2, and 3 of the *Handbook on Semiconductors,* edited by Moss [1992, 1994]. Our very limited purpose here is to review the contributions high-pressure experiments have made to clarification of their properties. More detailed discussion is to be found in Chapters 1 and 2 of Volume 55 by Maude and Portal, and Klipstein, respectively.

The electronic states of heterostructures are determined by the band structures of the constituent materials, along with parameters characterizing the boundary conditions at the interfaces and the band offsets, that is, the band lineup discontinuities at the conduction- and valence-band extrema. Theories for the band structures and properties of heterostructures are easily tested by the predicted effects of pressure. Knowledge of the effect of pressure on the band structures and properties of the different constituents can be exploited in the investigation of their properties. Prominent in this regard is the empirical rule on the shifts with pressure of the conduction-band extrema.

2. BAND OFFSETS

The determination of band alignments by different methods has been discussed by Brillson [1992]. Many of the methods are indirect. Among the more direct and accurate methods, unfortunately applicable only to the important limited set of heterostructures such as GaAs/AlGaAs, has been the study of photoluminescence spectra under pressure as analyzed by Venkateswaren *et al.* [1986] and by Wolford *et al.* [1986]. The technique involves measuring the pressure at which the X_1-band of the AlGaAs barriers drops below the Γ_1-band of the GaAs wells, resulting in a transfer of the photogenerated electrons to the AlGaAs layers and thus a precipitous drop in the luminescence intensity. The method works when the decrease in intensity is *not* caused by the X_1-band of the *well* falling lower in energy than the Γ_1-band of the well. Then, the photon energy at which the intensity decreases is just the indirect gap of the barrier at the pressure of intensity decrease minus the valence band offset (VBO) energy between well and barrier at that pressure, with a minor correction for quantum confinement energy. In practice, the VBO may be determined with milli–electron volt precision from experimental plots of the photon energies of the $\{\Gamma_1$ (well)–Γ_{15} (well)$\}$ and $\{X_1$ (barrier)–Γ_{15} (well)$\}$ transitions versus pressure. Since no fits to models are needed, this method is possibly the most reliable of any.

Whitaker et al. [1996] have proposed an extension of this method which gives the offsets, under limiting conditions on the band structure they outline, between a larger set of 3–5 compounds and alloys. The method relies on the band offsets being transitive. It is supposed that an absolute valence band maximum energy may be assigned to each 3–5 compound or alloy. Then the VBO is just the difference between the final absolute energies of the valence-band maxima in the two components of the heterostructure. Thus, if the VBOs between materials A and B, and between A and C, are measured, then the VBO between B and C is the difference between the two measurements. Specifically, if A is AlGaAs, B GaAs, and C GaAsSb, then the absolute valence band maximum of GaSb may be determined; if C is GaInAsP, then the absolute valence band for InP may be found.

Other systems, all designed for specific optoelectronic applications, whose band offsets have been determined by photoluminescence measurements under pressure are GaInAs/InP [Lambkin et al., 1988], GaAs/AlInP [Nakayama et al., 1993], GaInP/AlGaInP [Prins et al., 1995], and GaAs/GaInP [Chen et al., 1991].

The method just outlined assumes a pressure-independent VBO. This has been verified by Lambkin et al. for the GaAs/AlGaAs [1989] and GaInAs/InP [1988] systems. However, for the GaInP/AlInP system, Patel et al. [1993] found that the VBO increases at about 1.8 meV/kbar. Similarly, Cheong et al. [1994] determined that the CBO of the GaAs/AlGaAs increased at 0.73 meV/kbar, and the same authors [Cheong et al., 1997] found even larger changes in the VBO for two InAs/Ga$_{1-x}$As$_x$Sb superlattices, 3.5 and 5.6 meV/kbar, respectively. These findings necessitate in each case a change in the deduced magnitude of the VBO. It therefore appears that to this point no general rule or systematic behavior of the VBO under pressure has been established.

3. INFLUENCE OF THE X-STATES ON PHENOMENA IN HETEROSTRUCTURES

Goñi et al. [1990] have discussed the Γ–X mixing occurring in GaAs attributed to electron–phonon interaction. This mixing effect is expected to be larger in heterostructures because of the potential discontinuities at interfaces. The degree of mixing and the perturbation on discrete energies in wells depends on the ratio of the interaction potential to the energy separation of the uncoupled states. The known effect of pressure on this separation provides a convenient test of theories of mixing.

Perlin *et al.* [1995] have searched for this (possibly small) effect in the energy of the direct photoluminescence lines of GaAs/AlGaAs quantum wells, in experiments where they paid great attention to the precision with which the photon energy of the lines and the hydrostatic pressure were determined. They report that up to the Γ–X crossover pressure, the shift of the Γ–Γ lines agrees with a calculation that takes account of the increase of the direct gap in GaAs with pressure, and corrections to the confinement energy of the electrons and the exciton binding energy, both of which are changed by the pressure increase of the effective masses. When the X_1 minima in the barriers become lower than the Γ_1 level in the well, they find that the pressure coefficient of the direct line is lowered, and they attribute this to Γ–X mixing caused by the well–barrier interface. It is not clear, however, how the Γ–Γ lines above the crossover are differentiated from the (weak) lines attributable to X (barrier) to Γ (well) transitions.

Li *et al.* [1994, 1995] have measured the low-temperature photoluminescence spectra of InAs quantum dots embedded in a GaAs crystalline matrix to pressures of 70 kbar. Below 42 kbar the spectra are dominated by electron–heavy-hole exciton transitions. Above this (crossover) pressure the spectra show two X-related lines attributed respectively to a transition from the X_1 state of the InAs well to the heavy-hole state of the well, and to a second transition from the X_1-state of the GaAs barrier to the InAs heavy-hole states. In the crossover regime, a detailed fit to the pressure variation of the line energies shows convincing evidence of strong Γ–X mixing effects and nonlinear behavior of the lines. The strong coupling is attributed to the three-dimensional restriction of the motion of the electrons in the quantum dot.

The effect of Γ–X mixing in GaAs/Al$_x$Ga$_{1-x}$As coupled double quantum wells has also been investigated by Burnett *et al.* [1993] by using hydrostatic pressure to bring the X_1 valleys of the barriers nearly equal in energy to the energies of the confined electron states of the wells: whereas, when the wells are uncoupled (thick barriers), the pressure coefficients of the energies of the allowed transitions between the valence- and conduction-band quantized states are equal to the pressure coefficient of the GaAs gap, for the coupled wells (thin barriers) the energies of the allowed transitions all showed a decrease near 20 kbar. This decrease is attributed to a drop in the conduction-band quantum-well confinement energy due to the onset of Γ–X mixing.

4. RESONANT TUNNELING

The first paper of Esaki and Tsu [1970] discussed the physics of the resonant tunneling of electrons through a superlattice structure. The subject of tunneling is a complex one which, however, is illuminated by use of pressure as a parameter, as is discussed by Professor Klipstein in Chapter 2 of Volume 55. In a typical structure, an external voltage controls the energy difference of states in different regions of the sample whose wave functions overlap, and gives rise to a current with a nonlinear dependence (on V) which may include negative differential resistance. Pressure is a significant variable not only because it can influence the extent of wavefunction overlap by changing effective masses, but also because it can affect the energy separation of the tunneling states.

5. HETEROSTRUCTURES AND DX CENTERS

Although DX centers are generally regarded as bad actors with regard to efficient devices, they have been exploited to help understand the physics of different magnetotransport phenomena in AlGaAs/InGaAs/GaAs quantum wells [Knap et al. (1996), Dmowski et al. (1996)]. It was desired to vary the 2-D electron gas density over a wide range for these experiments. This was accomplished by using hydrostatic pressure to decrease the free carrier density and infrared LED light to increase it. The reduction in 2-D carrier density was accomplished by exploiting the occupation density of DX centers. When pressure is applied at room temperature, the DX centers' energy falls below the conduction level and the free carrier density is reduced. Cooling to low temperature before releasing the pressure back to ambient at 77 K ensures that the DX center occupation is maintained. Increases in the carrier density may be arranged by infrared light which ionizes the DX centers.

In other experiments which may be regarded as exploitation of DX centers, Bosc et al. [1996] used their knowledge of DX behavior to put in evidence dopant segregation in short-period superlattices resulting from planar Si doping performed during molecular beam epitaxy.

This brief account by no means exhausts the scope of papers discussing the application of high-pressure techniques to the investigation of quantum wells and superlattices. Other papers are to be found in the Proceedings of the High-Pressure Conferences on Semiconductors at Tomaszow (see *Semiconductor Science and Technology*, Vol. 4, 1989), Porto Carras (see *Semiconductor Science and Technology*, Vol. 6, 1991), Kobe (*Jpn. J. Appl.*

Phys., Vol. 32, Suppl. 32-1), Vancouver (*J. Phys. Chem. Solids,* Vol. 56, 1995), and Schwäbisch-Gmünd (*Phys. Stat. Solidi* (*b*), 198, 1996).

VI. Structural Phase Transitions

1. INTRODUCTION

The study of phase diagrams and structural phase transitions in solids induced by high pressure is a venerable subject dating back to Bridgman's early experiments [Bridgman, 1949]. Within the past 10 years, however, there has been a considerable improvement both in the experimental techniques available and in methods used by theoreticians to calculate the total energies of crystalline phases and to predict phase diagrams. Among the more important experimental advances have been the continuous development of DACs suitable for structural work, the advent of powerful X-rays from synchrotron sources, the development of the image plate detector for X-rays, and the use of neutron sources both for structural studies and for inelastic scattering measurements. On the theoretical side, Professor N. E. Christensen gives in Chapter 2 an account of progress since total energy calculations were pioneered by Yin and Cohen [1982]. Our discussion will be confined therefore to a very brief overview of the more recent work.

Besson [1993] has argued persuasively that, in view of the fact that the same solid–solid transition may be observed at different pressures depending on the experimental method used, and in view of the possible occurrence of intermediate phases and hysteresis, the most reliable procedure should be the full determination of phase lines over the entire P–T plane, rather than simply the changes observed at ambient temperature. He describes the many ways in which misleading results may be obtained, which have clearly occurred in one form or another, over the whole span of such measurements. His advice should be followed.

2. THE FAMILY OF ZINCBLENDE STRUCTURES AT
   ATMOSPHERIC PRESSURE

The advent of synchrotron X-ray radiation and improved DACs has had a great impact in the past six years on investigations of the structural transitions in the semiconductors having the zincblende structure at atmospheric pressure. The Edinburgh group (Nelmes *et al.* [1993a,b]. McMahon and Nelmes [1995], Nelmes *et al.* [1995a,b], McMahon and Nelmes [1996]) has applied angle-dispersive powder diffraction using the image plate area detector [Amemiya *et al.*, 1988; Hatton *et al.*, 1992] to reveal new transitions in some dozen members of the zincblende family that significantly modify the formerly accepted structural systematics of these materials. They conclude that much of these systematics, which generally supposed a tendency for crystals to progress to more highly coordinated forms, such as 4-fold (zincblende) → 6-fold (β-Sn or NaCl) → 8-fold (simple hexagonal) → 12-fold (fcc or hcp), are either incomplete or incorrect. Lower symmetry phases are found to be dominant in most of the 3–5 and 2–6 compounds (not group 4) at the highest pressures, but equally significant, the pattern of structural transformations is very complex and may show (different in different materials) several intermediate phases not discernible in the earlier work because of limitations on resolution. Thus, for example, they suggest a complete sequence may be zincblende → distorted zincblende → cinnabar (one form of distorted NaCl) → NaCl → Cmcm (another form of distorted NaCl). Although this group suggests many changes are needed, they believe that the increasing similarities being found in the zincblende family ultimately simplify the overall systematics of phase changes. Nelmes and his collaborators [Nelmes *et al.*, 1993a,b] review the 30 years of work on the much-investigated InSb, which involved contradictions and uncertainties, as an example of conclusions changed by the new and better techniques available. They found a body-centered orthorhombic structure (space group Immm) to be the stable form at high pressure (see, however, Mezouar *et al.* [1996]).

The new results suggest a re-evaluation of the contribution of calculations of total energy to the prediction of phase transitions. The usual formulation for $A^N B^{8-N}$ forms envisages a zincblende to β-Sn structural transition at high pressure as a result of a comparative study of these two phases; there is a possibility of the existence of lower energy forms than those being compared, and the new results on these other forms will almost certainly lead to new calculations of relative phase stability. As an example, new first-principles total energy calculations for the cubic zincblende, β-Sn and body-centered orthorhombic structure of InSb, the last structure being that determined by Nelmes and collaborators, found that the β-Sn structure was unstable with respect to the body-centered orthorhombic structure at high pressures [Guo *et al.*, 1993].

The Paris group of Besson (see Mezouar *et al.* [1996]) has also studied the phase diagram of InSb by energy-dispersive synchrotron X-ray diffraction at pressures up to 8 GPa and temperature up to 800 K, in contrast to the ambient-temperature results of the Edinburgh group. They conclude from this more extensive view that an orthorhombic form (six-fold coordinated orthorhombic, space group Cmcm) is the stable high-pressure phase and that the phase (orthorhombic, Immm) studied by Nelmes and collaborators is a metastable form only observed for rapid kinetic transformations. It is evident that conclusions are being debated and sorted out in rapid order in this field (reader, beware), but that the overall thrust should reach definitive conclusions soon. Obviously, in view of the conclusions of Guo et al. [1993] regarding the orthorhombic phase Immm, new total energy calculations on orthorhombic Cmcm should be carried out.

3. S, Se, and Te

Structural studies on the group 6 elements, especially Se and Te, have a long history dating back 50 years. They all show numerous phase transitions. Some of these transitions are from semiconductor to metal, and Se and Te have been shown to be superconducting at high pressures and sufficiently low temperatures [Akahama et al., 1992]. Their phase diagrams, and the symmetry of the different phases, have been and are being progressively clarified as the available high pressure and X-ray energy sources improve. In Se and Te similar sequences of transformations from the hexagonal form stable under ambient conditions to a bcc form stable at the highest pressures are displayed. A similar progression has been found for S, from the orthorhombic form stable at atmospheric pressure to a β-Po type formed at 162 GPa [Luo et al., 1993]. See also Akahama et al. [1993a]. Detailed histories are to be found in the above papers and those of Holzapfel et al. [1993] and Akahama et al. [1993b].

Nishikawa et al. [1993, 1995, 1996] have investigated the band structures and total energies of the β-Po (rhombohedral) type structure and the bcc structure of Te within the local density-functional formalism, determining that the latter is the more stable form at the top pressures, in agreement with experiment. They have specifically investigated the β-Po phase appearing at 11 GPa for Te, 60 GPa for Se, and 162 GPa for S, and inquired which phase is the preceding one. They confirm that the β-Po phase is stable [Luo et al., 1993] and that it approaches bcc with increasing pressure. Apparently these calculations are hindered somewhat by a lack of precise structural data on the different phases. These studied represent a work in progress, with steady gains being made in both the experimental and calculational arenas.

4. Other Systems

Besson et al. [1996] have expanded on earlier studies of the HgX (X ≡ S,Se,Te) system by measuring the equation of state of HgTe up to 800 K and pressures up to 2.5 GPa using X-ray diffraction. References to the earlier work are given in their paper.

Several studies have been done on the 1–3–6_2 chalcopyrite semiconductors, such as $CuInS_2$, $CuInSe_2$, $AgGaS_2$ and $AgGaSe_2$, which are analogs of the 2–6 binary materials. They have been measured to about 30 GPa using energy-dispersive X-ray diffraction [Tinoco et al., 1995] and their phase transitions indexed. The temperature dependence of the Raman-active modes has also been studied to 18 GPa and the temperature dependence of the mode Grüneisen parameters found [González et al., 1995]. Again, references to earlier studies are given in the papers cited.

Phase transitions and equations of state have also been examined in the 3–5 nitrides, BN, AlN, and InN, up to 110 GPa using X-ray diffraction [Ueno et al., 1993]. GaN has been investigated by Perlin et al. [1992]. Excepting BN, all exhibit phase transitions from the ambient wurtzite to the rock-salt structure. It is asserted that there is a progression to lower transition pressure as the ionicity of the compound increases in the sequence BN, GaN, AlN, and InN.

Finally, the several phase transitions induced by pressure in orthorhombic black phosphorus have been examined using *ab initio* pseudopotential total energy techniques by Morita et al. [1993]. For a review of the experimental and theoretical status to 1986, see Morita [1986].

5. LATTICE DYNAMICS: INELASTIC NEUTRON SCATTERING

Klotz et al. [1996a] reported in 1996 a study of the transverse acoustic phonon modes of Ge along the [100], [110], and [111] directions to 9.7 GPa, close to a phase transition, using for the first time inelastic neutron scattering. The transverse acoustic phonons in cubic semiconductors have attracted much theoretical and experimental attention for a long time, since their dispersion curves away from Γ are low in energy and flat, their Grüneisen parameters $\gamma = -(d \ln \omega)/(d \ln V)$ are negative at the zone boundary points L and X, and it is thought that the negative γ at X is related to the occurrence of pressure-induced phase transitions to higher coordination. The lack of extensive experimental data on the volume dependence of the dispersion curves has been a hurdle to a definitive theory. This first measurement of the pressure dependence of the dispersion curves across the entire Brillouin zone by inelastic neutron scattering (see Pintschovius [1994], Klotz et al. [1995, 1996b]) — the only possible technique — is a truly major advance likely to change completely our knowledge of the lattice dynamics of all semiconductors. It is obviously far more effective than determinations of the effect of pressure on phonons at isolated points of the zone.

The frequencies of modes along the [100], [110], and [111] directions were shown to have nonlinear volume dependences in all cases. The present results accord reasonably well with measurements of ultrasonic velocity by McSkimin and Andreatch [1963] at the Γ-point and with Raman measurements by Olego and Cardona [1982] at X, but disagree with an earlier measurement using phonon-assisted tunneling at L [Payne, 1964]. The values of γ at X and L are found to be the same, in agreement with past results on group 4, 3–5, and 2–6 materials. Thus, the present measurements fit well with past ones and calibrate their usefulness.

The results have been compared with *ab initio* calculations using the density functional formalisms.[5] The agreement between theory and experiment is excellent for the structural equilibrium properties, the phonon frequencies, and the mode Grüneisen parameters, without need for adjusting or fitting any parameters.

Raman scattering has also been used extensively to map out certain portions of the phase diagram. An example is InSe, a layered semiconductor currently being investigated as a potential solar cell material. The pressure dependence of all Raman-active zone-center phonons has been determined to 11 GPa at ambient temperature. Information about zone-edge modes is also found from second- and third-order scattering processes. The phonon dispersion curves were calculated using the rigid-ion model with six adjustable force constants [Ulrich *et al.*, 1996].

The pressure dependence of Raman scattering in α-boron at ambient temperature has also been measured to 30 GPa, and complete agreement found with *ab initio* lattice dynamics calculations using density functional theory [Vast *et al.*, 1996].

Yoshioka and Naga [1995] have studied Raman scattering in S to 20 GPa at room temperature, and found two reversible phase transitions.

VII. Other Materials

1. LEAD CHALCOGENIDES

The lead chalcogenides have been studied at high pressures by Paul *et al.* [1992], Averkin *et al.* [1962], Prakash [1966], Besson *et al.* [1965, 1968], Martinez *et al.* [1970], Martinez [1973], Schluter *et al.* [1975], and Khoklov and Volkov [1996]. These materials all have energy gaps corresponding to

---

[5]References for the theory given in Klotz *et al.* [1996b].

the infrared region of the spectrum. They and their alloys with the Sn chalcogenides form a family of photovoltaic detectors, light emitters, and lasers of great utility. An interesting characteristic of the Pb chalcogenides is that their energy gaps all *increase* with increasing temperature, in contrast to the behavior of the majority of other semiconductors. Consistent with this (although not the sole reason, there being an explicit effect of temperature that is positive and thus anomalous), the effect of pressure is a gap decrease at a rate between 8 and 9 meV/kbar.

The maximum of the valence band has L_6^+ symmetry (s-like around the Pb atom) and the minimum of the conduction band L_6^- symmetry (p-type around the chalcogen). There is strong $p-d$ mixing of wave functions which is supposed to "push up" the energy of the valence band at L with respect to that at Γ. One can suppose that the fact of the s-like symmetry combined with strong mixing might result in a large increase in the energy of the valence band maximum, and a decrease in the energy gap to the conduction band, as the lattice constant is reduced.

PbSe, the compound with the smallest gap, has been the most investigated. Besson *et al.* [1965, 1968] succeeded in tuning the laser emission of a PbSe laser diode at 77 K from 7.5 microns to 22 microns at 14 kbar, the end of their available pressure range.

Alloy studies show that in the Sn chalcogenides the order of the valence and conduction bands is reversed. Thus, Martinez [1973], applying pressure to a diode of $Pb_{0.88}Sn_{0.12}Se$, showed that the energy gap decreased from 70 to 58 meV between 0 and 1.7 kbar, and then increased from 58 to 70 meV between 12.9 and 14.7 kbar (the region between could not be observed because the emission threshold of the diode was very high). It is notable that the region of extremely small gaps, comparable with phonon energies, has not yet been observed directly, although technically that can now be done.

On a different front, Khoklov and Volkov [1996] have investigated persistent photoconductivity in PbTe doped with In and Ga, in some, but not complete, analogy to the features caused by DX centers in the 3–5 and 2–6 compounds. Experience suggests that pressure measurements might give interesting results.

Other semiconductors with the NaCl structure of the Pb and Sn chalcogenides have been described in the review by Martinez [1980].

2. GROUP 6 ELEMENTS

The group 6 elements, S, Se, and Te, were studied at high pressures in the 1930s and 1950s but seem to have been neglected since then. Bridgman

[1938] found that the resistivity of intrinsic Te decreased under pressure, which Bardeen [1949] interpreted as a decrease of the forbidden energy gap. Confirmatory measurements of the decrease in gap were made by Long [1956], Robin [1959], and Neuringer [1957], although their coefficients did not agree very well. Se was measured by Kozyrev and Nasledov [1956], and S by Slykhouse and Drickamer [1958]. Uniformly, the intrinsic gaps of these elements were found to decrease with pressure. Much more recently, Clark *et al.* [1995] have carried out total energy calculations on the hexagonal form of Se. They determined that the fundamental gap decreased at high pressures, in agreement with experiment.

3. AMORPHOUS SEMICONDUCTORS

There has been a fairly large number of investigations on amorphous semiconductors, particularly the chalcogenide glasses, over the past 25 years. Davis [1993] has reviewed these. The situation regarding the tetrahedrally coordinated semiconductors appears to be the following. Because of the absence of long-range order, there is nothing that resembles the specificity and exactness of the pressure coefficients of the zincblende family. Nevertheless, there is evidence that remnants of a correspondence exist between the states characterizing the maximum of the valence band and the minimum of the conduction band of the crystalline and amorphous materials. For example, the optical bandgap of a-Si and a-Si:H appears to decrease with pressure with roughly the coefficient, -1 meV/kbar, of crystalline silicon. On the other hand, the preponderance of experimental evidence [Imai *et al.*, 1993] suggests that the gap in a-Ge and a-Ge:H increases at low pressures, and eventually decreases at high pressures, qualitatively as the crystalline gap does. These facts are already intriguing evidence in a field where correspondence between experiment and exact theory is hard to come by: not so much the evidence on a-Si, which matches the overall approach of the valence and conduction band densities of states in a manner one might expect, but more so in Ge, where the initial increase is a reminder of a feature associated with the L-states of the conduction band.

The effect of pressure on the tetrahedrally bonded materials is, however, much less dramatic than that on other materials of lower coordination number. Witness arsenic, where the conductivity increases by eight orders of magnitude in 40 kbar, and then precipitously increases by a further two orders to that for a metallic phase [Elliott *et al.*, 1977]. Consistent with this, the optical gap decreases at a rate 20 times larger than that of a-Si and a-Si:H [Tanaka, 1988]. Davis gives a convincing argument in qualitative terms to explain these effects in terms of the electronic and atomic structure of the material.

The effects on the chalcogenide films and glasses are also considerable. These materials have two p electrons in a lone pair forming the uppermost valence band. The pressure coefficients of the energy gaps are generally large and negative, probably because of an increase in overlap between the lone pairs which results in band broadening. A direct comparison between the effects of pressure on the optical gaps of 2D-GeS_2, 3D-GeS_2, and a-GeS_2 may be found in the work of Weinstein et al. [1982], illustrated in Davis's paper.

4. Nanocrystals and Porous Silicon

The optical properties of nanocrystals are expected to differ from those of the bulk material (see, for example, Efros and Efros [1982] or Brus [1994]). Changes in the electronic and vibrational states have been examined in several different types of nanocrystal embedded in solid matrices for some time. More recently there has been a flurry of activity on a form of silicon obtained by anodic etching (porous silicon) which displays strong room-temperature visible photoluminescence (PL) and offers the possibility of obtaining electroluminescence from a silicon-based structure integrated into an optoelectronic device. The mechanism of luminescence is, however, not established, and there are two principal candidates: (1) from the Si, modified by confinement effects in the nanocrystals, which both increase the photon energy of emission and break the selection rules normally forbidding radiative recombination in bulk Si, and (2) extrinsic effects related to defects on the nanocrystal surface, or chemical complexes surrounding them. Here we discuss some of the pressure experiments that have been done in an attempt to identify the mechanism of luminescence.

Zeman et al. [1996] have offered persuasive evidence against the first explanation. They examined the luminescence and Raman scattering from porous silicon placed under hydrostatic pressure up to 22 GPa, and found that the intensity was not much affected even after the material had supposedly converted to the β-Sn phase, nor after it was recovered at 1 bar in the BC8 phase. The phase of the Si was followed through its several transitions using Raman scattering, since the Td, β-Sn, and BC8 phases are all Raman-active with characteristic signatures. It would appear to be hard to rebut this kind of evidence.

Support for this view has come from two experiments carried out by Cheong et al. [1995, 1996]. In the first they examined the PL to 50 kbar of Si nanocrystals in an SiO_2 matrix, fabricated by Si ion implantation and subsequent annealing. The Si/SiO$_2$ combination would appear to be mechanically and chemically stable and to avoid any complication produced by H as in the etched material. The PL peak energy shifted to lower energy with pressure at rates of -0.4 and -0.6 meV/kbar for two differently prepared samples, which is less than half the rate expected on the basis of a quantum confinement model. The authors note, correctly, that repetition of their experiments on nanocrystals such as Ge or GaAs ion-implanted into SiO$_2$ would provide a more definitive test, in view of the much larger, positive pressure coefficient of the energy gap.

In Cheong's second experiment, the PL of porous silicon produced in the normal manner was measured to 60 kbar, one sample with He as the pressure medium, a second with the methanol–ethanol medium used by several earlier investigators. In these earlier experiments, a blue shift of the peak energy was often found at low pressures, followed by a red shift at higher pressures. The blue shift was not found with the nonreactive He as pressure medium, but it was with the other, despite the fact that the two films used were from the same Si wafer. This shows that the blue shift is not an intrinsic property of porous silicon. The overall results are said to favor a PL mechanism determined by states at or near the surface of the porous silicon.

Zhao *et al.* [1996] have reported different conclusions. They consider their PL and Raman results consistent with the presence of strained Si quantum dots residing in the etched Si. For two differently prepared samples they find positive shifts of the peak PL energy to about 20 kbar, and negative shifts at higher pressure; this behavior is reversible, and the authors offer a rationalization for it. It may be notable that their pressure medium was alcohol-based, suggesting that the experiments be repeated with an inert medium such as He.

Arai *et al.* [1993] have examined the effect of pressure on the absorption edge, the PL, and the Raman spectra of CdS nanocrystals of 7- to 12-nm diameter, formed inside germanate glasses such as GeO$_2$ and Na$_2$GeO$_3$. They found no effects different from those in the bulk CdS.

5. Semiconducting Nitrides

Chapter 6 of Volume 55 deals with the growth of semiconductors at high pressures, especially the 3–5 nitrides. We may expect increased activity in the characterization of these materials, experimentally and theoretically. Indeed, there are already some papers, viz. Perlin *et al.* [1992], Perlin *et al.* [1993], Teisseyre *et al.* [1995], Perlin *et al.* [1996], Teisseyre *et al.* [1996], and Wetzel *et al.* [1996].

VIII. Other Matters

Of necessity, this historical overview has been brief and selective, too much so to do justice to the wide variety of interesting topics now being studied on semiconductors at high pressures. I regret not having been able to discuss many references, but it simply was not possible to make this a compendium of all the published papers. I trust the succeeding chapters will fill in the blank spaces. Let me, however, draw attention to some topics I omitted entirely or discussed only briefly.

I have referred without much elaboration to the extensive use of Raman spectra in establishing phase diagrams. One might add also the use of resonant Raman scattering, where either a tunable laser frequency is altered to match an energy separation, or, as was done by Lyapin *et al.* [1996], pressure is the tuning variable, while the laser frequency stays fixed. This technique has the potential to explore the effect of pressure on parts of the band structure not accessible directly by other optical or electrical measurements. Raman scattering is a tool ideally suited to the use of diamond anvil cells.

I have said nothing about dilute magnetic semiconductors, a topic which is discussed at all the conferences on semiconductors at high pressures, but this topic will be dealt with at length in Professor Kuroda's chapter. Nor did I discuss the growth of semiconductors, such as the nitrides, at high pressures, this subject being the topic of the contribution in Volume 55 by Dr. Grzegory and Professor Porowski. Lattice vibration spectra were not discussed per se, except indirectly in my references to the use of Raman spectra, and of course directly in my report on the advent of the technique of inelastic neutron scattering to our arsenal of investigative tools.

Impurity band conduction, d-band conduction, ionic conduction, and metal–insulator transitions are all topics which have engaged attention in the past, but which do not appear to be at the forefront now.

The subject of the use of high pressure and uniaxial stress in devices must finally be mentioned. Fortunately, Professor Adams and his collaborators will contribute Chapter 5 of Volume 55 on this topic. Diode lasers and detectors tunable by high pressure should continue to increase in usefulness. Their marriage with small DACs is, like the use of synchrotron radiation, clearly a successful union.

References

Akahama, Y., Kobayashi, M., and Kawamura, H. (1992). Pressure-induced superconductivity and phase transition in Se and Te. *Solid State Commun.* **84,** 803–806.

Akahama, Y., Kobayashi, M., and Kawamura, H. (1993a). Pressure-induced structural phase transition in sulfur at 83 GPa. *Phys. Rev.* **B48**, 6862–6864.

Akahama, Y., Kobayashi, M., and Kawamura, H. (1993b). Structural phase transitions in Se to 150 GPa. *Jpn. J. Appl. Phys.* **32**, Supp. 32-1, 22–25.

Altarelli, M., and Iadonisi, G. (1971). Donor ground states of group 4 and 3–5 semiconductors. *Nuovo Cimento* **B5**, 21–35.

Amemiya, Y., Matsushita, T., Nakagawa, A., Satow, Y., Miyahara, J., and Chikawa, J. (1998). Design and performance of an imaging plate system for X-ray diffraction study. *Nuclear Instr. and Meth.* **A266**, 645–653.

Arai, T., Inokuma, T., Makino, T., and Onari, S. (1993). Pressure effects on CdS microcrystals embedded in germanate glasses. *Jpn. J. Appl. Phys.* **32**, Supp. 32-1, 297–299.

Averkin, A. A., Ilisavsky, U. V., and Regel, A. R. (1962). The effect of elastic strain on the electric properties of PbTe, PbSe, $Bi_2 Te_3$ and $Bi_2 Se_3$. In "Proc. 6th Int. Conf. Phys. Semicond., Exeter," pp. 690–695. Inst. Physics and Physical Society, London.

Babb, S. E., Jr., and Robertson, W. W. (1963). High pressure spectroscopy of solids. In "High Pressure Physics and Chemistry" (R. S. Bradley, ed.), Vol. 1, pp. 375–409. Academic, London.

Baj, M., and Suski, T. (1990). Metastable defect states in GaAs and AlGaAs. In "Proc. 4th Int. Conf. on High Pressure in Semiconductor Physics, Porto Carras," pp. 10–17. Aristotle Univ., Thessaloniki.

Baj, M., Dmowski, L. H., and Slupinski, T. (1993). Direct proof of two-electron occupation of the Ge-DX centers in GaAs codoped with Ge and Te. *Phys. Rev. Lett.* **71**, 3529–3532.

Baranowski, J. M., and Grynberg, M. (1992). Impurities in semiconductors: experimental. In "Basic Properties of Semiconductors" (P. T. Landsberg, ed.), pp. 161–196, Vol. 1 of "Handbook on Semiconductors" (T. S. Moss, ed.). North-Holland, Amsterdam.

Bardeen, J. (1949). Pressure change of resistance of Te. *Phys. Rev.* **75**, 1777–1778.

Bassani, F., Iadonisi, G., and Preziosi, B. (1969). Band structure and impurity states. *Phys. Rev.* **186**, 735–746.

Bastard, G. (1988). "Wave Mechanics Applied to Semiconductor Heterostructures." Editions du Physique, Paris.

Bastard, G., Brum, J. A., and Ferreira, R. P. (1991). Electronic states in semiconductor heterostructures. *Solid State Phys.* **44**, 229–415.

Bate, R. T. (1962). Evidence for a selenium donor level above the principal conduction band edge in GaSb. *J. Appl. Phys.* **33**, 26–28.

Besson, J. M. (1993). High pressure structural transitions in semiconductors. *Jpn. J. Appl. Phys.* **32**, Supp. 32-1, 11–15.

Besson, J. M., Butler, J. F., Calawa, A. R., Paul, W., and Rediker, R. H. (1965). Pressure-tuned PbSe diode laser. *Appl. Phys. Lett.* **7**, 206–208.

Besson, J. M., Paul, W., and Calawa, A. R. (1968). Tuning of PbSe lasers by hydrostatic pressure from 8 to 22 microns. *Phys. Rev.* **173**, 699–713.

Besson, J. M., Grima, P., Gauthier, M., Itié, J. P., Mézouar, M., Häuserman, D., and Hanfland, M. (1996). Pretransitional behavior in zincblende HgTe under high pressure and temperature. *Phys. Stat. Solidi (b)* **198**, 419–425.

Bosc, F., Sicart, J., and Robert, J. L. (1996). High temperature electrical properties of Si-planar doped GaAs–AlAs superlattices under hydrostatic pressure. In "High Pressure Science and Technology" (W. A. Trzeciakowski, ed.), pp. 600–663. World Scientific, Singapore.

Bridgman, P. W. (1921). Electrical resistance under pressure, including certain liquid metals: black phosphorus. *Proc. Amer. Acad. Arts Sci.* **56**, 126–131.

Bridgman, P. W. (1935). Electrical resistance under pressure, with special reference to intermetallic compounds. *Proc. Amer. Acad. Arts Sci.* **70,** 285–317.
Bridgman, P. W. (1938). Resistance of nineteen metals to 30,000 kg/cm^2. *Proc. Amer. Acad. Arts Sci.* **72,** 200–204.
Bridgman, P. W. (1949). "The Physics of High Pressure." G. Bell and Sons.
Bridgman, P. W. (1951). The effect of pressure on the electrical resistance of certain semiconductors. *Proc. Amer. Acad. Arts Sci.* **79,** 127–148.
Bridgman, P. W. (1952). The resistance of 72 elements, alloys and compounds to 100,000 kg/cm^2: Si and Ge. *Proc. Amer. Acad. Arts Sci.* **81,** 219–221.
Bridgman, P. W. (1953). Further measurements of the effect of pressure on the electrical resistance of germanium. *Proc. Amer. Acad. Arts Sci.* **82,** 71–82.
Brillson, L. J. (1992). Surfaces and interfaces: atomic-scale structure, band bending and band offsets. *In* "Basic Properties of Semiconductors" (P.T. Landsberg, ed.), pp. 281–417, Vol. 1 of "Handbook on Semiconductors" (T. S. Moss, ed.). North-Holland, Amsterdam.
Brus, L. (1994). Luminescence of silicon materials: chains, sheets, nanocrystals, nanowires, microcrystals and porous silicon. *J. Phys. Chem.* **98,** 3575–3581.
Burkey, B. C., Khosla, R. P., Fischer, J. R., and Losee, D. L. (1976). Persistent photoconductivity in donor-doped $Cd_{1-x}Zn_xTe$. *J. Appl. Phys.* **47,** 1095–1102.
Burnett, J. H., Cheong, H. M., Paul, W., Koteles, E. S., and Elman, B. (1993). $\Gamma-X$ mixing in GaAs/Al_xGa_{1-x}As coupled double quantum wells under hydrostatic pressure. *Phys. Rev.* **B47,** 1991–1997.
Camphausen, D. L., Connell, G. A. N., and Paul, W. (1971). Calculation of energy-band pressure coefficients from the dielectric theory of the chemical bond. *Phys. Rev. Lett.* **26,** 184–188.
Chadi, D. J., and Chang, K. J. (1988). Theory of the atomic and electronic structure of DX centers in GaAs and Al_xGa_{1-x}As alloys. *Phys. Rev. Lett.* **61,** 873–876.
Chadi, D. J., and Chang, K. J. (1989). Energetics of DX-center formation in GaAs and Al_xGa_{1-x}As alloys. *Phys. Rev.* **B39,** 10063–10074.
Chen, J., Sites, J. R., Spain, I. L., Hafich, M. J., and Robinson, G. Y. (1991). Band offset of GaAs/$In_{0.48}Ga_{0.52}$P measured under hydrostatic pressure. *Appl. Phys. Lett.* **58,** 744–746.
Chen, R. J., Jiang, Z. X., Weinstein, B. A., and McCombe, B. D. (1996). Unusual Si impurity states, and the electron band mass, in GaAs observed by high pressure FIR magnetospectroscopy. *In* "Proc. 23rd Int. Conf. Physics of Semiconductors" (M. Scheffler and R. Zimmermann, eds.), pp. 2753–2756. World Scientific, Singapore.
Cheong, H. M., Burnett, J. H., Paul, W., Hopkins, P. F., and Gossard, A. C. (1994). Hydrostatic-pressure dependence of band offsets in GaAs/Al_xGa_{1-x}As heterostructures. *Phys. Rev.* **B49,** 10444–10449.
Cheong, H. M., Wickboldt, P., Pang, D., Chen, J. H., and Paul, W. (1995). Effects of hydrostatic pressure on the photoluminescence of porous silicon. *Phys. Rev.* **B52,** 11577–11579.
Cheong, H. M., Paul, W., Withrow, S. P., Zhu, J. G., Budai, J. D., White, C. W., and Hembree, D. M., Jr. (1996). Hydrostatic pressure dependence of the photoluminescence of Si nanocrystals in SiO_2. *Appl. Phys. Lett.* **68,** 87–89.
Cheong, H. M., Paul, W., Flatté, M. E., and Miles, R. H. (1997). Pressure dependence of band offsets in InAs/$Ga_{1-x}In_x$Sb superlattices. *Phys. Rev.* **B55,** 4477–4481.
Clark, S. J., Ackland, G. J., and Akbarzadeh, H. (1995). A theoretical study of pressure effects on selenium. *J. Phys. Chem. Solids* **56,** 329–334.
Contreras, S., Lorenzini, P., Mosser, V., Robert, J. L., and Piotrzkowski, R. (1993). DX centers in AlGaAs:Si under pressure: electron statistics for negative-U and multilevel model. *Jpn. J. Appl. Phys.* **32,** Supp. 32-1, 197–199.
Craford, M. G., Stillman, G. E., Rossi, J.A., and Holonyak, N., Jr. (1968). Effect of Te and

S donor levels on the properties of GaAs$_{1-x}$P$_x$ near the direct–indirect transition. *Phys. Rev.* **168,** 867–882.
Davis, E. A. (1993). Studies of amorphous semiconductors under pressure. *Jpn. J. Appl. Phys.* **32,** Supp. 32-1, 178–184.
Dmochowski, J. E., and Stradling, R. A. (1993). Farinfrared magneto-optics of the shallow-deep impurity transitions using diamond anvil cells. *Jpn. J. Appl. Phys.* **32,** Supp. 32-1, 227–232.
Dmochowski, J. E., Sadlo, M. A., Jakiela, R. S., Stradling, R. A., Prins, A. D., Sly, J. L., Dunstan, D. J., and Singer, K. E. (1996). Shallow and deep states of group 6 donors in GaAs and Al$_x$Ga$_{1-x}$As at high pressures. *In* "High Pressure Science and Technology" (W. A. Trzeciakowski, ed.), pp. 621–623. World Scientific, Singapore.
Dmowski, L., Baj, M., Iller, A., and Porowski, S. (1977a). Low temperature light induced change of the chlorine center configuration in CdTe. *In* "Proc. Int. Conf. High Pressure and Low Temperature Physics, Cleveland, Ohio," pp. 515–521. Plenum Press, New York.
Dmowski, L., Baj, M., and Porowski, S. (1977b). New hydrostatic pressure results on sulfur impurity center in GaSb. *In* "Proc. Int. Conf. High Pressure and Low Temperature Physics, Cleveland, Ohio," pp. 505–514. Plenum Press, New York.
Dmowski, L. H., Zduniak, A., Litwin-Staszewska, E., Contreras, S., Knap, W., and Robert, J. L. (1996). Study of quantum and classical scattering times in pseudomorphic AlGaAs/InGaAs/GaAs by means of pressure. *Phys. Status solidi (b)* **56,** 283–288.
Drickamer, H. G. (1965). The effects of high pressure on the electronic structure of solids. *In* "Solid State Physics" (F. Seitz and D. Turnbull, eds.), Vol. 17, pp. 1–133.
Efros, Al. L., and Efros, A. L. (1982). Interband absorption of light in a semiconductor sphere. *Soviet Phys. Semicond.* **16,** 772–775.
Elliott, S. R., Davis, E. A., and Pitt, G. D. (1977). Effect of high pressure on the electrical properties of amorphous arsenic. *Solid State Commun.* **22,** 481–484.
Erskine, D., Yu, P. Y., and Martinez, G. (1987). Technique for high-pressure electrical conductivity measurements in diamond anvil cells at cryogenic temperatures. *Rev. Sci. Inst.* **58,** 406–411.
Esaki, L., and Tsu, R. (1970). Superlattice and negative differential conductivity in semiconductors. *IBM J. Res. and Dev.* **14,** 61–65.
Foyt, A. G., Halsted, R. E., and Paul, W. (1966). Evidence for impurity states associated with high-energy conduction-band extrema in *n*-CdTe. *Phys. Rev. Lett.* **16,** 55–58.
Fritzsche, H. (1965). Use of pressure in the study of impurity states in semiconductors. *In* "Physics of Solids at High Pressures" (C. T. Tomizuka and R. M. Emrick, eds.), pp. 184–195. Academic Press, New York.
Goñi, A. R., Cantarero, A., Syassen, K., and Cardona, M. (1990). Effect of pressure on the low-temperature exciton absorption in GaAs. *Phys. Rev.* **B41,** 10111–10119.
González, J., Calderón, E., Tinoco, T., Itié, J. P., Polian, A., and Moya, E. (1995). CuGa(S$_x$-Se$_{1-x}$)$_2$ alloys at high pressure: optical absorption and X-ray diffraction studies. *J. Phys. Chem. Solids* **56,** 507–516.
Goutiers, B., Dmowski, L., Ranz, E., Aristone, F., Portal, J. C., and Chand, N. (1993). Investigation of shallow states related to Si-*DX* centers in AlGaAs near the Γ–*X*–*L* crossover. *Jpn. J. Appl. Phys.* **32,** Supp. 32-1, 249–251.
Groves, S. H., and Paul, W. (1963). Band structure of gray tin. *Phys. Rev. Lett.* **11,** 194–197.
Guo, G. Y., Crain, J., and Temmerman, W. M. (1993). On the high pressure phases of InSb. *Jpn. J. Appl. Phys.* **32,** Supp. 32-1, 39–41.
Hall, H. H., Bardeen, J., and Pearson, G. L. (1951). The effects of pressure and temperature on the resistance of *p–n* junctions in germanium. *Phys. Rev.* **84,** 129–132.

Haller, E. E., Hsu, L., and Wolk, J. A. (1996). Far infrared spectroscopy of semiconductors at large hydrostatic pressures. *Phys. Stat. Solidi* (*b*) **198,** 153–165.
Hatton, P. D., McMahon, M. I., Pilz, R. O., Crain, J., and Nelmes, R. J. (1992). Crystal structure refinement at high pressures using angle-dispersive powder diffraction techniques. *High Press. Res.* **9,** 194–204.
Herman, F., Kortum, R., Kuglin, C. D., and Short, R. A. (1966). New studies of the band structure of silicon, germanium and grey tin. *In* "Quantum Theory of Atoms, Molecules and Solids. A tribute to J. C. Slater," p. 381–428. Academic Press, New York.
Holland, M. G., and Paul, W. (1962). High pressure effects on impurity levels in semiconductors. *Phys. Rev.* **128,** 30–38, 43–55.
Holonyak, N., Nuese, C. J., Sirkis, M. D., and Stillman, G. E. (1966). Effect of donor impurities on the direct–indirect transition in $Ga(As_{1-x}P_x)$. *Appl. Phys. Lett.* **8,** 83–85.
Holzapfel, W. B., Krüger, T., Sievers, W., and Vijayakumar, U. (1993). Structural studies on Se and Te with synchrotron radiation to megabar pressures. *Jpn. J. Appl. Phys.* **32,** Supp. 32-1, 16–21.
Hoo, K., and Becker, W. M. (1976). Resonant and bound impurity states in n-Gasb(Se) from pressure dependence of Hall effect and resistivity at 77 K. *Phys. Rev.* **B14,** 5372–5383.
Howard, W. E. (1961). Thesis, Harvard University. Rept. No. HP7, Division of Engineering and Applied Physics, Harvard University.
Huant, S., Mandray, A., Martinez, G., and Etienne, B. (1996). DX centers and D^- centers in Al-rich AlGaAs based quantum wells. *In* "Proc. 23rd Int. Conf. Physics of Semiconductors" (M. Scheffler, and R. Zimmermann, eds.), pp. 2789–2792. World Scientific, Singapore.
Hutson, A. R. Jayaraman, A., Chynoweth, A. G., Coriell, A. S., and Feldman, W. L. (1965). Mechanism of the Gunn effect from a pressure experiment. *Phys. Rev. Lett.* **14,** 639–641.
Hutson, A. R., Jayaraman, A., and Coriell, A. S. (1967). Effects of high pressure, uniaxial stress, and temperature on the electrical resistivity of n-GaAs. *Phys. Rev.* **155,** 786–796.
Imai, M., Tsuji, K., and Yagi, T. (1993). Pressure effect on optical absorption coefficients for amorphous silicon–germanium alloys. *Jpn. J. Appl. Phys.* **32,** Supp. 32-1, 191–193.
Iseler, G. W., Kafalas, J. A., Strauss, A. J., and MacMillan, H. F. (1972). Non-Γ donor levels and kinetics of electron transfer in n-type CdTe. *Solid State Commun.* **10,** 619–622.
Jayaraman, A. (1983). Diamond anvil cell and high-pressure physical investigations. *Rev. Mod. Phys.* **55,** 65–108.
Kaplan, H. (1963). Impurity states associated with subsidiary energy-band minima. *J. Phys. Chem. Solids* **24,** 1593–1599.
Keyes, R. W. (1960). The effects of elastic deformation on the electrical conductivity of solids. *In* "Solid State Physics" (F. Seitz and D. Turnbull, eds.), Vol. 11, pp. 149–221. Academic Press, New York.
Khoklov, D. R., and Volkov, B. A. (1996). Mixed valence, electrical activity and metastable states in doped 4–6 compounds: Theory and experiment. *In* "Proc. 23rd Int. Conf. Physics of Semiconductors" (M. Scheffler and R. Zimmermann, eds.), pp. 2941–2948. World Scientific, Singapore.
Klotz, S., Besson, J. M., Schwoerer-Böhning, M., Nelmes, R., Braden, M., and Pintschovius, L. (1995). Phonon dispersion measurements at high pressures to 7 GPa by inelastic neutron scattering. *Appl. Phys. Lett.* **66,** 1557–1559.
Klotz, S., Besson, J. M., Braden, M., Karch, K., Bechstedt, F., Strauch, D., and Pavone, P. (1996a). Transverse acoustic phonons of Ge up to 9.7 GPa by neutron inelastic scattering. *Phys. Stat. Solidi* (*b*) **198,** 105–113.
Klotz, S., Besson, J. M., Hamel, G., Nelmes, R. J., Loveday, J. S., and Marshall, W. G.

(1996b). High pressure neutron diffraction, using the Paris–Edinburgh cell: experimental possibilities and future prospects. *High Pressure Research* **14**, 249–255.

Knap, W., Zduniak, A., Dmowski, L. H., Contreras, S., and Dyakonov, M. I. (1996). Study of transport, phase, and spin relaxation times of 2D electrons by means of pressure. *Phys. Stat. Solidi (b)* **198**, 267–281.

Koenig, S. H., Nathan, M. I., Paul, W., and Smith, A. C. (1960). Effect of high pressure on some hot electron phenomena in n-type Ge. *Phys. Rev.* **118**, 1217–1221.

Kosicki, B. B., and Paul, W. (1966). Evidence for quasi-localized states associated with high-energy conduction band minima in semiconductors, particularly Se-doped GaSb. *Phys. Rev. Lett.* **17**, 246–249.

Kosicki, B. B., Jayaraman, A., and Paul, W. (1968). Conduction band structure of GaSb from pressure experiments to 50 kbar. *Phys. Rev.* **172**, 764–769.

Kozyrev, P. T., and Nasledov, D. N. (1956). Electrical conductivity of selenium polycrystals as dependent on pressure up to 30,000 atmospheres. *Doklady Akademiia Nauk SSSR* **110**, 207–208.

Lambkin, J. D., Dunstan, D. J., O'Reilly, E. P., and Butler, B. R. (1988). The pressure dependence of the band offsets in a GaInAs/InP multiple quantum well structure. *J. Cryst. Growth* **93**, 323–328.

Lambkin, J. D., Adams, A. R., Dunstan, D. J., Dawson, P., and Foxon, C. T. (1989). Pressure dependence of the valence band discontinuity in GaAs/AlAs and GaAs/Al$_x$Ga$_{1-x}$As quantum-well structures. *Phys. Rev.* **B39**, 5546–5549.

Lang, D. V. (1986). DX centers in 3–5 alloys. In "Deep Centers in Semiconductors" (S. T. Pantelides, ed.), pp. 591–641. Gordon and Breach, New York.

Lang, D. V., and Logan, R. A. (1977). Large-lattice relaxation model for persistent photoconductivity in compound semiconductors. *Phys. Rev. lett.* **39**, 635–639.

Lang, D. V., Logan, R. A., and Jaros, M. (1979). Trapping characteristics and a donor-complex (DX) model for the persistent-photoconductivity trapping center in Te-doped Al$_x$Ga$_{1-x}$As. *Phys. Rev.* **B19**, 1015–1030.

Lannoo, M. (1992). Deep and shallow impurities in semiconductors: Theoretical. In "Basic Properties of Semiconductors" (P. T. Landsberg, ed.), pp. 113–160, Vol. 1 of "Handbook on Semiconductors" (T. S. Moss, ed.). North-Holland, Amsterdam.

Li, M. F., Yu, P. Y., Weber, E. R., and Hansen, W. L. (1987). Photocapacitance study of pressure-induced deep donors in GaAs: Si. *Phys. Rev.* **B36**, 4531–4534.

Li, G. H., Goñi, A. R., Syassen, K., Brandt, O., and Ploog, K. (1994). State mixing in InAs/GaAs quantum dots at the pressure-induced Γ–X crossing. *Phys. Rev.* **B50**, 18420–18425.

Li, G. H., Goñi, A. R., Syassen, K., Brandt, O., and Ploog, K. (1995). High pressure study of Γ–X mixing in InAs/GaAs quantum dots. *J. Phys. Chem. Solids* **56**, 385–388.

Long, D. (1956). Effects of pressure on the electrical properties of semiconductors. *Phys. Rev.* **101**, 1256–1263.

Luo, H., Greene, R. G., and Ruoff, A. L. (1993). β-Po phase of sulfur at 162 GPa: X-ray diffraction study to 212 GPa. *Phys. Rev. Lett.* **71**, 2943–2946.

Lyapin, S. G., Lomsadze, A. V., Trojan, I. A., Klipstein, P. C., Mason, N. J., and Walker, P. J. (1996). Pressure-tuned resonance Raman scattering in InAs/GaSb superlattices. *Phys. Stat. Solidi (b)* **198**, 321–327.

Malloy, K. J., and Khachaturyan, K. (1993). *DX* and related defects in semiconductors. *Semiconductors and Semimetals* **38**, 235–291.

Martinez, G. (1973). Band inversion in Pb$_{1-x}$Sn$_x$Se alloys under hydrostatic pressure. *Phys. Rev.* **B8**, 4678–4707.

Martinez, G. (1980). Optical properties of semiconductors under pressure. In "Optical Properties of Solids" (M. Balkanski, ed.), pp. 181–222, Vol. 2 of "Handbook on Semiconductors" (T. S. Moss, ed.). North-Holland, Amsterdam.

Martinez, G., Chambouleyron, I., Besson, J. M., and Balkanski, M. (1970). Variation sous pression de l'émission radiative et de l'effet photovoltaique de PbSe. In "Les propriétés physiques des solides sous pression," Colloque No. 188 du CNRS, pp. 241–246. Editions du CNRS, Paris.

McCombe, B. D., and Petrou, A. (1994). Optical properties of semiconductor quantum wells and superlattices. In "Optical Properties of Semiconductors" (M. Balkanski, ed.), pp. 285–384, Vol. 2 of "Handbook on Semiconductors" (T. S. Moss, ed.). North-Holland, Amsterdam.

McMahon, M. I., and Nelmes, R. J. (1995). Structural studies of tetrahedrally-coordinated semiconductors at high pressure — new systematics. *J. Phys. Chem. Solids* **56**, 485–490.

McMahon, M. I., and Nelmes, R. J. (1996). New structural systematics in the 2–6, 3–5 and group 4 semiconductors at high pressure. *Phys. Stat. Solidi* (*b*) **198**, 389–402.

McSkimin, H. J., and Andreatch, P. (1963). Elastic moduli of germanium versus hydrostatic pressure at 25°C and $-195.8°C$. *J. Appl. Phys.* **34**, 651–655.

Melz, P. J. (1971). Energy band structure of strained crystals: Pseudopotential calculations for Ge and Si with trial calculations for GaAs and CdTe. *J. Phys. Chem. Solids* **32**, 209–221.

Melz, P. J., and Ortenburger, I. B. (1971). Volume dependence of the spin–orbit splitting in representative semiconductors from high-pressure electroreflectivity measurements and relativistic OPW calculations. *Phys. Rev.* **B3**, 3257–3266.

Meyer, J. R., Hoffman, C. A., Myers, T. H., and Giles, N. C. (1994). HgTe–CdTe superlattices. In "Materials, Properties and Preparation" (S. Mahajan, ed.), pp. 535–593, Vol. 3a of "Handbook on Semiconductors" (T. S. Moss, ed.). North-Holland, Amsterdam.

Mezouar, M., Besson, J. M., Syfosse, G., Itié, J. P., Häusermann, D., and Hanfland, M. (1996). Phase diagram of InSb at high pressures and temperatures. *Phys. Stat. Solidi* (*b*) **198**, 403–410.

Miller, P. H., and Taylor, J. H. (1949). Pressure coefficient of resistance in intrinsic semiconductors. *Phys. Rev.* **76**, 179.

Mizuta, M., Tachikawa, M., Kukimoto, H., and Minomura, S. (1985). Direct evidence for the *DX* center being a substitutional donor in AlGaAs system. *Jpn. J. Appl. Phys.* **24**, L143–146.

Mooney, P. M. (1990). Deep donor levels (*DX* centers) in 3–5 semiconductors. *J. Appl. Phys.* **67**, R1–R26.

Morita, A. (1986). Semiconducting black phosphorus. *Appl. Phys.* **A39**, 227–242.

Morita, A., Shibata, K., and Shindo, K. (1993). A pressure-induced isostructural transition in black P. *Jpn. J. Appl. Phys.* **32**, Supp. 32-1, 6–10.

Moss, T. S. (1992). "Handbook on Semiconductors," Vol. 1. North-Holland, Amsterdam.

Moss, T. S. (1994). "Handbook on Semiconductors," Vols. 2, 3, and 4. North-Holland, Amsterdam.

Nakayama, T., Minami, F., Nagao, S., Inoue, Y., and Gotoh, H. (1993). Pressure dependence of band discontinuity in GaAs/AlInP quantum well structures. *Jpn. J. Appl. Phys.* **32**, Supp. 32-1, 151–153.

Nathan, M. I., and Paul, W. (1962). Pressure dependence of the resistivity of gold-doped silicon. *Phys. Rev.* **128**, 38–42.

Nathan, M. I., Paul, W., and Brooks, H. (1961). Interband scattering in *n*-type germanium. *Phys. Rev.* **124**, 391–407.

Nelmes, R. J., McMahon, M. I., Hatton, P. J., Piltz, R. O., and Crain, J. (1993a). New structural results for the high-pressure phases of InSb. *Jpn. J. Appl. Phys.* **32**, Supp. 32-1, 1–5.

Nelmes, R. J., McMahon, M. I., Hatton, P. D., Crain , J., and Piltz, O. (1993b). Phase transitions in InSb at pressures up to 5 GPa. *Phys. Rev.* **B47,** 35–54.

Nelmes, R. J., McMahon, M. I., Wright, N. G., Allan, D. R., Liu, H., and Loveday, J. S. (1995a). Structural studies of 3–5 and group 4 semiconductors at high pressure. *J. Phys. Chem. Solids* **56,** 539–543.

Nelmes, R. J., McMahon, M. I., Wright, N. G., and Allan, D. R. (1995b). Structural studies of 2–6 semiconductors at high pressure. *J. Phys. Chem. Solids* **56,** 545–549.

Neuringer, L. (1957). Pressure-induced changes in the optical absorption of Ge, Si and Te. *Bull. Am. Phys. Soc.* **2,** 134.

Nishikawa, A., Niizeki, K., and Shindo, K. (1993). Structural phase transitions and equations of state of Se and Te under high pressure. *Jpn. J. Appl. Phys.* **32,** Supp. 32-1, 48–50.

Nishikawa, A., Niizeki, K., Shindo, K., and Ohno, K. (1995). Band structure and structural stability of the high-pressure phases of the group 6b elements. *J. Phys. Chem. Solids* **56,** 551–554.

Nishikawa, A., Niizeki, K., Shindo, K., and Ohno, K. (1996). Structural stability of the group 6b elements under high pressure. *Phys. Stat. Solidi (b)* **198,** 475–480.

Olego, D., and Cardona, M. (1982). Pressure dependence of Raman phonons of Ge and 3C–SiC. *Phys. Rev.* **B25,** 1151–1160.

Patel, D., Hafich, M. J., Robinson, G. Y., and Menoni, C. S. (1993). Direct determination of the band discontinuities in $In_xGa_{1-x}P/In_yAl_{1-y}P$ multiple quantum wells. *Phys. Rev.* **B48,** 18031–18036.

Paul, W. (1961). Band structure of intermetallic semiconductors from pressure experiments. *J. Appl. Phys.* **32,** Supplement, 2082–2095.

Paul, W. (1966). The effect of isotropic and anisotropic stress on the optical properties of solids. *In* "The Optical Properties of Solids" (J. Tauc, ed.), pp. 257–309. Academic Press, New York.

Paul, W. (1968). Impurity levels associated with multi-conduction bands. *In* "Proc. IX Int. Conf. Phys. Semiconductors," pp. 16–26. Nauka Press, Leningrad.

Paul, W., and Brooks, H. (1954). Pressure dependence of the resistivity of germanium. *Phys. Rev.* **94,** 1128–1133.

Paul, W., and Brooks, H. (1963). Effect of pressure on the properties of Ge and Si. *In* "Progress in Semiconductors" (A. F. Gibson and R. E. Burgess, eds.), Vol. 7, pp. 135–238. Wiley and Sons, Inc., New York.

Paul, W., and Pearson, G. L. (1955). Pressure dependence of the resistivity of silicon. *Phys. Rev.* **98,** 1755–1757.

Paul, W., and Warschauer, D. M. (1963). The role of pressure in semiconductor research. *In* "Solids under Pressure" (W. Paul and D. M. Warschauer, eds.), pp. 179–249. McGraw-Hill.

Paul, W., DeMeis, W. M., and Finegold, L. X. (1962). Effect of pressure on the properties of PbS, PbSe and PbTe. *In* "Proc. 6th Int. Conf. Phys. Semic., Exeter," pp. 712–721.

Payne, R. T. (1964). Shift of [111] phonon energies at the Brillouin zone boundary under uniaxial stress in germanium. *Phys. Rev. Lett.* **13,** 53–55.

Perlin, P., Jauberthie-Carillon, C., Itié, J. P., Miguel, A. S., Grzegory, I., and Polian, A. (1992). Raman scattering and X-ray absorption spectroscopy in gallium nitride under high pressure. *Phys. Rev.* **B45,** 83–89.

Perlin, P., Gorczyca, I., Porowski, S., and Suski, T. (1993). 3–5 semiconducting nitrides: physical properties under pressure. *Jpn. J. Appl. Phys.* **32,** Supp. 32-1, 334–339.

Perlin, P., Sosin, T. P., Trzeciakowski, W., and Litwin-Staszewska (1995). The effect of Γ–X mixing on the direct excitonic photoluminescence in GaAs/AlGaAs quantum wells. *J. Phys. Chem. Solids* **56,** 411–414.

Perlin, P., Knap, W., Camassel, J., Polian, A., Chervin, J. C., Suski, T., Grzegory, I., and Porowski, S. (1996). Metal–insulator transition in GaN crystals. *Phys. Stat. Solidi (b)* **198**, 223–233.
Peterson, G. A. (1964). Theory of shallow impurity states for subsidiary valleys. *In* Proc. 7th Int. Conf. Phys. Semicond.," pp. 771–776. Dunod Cie, Paris.
Pfeiffer, G., Prescha, Th., and Weber, J. (1993). Transition metal deep level defects in silicon. *Jpn. J. Appl. Phys.* **32**, Supp. 32-1, 239–241.
Phillips, J. C. (1968a). Covalent bond in crystals: partially ionic bonding. *Phys. Rev.* **168**, 905–911.
Phillips, J. C. (1968b). Dielectric definition of electronegativity. *Phys. Rev. Lett.* **20**, 550–553.
Phillips, J. C. (1969). Dielectric definition of ionicity. *Chem. Phys. Lett.* **3**, 286–288.
Phillips, J. C., and Kleinman, L. (1959). New method for calculating wave functions in crystals and molecules. *Phys. Rev.* **116**, 287–294.
Pintschovius, L. (1994). Performance of a three-axis neutron spectrometer using horizontally and vertically focussing monochromators. *Nuclear Inst. and Methods* **A338**, 136–143.
Pitt, G. D. (1969). Hall effect measurements to 65 kbar in n-type GaSb. *High Temp. High Press.* **1**, 111–118.
Pitt, G. D. (1977). Developments in high pressure physics. *Contemp. Phys.* **18**, 137–164.
Pitt, G. D., and Lees, J. (1970). Semiconductor mobilities at high pressures. *In* "Les propriétés physiques des solides sous pression," Colloque No. 188 du CNRS, pp. 225–234. Editions du CNRS, Paris.
Porowski, S., and Trzeciakowski, W. (1985). The effect of pressure on deep impurity states with large lattice relaxation. *Phys. Stat. Solidi (b)* **128**, 11–22.
Porowski, S. A., Smith, J. E., Jr., McGroddy, J. C., Nathan, M. I., and Paul, W. (1970). Effects of hydrostatic pressure on low and high field transport in n-type InSb. *In* "Les propriétés physiques des solides sous pression," Colloque No. 188 du CNRS, pp. 217–224. Editions du CNRS, Paris.
Porowski, S., Kończykowski, and Chroboczek, M. (1974). On the existence of two non-equivalent lattice positions of donors in n-type InSb. *Phys. Lett.* **48A**, 189–190.
Porowski, S., Kończykowski, M., and Chroboczek, J. (1974). Evidence from transport measurements at high pressures for donor ions occupying non-equivalent lattice positions in InSb. *Phys. Stat. Solidi (a)* **63**, 291–295.
Prakash, V. (1966). Thesis, Harvard University. ONR Tech. Rept. No. 13.
Prins, A. D., Sly, J. L., Meney, A. T., Dunstan, D. J., O'Reilly, E. P., Adams, A. R., and Valster, A. (1995). Direct measurement of band offsets in GaInP/AlGaInP using high pressure. *J. Phys. Chem. Solids* **56**, 423–427.
Robin, J. (1959). *J. Phys. Radium* **20**, 506.
Saravia, L. R., and Brust, D. (1969). Strain-split energy bands in semiconductors: Ge. *Phys. Rev.* **178**, 1240–1243.
Schluter, M., Martinez, G., and Cohen, M. L. (1975). Pressure and temperature dependence of electronic energy levels in PbSe and PbTe. *Phys. Rev.* **B12**, 650–658.
Shockley, W., and Bardeen, J. (1950a). Energy bands and mobilities in monatomic semiconductors. *Phys. Rev.* **77**, 407–408.
Shockley, W., and Bardeen, J. (1950b). Deformation potentials and mobilities in nonpolar crystals. *Phys. Rev.* **80**, 72–80.
Sladek, R. J. (1964). Stress-induced donor deionization in GaAs. *In* "Proc. 7th Int. Conf. Physics of Semiconductors," pp. 545–551. Dunod, Paris.
Slykhouse, T. E., and Drickamer, H. G. (1958). The effect of pressure on the absorption edge of S. *J. Phys. Chem. Solids* **7**, 275.

Stradling, R. A. (1985). The use of hydrostatic pressure and alloying to introduce deep levels in the forbidden gap of InSb, GaAs, and $Ga_{1-x}Al_xAs$. *Festkörperprobleme* **25**, 591–603.
Suchan, H. L., Wiederhorn, S., and Drickamer, H. G. (1955). Effect of pressure on the absorption edges of certain elements. *J. Chem. Phys.* **31**, 355–357.
Suski, T. (1994). Hydrostatic pressure investigations of metastable defect states. *Mater. Sc. Forum* **143/147**, 975–982.
Takarabe, K. (1996). Photoluminescence study of rare-earth doped semiconductors under pressure. *Phys. Stat. Solidi (b)* **198**, 211–222.
Tanaka, K. (1988). Studies of amorphous arsenic under pressure. *Phil. Mag.* **B57**, 473–481.
Teisseyre, H., Perlin, P., Suski, T., Grzegory, I., Jun, J., and Porowski, S. (1995). Photoluminescence in doped GaN bulk crystal. *J. Phys. Chem. Solids* **56**, 353–355.
Teisseyre, H., Kozankiewicz, B., Leszczynski, M., Grzegory, I., Suski, T., Bockowski, M., Porowski, S., Pakula, K., Mensz, P. M., and Bhat, I. B. (1996). Pressure and time-resolved PL studies of Mg-doped and undoped GaN. *Phys. Stat. Solidi (b)* **198**, 235–241.
Tinoco, T., Polian, A., Itié, J. P., Moya, E., and Gonzalez, J. (1995). Equation of state and phase transitions in $AgGaS_2$ and $AgGaSe_2$. *J. Phys. Chem. Solids* **56**, 481–484.
Tsuji, K., Katayama, Y., Koyama, N., and Imai, M. (1993). Amorphization from quenched high-pressure phase at low temperatures and high pressures in semiconductors. *Jpn. J. Appl. Phys.* **32**, Supp. 32-1, 185–187.
Ueno, M., Yoshida, M., Onodera, A., Shimomura, O., and Takemura, K. (1993). Structural phase transition of 3–5 nitrides under high pressure. *Jpn. J. Appl. Phys.* **32**, Supp. 32-1, 42–44.
Ulrich, C., Mroginski, M. A., Goñi, A. R., Cantarero, A., Schwarz, U., Muñoz, V., and Syassen, K. (1996). Vibrational properties of InSe under pressure: Experiment and theory. *Phys. Stat. Solidi (b)* **198**, 121–127.
van Vechten, J. A. (1969a). Quantum dielectric theory of electronegativity in covalent systems I: Electronic dielectric constant. *Phys. Rev.* **182**, 891–905.
van Vechten, J. A. (1969b). Quantum dielectric theory of electronegativity in covalent systems II: Ionization potentials and interband transition energies. *Phys. Rev.* **187**, 1007–1020.
Vast, N., Baroni, S., Zerah, G., Besson, J. M., Polian, A., Grimsditch, M., and Chervin, J. C. (1996). Lattice dynamics of α-B from *ab initio* calculation and Raman scattering. *Phys. Stat. Solidi (b)* **198**, 115–119.
Venkateswaren, V., Chandrasekhar, M., Chandrasekhar, H. R., Vojak, B. A., Chambers, F. A., and Meese, J. M. (1986). High-pressure studies of $GaAs-Ga_{1-x}Al_xAs$ quantum wells of widths 26 to 150 Å. *Phys. Rev.* **B33**, 8416–8423.
Vul, A. Ya., Bir, G. L., and Shmartsev, Yu. V. (1970a). *Fiz. Poluprovodn.* **4**, 2331.
Vul, A. Ya., Golobev, L. V., Shronova, L. V., and Shmartsev, Yu. V. (1970b). *Fiz. Tekh. Poluprovokn.* **4**, 234.
Voisin, P., and Bastard, G. (1992). Coherence in 3–5 semiconductor superlattices. In "Basic Properties of Semiconductors" (P. T. Landsberg, ed.) pp. 817–861, Vol. 1 of "Handbook on Semiconductors" (T. S. Moss, ed.) North-Holland, Amsterdam.
Wasik, D., Baj, M., Przybytek, J., Slupiński, T., and Kudyk, K. (1996). Coexistence of DX and A_1 states in highly doped GaAs:Ge,Te and GaAs:Si,Te. *Phys. Stat. Solidi (b)* **198**, 181–186.
Wasilewski, Z., and Stradling, R. A. (1986). Magneto-optical studies of n-GaAs under high hydrostatic pressure. *Semic. Sci. Tech.* **1**, 264–274.
Weinstein, B. A., Zallen, R., Slade, M. L., and Mikkelson, J. C. (1982). Pressure-optical studies of GeS_2 glasses and crystals: Implications for network topology. *Phys. Rev.* **B25**, 781–792.
Weinstein, B. A., Ritter, T. M., Strachan, D., Li, M., Luo, H., Tamargo, M., and Park, R.

(1996). Competition of deep and shallow impurities in wide-gap 2–6 semiconductors under pressure. *Phys. Stat. Solidi (b)* **198,** 167–180.

Wetzel, C., Chen, A. L., Suski, T., Ager, J. W. III, and Walukiewicz, W. (1996). Si in GaN — on the nature of the background donor. *Phys. Stat. Solidi (b)* **198,** 243–249.

Whitaker, M. F., Dunstan, D. J., Missous, M., and Gonzalez, L. (1996). A general approach to measurement of band offsets of near-GaAs alloys. *Phys. Stat. Solidi (b)* **198,** 349–353.

Wisniewski, P., van der Wel, P., Suski, T., Singleton, J., Skierbiszewski, C., Giling, L. J., Warburton, R., Walker, P. G., Mason, N. J., Nicholas, R. J., and Eremets, M. (1993). Unusual behavior of the DX-centre in GaAs:Ge. *Jpn. J. Appl. Phys.* **32,** Supp. 32-1, 218–220.

Wolford, D. J., Kuech, T. F., Bradley, J.A., Gell, M.A., Ninno, D., and Jaros, M. (1986). Pressure dependence of GaAs/Al$_x$Ga$_{1-x}$As quantum-well bound states: the determination of valence-band offsets. *J. Vac. Sci. Technol.* **B4,** 1043–1050.

Yamaguchi, E., Shiraishi, K., and Ohno, T. (1990). First principle calculation of the DX-centers in GaAs, Al$_x$Ga$_{1-x}$As alloys and AlAs/GaAs superlattices. *In* "Proc. 20th Intl. Conf. Physics of Semiconductors," Vol. 1 (E. M. Anastassakis and J. D. Joannopoulos, eds.), pp. 501–504. World Scientific, Singapore.

Yin, M. T., and Cohen, M. L. (1982). Theory of static structural properties, crystal stability, and phase transformations: Application to Si and Ge. *Phys. Rev.* **B26,** 5668–5687.

Yoshioka, A., and Nagata, K. (1995). Raman spectrum of sulfur under high pressure. *J. Phys. Chem. Solids* **56,** 581–584.

Yu, P. Y., and Li, M. F. (1990). Pressure dependence of deep centers by capacitance spectroscopy inside the diamond anvil cell. *In* "Proc. 4th Intl. Conf. on High Pressure in Semiconductor Physics, Porto Carras," pp. 1–8. Aristotle Univ., Thessaloniki.

Zallen, R., and Paul, W. (1964). Band structure of gallium phosphide from optical experiments at high pressure. *Phys. Rev.* **A134,** 1628–1641.

Zallen, R., and Paul, W. (1967). Effect of pressure on interband reflectivity spectra of germanium and related semiconductors. *Phys. Rev.* **155,** 703–711.

Zeman, J., Zigone, M., Rikken, G. L. J. A., and Martinez, G. (1996). Is light coming out of silicon in porous silicon? *In* "High Pressure Science and Technology" (W. A. Trzeciakowski, ed.) pp. 591–595. World Scientific, Singapore.

Zhang, H. I., and Callaway, J. (1969). Energy-band structure and optical properties of GaSb. *Phys. Rev.* **181,** 1163–1172.

Zhang, S. B., and Chadi, D. J. (1990). Stability of DX centers in Al$_x$Ga$_{1-x}$As alloys. *Phys. Rev.* **B42,** 7174–7177.

Zhao, X.-S., Persans, P. D., and Schroeder, J. (1996). Carrier-induced strain in silicon nanocrystals. *In* "High Pressure Science and Technology" (W. A. Trzeciakowski, ed.), pp. 600–602. World Scientific, Singapore.

CHAPTER 2

Electronic Structure Calculations for Semiconductors under Pressure

N. E. Christensen

INSTITUTE OF PHYSICS AND ASTRONOMY
UNIVERSITY OF AARHUS
AARHUS, DENMARK

I. INTRODUCTION	49
II. BASIC THEORY	50
1. Zeroth Born–Oppenheimer Approximation	51
2. Density-Functional Theory	53
3. Self-Interaction Correction and Optimized Effective Potentials	57
4. Gradient Corrections	58
5. Excitation Energies: "GW"	60
6. "LDA Gap Error" and Gap Deformation Potentials	64
7. Ad Hoc LDA Gap Corrections	66
III. "ASA-LDA," "FROZEN-POTENTIAL METHOD"	69
1. Atomic-Spheres Approximation	71
IV. DEFORMATION POTENTIALS	79
1. Shear Deformation Potentials in Cubic Compound Semiconductors	80
2. Hydrostatic Deformation Potentials	86
V. PRESSURE-INDUCED STRUCTURAL CHANGES	98
1. $B3 \to B1$ Pressure-Induced Transitions	99
2. $B3 \to Imma$	105
3. Transitions in and from Wurtzite Structures	106
4. Pressure-Induced Transformations of Group IV Semiconductors	113
5. d-States	122
VI. PRESSURE DEPENDENCE OF PHONON FREQUENCIES	130
VII. CONCLUDING REMARKS	134
References	136

I. Introduction

Equations of state (EOSs) for condensed matter link temperature, pressure, density, magnetization, and other thermodynamic variables that specify the *state* of the system considered. Theoretical and experimental studies in general of the EOSs are therefore the most important, and — if "general" is taken seriously — the only goal of condensed matter physics. Knowing in detail the EOS of a material, we also know elementary interactions, forces, phases, and responses of the system to all kinds of external stimuli. Such a complete picture can never be formed for any system, and the physical and chemical researchers must be satisfied with exploring limited sets of possible thermodynamic variables and limited ranges of their values. Temperature, T, and pressure, P, belong to the most important variables used to describe the state. The intensive variable T forms together with the extensive variable S, the entropy, an "energy couple," and P (intensive) has as its partner the extensive variable V, the volume.

Microscopic models developed in theoretical condensed matter physics have now reached a level where it is possible, from parameter-free quantum mechanical calculations, to predict some of the interrelations between the macroscopic state variables. Total energies, forces, vibrational frequencies, magnetic moments, refractive indices, and so forth, can be calculated with a precision that is sufficient for predictions, and calculation of such properties are important in the analyses of experimental data.

The purpose of the present article is to present theoretical calculations showing the effects of applying external pressure to semiconducting materials. The presentation contains a description of the theory, including basic concepts (Sections II and III), and application of the theory to some selected examples: deformation potentials (Section IV), pressure-induced structural changes (Section V) and phonons (Section VI). The selection of examples is mainly related to the research of the author and does not attempt even a fairly complete presentation of results obtained by the many research groups in this field. Although the list of references is long, it is often necessary to seek more detailed information if specific subjects are studied.

II. Basic Theory

This section presents a review of the background for *ab initio* electronic structure calculations. The purpose is mainly to define the basic theoretical concepts, "density-functional theory," LDA, GGA, "LDA gap errors," GW, and so on, and to provide the reader who needs to go into more detail with a list of key references.

First (Subsection II.1) we give one of several reasons for considering the electronic structure at all. Within the Born–Oppenheimer approximation we can determine the structure of matter from the knowledge of the electronic energy as a function of nuclear coordinates. Subsection II.2 gives the important features of the density-functional theory, and it defines the crude, but useful, local approximation, LDA. Subsections II.3, II.4, and II.5 mention corrections to LDA: the self-interaction correction (SIC), gradient corrections and generalized gradient approximations (GGAs), and the calculations of quasiparticle spectra by, for example, the GW method. The difference between the quasiparticle energies, which give excitation energies, and the formal LDA eigenvalues is often (somewhat misleadingly) referred to as the "LDA gap error," and we discuss the reason for the LDA gaps being too *small*. Subsection II.6 comments on this "error," and ad hoc "corrections" are described in Subsection II.7.

Some references are given in the subsections, but in addition the reader is referred to the excellent reviews of density functional theory in Refs. [1–3].

1. ZEROTH BORN-OPPENHEIMER APPROXIMATION

A quantum theory of solids should be able to predict the *state* of the system, that is, it should predict structural, cohesive, and magnetic properties of the solid. The system (liquid, molecule, or crystal) contains many nuclei and many (N) electrons. Its Hamiltonian may be written as (using Ry atomic units, $e^2 = 2$, $m = \frac{1}{2}$, $\hbar = 1$)

$$H = H_e - \sum_\alpha \frac{m}{M_\alpha} \nabla_\alpha^2, \tag{II.1}$$

where m is the electronic mass, M_α the mass of ion α, and ∇_α the gradient operator acting on the coordinates of nucleus α. The last term in Eq. (II.1) represents the total nuclear kinetic energy, and

$$H_e \equiv \sum_{i=1}^N \left\{ -\nabla_i^2 - \sum_\alpha \frac{2Z_\alpha}{|\bar{r}_i - \bar{R}_\alpha|} + \frac{1}{2} \sum_{i \neq j} \frac{2}{|\bar{r}_i - \bar{r}_j|} \right\} + \frac{1}{2} \sum_{\alpha \neq \beta} \frac{2Z_\alpha Z_\beta}{|\bar{R}_\alpha - \bar{R}_\beta|}. \tag{II.2}$$

This *electronic Hamiltonian* does contain nuclear parameters such as the atomic numbers, Z_α, and the nuclear position vectors, $\bar{R}_\alpha$.

If we had $m/M_\alpha = 0$ for all α, then $H = H_e$, and solution of

$$H_e \Phi_e = \mathscr{E}_e \Phi_e \tag{II.3}$$

would give the electronic state for a given *structure*, that is, a given set of $\overline{R}_\alpha$-vectors:

$$\mathscr{E}_e = \mathscr{E}_e(\{\overline{R}_\alpha\})$$

$$\Phi_e = \Phi_e(\overline{r}_i s_i; \{\overline{R}_\alpha\}),$$
(II.4)

where $\overline{r}_i s_i$ symbolizes all electronic spatial coordinates and spins.

In the zeroth Born–Oppenheimer approximation one writes the wave function of the total system as a product of an electronic and a nuclear wave function:

$$\Phi(\overline{r}_i s_i, \overline{R}_\alpha) = \Phi_e(\overline{r}_i s_i; \{\overline{R}_\alpha\}) \cdot \Phi_n(\{\overline{R}_\alpha\}).$$
(II.5)

Applying the operator, Eq. (II.1), to this ansatz wave function, we get

$$\begin{aligned}
H\Phi &= \left\{ H_e - \sum_\alpha \frac{m}{M_\alpha} \nabla_\alpha^2 \right\} \Phi_e \Phi_n \\
&= \Phi_e \mathscr{E}_e(\{\overline{R}_\alpha\}) \Phi_n - \sum_\alpha \frac{m}{M_\alpha} \nabla_\alpha^2 \Phi_e \Phi_n \\
&= \Phi_e \left\{ -\sum_\alpha \frac{m}{M_\alpha} \nabla_\alpha^2 + \mathscr{E}_e(\{\overline{R}_\alpha\}) \right\} \Phi_n \\
&\quad - \sum_\alpha \frac{M}{M_\alpha} [\Phi_n \nabla_\alpha^2 \Phi_e + 2\nabla_\alpha \Phi_e \cdot \nabla_\alpha \Phi_n].
\end{aligned}$$
(II.6)

The last terms in Eq. (II.6) can be neglected, provided that

$$\nabla_\alpha \Phi_e \cdot \nabla_\alpha \Phi_n / (\Phi_e \nabla_\alpha^2 \Phi_n)$$

and

$$\Phi_n \nabla_\alpha^2 \Phi_e / (\Phi_e \nabla_\alpha^2 \Phi_n)$$

are small, that is, Φ_e should vary much less rapidly with $\overline{R}_\alpha$ than does Φ_n. The nuclear wave function vanishes essentially if $\overline{R}_\alpha$ is changed by a distance of the order of the amplitude in the zero-point motion. The length scale over which Φ_e changes significantly is of the order of interatomic distances. At "not too high" temperatures the ratio between these two distances is $\approx (m/M)^{1/4}$, which for a typical case is ≈ 0.1. In cases where conditions for neglecting the last terms are considered to be fulfilled, we then find a simple equation of motion (wave equation) for the nuclei.

$$\left\{ -\sum_\alpha \frac{m}{M_\alpha} \nabla_\alpha^2 + \mathcal{E}_e(\{\overline{R}_\alpha\}) \right\} \Phi_n = \mathcal{E} \Phi_n. \tag{II.7}$$

This is a Schrödinger equation with a *potential* which is the *electronic energy* $\mathcal{E}_e(\{\overline{R}_\alpha\})$ obtained by solving Eq. (II.3) for varying sets of $\{\overline{R}_\alpha\}$. This can, for example, be used to determine the equilibrium configuration of the atoms, that is, the structure of the system, and to derive pressure–volume relations (isotherms of the equation of state). Another example, which we shall illustrate later, is the calculation of phonon frequencies using the so-called frozen-phonon approach.

2. Density-Functional Theory

Provided that the approximations leading to Eq. (II.7) can be justified, we "only" need to solve the many-electron problem in order to arrive at an equation of motion for the atoms. A way to solve the electronic Schrödinger equation was suggested by the density-functional theory (DFT) [4], which maps the true many-electron problem onto an effective one-electron theory. According to this theory the ground-state properties of a system are functionals of the electron ground-state density. This includes the total energy, $E[n]$, and this functional satisfies the variational principle; it is minimized by the true ground-state density. Proof of the basic theorem is given in Refs. [4] and [5].

The Hamiltonian, H_e, of the N electrons is rewritten as

$$H = T + V_{ee} + \sum_{i=1}^{N} V_{ext}(\overline{r}_i), \tag{II.8}$$

where the index on H_e is omitted, $T = -\sum_{i=1}^{N} \nabla_i^2$ is the operator for the total electronic kinetic energy, V_{ee} represents the electron–electron interaction,

$$V_{ee} = \frac{1}{2} \sum_{\substack{i,j=1 \\ i \neq j}}^{N} \frac{2}{|\overline{r}_i - \overline{r}_j|}, \tag{II.9}$$

and $V_{ext}(\overline{r}_i)$ gives the "external" potential felt by an electron on the position $\overline{r}_i$. The latter includes the potential from the nuclei. A functional, $F[n]$, is defined as

$$F[n] = \min_{\Phi \to n} \langle \Phi | T + V_{ee} | \Phi \rangle, \tag{II.10}$$

meaning that the expectation value of $T + V_{ee}$ is minimized over the set of many-electron functions, Φ, which produce the density $n(\bar{r})$. The variational principle implies that the ground-state energy, E_0, and the ground-state density, $n_0 = n_0(\bar{r})$, fulfil

$$E_0 \leq E[n] \equiv \int d^3r V_{ext}(\bar{r}) n(\bar{r}) + F[n] \tag{II.11}$$

and

$$E_0 = \int d^3r V_{ext}(\bar{r}) n_0(\bar{r}) + F[n_0]. \tag{II.12}$$

No prescription is given for finding the energy functional, but the task of finding good approximations is simplified by splitting $F[n]$ into three terms as suggested by Kohn and Sham [6]:

$$F[n] = T_0[n] + \iint \frac{n(\bar{r})n(\bar{r}')}{|\bar{r} - \bar{r}'|} d^3r' d^3r + E_{xc}[n], \tag{II.13}$$

where $T_0[n]$ is the kinetic energy in a model system of noninteracting electrons with the density $n(\bar{r})$.

The difference between the true kinetic energy $T[n]$ and T_0 is included in the (unknown) functional $E_{xc}[n]$, the exchange-correlation energy. Equation (II.13) defines this quantity.

The variational principle for $E[n]$ yields:

$$\frac{\delta T_0[n]}{\delta n(\bar{r})} + V_{ext}(\bar{r}) + 2\int \frac{n(\bar{r}')}{|\bar{r} - \bar{r}'|} d^3r' + \frac{\delta E_{xc}[n]}{\delta n(\bar{r})} = \mu, \tag{II.14}$$

where μ is a Lagrange multiplier introduced by the requirement of particle conservation. For the model system (where the effective potential is $V(\bar{r})$), a similar variation implies that

$$\frac{\delta T_0[n]}{\delta n(\bar{r})} + V_{eff}(\bar{r}) = \mu, \tag{II.15}$$

and it follows that the two mathematical problems are identical, provided that

$$V_{eff}(\bar{r}) = V_{ext}(\bar{r}) + 2\int \frac{n(\bar{r}')}{|\bar{r} - \bar{r}'|} d^3r' + \frac{\delta E_{xc}[n(\bar{r})]}{\delta n(\bar{r})}. \tag{II.16}$$

Consequently, the density $n(\bar{r})$ in the ground state of the real system is obtained as

$$n(\bar{r}) = \sum_{i=1}^{N} |\psi_i(\bar{r})|^2, \qquad (\text{II}.17)$$

where the wave functions $\psi_i(\bar{r})$ are found by solving the one-particle equations

$$[-\nabla^2 + V_{\text{eff}}(\bar{r})]\, \psi_i(\bar{r}) = \epsilon_i \psi_i(\bar{r}). \qquad (\text{II}.18)$$

Here, the potential still contains an unknown exchange-correlation potential, and the ϵ_i are Lagrange multipliers that, in principle, are without a clear physical meaning. In particular, the ϵ_i's cannot be interpreted as excitation energies, and thus be used in connection with the interpretation of optical spectroscopy experiments. Surprisingly enough, it has nevertheless been found in several cases that they seem to be close to experimental "one-electron energies," and we shall return to this in the next subsection.

Practical calculations within the density-functional theory must be based on approximations to the functional $E_{\text{xc}}[n]$. However, a crude approximation, which has turned out to be useful and surprisingly accurate, is the so-called local-density approximation (LDA). In the LDA, the exchange-correlation energy is given by

$$E_{\text{xc}}[n] = \int n(\bar{r}) \epsilon_{\text{xc}}(n(\bar{r})) d^3 r, \qquad (\text{II}.19)$$

where $\epsilon_{\text{xc}}(n)$ is the exchange-correlation energy density in a homogeneous electron gas with the particle density n. The use of Eq. (II.19) then implies that we generate a local function of $\bar{r}$ by inserting, into the $\epsilon_{\text{xc}}(n)$ for the homogeneous gas, the density, $n(\bar{r})$, of the inhomogeneous gas at the position $\bar{r}$. The LDA can be extended to take spin-polarization into account: the local spin-density approximation, LSDA.

The exchange-correlation density of the uniform electron gas is obtained by first separating it into an "exchange-only" term, ϵ_x, and a correlation contribution:

$$\epsilon_{\text{xc}}(n_\uparrow, n_\downarrow) = \epsilon_x(n_\uparrow, n_\downarrow) + \epsilon_c(n_\uparrow, n_\downarrow), \qquad (\text{II}.20)$$

where (still using Ry atomic units)

$$\epsilon_x(n_\uparrow, n_\downarrow) = -6 \left(\frac{3}{4\pi}\right)^{1/3} \frac{1}{n} (n_\uparrow^{4/3} + n_\downarrow^{4/3}). \qquad (\text{II}.21)$$

The correlation contribution is obtained from numerical calculations followed by parameterization. Several parameterizations have been given, for example by von Barth and Hedin [7], Gunnarsson and Lundqvist [8], Vosko et al. [9], Perdew and Zunger [10], and Moruzzi et al. [11].

As an example, we here give the Barth–Hedin version where, with $r_s^3 = 3/(4\pi n)$ and $n = n_\uparrow + n_\downarrow$,

$$\epsilon_x(n_\uparrow, n_\downarrow) = \epsilon_x^P(r_s) + (\epsilon_x^F(r_s) - \epsilon_x^P(r_s))f(n_\uparrow, n_\downarrow) \tag{II.22}$$

$$\epsilon_c(n_\uparrow, n_\downarrow) = \epsilon_c^P(r_s) + (\epsilon_c^F(r_s) - \epsilon_c^P(r_s))f(n_\uparrow, n_\downarrow). \tag{II.23}$$

The function $f(n_\uparrow, n_\downarrow)$ is defined as

$$f(n_\uparrow, n_\downarrow) = \frac{1}{2^{4/3}-2}\left[\left(\frac{2n_\uparrow}{n}\right)^{4/3} + \left(\frac{2n_\downarrow}{n}\right)^{4/3} - 2\right]. \tag{II.24}$$

Further:

$$\epsilon_x^P(r_s) = -\frac{3}{2\pi}\left(\frac{9\pi}{4}\right)^{1/3} \cdot \frac{1}{r_s} (Ry), \tag{II.25}$$

and

$$\epsilon_x^F(r_s) = 2^{1/3}\epsilon_x^P(r_s).$$

Defining a function $F(z)$ as

$$F(z) = (1+z^3)\ln\left(1+\frac{1}{z}\right) + \frac{z}{2} - z^2 - \frac{1}{3}, \tag{II.26}$$

the correlation functions in Eqs. (II.22) and (II.23) are

$$\epsilon_c^P(r_s) = -c_P F\left(\frac{r_s}{r_P}\right) \tag{11.27}$$

$$\epsilon_c^F(r_s) = -c_F F\left(\frac{r_s}{r_F}\right). \tag{11.28}$$

The best set of parameters obtained in Ref. [7] is

$$c_P = 0.0504, \quad c_F = 0.0254$$
$$r_P = 30, \quad r_F = 75. \tag{II.29}$$

A large number of our LDA calculations are performed with this parameterization, but we also frequently use the Perdew–Zunger [10] expressions which fit the Monte-Carlo calculations by Ceperley and Alder [12].

3. SELF-INTERACTION CORRECTION AND OPTIMIZED EFFECTIVE POTENTIALS

The Coulomb interaction in the formalism reviewed in Subsection II.2 includes the electrostatic interaction of each electron with itself, the self-interaction. In the exact density-functional formalism this is cancelled by a contribution from the exchange-correlation term. But when the local approximations, LDA and LSDA, are made, this cancellation is no longer perfect. The error due to this is negligible for "usual" band states, that is, states which are spatially very extended (wide bands). But a proper treatment of *localized* states requires that corrections be made. An apparently simple way to do this consists of explicit subtraction of the self-interaction, that is, by replacing the LSDA potential for the orbital, α, by the self-interaction-corrected (SIC) potential [13]:

$$V_\alpha^{SIC}(\bar{r}) = V^{LSDA}(\bar{r}) - 2\int \frac{n_\alpha(\bar{r}')}{|\bar{r}-\bar{r}'|} d^3r' - V_{xc}^{LSDA}(n_\alpha(\bar{r})). \tag{II.30}$$

Note, that the potential now depends on the orbital (α). This makes a practical implementation into a self-consistent scheme for solid-state calculations very complex, but it has been done by Svane and Gunnarsson [14, 15]. The effects of SIC are very clear in, for example, transition-metal monoxides. Standard LDA predicts, for example, NiO to be a metal or a small-gap antiferromagnet if spin polarization is included, whereas it really is an insulator with a 4-eV gap. The SIC calculations predict the insulating state [16, 17], but the gap is somewhat small, 2.54 eV. Svane [18] also applied the SIC scheme to hcp Zn. The results are discussed in Subsection V.5.

Another way of avoiding the self-interactions (apart from performing Hartree–Fock calculations, of course) is to use an effective potential that leads to eigenfunctions, which minimize the Fock operator plus other potentials, such as the Coulomb potential and the Kohn–Sham correlation potential. The use of such optimized effective potentials (OEP) in solid state physics was recently suggested by Krieger et al. [19]. Atomic calculations have used them for quite a while, and the technique is in fact based on the work by Sharp and Horton [20] published in 1953. Bylander and Kleinmann [21] used the method in connection with pseudopotentials to study the electronic and cohesive energies of semiconductors, and Kotani and Akai [22] implemented it in LMTO-ASA and KKR-ASA band structure schemes. It is interesting that the eigenvalues, obtained by the OEP (or "exact exchange potentials") (+ correlation) to semiconductors, exhibit fundamental gaps which are close to experiments [21, 22].

4. Gradient Corrections

Hohenberg and Kohn suggested in their original DFT paper [4] that expansions in the gradient of the electron density be used to obtain useful expressions for the exchange-correlation energy. Early applications were not very succesful, mainly because computational problems prevented the inclusion of high-order terms. A second-order expansion used in the "$X_{\alpha\beta}$" exchange by Herman et al. [23] yielded potentials with divergences when applied to atoms, and it needs the adjustment of two parameters, α and β. The authors of Ref. [23] did, however, suggest the use of a fourth-order gradient expansion for the exchange,

$$E_x^{(4)}[n] = \int d^3r\, n(\bar{r})\epsilon_x^{\text{LDA}}(n(\bar{r}))\{1 + c_2\zeta + c_4[\eta^2 + c_{\eta\zeta}\eta\zeta + c_{\zeta^2}\zeta^2]\}, \quad (\text{II}.31)$$

with

$$\zeta(\bar{r}) = \left[\frac{\nabla n(\bar{r})}{2k_F(\bar{r})n(\bar{r})}\right]^2 \quad (\text{II}.32)$$

$$\eta(\bar{r}) = \frac{\nabla^2 n(\bar{r})}{4k_F^2(\bar{r})n(\bar{r})} \quad (\text{II}.33)$$

$$k_F(\bar{r}) = [3\pi^2 n(\bar{r})]^{1/3}. \quad (\text{II}.34)$$

The exchange potential corresponding to this is of a rather complicated form [24], and sufficient numerical accuracy is difficult to obtain. $E_x^{(4)}$ also has divergences, but they only show up very close to the nuclei and are easily eliminated by using finite-sized nuclei.

More recently, *generalized gradient approximations* (GGAs) have been developed [25–29]. Within the the GGA the exchange-correlation energy is expressed as

$$E_{xc}^{GGA}[n_\uparrow, n_\downarrow] = \int d^3r f(n_\uparrow, n_\downarrow, \nabla n_\uparrow, \nabla n_\downarrow). \qquad (II.35)$$

The approximation is called "generalized" in the sense that density gradients are included in f, not only to a certain order, but in summations of (infinite) series of important ∇n terms. Furthermore, several of the GGAs now also include correlation; hence the index "xc" in Eq. (II.35). These may be called "semi-local" [29] because they include, through the ∇n dependencies, some of the nonlocality, but nevertheless have a form which allows them relatively easy to be implemented as corrections in usual LSDA schemes.

One of the simplest, but nevertheless quite successful GGAs to exchange was suggested only by Becke [26]:

$$E_x = E_x^{LDA} - \beta \sum_\sigma \int n_\sigma^{4/3} \frac{x_\sigma^2(\bar{r})}{1 + 6\beta x_\sigma(\bar{r}) \sinh^{-1} x_\sigma(\bar{r})} d^3r. \qquad (II.36)$$

Here, $\sigma = \uparrow$ or $\downarrow$, and

$$x_\sigma(\bar{r}) \equiv \frac{|\nabla n_\sigma(\bar{r})|}{(n_\sigma(\bar{r}))^{4/3}}. \qquad (II.37)$$

It appears that a single value of β, obtained by least-squares fit to selected atomic data, can be used for many systems [26]. Further, Eq. (II.36) is not very different from the exchange part of the functional GGA-II in Ref. [29].

It is interesting to note that application of GGA to NiO yields a gap which is much larger [30], 1.2 eV, than the LSDA value, 0.2 eV. However, it is still smaller than that obtained by SIC [14] and considerably below the experimental value, around 4 eV.

Total-energy calculations, in particular for molecules [31, 32] and surfaces [33], are often considerably improved by means of the GGA. For the sake of simplicity, the GGA is frequently added to the total energy in a final step, whereas self-consistency has been obtained within the LSDA alone. This may lead to errors in some cases, since the structure obtained by LSDA for some systems (hydrogen-bonded) may be substantially in error. This is, for example, stressed in the work by Barnett and Landman [34].

We shall later discuss further examples in some detail (Section V).

5. EXCITATION ENERGIES: "GW"

We have already (see the discussion of NiO in the preceding subsection) started to compare the DFT eigenvalues, ϵ_i of Eq. (II.18), to experiments, although it was stressed that such a comparison in fact cannot be justified. The DFT yields ground-state properties, and the ϵ_i are Lagrange multipliers ensuring that the one-particle wave functions in the model system of noninteracting electrons are normalized.

In this work we are mainly interested in electronic properties of semiconductors, and for these the single, most important feature is the energy gap. To study this we need to calculate excitation energies. The fundamental gap is the difference between the ionization potential, I, and the affinity, A,

$$E_g = I - A = \{E_0(N-1) - E_0(N)\} - \{E_0(N) - E_0(N+1)\}, \qquad \text{(II.38)}$$

where $E_0(N)$ is the ground-state energy of the N-electron system. This gives [35, 36]

$$E_g = \Delta\epsilon + \Delta, \qquad \text{(II.39)}$$

where $\Delta\epsilon$ is the eigenvalue gap calculated for the N-electron system and Δ is a discontinuity in the exchange-correlation potential; see also Ref. [37].

Borrmann and Fulde [38] calculates the Hartree–Fock (HF) exchange contributions to the valence and conduction bands, and correlation is added using an ansatz [38] for the excited state that is analogous to that used earlier [39] for the ground-state wave function. The model, which is based on the bond orbital approximation, allows relatively simple analytical calculations, and at the same time a transparent physical picture can be obtained. It predicts that at the zone center (Γ), the exchange part of the quasi-particle energy has a discontinuity

$$\Delta_x(\Gamma) \approx K - 4V_{1114}, \qquad \text{(II.40)}$$

where $K = V_{1122}$ is the Coulomb interaction between two electrons in two different hybrid orbitals, 1 and 2, that form a bond, and V_{1114} relates to interactions between bond orbitals in different bonds. In sp^3-bonded materials this seems very plausible, and actual calculations confirm that K is much larger than $4V_{1114}$, and thus $\Delta_x(\Gamma)$ is >0. The exchange term increases the separation between the bonding and the antibonding states. Therefore, the HF gaps are very large in the covalently bonded semiconductors. The correlations were shown (consistent with GW) within the model to reduce the gaps since they give rise to two competing effects: (1) *gain* in polarization energy and (2) *loss* in ground-state correlation energy. The first contribution dominates, and if we examine the effects of adding an extra electron and a hole, we see that the buildup of a polarization cloud causes the ionization energy to decrease while the electron affinity increases. Consequently, a discontinuity across the gap exists, and its sign is opposite that in the exchange contribution. Numerical calculations made so far have always yielded exchange discontinuities larger in magnitude than the jump in correlation contribution. In an extreme tight binding model the exchange contribution to the gap becomes $\Delta_x = K$ (Eq. (II.40)), and following Ref. [40] this Coulomb interaction may be estimated by a classical contribution so that

$$\Delta_x \simeq \frac{e^2}{l_x}, \qquad (II.41)$$

where the characteristic "exchange length," l_x, is somewhat smaller than the bond length, d. The contributions from correlations to the gap correction may be [40] estimated as the classical polarization energy gain in the continuum approximation. Adding the contributions of an extra electron and an extra hole, this gives a correction

$$\Delta_c = -\frac{\epsilon_0 - 1}{\epsilon_0} \frac{e^2}{l_c}, \qquad (II.42)$$

where ϵ_0 is the static dielectric constant and l_c a correlation length. The value of l_c is [40] $l_c \simeq a/2 = 1.16d$, where a is the lattice constant. Equations (II.43) and (II.44) show that $\Delta_x > 0$, $\Delta_c < 0$, and $\Delta_x > -\Delta_c$, as mentioned earlier. For Si, Borrmann and Fulde found [40] $\Delta_x = 6.6$ eV and $\Delta_c = -3.4$ eV. Adding these to the Hartree gap, 1.6 eV, one gets a gap of 3.4 eV, which should correspond to the Γ_{15}–Γ'_{25} gap in Si. The experimental value is 3.4 eV, that is, in agreement with the theoretical estimate. For C and Ge, Γ_{15}–Γ'_{25} gaps equal to 7.4 and 3.2 eV, similar agreement with experiments were obtained. LDA values of these gaps are Si: 2.6 eV, C: 5.6 eV, and Ge: 2.6 eV.

The LDA eigenvalues are calculated from a one-electron equation where the exchange-correlation potential is energy independent. It does not, therefore, reproduce the discontinuities at the gap, and the semiconductor bond gaps are therefore underestimated when deduced from the LDA eigenvalues. The Kohn–Sham eigenvalue spectrum is also obtained (see Eqs. (II.16)–(II.18)), using potentials which are without the explicit energy dependence, suggesting that the "gap problem" is a "DFT gap problem" rather than an "LDA gap error."

Quasiparticle band structures can be obtained by means of Green's-function techniques to solve the equation

$$\{T + V_{\text{ext}}(\bar{r}) + V_{\text{H}}(\bar{r})\}\psi_k(\bar{r}) \quad \text{(II.43)}$$
$$+ \int d^3 r' \sum (\bar{r}, \bar{r}', \epsilon_k)\psi_k(\bar{r}') = \epsilon_k \psi_k(\bar{r}).$$

Here, V_{ext} and V_{H} are the "usual" external and Hartree potentials, whereas Σ is a complicated function, the "self-energy," which can only be evaluated within certain approximations, and even then it requires substantial computational effort. A first approximation is the "GW" approximation [41, 42]:

$$\sum (\bar{r}, \bar{r}', \epsilon) = \int d\omega \frac{ie^{-i\omega 0^+}}{2\pi} G(\bar{r}, \bar{r}', \epsilon - \omega) W(\bar{r}, \bar{r}'; \omega). \quad \text{(II.44)}$$

The name "GW" stems from the presence of G, the Green's function, and W, the screened interaction. In Eq. (II.44), $W(\bar{r}, \bar{r}'; \omega) = e^2 \int d^3r'' \epsilon^{-1}(\bar{r}, \bar{r}''; \omega)/|\bar{r}'' - \bar{r}'|$.

Several GW calculations [43–46] have been performed, and in general good agreement with optical data is obtained. It should be noted, however, that GW calculations so far have been made only as correction to, for example, LDA band structures, that is, they use the LDA wave functions.

Figure 1 compares the LDA [47, 48], GW [46], and experimental [49] bands of Ge, and in Table I specific energy levels are compared. The same table includes results of three GW calculations [46, 44, 43].

The GW calculations further show, in agreement with the calculations by Borrmann and Fulde discussed earlier, that the exchange contribution to the quasiparticle energies varies smoothly, except at the gap where a large discontinuity is found, this being positive in the sense that the empty conduction states are shifted upwards in energy relative to the valence bands. Even if the corresponding LDA contribution is subtracted, an ≈ 4-eV discontinuity remains [46]. The GW correlation contribution exhibits a negative step at the gap, but the magnitude is smaller. Since the LDA correlation energy contribution varies smoothly with ϵ, the GW calculations yield a net eigenvalue upshift of the empty conduction states as compared to the LDA.

The use of plane waves as basis functions is convenient in many respects, but applications of the GW method to large systems, including surfaces, may be more efficient, considering computational speed, if other types of basis functions are used, such as muffin-tin orbitals [50] or Gaussian orbitals as applied to Si and the Si(001)–(2 × 1) surface by Rohlfing et al. [51].

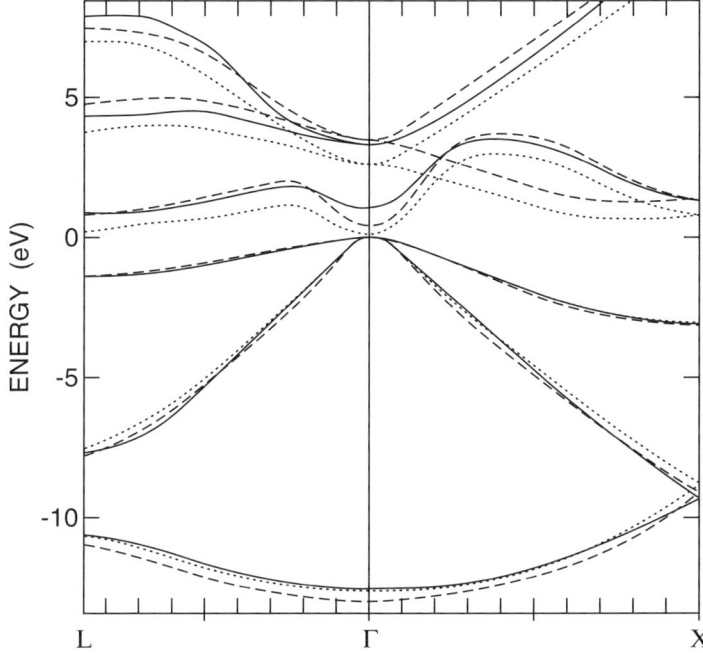

FIG. 1. Band structure of Ge. —Experiment, ----GWA (Ref. [46]), ··· LDA (Refs. [47, 48]). (The LDA calculation used here differs slightly from that referred to in Table I, which is more accurate and yields a negative gap at Γ.)

The gap correction, as calculated for GaN by Palummo *et al.* [53], is 0.88 eV at the experimental equilibrium volume. Adding this to our "best" [52] LDA eigenvalue gaps gives $E_g \simeq 3.0$ eV for cubic and $E_g \simeq 3.2$ eV for hexagonal GaN at zero pressure. These gaps are 0.2–0.4 eV too small as compared to experiments (see Ref. [52]). The GW calculations by Rubio *et al.* [54] yield a gap, 3.5 eV, which agrees even better with experiments. Also, ultra-short-period superlattices and alloys of AlN and GaN were examined by Rubio *et al.* [55] using a plane-wave basis.

Bechstedt and del Sole [56] introduced an approximate, analytic method of deducing the GW corrections to the gaps in sp^3-bonded zincblende-type semiconductors. For GaN, their model yields [52] a 1.3-eV correction, which is too large. It should be noted that the scheme as applied here uses several approximations, the worst being that the gap correction is derived from the shift of *p*-states alone. This is not correct for GaN, where the conduction band edge is *s*-like. This error is discussed in Ref. [56].

TABLE I

QUASIPARTICLE ENERGIES (eV) IN Ge.

$\vec{k}$	m	LDA $\epsilon_{\vec{k}m}$	GW(a) $\epsilon_{\vec{k}m}$	GW(b) $\epsilon_{\vec{k}m}$	GW(c) $\epsilon_{\vec{k}m}$	Exp $\epsilon_{\vec{k}m}$
	6–8	2.57	3.43	3.43	3.26	3.25(1)
	5	−0.09	0.39	1.11	0.71	1.0(1)
Γ	2–4	0.00	0.00	0.00	0.00	0.00
	1	−12.70	−13.07	−13.06	−12.86	−12.6
	8	7.84	8.26	8.00	7.61	7.8(6)
	6–7	3.73	4.70	4.54	4.33	4.3(2)
	5	0.04	0.61	0.77	0.75	0.8(2)
L	3–4	−1.38	−1.37	−1.42	−1.43	−1.4(3)
	2	−7.56	−7.87	−7.94	−7.82	−7.7(2)
	1	−10.62	−10.95	−10.94	−10.89	−10.6(5)
	7–8	9.46	10.62	10.60		13.8(6)
	5–6	0.70	1.33	1.23	1.23	1.3(2)
X	3–4	−3.05	−3.21	−3.20	−3.22	−3.2(0)
	1–2	−8.84	−9.15	−9.17	−9.13	−9.3(2)

LDA [47, 48] (note: "negative" gap at Γ). GW(a) [46]. GW(b) [44]. GW(c) [43]. Exp, experiments [49].

The gap correction suggested in Ref. [56] for the tetrahedrally bonded semiconductors is, in its crudest approximation,

$$\Delta = \frac{e^2}{\epsilon_0} q_{TF}/(1 + cx), \quad (II.45)$$

where ϵ_0 is the (static) dielectric constant, q_{TF} the Thomas–Fermi screening wave number, c a constant (≈ 7.62), and

$$x = q_{TF} \left[\frac{1}{2}(1 - \alpha_p)r_A + \frac{1}{2}(1 + \alpha_p)r_B \right]. \quad (II.46)$$

The quantities r_A and r_B are half the decay lengths of the Slater-type orbitals, assumed here to be of p-type, for the cation (A) and anion sites, and α_p is a bond polarity as defined in Harrison's tight-binding scheme [57].

6. "LDA GAP ERROR" AND GAP DEFORMATION POTENTIALS

In the preceding subsection it has been mentioned how one (in principle) can calculate quasiparticle band structures which may be compared to optical experiments. Still, the interesting question remains whether the "error" in the LDA eigenvalues is entirely due to the discontinuity in the exchange-correlation potential, or whether part of it is due to the *local* approximation itself. In other words, how should the exact Kohn–Sham eigenvalue gap compare to the correct gap E_g (Eq. (II.38))? This has been discussed by Gunnarsson and Schönhammer [58]. This, and the discussion in Subsection II.3, shows that the "gap error" is not as a rule created by the LD approximation itself. However, eigenvalues obtained by the OEP method (see II.3) seem to yield gaps that are not necessarily smaller than the experimental values. In what follows we will use the customary terminology by referring to the "LDA gap error."

An important quantity in connection with high-pressure studies of semiconductors is the pressure coefficient (dE_g/dP) of the gap or the deformation potential, $a_g = dE_g/d \ln V$, where V is the volume. Experience shows that even if the gaps themselves, when extracted from LDA band structures, may be in error by 50–100% [47], the deformation potentials are nevertheless quite accurately predicted. This has been further supported [59, 60] by quasiparticle calculations. But it cannot be generally assumed that LDA errors do not show up at all in the gap deformation potentials. As an example, consider GaN where we calculate a_g to be -7.4 eV (LDA) [52]. Palummo *et al.* obtained, also within the LDA, a similar value, -6.95 eV. In addition, they included GW corrections which yielded $a_g = -8.96$ eV, that is, the quasiparticle correction to the gap deformation potential is in the case not negligible, at least as calculated in this GW approximation. It is interesting to note that in the case of GaN, the LDA deformation potential agrees [61] with *experiment* (see also Fig. 2).

It would be interesting to examine whether the model proposed by Bechstedt and del Sole [56] could be refined so that volume derivatives of the gap correction could be estimated from analytical expressions. We note that the range parameters, r_A and r_B, in Eq. (II.46) essentially scale with the lattice constant (a). The Thomas–Fermi wave number $q_{TF} \sim 1/\sqrt{r_s}$, that is, it scales as $a^{-1/2}$. The volume dependencies of α_p and ϵ_0 can be obtained from *ab initio* LDA calculations (which is sufficient in this connection) [52, 62, 63], or Harrison's scalings [57] may be used (where appropriate). But, in spite of this, the resulting estimate of the GW correction may be of little use because the approximations made in order to obtain Eq. (II.45) are too crude for this purpose. It is necessary at least to go back to the level where realistic wave functions are used for initial as well as final states [64].

The crude estimates, Eqs. (II.43)–(II.44), would suggest that the correction to the gap deformation potential should be $d \Delta/d \ln V \simeq -\Delta/3$, and with $\Delta \cong 1.2$ eV for GaN, this would give a -0.4 eV correction to the

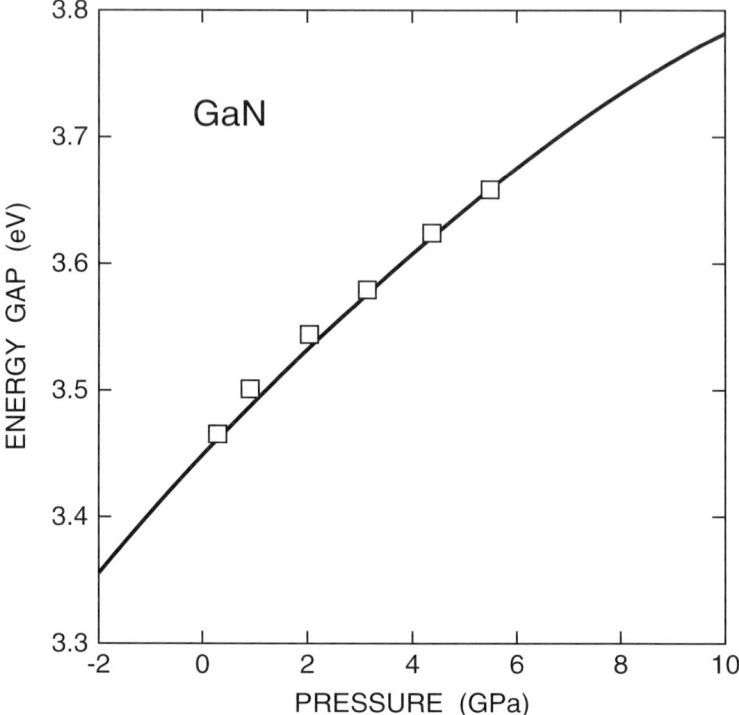

FIG. 2. Pressure dependence of the energy gap in GaN. Squares: Experiment [61]. Solid curve: LDA calculation, rigidly upshifted by 0.82 eV [52].

deformation potential. The sign of the correction is the same as found by the GW calculation (see earlier discussion), but the magnitude is smaller (and in better agreement with experiment).

7. *Ad Hoc* LDA GAP CORRECTIONS

Realizing that several applications of electronic structure calculations for semiconductors depend on the accuracy of the bandgap value, and that full quasiparticle calculations are extremely cumbersome, it is often decided to apply *ad hoc* gap corrections. The simplest approach consists of shifting all unoccupied states upwards on the energy scale so that the minimum gap agrees with experiments. This is often referred to as the application of a "scissors operator" [65, 66]. This procedure leaves the conduction-band effective masses unchanged from their LDA values. For some applications this may lead to serious errors. Even in the lowest conduction band, one usually finds that states at different points of the Brillouin zone may have quite different LDA errors. The *dispersion* of the conduction bands is then not given correctly by the LDA-eigenvalue band structure. As an example we consider the energetic position of the lowest GaAs conduction band at three symmetry points, Γ, X, and L, as measured from the valence band maximum at Γ_4 (Table II). Also (last column of Table II) the effective mass at the conduction-band minimum at Γ is included [64, 67, 47]. First, we see that the apparent gap errors depend on the choice of method to solve the one-electron problem, in particular on whether relativistic corrections [47] are included or not. The "best" calculations yield the results at the bottom of the table, and these are the same time the gaps that differ most from the experiments. A nonrelativistic calculation strongly underestimates the "LDA group errors" in GaAs. Secondly, we note that the errors are different at Γ, X, and L, and that the effective mass, as a consequence of the wrong dispersion, in the "best" LDA calculation is only $\approx \frac{1}{5}$ (!) of the experimental value. A convenient way of examining sensitive details of the band structure of compound semiconductors consists of comparing experimental and theoretical values for the *crossover pressure*, P_0. This is the pressure above which the X_6^c state (conduction-band minimum) has moved below the Γ_6^c state. For GaAs the value of P_0 is 73 kbar [67] when derived from an LDA calculation. This differs substantially from the experimental values, 35 kbar [70], 30 kbar [71], and 37 kbar [72]. In order to "correct" these errors, we first noted [67] that mainly s-like conduction states had

TABLE II

Lowest Conduction Band of GaAs (Energies in eV Relative to the Valence Band Top at Γ) and the Effective Mass at the Conduction-Band Minimum

	$\Gamma(E_0)$	X	L	m_c^*/m_0
Exper. (300 K)	1.42	1.81	1.72	—
Exper. (0 K)	1.52	1.98	1.82	0.066
Non-rel. LDA	1.26	1.37	1.38	0.052
Scalar-rel[a]	0.64	1.35	1.07	0.027
Fully rel.[b]	0.25	1.05	0.67	0.012

[a] The nonrelativistic and the scalar-relativistic (no spin–orbit) calculations were carried out by treating the Ga-3d states as frozen core states.
[b] This calculation includes spin–orbit coupling, *and* the Ga-3d states are treated as self-consistently relaxed band states (two-panel calculation).

to be upshifted, and we therefore applied "artificial Darwin shifts" by adding rather localized external potentials,

$$V_W(r) = V_0 \frac{r_0}{r} e^{-(r/r_0)^2}, \qquad (II.47)$$

to each atomic as well as interstitial site in the zincblende structure. The range parameters, r_0, are chosen to be so small that the potentials are almost δ-function-like, and they therefore act mainly on the states with nonvanishing wave-function amplitudes on the nucleus, that is, s-states. By choosing the parameters in Eq. (II.47) properly, we can reproduce the experimental minimum gaps at Γ, X, and L, which simultaneously yields very good dispersions in general, and causes the effective mass at Γ to agree with experiment. The hydrostatic pressure coefficient, $m_c^{-1} dm_c/dP$, of the conduction-band mass in GaAs as calculated [67] 6.8 Mbar^{-1}, with inclusion of the adjusting potentials agrees extremely well with experiments by Pitt et al. [68], 7.0 Mbar^{-1}, and DeMeis [69], 6.5 Mbar^{-1}. Also, cyclotron resonance experiments [72] give a similar pressure coefficient, 6.1 ± 0.3 Mbar^{-1}. The value of the same quantity calculated without adjustment is 35 Mbar^{-1}, but the large error is mainly due to the error in the value of m_c^* itself, m_c^* (adjusted)/m_c^* (unadjusted) = 0.068/0.012 = 5.6. The absolute pressure coefficient $dm_c^*/dP \simeq 0.46\, m_0$ Mbar^{-1}, is quite well predicted even by the unadjusted band model. The adjusting potentials (II.47) only slightly affect the occupied states. They are included in the full self-consistency cycle. An important property of the adjusting potentials is the fact that they are *transferable*. Once the adjustment has been made for a given compound at equilibrium volume, the potentials can be used in self-consistent calculations at varying volumes without any change in parameter values. Figure 3 illustrates this by showing for GaAs the Γ and X conduction-band edges as functions of (theoretical) pressure. The crossover pressure is now 30.5 kbar, that is, in agreement with experiment. Figure 4 shows that this improvement of the predicted crossover pressure is not restricted to one compound, GaAs. The dashed curves in Fig. 4 show the conduction-band edges at Γ and X in InP as calculated within the LDA, whereas the solid lines (also including the L point) give the band edges versus pressure when the external potentials are added [73]. Experimental values of the crossover pressure range from 8.0 GPa [74] to 10.4 GPa [75], in good agreement with our calculation, 10.4 GPa. This transferability is also essential for the calculation of optical properties of semiconductor superlattices [76].

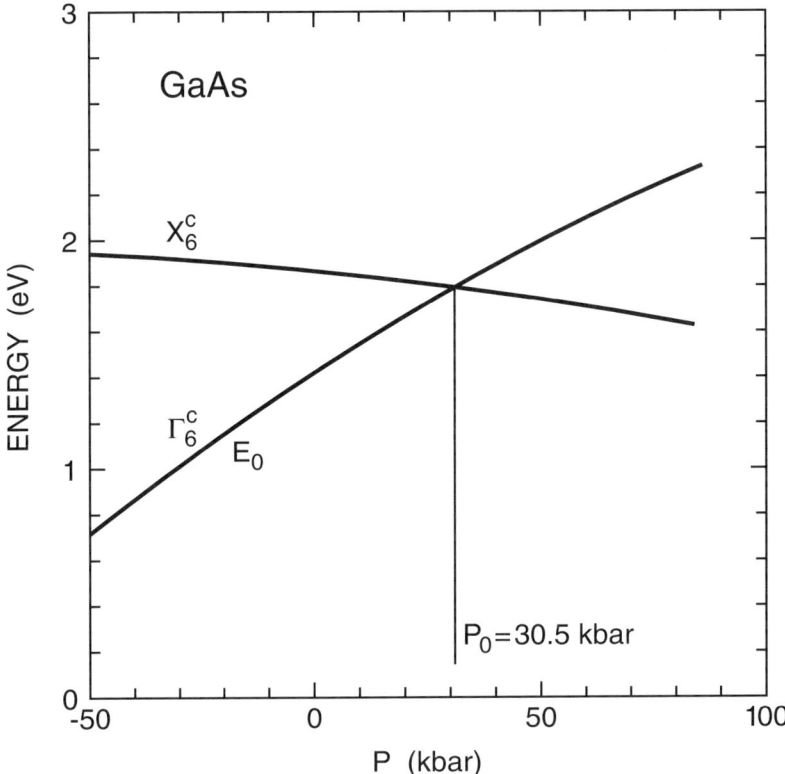

FIG. 3. Conduction-band edges at X and Γ of GaAs as functions of (theoretical) pressure. The gaps are calculated from the adjusted band structure for varying volume (self-consistent at each volume). The parameters V_0 and r_0 in Eq. (II.47) were determined at the experimental equilibrium volume.

III. "ASA-LDA," "Frozen-Potential Method"

This section describes a simple, but physically very transparent, way of studying pressure-induced structural phase transitions. It is called the "frozen-potential method" [77] and is particularly simple when applied within the atomic-spheres approximation (ASA) [78]. The ASA is usually considered as referring to the simplified version of the LMTO (linear muffin-tin orbital) method [78], consisting of making the crystal potential spherically symmetric within atomic spheres. But ASA is more than just that, since it defines, by the use of atomic spheres, a *total-energy functional*, the ASA-LDA functional, which is minimized by the ASA effective potential.

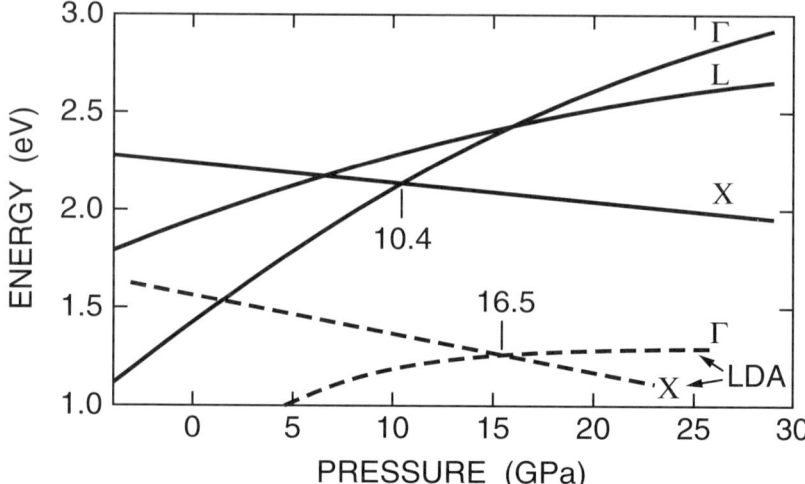

FIG. 4. Pressure dependence of the Γ_{6c}, X_{6c}, and L_{6c} minima of the conduction band of InP. The dashed curves show the X and Γ gaps as calculated in the LDA approximation (i.e., without the external potentials).

The "force theorem" [79] states that the change, δE, in total energy of a system following an infinitesimal displacement of a part of the system (for example, a group of atoms) can be calculated as

$$\delta E = \delta \sum_{occ} \epsilon_i + \delta U_{el}. \tag{III.1}$$

This theorem, which holds to first order in the displacement [80], was derived within the LDA. The first term, $\delta \sum \epsilon_i$, is the change in the sum of one-electron energies for a virtual displacement with the following restrictions. First, the electronic structure is calculated self-consistently for the undistorted system. This calculation yields a sum of one-particle energies, $\sum \epsilon_i(1)$, and a self-consistent effective potential. The part of the system that has to be moved is virtually "cut loose" and displaced to its new position with all its potentials kept fixed as functions of local variables but moved with the parts of the system. For this distorted system with the "frozen potentials," one single new electronic structure calculation is done, that is, there is no iteration toward self-consistency. This second calculation yields $\sum \epsilon_i(2)$, and the first term in Eq. (III.1) is

$$\delta \sum \epsilon_i = \sum \epsilon_i(2) - \sum \epsilon_i(1).$$

2 ELECTRONIC STRUCTURE CALCULATIONS FOR SEMICONDUCTORS

The second term, δU_{el}, is the change in classical electrostatic interaction energy between the piece that has been cut loose and the rest of the system following the restricted distortion. As an application of the force theorem, we may mention the calculation of elastic shear constants [81]. In that case the calculation of the second term was very complicated, since the full nonsphericity of the charge distributions, as well as the shear dependence of the cellular multipole moments, had to be calculated accurately. Within the ASA (see later discussion), the use of the force theorem becomes much simpler. For an elemental solid, the atomic spheres are neutral and the δU_{el} term vanishes. In that case δE is simply the change in the one-electron-energy sum. Although the derivation of Eq. (III.1) assumes the distortions to be small, Skriver [82] assumed that it could be used to predict structural energy differences. The method yielded good results, and the trends over the transition-metal series agree essentially with observations. In the following, we shall show how — within the ASA — the more rigorous frozen-potential method can be applied to calculations of structural energy differences; an analysis of a zincblende → rocksalt structural transformation of a compound semiconductor under pressure illustrates the transparency of the method.

1. ATOMIC-SPHERES APPROXIMATION (ASA)

Equations (II.13) and (II.19) define the LDA total-energy functional. The so-called ASA-LDA functional is defined as

$$E = T_0 + \sum_{\substack{R,R' \\ R \neq R'}} \frac{q_R q_{R'}}{|\overline{R} - \overline{R'}|} + U, \tag{III.2}$$

with

$$T_0 = \sum_{\text{occ}} \epsilon_i - \sum_R \int_R V_R(r) n_R(r) \, d^3r \tag{III.3}$$

and

$$U = \sum_R \int_R n_R(r) \left\{ \int_R \frac{n_R(r')}{|\vec{r} - \vec{r'}|} d^3r' - \frac{2Z_R}{r} + \epsilon_{\text{xc}}(n(r)) \right\} d^3r. \tag{III.4}$$

Here the electron densities, $n_R(r)$, and the potentials, $V_R(r)$, are spherically symmetric within the atom spheres centered at the points $\overline{R}$. The radii of the atomic spheres are chosen such that the total volume enclosed by spheres equals the volume of the crystal. Therefore, the atomic spheres (slightly) overlap. The quantity q_R is the number of electrons in sphere $\overline{R}$ minus the nuclear atomic number, Z_R. Integrals $\int_R$ are performed over the atomic sphere centered at $\overline{R}$. The sum of eigenvalues is obtained by solving the Schrodinger-like one-electron equation, and it is straightforward to prove that the functional Eq. (III.2) attains its minimum for a density generated by the *ASA effective potentials*,

$$V_R(r) = \sum_{\overline{R} \neq \overline{R}'} \frac{2q_{R'}}{|\overline{R} - \overline{R}'|} + \int \frac{2n_R(r')}{|\vec{r} - \vec{r}'|} d^3r' - \frac{2Z_R}{r} + \mu_{xc}(r), \qquad \text{(III.5)}$$

with the local exchange potential μ given by

$$\mu_{xc}(r) = \frac{\delta \epsilon_{xc}(n(r))}{\delta n}. \qquad \text{(III.6)}$$

The important feature of ASA is that the ASA functional satisfies the variational principle so that among the spherically symmetric potentials, those given by Eq. (III.6) minimize the ASA functional when they are used to produce ASA densities by insertion in the one-electron equation.

Now, consider a compound in two structures, two different topological arrangements, I and II, of the same atoms. We wish to calculate, at fixed volume, the energy difference

$$\Delta E = E_{II} - E_I. \qquad \text{(III.7)}$$

We use the same trial ASA potentials for the two structures, namely the $V_R(r)$, which were obtained from a self-consistent calculation with the atoms placed in structure I. Clearly, we here take advantage of the variational property of the ASA functional. Note the simplicity of the ASA: The "crystal potential" is

$$V(\vec{r}) = \sum_{\overline{R}} V_R(|\vec{r} - \overline{R}|) \, \theta_R(|\vec{r} - \overline{R}|), \qquad \text{(III.8)}$$

where

$$\theta(x) = \begin{cases} 1 & \text{for } x < S_R \\ 0 & \text{for } x > S_R \end{cases}. \tag{III.9}$$

We can therefore easily move a potential, $V_R(r)$, rigidly along with the atom originally at position $\overline{R}$ when the structure is altered. This is the "frozen-potential" procedure. With the potentials placed in the new structure, II, we now perform *one* single electronic structure calculation. The electron density in the sphere, R, *resulting* from this calculation is called $n_{IIR}(r)$, whereas $n_{IR}(r)$ refers to the (self-consistent) calculation for structure I. We then have, with $\Delta n_R(r) = n_{IIR}(r) - n_{IR}(r)$ and $\Delta q_R = q_{IIR} - q_{IR}$,

$$\Delta E = \Delta T_0 + \Delta \left\{ \sum_{\substack{\overline{R},\overline{R}' \\ \overline{R} \neq \overline{R}'}} \frac{q_R q_{R'}}{|\overline{R} - \overline{R}'|} \right\} + \Delta U, \tag{III.10}$$

where

$$\Delta T_0 = \Delta \left\{ \sum_{\text{occ}} \epsilon_i \right\} - \sum_R \int_R V_R(r) \Delta n_R(r) \, d^3r \tag{III.11}$$

and

$$\Delta U = \sum_R \int_R \Delta n_R(r) V_R(r) \, d^3r - \sum_{\substack{\overline{R},\overline{R}' \\ \overline{R} \neq \overline{R}'}} \frac{2 q_{IR} \Delta q_R}{|\overline{R} - \overline{R}'|_I}$$
$$+ \sum_R \int_R \int_R \Delta n_R(r) u_R (\overline{r} - \overline{r}') \Delta n_R(r') \, d^3 r d^3 r'. \tag{III.12}$$

Here,

$$u(\overline{r} - \overline{r}') = \frac{1}{|\overline{r} - \overline{r}'|}$$
$$+ \delta(\overline{r} - \overline{r}') \cdot \frac{n_{IIR}(r)\epsilon_{xc}(n_{IIR}(r)) - n_{IR}(r)\epsilon_{xc}(n_{IR}(r)) - \Delta n_R(r)\mu_{xc}(n_{IR}(r))}{[\Delta n_R(r)]^2}.$$
$$\tag{III.13}$$

From Eqs. (III.10)–(III.14), we then have

$$\Delta E = \Delta \left\{ \sum_{\text{occ}} \epsilon_i \right\}$$
$$+ \left[\sum_{\overline{R}} \int_R \int_R \Delta n_R(r) u(\overline{r} - \overline{r}') \Delta n_R(r') \, d^3r \, d^3r' + \sum_{\substack{\overline{R},\overline{R}' \\ \overline{R} \neq \overline{R}'}} \frac{\Delta q_R \Delta q_{R'}}{|\overline{R} - \overline{R}'|_{\text{I}}} \right]$$
$$+ \sum_{\substack{\overline{R}',\overline{R} \\ \overline{R} \neq \overline{R}'}} q_{\text{IIR}} q_{\text{IIR}'} \left\{ \frac{1}{|\overline{R} - \overline{R}'|_{\text{II}}} - \frac{1}{|\overline{R} - \overline{R}'|_{\text{I}}} \right\}. \qquad (\text{III}.14)$$

The indices I, II, on $1/|\overline{R} - \overline{R}'|_{\text{I,II}}$ indicate the structure for which these quantities are to be evaluated. The "frozen-potential" expression for the structural energy difference, Eq. (III.14), thus has these three contributions: (1) the change in one-electron energy sum, (2) the "self-interaction" of Δn (the two terms in the square bracket), and (3) the change in electrostatic energy (Madelung) for fixed values of q. The contribution (2) may also be referred to as a "charge rearrangement" contribution.

In a monoatomic solid, q and Δq are zero and ΔE only differs from the one-particle energy contribution by the first term in the square bracket. This is in most cases small. For Mo, for example, it only amounts to ≈ 1mRy, when we consider the $bcc - fcc$ energy difference. This essentially justifies the calculation [82, 83] of structural energy differences for elemental solids using just the force theorem, Eq. (III.1).

As an example of the use of the frozen-potential method we examine the pressure-induced zincblende → rock-salt transformation of a compound semiconductor, in this case ZnTe (see Fig. 5). First, we see that ΔE is positive for large volumes, that is, the zincblende structure is stable. But for compressed lattices, $S < 2.53$ a.u., the rock-salt structure has the lower energy. Secondly, the charge rearrangement term (2) is very small over the volume range considered. The eigenvalue sum is lower in the B3 than in the B1 structure. This term is strongly influenced by the formation of covalent bonds, as in the sp^3-bonded B3 structure. The bonding states are below the antibonding states, and therefore the $\Delta \sum \epsilon_i$ term will favor the B3 structure. The other term of significant magnitude, $\Delta E_{\text{MAD}}^{\text{STR}}$, is the Madelung contribution, the last term in Eq. (III.14). This favors the B1 structure, a simple consequence of the larger coordination number in B1 than in B3. The magnitude of this ionic Coulomb energy increases as the interatomic distances are reduced, and eventually it exceeds that of the (positive) covalent term, $\Delta \sum \epsilon_i$.

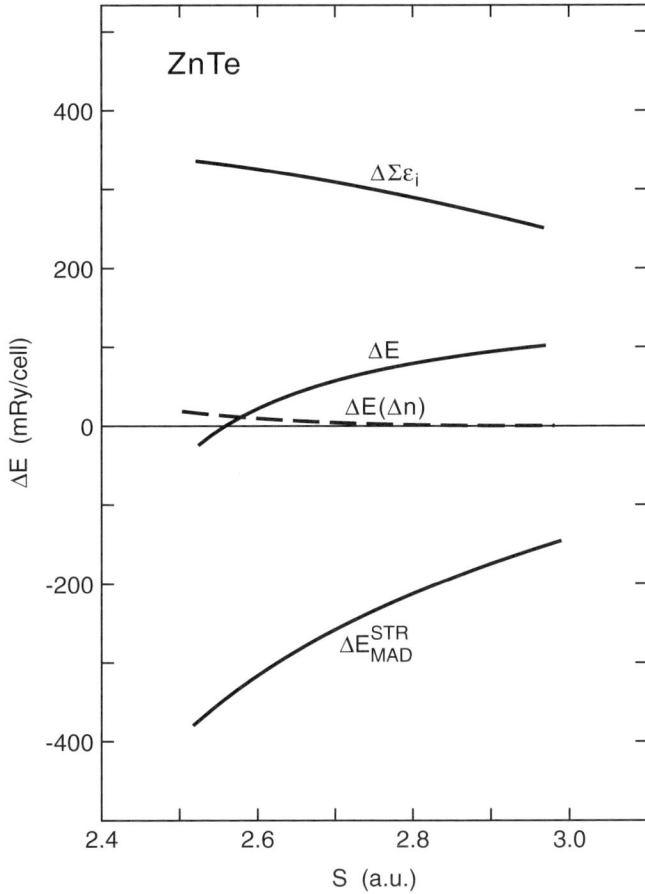

FIG. 5. Difference, ΔE, in Helmholz free energy (at $T = 0$) between the rock-salt (B1) and zincblende (B3) phases of ZnTe as calculated by means of the frozen-potential method (Eq. (III.14)) as a function of S, the (average) atomic-sphere radius ($4 \times 4\pi/3\, S^3 = a^3/4$, a being the lattice constant). The basis contains two atoms plus two "empty spheres."

The frozen-potential method therefore allows us to understand this type of phase transformation in very simple terms, although the full density-functional total-energy calculation involves the calculation of several very complex quantities, each of which may be difficult to give a simple physical meaning.

Theoretical calculations of the pressure may be performed by means of Pettifor's pressure relations [85], which were rederived in an elegant way by means of the force theorem [79]. Alternatively, the pressure can be obtained by numerical volume differentiation of the total energy. But the ASA pressure relations often allow a separation into terms of clear physical meaning, and they are therefore very useful for the analysis of the interactions in solids [79, 84]. Figure 6 shows the pressure as calculated [86] for GaAs as a function of the lattice constant. The agreement with experiment [87] is excellent. The theoretical equilibrium lattice constant is determined as the one for which the theoretical pressure vanishes. We have compared (see Fig. 7) the theoretical (ASA) and experimental data for a large number

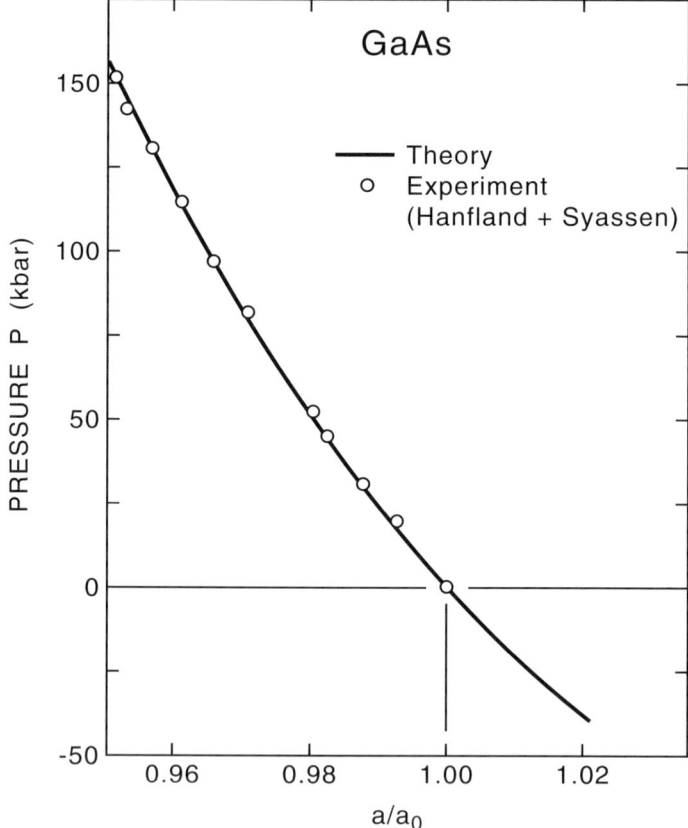

FIG. 6. Pressure P versus lattice constant a for GaAs as obtained from the ASA pressure formula [79], using the LDA and LMTO [78] methods. Experimental data (open circles) are from Ref. [87]. (The experimental lattice constant, a_0, is 6.654 Å.)

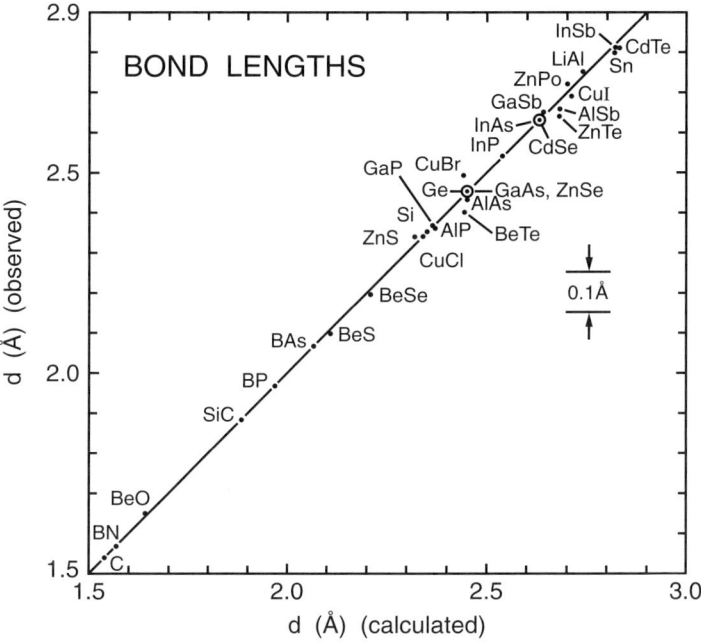

FIG. 7. Observed equilibrium bond lengths plotted against the values obtained from ASA pressures calculated within the LDA. The straight line corresponds to full agreement between theory and experiment.

of compound semiconductors. In general, the calculated values are very close to experiment.

The results represented by Figs. 6 and 7 may seem surprising, since it usually is found that the LDA leads to some overbinding, that is, it would be expected to yield lattice parameters that are somewhat small (by ~1%, typically) when compared to experiment. The absence of this effect here is an artifact due to the ASA. Full-potential (FP) calculations with the LDA often yield smaller equilibrium volumes. As examples we compare (Figs. 8 and 9) pressure–volume relations for AlN, wurtzite as well as rock-salt phases, calculated within ASA and obtained without making the shape approximations, FP-LMTO [91]. Figure 8 compares the shapes of isotherms, that is, the theoretical curves use V_0 as the theoretical equilibrium volume, $V_0^{th}(\text{ASA}) = 0.994 V_0^{exp}$, V_0^{th} (FP-LMTO) $= 0.961 V_0^{exp}$. It is seen that the ASA PV relation also has the better overall agreement with experiment [92]. Similar conclusions can be drawn from Fig. 9. The pressures in these

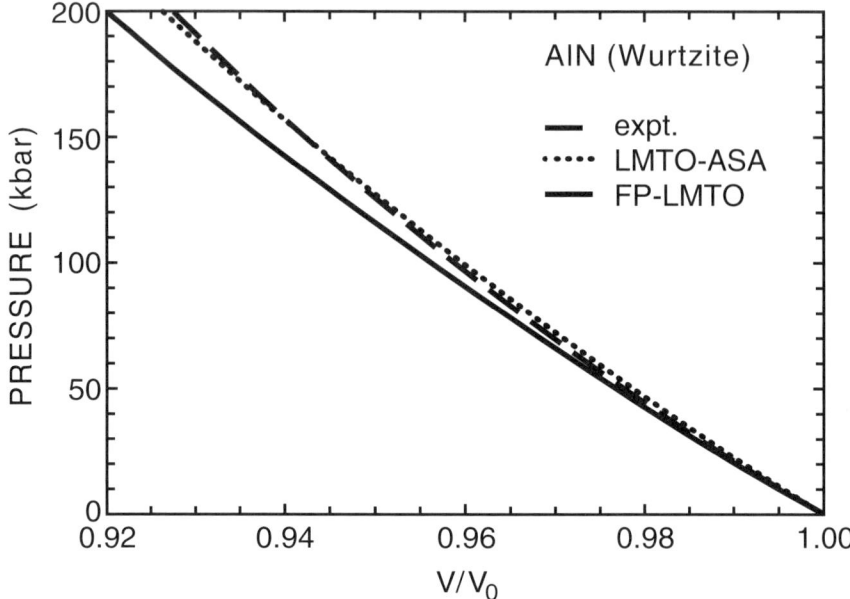

FIG. 8. Wurtzite AlN, see Ref. [91]: Calculated (FP-LMTO, solid line) pressure–volume relation (here V_0 is the theoretical equilibrium volume). Dashed curve: Murnaghan fit to experimental data (Ref. [92]) with B_0 = 2079 kbar, B'_0 = 6.3. Dotted: Theoretical relation obtained with the LMTO-ASA.

ASA as well as FP calculations are obtained by numerical differentiation of the total energy with respect to volume.

We return again to the form of the crystal potential, Eq. (III.8) with Eq. (III.5), in the ASA. The equations show that the potential consists of individual atomic potentials that are spherically symmetric, and where the boundary conditions are chosen so that the sphere containing q_R electrons contribute $2q_R/r$ to the potential (potential energy) outside the sphere. This means that we have defined an "ASA energy scale," and the eigenvalues will be obtained with the energy reference specified by this ASA scale. Therefore, an estimate of, for example, heterojunction band offsets can simply be obtained by comparing the relevant eigenvalues obtained by separate bulk ASA band-structure calculations for the two compounds. These *zero-order* offsets differ from the correct discontinuities because they do not take into account the specific interface effects causing charge redistribution when the two solids are joined [93]. Although these dipole effects may be large, the ASA zero-order discontinuities are still useful as initial values to which physically meaningful corrections may be added. Also, acoustic-phonon deformation potentials may be derived from zero-

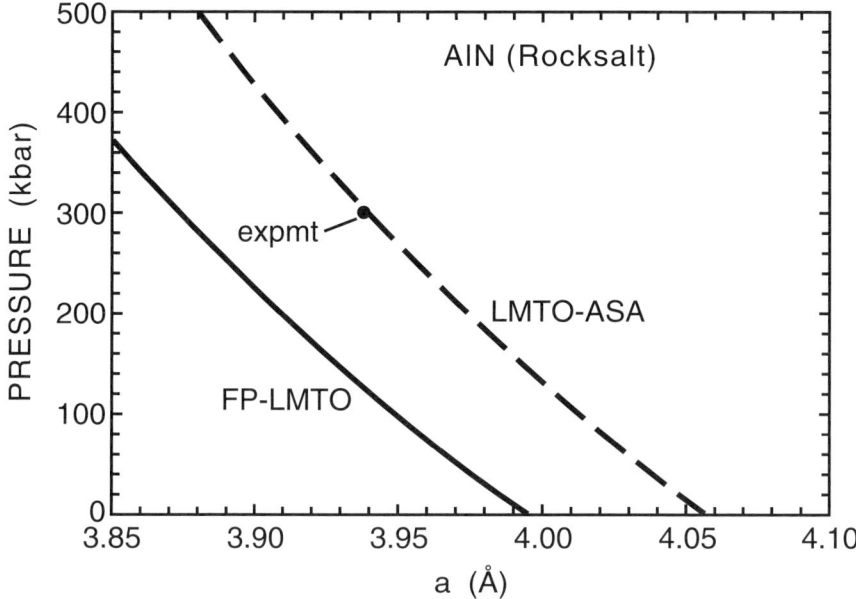

FIG. 9. Rock-salt AlN: Pressure as function of lattice constant. The solid line represents the FP-LMTO calculation ("best LDA"); the dashed curve is the result obtained with LMTO-ASA; and a single experimental point (Ref. [92]) has been marked by a dot.

order "absolute" deformation potentials calculated as volume derivatives of edge eigenvalues given on the ASA energy scale. The bare deformation potentials need, however, to be modified by inclusion of screening effects (see Subsection IV.2).

IV. Deformation Potentials

Deformation potentials describe the response of the band structure of a material to applied (small) pressures. The first-order change in energy of a specific level caused by a uniform strain is related to the coupling between the electrons and long-wavelength acoustic phonons [94]. Changes in the electronic levels caused by nonhydrostatic deformations are also examined; particularly for semiconductors, deformation potentials have played important roles in the analyses of many different phenomena, such as electrical conductivity, Raman and Brillouin scattering, splittings and shifts of excitons and band edges under stress [95], temperature dependence of band edges [98–100], stress dependence of effective masses [101], and electronic properties of strained-layer superlattices [96, 97]. Therefore, determination of the deformation potentials is one of the most important tasks of experimental and theoretical studies of semiconductors under pressure.

1. SHEAR DEFORMATION POTENTIALS IN CUBIC COMPOUND SEMICONDUCTORS

The valence-band maximum (VBM) in cubic, zincblende-type semiconductors is, when spin–orbit coupling is omitted, the triply (not counting Kramers degeneracy) degenerate Γ_{15}^v states. We shall here consider the splitting of this state caused by volume-conserving tetragonal shear ((001) axis) and trigonal ((111) axis) shear.

A Bravais lattice point, which in the undeformed crystal is at $\overline{R}$ moves, when the strain is applied, to $\overline{R}'$, given by

$$\overline{R}' = \overline{\overline{E}} \cdot \overline{R}, \tag{IV.1}$$

where

$$\overline{\overline{E}} = \begin{Bmatrix} e^{-\gamma/2} & 0 & 0 \\ 0 & e^{-\gamma/2} & 0 \\ 0 & 0 & e^{\gamma} \end{Bmatrix} \tag{IV.2}$$

for the tetragonal shear, and

$$E_{ij} = \begin{cases} \dfrac{1}{3}(e^{\gamma} + 2e^{-\gamma/2}) & \text{for } i = j \\ \dfrac{1}{3}(e^{\gamma} - e^{-\gamma/2}) & \text{for } i \neq j \end{cases} \tag{IV.3}$$

in the trigonal case. In the trigonal case, however the atomic positions do not necessarily transform in this way. The two sublattices in the zincblende structure are not fixed relative to each other by symmetry. Assuming that Zn is at (0, 0, 0) and S at $\overline{R}_2 = (1, 1, 1)a/4$ in the undistorted crystal, then S in the ZnS crystal subject to trigonal strain will be at

$$\overline{R}'_2 = [\overline{\overline{E}} - \zeta(\overline{\overline{E}} - \overline{\overline{1}})] \cdot \overline{R}_2, \tag{IV.4}$$

where ζ is the "internal strain parameter" defined by Kleinman [102]. The deformation potentials, b_1 and d_1, are defined for $\gamma \to 0$ through

$$\delta\omega_0 \equiv E_1 - \frac{1}{3}(E_1 + 2E_2) = 3b_1\gamma \text{ (tetragonal)} \qquad \text{(IV.5)}$$

$$\delta\omega_0 \equiv E_1 - \frac{1}{3}(E_1 + 2E_2) = \sqrt{3}d_1\gamma \text{ (trigonal)}. \qquad \text{(IV.6)}$$

Here E_1 and E_2 are the singly and doubly degenerate states into which the shear has split the Γ_{15}^v state. Spin orbit coupling splits the Γ_{15}^v VBM of the undistorted crystal into Γ_7^v (singly degenerate) and Γ_8^v (doubly degenerate). The shears split Γ_8^v into two states separated by $\delta\omega$, and the deformation potentials, b and d, are defined by $\delta\omega = 3b\gamma$ (tetragonal shear) and $\delta\omega = \sqrt{3}d\gamma$ (trigonal). The effect of the spin–orbit coupling is then characterized by b_2 and d_2:

$$b = b_1 + 2b_2; d = d_1 + 2d_2. \qquad \text{(IV.7)}$$

These spin–orbit contributions are small in magnitude. For Si and GaAs we found $b_2 = 0.00$ eV and $d_2 \simeq -0.05$ eV [67, 103]. The trigonal shear deformation potentials depend [104] on ζ:

$$d = d' - \frac{1}{4}d_0\zeta. \qquad \text{(IV.8)}$$

Here, d_0 is the optical-phonon deformation potential. The optical phonon at Γ (Γ_{15} symmetry) displaces the atoms relative to each other by $|u_{\text{rel}}|$ along [111]. This causes a splitting of Γ_8^v by

$$\delta\omega = \frac{|u_{\text{rel}}|}{a}d_0, \qquad \text{(IV.9)}$$

where a is the lattice constant.

The internal-strain parameter can be obtained directly from total-energy calculations. These must, however, be performed without the shape approximations of the ASA, that is, by means of FP-LMTO, FP-LAPW, or *ab initio* norm-conserving pseudopotential methods. The last were used by Nielsen and Martin [105].

Our early calculations [67, 103, 106] used the ASA, and therefore ζ was determined as the value for which d — satisfying Eq. (IV.8) — agrees with the deformation potential determined experimentally. This leads to quite reasonable values of ζ, ≈0.53 for Si and GaAs, but the spread in experimental values for d was also reflected in the accuracy of the ζ values. Further, it became clear [107] that d_0, in particular, is very sensitive to the omission of nonspherical components in the potential. The calculated values of d_0, d', as well as the corrections due to the nonsphericity of the potentials, and the deduced internal strain parameter are listed for 22 compounds in Ref. [107]. The nonspherical corrections are discussed in detail in that reference, and we shall just present the final results and compare them to other calculations and experiments (see Table III and the "chemical trend" plot, Fig. 10). The abcissa values in this figure represent the sp^3-bond order, $b \equiv \mathcal{B} - \mathcal{A}$, the difference between the bonding and the antibonding content in the occupied states. This quantity was calculated from first principles [62] and is a measure of the strength of the bond. The force constant determining the frequency of the $\Gamma(TO)$ phonon mode scales linearly with the bond order [109, 110]. Figure 10 would suggest that the calculated values for d_0 for BeO and SiC are too large. Lambrecht et al. [112] calculated elastic constants and deformation potentials for cubic SiC. They used self-consistent FP-LMTO calculations, and found $d_0 = 24.5$ eV, that is, smaller than our 30.8 eV. Their value is thus in better agreement with the general trend in Fig. 10 than our result for this compound. The same authors also calculated ζ for SiC and found 0.49. In the cases of C, Si, Ge, and GaAs, we see that the d_0 values obtained by full-potential calculations are very close to our "warped ASA" calculations (d_0 in Table III). The experimental data, d_0^{exp}, exhibit substantial scatter in some cases where more experiments were made.

The internal-strain parameter is strongly pressure dependent. This is illustrated by the calculations [120] of the ζ parameters of ZnS and ZnSe as derived by total-energy minimization using the FP-LMTO scheme.

The calculations of the optical-phonon deformation potentials have caused particular problems in the cases of ZnS and ZnSe. The values which we obtained [62, 107] by means of the LMTO-ASA for d_0 were very small, 5.2 eV for ZnS and 7.1 eV for ZnSe. The value for ZnS was in fact the lowest among all the semiconductors examined. Surprisingly enough, these two compounds were also those for which the nonspherical corrections to d_0 were found to be the *largest*, 18.37 eV for ZnS and 14.35 eV for ZnSe. Thus, in these cases the corrections are much larger than the ASA values of d_0 themselves, and the validity of the perturbative approach is definitely questionable. Indeed, as follows from the numbers in the fourth column of Table III, full-potential LMTO calculations [120] give quite different results, 6.6 and 8.2 eV. Presumably, the "warped ASA" calculations in these cases are in error because the nonspherical corrections do not take

TABLE III

OPTICAL-PHONON DEFORMATION POTENTIAL d_0 (eV), d'(eV) (SEE EQ. (IV.8)) AS CALCULATED FROM FIRST PRINCIPLES WITH INCLUSION OF NONSPHERICAL CORRECTIONS TO LMTO-ASA.

Compound	d_0	$d_0^{\text{FP-TB}}$	d_0^{TB}	d_0^{other}	d_0^{expt}	d'	d^{expt}	ζ	ζ^{other}	ζ^{expt}
C	61.3	68.7	106.2	63.1[a]	90, 69[d]	−1.80			0.108[a]	
Si	27.1	36.0	45.8	29.8[b]	40, 27[e,f]	−2.28	−5.3	0.45	0.53[b] 0.53[g]	0.54[h]
Ge	29.3	33.9	42.3	28.9[l]	34, 39[e]	−1.30	−5.0	0.51	0.44[b] 0.50[g]	0.54[h]
AlAs	22.0	30.5	36.2			−2.86				
AlP	22.1	29.5	37.9			−2.16				
AlSb	21.3	29.0	31.8		37	−2.25	−4.3	0.38		
BN	40.0	55.9	95.9			−1.15				
BP	35.8	46.4	65.1			1.80				
BeO	21.4	31.8				1.34				
CdTe	18.8	14.8	19.4		22	−0.51	−4.8	0.91		
GaAs	25.0	31.4	36.3	23.5[b]	48[g], 41[e]	−0.99	−4.5	0.56	0.48[b]	0.76[l]
GaP	24.3	31.1	38.1		44, 47[e]	−0.82	−4.5	0.88		
GaSb	23.4	30.5	32.2		32	−1.17	−4.6	0.99		
InAs	20.8	21.3	29.5		42	−0.57	−3.6	0.58		
InP	20.1	24.8	30.5		35	0.65	−5.0	0.87		
InSb	19.7	25.3	26.9		39	−0.55	−5.0	0.90		
MgS	15.2	16.0				−2.26				
SiC	30.8	41.5	66.5	24.5[c]		−2.39			0.49[c]	
SiGe	28.2	32.7				−2.25				
ZnS	23.5	17.9	28.7	6.6[j]	4[k]	1.79	−3.7	0.93	0.66[j]	
ZnSe	19.2	23.7	27.0	8.2[j]	12[k], 27[e]	0.30	−3.8	0.73	0.65[j]	
ZnTe	13.8	22.2	23.6		23	−1.12	−4.6	1.01		

Tight-binding (TB) results $d_0^{\text{FP-TB}}$ and d_0^{TB} are also given. The former is a TB-scheme using first-principles TB term energies and matrix elements, whereas d^{TB} are sp^3 TB results from Ref. [111]. The results d_0^{other} are all derived from self-consistent full-potential calculations, pseudopotentials, and FP-LMTO. Experimental values d_0^{expt} and d^{expt} are taken from Landolt–Bornstein [113] in the cases where no specific reference is made. The internal-strain parameters ζ are determined, as the values that through Eq. (IV.8) reproduce the experimental values of d, whereas ζ^{other} are obtained by direct total-energy minimizations.

[a] O. H. Nielsen [108].
[b] O. H. Nielsen and R. M. Martin [105].
[c] W. R. L. Lambrecht et al. [112].
[d] C. Canali et al. [114].
[e] P. Lawaetz [115].
[f] C. Jacoboni et al. [116].
[g] S. Wei [117].
[h] C. S. G. Cousins et al. [118].
[i] C. N. Koumelis et al. [119].
[j] R. A. Casali and N. E. Christensen [120].
[k] J. M. Calleja et al. [121].
[l] FP-LMTO (present).

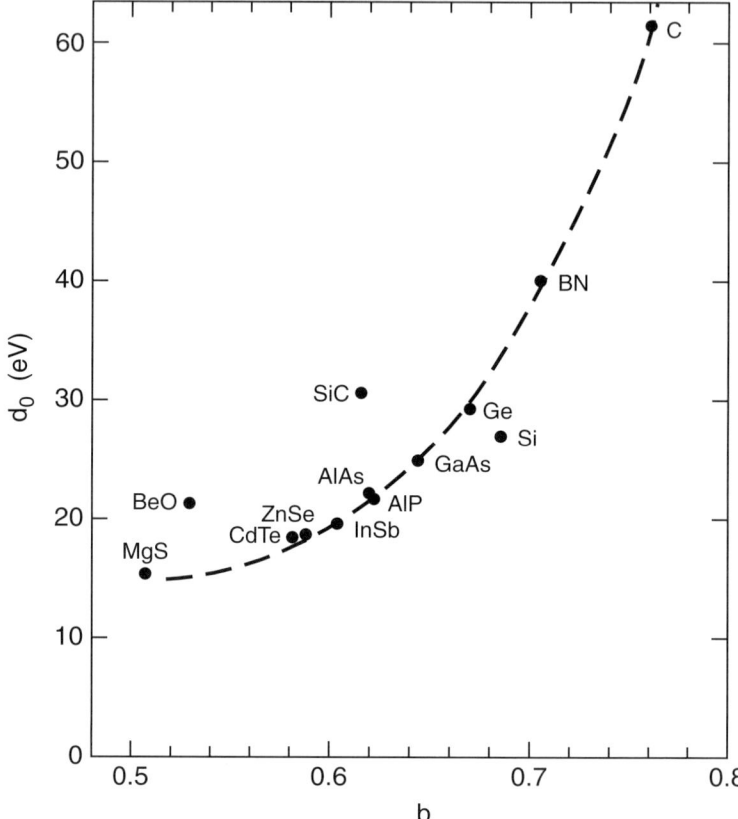

FIG. 10. "Chemical trend" of d_0 illustrated by plotting it against the sp^3 bond order b (see Ref. [62]); $b \equiv \mathscr{B} - \mathscr{A}$, where $\mathscr{B}$ and $\mathscr{A}$ are the bonding and antibonding sp^3 projections of the occupied states integrated over the Brillouin zone. The dashed curve is just a guide to the eye.

the Zn 3d-states properly into account. The FP-LMTO results agree very well with the data extracted from Raman experiments by Calleja et al. [121]. This excellent agreement may be somewhat fortuitous, however, because the determination of d_0 from Raman intensities is difficult and involves crude assumptions of magnitudes of parameters.

The two columns, $d_0^{\text{FP-TB}}$ and d_0^{TB}, represent d_0 values calculated from tight-binding models. They were both calculated from the expression [57, 122]

$$d_0^{\text{TB}} = \frac{32}{3\sqrt{3}} \frac{|H_{xx}V_{xy}|}{\{[(E_{\text{pa}} - E_{\text{pc}})/2]^2 + H_{xx}^2\}^{1/2}}, \quad \text{(IV.10)}$$

with the matrix elements as defined in Ref. [122]. E_{pa} and E_{pc} are the anion and cation p-level energies. The values d_0^{TB} are obtained by Blacha et al. [111] using the "standard" sp^3 tight-binding scheme [57, 122], whereas d_0^{FP-TB} are calculated from (IV.10) with matrix elements and term energies as derived from first-principles calculations [62]. Both sets of TB calculations tend to give larger d_0 values than those obtained from the full calculations (d_0 and d_0^{other} in Table III). Experiments [123] examining femtosecond interband hole scattering in Ge are consistent with a value of d_0 close to 30. This compares favorably to our FP-LMTO result, 28.9 eV.

The direct measurement of the internal strain parameter ζ is difficult since it requires the measurement of the intensity of a reflection, which is forbidden without strain. There are some possibilities, though, for independent control of the value of ζ. Some were referred to in Ref. [67], such as comparison to piezo resistance measurements, which can give information about the shear deformation potential, $\mathscr{E}_2^L$, of the L conduction band edge [124]. In the case of GaAs the experiment yielded $\mathscr{E}_2^L = 19.6 \pm 3$ eV, and by comparing to calculations of $\mathscr{E}_2^L$ as a function of ζ we find that this would correspond to $\zeta \simeq 0.59 \pm 0.25$, that is, a reasonable value, but with a very large error bar.

An interesting study of the changes in the strain in ZnSe/GaAs junctions, when external pressure was applied, was performed by Rockwell et al. [125]. ZnSe and GaAs have at zero pressure different bulk lattice constants, and ZnSe pseudomorphically grown on a GaAs epilayer by molecular-beam epitaxy will be under compressive biaxial strain. However, because of the differences in compressibility of ZnSe and GaAs, a tensile strain is generated when pressure is applied, and this progressively compensates for the lattice mismatch. A certain pressure (experiment Ref. [125]: 36.2 kbar) makes the heterostructure strain-free. The changes in the electronic structure were observed by means of modulation spectroscopy [125]. One of the results of the analysis was that the uniaxial shear deformation potential of ZnSe is clearly pressure dependent; $b = -1.14 \pm 0.03$ eV $+ b' \cdot P$, where the first-order pressure coefficient is $b' = -0.017 \pm 0.002$ eV/kbar.

Using the LCAO result,

$$b = -\frac{\sqrt{3}}{16} d_0, \qquad (IV.11)$$

with d_0 given by Eq. (IV.10), we get for ZnSe $b' \approx -0.005$ eV/kbar, provided that we use Harrison's d^{-2} bond-length (d) scaling law for the matrix elements [57] and the bulk modulus 684 kbar. The sign of b' agrees with that inferred from the experiment, but the magnitude is too small. It is interesting to note that, particularly for II–VI compounds, we find that the d^{-2} scaling is not well supported by *ab initio* calculations (see Subsection IV.2). Nevertheless, FP-LMTO calculations have yielded [120] $b' = -0.004$ eV/kbar for ZnSe, very close to the TB estimate just given. We examined [120] the effects of the pressure dependence of the elastic constants on the analysis of the experiments of Ref. [125] and found that these would reduce the magnitude of b' deduced from these data by only 3 meV/kbar.

A multitude of calculations of shear deformation potentials (at zero pressure) exists, also for other levels than the band edges and for other points in the Brillouin zone. For specific details the reader is referred to reviews, such as Ref. [111], and the original papers.

2. Hydrostatic Deformation Potentials

The deformation potentials originally introduced by Bardeen and Schockley [94] were defined as the shift in energy levels due to small volume changes, $dE/d \ln V$, because these quantities were needed in the theory of electron–phonon interactions. It is now clear that such *absolute* deformation potentials, ADPs, cannot be derived from electronic structure calculations for infinite crystals, because an energy reference level cannot be defined unambiguously for such a system. (Calculations of volume derivatives of energy gaps do not lead to similar difficulties, of course.) It is reasonable to try to calculate ADPs by means of strained "homojunction" geometries [126, 97, 112]. In this a superlattice is formed with layers of the compound semiconductor in strained and unstrained states. For this system, band offsets are derived in the same way that band-edge discontinuities are calculated for heterojunctions [127]. After subtraction of shear contributions, the hydrostatic ADP is obtained. The argument for this procedure is that although the system as such, still being an infinite crystal, does not have a well-defined energy reference, the strained and unstrained portions do have a common reference because they are in contact. Resta *et al.* [128, 129] showed that for given strain direction, determined by the "growth direction" of the superlattice referred to earlier, the ADPs can be defined in terms of bulk properties for homopolar semiconductors.

The calculation of ADPs for real systems, such as finite crystals, would in principle not suffer from the problem just mentioned. But the theoretical calculation of the electronic energy spectrum, including its volume dependence, relative to the vacuum level is a formidable problem, since all effects of the surfaces must be included. Instead of trying to calculate the "genuine" ADPs, it has been suggested that the acoustic-phonon deformation potentials needed to evaluate the interaction between electrons and phonons with wavelengths much shorter than the sample size should be bulk properties.

Vergés *et al.* [130] suggested using the energies defined on the ASA-energy scale (see Section III). These "bare" deformation potentials did not lead to very good electron–phonon coupling constants. The reason was that the effects of screening had been overlooked [131]. This was corrected in the DME (dielectric midgap energy) model [131].

The DME level is similar to the charge neutrality levels introduced by Tejedor, Flores, and Louis [132, 133] and by Tersoff [134, 135] in connection with model theories for band offsets at semiconductor interfaces and Schottky barriers. Within the DME model the screening potential which accompanies the LA phonon is obtained from the "bare" deformation potential, a_D, of the DME level as

$$\Delta a_D = a_D [\epsilon^{-1}(\overline{q}, \omega) - 1], \qquad (IV.12)$$

where $\epsilon(\overline{q}, \omega)$ is the dielectric constant with arguments defined by the phonon mode $(\overline{q}, \omega)$. For the acoustic phonons we chose to take $\omega \approx 0$ in Eq. (IV.12), and since we only considered intraband scattering, the limit, $q \to 0$, was also taken. The dielectric constants inserted in the screening formula were then the static values, $\epsilon(0, 0)$. The band-edge deformation potentials, $\overline{a}_v$ and $\overline{a}_c$, were than calculated from

$$\overline{a}_{v,c} = a_{v,c} + \Delta a_D, \qquad (IV.13)$$

where $a_{v,c}$ are the unscreened deformation potentials of the valence/conduction band edges as calculated directly from the band structures on the ASA energy scale.

The midgap energy, the DME level, was calculated as the level in the middle of the average gap, the Penn gap, and we took this to be limited by the band edges at the first Baldereschi point [136]. The reason for this choice is that $\overline{k}$-space integrations performed to calculate the imaginary parts of the dielectric function with good precision can be replaced by the summation of a few special points. In the extreme limit where only one point is included, this should be the Baldereschi point, which in the *fcc* case is $(0.622, 0.295, 0)2\pi/a$. The gap defined in that way is very close to the spectral position of the E_2-peak in $\epsilon_2(\omega)$, and we therefore call this gap "E_2." Figure 1 of Ref. [137] clearly shows that the gaps at the Baldereschi points are very close to the E_2 peak positions as listed by Phillips [138].

The acoustic-phonon deformation potentials as calculated from the DME model for several compounds are listed in Ref. [131]. Later calculations [97] yielded $\overline{a}_v = -0.70$ eV for ZnSe and -4.10 eV for ZnS. Strained (110) homojunction calculations yield -1.20 eV (ZnSe) and -3.50 eV. Thus, these results, obtained from quite different calculation schemes, are very close. For Si, the DME calculation gave $\bar{a}_v = -1.6$ eV, which is rather far from the homojunction calculations by Lambrecht et al. [112], 2.3 eV (110 strain axis), and 2.2 eV (110) by Resta et al. [128].

The Penn gap, E_2, also enters in the simplest model of the energy (E)-dependent dielectric constant:

$$\epsilon = 1 + \frac{A}{E_2^2 - E^2}, \tag{IV.14}$$

where A is inversely proportional to the volume ($\sqrt{A} \propto \omega_p$, the plasma frequency). It follows that, within this model with $E = 0$, the volume and pressure dependencies of the refractive index, n, will be given by

$$\frac{d \ln n}{d \ln V} = \frac{1 - \epsilon}{2\epsilon} \left\{ 1 + 2 \frac{d \ln E_2}{d \ln V} \right\} \tag{IV.15}$$

and

$$\frac{1}{n} \frac{dn}{dP} = \frac{\epsilon - 1}{2\epsilon} \left\{ K - 2E_2^{-1} \frac{dE_2}{dP} \right\}, \tag{IV.16}$$

where K is the compressibility. Tetrahedrally bonded semiconductors have in general positive gap pressure coefficients, and the second term in the curly bracket in Eq. (IV.16) may exceed K. Consequently, dn/dP becomes negative, that is, the refractive index *decreases* with pressure. If a tight-binding model is used with Harrison's [57] d^{-2} scaling of matrix elements with band length, all semiconductors would have $d \ln E_2/d \ln V = -2/3$, and $1 d \ln n/d \ln V = (\epsilon - 1)/6\epsilon$. Taking, as an example, $\epsilon = 5.4$ to match reasonable experimental values of GaN, we should for this material have $d \ln n/d \ln V = 0.136$. Using the value 200 GPa for the bulk modulus, the pressure coefficient for GaN becomes $\sim -0.07 \times 10^{-2}$ GPa^{-1}. This is too small when compared to the experimental value [61], -0.2×10^{-2} GPa^{-1}. The calculation is off, not only because the model is crude, but also because the d^{-2} scaling is not adequate in general.

Average gaps, E_2, as well as ionic gap components, C, as derived from first-principles LDA calculations are listed in Table IV, together with their volume derivatives for selected semiconductors. The values obtained for $d \ln E_2/d \ln V$ clearly differ from -0.67, Harrison's d^{-2} scaling. The deviations are particularly pronounced for the compounds with cation d-states in, or close to, the valence-band energy regime (for example, the II–VI compounds). Camphausen et al. [139] based their theory on the Phillips–van

TABLE IV

Average Gap, E_2 (eV), $d \ln E_2/d \ln V$, Ionic Gap, C (eV), $d \ln C/d \ln V$, Dielectric Constant ϵ, and $d \ln n/d \ln V$, Where n is the Refractive Index, for Selected Semiconductors

	E_2	$\dfrac{d \ln E_2}{d \ln V}$	C	$\dfrac{d \ln C}{d \ln V}$	ϵ	$\dfrac{d \ln n}{d \ln V}$	f	f_P
C	11.66	−0.81	—		5.7	0.25	0	0
Si	4.73	−0.54	—		12	0.04	0	0
Ge	4.85	−0.73	—		16	0.22	0	0
α-Sn	4.03	−0.88	—		20	0.36	0	0
AlP	5.22	−0.40	3.39	−0.44	8	−0.09	0.182	0.307
AlAs	5.18	−0.47	3.15	−0.26	9.1	−0.03	0.370	0.274
AlSb	4.38	−0.57	1.53	0.46	10.2	0.08	0.122	0.250
GaP	5.29	−0.71	3.18	−0.47	9.1	0.19	0.361	0.327
GaAs	5.12	−0.81	2.85	−0.34	10.9	0.27	0.310	0.310
GaSb	4.31	−0.95	1.42	0.77	14.4	0.42	0.109	0.261
InP	4.94	−1.07	3.63	−0.72	9.6	0.51	0.540	0.421
InAs	4.80	−1.09	3.57	−0.62	12.3	0.54	0.216	0.357
InSb	4.09	−1.17	2.26	−0.21	15.7	0.62	0.305	0.321
ZnSe	5.94	−1.13	5.11	−0.47	5.9	0.53	0.740	0.630
ZnTe	4.98	−1.02	3.73	−0.05	7.3	0.45	0.566	0.609
CdTe	4.59	−1.20	3.95	−0.30	7.2	0.60	0.741	0.717
HgTe	3.92	−1.36	3.24	−0.51	9.3	0.76	0.683	0.65

The last two columns give the calculated ionicity, $f = (C/E_2)^2$ and the Phillips ionicity, f_P.

Vechten theory. They were at that time forced to make assumptions for the volume dependence of the gaps, in particular for those of the ionic components, and for interactions with d-states, that could not be justified by later *ab initio* calculations. Their calculations suggested that $d \ln n/dP$ is negative for (all) III–V and positive for (most) II–VI tetrahedrally bonded component semiconductors. Our model combined with calculated volume coefficients does not predict similar trends (Table IV).

The calculation of Ref. [139] yields for GaN $d \ln n/dP = -0.05 \times 10^{-2}$ GPa^{-1}, i.e., very similar to the LCAO estimate presented earlier, and thus also far too small in magnitude. The value which we derive [52] from the $\omega \to 0$ limit of $\epsilon_1(\omega)$ calculated directly for GaN under pressure is $d \ln n/dP = -0.19 \times 10^{-2}$ GPa^{-1}, i.e., in excellent agreement with experiment.

The model used by Camphausen *et al.* [139] does, however, give excellent results for Ge, $d \ln n/dP = -(1.0 \pm 0.2) \times 10^{-2}$ GPa^{-1}, and GaAs, $d \ln n/dP = -0.5 \times 10^{-2}$ GPa^{-1}. The FP-LMTO calculations by Alouani and Wills [140] give -1.0×10^{-2} and -0.5×10^{-2} GPa^{-1}; experimental values are -1.1×10^{-2} and -0.4×10^{-2} GPa^{-1}.

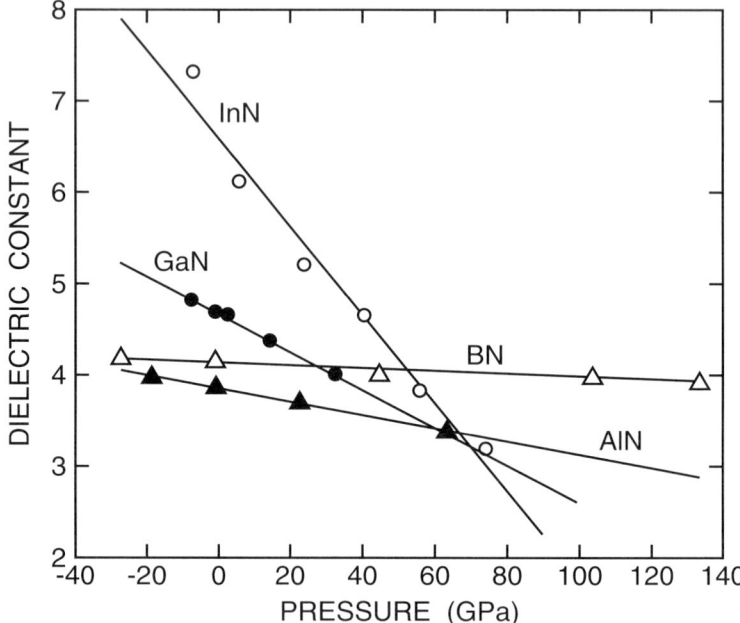

FIG. 11. Pressure dependence of $\epsilon(0)$ for the III–V nitrides (Ref. [77]). The lines represent least-squares fits to the calculated points.

In Fig. 11 we show for III–V nitrides the calculated [62] $\epsilon_1(\omega)$ $\omega \to 0$, that is, dielectric constants as functions of pressure up to $\simeq 100$ GPa. They all decrease with pressure, but $\epsilon_1(0)$ for BN is almost constant. In Fig. 12 we reproduce similar results for Ge, Si, and GaAs as obtained in Ref. [140].

Table IV also contains the calculated ionicities

$$f = (C/E_2)^2 \qquad (IV.17)$$

as well as the Phillips [138] ionicities, f_P, deduced from experiments. The general trends in the calculated ionicity scale [141] are very similar to those of the Phillips scale. The predicted dependence of the ionicity on volume is easily obtained as $2 \times \{d \ln C/d \ln V - d \ln E_2/d \ln V\}$, and it follows that, except perhaps for the marginal case of AlP, the ionicity for all compound semiconductors listed in Table IV is predicted to *decrease* when pressure is applied (at least considering the first-order pressure coefficient). Among all cases that we examined, only one, namely SiC, is clearly predicted to have $df/dP > 0$ [62]. This behavior is related to the fact that in SiC the anion (C) has no core p-states [62]. Therefore, there are no orthogonality requirements keeping valence states from being transferred

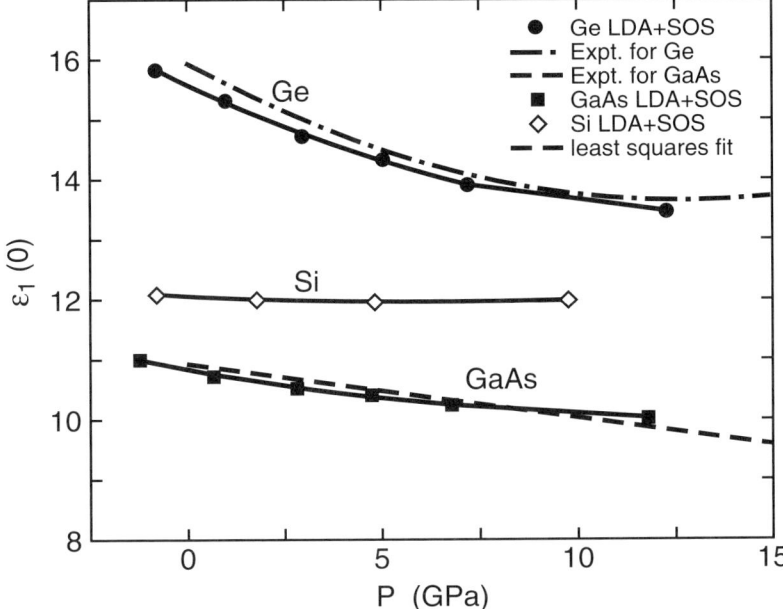

FIG. 12. Dielectric constants of Si, Ge, and GaAs as obtained from the optical spectra in Ref. [140] using gap-corrected density functional theory (SOS: "scissors operator"; see Section III). The experimental data for GaAs are from Ref. [142].

to C from Si under compression of the lattice. The same effect is also manifested in the large value of f for SiC ($f = 0.394$ in Ref. [62], whereas f_P is only 0.177).

The optical Γ phonons in the zincblende-type semiconductors are split into longitudinal (LO) (vibrating along $\overline{k}$, $|\overline{k}| \to 0$) and transverse (TO) (vibrating perpendicular to $\overline{k}$) modes. The frequency splitting is related to the "transverse effective charge" (or Born effective charge), e_T^*, through

$$\omega_{LO}^2 - \omega_{TO}^2 = \frac{4\pi(e_T^*)^2}{\mu \Omega_c \epsilon_\infty}, \qquad (IV.18)$$

where μ is the reduced mass of the two atoms, Ω_c the unit cell volume, and ϵ_∞ the infrared dielectric constant. The model used by Harrison [57] suggests that the effective charge is related to the polarity, α_p, through

$$e_T^* = Z - 4 + (\alpha_p/3)(20 - 8\alpha_p^2). \qquad (IV.19)$$

Z is the cation core charge, that is, 3 and 2 for III–V and II–III compounds, respectively.

We can estimate the pressure or volume dependence of e_T^* from (IV.19):

$$\frac{de_T^*}{d \ln V} = 8(5/6 - \alpha_p^2) \frac{d\alpha_p}{d \ln V}. \qquad \text{(IV.20)}$$

For all semiconductor compounds examined, we [62] find from *ab initio* calculations that $\alpha_p^2 < 5/6$ (CuBr is an extreme case having α_p as large as 0.91). This also agrees with the TB results of Ref. [57]. Consequently, the sign of $de_T^*/d \ln V$ is given by $d\alpha_p/d \ln V$. For all compound semiconductors, *except* SiC, we found that $d\alpha_p/d \ln V$ is positive when calculated at the equilibrium volume. This agrees with experimental observations [143, 144]. (See also Ref. [145].)

The pressure coefficients of particular gaps have been calculated and compared to experiments for several compounds and several interband transitions. First, in Table V we give the linear pressure coefficients for Si and Ge. Results obtained in three different calculations [130, 146, 147] are seen to agree well. For GaAs we calculated, using LDA + adjusting potentials, first- and second-order pressure coefficients for the fundamental gap, E_0, that is, b and c in

$$E_0 = E_0^0 + b \cdot P + c \cdot P^2. \qquad \text{(IV.21)}$$

The values $E_0^0 = 1.464$ eV, $b = 12.59$ eV/Mbar, and $c = -31.02$ eV/Mbar^2 obtained [67] agree well with experiments [87]. The pressure values used in that calculation are given by the theoretical equation of state [67]. Note, however, that our linear pressure coefficient is somewhat large when compared to $b = 10.8$ eV/Mbar as derived [88] by Goñi *et al.* from experiments. Perlin *et al.* [89] got a larger value, 11.6 eV/Mbar, and Prins *et al.* [90]

TABLE V

Linear Pressure Coefficients for Gaps in Si and Ge (LDA Calculations) in eV/Mbar

	Si			Ge		
	Ref. [130]	Ref. [146]	Ref. [147]	Ref. [130]	Ref. [146]	Ref. [147]
$\Gamma_{2'}-\Gamma_{25'}$	12.6	11.6	—	13.8	12.8	16.2
$\Gamma_{15}-\Gamma_{25'}$	0.3	0.5	0.5	0.3	0.8	—
$L_1-\Gamma_{25'}$	3.5	3.8	3.2	4.1	4.6	4.9
$X_1-\Gamma_{25'}$	−2.4	(−1.6)[a]	−1.3	−2.5	−1.4	−1.1
$L_1-L_{3'}$	4.3	—	4.4	6.3	—	6.3

[a] At $k = 0.84 \cdot (\Gamma X)$.

showed that it may lead to some systematic errors if the linear coefficient is extracted by fitting the experimental data to a quadratic expression such as Eq. (IV.21). Tables VI and VII summarize linear and quadratic pressure terms for important gaps in the III–V nitrides in the wurtzite structure. Similar calculations were performed for the same compounds in the cubic zincblende and rock-salt structures [52]. We compare, in Table VI, our results to other calculations [53, 148–154] and experiments [161–165].

The calculations represented in these tables for the nitrides in the wurtzite structure were, at least as far as our work is concerned, made by assuming the structure to be ideal: the axial ratio is $c/a = \sqrt{8/3}$, and the c-axis bond-length scale parameter u is taken to be 3/8. But the real structural parameters differ somewhat from their ideal values, in particular for AlN, where $c/a = 1.600$, $u = 0.3821$ (experiment: see Ref. [166] and theory: Ref. [91]). Further, the parameters vary with pressure [91], and this influences the value of the deformation potential. For the fundamental gap, E_g, at Γ in AlN, we find [167] $dE_g/d \ln V = -9.1$ eV (to be compared to -9.0 eV in Table VI) without structural optimization, and -7.6 eV with optimization at each volume. The change thus amounts to $\simeq 20\%$ in this case.

Structural optimization in connection with calculations of the pressure-dependent electronic properties is particularly important for heterojunctions and superlattices — as we also discussed in the case of shear deformation potentials. This will be illustrated by calculations for ZnS/ZnSe heterojunctions. The linear pressure coefficient, dE_g/dP, for the gaps in ZnS and ZnSe are very similar, 62 and 63 meV/GPa according to the calculation in Ref. [167]. These values are in good agreement with experiments (ZnS: 63.5 [168] and 64 [169] and ZnSe: 71.7 [170], 75 [169], 60 [171], all in meV/GPa). In a heterojunction, assumed to be grown by molecular-beam epitaxy, application of hydrostatic pressure will affect the positions of the atoms, and the strain state of each of the constituent materials depends on the elastic properties of both components. As a consequence, the pressure response observed in the electronic structure will critically depend on the growth conditions. Figure 13 shows the variation of the gaps in ZnS and ZnSe in heterojunctions all "grown" in a (110) direction (nonpolar interface). Three strain modes have been considered, as explained in the caption.

Calculations like the one represented by Fig. 13 are considerably more involved than a calculation of pressure effects in a bulk compound. For the latter type of investigation the volume is varied, and pressures are then derived from the slope of the energy-versus-volume curve. Here the total energy depends not solely on the volume, but also on the internal arrangement of the atoms in the interface region as well as the nonhydrostatic strains. We performed the microscopic structure optimization by means of a valence-force model [172, 173].

TABLE VI

ENERGY GAPS IN EV AND DEFORMATION POTENTIALS, a, OF NITRIDES IN THE WURTZITE STRUCTURE

	Gaps (eV)				a (eV)		
	Present work			Other calculations	Experiment	Present work	Other results
	LDA		Adjusted				
InN	V_{expmt}	V_{th}					
$\Gamma_v-\Gamma_c$	0.26	0.43	2.04	2.04^a, 1.02^b $0.30(2.05)^c$	1.90^d, 2.05^e 2.2^f	−4.1	-3.3^f
Γ_v-K_c	5.25	5.26	5.72	$5.41(7.16)^c$	—	−0.23	—
Γ_v-M_c	4.09	4.34	5.33	$4.18(5.93)^c$	—	−5.4	—
Γ_v-A_c	2.62	2.81	3.85	—	—	−4.1	—
Γ_v-L_c	3.66	3.79	4.70	—	—	−2.8	—
Γ_v-H_c	5.42	5.66	6.77	—	—	−5.1	—
GaN							
$\Gamma_v-\Gamma_c$	2.45 [2.34]	2.64	3.44	3.0^a, 2.71^b, $2.42(3.65)^c$ 1.63^g, 2.76^h, 3.0^i	3.50^j, 3.60^k 3.457^l	−7.8	-11.5^l
Γ_v-K_c	5.45	5.44	6.35	$5.74(6.97)^c$, 4.57^g, 4.93^h	—	0.36	—
Γ_v-M_c	5.09	5.17	5.94	$5.22(6.45)^c$, 4.63^g, 5.02^h	—	−2.8	—
Γ_v-A_c	4.59	4.77	5.31	4.28^g, 5.00^h	—	−7.2	—
Γ_v-L_c	4.67	4.75	5.49	3.99^g, 4.54^h	—	−3.4	—
Γ_v-H_c	6.73	6.85	7.47	6.62^h	—	−3.2	—
AlN							
$\Gamma_v-\Gamma_c$	4.73	4.78	6.05	4.64^b, $4.56(6.28)^c$	6.28^n	−8.8	-7.1^g
Γ_v-K_c	5.44	5.44	6.29	$5.58(7.30)^c$, 4.36^g	—	−0.6	1.2^g
Γ_v-M_c	6.10	6.11	6.82	$5.90(7.62)^c$, 4.93^g	—	−2.8	-1.5^g
Γ_v-A_c	6.83	6.88	7.78	5.57^g	—	−8.8	-7.0^g
Γ_v-L_c	5.69	5.71	6.60	4.59^g	—	−3.4	-1.6^g
Γ_v-H_c	7.80	7.81	8.56	7.14^g	—	−3.5	-1.6^g
BN							
$\Gamma_v-\Gamma_c$	8.50	8.52	—	8.0^g, $8.89(11.0)^c$	—	−4.3	—
Γ_v-U_c	6.72	6.72	—	5.81^b	—	−3.7	—
Γ_v-K_c	5.45	5.44	—	$5.70(7.8)^c$	—	−1.4	—
Γ_v-M_c	6.66	6.67	—	$6.65(8.7)^c$	—	−3.7	—
Γ_v-A_c	9.63	9.66	—	—	—	−9.8	—
Γ_v-L_c	6.75	6.76	—	—	—	−3.8	—
Γ_v-H_c	9.84	9.86	—	—	—	−8.0	—

TABLE VI continued

Our calculated gaps are given for two volumes, V_{expmt} and V_{th}, the "experimental" and "theoretical" equilibrium volumes, respectively. "Other results" includes experiments as well as theory. (The gap value [2.34] given to GaN is discussed in the text.)
[a] Model pseudopotential [155].
[b] OLCAO [154].
[c] LMTO; values in parentheses include bandgap correction [148].
[d] Absorption at 300 K [156].
[e] Absorption at 300 K [157].
[f] P. Perlin, private communication.
[g] Nonlocal pseudopotential [149].
[h] Norm-conserving pseudopotential [211].
[i] Norm-conserving pseudopotential [151].
[j] Photoluminescence excitation spectra [158].
[k] Reflectivity [159].
[l] Absorption at 20 K [212].
[m] LCAO [209].
[n] Absorption at 300 K [160].

The electronic structure calculation for a given structure was performed by means of the supercell method, that is, an infinite, periodic structure consisting of alternating ZnS and ZnSe layers was used to represent the heterojunction. The representation of the junction by a superlattice makes sense, provided that the layers of the constituents, here ZnS and ZnSe, are chosen so thick that the interior of each layer is "bulklike." By this we mean that the shapes of the potentials in a central layer are identical to those of the potentials in a bulk sample of the same material with the same lattice parameters. The superlattice potentials will differ from those of the self-consistent bulk by a constant value. The results presented in Fig. 13 were obtained by "exporting" the central-layer potentials to the proper bulk crystal structures and then performing (a single) band calculation. This method is possible (see Section III) because we used ASA potentials, and it ensures that the wave functions satisfy boundary conditions of a bulk crystal as they should. The band offsets are conveniently calculated in a similar way [127]. Figure 14 shows how the valence-band offsets in ZnS/ZnSe heterojunctions (superlattices) grown along (110) vary with applied pressure. The magnitudes and the pressure variation depend sensitively on the strain mode (the modes A, B, and C are defined in the caption of Fig. 13). Knowing the band offsets as well as the pressure dependence of the eneregy gaps (Fig. 13), we can then, after having corrected for the "LDA gap error" (Section II), also deduce the variation of the conduction-band edges in the heterostructure. The conduction-band edges, as they are lined up in the supercell calculations, are shown in Fig. 15 as functions of pressure. We see that they coincide at the pressure $P_C \simeq 4.5$ GPa. This

TABLE VII

The Coefficients α, β, γ, and δ for the Nitrides in the Wurtzite Structure Related to $E = E(a_0) + \alpha(\Delta a/a_0) + \beta(\Delta a/a_0)^2$ and $E = E(0) + \gamma P + \delta P^2$

	α (eV)	β (eV)	γ (meV/GPa)	δ (meV/GPa^2)
InN				
Γ_v–Γ_c	−12	49	33	−0.55
Γ_v–K_c	−0.63	−11	11	−0.08
Γ_v–M_c	−16	−44	−43	−1.1
Γ_v–A_c	−12	15	33	−0.87
Γ_v–L_c	−8.5	−6.5	23	−0.70
Γ_v–H_c	−15	−49	41	−1.6
GaN				
Γ_v–Γ_c	−23	23	39	−0.32
Γ_c–K_c	1.1	−16	−1.8	−0.03
Γ_v–M_c	−8.3	−6.9	14	−0.17
Γ_v–A_c	−21	25	36	−0.30
Γ_v–L_c	−10	−7.6	17	−0.21
Γ_v–H_c	−10	30	16	−0.16
AlN				
Γ_v–Γ_c	−27	33	40	−0.32
Γ_v–K_c	−1.8	−10	2.7	−0.05
Γ_v–M_c	−8.0	−29	13	−0.24
Γ_v–A_c	−26	28	40	−0.33
Γ_v–L_c	−10	−3.8	16	−0.17
Γ_v–H_c	−11	−2.4	16	−0.17
BN				
Γ_v–Γ_c	−13	3.0	11	−0.05
Γ_v–K_c	4.1	−30	−3.5	−0.00
Γ_v–M_c	−11	−0.6	9.4	−0.04
Γ_v–A_c	−29	−45	25	−0.12
Γ_v–L_c	−11	−7.8	9.7	−0.03
Γ_v–H_c	−24	−18	21	−0.06

E is the relevant gap and a_0 the equilibrium lattice constant. LDA calculation using the LMTO method.

means that the calculations predict that at P_C the superlattice is converted from a type I superlattice to type II (see also Fig. 16). This calculation is performed for strain A. In the case of B, the estimated value of P_C is much higher, and in case C we found that conversion should not occur. Experimentally [174] it has been found that I → II conversion takes place at $P = 3$ GPa in ZnS/ZnSe (110) grown so that the strain mode is very similar to our case A.

2 ELECTRONIC STRUCTURE CALCULATIONS FOR SEMICONDUCTORS 97

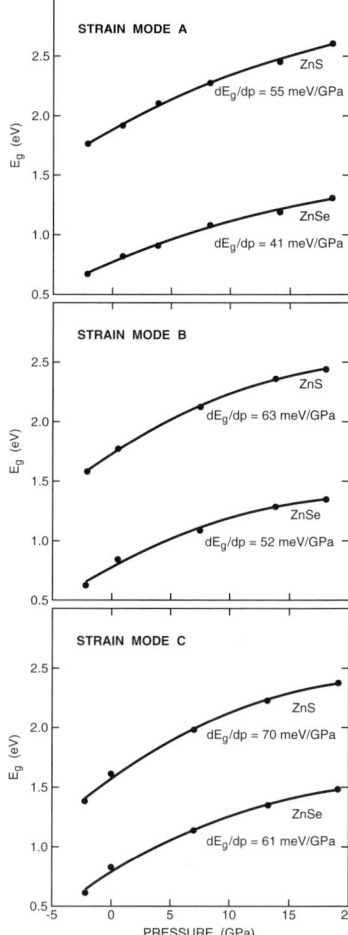

FIG. 13. Pressure dependence of the fundamental energy gaps in ZnS and ZnSe when these compounds are joined in a heterostructure. Three growth conditions are considered: A is the pseudomorphic growth of ZnSe on a ZnS substrate; B corresponds to a "free-standing" superlattice; and C is the pseudomorphic growth of ZnS on a ZnSe substrate. The pressure coefficients give the slopes at $P = 0$.

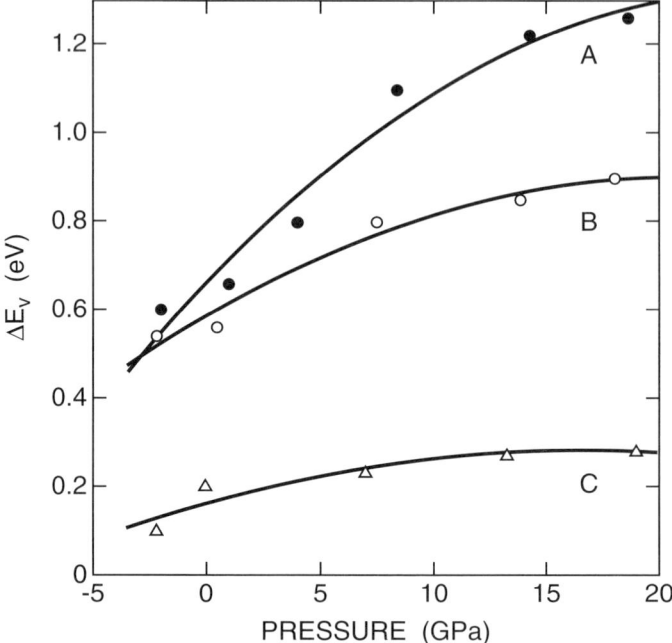

FIG. 14. Calculated pressure dependence of the valence-band offset in ZnS/ZnSe heterojunctions (ZnSe higher). The strain modes A, B, and C are defined in the caption of Fig. 13.

V. Pressure-Induced Structural Changes

Total energy calculations based on the density functional theory allow us to derive free energies, and in that way determine the atomic positions in a solid such that the structure is the most stable at given conditions. In the description of the basic theory we have already shown examples: the calculation of the equilibrium volume of GaAs, and the $P-V$ relation at zero temperature. Also, the frozen-potential approach was discussed, and an example showed that bond energy and structural difference in Madelung energy are competing terms with simple physical meanings that account for the structural energy difference at fixed volume. This chapter discusses further applications of the first-principles structural optimization for semiconductors under pressure. First, we examine a few simple examples of pressure-induced transformations from zincblende to rock-salt structures (Subsection V.1).

Subsection V.2 briefly describes the transition to the Imma structure, the monoatomic equivalent of the high-pressure phase GaAs III [189]. Then, in Subsection V.3, semiconductor compounds, which at ambient conditions

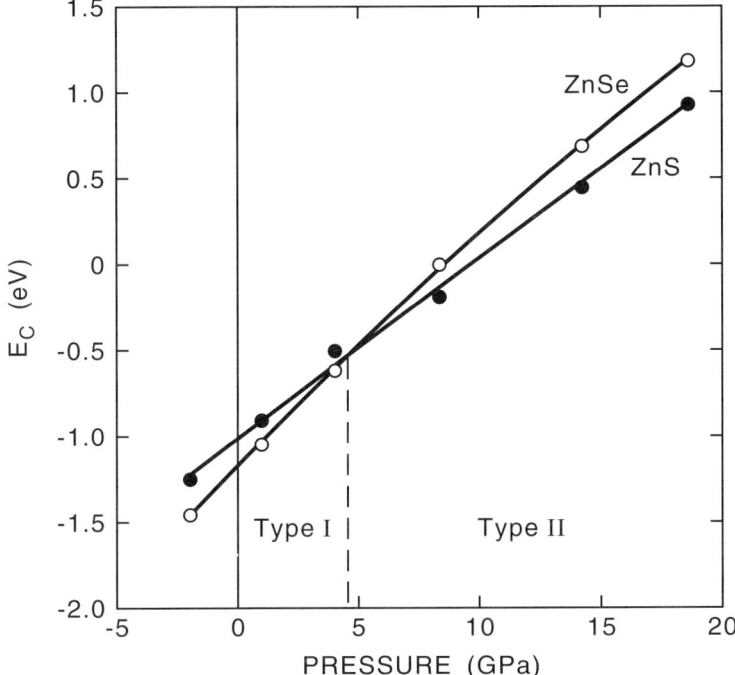

FIG. 15. Pressure dependence of the conduction-band edges related to the central layers of ZnS and ZnSe in the (110) ZnS/ZnSe superlattice. The strain mode is A.

crystallize in the wurtzite structure, are examined (pressure dependence of c/a and u, and structural phase transitions).

In the present study we also include Sn formally as a semiconductor, together with Si and Ge, and its pressure-induced transitions, α-Sn $\rightarrow$ β-Sn $\rightarrow$ bct $\rightarrow$ bcc, are analyzed in Subsection V.4.

In Subsection V.5 we discuss particular problems with LDA calculations of structural properties of the compounds which contain atoms with "semicore" states, in particular Ga-$3d$ states in GaAs, GaN, Zn-$3d$ states in ZnSe, ZnS, and Zn, and Sn-$4d$ in Sn.

1. B3 → B1 PRESSURE-INDUCED TRANSITIONS

The structure of the $A^N B^{8-N}$ compounds is determined by the competition between the energy gain obtained by formation of sp^3 bonds (as in the zincblende, B3 structure) and the gain in Madelung energy due to a larger coordination number. The B3 structure is fourfold coordinated (CN4), and

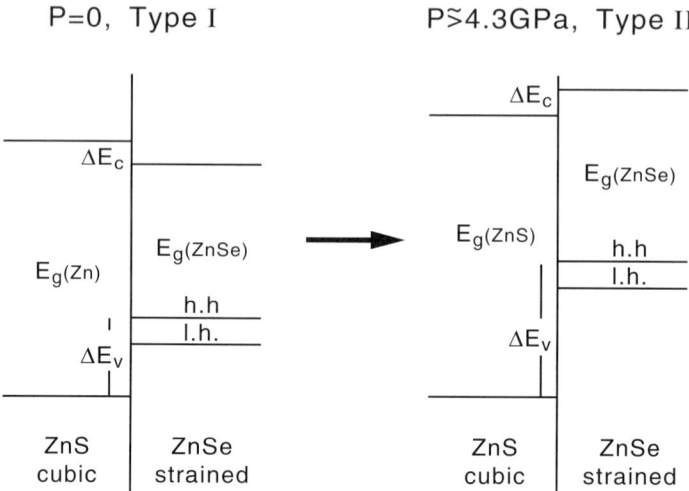

FIG. 16. Sketch of the I → II type conversion of the ZnS/ZnSe superlattice (strain mode A).

B1 (rock-salt) is sixfold coordinated (CN6). The more ionic compounds will therefore prefer B1 to B3, and according to the Phillips–van Vechten theory [138], a critical ionicity, $f_c = 0.785$, exists such that compounds with ionicities larger than this crystallize in the sixfold coordinated structure, whereas the others are stable in the tetrahedrally bonded structures, zincblende or wurtzite. This agrees well with our calculations [62] of trends in theoretical B3 → B1 transition pressures versus the calculated ionicity; see Fig. 17. Our critical ionicity is only slightly larger, 0.802, than that of the Phillips scale, and also we find MgS just at the border between the CN4 and CN6 stability regimes. Often these kinds of chemical trends offer useful general pictures, but they may fail in special cases. Therefore, we choose three cases, InP, ZnO, and BeO, and compare the transition pressures estimated from the trends (such as Fig. 17) to full calculations and experiments. The ionicity of InP is 0.54 (Table IV), ZnO: 0.65 (Phillips), BeO: 0.602 (Phillips). Rough estimates of P_t would then be InP: ≈ 18 GPa, ZnO: ≈ 10 GPa, and BeO: 20–30 GPa.

Figure 18 shows the experimental [175] and calculated [73] pressure–volume relations for InP. The calculated value $P_t = 10$ GPa agrees well with the 10.9 GPa of Ref. [175] and other experimental [75, 176] values ranging from 10.1 to 11 GPa [75, 176]. Pseudopotential calculations by Soma et al. [178] predicted that the transition pressure is between 11 and 14 GPa. InP does not seriously violate the ionicity "trend rule" mentioned earlier.

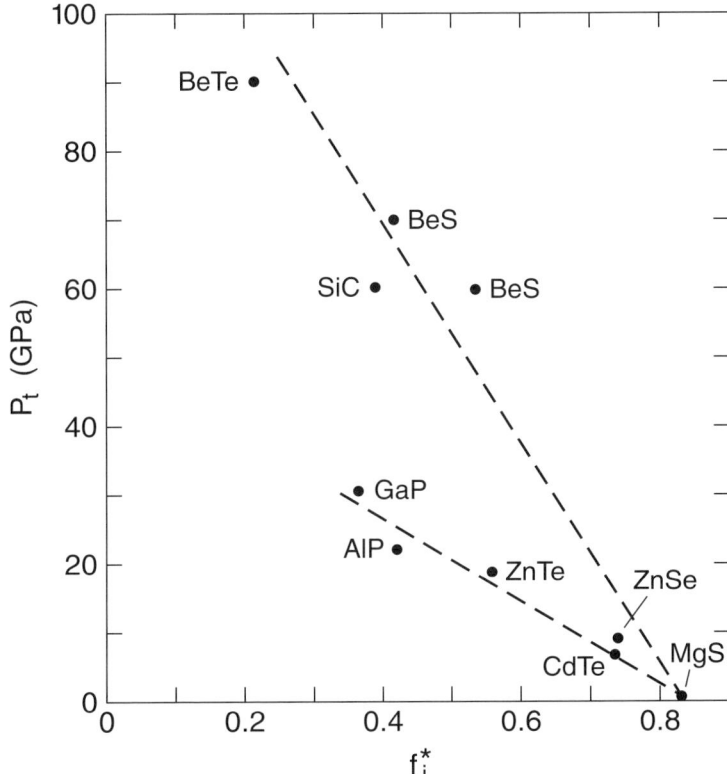

FIG. 17. Theoretical transition pressure versus ionicity (theory: Ref. [62]).

It should be noted that our results for the B3 → B1 transition in InP do not agree with the pseudopotential calculation by Mujica and Needs [179], who find that it should occur at 5.6 GPa. This calculation further predicts a transition to the Cmcm structure at 11–12 GPa, followed by a change to Immm at 50.2 GPa. A final transition to the cesium chloride structure is calculated to occur at 102 GPa. Our current FP-LMTO simulations [180] predict the B1 → Cmcm transition to occur near 26 GPa, higher than found in Ref. [179]. But even very small changes in the calculated total energies are needed to shift appreciably the calculated transition pressure.

ZnO crystallizes at ambient conditions in the wurtzite structure, but energetically this is so close to the cubic B3 structure that for the present purpose we can equally well take that as the representative for the CN4 structures. Within the LDA, and using the FP-LMTO method, we have calculated total energies and pressures as well as the enthalpies. The results were interpolated to yield the enthalpies ($H = E + PV$) as functions of

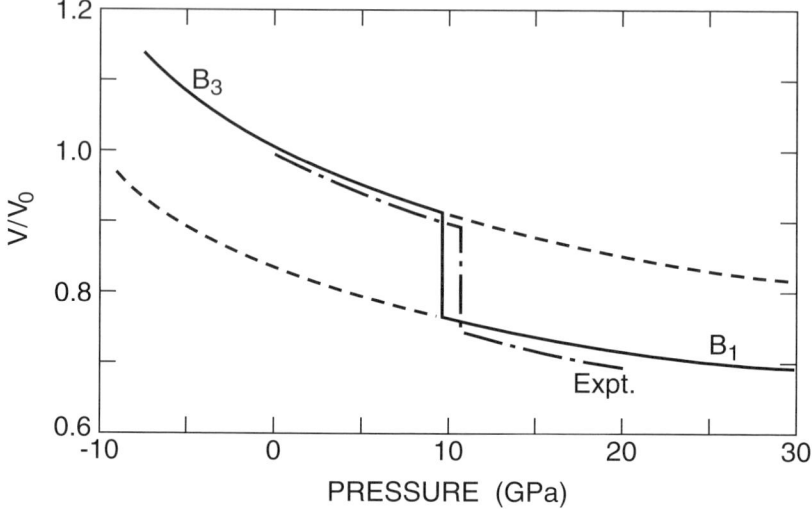

FIG. 18. Volume–pressure relation of InP in the B3 and B1 structures. Solid line: theory; dash-dotted: experiment, Ref. [175].

pressure. Figure 19 shows that this calculation [177] yields $P_t \simeq 65$ kbar (6.5 GPa). The experiment by Bates et al. [181] yielded a somewhat higher value, $P_t = 95$ kbar, and later Yu et al. [182] found $P_t = 80 \pm 3$ kbar. All three values, however, follow the ionicity trend that suggests $P_t \approx 100$ kbar.

For BeO the situation is quite different. As mentioned, the dielectric theory would suggest $P_t \approx 20$–30 GPa, and that agrees in fact with the pseudopotential calculation [183] yielding $P_t = 22$ GPa. The more recent calculation [184] yields a much higher value, $P_t = 136$ GPa. Further, no experimental evidence seems to have been found for CN 4 → CN 6 transitions in BeO below 55 GPa [185]. In order to examine this in more detail, Boettger and Wills [186] performed two all-electron full-potential calculations, one using linear combinations of Gaussian orbitals, the other FP-LMTO. Both calculations yield high values of P_t: 95 GPa. (Actually, they calculate the series of transitions wurtzite → zincblende → rock-salt.) Using the ASA functional and the LMTO method we get $P_t = 92$ GPa, that is, in agreement with Ref. [186]. We also performed an FP-LMTO calculation similar to the one for ZnO with the result shown in Fig. 20, and in that case we get $P_t = 110$ GPa. We thus agree with Ref. [186] and Ref. [184] in finding a very high value of P_t. This value clearly falls far outside the error bars of the ionicity trend scheme.

Why then is BeO such a peculiar case? The answer may be quite simple. Apart from the "ionicity rule," there are other rules of thumb which in some cases may be useful. One is based on a "rigid ion model" and states

2 Electronic Structure Calculations for Semiconductors   **103**

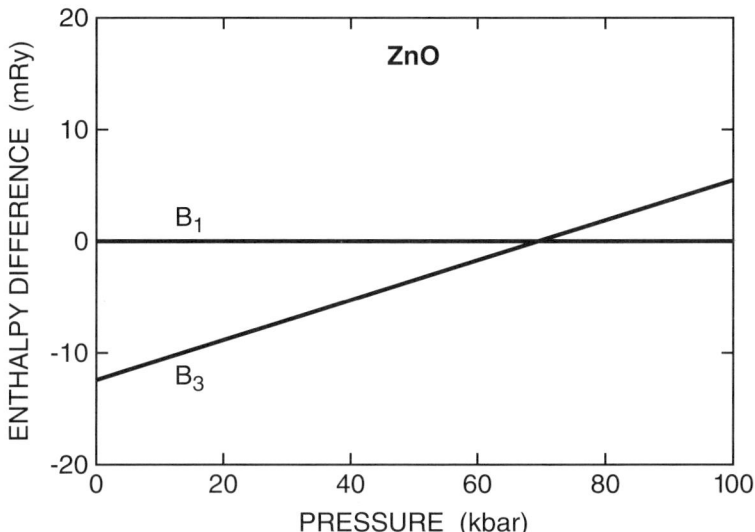

Fig. 19. B1–B3 enthalpy difference calculated for ZnO. The predicted transition pressure, $P_t \approx 65$ kbar, is given by the crossing point (FP-LMTO).

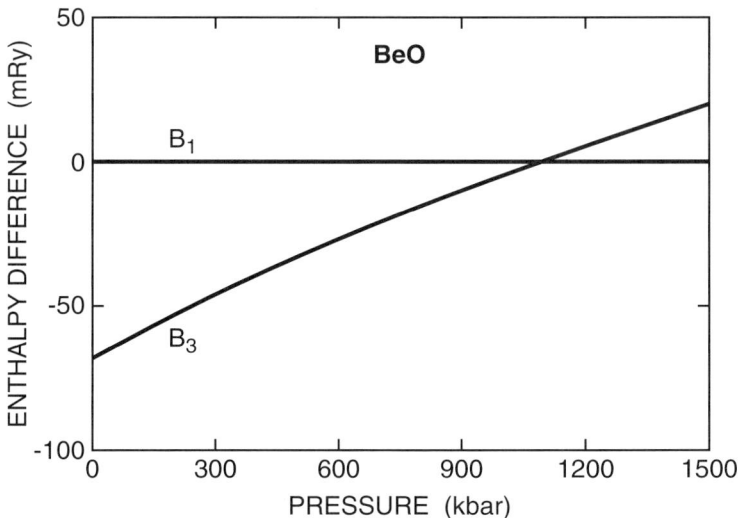

Fig. 20. B1–B3 enthalpy difference for BeO. $P_t \simeq 110$ GPa (FP-LMTO).

that the NaCl cannot be stable if the ionic radii have a ratio smaller than $r_c/r_a = 0.33$. This rule is not valid in general, as pointed out in Ref. [187]. If, however, from our calculated charge distributions in BeO we determine this ratio, we get $r_{Be^{2+}}/r_{O^{2-}} = 0.23$, that is, far below the critical ratio 0.33. Thus, it would seem that BeO, irrespective of its large ionicity, is far from "wanting" to go to the B1 structure. The O^{2-} ions are "too large" and they overlap. This picture is supported by real calculations using the frozen-potential approach. Figures 21 and 22 show the B1–B3 structural energy differences versus volumes for ZnO and BeO. Here $\Delta E_{tot} \equiv E_{tot}(B1) - E_{tot}(B3)$, and a positive value corresponds to stability of the zincblende structure. As seen earlier (Section III), the one-electron energy sum favors the B3 structure, but its magnitude is much larger for BeO than for ZnO. Further, it is seen that ΔE_{one-el} *increases* for BeO under compression, whereas it decreases in the case of ZnO. This is caused by the stronger O–p overlap in BeO, which we also may interpret as a "hard-core repulsion." Consequently, the simple hard-sphere model is in some sense here supported by the detailed quantum-mechanical calculation, and we understand why BeO must have a value of P_t that by far exceeds the prediction of the "ionicity rule."

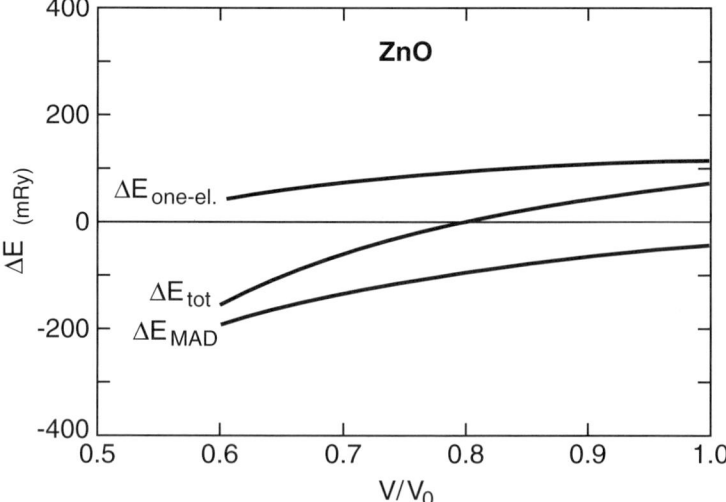

FIG. 21. Frozen-potential calculation for ZnO of the B1–B3 structural energy difference. The reference volume, V_0, is the experimental equilbrium volume (wurtzite structure), ΔE_{one-el} is the difference between the sum of the one-electron energies as calculated in the B1 structure and the B3 structures. ΔE_{MAD} is the structural difference in Madelung energy, and ΔE_{tot} is the sum of these two contributions *plus* the "charge rearrangement term." The latter is very small at all volumes.

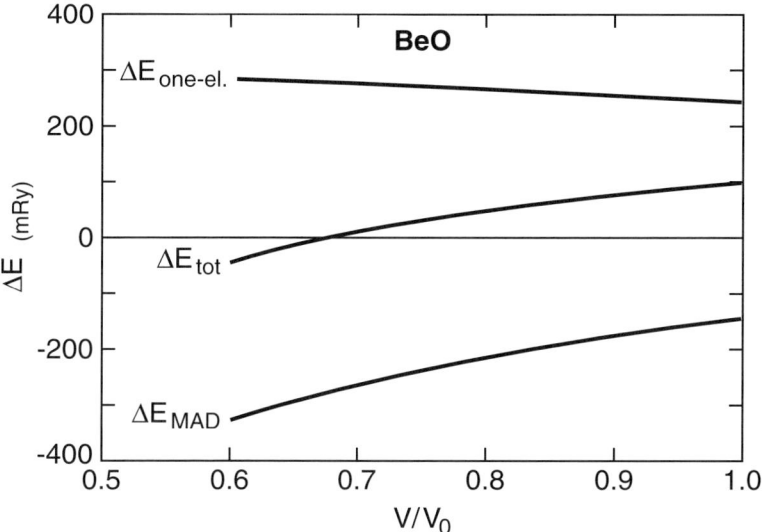

FIG. 22. Same as Fig. 21, but for BeO.

2. B3 → IMMA

As follows from Table IV, GaSb has a very low ionicity value, and it might then be expected to be the compound to be closest in behavior under pressure to Si and Ge [188]. Its structural transformation attracted special attention in several studies, and it appeared that the sequence diamond → β-tin → simple hexagonal (SH) found in Se and Ge [190, 191] also occurred in GaSb. The early diffraction experiments [176, 193, 194, 182] seemed to confirm this with a transition from the zincblende structure to the β-tin (GaSb II) phase around 7 GPa, followed by a transition to an SH structure (GaSb-III) at 27.8 GPa [192]. Further, a transition was reported [192] to a fourth, perhaps orthorhombic, structure at 61 GPa. It is interesting to notice that the pressure determined theoretically [188] for the β-tin → SH transition was 52.8 GPa, that is, significantly different from the experimental value, 27.8 GPa. In their new, systematic study of the pressure-dependent structural properties, McMahon et al. [195] found that the transition at 7 GPa is to a disordered structure which is the same as the new Imma phase found in Si between the β-tin and SH phases [196]. (The Imma structure is named after its space group. Imm2 is a related, but site-ordered, structure.) In GaSb the Imma structure was found to be stable up to at least 35 GPa [195]. In the case of Si and Ge, the Imma pressure dependence of the c/a and b/a ratios, as well as the internal structural parameter, u, was examined theoretically by *ab initio* calculations [197, 198] (see also

Subsection V.4). In fact, the calculations in 1984 by Needs and Martin [199] of total energies in Imma structures already suggested that such an orthorhombic structure might have a stability regime for Si between the β-tin and SH phases.

Total-energy calculations within the LDA and performed by means of the LMTO-ASA method showed [200] that for InAs one should expect a B3–B1 transition around 7.8 GPa and a further transition to the diatomic (site-ordered) β-tin structure around 19.5 GPa. These results, and the calculated pressure–volume variations, agreed extremely well with the conclusions drawn from the experiments by Vohra *et al.* [201]. Evaluating the calculations, one should remember that they do not represent a full structural simulation; they only compare energies (enthalpies) of a limited set (here, three) of selected structures. Secondly, the LDA is an approximation, although not seriously bad, and the ASA is less accurate than more recent full-potential methods. The numerical agreement between experiment and theory is therefore considerably better than could be expected. A more recent FP-LMTO calculation [202] yields a lower B3 → B1 transition pressure, ≈6 GPa. Even more interesting is the observation [203, 204, 205] that *no II–VI or III–V compound semiconductor* transforms to the diatomic β-tin structure. The high-pressure phases are more complex and with a lower degree of symmetry than previously assumed. As a consequence, new theoretical calculations of pressure-induced structural changes have been initiated; see, for example, Refs. [179] and [180]. These include full relaxations of lattice constants and internal atomic coordinates, and they are performed within the usual LDA as well as by using GGA. There are cases where the diatomic β-tin structure appears to have a stability range, in contrast to the experimental observation. However, several structures are separated in energy (enthalpy) by only a few mRy/f.u. (see, for example, Ref. [179]). Different numerical methods (FP-LMTO, FP-LAPW, or pseudopotentials in a plane-wave basis) used to solve the one-electron equations may then easily predict different energetic orderings of the various phases, even if the same density-functional formulation is used. Also, it should be recalled that most experiments are carried out at room temperature. For a better comparison, the calculations should therefore in fact include entropy terms. The phonon contributions to these may be significantly different in various structures (example: α- and β-Sn). (Configurational entropy contributions are of course also important when ordered and disordered phases are compared.)

3. TRANSITIONS IN AND FROM WURTZITE STRUCTURES

Some of the compounds discussed in Subsection V.1 (BeO, ZnO) do in fact form in the wurtzite structure under ambient conditions, but for the purpose of the study there we chose to consider these compounds in the zincblende structure. In this subsection we first examine some structural details of wurtzite compounds, and as examples we take AlN, GaN, and InN. It was mentioned earlier (Subsection IV.2) that the gap deformation potentials for wurtzite-type semiconductors are affected by the pressure dependence of the structural parameters, c/a and u. The reason is that these affect the crystal-field splitting (see also Ref. [206]) at the valence-band top at Γ. In our FP-LMTO calculations for AlN [91], we performed the total-energy optimization of c/a and u at eight volumes, and deduced the volume dependence of these parameters as shown in Figs. 23 and 24. If all bond lengths were equal, the relation between u and c/a would be

$$u = \frac{1}{4}\left[1 + \frac{4}{3}(c/a)^{-2}\right]. \tag{V.1}$$

For $c/a = \sqrt{8/3}$, this gives $u = \frac{3}{8}$ ("ideal ideal" values). The zero-pressure values of c/a and n are 1.596 and 0.3820, very close to the experimental values, 1.601 and 0.3821 [166]; Satta *et al.* [207] measured for GaN, AlN, and InN, respectively, $(c/a, u)$ = (1.634, 0.375), (1.619, 0.380), and (1.627, 0.377) at zero pressure. This shows that, at ambient pressure, AlN is the

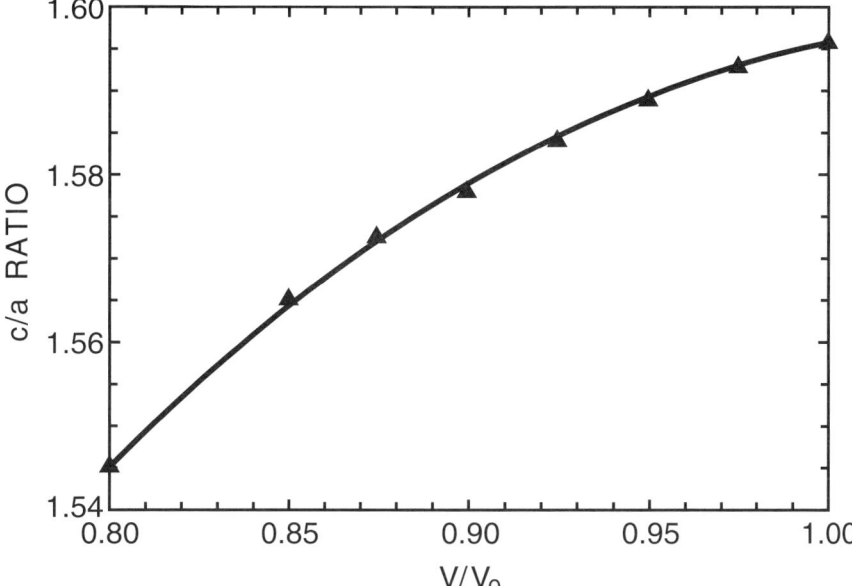

FIG. 23. AlN (wurtzite structure): Optimized c/a ratio as a function of volume (both c/a and u are varied independently to minimize the total energy) (FP-LMTO).

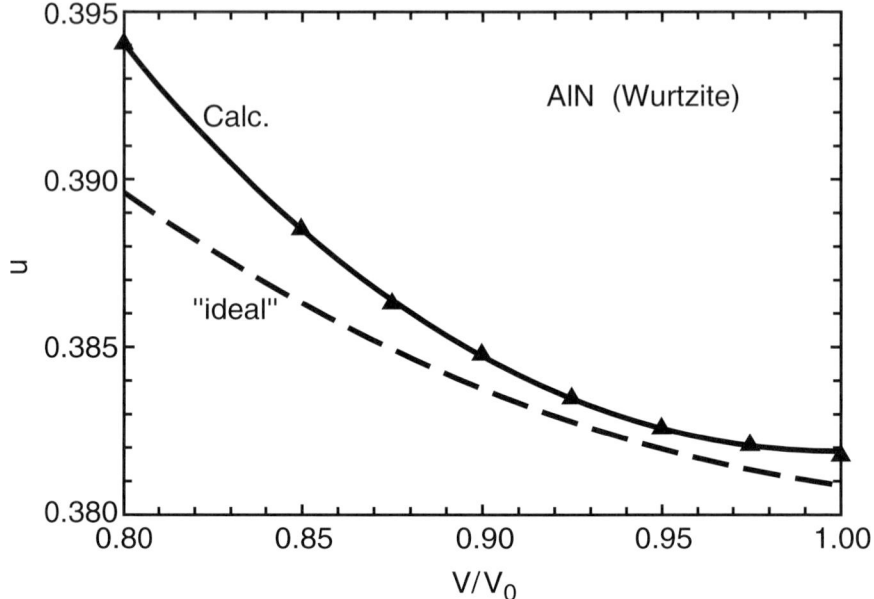

FIG. 24. AlN (wurtzite structure): Optimized u parameter versus volume. The dashed curve represents the "ideal" case where all bond lengths are equal (Eq. (V.1)), but where the c/a ratio is obtained by the optimization.

III–V nitride among those three which has hexagonal structural parameters that differ most from the ideal values.

Bellaiche et al. [208] performed first-principles structural optimization of wurtzite InN under pressure. They suggested that an isostructural phase-transition occurs close to the pressure regime of the transition to the B1 structure. It is interesting to note that the reduction in c/a, which they calculate for a volume compression of $\approx 7\%$, is almost the same as what we found for AlN, but the kink in c/a found close to 16 GPa in InN [208] has no counterpart in our AlN calculation.

Table VIII contains, for InN, GaN, and AlN theoretical and experimental values of equilibrium volume, bulk modulus B, and its pressure derivative, B'. The results "Present (Ref. [52])" in this table were all calculated within the LDA using the ASA total energies calculated by means of the LMTO method (including the "combined correction term") [78]. The results using the FP-LMTO differ only slightly, as may be seen in the case of AlN [91]. The wurtzite form the InN, which is stable at ambient conditions, appears to have an unusually large value of B'. Our calculation yields 8.1, approximately twice the value found for many compound semiconductors. The experimental value [216] is even higher, 12.7. The calculation of Ref. [154],

TABLE VIII

CALCULATED EQUILIBRIUM LATTICE CONSTANTS a, BULK MODULI B, AND PRESSURE DERIVATIVES OF THE BULK MODULI B', FOR THREE III–V NITRIDES IN THE WURTZITE, ZINCBLENDE, AND ROCK-SALT STRUCTURES

	Wurtzite			Zincblende		Rock-salt	
	Present (Ref. [52])	Other calc.	Exp.	Present (Ref. [52])	Other results	Present (Ref. [52])	Other results
InN							
a (Å)	4.922	—	4.98[a]	4.95	—	4.65(4.49)	—
B (GPa)	125	166[b], 212[c]	125.5[d]	137	—	154(251)	—
B'	8.1	3.40[c]	12.7[d]	4.3	—	8.8	—
GaN							
a (Å)	4.433	—	4.50[a]	4.46	4.419[b], 4.54[j]	4.18(3.96)	4.098[b] 4.22[e]
B (GPa)	200	190[b], 203[c], 179[e] 176[f], 240[g]	237[d], 245[h] 195[i], 188[p], 237[q] 245[r], 207 ± 3[s]	184	173[b]	227(397)	223[b]
B'	3.8	2.92[b], 3.98[c] 3.93[e], 2.66[f]	4.3[d] 3.2[p]	4.6	3.64[b]	4.0	3.69[b]
AlN							
a (Å)	4.357	—	4.37[a]	4.37	—	4.06(3.99)	4.032[b]
B (GPa)	220	195[b], 207[c,k] 205[l]	207.9[m]	215	216[l]	281(348)	270[l]
B'	3.9	3.74[b], 5.60[c] 3.98[k]	6.3[m]	4.0	—	4.0	—

The a given for the wurtzite structures are the "effective lattice constants" ($a_{\text{eff}}^3 = \sqrt{3}a^3c$). For the rock-salt structure are also given (in parentheses) values corresponding to the volume to which the phase transition occurs. "Other results" includes experiment and theory.

[a] Landolt–Börnstein [113].
[b] Nonlocal pseudopotential [149].
[c] OLCAO [154].
[d] X-ray diffraction [216].
[e] Norm-conserving pseudopotential [210].
[f] Norm-conserving pseudopotential [211].
[g] Norm-conserving pseudopotential [151].
[h] EXAFS [212].
[i] From values of atomic displacements [214].
[j] Transmission electron microscopy [213].
[k] LCAO [209].
[l] Full-potential LMTO [91].
[m] X-ray diffraction [92].
[n] Pseudopotential with nonlocal exchange potential [152].
[o] Nonlocal pseudopotential [150].
[p] X-ray diffraction (powder) [215].
[q] X-ray diffraction (powder) [216].
[r] X-ray diffraction [217].
[s] X-ray diffraction (monocrystals) [218].

on the other hand, yielded only 3.4. This may be due to the less accurate method. One also notices that this calculation yields a value $B = 212$ GPa, which is far too large. Our value, $B = 125$ GPa, agrees with the experiment [216]. In the case of GaN (wurtzite), however, our calculated values of B and B' agree well with those of Ref. [154]. The spread in the experimental values for B, 188–245 GPa, is rather large. This reflects the fact that the experimental result depends sensitively on the sample preparation. Also, the large number of experiments reflects the fact that GaN is the technologically most interesting compound among the III–V nitrides. The latest experimental result [218] was obtained by means of high-resolution X-ray diffraction on single crystalline samples. Presumably this, $B = 207 \pm 3$ GPa, is the value which we should compare to the theoretical calculations. For AlN (wurtzite) all calculated values of B are in reasonable agreement with experiment [92]. The FP-LMTO calculation [91] gave $B = 205$ GPa, a little smaller than the ASA result, 220 GPa, and even closer to experiment. The calculated B' values are somewhat lower than the (single) experimental results, 6.3, from Ref. [92].

The III–V nitrides are characterized by having high ionicity values. The Phillips ionicities of InN, GaN, AlN, and BN are 0.578, 0.500, 0.449, and 0.256, respectively. The ionicities derived from *ab initio* calculations by Garciá et al. [219] are (in the same order): 0.853, 0.778, 0.754, and 0.484, rather close to our results [52], 0.859, 0.770, 0.775, 0.380. If we then, as in the case of the B3 → B1 transitions, use the ionicity trend plot to estimate the pressure, P_t, at which a CN4 → CN6 transition should occur, we would get $P_t \approx 0$ for InN, $P_t \approx 10$ GPa for AlN and GaN, and $P_t \approx 70$ GPa for BN. Again this simple rule fails. In the case of BN (which in fact at $P = 0$ assumes the B3 structure), the error is at least as spectacular as the one we saw for BeO. The (theoretical) transition pressure is ≈ 850 GPa, which is 12(!) times the value predicted from the ionicity. The explanation for this will be similar to that for BeO. The bond length of the nitrides are small, but the nitrogen ion is "large." The ionicity value of InN is larger than the critical value, which would suggest that at zero pressure this compound should already prefer the rock-salt structure. Our first-principles calculations (see Table IX) [52], $P_t = 21.6$ GPa, as well as experiments, 23 GPa (Ref. [220]) and 12.1 GPa (Ref. [216]), show that a substantial external pressure is required to induce the transformation. For AlN the prediction $P_t \approx 10$ GPa is good (see Table VII and the detailed discussion later). However, the prediction that GaN and AlN, which have almost the same ionicity values, would thus have nearly the same values of P_t is not supported by the total energy calculations. These yield [91, 52] $P_t = 12.5$ GPa (AlN) and 51.8 GPa (GaN). The volumes per formula unit are quite similar for GaN and AlN (with that of AlN being the smaller), and the anion (N) is

TABLE IX

CALCULATED (LDA-LMTO) STRUCTURAL PHASE TRANSITION (1 → 2) PRESSURES P_t AND RELATIVE VOLUMES V_1/V_0 AND V_2/V_0 OF THE TWO PHASES AS THE TRANSITION PRESSURE

	Present work	Other calculations	Experiment
InN			
P_t (GPa)	21.6	5.0[a]	23.0[b], 12.1[c]
V_1/V_0	0.85	—	—
V_2/V_0	0.72	—	—
GaN			
P_t (GPa)	51.8	55[a], 51.9[i] 35.3[j], 32.4[k], 40[l]	52.2[c], 47–50[d], 37[m]
V_1/V_0	0.81	0.82[a]	0.86[d]
V_2/V_0	0.69	0.71[a]	0.73[d]
AlN			
P_t (GPa)	16.6	12.9[e], 12.5[f]	22.9[c], 14–16.5[d]
V_1/V_0	0.93	0.95[e]	0.93[d], 0.92[g]
V_2/V_0	0.76	0.77[e]	0.74[g]
BN			
P_t (GPa)	850	1110[h]	—
V_1/V_0	0.51	0.45[h]	—
V_2/V_0	0.47	0.42[h]	—

V_0 is the experimental equilibrium volume of phase 1 (wurtzite for InN, GaN, AlN, and zincblende for BN).
[a] Norm-conserving pseudopotentials [210].
[b] Visual observation [220].
[c] X-ray diffraction [216].
[d] EXAFS [212], visual observation [220].
[e] Nonlocal pseudopotentials [149].
[f] Full-potential LMTO [91].
[g] X-ray diffraction [92].
[h] Nonlocal pseudopotential [221].
[i] Hartree–Fock–LCAO [222].
[j] HF + correl. [223]
[k] LDA-LCAO [223].
[l] GGA-LCAO [223].
[m] x-ray diffraction [215].

the same; the ionicity is also the same. Thus, the large difference in transition pressure is due to the different cations. In Ref. [52] we summarized an analysis in terms of partial pressures referring to the decomposition of terms in the pressure formula [84]. This showed that the presence of the $3d$ (semi-) core states on Ga in GaN is responsible for the difference.

Their presence pushes, because of orthogonality requirements, the empty $4d$ states up on the energy scale, whereas the (empty) $3d$ states on Al lie much lower in AlN. This leads to a net positive (though rather small) Ga-d pressure in GaN, whereas the Al-3d "tail pressure" is large in magnitude and *negative* in AlN. It was further shown that this led to a larger relative volume change at the transition in AlN than in GaN and consequently a smaller value of P_t for AlN (P_t being determined by the slope of the common tangent to the energy–volume curves of the two phases).

The summary in Table VII shows that, in particular for GaN, where the largest data base exists, the scatter in theoretical as well as experimental values for P_t is rather large, in both cases, however, essentially covering the same range, 33–55 GPa. As far as theory is concerned this in part reflects that quite different methods with different levels of accuracy are used. A comparison of LDA results obtained by using a linear combination of atomic orbitals (LCAO) method and, for example, LDA combined with the LMTO method is sufficient to illustrate this. The former yields P_t = 32.4 GPa (the lowest theoretical value in Table VII), whereas the latter predicts P_t = 51 GPa. It is further to be noted that the LDA-LMTO result is close to that obtained (55 GPa) within the LDA [210] but by using the norm-conserving pseudopotential method.

The calculations of Ref. [223] using the LCAO–Hartree–Fock scheme without and with addition of correlation (from the generalized gradient corrections to LDA (!); see Section II) show that this correction reduces P_t from 52 to 35 GPa, whereas the GGA, when added to the LDA, increases P_t. Further, the equilibrium volume, V_0^{th}, obtained from HF-LCAO is 154.9 a.u., very close to experiment [215], 154.4 a.u. When "correlations" are added as described earlier, V_0^{th} is too low, 145.1 a.u., by $\approx 6\%$. The GGA, when added to LDA, usually increases the equilibrium volume. The GGA-LCAO calculation of Ref. [223] also shows this by giving V_0^{th} = 156.2 a.u. The large range of experimental values for P_t appears to be more surprising, since the extreme values, 37 and 52.2 GPa, are obtained by very similar techniques, high-resolution X-ray diffraction.

Several possible high-pressure phases of the III–V nitrides were examined theoretically in Ref. [52] and in other calculations referred to in that paper. For all four compounds examined, one finds that the β-Sn structure has a total energy which is so far above those of the wurtzite and zincblende structures that this cannot be a stable high-pressure phase within a range of some megabars. Using AlN as an example, we illustrate these calculations in Fig. 25. As for the wurtzite structure, we also optimized the structural parameters for the NiAs structure [91]. The u-parameter, however, did not differ from the ideal values, 0.25, in the pressure regime examined, but c/a increases slightly with pressure. At the volume $V/V_0 = 0.65$ we find $c/a = 1.72$. Our calculations predict (Fig. 26) that the wurtzite $\to$ rock-salt

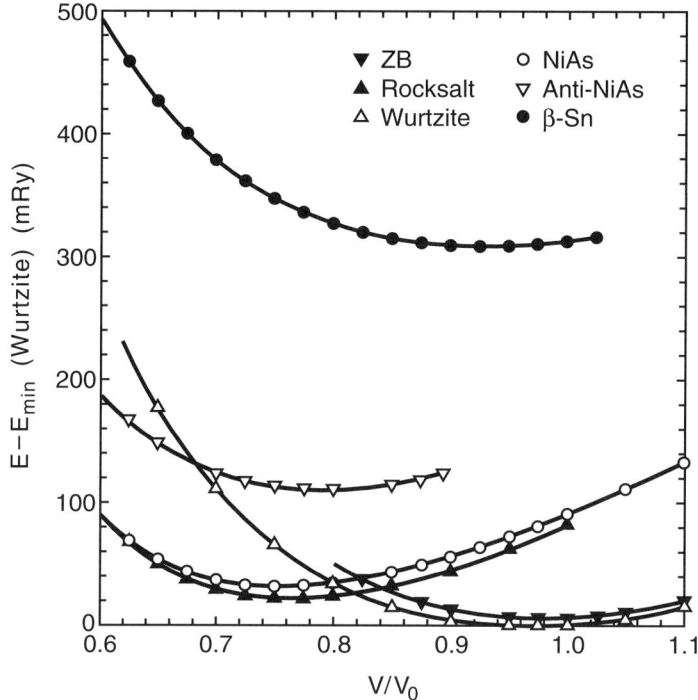

FIG. 25. Total enegies as calculated, within the LDA using the FP-LMTO method, for AlN in the wurtzite (c/a and u optimized), zincblende (ZB), rock-salt, NiAs ($c/a = 1.633$, $u = 0.25$), anti-NiAs ($c/a = 1.633$, $u = 0.25$), and β-Sn (ideal c/a ratio) structures.

transition occurs at 12.5 GPa, and that a further transition to the NiAs structure may occur in the pressure range 26–30 GPa. Experimentally, this has not been observed.

As follows from Table VII, the value of $P_t = 12.5$ GPa agrees extremely well with the LDA result, 12.9 GPa, obtained in Ref. [149] also within the LDA, but using the norm-conserving pseudopotential method. The LMTO-ASA calculation [52] gave a higher value, 16 GPa. This last value, in fact, agrees best with experiments (Table VII). This was discussed in the theory sections (Section III).

4. PRESSURE-INDUCED TRANSFORMATIONS OF GROUP IV SEMICONDUCTORS

Free atoms of the group-IV elements, C, Si, Ge, Sn, and Pb, all have the (ns^2, np^2) outer electron configuration, and in the solid state these elements

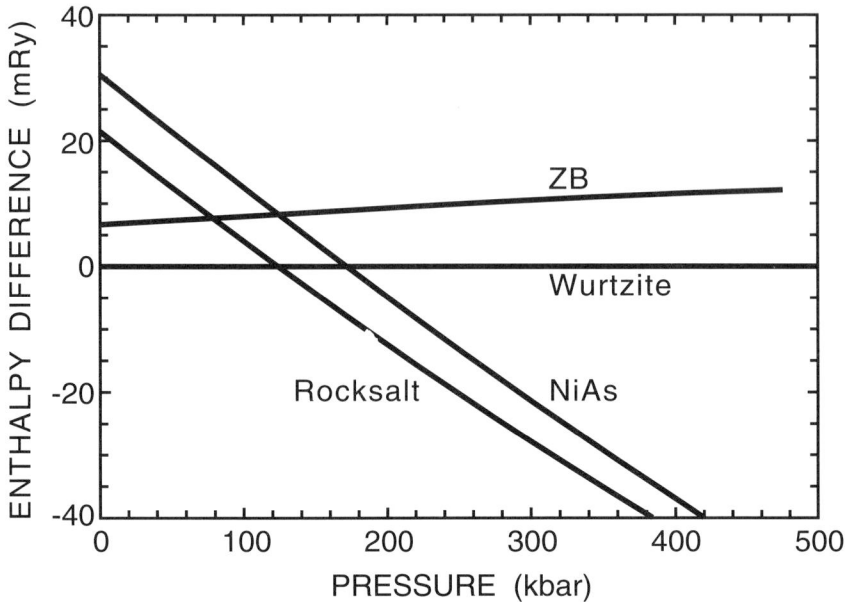

FIG. 26. Calculated (FP-LMTO) enthalpies of AlN in the NiAs (c/a = 1.633 and 1.72), zincblende (ZB), rock-salt, and wurtzite structures versus pressure (theoretical). (The enthalpy function of AlN in the wurtzite structure, fully optimized, is used as a reference.)

all, except lead, crystallize in the diamond structure at zero pressure and room temperature. For C, Si, Ge, and Sn, the $s \to p$ promotion energy cost is smaller than the gain in energy due to formation of sp^3 bonds. In Pb this is not the case because the separation between the 6s and 6p levels (in the solid) is too large, due to the relativistic effects [224]. This is demonstrated in Fig. 27, which shows the ratio between the bond formation energy, $-4h$, and the $s \to p$ promotion energy plotted against b, the sp^3 bond order (see Subsection IV.A). These quantities are all derived from first-principles calculations for the various elements in the (assumed) diamond structure. E_p, E_s, and $-4h$ are deduced by transforming the LMTO into its tight-binding version [225, 226]. The fact that Pb is below the horizontal line, that is, that δ(Pb) < 1, means that the energy accounting mentioned earlier does not favor the formation of the tetrahedral bonds in this case. The point "NR Pb" is obtained from a calculation for Pb with all relativistic effects switched off (speed of light set to infinity). In this case δ is just at the border δ = 1, very close to the value for Sn. The reason Pb does not want to assume the diamond structure is then what Phillips called "relativistic dehybridization" [138]. This is well known, but here the effect is clearly demonstrated by *ab initio* calculations. Further, through the abscissa values

2 ELECTRONIC STRUCTURE CALCULATIONS FOR SEMICONDUCTORS

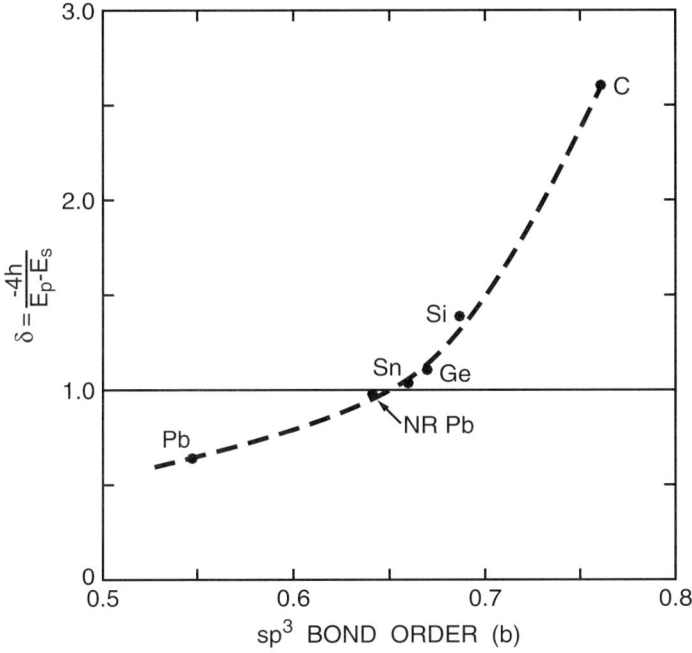

FIG. 27. Group VI: Ratio, δ, between the bond-formation energy $(-4h)$ and the $s \to p$ promotion energy plotted against b, the bond order calculated as the difference between the sp^3 bonding and antibonding character summed over all occupied states in the (assumed) diamond structure. Pb forms in reality in the *fcc* structure because δ does not favor sp^3 bonding in that case. NR Pb is a fully nonrelativistic calculation.

of b, Fig. 27 also serves to characterize the relative strength of the covalent bonds [110] in the Group-IV semiconductors.

The behavior of the Group-IV semiconductors under pressure is complex and has been studied in detail. This is in particular the case for Si and Ge, still the most important materials for electronic and micromechanic device applications.

The understanding of the high-pressure phases of Si, Ge, and Sn has, especially for Si and Ge, been subject to several modifications over the past couple of decades. The complexity is surprising. At the present we believe that the pressure-induced transitions in Si are, with increasing pressure,

$$\text{Diamond} \to \beta\text{-Sn} \to \text{Imma} \to \text{SH} \to \text{Si-VI} \to hcp \to fcc,$$

where SH means simple hexagonal, the Imma is the "generalized β-Sn" structure mentioned earlier (see also Fig. 28), and Si-VI is an intermediate

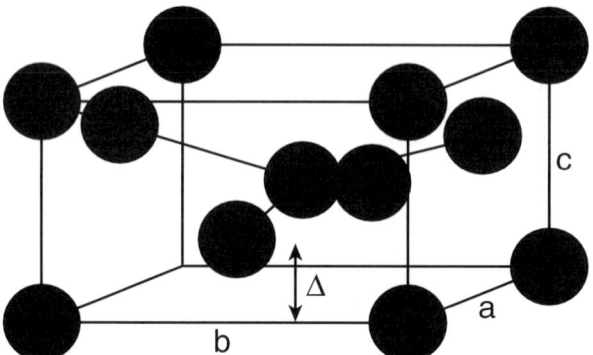

FIG. 28. The orthorhombic Imma structure. The special case $a = b$, $\Delta = c/4$ is the β-Sn structure. Also, the simple hexagonal structure is generated by a suitable choice of lattice parameters: $b/c = \sqrt{3}$ and $\Delta = c/2$.

phase. The pressure, $P_{t\beta}$, at which the transformation to the β-Sn structure occurs was first [176] found to be near 150 kbar, whereas later experiments [227, 228] found the transition to occur at a somewhat lower pressure, 125 kbar. Although the theoretical calculations based on *ab initio* theory presented by Yin and Cohen [229] did not include all relevant high-pressure structures, this paper is still a central reference in the theory of pressure-induced transformations in semiconductors. The calculation of Ref. [229] yielded $P_{t\beta} = 99$ kbar for Si. The theoretical value of the c/a ratio was $0.55 \pm 5\%$, in good agreement with experiments [193], 0.552. FP-LMTO calculations [230] give $P_{t\beta} = 132$ kbar. The pressure-dependent c/a ratio was also calculated, but in the pressure range 132 to 170 kbar, where Si was found (theoretically) to exist in the β-Sn structure, c/a stays fixed at the value 0.5516. Experimentally it was found [196] that the Imma phase of Si is stable between $P_{\text{Imma}} = 132$ kbar and 156 kbar (P_{SH}). The first-principles pseudopotential calculation by Lewis and Cohen [197] places the Imma below β-Sn in energy for all volumes, presumably because the energy differences are smaller than the computational uncertainty. This is not the case in our current FP-LMTO calculations, which yield the correct sequence, diamond $\rightarrow$ β-Sn $\rightarrow$ Imma, but our transition pressures $P_{t\beta} = 132$ and $P_{\text{Imma}} = 170$ kbar are too high.

The primitive hexagonal phase of Si exists between $P_{\text{SH}} = 156$ kbar and $P_{\text{VI}} \simeq 370$ kbar [231, 232], and the intermediate Si-VI is found up to $P_{\text{hcp}} \simeq 400$ kbar [231, 232]. *SH*-Si is metallic and a combined theoretical and experimental investigation shows that the band structure can be rather well accounted for by a nearly free-electron model [235]. This is also seen from the band plots in Fig. 29.

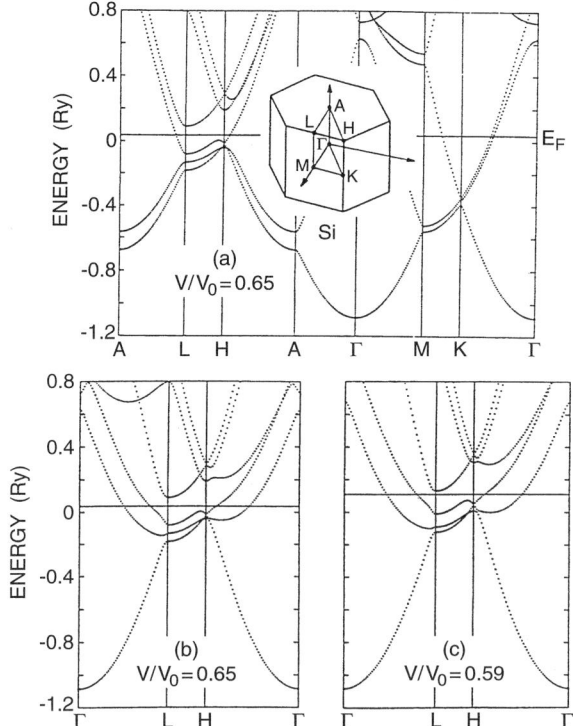

FIG. 29. (a) LDA bands of *SH*-Si at $V/V_0 = 0.65$ (theoretical pressure = 22.3 GPa). (b) Same as (a) but for other symmetry lines in the Brillouin zone. (c) Same as (b) but for $V/V_0 = 0.59$ ($P = 42.6$ GPa). The one-particle equation was solved by means of the LMTO method.

Si is stable in the *hcp* structure up to $P_{fcc} \approx 780$ kbar, where it transforms [237] to the *fcc* structure. The so-called "intermediate phase," Si-VI, is presumably [190] similar in structure to the X-phase of $Bi_{0.8}Pb_{0.2}$. The structural behavior of Si under pressure is apparently even more complicated than would appear from the preceding discussion. Four additional phases are found under slow pressure decrease. The BC8 (Si-III) structure has a *bcc* Bravais lattice [233, 234]. The BC8 and ST12 structures have been examined theoretically by Crain *et al.* [238] and by Clark *et al.* [239]. Upon heating, Si-III transforms to Si-IV, which has a hexagonal (wurtzite-type) structure. A very fast decompression of Si-II (the β-Sn structure) can produce two new tetragonal phases [236], Si-VIII and Si-IX.

The structural sequence suggested for Ge is diamond → β-Sn →

Imma → SH → dhcp. For Ge the experimental value of $P_{t\beta}$ is ≈ 110 kbar, and the calculation of Ref. [229] gave 96 kbar. The Imma structure of Ge was studied by means of theoretical total energy calculations in Ref. [198], and it was found, as in the calculations for Si, that this structure seems to have the lowest energy at all volumes. Again, however, the energy differences are small and structural predictions are difficult to make. Experimentally, the Imma structure of Ge was observed at Daresbury [240]. The β-Sn phase of Ge appears to exist over a much larger pressure range, up to P_{SH} ≈ 750 kbar [241], and above ≈ 1 Mbar Ge assumes the *dhcp* structure [241].

The third group-IV semiconductor, tin, behaves in yet another way under pressure; see Figs. 30 and 31 (and further comments in Subsection V.5). At low temperatures (less than 286 K) tin is stable in the diamond structure (α-Sn phase) but at "room temperature" (T > 286 K), the β-Sn structure is the stable form. At low T only a very small external pressure is required to produce the α → β-Sn transition. At a pressure close to 100 kbar (104

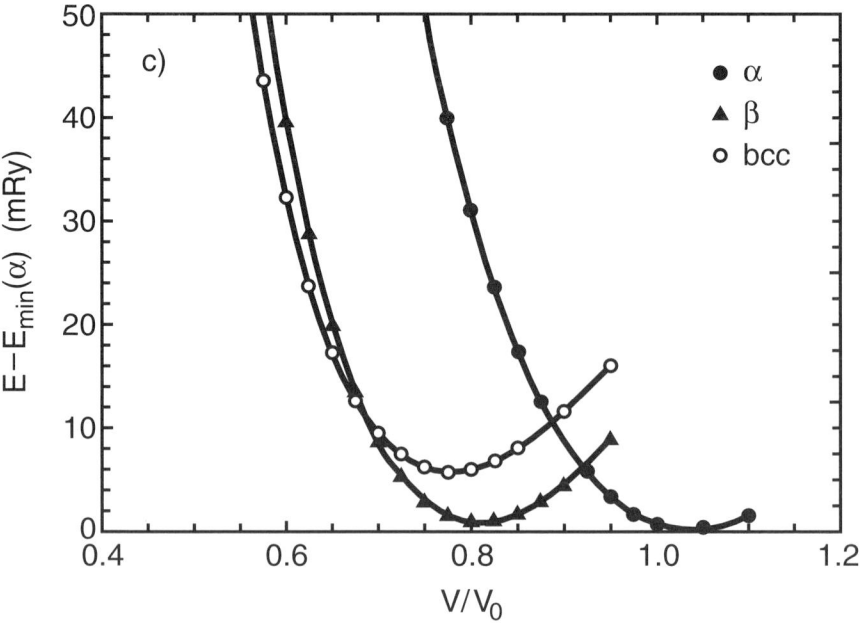

FIG. 30. Total energies of Sn in the diamond (α), β, and *bcc* structures calculated by means of the FP-LMTO method as functions of volume V. V_0 is the experimental equilibrium volume of α-Sn [242]. (These calculations are not straight LDA, but the 4d-states have been downshifted in energy by 1.5 eV; see Subsection V.5.)

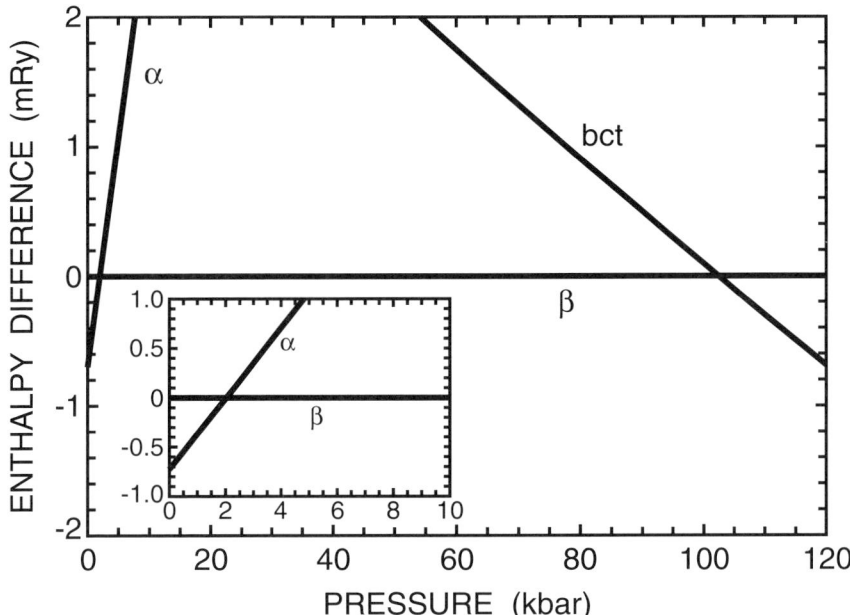

FIG. 31. Enthalpies of α-, β-, and bct-Sn (c/a optimized) as functions of (theoretical pressure (Ref. [242]). The insert shows the regime close to $P = 0$ on an expanded scale. (See also comments in Subsection V.5.)

from the calculation of Ref. [242], 95 kbar as measured at room temperature [243]), the c/a ratio of this bct structure is ≈ 0.90, and it increases slightly with pressure. At high pressures, $P \approx 500$ kbar, a first-order transition to the bcc structure occurs. Thus, c/a does not increase continuously to the value 1.00 that corresponds to the bcc phase.

The LDA calculations performed by means of the FP-LMTO method also yield this result, as can be seen from the enthalpy plots in Fig. 32. The peculiarity that tin does not transform directly into the bcc phase, but rather prefers the bct structure in a certain energy range, was explained as being due to the splitting of the level H_{15}, which in the bcc structure is triply degenerate (see Fig. 33). At sufficient compression the upper level ("z" in Fig. 33) is above E_F and thus empty. The electrons that otherwise were contained in states around this level are transferred to states near the Fermi level, and this increases slightly, $E_F \to E_F + \Delta E_F$. If the splitting of the H_{15} level were the only effect of the tetragonal distortion, we see that energy would be gained. A smaller amount, however, is spent to increase E_F. But in addition to these effects, the center of gravity of the level also moves. This shift, $\Delta E_{xyz}^{av} = (2E_{xy}^{bct} +$

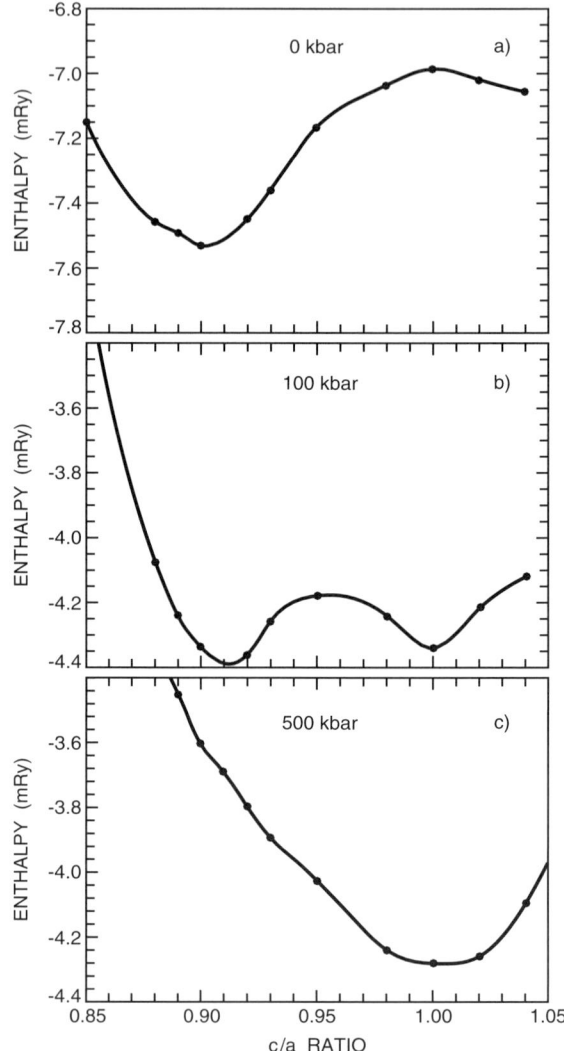

FIG. 32. BCT-Sn. Enthalpy versus axial ratio c/a at various pressures P. (a) $P = 0$; (b) 100 kbar; (c) 500 kbar. The numerical scatter is $\pm\ 0.02$ mRy. (The three figures have no common energy reference.)

$E_t^{bct} - 3E_{xyz}^{bcc})/3$, is zero to first-order in the strain parameter, but the change in c/a from 1.0 to 0.9 is so large that higher-order terms become important. Our simple picture then suggests that the *bct–bcc* total energy difference is

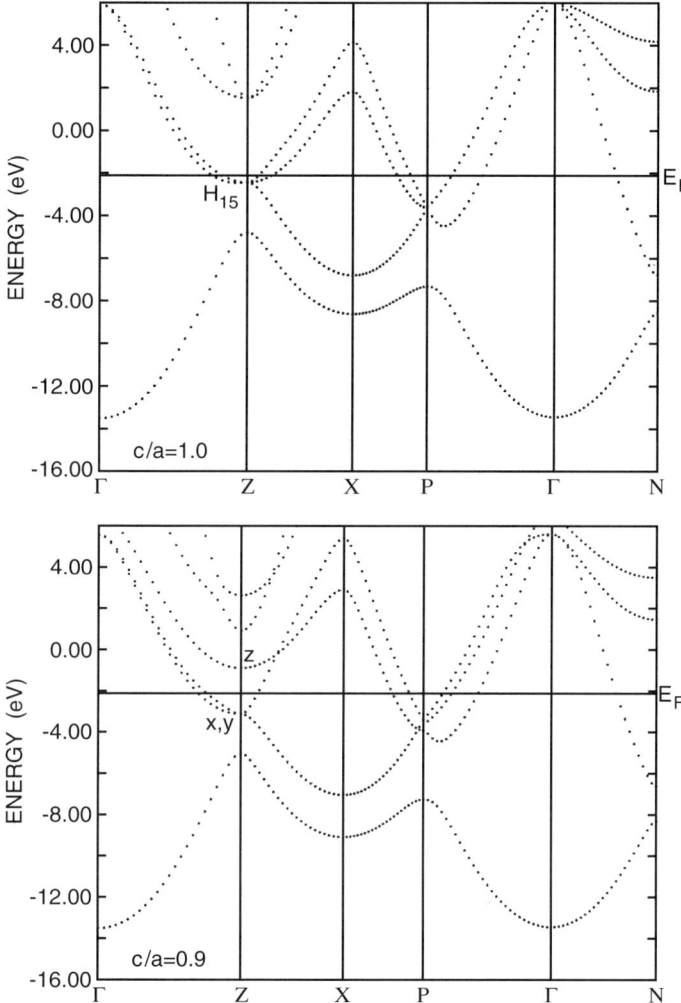

FIG. 33. Band structure of tin in the *bcc* (upper figure) structure and *bct* structures with $c/a = 0.9$ (lower figure). Symmetry points are labeled as in the *bct* structure.

$$\Delta E = E(bct) - E(bcc) \\ = n(3\Delta E_{xyz}^{av} - (E_z^{bct} - E_F^{bct}) + \Delta E_F/2), \quad \text{(V.2)}$$

where n is the number of electrons shifted above E_F with E_z and subsequently released to the Fermi surface. With slight modifications this model could explain qualitatively as well as quantitatively the occurrence of the *bct* intermediate phase in Sn under pressure [244].

5. d-States

As mentioned in Section II, there are cases where the LDA is not sufficiently accurate, even disregarding the rather trivial "LDA gap problem" in semiconductors. Even ground-state properties in some cases require methods that go beyond the LDA, although correlation effects are not as pronounced as in f-electron systems (heavy-fermion systems). Already in GaAs we noticed [47] that the Ga-$3d$ states need to be included as "relaxed band states," although they form a rather low-lying, narrow band. Their effect was included in the actual [47] LMTO band calculation by making a so-called two-panel calculation, that is, in each iteration the total valence charge was obtained by summation of contributions from two LMTO calculations, one covering a lower energy range including the Ga-$3d$ states, the other covering the usual s–p valence bands, but not including the $3d$'s. Rather, the upper panel used a coupling to the (empty) Ga-$4d$ states. The importance of including cation semi-core d-states became very clear in connection with calculation of heterojunction band offsets [93]. The valence-band maximum (VBM) in ZnSe is 0.6 eV higher in a band calculation, which treats the Zn-$3d$ states as band states, than in one where they are frozen as core states. If *hybridization* to the Zn-$3d$ states is included in a band circulation for the upper (valence) panel, the VBM is pushed further up by 0.1 eV. These shifts strongly influence the calculated VBM offsets in ZnSe/Ge heterojunctions, whereas the discontinuities in ZnS/ZnSe junctions are less affected since the Zn-$3d$ states are present on both sides of the interface. The two-panel calculations did include the cation d-states in the calculation of the self-consistent potentials. But the (smaller) effects of the hybridization between these states had to be treated separately; once the potentials were obtained, a single band calculation was made where coupling to the Ga-$3d$ states was included. This appeared to overestimate the effect (see also the discussion of calculated LDA gaps in GaN in Ref. [52]). In the VBM-offset calculations we suggested compensating for this either by making an eigenvector-dependent averaging [52] of the VBM values obtained with and without including the hybridization to the Ga-$3d$'s, or by downshifting the Ga-$3d$ states in energy so that they correspond roughly to the binding energies as obtained from photoemission experiments. Although comparison of the measured binding energies to DFT eigenvalues is not quite simple, our analyses for several compounds suggested that the valence bands were improved when the (d-like) semicore states were downshifted. The Sn-$4d$ states were found to give rise to some difficulties in LDA calculations [245] for SnGe. More dramatic effects were detected during the calculation of the total energies of time in various structures [242]. First, we performed FP-LMTO calculations using two energy panels (in analogy to the GaAs calculations described earlier). This led to total energies of the α- and β-phases which have minima differing by less than 1 mRy, α-Sn being lowest. This is correct when compared to experiments, but the *bct* and *bcc* phases of Sn have energies far too low when calculated in this way (see Fig. 6a of Ref. [242]). The reason for this error is that the two-panel calculation underestimates the effects of hybridization between the 4d states and the top of the 5p band. We therefore performed another calculation where the 4d states were included in the valence panel (a one-panel calculation). This gave a correct sequence of the α-, β-, *bct* energy minima (see Fig. 6b of Ref. [242]), but the energy differences were far too large and the pressures for transition to the *bct* phase much too high. Thus, in this case the hybridization effects are *overestimated*. This was interpereted as a manifestation of the LDA yielding 4d states that are too high-lying due to incomplete cancellation of self-interaction in the exchange term for these rather localized states. We did not perform a complete SIC calculation (see Subsection II.3 and references given therein), but decided rather to simulate the small changes in the total-energy differences by performing LDA calculations, which were modified by having (in each *scf* iteration) the 4d states somewhat downshifted. (This is particularly easy to do when the LMTO method is used.) Thus, we do not include a SIC potential, and therefore the shift that has to be applied is given by the orbital exchange-correlation energy corrections, $\delta_{\alpha\sigma}$, as defined by Perdew and Zunger [10]. The shift can be estimated by integration of the orbital density, $n_{\alpha\sigma}(\bar{r})$:

$$\delta_{\alpha\sigma} \approx -0.2 \int n_{\alpha\sigma}(\bar{r})^{4/3} d^3r \tag{V.3}$$

(Rydberg atomic units).

Using a spherically symmetrized density, the Sn-4d shift was estimated to be $\delta_{4d} \approx -2$ eV. We had, before performing this estimate, decided to apply a 1.5 eV downshift. The actual choice of the value is not very critical. A 2-eV shift leads to essentially the same result, whereas downshifting the 4d's by 3 eV appears to overestimate the correction.

Obviously, such a correction must be pressure- (or rather volume-) dependent. When the Sn interatomic distances become sufficiently small, the 4d band will have broadened so much that the SIC corrections disappear. However, this does not happen in the pressure ranges that are discussed in Ref. [242] and the Sn results described in Subsection V.4. The first-order volume coefficient can be estimated to be $d \ln|\delta_{4d}|/d \ln V \approx 4 \times 10^{-3}$. We then used a volume-independent shift. The results shown in Figs. 30–34 were all obtained by applying this downshift, $\delta_{4d} = -1.5$ eV. This gave a

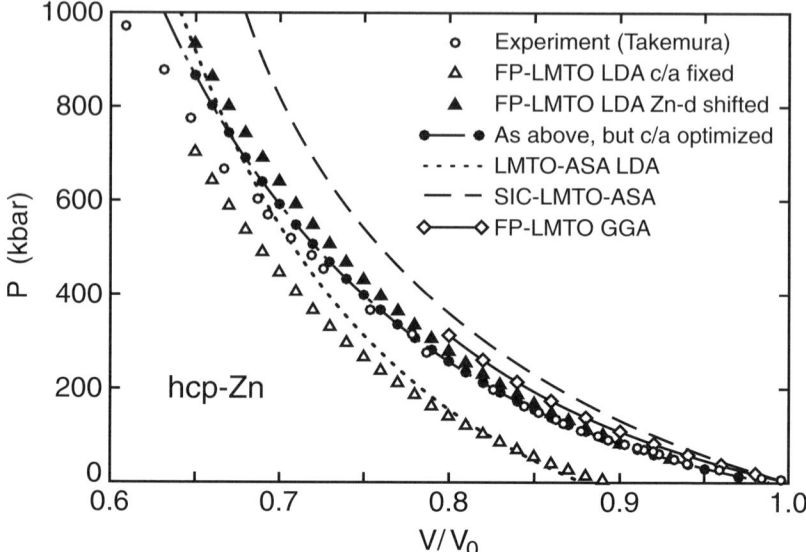

FIG. 34. Pressure–volume relations for (hcp) Zn calculated within the LDA and with inclusion of various corrections. V_0 is the experimental equilibrium volume. Experimental data are from Ref. [249].

consistent set of ground-state properties in quantitative agreement with experiments.

Zinc compounds, such as ZnSe, ZnS, and ZnTe, may also require that similar corrections be applied to the Zn-$3d$ states. Calculations [246, 247] for metallic (hcp) Zn clearly show this effect. Although the present article should concentrate on semiconductor properties, it is nevertheless relevant to analyze these calculations of structure (c/a ratio) and volume versus pressure for Zn. This will further substantiate the preceding discussion, and also illustrate the effects of some of the LDA corrections described in Section II.

The main reason for performing total-energy optimizations for an otherwise rather uninteresting metal such as Zn was the observation made by Takemura [248] of a discontinuity in the axial c/a ratio near $\sqrt{3}$, when pressure is applied to Zn. A detailed discussion of this is beyond the scope of the present article. We performed the optimizations, first using conventional LDA and full-potential LMTO. The calculated pressure–volume relation differs substantially from experiment [248, 249]. This is seen from Fig. 34. The LDA yields an equilibrium volume which is too small by 10–11%, indicating overbinding. First it was examined whether this error could be caused by numerical inaccuracies in the band-structure calculations or by application of too-small basis sets. LMTO-ASA, dotted curve in Fig. 34, gives essentially the same equilibrium volume, but the pressure increases more rapidly than those derived from the FP-LMTO calculations (open triangles in Fig. 34). This difference is a result of the ASA. A new FP-LMTO calculation using a completely different code [250] gave results which agree with those given by the open triangles in Fig. 34. One might suspect that an extension of the basis set could affect the total energies. Especially in a case such as this, where the $3d$ states obviously must be included in the band structure, a simultaneous inclusion of $4d$ states (in the same panel) could be important. This option is included in the codes of Ref. [250]. Expanding the basis set in this way changed the theoretical equilibrium volume by less than 0.5%. Finally, the application of a full-potential linear augmented-plane-wave method [78, 251] (FP-LAPW) [252] did not change the calculated value of the equilibrium volume either. Consequently, the error is ascribed to the LDA, and several ways of correcting for this were examined. The dashed curve represents the self-consistent self-interaction-corrected LDA calculations using the LMTO-ASA band-structure method (hence the label SIC-LMTO-ASA) performed by Svane [18]. The theoretical equilibrium volume is now in perfect agreement with experiment. The fact that the P–V curve is too steep is due to the atomic-sphere approximation (ASA): the dashed and dotted curves in Fig. 34 are parallel. The calculations represented by diamonds and labelled FP-LMTO-GGA include the generalized gradient corrections [29]. Also, the GGA yields an equilibrium volume in perfect agreement with experiment. FP-LAPW-GGA calculations performed with the WIEN95 code [252] gave the same results. We compared the band structures of Zn as calculated with and without inclusion of GGA and found no differences. Therefore, the GGA corrections in this case are exclusively due to changes in the Coulomb interactions in the energy functional. The correction mechanism is then very different from that of the SIC as well as from the approach leading to the results represented by the filled triangles in Fig. 34. This was performed in the same way as the corrected Sn calculations described earlier, namely by downshifting the Zn-$3d$ states, but otherwise performing the usual LDA calculation. The downshift was taken to be 2 eV. This gives an equilibrium volume which is 1% smaller than experiment, that is, a small overbinding. This error is well inside the limits defining "good" LDA ground-state calculations. All the calculations described here so far for Zn were made with fixed c/a ratio. Now, consider the black dots connected by the full-line curve in Fig. 34, which are the results obtained when at each volume we perform a total-energy optimization of c/a. On the pressure scale of the figure the effects of this optimization (compare to the filled triangles) is not dramatic, but still the agreement with experiment is clearly improved. This is the calculation, among those discussed here, which agrees best with experiments in the pressure range 0–400 kbars. At higher pressures the experiments gradually approach the standard LDA results. This means that the downshift of Zn-$3d$ in the calculation should be reduced as the compression widens the bands. This is not surprising, as discussed earlier.

The LDA corrections discussed here for Zn considerably improve the calculated mechanical properties, but the ways they do so are different. One type (e.g., GGA) affects essentially the Coulomb interactions in the total energy functional, whereas another (e.g., $3d$ downshift) reduces the hybridization between the localized and the truly itinerant band states. This narrows the $3d$ band and reduces the binding (see also the discussions in Ref. [84]). Perhaps LSDA corrections that reduce the width of the Fe-$3d$ band will be able to explain quantitatively as well as give a clear physical picture of the LSDA errors in predicting the ground state of bulk iron [253].

Calculations for the III–V nitrides by Vogel, Krüger, and Pollmann [254] use self-interaction-corrected pseudopotentials, and the results show that the semi-core states are shifted with respect to the spectral positions obtained by straight LDA calculations, in close agreement with our discussion here.

Several important semiconductor compounds, some of which were mentioned earlier, have electronic and cohesive properties and cohesive properties which are influenced by d-states. The II–VI compounds (CdS, CdSe, MnTe and CdMnTe$_2$ are other examples) and the copper and silver halides, such as CuBr, CuCl, and CuI, are compounds where the bonding is very strongly influenced by the interaction between the Cu (or Ag) d-states and the halogen p-states. For example, this strongly affects the spin–orbit splitting, Δ_0, at the valence-band maximum. In CuCl, Δ_0 is even negative. This, and the pressure dependence of Δ_0, was examined in Ref. [255]. Measurements and calculations of $d\Delta_0/d \ln V$ allowed an examination of the pressure-dependent p–d hybridization in these compounds. Also, the deformation potentials of the VBM states, like d_0, are strongly influenced by this hybridization [256, 257].

Even more exotic semiconductors, such as β-FeSi$_2$ and ϵ-FeSi, have gaps which are formed in the Fe-$3d$ band complex. NiSi$_2$ and CoSi$_2$ form in the fluorite structure under ambient conditions, and they are metals. FeSi$_2$ would also be a metal in this cubic phase (γ-FeSi$_2$), but it is not stable in bulk form. A Jahn–Teller-like distortion into a complex orthorhombic structure with 24 atoms in the primitive cell creates a gap ≈ 0.9 eV in the band structure [258]. The gap edge states interact strongly with the phonons, and from calculated acoustic-phonon deformation potentials it was concluded [258] that electron–phonon scattering severely limits the mobilities in this compound. FP-LMTO calculations have been used to examine the changes in the gap with pressure [259]. Tight-binding calculations [260] suggested that the nature of the gap, direct/indirect, would change when pressure is applied, but our calculations, at their present stage, do not confirm this.

The monosilicide in the ϵ-phase is a semiconductor with a small gap, ≈ 0.1 eV. Our LDA calculation also predicts this (see Fig. 35), but closer examination shows that the LDA band structure fails to explain measured temperature variations of the specific heat, the magnetic susceptibility, and Fe quadrupole splittings (see Refs. [261–264] and references given therein). The density of states function, Fig. 35, derived within the LDA has peaks which are too low and too wide by a factor of ≈ 10 when compared to photoemission data [264]. This, as well as the clear discrepancies with experiment as mentioned above, shows that a renormalization of (part) of the Fe-d states beyond the LDA should be invoked in the theory. The d-states at the gap edges are more localized than predicted by the simple band picture, and ϵ-FeSi is presumably a Kondo insulator.

The LDA calculation yields an equilibrium volume which is 7% smaller than the experimental value. Thus, this strong overbinding also indicates that the LDA produces Fe-$3d$ bands which are too wide.

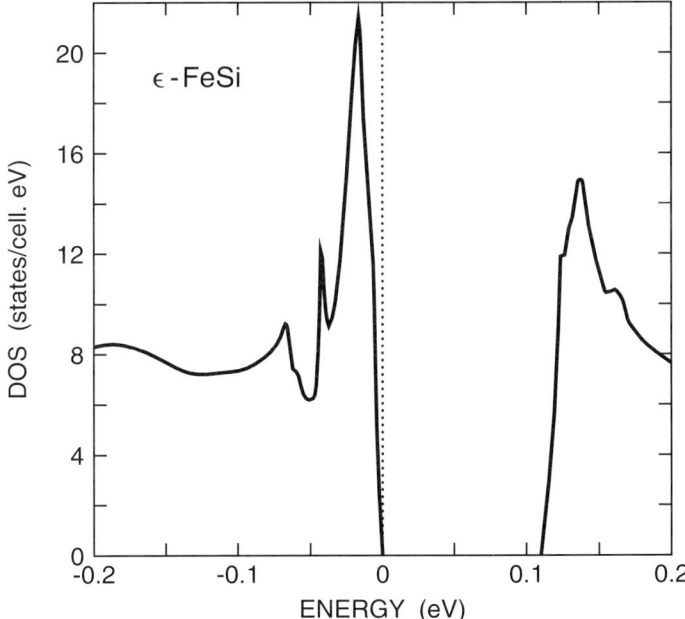

FIG. 35. Density of states (LDA) of ϵ-FeSi around the energy gap. The vertical, dotted line marks the top of the valence band.

Mössbauer spectroscopy provides information on the interaction of a nucleus with the surrounding charge distribution. One of the measured quantities, the quadrupole splitting Δ (in velocity units), depends on the asymmetry of the electrostatic potential V at the nuclear site. The relation between these two quantities is given by

$$\Delta = \frac{eV_{zz}Qc}{2E_0}\left(1 + \frac{1}{3}\eta^2\right)^{1/2}, \tag{V.4}$$

where $E_0 = 14.41$ keV is the γ-ray energy. V_{zz} is the electric field gradient (EFG), the largest element of the diagonalized tensor of second spatial derivatives of the ($l = 2$) component of the crystalline Hartree potential evaluated at the nucleus. Q is the nuclear quadrupole moment for ^{57}Fe, and η, the asymmetry parameter, is defined as

$$\eta = \frac{V_{xx} - V_{yy}}{V_{zz}}. \tag{V.5}$$

Because of the trigonal symmetry of ϵ-FeSi, $\eta = 0$. The major axis, the z-axis, of the EFG tensor is along [111].

The quadrupole splitting of Fe in FeSi varies strongly with temperature (Fig. 36). Our calculated splitting is 0.7 mm/sec, which in magnitude is

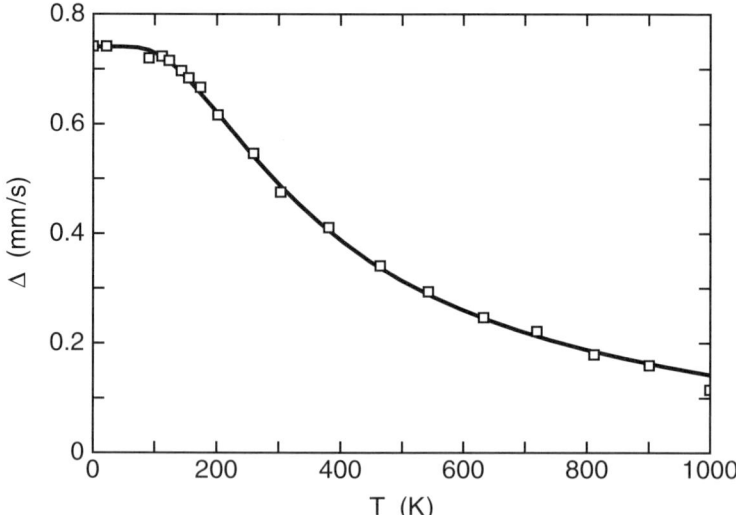

FIG. 36. Temperature variation of the quadrupole splitting (Fe) in ϵ-FeSi. The squares are experimental results (Ref [262]); the curve represents a simple model (see text).

in excellent agreement with experiment, 0.5 mm/sec, recalling that the measurement [263] was carried out at room temperature, and our calculation refers to $T = 0$. Further, the calculation agrees with the earlier [262] measurement at very low temperatures.

The conversion electron Mössbauer spectroscopy measurements [263] were carried out with varying externally applied magnetic fields, and this allowed also the sign of the EFG to be determined from experiment. It agrees with the calculation. Further, an asymmetry between the intensities of the lower- and higher-energy transitions was explained as being due to strain in the sample which could be estimated from the curvature. A calculation for strained ϵ-FeSi shows that the quadrupole splitting at one of the Fe nuclei decreases (linearly) with the strain in the actual deformation mode, whereas the splitting on the other three Fe atoms in a group of four nearest neighbors to Si is essentially strain independent (Fig. 37). The calculation agrees very well with experiment, and this provided an additional check of the sign of the EFG. It would nevertheless be interesting to perform measurements on single crystals subject to externally controlled strain.

The temperature dependence of the quadrupole splitting (Fig. 36) is explained in terms of a change in symmetry of the charge distribution due to the thermal excitations across the gap of electrons from Fe-d_{3z^2-1} states to $d_{x^2-y^2}$ states which are split by the distorted tetrahedral field. The squares

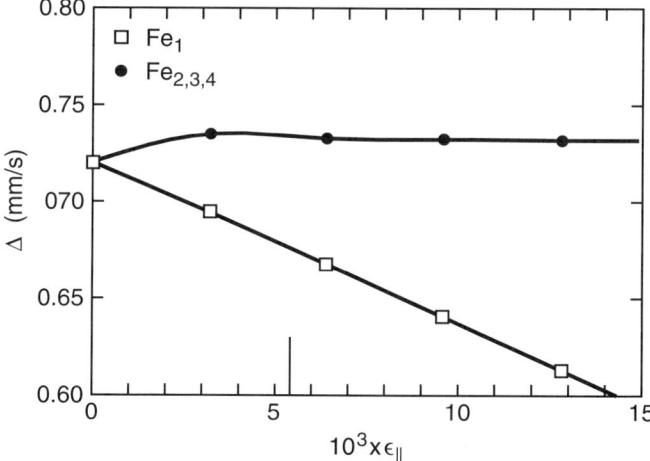

FIG. 37. Calculated strain dependence of the quadrupole splitting in ϵ-FeSi; $\epsilon_\|$ is the component of the strain parallel to the [111] direction, and $\epsilon_\|/\epsilon_\perp$ was taken as given by experimental data ($-5.4/3.2$). Fe$_1$ is on the [111] axis. The vertical bar indicates the strain in the experiment, Ref. [263].

in Fig. 36 are experimental data, whereas the curve is obtained from a simple model based on just this picture (see Eq. 3 of Ref. [265]) with parameters chosen to fit the experiment at $T = 0$ and a very high temperature. Calculations using thermal broadening in the LDA band structure to calculate the FP-LMTO charge density are in fact able to reproduce the experiment, *provided* that the temperature, T, is replaced by $10 \times T$. Thus, it appears that the aforementioned renormalization would improve the theoretical model. Our picture is different from those based on spin-fluctuation theories.

Under pressure the Fe-d states will broaden, and eventually all d-states become "good band states." It would be very interesting to try to observe this transition experimentally, and, if possible, determine the specific heat, magnetic susceptibility, and quadrupole splittings when the pressure is varied. Also, measurements of the electrical resistivity as a function of T and P will be most useful to the study of this compound.

VI. Pressure Dependence of Phonon Frequencies

Discussing the Born–Oppenheimer approximation in Subsection II.1, we mentioned its application to calculation of phonon frequencies using the "frozen-phonon approach." This method consists of calculating the total energy changes due to a series of distortions defined by a chosen phonon mode. From the coefficient to the term varying with the distortion parameter in second order, one then extracts the vibration frequency. In general this is best done by diagonalizing a dynamical matrix, and in this way the mode eigenvectors are also obtained.

First, consider the zone-center phonon modes, Γ_3 and Γ_5, of β-tin. These modes (see, for example, Fig. 6 of Ref. [266]) are Raman active, and the pressure dependence of the frequencies was measured by Raman spectroscopy by Olijnyk [267]. Figure 38 shows a comparison between the experimental data and our [242] calculations. The calculated frequencies are at $P = 0$, $\nu(\Gamma_3) = 45$ cm^{-1}, and $\nu(\Gamma_5) = 130$ cm^{-1}. Both are higher than the experimental values, 42.44 and 126.60 cm^{-1}, but slightly lower than the neutron scattering results [268], 46.5 and 132.5 cm^{-1}. It was found [267] that within experimental error, the pressure dependence could be expressed (with P in kbar and ν in cm^{-1}) as

$$\nu(\Gamma_3) = 42.44 + 0.1179 P - 0.436 \times 10^{-3} P^2$$

$$\nu(\Gamma_5) = 126.60 + 0.5517 P - 1.219 \times 10^{-3} P^2. \tag{VI.1}$$

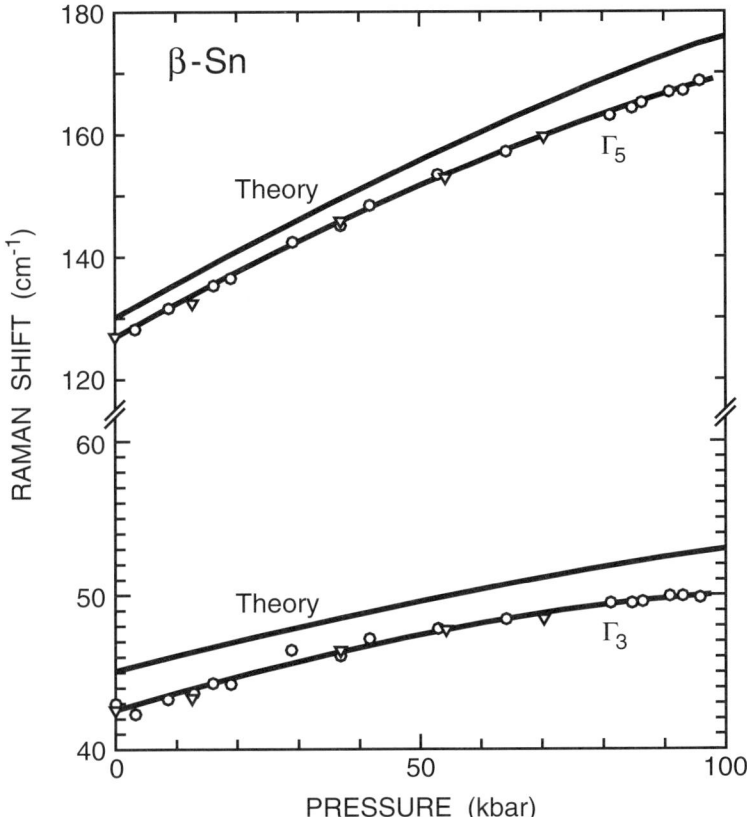

Fig. 38. Frequencies of the Γ_3 and Γ_5 phonon modes in β-Sn. Data points (and fitted curves) are from the Raman experiments of Ref. [267].

Similar fits made to the calculations yield

$$\nu(\Gamma_3) = 45.08 + 0.1460 P - 0.538 \times 10^{-3} P^2 \qquad \text{(VI.2)}$$

$$\nu(\Gamma_5) = 129.90 + 0.5349 P - 0.512 \times 10^{-3} P^2.$$

The pressure variation of the mode Grüneisen parameters is shown in Fig. 16 of Ref. [242]. For both modes they are positive, but decreasing with pressure, in the entire pressure range where the β-phase of Sn is stable.

Germanium in the β-Sn structure also has Γ_3 and Γ_5 frequencies that increase with pressure [267], but in Si they decrease. This is presumably related to the transition to the Imma structure and the fact that in Si the β-Sn structure only exists in a narrow pressure range. The calculations by

Lewis and Cohen [269] agree well with the experiments of Olijnyk [270, 267] and — in the case of Sn — with our calculations.

As another example of application of the frozen-phonon method to calculation of pressure dependence of phonon frequencies, we consider the zone-center phonons of wurtzite GaN and AlN. The small group of $\vec{k} = 0$ is the C_{6v} point group. The full reducible representation may be decomposed into irreducible representations according to $\Gamma_{\text{opt}} = A_1 + 2B_1 + E_1 + 2E_2$, using the notation of Ref. [271]. The mode patterns are illustrated in Fig. 39. Only optical modes are considered. The acoustic modes are, since we here set $\vec{k} = 0$, just translations. The E_2 modes are Raman active; A_1 and E_1 are both Raman and infrared active, whereas the B_1 modes are "silent" in the sense that they are forbidden in Raman as well as infrared excitation modes. Figures 40 and 41 summarize our [272] calculations and compare them to experiments [212, 273]. The mode Grüneisen parameters (γ) calculated for E_1, E_2^1, E_2^2, A_1, B_1^1, and B_1^2 are 1.48, −0.28, 1.66, 1.50, 1.08, and 1.15 in the case of AlN, and, in the same ordering, 1.48, −0.20, 1.60, 1.52, 1.04, and 1.29 for GaN. In the latter case, experiments yield $\gamma(E_1) = 132$, $\gamma(E_2^1) = -0.34$, $\gamma(E_2^2) = 1.47$, and $\gamma(A_1) = 1.53$, showing good agreement between theory and experiment. If one compares the results for GaN and AlN, it is seen that the E_2^1 mode is softer (γ negative and larger in magni-

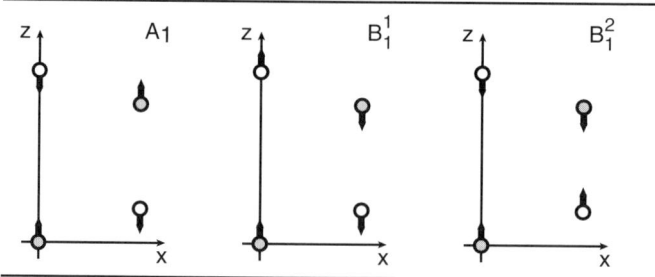

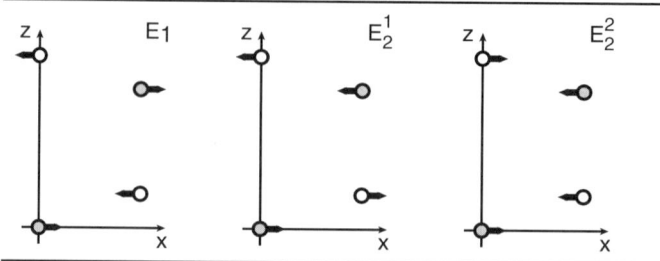

FIG. 39. Optical phonon modes for $k = 0$ in the wurtzite structure.

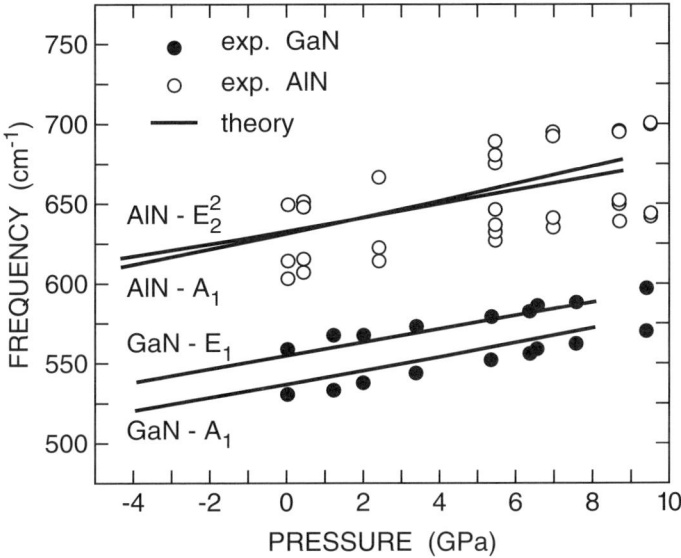

FIG. 40. Optical zone-center phonon frequencies of wurtzite GaN and AlN (high-frequency modes) as functions of hydrostatic pressure. Calculated results (solid lines) are compared with experimental data (Refs. [212, 273]).

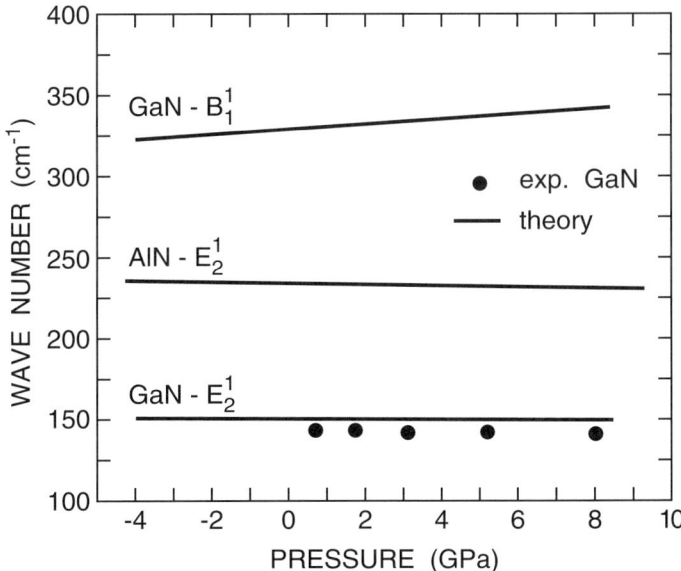

FIG. 41. Same as Fig. 40, but for the low-frequency modes.

tude) for AlN. This is probably related to the fact that in AlN the transformation to the rock-salt structure occurs at a lower pressure than in GaN. The difference is then caused by the presence of the $3d$ (semi-) core states on Ga.

The phonon frequencies calculated [272] by us at zero pressure agree well with those obtained by Miwa and Fukumoto [274]. We also calculated the TO phonon frequencies for AlN and GaN in the zincblende structures, finding 652 cm^{-1} for AlN and 551 cm^{-1} for GaN at $P = 0$. The results of Ref. [274] are very close, 648 and 558 cm^{-1}. Also, the TO frequencies, 603 and 600 cm^{-1}, both for cubic GaN, obtained in Ref. [148] and [153] are in good agreement with our theoretical values. No experimental data seem to be available for the cubic phases.

The pressure dependence of the phonon modes with wave vectors different from zero have also been extensively discussed. In particular the transverse acoustic (TA) modes in the cubic semiconductors are interesting, since their dispersion curves are flat away from the zone center and thus make a large contribution to the density of states. Furthermore, their energy is low, and the mode Grüneisen parameters at the zone faces (X and L) are negative [275]. These features cause the peculiar temperature dependence of the thermal expansion coefficient, α, which in a certain temperature range, in Si around 80 K, becomes *negative* [276]. Biernacki and Scheffler [277] calculated, from first principles, the total electronic energy and the phonon frequencies (and were thus able to derive the vibrational entropy) as functions of crystal volume. From this they calculated the Gibbs free energy and obtained the temperature dependence of α for Si. The quantitative agreement with experiment [276] was very good, and the occurrence of the temperature region where α is negative was explained. The negative Grüneisen parameter at the X-point has also been related [278] to the diamond → β-tin transition in Ge and Si. Klotz *et al.* [279] measured the TA phonons along the [100], [110], and [111] directions in Ge as functions of pressure up to 9.7 GPa by inelastic neutron scattering. The experimental results were compared to LDA calculations, and very good agreement was found. The Grüneisen parameters, $\gamma_{TA}(L)$ and $\gamma_{TA}(X)$, were measured to be -1.51 ± 0.15 and -1.31 ± 0.15, respectively. The theoretical values as obtained [279] by the pseudopotential method are -1.18 and -1.36. The value of $\gamma_{TA}(X)$ obtained [280] from Raman experiments is -1.53 ± 0.05.

VII. Concluding Remarks

The density-functional theory of electronic states has been reviewed briefly and applied in actual *ab initio* calculations of some physical properties of semiconductors under pressure. These include pressure-induced structural phase transitions, hydrostatic and uniaxial strain deformation potentials, internal strain parameters, elastic constants, phonon frequencies, mode Grüneisen parameters, pressure dependence of refractive indices, polarities, ionicities, and dynamical effective masses. Too little attention has been paid to pressure effects on superlattice electronic structures; we only discussed one case, ZnS/ZnSe, and showed that a type I → type II conversion occurs. We completely left out the extremely important and large field of defect states and their pressure dependencies. A separate review would be needed to discuss just some of the important results in this area obtained by several theoretical as well as experimental groups. Some recent results as well as references to relevant papers can be obtained, for example, from the Proceedings of the Conference on High Pressure Semiconductor Physics [281].

The crudest, but most frequently used, approximation — the local (LDA) approximation — to the density-functional theory underestimates the gaps when these are derived from the eigenvalue spectrum. Corrections to this well-known gap problem were discussed, and it was shown that the pressure coefficients of the gaps are only slightly affected by the gap error. Effective masses of the conduction bands are usually not given correctly in the LDA, since the gap corrections are $\overline{k}$ dependent. Si may be an example where essentially a rigid upshift of the conduction bands is sufficient, but this is fortuitous.

Apart from this "LDA gap problem," we are perhaps not used to encountering problems when applying the LDA in semiconductor physics, if we disregard systems like the transition-metal monoxides. Considering the cohesive properties, P–V relations, elastic constants, and so forth, this is also true as long as these properties are described essentially by the contributions to the forces from the s- and p-states alone. In compounds with localized d-states, like the Ga-3d in GaN, Zn-3d in ZnSe, 4d in tin, and Fe-3d in ϵ-FeSi, this is no longer the case. Corrections are required, and we examined a few cases by applying GGA, SIC as well as a reduction of hybridization (downshifting of d-states) and mass-renormalization (ϵ-FeSi). The latter corrections reduce the LDA overbinding, and the method resembles the kind of renormalization (of f-states) applied to heavy-fermion systems [282, 283]. Boebinger [284] argues that the study of Kondo insulators in strong magnetic fields will provide new information about these systems. It has been suggested in the present article that pressure studies, for example of ϵ-FeSi, would also be very interesting.

Future theoretical research in the field discussed here will be concerned with refined density-functional theory making possible a still better description for (strongly) correlated electron systems amenable to practical numerical calculations. the variation of pressure allows testing of these methods, but the theoretical calculations must also, to a larger extent than done at the present, include variation of other thermodynamic parameters. Inclusion of finite temperatures and entropy contributions — configurational as well as vibrational — can be made at the present. This will become even more important, and it implies that larger parts of the equation-of-state of the materials can be explored. The combination of experimental and theoretical data will allow us to form an increasingly detailed physical picture of the basic interactions that determine the physical and chemical properties of solids, nanoparticles, clusters, and molecules. Apart from the gain in basic scientific knowledge, this will also help us to proceed towards the goal of designing new materials with properties optimized to technological device applications. Some of these materials may, in fact, be metastable structures produced by relaxing pressure after structural phase transitions have been induced.

Acknowledgments

A large part of the work described here was performed in collaboration with colleagues and inspired by several discussions. In particular the author wishes to thank O. K. Andersen, M. Cardona, M. Alouani, I. Gorczyca, C. O. Rodriguez, A. Svane, K. Syassen, S. Satpathy, I. Wenneker, M. Fanciulli, and G. Weyer. A. Heiring is thanked for her expert assistance in the composition of the manuscript.

REFERENCES

[1] Jones, R. O., and Gunnarsson, O. (1989). *Rev. Mod. Phys.* **61**, 689.
[2] Parr, R. G. W., and Yang, W. (1989). "Density Functional Theory of Atoms and Molecules." Oxford.
[3] Dreizler, R. M., and Gross, E. K. U. (1990). "Density Functional Theory." Springer Verlag, Berlin.
[4] Hohenberg, P., and Kohn, W. (1964). *Phys. Rev.* **136**, B864.
[5] Levy, M. (1979). *Proc. Natl. Acad. Sci. (USA)* **76**, 6062.
[6] Kohn, W., and Sham, L. J. (1965). *Phys. Rev.* **140**, A1133.
[7] von Barth, U., and Hedin, L. (1972). *J. Phys.* **C5**, 1629.
[8] Gunnarsson, O., and Lundqvist, B. I. (1976). *Phys. Rev.* **B13**, 4274.
[9] Vosko, S. H., Wilk, L., and Nusair, M. (1980). *Cand. J. Phys.* **58**, 1200.
[10] Perdew, J. P., and Zunger, A. (1981). *Phys. Rev.* **B23**, 5048.
[11] Moruzzi, V. L., Janak, F. F., and Williams, A. R. (1978). "Calculated Electronic Properties of Metals." Pergamon Press, New York.
[12] Ceperley, D. M., and Adler, B. J. (1980). *Phys. Rev. Lett.* **45**, 566.
[13] Perdew, J. P., and Zunger, A. (1981). *Phys. Rev.* **B23**, 5048.

[14] Svane, A., and Gunnarsson, O. (1990). *Phys. Rev. Lett.* **65,** 1148.
[15] Svane, A. (1996). *Phys. Rev.* **B53,** 4275.
[16] Braicovich, L., Ciccacci, F., Puppin, E., Svane, A., and Gunnarsson, O. (1992). *Phys. Rev.* **B46,** 12165.
[17] Svane, A. (1992). *Phys. Rev. Lett.* **68,** 1900.
[18] Svane, A. (1996). Private communication.
[19] Krieger, J. B., Li, Y., and Iafrate, G. J. (1990). *Phys. Lett.* **A146,** 256.
[20] Sharp, R. T., and Horton, G. K. (1953). *Phys. Rev.* **90,** 317.
[21] Bylander, D. M., and Kleinman, L. (1995). *Phys. Rev.* **B52,** 14566.
[22] Kotani, T. (1995). *Phys. Rev. Lett.* **74,** 2989; Kotani, T., and Akai, H. (1996). *Phys. Rev.* **B54,** 16502.
[23] Herman, F., Van Dyke, J. P., and Ortenburger, I. B. (1969). *Phys. Rev. Lett.* **22,** 807.
[24] Engel, E., and Vosko, S. H. (1994). *Phys. Rev.* **B50,** 10498.
[25] Langreth, D. C., and Mehl, M. J. (1983). *Phys. Rev.* **B28,** 1809.
[26] Becke, A. D. (1988). *Phys. Rev.* **A38,** 3098.
[27] Perdew, P., and Wang, Y. (1986). *Phys. Rev.* **B33,** 8800.
[28] Perdew, P. (1986). *Phys. Rev.* **B33,** 8822; (1986) **34,** 7406 (E).
[29] Perdew, P., Chevary, J. A., Vosko, S. H., Jackson, K. A., Pederson, M. R., Singh, D. J., and Fiolhais (1992). *C. Phys. Rev.* **B46,** 6671.
[30] Dufek, P., Blaha, P., Sliwko, V., and Schwarz, K. (1994). *Phys. Rev.* **B49,** 10170.
[31] Fan, L., and Ziegler, T. (1991). *J. Chem. Phys.* **94,** 6057.
[32] Ortiz, G., and Ballone, P. (1991). *Phys. Rev.* **B43,** 6376.
[33] Hammer, B., and Scheffler, M. (1995). *Phys. Rev. Lett.* **74,** 3487.
[34] Barnett, R. N., and Landman, U. (1993). *Phys. Rev.* **B48,** 2081.
[35] Perdew, J. P., and Levy, M. (1983). *Phys. Rev. Lett.* **51,** 1884.
[36] Sham, L. J., and Schlüter, M. (1983). *Phys. Rev. Lett.* **51,** 1888.
[37] Fritsche, L. (1995). "Density Functional Theory" (Gross, E. K. U., and Dreizler, R. M., eds.). Plenum Press, New York, p. 119.
[38] Borrmann, W., and Fulde, P. (1987). *Phys. Rev.* **B35,** 9569.
[39] Borrmann, W., and Fulde, P. (1985). *Phys. Rev.* **B31,** 7800.
[40] Borrmann, W., and Fulde, P. (1986). *Europhys. Lett.* **2,** 471.
[41] Hedin, L. (1965). *Phys. Rev.* **139,** A796.
[42] Hedin, L., and Lundqvist, S. (1969). *In* "Solid State Physics" (Ehrenreich, H., Turnbull, and Seitz, F., eds.). Academic Press, New York, Vol. 23, p. 1.
[43] Hybertsen, M. S., and Louie, S. G. (1986). *Phys. Rev.* **B34,** 5390.
[44] von der Linden, W., and Horsch, P. (1988). *Phys. Rev.* **B37,** 8351.
[45] Godby, R. W., Schlüter, M., and Sham, L. J. (1986). *Phys. Rev. Lett.* **56,** 2415.
[46] Hott, R. (1990). "Ab Initio Bandstrukturrechnung in der GW-Näherung." Thesis, Max-Planck-Institut für Festkörperforschung, Stuttgart; *Phys. Rev.* **B44,** 1057 (1991).
[47] Bachelet, G. B., and Christensen, N. E. (1984). *Phys. Rev.* **B31,** 879.
[48] Christensen, N. E. (1991). Unpublished.
[49] Landolt-Börnstein. (1982). "Zahlenwerte und Funktionen aus Naturwissenschaft und Technik," Vol. III, (Madelung, O., ed.). Springer Verlag, Berlin.
[50] Aryasetiawan, F., and Gunnarsson, O. (1994). *Phys. Rev.* **B49,** 7219; *Phys. Rev.* **B49,** 16214 (1994).
[51] Rohlfing, M., Krüger, P., and Pollmann, J. (1995). *Phys. Rev.* **B52,** 1905.
[52] Christensen, N. E., and Gorczyca, I. (1994). *Phys. Rev.* **B50,** 4397.
[53] Palummo, M., Reining, L., Godby, R. W., Bertoni, C. M., and Börnsen, N. (1994). *Europhys. Lett.* **26,** 607.

[54] Rubio, A., Corkill, J. L., Cohen, M. L., Shirley, E. L., and Louie, S. G. (1993). *Phys. Rev.* **B48,** 11810.
[55] Rubio, A., Corkill, J. L., and Cohen, M. L. (1994). *Phys. Rev.* **B49,** 1952.
[56] Bechstedt, F., and del Sole, R. (1988). *Phys. Rev.* **B38,** 7710.
[57] Harrison, W. A. (1980). "Electronic Structure." Freeman, San Francisco.
[58] Gunnarsson, O., and Schönhammer, K. (1986). *Phys. Rev. Lett.* **56,** 1968.
[59] Zhu, X., Fahy, S., and Louis, S. G. (1989). *Phys. Rev.* **B39,** 7840.
[60] Hanke, W., and Sham, L. J. (1989). *Solid State Commun.* **71,** 211.
[61] Perlin, P., Gorczyca, I., Christensen, N. E., Grzgory, I., Tesseyre, H., and Suski, T. (1996). *Phys. Rev.* **B45,** 13307.
[62] Christensen, N. E., Satpathy, S., and Pawlowska, Z. (1987). *Phys. Rev.* **B36,** 1032.
[63] Christensen, N. E. (1996). *Phys. Stat. Sol.* (*b*) **198,** 23.
[64] Christensen, N. E. (1996). Unpublished.
[65] Baraff, G. A., and Schlüter, M. (1984). *Phys. Rev.* **B30,** 3460.
[66] Gygi, F., and Baldereschi, A. (1989). *Phys. Rev. Lett.* **62,** 2160. (This work describes a useful model self-energy correction method that approximates complete GW calculations well.)
[67] Christensen, N. E. (1984). *Phys. Rev.* **B30,** 5753.
[68] Pitt, G. D., Lees, J., Hoult, R. A., and Stradling, R. A. (1973). *J. Phys.* **C6,** 3282.
[69] DeMeis, W. M. (1965). Technical Report HP-15, Harvard University (unpublished).
[70] Welber, B., Cardona, M., Kim, C. K., and Rodriguez, S. (1975). *Phys. Rev.* **B12,** 5729.
[71] Yu, P. Y., and Welber, B. (1978). *Solid State Commun.* **25,** 209.
[72] Jiang, Z. X., Chen, R. J., Tischler, J. G., Weinstein, B. A., and McCombe, B. D. (1996). *Phys. Stat. Sol.* (*b*) **198,** 41.
[73] Gorczyca, I., Christensen, N. E., and Alouani, M. (1989). *Phys. Rev.* **B39,** 7705.
[74] Kobayashi, T., Tei, T., Aoki, K., Yamamoto, K., and Abe, K. (1989). *In* "Physics of Solids under High Pressure" (Schilling, J. S., and Shelton, R. N., eds.). North-Holland, Amsterdam.
[75] Müller, H., Trommer, R., Cardona, M., and Vogl, P. (1980). *Phys. Rev.* **B21,** 4879.
[76] Alouani, M., Gopalan, S., Garriga, M., and Christensen, N. E. (1988). *Phys. Rev. Lett.* **61,** 1643.
[77] Christensen, N. E., and Andersen, O. K. (1985). Unpublished.
[78] Andersen, O. K. (1975). *Phys. Rev.* **B12,** 3060.
[79] Mackintosh, A. R., and Andersen, O. K. (1979). *In* "Electrons at the Fermi Surface" (Springford, M., ed.). Cambridge University Press, Cambridge, UK, Sect. 5.3.11. See also Heine, V. (1980). *Solid State Physics* **35,** 1.
[80] The fact that, for example, (III-1) is a first-order theory implies that it correctly gives the slope of the energy-versus-displacement curve, i.e., the "force." Hence the name "force theorem."
[81] Christensen, N. E. (1984). *Solid State Commun.* **49,** 701.
[82] Skriver, H. L. (1982). *Phys. Rev. Lett.* **49,** 1768.
[83] Skriver, H. L. (1985). *Phys. Rev.* **B31,** 1090.
[84] Christensen, N. E., and Heine, V. (1985). *Phys. Rev.* **B32,** 6145.
[85] Pettifor, D. G. (1976). *Commun. Phys.* **1,** 141.
[86] Christensen, N. E. (1987). *Physica Scripta* **T19,** 298.
[87] Hanfland, M., and Syassen, K. (1984). *J. Phys. Coll. Suppl. 11,* **45,** C8–57.
[88] Goñi, A. R., Strössner, K., Syassen, K., and Cardona, M. (1987). *Phys. Rev.* **B36,** 1581.
[89] Perlin, P., Trzeciakowski, W., Litwin-Staszewska, E., Muszalski, J., and Micovic, M. (1994). *Semicond. Sci. Technol.* **9,** 2239.
[90] Prins, A. D., Sly, J. L., and Dunstan, D. J. (1996). *Phys. Stat. Sol.* (*b*) **198,** 57.

[91] Christensen, N. E., and Gorczyca, I. (1993). *Phys. Rev.* **B47**, 4307.
[92] Ueno, M., Onodera, A., Shimomura, D., and Takemura, K. (1992). *Phys. Rev.* **B45**, 10123.
[93] Christensen, N. E. (1988). *Phys. Rev.* **B37**, 4528.
[94] Bardeen, J., and Shockley, W. (1959). *Phys. Rev.* **80**, 72.
[95] Chandrasekhar, M., and Pollak, F. H. (1977). *Phys. Rev.* **B15**, 2127.
[96] Willatzen, M., Lew Yan Voon, L. C., Santos, P. V., Cardona, M., Munzar, D., and Christensen, N. E. (1995). *Phys. Rev.* **B52**, 5070.
[97] Christensen, N. E., and Gorczyca, I. (1991). *Phys. Rev.* **B44**, 1707.
[98] Fan, H. Y. (1951). *Phys. Rev.* **82**, 900.
[99] Cardona, M. (1987). *Phys. Rev.* **B35**, 9174.
[100] Gopalan, S., Lautenschlager, P., and Cardona, M. (1987). *Phys. Rev.* **B35**, 5577.
[101] Aspnes, D. E., and Cardona, M. (1978). *Phys. Rev.* **B17**, 741.
[102] Kleinman, L. (1962). *Phys. Rev.* **128**, 2614.
[103] Christensen, N. E. (1984). *Solid State Commun.* **50**, 177.
[104] Balslev, I. (1967). *Solid State Commun.* **5**, 375.
[105] Nielsen, O. H., and Martin, R. M. (1993). *Phys. Rev. Lett.* **50**, 697 (1993); *Phys. Rev.* **B32**, 3780 (1985); *Phys. Rev.* **B32**, 3792 (1985).
[106] Christensen, N. E. (1984). *Phys. Stat. Sol.* (*b*) **125**, K59.
[107] Brey, L., Christensen, N. E., and Cardona, M. (1987). *Phys. Rev.* **B36**, 2638.
[108] Nielsen, O. H. (1986). *Phys. Rev.* **B34**, 5808.
[109] Weyrich, K. H., Brey, L., and Christensen, N. E. (1988). *Phys. Rev.* **B38**, 1392.
[110] As a consequence of this linear relation between the force constant, A, and the bond order, $b(A = b_0 + \text{cst} \cdot b)$, we were able to calculate the Γ(TO)-phonon frequencies with very good precision for many compounds by using theoretical values for b and ω_{TO} taken either from experiment or from full-potential LMTO calculations for *two* semiconductors (Si and Ge) only. As a curiosity it is mentioned that the b values were obtained from LMTO-ASA calculations. Calculation of phonon frequencies directly from total energies or forces requires that a full-potential method be used.
[111] Blacha, A., Presting, H., and Cardona, M. (1984). *Phys. Stat. Sol.* (*b*) **126**, 11.
[112] Lambrecht, W. R. L., Segall, B., Methfessel, M., and van Schilfgaarde, M. (1991). *Phys. Rev.* **B44**, 3685.
[113] Cardona, M., and Harbeke, G. (1982). *In* "Landolt–Börnstein Numerical Data and Functional Relationships in Science and Technology, New Series" (Madelung, O., Schulz, H., and Weiss, K., eds.). Springer, Berlin, Vol. 17a.
[114] Canali, C., Jacoboni, C., Nava, F., and Reggiana, L. (1978). "Proc. of the International Conference on the Physics of Semiconductors, Edinburgh" (Wilson, B. L. H., ed.). Bristol, New York, p. 327.
[115] Lawaetz, P. (1978). "The Influence of Holes on the Phonon Spectrum of Semiconductors." Thesis, The Technical University of Denmark.
[116] Jacoboni, J., Gagliani, G., Reggiani, L., and Turci, V. (1978). *Solid State Electron.* **21**, 315.
[117] Wei, S. (1991). "Structural, Dynamical and Optical Properties of the Semiconductors Si and Ge and Their Superlattice." Thesis, Ohio State University, Columbus, OH.
[118] Cousins, L. S. G., Gerward, L., Staun Olsen, J., Selsmark, B., and Sheldon, B. J. (1987). *J. Phys.* **C20**, 29.
[119] Koumelis, C. N., Zardas, G. E., Landos, C. A., and Leventuri, D. K. (1975). *Acta Crystalogr. Sect. A* **32**, 84.
[120] Casali, R. A., and Christensen, N. E. (1998). In press, *Solid State Commun.* Unpublished, and *Bull. Am. Phys. Soc.* **42**, 578.
[121] Callega, J. M., Vogt, H., and Cardona, M. (1982). *Phil. Mag.* **45**, 239.
[122] Pötz, W., and Vogl, P. (1981). *Phys. Rev.* **B24**, 2025.

[123] Zollner, S., Myers, K. D., Jensen, K. G., Dolan, J. M., Bailey, D. W., and Stanton, C. J., *Solid State Commun.* (1997). In press.
[124] Aspnes, D. E., and Cardona, M. (1978). *Phys. Rev.* **B17,** 741.
[125] Rockwell, B., Chandrasekhar, H. R., Chandrasekhar, M., Ramdas, A. K., Kobayashi, M., and Gunshor, R. L. (1991). *Phys. Rev.* **B44,** 11307.
[126] Van de Walle, C. G., and Martin, R. M. (1989). *Phys. Rev. Lett.* **62,** 2028.
[127] Christensen, N. E. (1988). *Phys. Rev.* **B37,** 4528.
[128] Resta, R., Colombo, L., and Baroni, S. (1990). *Phys. Rev.* **B41,** 12358 (1990); *Phys. Rev.* **B41,** 14273(E) (1990).
[129] Resta, R. (1991). *Phys. Rev.* **B44,** 11035.
[130] Vergés, J. A., Glötzel, D., Cardona, M., and Andersen, O. K. (1982). *Phys. Stat. Sol.* (*b*) **113,** 519.
[131] Cardona, M., and Christensen, N. E. (1987). *Phys. Rev.* **B35,** 6182 (1987); *Phys. Rev.* **B36,** 2906 (E) (1987).
[132] Tejedor, C., Flores, F., and Louis, E. (1977). *J. Phys.* **C10,** 2163.
[133] Flores, F., and Tejedor, C. (1979). *J. Phys.* **C12,** 731.
[134] Tersoff, J. (1984). *Phys. Rev. Lett.* **52,** 465.
[135] Tersoff, J. (1984). *Phys. Rev.* **B30,** 4870.
[136] Baldereschi, A. (1973). *Phys. Rev.* **B7,** 5212.
[137] Christensen, N. E. (1994). *Phil. Mag.* **B70,** 567.
[138] Phillips, J. C. (1973). "Bonds and Bands in Semiconductors." Academic Press, New York.
[139] Camphausen, D. L., Neville Conell, G. A., and Paul, W. (1971). *Phys. Rev. Lett.* **26,** 184.
[140] Alouani, M., and Wills, J. M. (1996). *Phys. Rev. B,* in press and private communication.
[141] The *f*-values given here differ a little from those presented earlier in Ref. [62] because of some calculational details.
[142] Goñi, A. R., Syassen, K., and Cardona, M. (1989). *Phys. Rev.* **B41,** 10104.
[143] Wickboldt, P., Anastassakis, E., Sauer, R., and Cardona, M. (1987). *Phys. Rev.* **B35,** 1362.
[144] Anastassakis, E., and Cardona, M. (1984). *Phys. Stat. Sol.* (*b*) **126,** 11.
[145] Cardona, M. (1996). *Phys. Stat. Sol.* (*b*) **198,** 5.
[146] Chang, K. J., Froyen, S., and Cohen, M. L. (1984). *Solid State Commun.* **50,** 105.
[147] Lee, S., Sanchez-Dehesa, J., and Dow, J. D. (1985). *Phys. Rev.* **B32,** 1152.
[148] Lambrecht, W. R. L., and Segall, B. (1994). *In* "Properties of Group III Nitrides" (Edgar J. H., ed.), Electronic Materials Information Service (EMIS) Data Reviews Series No. 11. INSPEC, the Institution of Electrical Engineers, London, p. 124.
[149] Van Camp, P. E., Van Doren, V. F., and Devreese, J. T. (1991). "Proc. XIII AIRAPT Int. Conf. on High Pressure Sci. and Technol., Bangalore, India," p. 237; *Phys. Rev.* **B44,** 9056 (1991); *Solid State Commun.* **81,** 23 (1992).
[150] Wentzcovitch, R. M., Chang, K. J., and Cohen, M. L. (1986). *Phys. Rev.* **B34,** 1071.
[151] Min, B. J., Chan, C. T., and Ho, K. M. (1992). *Phys. Rev.* **B45,** 1159.
[152] Kikuchi, K., Uda, T., Sakuma, A., Hirao, M., and Murayama, Y. (1992). *Solid State Commun.* **81,** 653.
[153] Fiorentini, V., Methfessel, M., and Scheffler, M. (1993). *Phys. Rev.* **B48,** 13353.
[154] Yong-Nian, Xu, and Ching, W. Y. (1993). *Phys. Rev.* **B48,** 4335.
[155] Grinyaev, S. N., Ya. Malakhov, V., and Chaldyshev, V. A. (1986). *Izv. Vyssh. Uchebn. Zaved. Fiz.* **4,** 69.
[156] Zetterstrom, R. B. (1970). *J. Mater. Sci.* **5,** 1102.
[157] Tyagay, W. A., and Yevstigneev, A. M. (1977). *Fiz. Tech. Polupr.* **11,** 2142.
[158] Monemar, B. (1974). *Phys. Rev.* **B10,** 676.
[159] Bloom, S., Harbeko, G., Meier, E., and Ortenburger, I. B. (1974). *Phys. Stat. Sol.* (*b*) **66,** 161.

[160] Perry, B., and Rutz, R. F. (1978). *Appl. Phys. Lett.* **33,** 319.
[161] Fomichev, V. A., and Rumsh, M. A. (1968). *J. Phys. Chem. Solids* **29,** 1015.
[162] Strite, S., Ruan, J., Li, Z., Salvador, A., Chen, H., Smith, D. J., Choyke, W. J., and Morkoç M. (1991). *J. Vac. Sci. Technol.* **B9,** 192.
[163] Chrenko, R. M. (1974). *Solid State Commun.* **14,** 511.
[164] Lei, T., Moustakas, T. D., Graham, R. J., He, Y., and Berkowitz, S. J. (1992). *J. Appl. Phys.* **71,** 4933.
[165] Miyata, M., and Moriki, K. (1989). *Phys. Rev.* **B40,** 12028.
[166] Schultz, H., and Thiemann, K. H. (1977). *Solid State Commun.* **23,** 813.
[167] Gorczyca, I., and Christensen, N. E. (1993). *Phys. Rev.* **B48,** 17202.
[168] Ves, S., Schwarz, U., Christensen, N. E., Syassen, K., and Cardona, M. (1990). *Phys. Rev.* **B42,** 9113.
[169] Jaszcyn-Kopeç, A., Canny, B., and Sysoffe, C. (1983). *J. Lumin.* **28,** 319.
[170] Ves, S., Strössner, K., Christensen, N. E., Chul Koo, Kim, and Cardona, M. (1985). *Solid State Commun.* **56,** 479.
[171] Cardona, M. (1963). *J. Phys. Chem. Solids* **24,** 1543.
[172] Keating, P. N. (1966). *Phys. Rev.* **145,** 637.
[173] Martin, R. M. (1970). *Phys. Rev.* **B1,** 4005.
[174] Yamada, Y., Masumoto, Y., Taguchi, T., and Takemura, K. (1991). *Phys. Rev.* **B44,** 1801.
[175] Menoni, C. S., and Spain, I. L. (1987). *Phys. Rev.* **B35,** 7520.
[176] Minomura, S., and Drickamer, H. G. (1962). *J. Phys. Chem. Solids* **23,** 451.
[177] Christensen, N. E. (1996). Unpublished.
[178] Soma, T., and Matsuo Kasaya, M. (1984). *Solid State Commun.* **49,** 261.
[179] Mujica, A., and Needs, R. (1997). *Phys. Rev.* **B55,** 9659.
[180] Christensen, N. E., Novikov, D. L., Alonso, R., Weht, R., and Rodriguez, C. O. (1998). Work in progress.
[181] Bates, C. H., White, W. B., and Ray, R. (1962). *Science* **162,** 993.
[182] Yu, S. C., Spain, I. L., and Skelton, E. F. (1978). *Solid State Commun.* **25,** 49.
[183] Chang, K. J., and Cohen, M. L. (1984). *Solid State Commun.* **50,** 487.
[184] Van Camp, P. E., and Van Doren, V. E. (1996). *J. Phys. Condens. Matter* **8,** 3385.
[185] Jephcoat, A. P., Hemley, R. J., Mao, H. K., Cohen, R. E., and Mehl, M. J. (1988). *Phys. Rev.* **B37,** 4727.
[186] Boettger, J. C., and Wills, J. M. (1996). *Phys. Rev.* **B54,** 8965.
[187] Tosi, M. P. (1964). *Solid State Phys.* **16,** 1.
[188] Zhang, S. B., and Cohen, M. L. (1987). *Phys. Rev.* **B35,** 7604.
[189] Weir, S. T., Vohra, Y. K., Vanderborgh, C. A., and Ruoff, A. L. (1989). *Phys. Rev.* **B39,** 1280.
[190] Duclos, S. J., Vohra, Y. K., and Ruoff, A. L. (1990). *Phys. Rev.* **B41,** 12021.
[191] Vohra, Y. K. (1986). *Phys. Rev. Lett.* **56,** 1944.
[192] Weir, S. T., Vohra, Y. K., and Ruoff, A. L. (1987). *Phys. Rev.* **B36,** 4543.
[193] Jamieson, J. C. (1963). *Science* **139,** 845.
[194] Jayaraman, A., Klement, W., and Kennedy, G. C. (1963). *Phys. Rev.* **130,** 540.
[195] McMahon, M. I., Nelmes, R. J., Wright, N. G., and Allan, D. R. (1994). *Phys. Rev.* **B50,** 13047.
[196] McMahon, M. I., and Nelmes, R. J. (1993). *Phys. Rev.* **B47,** 8331; McMahon, M. I., Nelmes, R. J., Wright, N. G., and Allan, D. R. (1994). *Phys. Rev.* **B50,** 739.
[197] Lewis, S. P., and Cohen, M. L. (1993). *Phys. Rev.* **B48,** 16144.
[198] Lewis, S. P., and Cohen, M. L. (1994). *Solid State Commun.* **89,** 483.
[199] Needs, R. J., and Martin, R. M. (1984). *Phys. Rev.* **B30,** 5390.
[200] Christensen, N. E. (1986). *Phys. Rev.* **B33,** 5096.

[201] Vohra, Y. K., Weir, S. T., and Ruoff, A. L. (1985). *Phys. Rev.* **B31,** 7344.
[202] Christensen, N. E. (1997). Unpublished.
[203] Nelmes, R. J., McMahon, M. I., Wright, N. G., and Allan, D. R. (1995). *Phys. Rev.* **B51,** 15723. (And references therein).
[204] Nelmes, R. J., McMahon, M. I., and Belmonte, S. A. (1997). To be published.
[205] Nelmes, R. J., McMahon, M. I., Belmonte, S. A., and Allan, D. R. (1997). *High Pressure Science and Technology* **6,** 136.
[206] Lambrecht, W. R. L., Kim, K., Rashkeev, S. G., and Segall, B. (1996). *Mat. Res. Symp. Proc.* **395,** 455, "Gallium Nitride and Related Materials," (Pouce, F. A., Dupuis, R. D., Nakamura, S., and Edmond, J. A., eds.) MRS, Pittsburgh.
[207] Satta, A., Fiorentini, Bosin, A., and Meloni, F. (1996). *MRS Proc.* **395,** 515.
[208] Bellaiche, L., Kunc, K., and Besson, J. M. (1996). *Phys. Rev.* **B54,** 8945.
[209] Ching, W. Y., and Harmon, B. N. (1986). *Phys. Rev.* **B34,** 5305.
[210] Muñoz, A., and Kunc, K. (1991). *Phys. Rev.* **B44,** 10372.
[211] Palummo, M., Bertoni, C. M., Reining, L., and Finochi, F. (1993). *Physica* **B185,** 404.
[212] Perlin, P., Jauberthie-Carillon, C., Itie, J. P., Miguel, A. S., Grzegory, I., and Polian, A. (1991). *High Pressure Research* **71,** 96.
[213] Paisley, M., Sitar, Z., Posthill, J., and Davis, R. (1989). *J. Vac. Sci. Tech.* **A7,** 701.
[214] Sheleg, A., and Savastenko, V. (1979). *Inorg. Mater.* **15,** 1257.
[215] Xia, H., Xia, Q., and Ruoff, A. L. (1993). *Phys. Rev.* **B47,** 12925.
[216] Ueno, M., Yoshida, M., Onodera, A., Shimomura, O., and Takemura, K. (1994). *Phys. Rev.* **B49,** 14.
[217] Perlin, P., Jauberthie-Carillon, C., Itié, J. P., Miguel, A. S., Grzegory, I., and Polian, A. (1992). *Phys. Rev.* **B45,** 83.
[218] Leszczynski, M., Suski, T., Perlin, P., Teisseyre, H., Grzegory, I., Bokowski, M., Jim, J., Porowski, S., and Major, J. (1995). *J. Phys. D: Appl. Phys.* **28,** 1.
[219] Garciá, A., and Cohen, M. L. (1993). *Phys. Rev.* **B47,** 4215.
[220] Perlin, P., Gorczyca, I., Porowski, S., Suski, T., Christensen, N. E., and Polian, A. (1993). *Jpn. J. Appl. Phys.* **32,** 334.
[221] Wentzcovitch, R. M., and Cohen, M. L. (1987). *Phys. Rev.* **B36,** 6058.
[222] Pandey, R., Jaffe, J. E., and Harrison, N. M. (1994). *J. Phys. Chem. Solids* **55,** 1357.
[223] Pandey, R., Causá, M., Harrison, N. M., and Seel, M. (1996). *J. Phys.: Condens. Matter* **8,** 3993.
[224] Christensen, N. E., Satpathy, S., and Pawlowska, Z. (1986). *Phys. Rev.* **B34,** 5977.
[225] Andersen, O. K., and Jepsen, O. (1984). *Phys. Rev. Lett.* **53,** 2571.
[226] Andersen, O. K., Jepsen, O., and Glötzel, D. (1985). Highlights in condensed matter theory." *In* "Proc. of the Enrico Fermi International School of Physics, Course LXXXIX" (Bassani, F., Fumi, F., and Tosi, M. P., eds.). North-Holland, Amsterdam, p. 59.
[227] Weinstein, B. A., and Piermarini, G. J. (1975). *Phys. Rev.* **B12,** 1172.
[228] Piermarini, G. J., and Block, S. (1975). *Rev. Sci. Instrum.* **46,** 973.
[229] Yin, M. T., and Cohen, M. L. (1982). *Phys. Rev.* **B26,** 5668.
[230] Christensen, N. E. (1996). Unpublished.
[231] Olijnyk, H., Sikka, S. K., and Holzapfel, W. B. (1984). *Phys. Lett.* **103A,** 137.
[232] Hu, J. Z., and Spain, I. L. (1984). *Solid State Commun.* **51,** 263.
[233] Kasper, J. S., and Richards, S. M. (1963). *Acta Cryst.* **17,** 752.
[234] Verges, J. A., Alouani, M., and Christensen, N. E. (1988). *Phys. Rev.* **B38,** 1378.
[235] Hanfland, M., Alouani, M., Syassen, K., and Christensen, N. E. (1988). *Phys. Rev.* **B38,** 12864.
[236] Zhao, Y. X., Buchler, F., Sites, J. R., and Spain, I. L. (1986). *Solid State Commun.* **59,** 679.

[237] Duclos, S. J., Vohra, Y. K., and Ruoff, A. L. (1987). *Phys. Rev. Lett.* **58**, 775.
[238] Crain, J., Clark, S. J., Ackland, G. J., Payne, M. C., Milman, V., Hatton, P. D., and Reid, B. J. (1994). *Phys. Rev.* **B49**, 5329.
[239] Clark, S. J., Ackland, G. J., and Crain, J. (1994). *Phys. Rev.* **B49**, 5341.
[240] McMahon, M. I., and Nelmes, R. J. (1996). *Phys. Stat. Sol.* (*b*) **198**, 389.
[241] Vohra, Y. K., Brister, K. E., Desgreniers, S., Ruoff, A. L., Chang, K. J., and Cohen, M. L. (1986). *Phys. Rev. Lett.* **56**, 1944.
[242] Christensen, N. E., and Methfessel, M. (1993). *Phys. Rev.* **B48**, 5797.
[243] Olijnyk, H., and Holzapfel, W. B. (1984). *J. Physique* (*Paris*) *Colloq.* **45**, Suppl. 11, C8-153.
[244] Christensen, N. E. (1985). *Solid State Commun.* **85**, 151.
[245] Brudevoll, T., Citrin, D. S., Christensen, N. E., and Cardona, M. (1993). *Phys. Rev.* **B48**, 17128.
[246] Christensen, N. E. (1996). Unpublished.
[247] Christensen, N. E., Svane, A., and Brudevoll, T. (1996). *Bull. Am. Phys. Soc.* **41**, 717.
[248] Takemura, K. (1995). *Phys. Rev. Lett.* **75**, 1807.
[249] Takemura, K. (1996). Private communication.
[250] Full-potential LMTO code developed by Wills, J., Los Alamos Natl. Lab., New Mexico.
[251] Wimmer, E., Krakauer, H., Weinert, M., and Freeman, A. J. (1981). *Phys. Rev.* **B24**, 864, and references therein.
[252] Blaha, P., Schwarz, K., Dufek, P., and Augustyn, R. (1995). WIEN95, Technical University of Vienna. (Improved and updated Unix version of the original copyrighted WIEN code, which was published by Blaha, P., Schwarz, K., Soratin, P., and Trickey, S. B. in *Comput. Phys. Commun.* **59**, 399 (1990).)
[253] Christensen, N. E., Gunnarsson, O., Jepsen, O., and Andersen, O. K. (1988). *Journ. de Phys., Coll. C8,* Suppl. au #12, **49**, 17.
[254] Vogel, D., Krüger, P., and Pollmann, J. (1996). *Phys. Rev. B* **54**, 5495.
[255] Blacha, A., and Cardona, M., Christensen, N. E., Ves, S., and Overhof, H. (1982–83). *Physica* **117B** & **118B**, 63 (1983), and *Sol. Stat. Commun.* **43**, 183 (1982).
[256] Christensen, N. E. (1984). *Phys. Stat. Sol.* (*b*) **125**, K59.
[257] Christensen, N. E. (1984). *Phys. Stat. Sol.* (*b*) **123**, 281.
[258] Christensen, N. E. (1990). *Phys. Rev.* **B42**, 7148.
[259] Christensen, N. E., Wenneker, I., Svane, A., and Fanciulli, M. (1996). *Phys. Stat. Sol.* (*b*) **198**, 23.
[260] Miglio, L. (1995). *Phys. Rev.* **B52**, 1448.
[261] Wenneker, I., Christensen, N. E., and Svane, A., to be published (1998).
[262] Jaccarino, V., Wertheim, G. K., Wernick, J. H., Walker, L. R., and Arash, S. (1967). *Phys. Rev.* **160**, 476.
[263] Fanciulli, M., Zankevich, A., Wenneker, I., Svane, A., Christensen, N. E., and Weyer, G. (1996). *Phys. Rev.* **B54**, 15985.
[264] Park, C.-H., Shen, Z.-X., Loeser, A. G., Dessau, D. S., Mandrus, D. G., Migliori, A., Sarrao, J., and Fisk, Z. (1995). *Phys. Rev.* **B52**, 16981.
[265] Edwards, P. R., Johnson, C. E., and Williams, R. J. P. (1967). *J. Chem. Phys.* **47**, 2074.
[266] Peltzer y Blancá, E. L., Svane, A., Christensen, N. E., Rodriguez, C. O., Cappanini, O. M., and Moreno, M. S. (1993). *Phys. Rev.* **B48**, 15712.
[267] Olijnyk, H. (1992). *Phys. Rev.* **B46**, 6589.
[268] Rowe, J. M. (1967). *Phys. Rev.* **163**, 547.
[269] Lewis, S. P., and Cohen, M. L. (1993). *Phys. Rev.* **B48**, 3646.
[270] Olijnyk, H. (1992). *Phys. Rev. Lett.* **68**, 2232; *Phys. Rev.* **B46**, 6589 (1992).

[271] Tinkham, M. (1964). "Group Theory and Quantum Mechanics." McGraw-Hill, New York.
[272] Gorczyca, I., Christensen, N. E., Peltzer y Blancá, E. L., and Rodriguez, C. O. (1995). *Phys. Rev.* **B51,** 11936.
[273] Perlin, P., Polian, A., and Suski, T. (1993). *Phys. Rev.* **B47,** 2874.
[274] Miwa, K., and Fukomoto, A. (1993). *Phys. Rev.* **B48,** 7897.
[275] Weinstein, B. A., and Zallen, R. (1984). *In* "Light Scattering in Solids IV" (Cardona M., and Gühtherodt, G., eds.), Topics in Applied Physics, Vol. 54. Springer-Verlag, Berlin.
[276] Ibach, H. (1969). *Phys. Stat. Sol.* **31,** 625.
[277] Biernacki, S., and Scheffler, M. (1989). *Phys. Rev. Lett.* **63,** 290.
[278] Weinstein, B. A. (1977). *Solid State Commun.* **24,** 595.
[279] Klotz, S., Besson, J. M., Braden, M., Karch, K., Bechstedt, F., Strauch, D., and Pavone, P. (1996). *Phys. Stat. Sol.* (*b*) **198,** 105.
[280] Olego, D., and Cardona, M. (1982). *Phys. Rev.* **B25,** 1151.
[281] "Proc. 7th International Conference of High Pressure Semiconductor Physics," Schwäbisch Gmünd, Germany, July 28–31 (1996) (Syassen, K., Stradling, R. A., and Goñi, A. R., eds.); *Phys. Stat. Sol.* (*b*) **198** (1996), Akademie Verlag, Berlin.
[282] Zwicknagl, G., Runge, E., and Christensen, N. E. (1990). *Physica* **163B,** 97.
[283] Zwicknagl, G. (1992). *Adv. Phys.* **41,** 203.
[284] Boebinger, G. (1996). *Physics Today* **49**(6), 361.

CHAPTER 3

Structural Transitions in the Group IV, III–V, and II–VI Semiconductors under Pressure

R. J. Nelmes

DEPARTMENT OF PHYSICS AND ASTRONOMY, THE UNIVERSITY OF EDINBURGH,
EDINBURGH, UK

M. I. McMahon

DEPARTMENT OF PHYSICS, THE UNIVERSITY OF LIVERPOOL, LIVERPOOL, UK

I. INTRODUCTION	146
II. STRUCTURES	150
1. Introduction	150
2. Diamond and Zincblende	152
3. Wurtzite and Lonsdaleite	154
4. β-Tin and Diatomic β-Tin	155
5. NaCl	157
6. NiAs	157
7. Imma, Imm2, and Immm	159
8. Simple Hexagonal	161
9. Cinnabar	163
10. Cmcm	164
11. $C222_1$	168
12. CsCl and Body-Centered Cubic	169
13. ST12	169
14. BC8, SC16, and R8	171
III. INDIVIDUAL SYSTEMS	175
1. Introduction	175
2. Silicon	177
3. Germanium	181
4. Boron-V Compounds	183
5. Aluminum Nitride	183
6. Aluminum Phosphide	184
7. Aluminum Arsenide	186
8. Aluminum Antimonide	187
9. Gallium Nitride	189
10. Gallium Phosphide	190
11. Gallium Arsenide	191
12. Gallium Antimonide	198

13. Indium Nitride	200
14. Indium Phosphide	202
15. Indium Arsenide	203
16. Indium Antimonide	205
17. Zinc Oxide	211
18. Zinc Sulfide	212
19. Zinc Selenide	214
20. Zinc Telluride	215
21. Cadmium Oxide	218
22. Cadmium Sulfide	218
23. Cadmium Selenide	220
24. Cadmium Telluride	222
25. Mercury Oxide	225
26. Mercury Sulfide	226
27. Mercury Selenide	228
28. Mercury Telluride	230
IV. Discussion	233
V. Concluding Remarks	244
Acknowledgments	245

I. Introduction

Structural studies of the group IV, III–V, and II–VI semiconductors date from the pioneering work of Jamieson in 1963 on silicon and germanium (Jamieson, 1963a) and on the III–V's AlSb, GaSb, InP, InAs, and InSb (Jamieson, 1963b). The following three decades saw an extraordinary amount of interest and work, with structural studies stimulating — and stimulated by — numerous spectroscopic and transport-property measurements, and a growing level and sophistication of theoretical calculations. The general picture to emerge from these structural studies was one in which increasing pressure brings about a sequence of phase transitions to high-symmetry structures of increasing coordination: fourfold (diamond, zincblende, or wurtzite) → sixfold (NaCl and β-tin) → eightfold (simple hexagonal) → eightfold with six close second-nearest neighbors (bcc or CsCl) or 12-fold (fcc or hcp). Nearly all of the structural work had to be carried out by powder-diffraction methods because single-crystal samples could not survive intact through the abrupt density increases that accompany many of these transitions.

Much of this picture was obtained in the period following the start of wide access to synchrotron sources in the early 1980s. The technique of choice was always energy-dispersive diffraction (EDX), in which the full white beam of the synchrotron falls on the sample and the diffraction pattern is recorded as a function of X-ray energy at a fixed scattering angle. This gave huge gains over previous laboratory-source techniques in flux on the sample, and thus permitted a large increase in the accessible pressure range (which necessarily entails smaller sample volumes), allowed much more rapid surveying in P and T, and made it possible to carry out studies of more weakly scattering systems. Such clear advantages for the demanding requirements of high-pressure powder diffraction stimulated much new structural work and greatly extended its range and power. However, there were also disadvantages. The resolution of EDX powder patterns is limited to that of the solid-state detector, patterns are unavoidably contaminated with fluorescence lines from the sample, and the tight collimation of the diffracted beam required to define the scattering angle often leads to poor powder averaging with the effect that peak intensities are unreliable. Thus, while ambient-pressure *angle*-dispersive (ADX) powder diffraction studies on synchrotron sources were developing and applying the techniques of full profile (Rietveld) refinement (Rietveld, 1969), EDX studies of high-pressure phases remained at a relatively approximate level of crystallography, mostly limited to the identification of structures through a tabulated comparison of observed and calculated d-spacings and intensities of reflections. Though EDX techniques made a major creative impact on the field through the mapping out of numerous transitions to very high pressures and stimulated a steadily growing effort in theoretical studies, they set ultimately unacceptable limits on the level of detail and precision attainable in high-pressure work.

The way out of this state of affairs was found by workers in Japan who recognized the potential benefits of the so-called image-plate detector for high-pressure powder diffraction (Shimomura *et al.*, 1992). This is a storage phosphor in which incident X-rays create trapped electron states which can be "read out" as emitted light, stimulated by a scanning laser beam, to give a digital image of the powder pattern. Image plates have ideal characteristics for recording powder diffraction patterns — a low intrinsic noise level, a wide dynamic range, a size large enough (20 cm across or more) to record a large part of complete 2-D patterns without loss of resolution, and ease of use. It is quite simple to integrate the 2-D pattern around the rings of the powder pattern to produce the 1-D profile required for structure solution and refinement. The introduction of this well-adapted area detector allowed a move away from EDX techniques to ADX, using a monochromatic beam. By recording all the lines in a powder pattern in parallel, a sufficient signal is still obtained from the small sample volume despite the much reduced flux of the monochromatic beam, and — moreover — the integration gives greatly improved powder averaging. The advent of image-plate ADX techniques has thus brought a remarkable breakthrough in data quality — higher resolution, freedom from contaminant features, reliable peak intensities, and high sensitivity to weak reflections.

Much other technique development has been necessary, and still remains to be done, to make the best use of 2-D data. Substantial progress has already been made in the implementation of image-plate techniques at SRS Daresbury (Nelmes and McMahon, 1994). The first step was to introduce diamond-anvil pressure cells (DACs) better matched to 2-D data collection. In EDX work, the diffracted beam passes through a narrow aperture at a 2θ angle between $\sim 10°$ and $\sim 20°$, and so the DAC requires only a slot in its steel housing for the exiting beam. This was the type of cell used in the initial development of image-plate methods (Shimomura et al., 1992), and the diffraction patterns obtained were thus limited to a quite narrow strip through the full 2-D patterns. The DACs used at SRS have a wide-angle conical aperture which allows complete Debye–Scherrer powder rings to be recorded up to $2\theta \sim 40°$. This brings large further gains in signal-to-noise and in the quality of powder averaging. In addition, much effort has been directed toward minimizing the background levels and eliminating all nonsample sources of powder pattern contamination (Nelmes and McMahon, 1994). The result has been a whole new level of accuracy and sensitivity which has made it possible to record reflections previously too weak to detect. One consequence of this enhanced sensitivity has been the ability to use anomalous dispersion effects successfully (for example, McMahon et al., 1993; Nelmes et al., 1993). It has also now become entirely routine to apply full profile (Rietveld) refinement techniques to high-pressure data analysis. On the one hand this has led to much greater precision in results, while on the other hand it has made misfits between structural models and data much more evident than they were when presented in terms of tabulated intensities and d-spacings. Similar technique enhancements based on ADX methods are now in place at many other synchrotron centers.

Access to full 2-D data has also brought other (often unexpected) benefits. An important example has been the ability to detect and characterize sample microstructure effects, which have long been known to cause serious problems in high-pressure work. Preferred orientation (PO) can alter peak intensities dramatically, sometimes even causing a quite strong reflection to be apparently absent from a powder pattern. This can introduce large uncertainties into structure solution and refinement. Other effects such as nonhydrostatic stress and stacking faults can also displace reflection d-spacings, which may make it difficult even to identify a unit cell with confidence. The recognition of such problems is not at all new. But they have been made much more evident by Rietveld fitting of ADX data and they pose a serious impediment to accurate crystallography, however good the data quality in terms of resolution and signal-to-noise. Significant progress has already been made in detecting and characterizing PO from its effects on 2-D data (Wright *et al.*, 1996). This new approach to PO is now at the stage where it can be modeled, entirely separately from the crystal-structure parameters, by fitting to the 2-D data, and it is very probable that a similar approach will prove effective in handling the causes of d-spacing displacement (Belmonte, 1997). Another pertinent example of the value of full 2-D data is the possibility of recognizing a sample to be made up of more than one phase because the diffraction patterns of the different phases are distinguishable by their appearance — for example, one giving spotty lines and the other giving smooth lines (Nelmes *et al.*, 1993).

All these recent and large advances in techniques and data quality have had a dramatic impact on the picture of the structural systematics of the group IV, III–V, and II–VI semiconductors. Many phases have been found to have more complex, lower-symmetry structures than previously thought; some supposed phase transitions have been found not to exist or to occur at quite different pressures; some previous long-standing problems have been solved; several new phenomena — such as intermediate phases — have been found; in some cases it has been possible to determine in detail how structures vary with pressure; and it has been possible to distinguish site-ordered and non-site-ordered phases. Alongside these various changes and clarifications, the results have also revealed a significant number of phases for which decisive structure solution appears likely to remain elusive until microstructural effects can be modeled or reduced (perhaps by annealing). Many of the revised interpretatons and new results are based on the detection of very weak reflections, some have required the measurement of anomalous dispersion effects (for example, to clarify site ordering), and others have depended on new experimental techniques to detect effects such as PO.

Though a large part of the new results have come from ADX studies, there have also been significant contributions from recent EDX work, particularly on the aluminum-V systems. And new EDX methods are very effective in mapping out $P–T$ phase diagrams.

In Section III, we review each of the group IV, III–V, and II–VI systems in turn, summarize previous work on each system, and present the best present understanding of its structures and transitions, with a critical review of remaining uncertainties. Where studies have also been done on pressure decrease, we include the results. We also include a significant amount of unpublished work of our own where this adds usefully to what can be said. Most of the structural results to date have been obtained at room temperature, but where studies have been done at nonambient temperature, they are included. Some theoretical and nonstructural experimental work is discussed where the results bear directly on the clarification of the structural behavior or the interpretation of the possible significance of structural features. To assist in the presentation of Section III, we first provide, in Section II, a summary and description of the key structures that will be referred to in the rest of the review. The principal results and overall systematics are summarized and discussed in Section IV, along with an assessment of important remaining problems and uncertainties. Some significant questions and challenges for future work are summarized in Section V.

The reviews of individual systems in Section III refer to Section II for details of the structures presented and are also illustrated by a few additional figures, particularly where we are presenting new and previously unpublished results. In addition, it is useful throughout to make reference to specific figures and tables in the literature to clarify the presentation. To avoid any confusion, such references to figures and tables in other papers are given in italics.

As far as we are aware, this is the first comprehensive review of all the group IV, III — V, and II–VI semiconductors. Much briefer versions, at earlier stages of the Daresbury program can be found in McMahon and Nelmes (1995) and McMahon and Nelmes (1996). A review by Ruoff and Li (1995) covers the III–V systems in detail, and summarizes results for Si and Ge; it also includes a useful overview of factors such as sample conditions, different experimental techniques, and amorphization.

REFERENCES

Belmonte, S. A. (1997). "2-D Data Analysis in High-Pressure Powder Diffraction." Ph.D. Thesis, University of Edinburgh.
Jamieson, J.C. (1963a). *Science* **139,** 762.
Jamieson, J. C. (1963b). *Science* **139,** 845.
McMahon, M. I. and Nelmes R. J. (1995). *J. Chem. Phys. Sol.* **56,** 485.
McMahon, M. I. and Nelmes R. J. (1996). *Phys. Stat. Sol. (b)* **198,** 389.
McMahon, M. I., Nelmes R. J., Wright N. G., and Allan D. R. (1993). *Phys. Rev. B* **48,** 16246.
Nelmes, R. J., McMahon, M. I., Hatton, P. D., Crain, J., and Piltz, R. O. (1993); *Phys. Rev. B* **47,** 35; *Phys. Rev. B* **48,** 9949 (Corrigendum).
Nelmes, R. J. and McMahon, M. I. (1994). *J. Synchrotron Rad.* **1,** 69.
Rietveld, H. M. (1969). *J. Appl. Cryst.* **2,** 65.
Ruoff, A. L. and Li, T. (1995). *Annu. Rev. Mater. Sci.* **25,** 249.
Shimomura, O., Takemura, K., Fujihisa, H., Fujii, Y., Ohishi, Y., Kikegawa, T., Amemiya, Y., and Matsushita, T. (1992). *Rev. Sci. Instrum.* **63,** 967.
Wright, N. G., Nelmes, R. J., Belmonte, S. A., and McMahon, M. I. (1996). *J. Synchrotron Rad.* **3,** 112.

II. Structures

1. INTRODUCTION

This section gives brief descriptions and guides to all the structures found so far in the group IV, III–V, and II–VI semiconductors, except some that are peculiar to particular systems (such as Pnma HgO), some that are simple, standard structures (such as hexagonal-close-packed (hcp) and face-centered-cubic (fcc) silicon), and — of course! — some that have yet to be solved (such as the monoclinic "intermediate" phase of silicon). Details of any structures discussed in Section III but not given here — including some previously proposed structures that are now thought to be incorrect — are presented in the relevant parts of Section III. Comprehensive descriptions of many of the structures can be found in the textbooks by Megaw (1973) and Donohue (1974).

The structures have come to be referred to in a variety of different ways — by name (diamond, zincblende, etc), by chemical formula of the prototype of the structure (NaCl, CsCl, NiAs), by phase diagram labelling (β-tin), by symmetry class (monoclinic Si), by space group symmetry (Imma, Cmcm, etc.), by standard structure type (hcp, fcc, etc.), and by lattice type plus the number of atoms in the primitive volume (BC8, ST12, R8). This is undesirably nonuniform, and it is now recommended by the International Union of Pure and Applied Chemistry (1990) that a standard description of structures be adopted in terms of their symmetry class, lattice type, and number of atoms per unit cell: For example, the structure of fcc silicon would be denoted "cF4" — for cubic, with an F lattice (fcc), and four atoms per cell. But it seems too soon to make such a strong break with familiar names.

In some cases in the text, in Table I, and in the figures and their captions, it is useful to adopt abbreviations. These are "diam" for diamond; "SH" for simple hexagonal; "wurtz" for wurtzite; "ZB" for zincblende; "cinn" for cinnabar; "mono" for monoclinic; "bcc" for body-centered cubic; "dhcp" for double hexagonal close-packed; and "hcp" and "fcc," which were spelled out earlier.

In the following descriptions of structures, the settings of the space groups and the atomic sites referred to — such as the 8(a) site in diamond — are taken from the *International Tables for Crystallography*, Vol. A (1983). It is important to note that it is sometimes useful to adopt a nonstandard space group setting (such as *Bmmb* instead of *Cmcm*) or a nonstandard origin (as in the description below of the Imma structure). As here, space groups are in italics when referred to as the symmetry of a structure, but not when used as a name only. All atomic positions are given as fractional coordinates. Lattice parameters are designated a, b, and c, with angles α, β, and γ. For each space group, the "reflection conditions" are specified — that is, the conditions on h, k and ℓ that have to be satisfied for the ($hk\ell$) reflection to be allowed. (Some conditions are not exact — for example, the $h + k + \ell = 4n$ condition for diamond. The tetrahedral bonding electron densities around nearest-neighbor atoms are differently oriented, and this gives rise to extremely weak $h + k + \ell \neq 4n$ reflections, such as (222). The (222) reflection in Si has been examined under pressure by Yoder-Short *et al.* (1982) to look for changes in bonding electron density approaching the transition to the β-tin phase, but using single-crystal samples. So far reflections of this type have been much too weak to detect in high-pressure powder diffraction data and will not be considered further.) For economy of presentation, only the ordered diatomic form of each structure is illustrated in most cases, and the monatomic form (Si, Ge, or site-disordered III–V or II–VI) is described as having the illustrated structure "but with all atoms identical."

Reference is made later to "difference" reflections in discussing several of the binary systems. These are reflections in which the two different atomic species scatter in exact anti-phase, with the result that the reflection intensity depends solely on the difference in scattering power of the atoms. The difference reflections in InSb, for example, are thus extremely weak (since In has $Z = 49$ and Sb has $Z = 51$), while in HgO, for example, these reflections are nearly as strong as those in which the two species scatter in phase. The observation of difference reflections is required to demonstrate site ordering, which is thus difficult to do in systems such as InSb and GaAs where the two atomic species have very similar scattering powers.

Structures described as site-disordered cannot be shown by diffraction techniques to be *randomly* site-disordered. It is more strictly correct to say that they lack long-range site ordering. The length scale of the ordering may be anywhere between the nearest-neighbor distance (random site disorder) and several unit cells. The difference reflections which signal site ordering may also be apparently absent because of anti-phase domains, which have the effect of broadening those reflections (Warren, 1990, Section 12.3). (It has been proposed that such domains, some 10 unit cells across, account for the absence of difference reflections in the β-tin-like P2 phase of InSb (Crain *et al.* 1993). However, equivalent reflections are absent in the Imma phase of GaSb (Section III.12). In this case the difference reflections are very much stronger, and the domains would need to be so small as to become similar to general site disordering. As discussed further in Section IV, this — and other evidence for very short length scales of any ordering — probably casts some doubt on the interpretation of the InSb case in terms of ordered domains extending over many unit cells.)

2. DIAMOND AND ZINCBLENDE

The diamond structure is found in Si and Ge at ambient conditions. The structure is as shown for zincblende in Fig. 1a, but with all the atoms identical. It is cubic, spacegroup *Fd3m*, with atoms on the 8(a) site at

3 GROUP IV, III–V, AND II–VI SEMICONDUCTORS

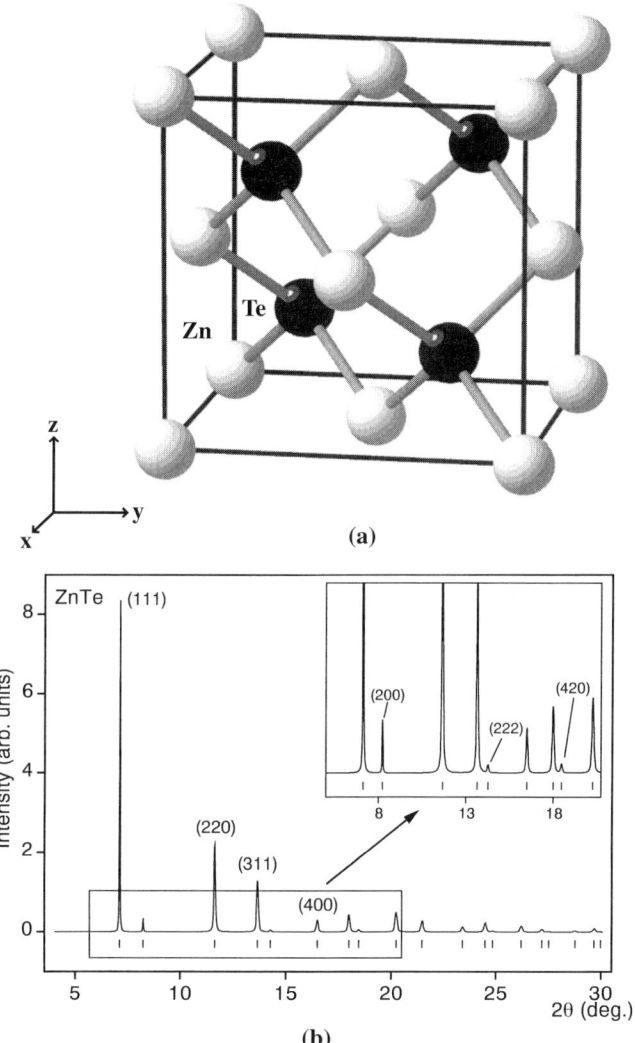

FIG. 1. (a) Zincblende structure of ZnTe. Zn is at $(0, 0, 0)$ and Te is at $(1/4, 1/4, 1/4)$. If all the atoms are made identical (a site-disordered or monatomic system), this becomes the diamond structure. (b) Calculated diffraction pattern for ZnTe in the zincblende structure. The (200), (222), and (420) difference reflections are shown in the part of the pattern enlarged in the inset.

(1/8, 1/8, 1/8) etc. The structure is centrosymmetric, and the standard origin is at a center of inversion symmetry — midway between neighboring atoms. However, the structure is usually drawn with a nonstandard origin, shifted to (1/8, 1/8, 1/8), for convenience and ease of comparison with zincblende. The atoms are then at (0, 0, 0) and (1/4, 1/4, 1/4), etc., as in Figure 1a. The structure is tetrahedrally coordinated with four nearest neighbors at a distance of $\sqrt{3}a/4$. The general reflection conditions are h, k, and ℓ all even or all odd in $hk\ell$; and $k + \ell = 4n$, and k and $\ell = 2n$, in $0k\ell$ — and likewise for cyclic permutations. The location of all atoms on the 8(a) special positions gives the further condition $h + k + \ell = 2n + 1$ or $4n$ in $hk\ell$.

Zincblende (ZB) is the dominant ambient-pressure structure in the III–V and II–VI semiconductors and is the binary analog of the diamond structure. It is cubic (spacegroup $F\bar{4}3m$) with one atom species on the 4(a) site at (0, 0, 0) and the other on the 4(c) site at (1/4, 1/4, /4), as shown in Fig. 1a. Each atom is therefore tetrahedrally coordinated with four *unlike* nearest neighbors. The reflection conditions are h, k, and ℓ all even or all odd in $hk\ell$. The diffraction pattern (Fig. 1b) is very similar to that of diamond, but contains extra reflections, such as those labeled (200), (222) and (420), which are difference reflections.

3. WURTZITE AND LONSDALEITE

The wurtzite structure (Fig. 2) is adopted by AlN, GaN, InN, ZnO, ZnS, CdS, and CdSe at ambient conditions. It is hexagonal (spacegroup $P6_3mc$) and is related to ZB in the same way that hcp is related to fcc: that is, the stacking of puckered layers along the (111) axis of ZB is $ABCABC$ while in wurtzite it is $ABABAB$. The "ideal" wurtzite structure has $c/a = \sqrt{(8/3)} = 1.633$ and has one atomic species on a 2(b) site at (1/3, 2/3, 0) and the second on a 2(b) site at (1/3, 2/3, u), where $u = 3/8$. The structure is noncentrosymmetric and has an arbitrary origin along z; the choice of $z = 0$ for one species fixes the origin. The difference in z between the two species can vary from 3/8 — as occurs in ZnO, for example (Megaw, 1973). The only general reflection condition is $\ell = 2n$ in $hh\ell$. The location of all atoms on the 2(b) special positions gives the further condition that $\ell = 2n$ if $h - k = 3n$ in $hk\ell$.

The monatomic equivalent of wurtzite is the so-called hexagonal diamond structure found under ambient conditions as the mineral lonsdaleite. The BC8 forms of Si and Ge (see Section II.14) are believed to transform to this structure — at room temperature (Ge) or on heating (Si). The (centrosymmetric) structure is as in Fig. 2 but with all the atoms identical. It has

3 Group IV, III–V, and II–VI Semiconductors

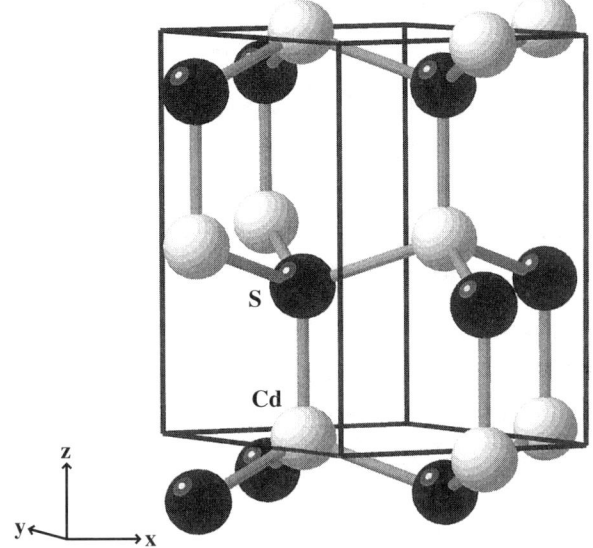

Fig. 2. Wurtzite structure of CdS. Cd is at $(1/3, 2/3, 0)$ and S is at $(1/3, 2/3, u \sim 3/8)$. If all the atoms are made identical (a site-disordered or monatomic system), this becomes the hexagonal diamond structure of lonsdaleite.

space group $P6_3/mmc$, with atoms on the 4(f) site at $(1/3, 2/3, z \sim 1/16)$ etc., relative to the standard origin with a center of inversion symmetry at $(0, 0, 0)$. In Fig. 2 (with all atoms identical), this origin is shifted to $(0, 0, -1/16)$. The only general reflection condition is $\ell = 2n$ in $hh\ell$. The location of all atoms on the 4(f) special positions gives the further condition $\ell = 2n$ if $h - k = 3n$ in $hk\ell$.

4. β-Tin and Diatomic β-Tin

β-tin, or the white tin structure, is found in elemental Sn above 291 K at atmospheric pressure. The structure is as shown in Fig. 3a, but with all the atoms identical. It is tetragonal, spacegroup $I4_1/amd$, with atoms on the 4(a) positions at $(0,0,0), (0, 1/2, 1/4)$, etc. The structure is centrosymmetric, but in this standard setting the origin is taken at the $\bar{4}$ inversion point; the centers of inversion symmetry are midway between neighboring atoms. The general reflection conditions are $h + k + \ell = 2n$ in $hk\ell$; h and $k = 2n$ in $hk0$; and $2h + \ell = 4n$, and $\ell = 2n$, in $hk\ell$. The location of all atoms on the 4(a) special positions gives the further condition $2k + \ell = 2n + 1$ or $4n$ in $hk\ell$. In Sn at 298 K, $a = 5.8316(2)$ Å, $c = 3.1815(2)$ Å, and

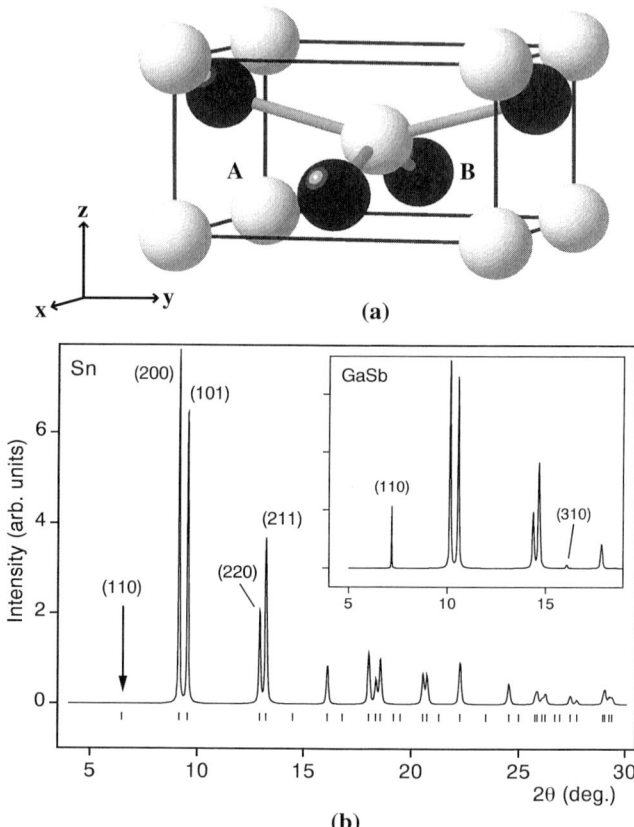

FIG. 3. (a) Diatomic β-tin structure — not yet shown to exist in any binary AB system. A is at (0, 0, 0) and B is at (0, 1/2, 1/4). If all the atoms are made identical (a site-disordered or monatomic system), this becomes the β-tin structure. (b) Calculated powder pattern for Sn in the β-tin structure. The arrow marks the position of the absent (110) reflection. The inset shows a calculated pattern for Ga and Sb in the diatomic β-tin structure; the additional reflections, such as (110), can be seen.

$c/a = 0.54556(4)$, and the structure therefore has four nearest neighbors at 3.022 Å and two slightly further away at 3.182 Å (Donohue, 1974). The structure would have six equidistant nearest-neighbors if $c/a = \sqrt{(4/15)} = 0.5164$; this has been termed the "perfect" β-tin structure.

The diffraction pattern from the β-tin structure is dominated at low angles by two strong doublets formed by the (200)/(101) and (220)/(211) pairs of reflections, as shown by the pattern for Sn in Fig. 3b. The 2θ

splitting of the second doublet is completely determined by the splitting of the first.

The diatomic equivalent of the β-tin structure — d-β-tin — has long been thought to exist as a high-pressure phase in many of the binary semiconductors. However, as discussed in Section IV, no example of this structure has in fact yet been found in these materials, and, indeed, it has not been reported in *any* binary material under any conditions.[1] Diatomic β-tin has spacegroup $I\bar{4}m2$, with one species of atom on the 2(a) site at (0, 0, 0) and the other on the 2(c) site at (0, 1/2, 1/4). The structure is shown in Fig. 3a; it is noncentrosymmetric. Of the β-tin reflection conditions (given earlier), only "$h + k + \ell = 2n$ in $hk\ell$" applies to d-β-tin. The β-tin and d-β-tin powder patterns are thus distinguished by the presence of difference reflections, the most intense of which is (110). The inset to Fig. 3b shows the pattern that would be given by Ga and Sb in the d-β-tin structure, and the difference reflection (110) is very clearly visible. The position of the absent (110) reflection is marked in the β-tin profile. Observation of the difference reflections is required to demonstrate the existence of true, site-ordered d-β-tin.

5. NaCl

The NaCl or rock-salt structure is found in CdO at ambient pressure and in AlN, GaN, InN, InP, InAs, ZnO, HgO, and ZnX, CdX and HgX (X = S, Se, Te) at high pressure — at high temperature and pressure in the case of ZnTe. The structure is shown in Fig. 4a. It is cubic and centrosymmetric (spacegroup $Fm3m$) with atoms on the 4(a) and 4(b) sites at (0, 0, 0) and (1/2, 1/2, 1/2), respectively. The origin is at a center of inversion symmetry. The reflection conditions are h, k, and ℓ all even or all odd in $hk\ell$. Each atom has perfect octahedral coordination with six unlike nearest neighbors at a distance $a/2$. The diffraction pattern from ZnTe in the NaCl structure is shown in Fig. 4b. Reflections with h, k, and ℓ all odd, such as (111), are difference reflections which have to be detected to show that the structure is site-ordered.

6. NiAs

The NiAs structure is found in AlP and AlAs at high pressure. It is hexagonal with the same spacegroup as wurtzite, $P6_3mc$. The structure is

[1]Search on spacegroups $I\bar{4}m2$ and $I4_1/amd$ in the U.K. Chemical Database Service, D. A. Fletcher, R. F. McMeeking and D. J. Parkin, *Chem. Inf. Comput. Sci.* **36**, 746 (1996). (Note that in Nelmes *et al.* (1997) spacegroup $I\bar{4}m2$ was incorrectly described as being $I4md$.)

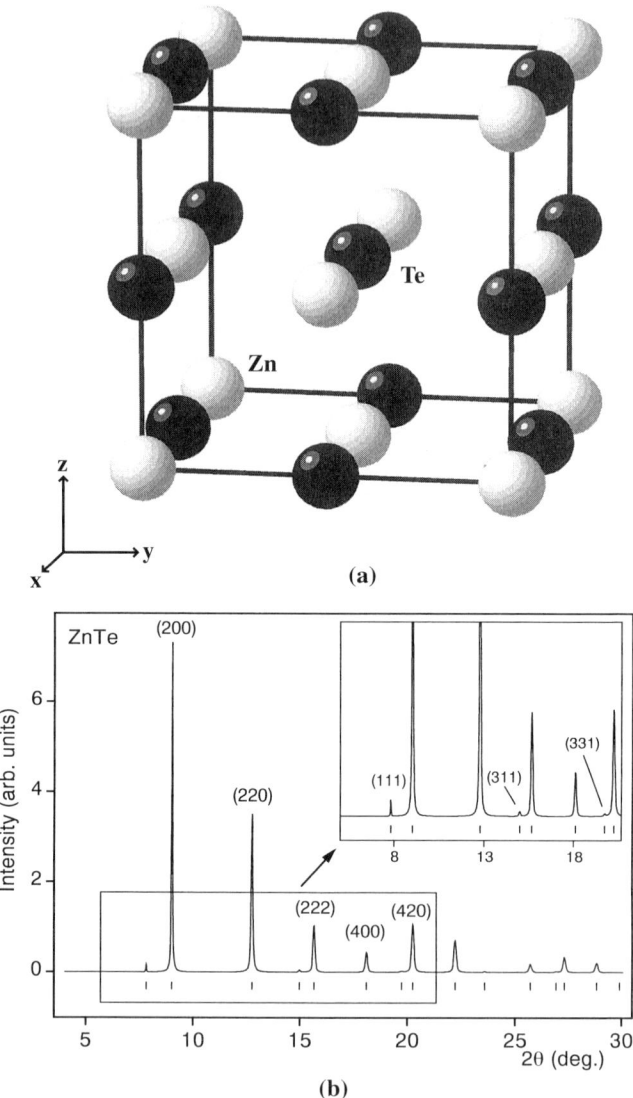

FIG. 4. (a) Sodium chloride structure of ZnTe. Zn is at (0, 0, 0) and Te is at (1/2, 1/2, 1/2). (b) Calculated powder diffraction pattern for ZnTe in the NaCl structure. The (111), (311), and (331) difference reflections are shown in the part of the pattern enlarged in the inset.

3 Group IV, III–V, and II–VI Semiconductors

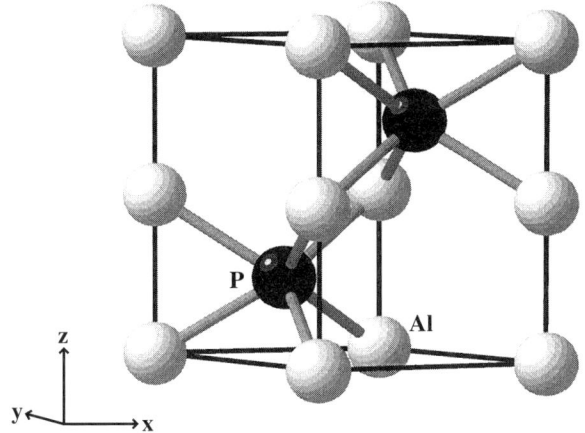

FIG. 5. Nickel arsenide structure for AlP. Al is at (0, 0, 0) and P is at (1/3, 2/3, $u \sim 1/4$). The Al and P sites have quite different coordination.

shown in Fig. 5. The two atomic species are on the 2(a) site at (0, 0, 0) and the 2(b) site at (1/3, 2/3, u) with $u \sim 1/4$. As discussed for wurtzite, the position of the origin is arbitrary in z and is fixed by setting $z = 0$ for one of the sites. The only general reflection condition is $\ell = 2n$ in $hh\ell$. The location of all atoms on 2(a) and 2(b) special positions gives the further condition that $\ell = 2n$ if $h - k = 3n$ in $hk\ell$. When $u = 1/4$, each anion (P or As) is surrounded by six equidistant Al atoms, while each cation (Al) has eight close neighbors, two of which are like atoms. The anions and cations thus have very different environments in this structure.

7. Imma, Imm2, and Immm

The Imma structure is an orthorhombic distortion of both tetragonal β-tin and the simple hexagonal structure and is found in Si, Ge, and site-disordered GaSb and InSb. All the atoms are on the 4(e) site at (0, 1/4, Δ/2) etc. of spacegroup *Imma*. The structure is centrosymmetric, and the standard origin is at a center of inversion symmetry — midway between neighboring atoms. However, to facilitate comparison with the β-tin, simple hexagonal, Imm2 and Immm structures, the structure is usually described with the origin shifted to (0, 1/4, Δ/2). The atoms are then at (0, 0, 0), (0, 1/2, Δ) etc. This structure is as shown in Fig. 6, but with all the atoms identical. The reflection conditions are $h + k + \ell = 2n$ in $hk\ell$, and $h = 2n$ in $hk0$. When $a = b$ and $\Delta = 1/4$, Imma becomes the same as β-tin, while if $b = \sqrt{3}c$ and $\Delta = 1/2$, then the structure is simple hexagonal. A

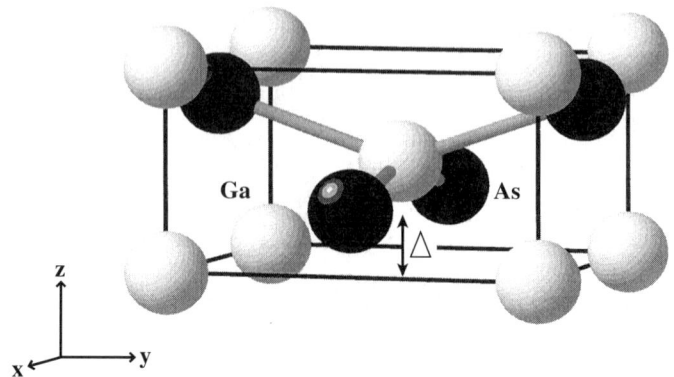

FIG. 6. Imm2 structure proposed for GaAs (Weir *et al.*, 1989). Ga is at (0, 0, 0) and As is at (0, 1/2, Δ). If all the atoms are made identical (a site-disordered or monatomic system), this becomes the Imma structure.

continuous distortion from β-tin to simple hexagonal is thus possible through the Imma structure. Figure 7 shows a diffraction pattern recorded from GaSb in its Imma phase. The complete absence of the (110) reflection establishes the site-disordered nature of the structure. (Compare the quite large intensity of (110) in the inset to Fig. 3B for Ga and Sb in a site-ordered d-β-tin structure — which is very similar to Imma apart from the site ordering.)

The diatomic equivalent of the Imma structure has spacegroup *Imm2*,

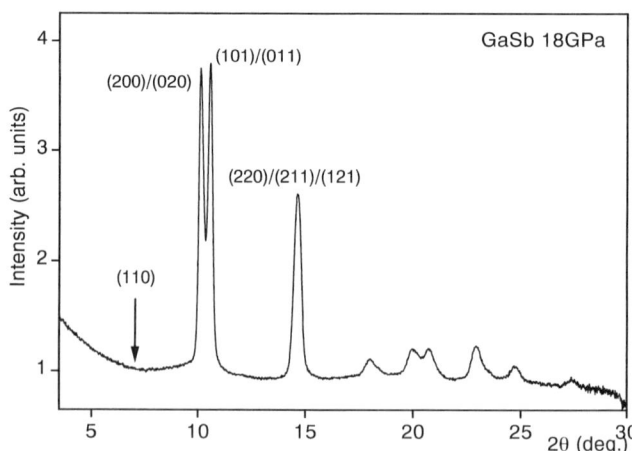

FIG. 7. Observed powder diffraction pattern obtained from GaSb at 18 GPa. The arrow marks the position of the absent (110) reflection. Copyright 1994 American Physical Society.

with one atomic species on the 2(a) site at (0, 0, 0) and the other on the 2(b) site at (0, 1/2, Δ), as shown in Fig. 6. The P3 phase of InSb (see Section III.16) was at first thought to have this structure, but has since been found to have *Immm* symmetry (see following discussion). The structure has also been proposed in GaAs (Weir et al., 1989). It is noncentrosymmetric and has an arbitrary origin along z; the choice of $z = 0$ for one species fixes the origin. If $\Delta = 1/4$, then Imm2 becomes the diatomic β-tin structure, whereas if $\Delta \neq 1/4$ but the two atomic species are the same, or are 50:50 site-disordered (as in GaSb earlier), the structure becomes Imma. The only reflection condition for *Imm2* is $h + k + \ell = 2n$ in $hk\ell$. So, as with β-tin and d-β-tin, Imma and Imm2 can be distinguished by the absence or presence of difference reflections ($hk0$) with $h \neq 2n$, such as (110) — see Fig. 7.

If $\Delta = 1/2$ in Fig. 6, then the Imm2 structure becomes Immm, with the two atomic species on the 2(a) and 2(b) sites at (0, 0, 0) and (0, 1/2, 1/2), respectively. This structure is centrosymmetric, and the standard origin — which has been adopted here — is at a center of inversion symmetry. The only reflection condition is $h + k + \ell = 2n$ in $hk\ell$. This structure is found in the P3 phase of InSb. Distinguishing Immm from Imm2 with Δ close to 1/2 is difficult. However, in Immm, those reflections with $h =$ odd in ($hk0$) and $k + \ell =$ odd in ($hk\ell$) are difference reflections — the latter arising because for $\Delta = 1/2$ the structure becomes pseudo-A-face-centered; in Imm2 only the reflections with $h =$ odd in ($hk0$) are difference reflections. A distinction between the two structures can therefore be made by an accurate measurement of the relative intensities of groups of reflections such as (110)/(101) and (310)/(130)/(301), since in Imm2 ($\Delta \neq 1/2$), those reflections with $\ell \neq 0$ will be more intense because of an extra "structural" component in their structure factors.

An important feature of the Immm structure (or Imm2 with Δ near 1/2) is that it can be considered as a stacking of NaCl-like planes — parallel to (011). The atoms labeled A, B, C, and D mark one of these planes in the Immm structure of Fig. 8.

8. SIMPLE HEXAGONAL

The primitive, or simple, hexagonal structure (SH) is found in Si and Ge at high pressure. It has the centrosymmetric spacegroup *P6/mmm* with atoms on the 1(a) site at (0, 0, 0) etc. — at centers of inversion symmetry. The structure is as shown in Fig. 8, but with all atoms identical. There are no reflection conditions. Since $c/a \sim 1$, each atom is eightfold coordinated, with six nearest-neighbor atoms within the same hexagonal layer and two atoms in the adjacent layers. If $c/a = 1$, then these eight nearest-neighbor

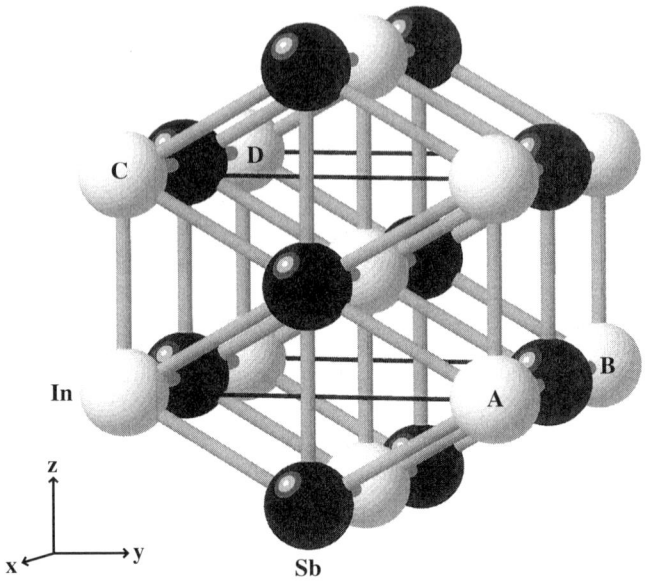

FIG. 8. Immm structure of InSb, with In at (0, 0, 0) and Sb at (0, 1/2, 1/2), for the special case that $b = \sqrt{3}\,c$. The spatial arrangement of atomic sites is then simple hexagonal, as illustrated. However, the site ordering of In and Sb breaks the hexagonal symmetry, and the structure remains orthorhombic (see text). If all the atoms are made identical (a site-disordered or monatomic system), the structure becomes true simple hexagonal.

distances are all the same. As discussed earlier, the SH structure can be obtained from β-tin by a continuous distortion through the Imma structure. However, there is no diatomic equivalent of SH to be obtained from d-β-tin by distortion through Imm2 — the resulting structure, as shown in Fig. 8, does not possess true hexagonal symmetry. The ordered arrangement of two different atomic species rules out any true six-fold (or three-fold) rotation axes along x, and the true symmetry remains orthorhombic *Immm*. The reported simple-hexagonal phases of InSb (Vanderborgh *et al.*, 1989), GaSb (Weir *et al.*, 1987), and GaAs (Weir *et al.*, 1989) could therefore be truly hexagonal only if they have no long-range order and are thus quasi-monatomic.

Another example of a quasi–simple-hexagonal arrangement in a binary system can be obtained in structures with the Cmcm structure, as was observed in CdTe (Nelmes *et al.*, 1995). However, as illustrated in *Figure 6* of Nelmes *et al.* (1995), this arrangement also cannot possess true six-fold or three-fold symmetry and remains orthorhombic.

9. CINNABAR

The cinnabar structure is found under ambient conditions in HgS and as a metastable phase is HgO. It is found at high pressure in HgSe, HgTe, CdTe, ZnTe, and GaAs. The structure is trigonal (spacegroup $P3_121$) with the two atomic species on the 3(a) and 3(b) sites at $(u, 0, 1/3)$ and $(v, 0, 5/6)$, respectively. (The cinnabar structure can be described in $P3_121$ or the enantiomorphic spacegroup $P3_221$. A single-crystal study of HgS at ambient pressure shows that it is correctly described by $P3_221$ in that case (Auvray and Genet, 1973). However, it is impossible to distinguish enantiomorphic structures using powder data, and the $P3_121$ description has generally been adopted.) The only reflection condition is $\ell = 3n$ in (00ℓ). The (noncentrosymmetric) structure can be viewed as a trigonal distortion of NaCl — to which it would be identical if $c/a = \sqrt{6} = 2.449$ and $u = v = 2/3$. In HgS under ambient conditions $c/a = 2.291(2)$, $u(\text{Hg}) = 0.7198(5)$, and $v(\text{S}) = 0.4889(28)$ (Auvray and Genet, 1973).

The HgS structure is strongly distorted from the six-fold coordination of NaCl, having 2 + 4 coordination in which the two nearest neighbors are (unlike) atoms at 2.368 Å, much closer than the next four (unlike) neighbors at 3.094 Å and 3.287 Å. The closest near-neighbor distances link the atoms into –S–Hg–S– spiral chains running along the z-axis, as shown in Fig. 9a. However, the detailed coordination of the cinnabar structure depends critically on the values of u and v. In HgSe, HgTe, and CdTe, the u and v values are more similar than in HgS and HgO, with $u < 2/3$. This leads to a 2 + 2 + 2 (HgSe) or 4 + 2 coordination (HgTe and CdTe) in which interchain nearest neighbor distances become comparable with the in-chain distances. In the cases of ZnTe and GaAs, u and v are even closer in value (~0.54 and ~0.50, respectively) and well away from the NaCl value of 2/3. The structure is then almost exactly four-fold coordinated, as shown in Fig. 9b for the cinnabar phase of GaAs. This has four nearest neighbors, two at ~2.37 Å and two at ~2.48 Å, and two like-atom next-nearest neighbors much farther away at ~3.48 Å. (Mujica et al. (1998) have questioned whether this structure is even properly described as "cinnabar" since it is so different from the prototype cinnabar structure of HgS (Figs. 9a and 9b). This is discussed further in Sections III.11 and IV.)

The diffraction pattern from HgS at ambient conditions is shown in Fig. 9c. The (003) difference reflection is labeled. The presence of this reflection shows the structure to be site-ordered. The inset illustrates the much greater sensitivity required to detect site-ordering in the case of CdTe.

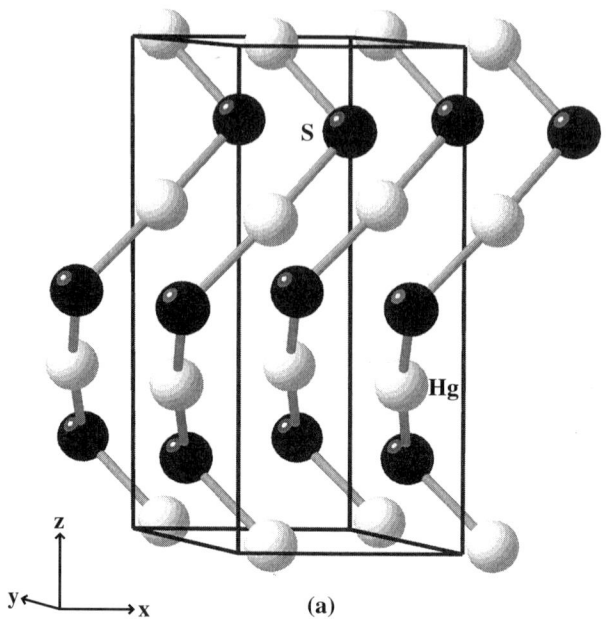

FIG. 9. Cinnabar structure of (a) HgS and (b) GaAs. Hg and Ga are at $(u, 0, 1/3)$, and S and As are at $(v, 0, 5/6)$, with $u \sim 0.72$ for Hg and ~ 0.54 for Ga, and $v \sim 0.50$ in both cases. Only the short nearest-neighbor (<2.75 Å) bonds are shown. In (a) the four S–Hg–S chains spiral around each of the four vertical (z-axis) edges of the cell. (c) Calculated powder diffraction pattern for HgS in the cinnabar structure. The (003) difference reflection is marked. The inset shows the calculated pattern for CdTe in the same structure; the relative intensity of the (003) reflection is much weaker in this case. (b) Copyright 1998 The Japan Society of High Pressure Science and Technology.

10. Cmcm

Cmcm is a structure that was previously unobserved in any binary material but has become increasingly important in recent studies of semiconductors at high pressure. It, or a very similar structure, has been found so far in AlSb, GaP, GaAs, InP, InAs, InSb (as a superstructure), ZnSe, ZnTe, CdS, CdSe, CdTe, HgSe, and HgTe. It has also been found at high pressure in CuBr and CuCl (Hull and Keen, 1994). The Cmcm structure can be viewed as an orthorhombic distortion of NaCl with the two atomic species on 4(c) sites of spacegroup $Cmcm$ at $(0, y_1, 1/4)$ and $(0, y_2, 1/4)$, respectively. The reflection conditions are $h + k = 2n$ in $hk\ell$, and h and $\ell = 2n$ in $h0\ell$. If $a = b = c$, $y_1 = 3/4$, and $y_2 = 1/4$, then the structure is

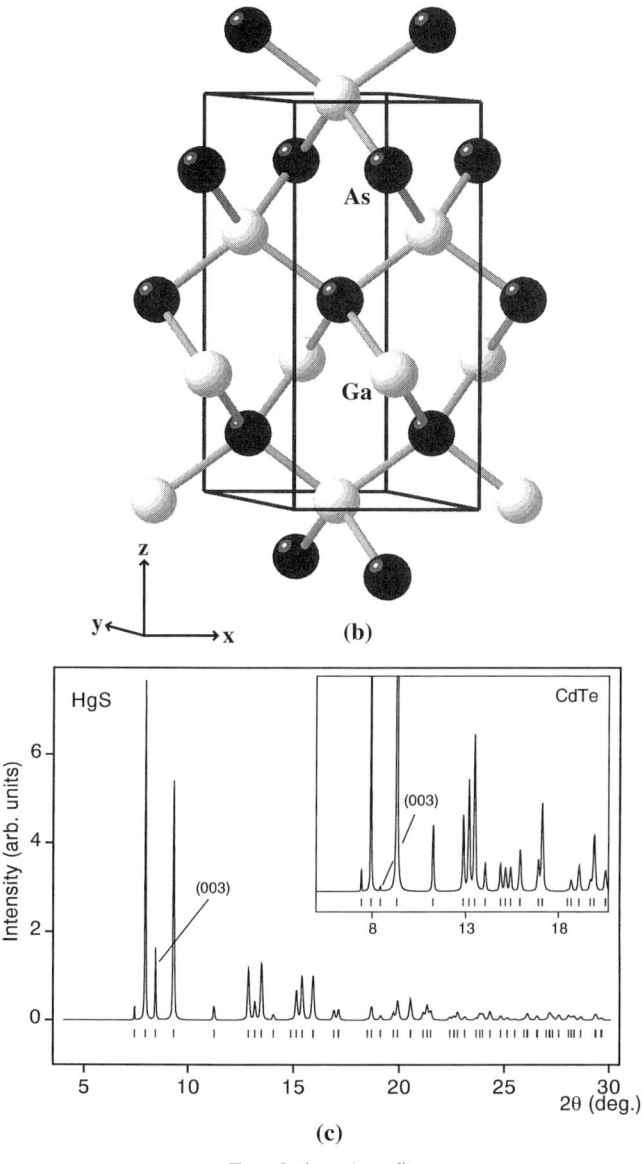

FIG. 9. (*continued*)

identical to NaCl. The principal distortion from NaCl is a displacement along $\pm y$ of alternate x–y NaCl planes. There is a second, smaller distortion if $\Delta y = y_1 - y_2 \neq 0.5$ which results in a $\pm(0.5 - \Delta y)/2$ zigzag along the x-axis chains. The Cmcm structure is shown in Fig. 10 for typical values of

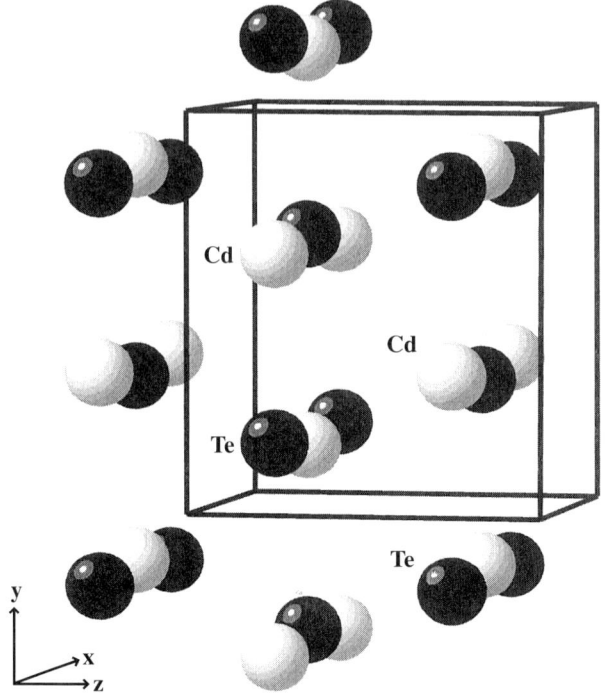

FIG. 10. Cmcm structure of CdTe. Cd is at $(0, y_1 \sim 0.65, 1/4)$ and Te is at $(0, y_2 \sim 0.18, 1/4)$. $y_1 - y_2 = \Delta y \sim 0.47$. The $\pm(0.5 - \Delta y)/2$ zigzag can be seen along the x-axis chains. The shortest Cd–Cd and Te–Te distances are between the labeled atoms. For $\Delta y < 0.5$, the shortest Cd–Cd distance is less than the shortest Te–Te distance. Copyright 1996 WILEY-VCH Verlag Berlin GmbH.

$y_1 = 0.65$ and $y_2 = 0.18$ ($\Delta y = 0.47$). $\Delta y < 0.5$ makes the shortest cation–cation distance (between the atoms labeled "Cd," in Fig. 10) less than the shortest anion–anion distance (between the atoms labeled "Te" in Fig. 10); $\Delta y > 0.5$ has the opposite effect.

The way in which the coordination of the Cmcm structure differs from NaCl depends critically on the values of the lattice parameters and the variable atomic coordinates. The continuous NaCl-to-Cmcm transition in CdTe at 10 GPa has allowed the evolution of the Cmcm coordination to be followed in detail to 28 GPa (Nelmes et al., 1995). The results reveal that five of the (unlike-atom) nearest-neighbor distances remain almost constant with pressure, while the sixth increases sharply as a consequence of the change in Δy from 1/2. At the same time, the displacement along $\pm y$ of alternate NaCl planes causes a large reduction in two of the next-nearest-neighbor (like-atom) distances — the more so for the cation–cation distance if $\Delta y < 0.5$ as discussed previously — such that the coordination can be viewed as (5 + 3)-fold. This is shown in *Figure 12* of Nelmes *et al.* (1995). The same 5 + 3 coordination is observed in the Cmcm phase of ZnTe (Nelmes *et al.*, 1994). However, in Cmcm-HgTe, the atomic coordinates are different from those observed in CdTe and ZnTe, and although the coordination is still five-fold it is less clearly so, especially around the Hg atom. In all cases the Cmcm structure is quite different from the diatomic β-tin structure for which it has previously been frequently mistaken. Among other things, d-β-tin (Fig. 3a) does not have the NaCl-like planes of Cmcm.

The diffraction pattern from Cmcm differs from that of NaCl in two major respects:

1. The orthorhombic distortion results in a splitting of the NaCl peaks — for example, the NaCl (200) splits into (200), (020), and (002) of Cmcm.
2. The displacements of the alternate NaCl layers along $\pm y$ result in the appearance of additional reflections — principally (021) and (221) — the intensities of which are directly related to the degree of displacement.

The relative prominence of these features varies from system to system (see, for example, *Figures 1, 2,* and *3* of Nelmes *et al.*, 1997). In particular, the displacement of the layers, coupled with only a small orthorhombic distortion, results in what is essentially an NaCl diffraction pattern with two additional strong lines — (021) and (221) — leading to two low-angle doublets. The apparent resemblance of the strong peaks in this pattern to those of the β-tin pattern (Fig. 3b) has played a misleading role in past work. However, the orthorhombic splitting of Cmcm does allow the β-tin and Cmcm patterns to be distinguished, while a comparison of weaker peaks reveals that the two diffraction patterns are very different in detail.

Even when distinguished from d-β-tin, the Cmcm structure has been the subject of some misidentification. Zhang and Cohen (1989) described an orthorhombically distorted NaCl-like structure for GaAs-II, with a single variable atomic coordinate, α, defining the distortion. This structure does not have the *Amm2* symmetry identified for it by Vanderborgh *et al.* (1989). First, the structure drawn by Zhang and Cohen (1989) — their *Figure 1(a)* — is B-face-centered rather than A-face-centered. In fact, it is *Bmmb* with z_1 and z_2 constrained to differ by 1/2, which can be redescribed as *Cmcm* with $\Delta y = 1/2$ by interchange of the y and z axes. The $\Delta y = 1/2$ constraint does not represent any change of symmetry: It remains *Cmcm*. The B-face-centered setting of the space group has also been adopted in the figures of recent papers by Mujica *et al.* (1996) and Mujica and Needs

(1996, 1997), though the symmetry is referred to as *Cmcm* in the text. (It is perhaps a danger of using the space group symmetry as the name of a structure type that the name and the precise meaning of the symmetry may become separated.)

11. $C222_1$

The $C222_1$ structure is found in HgSe and HgTe as a "hidden" intermediate phase between the ZB and cinnabar phases. The structure is orthorhombic and noncentrosymmetric, with the Hg and Se/Te atoms on the 4(a) site at $(x, 0, 0)$ and on the 4(b) site at $(0, y, 1/4)$, respectively. The general reflection conditions are $h + k = 2n$ in $hk\ell$, and $\ell = 2n$ in 00ℓ. The structure (Fig. 11) can be viewed as an orthorhombic distortion of ZB — they would be identical if $a = b = c$, $x = 0.25$, and $y = 0.25$. In HgSe at 2.25GPa (McMahon et al., 1996), $a = 5.992(1)$ Å, $b = 5.879(1)$ Å, and $c = 6.045(2)$ Å, with $x(\text{Hg}) = 0.302(1)$ and $y(\text{Se}) = 0.207(2)$. The structure thus appears only slightly distorted from ZB, and it is four-fold coordinated. However, the bond angles, which vary from 94° to 125°, are distorted considerably from the ideal tetrahedral value. The range of values is similar to those observed in the ST12 structure.

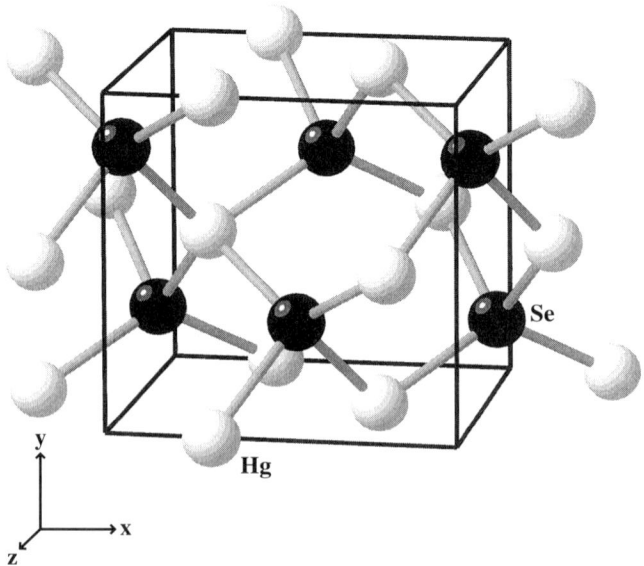

FIG. 11. $C222_1$ structure of HgSe. Hg is at $(x \sim 0.3, 0, 0)$ and Se is at $(0, y \sim 0.2, 1/4)$.

The C222$_1$ structure has a close structural relationship with the low-cristobalite forms of AlPO$_4$ and GaPO$_4$, which have the same spacegroup and a similarly pseudo-cubic unit cell (Mooney, 1956). Moreover, the fractional coordinates x(Al) = 0.306 and y(P) = 0.198 of Al and P in AlPO$_4$, and x(Ga) = 0.327 and y(P) = 0.183 of Ga and P in GaPO$_4$ (Mooney, 1956) are similar to those of Hg and Se/Te here. Since the ZB phases of HgSe and HgTe have the same spatial arrangement of atoms as the Al and P, and Ga and P atoms in the (idealized) high-cristobalite phases of AlPO$_4$ and GaPO$_4$, the ZB-to-C222$_1$ transition in HgSe and HgTe is analogous to a high–low cristobalite transition. This relationship, and that between the ST12 and keatite structures (Section II.13), suggests that the structures of SiO$_2$ and its polymorphs may provide a fruitful source of structural models for other high-pressure or metastable phases of tetrahedrally coordinated semiconductors.

12. CsCl and Body-Centered Cubic

The CsCl structure has not yet been found in any of the II–VI or III–V systems, but is expected to occur as a very high-pressure phase. It is centrosymmetric cubic (spacegroup $Pm3m$) with the two atomic species on the 1(a) and 1(b) sites at (0, 0, 0) and (1/2, 1/2, 1/2), respectively. The origin is at a center of inversion symmetry. The structure has perfect eight-fold coordination, with six next-nearest neighbors less than 15% farther away, and is shown in Fig. 12. There are no reflection conditions, but reflections with $h + k + \ell \neq 2n$ are difference reflections. If there is no long-range order, then only reflections with $h + k + \ell = 2n$ are present and the structure is identical to body-centered cubic. The spacegroup is then $Im3m$, with atoms on the 2(a) site at (0, 0, 0) etc. This bcc site-disordered form has been found in both InSb and HgTe, and probably GaSb, to date.

13. ST12

The ST12 structure, shown in Fig. 13, is found in Ge on slow pressure decrease from the β-tin phase. It has been sought but never yet found in Si. It is noncentrosymmetric and tetragonal with spacegroup $P4_32_12$. (As in the case of cinnabar (Section II.9), the ST12 structure can also be described using an enantiomorphic spacegroup, $P4_12_12$. However, since the ST12 phase has only ever been obtained in powdered samples, no distinction between the two spacegroups has been possible. It has been standard, however, to adopt the $P4_32_12$ description.) The atoms are on the 4(a) and

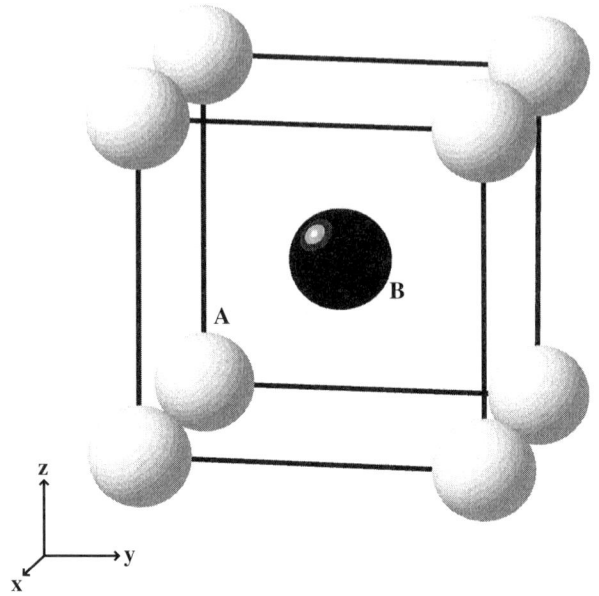

Fig. 12. Cesium chloride structure for a binary AB system, with A at (0, 0, 0) and B at (1/2, 1/2, 1/2). If all the atoms are made identical (a site-disordered or monatomic system), this becomes the body-centered cubic structure.

8(b) sites at $(u, u, 0)$ and (x, y, z), respectively. At ambient pressure $a = 5.93$ Å and $c = 6.98$ Å, with $u = 0.0912(60)$, $x = 0.1730(37)$, $y = 0.3784(51)$, and $z = 0.2486(48)$ (Donohue, 1974). The general reflection conditions are $\ell = 4n$ in 00ℓ, and $h = 2n$ in $h00$.

The ST12 structure has fourfold coordination and contains five- and seven-membered rings. Therefore, unlike the BC8 structure (see next section), which contains only even-membered rings, any binary equivalent of ST12 would have to contain energetically unfavorable like-atom bonds. The average bond length in the ST12 structure is 2.485 Å, slightly *greater* than the value of 2.450 Å observed in the diamond phase at ambient pressure. However, the bond angles vary from 98° to 133° and are thus distorted considerably from their tetrahedral values, such that ST12 has an atomic volume 9.6% less than the diamond phase at ambient pressure and 1.6% less than the BC8 phase.

It is interesting to note — especially with regard to the close relationship between the $C222_1$ (see Section II.11) and cristobalite structures — that the arrangement of the Si atoms in the keatite polymorph of SiO_2 is almost identical to ST12 Ge: Both have the same spacegroup and the Si atoms in

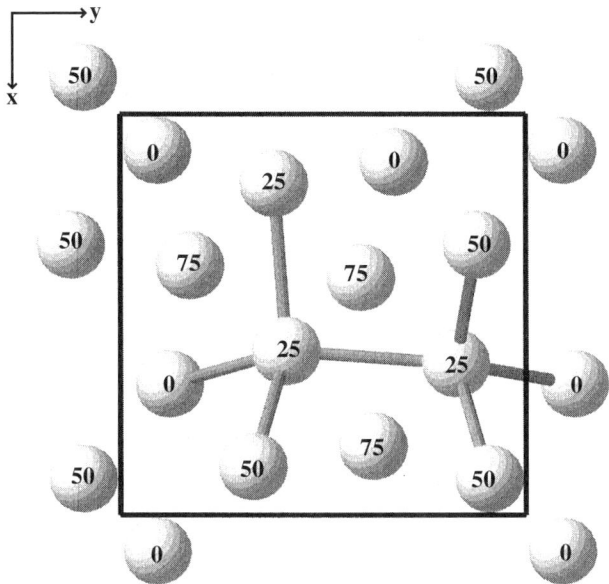

FIG. 13. ST12 structure of Ge, shown in projection down the (unique) z-axis. Numbers on each atom give the z coordinates in units of 0.01. Some short nearest-neighbor Ge–Ge bonds are shown. Adapted with permission from *Figure 2* of Bundy, F. P., and Kasper, J. S. (1963) *Science* **139**, 340. Copyright 1963 American Association for the Advancement of Science.

keatite are on the 4(a) and 8(b) sites at (0.09, 0.09, 0) and (0.174, 0.38, 0.252) (Shropshire *et al.*, 1959).

14. BC8, SC16 AND R8

The BC8 structure is found in Si and Ge at ambient conditions on pressure release from their high-pressure β-tin phases. The structure is body-centered cubic, spacegroup $Ia\bar{3}$, with atoms on the 16(c) site at (u, u, u) etc. In BC8-Si at ambient pressure, $a = 6.636(5)$ Å and $u = 0.1003(8)$ (Kasper and Richards, 1964). In BC8-Ge at ambient pressure, $a = 6.932(1)$ Å and $x = 0.1004(3)$ (Nelmes *et al.*, 1993). The reflection conditions are $h + k + \ell = 2n$ in $hk\ell$, and k and $\ell = 2n$ in $0k\ell$ — and cyclic permutations. The structure is tetrahedrally coordinated, with each atom having one short "A" bond and three slightly longer "B" bonds to nearest neighbors. Although the bond lengths are similar to those found in the diamond phases of Si and Ge, the bond angles are distorted considerably from their ideal tetrahedral values, with the result that at ambient pressure the BC8 structure has an atomic volume 8.0% (Ge) or 8.8% (Si) less than the diamond form.

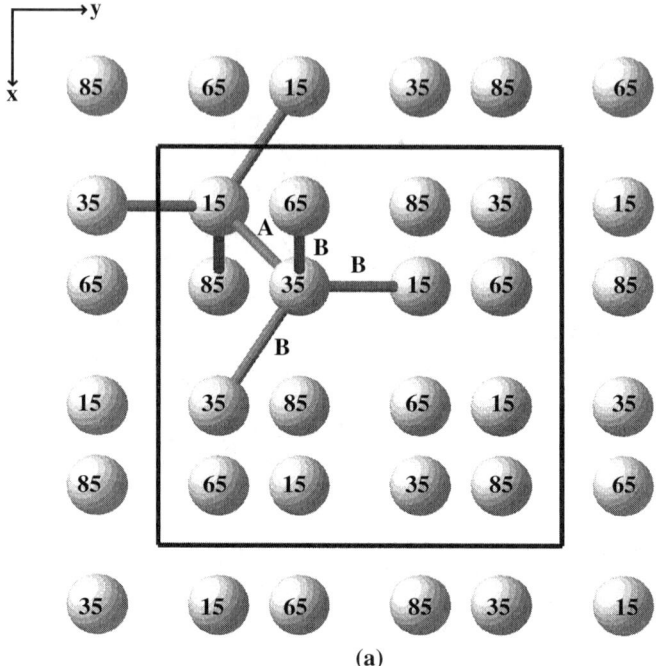

FIG. 14. (a) BC8 structure of Si, shown in projection down the z-axis. Si is at $(u \sim 0.15, u, u)$. Numbers on each atom give the z coordinates in units of 0.01. Some of the short nearest-neighbor A and B bonds are shown. (b) SC16 structure of GaAs, shown in projection down the z-axis. Ga is at $(u \sim 0.15, u, u)$ and As is at $(v \sim 0.64, v, v)$. The z coordinates are as given in (a) to within 0.01. A and B bonds are shown as in (a). (b) Copyright 1998 American Physical Society.

Since the first observation of the BC8 structure in 1963, it has been described with the single variable atomic coordinate $u \sim 0.1$, as above. However, an *identical* description of the structure can be made using a value of $u \sim 0.15$. This description of the structure is drawn in Fig. 14a. In both cases, the structure has centers of inversion symmetry at $(0, 0, 0)$ and $(1/4, 1/4, 1/4)$. However, with the $u \sim 0.1$ description, the center of inversion symmetry at $(0, 0, 0)$ must disappear in the binary analog of BC8, the SC16 structure, to allow the A-bonded atoms at $(0.1, 0.1, 0.1)$ and $(-0.1, -0.1, -0.1)$ to be of different species. This appears to have misled Crain *et al.* (1994b) into concluding that SC16 has to be noncentrosymmetric, space-group $P2_13$. However, the SC16 structure they describe still contains a center of inversion symmetry at $(1/4, 1/4, 1/4)$, and the structure actually has $Pa\bar{3}$ symmetry. The other set of inversion centers was noted by Mujica

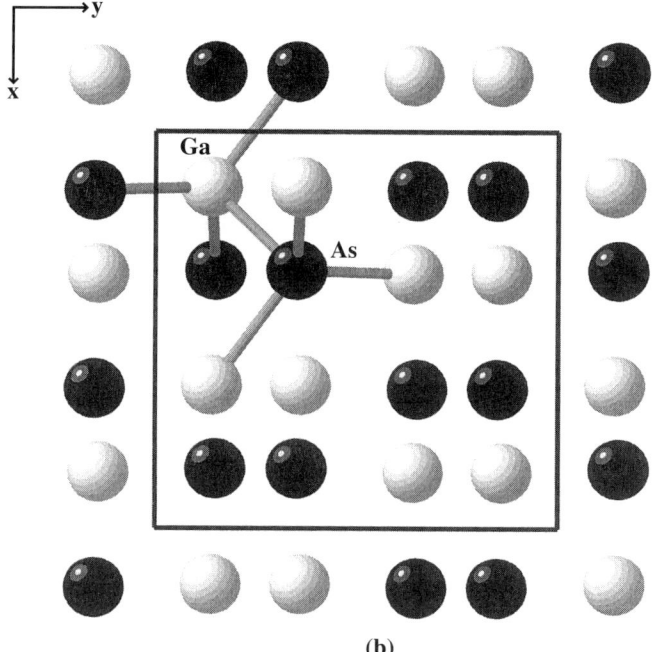

(b)

FIG. 14. (*continued*)

et al. (1995), who described the SC16 structure as being centrosymmetric with $Pa\bar{3}$ symmetry. This symmetry is supported experimentally by Hull and Keen (1994) in their high-pressure diffraction studies of the SC16 structure of copper halides. In spacegroup $Pa\bar{3}$, one atomic species is on an 8(c) site at (u, u, u) with $u \sim 0.15$, and the other is on another 8(c) site at (v, v, v) with $v \sim 0.65$, as shown in Fig. 14b. The reflection conditions are $k = 2n$ in $0k\ell$, $h = 2n$ in $h00$, and cyclic permutations. Until very recently, the copper halides were the only systems in which the SC16 structure had been observed experimentally. Calculations have shown that the structure is energetically favored over the known phases of several III–V systems (Crain *et al.*, 1994b; Mujica *et al.*, 1995; Mujica and Needs, 1997) but no experimental evidence for its existence could be found. Now an SC16 phase has been obtained in GaAs by heating to ~450K just below 14GPa (McMahon *et al.*, 1998). At 18.9 GPa, $u(\text{Ga}) = 0.152(1)$ and $v(\text{As}) = 0.640(1)$ with $a = 6.594(1)$ Å. The structure is 5.1(1)% denser than the ZB phase and 7.4(1)% less dense than the Cmcm phase at the same pressure.

Although the $u \sim 0.1$ and $u \sim 0.15$ descriptions of the BC8 structure are completely equivalent, the relationship between the BC8 and SC16 structures is greatly simplified if the $u \sim 0.15$ description of BC8 is adopted because it allows both structures to be described with respect to a common origin — as shown in Figs. 14a and 14b. The center of inversion symmetry at (0, 0, 0) is then retained on transforming from $Ia\bar{3}$ to $Pa\bar{3}$, and the center at (1/4, 1/4, 1/4) is the one that is lost. The A bond is between atoms at (0.15, 0.15, 0.15) and (0.35, 0.35, 0.35) in both structures when drawn this way.

The R8 phase, which has a small rhombohedral distortion of the BC8 structure, is found in Si on pressure decrease from the high-pressure β-tin phase and is stable from 10 GPa down to ~2 GPa. It has spacegroup $R\bar{3}$, with atoms on the 2(c) and 6(f) sites at (w, w, w) and (x, y, z). There are no reflection conditions. At 6.3 GPa, $w = 0.2836(5)$, $x = 0.4620(5)$, $y = -0.0322(4)$, and $z = 0.2667(4)$, with $a = 5.630(3)$ Å and $\alpha = 110.00(3)°$ (see Section III.2). This description of the structure relates to the $u \sim 0.1$ description of the BC8 structure and the coordinates should be transformed to $w' = 0.5 - w$, $x' = 0.5 - y$, $y' = 0.5 - x$, and $z' = 0.5 - z$ for comparison with the recommended $u \sim 0.15$ description of BC8. At 6.3 GPa, $w' = 0.216(1)$, $x' = 0.532(1)$, $y' = 0.038(1)$, and $z' = 0.234(1)$. To facilitate a comparison with BC8, the R8 structure can also be described using a pseudo-BC8 unit cell with $a = 6.460$ Å and $\alpha = \beta = \gamma = 91.1°$. In this description, the atom at (w', w', w') in the standard setting is transformed to $(w'/2, w'/2, w'/2) = (0.108, 0.108, 0.108)$. The displacement of this atom from the position it would occupy in the BC8 structure — (0.15, 0.15, 0.15) — is a large movement along the [111] direction, and results in the bond breaking and reforming along this direction at the BC8-to-R8 transition noted Crain et al. (1994a). The A-bond between the atoms at (0.15, 0.15, 0.15) and (0.35, 0.35, 0.35) in BC8 is broken, these atoms move apart along [111], and new bonds are formed to the atoms at (−0.15, −0.15, −0.15) and (0.65, 0.65, 0.65), which move closer together along [111] — see Fig. 14a. The resulting R8 structure contains odd-membered rings. Therefore, as with the ST12 structure, a binary equivalent of the R8 structure is very unlikely, as it would have to contain energetically unfavorable like-atom bonds.

References

Auvray, P. and Genet, F. (1973). *Bull. Soc. Minéral Cristallogr.* **96**, 218.
Crain, J., Ackland, G. J., Piltz, R. O., and Hatton, P. D. (1993). *Phys. Rev. Lett.* **70**, 814.
Crain, J., Ackland, G. J., Maclean, J. R., Piltz, R. O., Hatton, P. D., and Pawley, G. S. (1994a). *Phys. Rev. B* **50**, 13043.
Crain, J., Piltz, R. O., Ackland, G. J., Clark S. J., Payne, M. C., Milman, V., Lin, J. S., Hatton, P. D., and Nam, Y. H. (1994b). *Phys. Rev. B* **50**, 8389.
Donohue, D. (1974). "The Structures of the Elements." Wiley, New York.
Hull, S., and Keen, D. A. (1994). *Phys. Rev. B* **50**, 5868.

"International Tables for Crystallography," Volume A (T. Hahn, Ed.). Reidel, Dordrecht, 1983.
International Union of Pure and Applied Chemistry (1990). "Nomenclature of Inorganic Chemistry," Recommendations 1990. Blackwell Scientific Publication, Oxford.
Kasper, J. S., and Richards, S. M. (1964). *Acta Crystallogr.* **17,** 752.
McMahon, M. I., Nelmes, R. J., Liu, H., and Belmonte S. A. (1996). *Phys. Rev. Lett.* **77,** 1781.
McMahon, M. I., Nelmes, R. J., Allan, D. R., Belmonte, S. A., and Bovornratanaraks, T. (1998). *Phys. Rev. Lett.*, in press.
Megaw, H. D. (1973). "Crystal Structures: A Working Approach." W.B. Saunders, Philadelphia.
Mooney, R. C. L. (1956). *Acta Crystallogr.* **9,** 728.
Mujica, A., and Needs, R. J. (1996). *J. Phys.: Condens. Matter* **8,** L237.
Mujica, A., and Needs, R. J. (1997). *Phys. Rev. B* **55,** 9659; *Phys. Rev. B* **56,** 12653E (Erratum).
Mujica, A., Muñoz, A., and Needs, R. J. (1998). *Phys. Rev. B* **57,** 1344.
Mujica, A., Needs, R. J., and Muñoz, A. (1995). *Phys. Rev. B* **52,** 8881.
Mujica, A., Needs, R. J., and Muñoz, A. (1996). *Phys. Stat. Sol. (b)* **198,** 461.
Nelmes, R. J., McMahon, M. I., Wright, N. G., Allan, D. R., and Loveday, J. S. (1993). *Phys. Rev. B* **48,** 9883.
Nelmes, R. J., McMahon, M. I., Wright, N. G., and Allan, D. R. (1994). *Phys. Rev. Lett.* **73,** 1805.
Nelmes, R. J., McMahon, M. I., Wright, N. G., and Allan D. R. (1995). *Phys. Rev. B* **51,** 15723.
Nelmes, R. J., McMahon, M. I., and Belmonte, S. A. (1997). *Phys. Rev. Lett.* **79,** 3668.
Shropshire, J., Keat, P. P., and Vaughan, P. A. (1959). *Z. Kristalogr.* **112,** 409.
Vanderborgh, C. A., Vohra, Y. K., and Ruoff, A. L. (1989). *Phys. Rev. B* **40,** 12450.
Warren, B. E. (1990). "X-ray Diffraction." Dover, New York.
Weir, S. T., Vohra, Y. K., and Ruoff, A. L. (1987). *Phys. Rev. B* **36,** 4543.
Weir, S. T., Vohra, Y. K., Vanderborgh, C. A., and Ruoff, A. L. (1989). *Phys. Rev. B* **39,** 1280.
Yoder-Short, D., Colella, R., and Weinstein, B. A. (1982). *Phys. Rev. Lett.* **49,** 1438.
Zhang, S. B., and Cohen, M. L. (1989). *Phys. Rev. B* **39,** 1450.

III. Individual Systems

1. INTRODUCTION

In this section we give a detailed presentation of each of the 30 group IV, III–V, and II–VI systems in turn, except for the boron-V systems, on which a relatively small amount of work has been done until recently, work that has been fully summarized in a review by Ruoff and Li (1995).

The results presented have all been obtained from X-ray powder diffraction techniques — variously energy-dispersive diffraction (EDX) and angle-dispersive diffraction (ADX) on synchrotron sources (usually with an image-plate detector for recent ADX), or ADX on laboratory sealed or rotating-anode sources. The abbreviations EDX and ADX will be used hereafter. Other details of experimental techniques — such as the use or not of a pressure-transmitting medium — will be discussed wherever they appear to affect the results obtained.

In several of the more recent studies, anomalous dispersion effects have been used. Monochromatic ADX techniques allow the relative scattering power of the constituent species to be changed by exploiting these effects. The effective number of scattering electrons is modified at absorption edges — for example, the difference in scattering power of In and Sb is changed from ~2 electrons away from edges to over 5 electrons close to the In K-edge. This effect can greatly aid the detection of site ordering in systems where the atoms are close in Z, and even aid structure solution. However, the technique can be used only for edges that lie within the range between the high-energy flux cutoff of the particular synchrotron source and the low-energy limit of ~10 keV for passing through the diamond anvils of the pressure cell without undue attenuation.

Where we have included as yet unpublished work of our own, it is distinguished by being described in the first person — "we have . . . ," "work done by us," etc. All such work has been carried out with our standard image-plate ADX techniques at SRS, Daresbury (Nelmes and McMahon, 1994), as outlined in Section I.

The large majority of *in situ* diffraction studies of the semiconductors have been carried out at room temperature (RT) so far, and the results summarized from this section in Table I (of Section IV) are those obtained at RT only. However, in the presentations that follow, available results from high-temperature studies are also included.

There have been only a few attempts until recently to determine the pressure dependence of structures, and we have not generally included any detailed discussion of it in what follows. However, reference to work on structural pressure dependence is included wherever it has been studied.

Lattice parameters are given for all phases — at one pressure. For ambient-pressure phases, all values have been taken from Landoldt–Börnstein (1982a, 1982b), except where otherwise indicated. In many cases, the values tabulated in Landoldt–Börnstein (1982a, 1982b) show quite wide variations. We have quoted a value that is representative of the most recent or accurate results.

The volume changes between phases are given wherever they are known. The volume change is defined as the difference, ΔV, between two phases in the volume per formula unit in their structures. This is sometimes quoted in the literature as $\Delta V/V_{trans}$, where V_{trans} is the volume of the lower-pressure (less dense) phase at the transition, and sometimes at $\Delta V/V_0$, where V_0 is the volume for the ambient-pressure phase at ambient pressure. In some cases authors have not specified which of these has been used. Where possible we have clarified this from other data given in the paper concerned. If there is insufficient information to be certain, we simply quote the percentage volume change given in the paper without any definition of its meaning. Values of volume change are sometimes given at a pressure close to but not identical with the transition pressure because this allows an accurate value to be obtained from refinement of a mixed-phase sample.

It is important to note that a reliable value of volume change can be obtained only if the unit cell has been correctly identified and accurate lattice parameters have been measured — and the structure is understood well enough that at least the correct number of atoms in the unit cell is known. Many previous values for volume change in the literature are incorrect for want of satisfying these conditions. We believe the values given herein are all correctly related to the current best interpretations of the structures involved. Nonetheless, there are often quite large variations in the values reported in different studies.

We have endeavored to make a critical assessment of the reliability attached to the structural characterization of all phases — including apparently straightforward cases such as NaCl phases. Where we make no comment, it can be taken that we believe the structure to be correct within the minimum level at which we note uncertainty in other cases. It is important to stress that the standards applied have been forced to be considerably more rigorous following the introduction of full-profile (Rietveld) analysis of powder diffraction data — see Section I.

We have noted all cases where it is possible to determine whether or not a phase is site-ordered — either from data in the literature, or our own data (sometimes unpublished).

REFERENCES

Landolt–Börnstein (1982a). "Semiconductors, Physics of Group IV Elements and III–V Compounds" (O. Madelung, Ed.), New Series, Group II, Vol. 17, Pt. A. Springer-Verlag, Berlin.

Landolt–Börnstein (1982b). "Semiconductors, Physics of III–VI and I–VI Compounds, Semimagnetic Semiconductors" (O. Madelung, Ed.), New Series, Group II, Vol. 17, Pt. B. Springer-Verlag, Berlin.

Nelmes, R. J. and McMahon, M. I. (1994). *J. Synchrotron Rad.* **1,** 69.

Ruoff, A. L. and Li, T. (1995). *Annu. Rev. Mater. Sci.* **25,** 249.

2. SILICON

Si has the diamond structure under ambient conditions with $a = 5.4310$ Å. Its high-pressure behavior has been the subject of intense experimental and theoretical interest for over 35 years (Minomura and Drickamer, 1962). Comprehensive reviews of previous diffraction work on Si have been given by Hu *et al.* (1986) and Ruoff and Li (1995).

A high-pressure transition in Si was first reported by Minomura and Drickamer, who determined a transition pressure of ~20 GPa (Minomura and Drickamer, 1962). A subsequent diffraction study of the transition reported the transition pressure as ~16 GPa and found the high-pressure phase to have the β-tin structure (Jamieson, 1963). More recent studies have found the transition at 11.3(2) GPa (Hu et al., 1986) and 11.7 GPa (McMahon et al., 1994), with a volume decrease at the transition ($\Delta V/V_{trans}$) of 20.4(4)% (Hu et al., 1986). The transition pressure is known to be sensitive to the effects of nonhydrostatic conditions, and a transition pressure of ~8.5 GPa is obtained if no pressure transmitting fluid is used (Hu et al., 1986). However, a transition pressure of 8.8 GPa has also been reported in a study that did use a methanol:ethanol pressure medium (Olijnyk et al., 1984).

At 11.3 GPa, the lattice parameters of the β-tin phase (Si-II) are $a = 4.690(6)$ Å and $c = 2.578(5)$ Å (Hu et al., 1986). At 13–16 GPa, Si-II has long been reported to transform to Si-V with the simple hexagonal (SH) structure (Hu and Spain, 1984; Hu et al., 1986; Ruoff and Li, 1995). At ~16 GPa the lattice parameters of the SH phase are $a = 2.551(6)$ Å and $c = 2.387(7)$ Å (Hu et al., 1986). However, the improved resolution afforded by image-plate ADX techniques revealed a new orthorhombic phase, with Imma symmetry, between the β-tin and SH phases (McMahon and Nelmes, 1993). At 15 GPa, the Imma structure has $a = 4.737(1)$ Å, $b = 4.502(1)$ Å, $c = 2.550(1)$ Å, with $\Delta = 0.386(2)$. The structure is as shown in Fig. 3a, but with all the atoms identical. The possible existence of such a phase had been considered previously in energy calculations (Needs and Martin, 1984). Further detailed diffraction studies (McMahon et al., 1994) determined the stability range of the Imma phase as being from 13.2(3) GPa to 15.6(3) GPa and revealed that both the β-tin-to-Imma and Imma-to-SH transitions are first-order, with volume discontinuities ($\Delta V/V_0$) of 0.2(1)% and 0.5(1)%, respectively. These volume changes are accompanied by significant discontinuities in Δ, which is found to vary from 0.3 to 0.4 over the stability range of the Imma phase (see *Figure 5* of McMahon et al., 1994). Calculations stimulated by the discovery of the Imma structure confirmed its stability relative to β-tin and SH (Lewis and Cohen, 1993; Ackland, 1994).

In light of Chang and Cohen's (1984, 1985) suggestion that phonons near the β-tin-to-SH transition would be soft, enhancing the superconductivity, McMahon and Nelmes (1993) noted that a discontinuity in the superconducting transition temperature (T_c) at ~13 GPa (Il'ina and Itskevich, 1980; Chang et al., 1985; Mignot et al., 1986), and a maximum in T_c found at ~15 GPa (Chang et al., 1985, Mignot et al., 1986) coincide with the β-tin-to-Imma and Imma-to-SH transition pressures.

On further pressure increase, Si-V transforms at 37.6(16) GPa to the so-called "intermediate phase" (Olijnyk et al, 1984; Duclos et al., 1990), Si-VI, the structure of which is as yet unknown, despite several attempts to solve this long-standing problem (Vijaykumar and Sikka, 1990; Duclos et al., 1990). The most recent study concluded that the structure is monoclinic, with the same structure as the X phase of the alloy $Bi_{0.8}Pb_{0.2}$ (Duclos et al., 1990). The reported lattice parameters of the monoclinic structure at 40.0 GPa are $a = 2.351$ Å, $b = 3.996$ Å, and $c = 5.279$ Å, with $\beta = 64.7°$ (Duclos et al., 1990). Si-VI transforms to hcp Si-VII at 41.8(5) GPa (Duclos et al., 1990), with $a = 2.524(9)$ Å and $c = 4.142(50)$ Å at 41–42 GPa (Hu et al., 1986). Si-VII subsequently transforms to fcc Si at 78(3) GPa (Duclos et al., 1987). The fcc phase has $a = 3.341$ Å at 87(7) GPa and is stable to at least 248 GPa (Duclos et al., 1990). Because the structure of Si-VI is unknown, no reliable value can be given for the volume change at the SH-to-Si-VI and Si-VI-to-hcp transitions. The volume change ($\Delta V/V_0$) at the hcp → fcc transition is 0.3(6)% (Duclos et al., 1990).

Five further phases have been found on pressure decrease. On slow decompression from the β-tin phase, Si has long been reported to transform to Si-III, also known as the BC8 phase, which is a hole semimetal (Besson et al., 1987). It has a body-centered cubic structure (Fig. 14a) with $a = 6.636(5)$ Å at ambient pressure (Kasper and Richards, 1964). The BC8 structure is discussed in detail in Section II.14, where it is suggested that a description with atoms at ($u \sim 0.15, u, u$), etc, in spacegroup $Ia\bar{3}$, rather than ($u \sim 0.1, u, u$) as chosen by Kasper and Richards (1964), is preferable because it simplifies the relationship to the SC16 structure (the diatomic equivalent of BC8). The improved resolution of ADX techniques has revealed that the β-tin phase does not transform directly to BC8, however, but rather transforms to the rhombohedral R8 phase at 10.1(1) GPa (Crain et al., 1994; Piltz et al., 1995). We have measured the volume increase ($\Delta V/V_0$) at the β-tin-to-R8 transition as 10.7(2)% at 9.4 GPa. At 8.2 GPa, the reported R8 structure has $a = 5.620(3)$ Å and $\alpha = 110.07(3)°$, with atoms at ($w = 0.2922(9), w, w$)) and ($x = 0.4597(8), y = -0.0353(7), z = 0.2641(7)$) (Piltz et al., 1995) — see Section II.14. Our own ADX studies of R8 silicon at 6.3 GPa gave $a = 5.630(3)$ Å and $\alpha = 110.00(3)°$, with $w = 0.2836(5), x = 0.4620(5), y = -0.0322(4)$, and $z = 0.2667(4)$, and we note that the lower w value is in closer agreement with the computational study of Pfrommer et al. (1997). On further pressure decrease to 2 GPa, the R8 phase transforms to BC8 with a volume increase ($\Delta V/V_0$) of 2.1(2)% (Piltz et al., 1995). The R8 phase is thus the densest of the tetrahedrally bonded phases of silicon. The BC8-to-R8 transition is found to be completely reversible on pressure increase, taking place at $\sim$2 GPa (Crain et *al.*, 1994). Total-energy calculations using the R8 and BC8 structures suggest that both phases may be stable, or very close to stable, at 8 GPa (Piltz *et al.*, 1995). Calculations of the R8 band structure indicate that it is a semimetal (Piltz *et al.*, 1995; Pfrommer *et al.*, 1997).

On heating at ambient pressure, Si-III (BC8) is reported to transform to Si-IV (Wentorf and Kasper, 1963) with the hexagonal diamond structure (Section II.3). The reported lattice parameters at ambient conditions are $a = 3.837$ Å and $c = 6.317$ Å (Besson *et al.*, 1987). However, the observed d-spacing of the (002) reflection in *Table 1* of Besson *et al.* (1987) is given as 3.189 Å, which implies $c = 6.378$ Å. In fact, a least-squares fit to the observed d-spacings reported by Besson *et al.* (1987) gives the significantly amended values of $a = 3.820$ Å and $c = 6.374$ Å. In addition, Nelmes *et al.* (1993) have questioned the correctness of the hexagonal diamond structure; neither they nor Besson *et al.* (1987) observe the strong (102) reflection which should be present, and Nelmes *et al.* (1993) find an orthorhombic unit cell to give a better fit to the observed d-spacings. It is perhaps conceivable that the absence of (102) and displacements of peak positions from hexagonal symmetry arise from stacking faults (Treacy *et al.*, 1991). Certainly more work is required on this phase.

On very fast decompression of Si-II to atmospheric pressure, two tetragonal phases have been reported (Zhao *et al.*, 1986) and designated Si-VIII and Si-IX. The structures of these phases are unknown. Si-VIII is reported to be tetragonal with $a = 8.627(3)$ Å and $c = 7.500(3)$ Å, and density considerations suggest that there are approximately 30 atoms per unit cell. The systematically absent reflections were found to be consistent with spacegroup $P4_12_1$. Si-IX is also tetragonal, with $a = 7.482(5)$ Å and $c = 3.856(5)$ Å, and density considerations suggest there are 12 atoms per unit cell. The systematically absent reflections are consistent with spacegroup $P4_22$ (Zhao *et al.*, 1986).

References

Ackland, G. J. (1994). *Phys. Rev. B* **50**, 7389.
Besson J. M., Mokhtari E. H., Gonzalez J., and Weill G. (1987). *Phys. Rev. Lett.* **59**, 473.
Chang, K. J. and Cohen, M. L. (1984). *Phys. Rev. B* **30**, 5376.
Chang, K. J. and Cohen, M. L. (1985). *Phys. Rev. B* **31**, 7819.
Chang, K. J., Dacorogna, M. M., Cohen, M. L., Mignot, J. M., Chouteau, G., and Martinez, G. (1985). *Phys. Rev. Lett.* **54**, 2375.
Crain, J., Ackland, G. J., Maclean, J. R., Piltz, R. O., Hatton, P. D., and Pawley, G. S. (1994). *Phys. Rev. B* **50**, 13043.
Duclos, S. J., Vohra, Y. K., and Ruoff, A. L. (1987). *Phys. Rev. Lett.* **58**, 775.
Duclos, S. J., Vohra, Y. K., and Ruoff, A. L. (1990). *Phys. Rev. B* **41**, 12021.
Hu, J. Z. and Spain, I. L. (1984). *Solid State Commun.* **51**, 263.

Hu, J. Z., Merkle, L. D., Menoni, C. S., and Spain, I. L. (1986). *Phys. Rev. B* **34**, 4679.
Il'ina, M. A. and Itskevich, E. S. (1980). *Sov. Phys. Solid State* **22**, 1833.
Jamieson, J. C. (1963). *Science* **139**, 762.
Kasper, J. S. and Richards, S. M. (1964). *Acta Crystallogr.* **17**, 752.
Lewis, S. P. and Cohen, M. L. (1993). *Phys. Rev. B* **48**, 16144.
McMahon, M. I. and Nelmes, R. J. (1993). *Phys. Rev. B* **47**, 8337.
McMahon, M. I., Nelmes, R. J., Wright, N. G., and Allan, D. R. (1994). *Phys. Rev. B* **50**, 739.
Mignot, J. M., Chouteau, G., and Martinez, G. (1986). *Phys. Rev. B* **34**, 3150.
Minomura, S. and Drickamer, H. G. (1962). *J. Phys. Chem. Solids* **23**, 451.
Needs, R. J. and Martin, R. M. (1984). *Phys. Rev B* **30**, 5390.
Nelmes, R. J., McMahon, M. I., Wright, N. G., Allan, D. R., and Loveday, J. S. (1993). *Phys. Rev. B* **48**, 9883.
Olijnyk, H., Sikka, S. K., and Holzapfel, W. B. (1984). *Phys. Lett.* **103a**, 137.
Pfrommer, B. G., Côté, M., Louie, S. G., and Cohen, M. L. (1997). *Phys. Rev. B* **56**, 6662.
Piltz, R. O., Maclean, J. R., Clark, S. J., Ackland, G. J., Hatton, P. D., and Crain, J. (1995). *Phys. Rev. B* **52**, 4072.
Ruoff, A. L. and Li, T. (1995). *Annu. Rev. Mater. Sci.* **25**, 249.
Treacy, M. M. J., Newsam, J. M., and Deem, M. W. (1991). *Proc. R. Soc. Lond. A* **433**, 499 (1991).
Vijaykumar, V. and Sikka, S. K. (1990). *High Pressure Research* **4**, 306.
Wentorf, R. H. Jr. and Kasper, J. S. (1963). *Science* **139**, 338.
Zhao, Y. X., Buehler, F., Sites, J. R., and Spain, I. L. (1986). *Solid State Commun.* **59**, 679.

3. GERMANIUM

Although the reported phase transition sequence in Ge is somewhat similar to that found in Si, the transition pressures are very different. Reviews of previous diffraction results have been given by Menoni *et al.* (1986) and Ruoff and Li (1995).

At ambient pressure, germanium has the diamond structure (Ge-I) with $a = 5.6579$ Å. Ge-I transforms to β-tin (Ge-II) at 10.6(5) GPa with a volume decrease ($\Delta V/V_{\text{trans}}$) of 18.9(7)% (Menoni *et al.*, 1986). The lattice parameters of Ge-II at 12 GPa are $a = 4.9585$ Å and $c = 2.7463$ Å (Qadri *et al.*, 1983). The diamond-to-β-tin transition pressure is known to be sensitive to the level of stress in the sample and transition pressures as low as 8.0(5) GPa (Baublitz and Ruoff, 1982) and 6.7 GPa (Qadri *et al.*, 1983) have been reported using, respectively, ice or no pressure-transmitting fluid. In contrast to the limited stability range of the β-tin phase of Si, Ge-II is reported to be stable to the very much higher pressure of 75(3) GPa where it was initially reported to transform to the simple-hexagonal (SH) structure (Vohra *et al.*, 1986) with a volume change (undefined) of less than 0.5% (Vohra *et al.*, 1986). The much greater stability range of the β-tin phase in Ge is due to the effects of the $3d$ core electrons present in Ge (Vohra *et al.*, 1986). As in Si, however, image-plate ADX data have revealed that the transition from β-tin is first to Imma (see Section II.7), at ~75 GPa (Nelmes *et al.*, 1996) — as predicted in the calculations of Lewis and Cohen (1994). At 81 GPa, the refined structure has $a = 4.572(1)$ Å, $b = 4.386(1)$ Å, $c = 2.461(1)$ Å, and $\Delta = 0.390(4)$. The orthorhombic splitting at 81 GPa is the same as that observed in Si at 14.4(2) GPa, while the value of Δ is the same as that observed in Si at the slightly higher pressure of ~15 GPa. If the behavior of Imma-Ge is the same as that of Imma-Si, an estimate of ~85 GPa can be obtained for the Imma-to-SH transition pressure. This would accord with the report of a single-phase SH diffraction pattern at 90 GPa by Vohra *et al.* (1986). The best available estimate of the stability range of the Imma phase is ~75 to ~85 GPa, but we believe from some subsequent (as yet unpublished) work that these values may need to be revised downward a little.

On further pressure increase, the simple-hexagonal phase is reported to transform to double hexagonal close packed (dhcp) at 102(5) GPa, with $a = 2.414$ Å and $c = 8.114$ Å at 125 GPa (Vohra *et al.*, 1986). However, the authors showed no experimental data from the dhcp phase and the identity of the structure remains to be decisively confirmed.

At 7.60(5) GPa on slow pressure decrease, β-tin transforms to the metastable ST12 structure (Ge-III), shown in Fig. 13, with a volume change ($\Delta V/V_{trans}$) of 9.2(1)% (Bundy and Kasper, 1963; Menoni *et al.*, 1986). Like the BC8 structure (Fig. 14a), ST12 comprises distorted tetrahedra, but with the atoms arranged in fivefold and sevenfold rings instead of six-fold rings (Bundy and Kasper, 1963). The angular distortions of the ST12-Ge tetrahedra are greater than in BC8-Ge, however, and the ST12 structure is 9.6(1)% denser than that of diamond at ambient pressure and 1.7(1)% denser than BC8 (Nelmes *et al.*, 1993). Although it has been claimed that a direct transition from diamond to ST12 is possible at pressures as low as 2.5 GPa (Bates *et al.*, 1965), attempts to repeat this result have been unsuccessful (Menoni *et al.*, 1986) and other studies have reported that ST12 forms only from β-tin (Qadri *et al.*, 1983). The β-tin-to-ST12 transition is reversible; repressurizing ST12 from ambient pressure to ~11 GPa results in a complete transition back to the β-tin phase (Qadri *et al.*, 1983).

The BC8 phase of Ge is obtained at ambient pressure on fast pressure release from the β-tin phase (Nelmes *et al.*, 1993). The lattice parameter at ambient pressure is 6.932(1) Å. As discussed in Section II.14, we now suggest that a description of BC8 with $u \sim 0.15$ should be used rather than the $u = 0.1004(3)$ description proposed initially (Nelmes *et al.*, 1993) — that is, with $u = 0.1496(3)$. Diffraction patterns had to be obtained from the BC8-Ge samples immediately on recovery as they rapidly transform to a different phase characterized by very broad reflections. Comparisons with both the broadened diffraction patterns obtained by Minomura

(1981) starting from amorphous Ge and with the patterns of Si-IV obtained on heating BC8-Si, show BC8-Ge to transform into a phase very similar to Si-IV (Nelmes et al., 1993). This suggests that both BC8-Si and BC8-Ge behave in a like manner, in transforming on heating to the supposed hexagonal-diamond structure, but that the activation energy is much lower for the transition in BC8-Ge such that it occurs quite readily at room temperature. As for diffraction patterns from Si-IV (see Section III.2), the (102) reflection, which should be quite strong, is unobserved, raising the same question as for Si-IV of whether hexagonal diamond is the correct structure assignment.

References

Bates, C. H., Dachille, F., and Roy, R. (1965). *Science* **147,** 860.
Baublitz, M. and Ruoff, A. L. (1982). *J. Appl. Phys.* **53,** 5669.
Bundy, F. P. and Kasper, J. S. (1963). *Science* **139,** 340.
Lewis, S. P. and Cohen, M. L. (1994). *Solid State Commun.* **89,** 483.
Menoni, C. S., Hu, J. Z., and Spain, I. L. (1986). *Phys. Rev. B* **34,** 362.
Minomura, S. (1981). *J. Phys. (Paris) Colloq.* **42,** C4-181 — and references therein.
Nelmes, R. J., McMahon, M. I., Wright, N. G., Allan, D. R., and Loveday, J. S. (1993). *Phys. Rev. B* **48,** 9883.
Nelmes, R. J., Liu, H., Belmonte, S. A., Loveday, J. S., McMahon, M. I., Allan D. R., Häusermann, D., and Hanfland M. (1996). *Phys. Rev. B* **53,** R2907.
Qadri, S. B., Skelton, E. F., and Webb, A. W. (1983). *J. Appl. Phys.* **54,** 3609.
Ruoff, A. L. and Li, T. (1995). *Annu. Rev. Mater. Sci.* **25,** 249.
Vohra, Y. K., Brister, K. E., Desgreniers, S., Ruoff, A. L., Chang, K. J., and Cohen, M. L. (1986). *Phys. Rev. Lett.* **56,** 1944.

4. Boron-V Compounds

The high-pressure behavior of the boron III–V compounds will not be discussed in detail in this review. A review by Ruoff and Li (1995) covers work on BN, BP, BAs, and BSb.

Reference

Ruoff, A. L. and Li, T. (1995). *Annu. Rev. Mater. Sci.* **25,** 249.

5. Aluminum Nitride

Under ambient conditions AlN has the wurtzite structure with $a = 3.110$ Å and $c = 4.980$ Å. As a result of the difficulty in manufacturing samples, and in marked contrast to many other of the binary semiconductors, the great majority of the experimental and theoretical high-pressure work on AlN has taken place only in the last decade.

While Van Vechten predicted a transition to a β-tin structure at high pressure (Van Vechten, 1973), an NaCl phase was obtained on quenching from 16.5 GPa and 1700–1900 K (Vollstädt *et al.*, 1990). The first *in situ* high-pressure diffraction measurement of AlN (Ueno *et al.*, 1992), using an image-plate system on a rotating-anode generator, found a transition to a site-ordered NaCl structure at 22.9 GPa. The volume change ($\Delta V/V_0$) was determined to be 17.9% at ~23 GPa. The measured lattice parameter of the NaCl phase at 30.0(8) GPa is 3.938 Å (Ueno *et al.*, 1992). A second *in situ* study to 65 GPa, using EDX methods, confirmed the site-ordered NaCl structure but found the transition to start at the lower pressure of 14 GPa and to be complete at 20 GPa (Xia *et al.*, 1993). This study, however, makes no mention of the use of a pressure-transmitting medium — unlike Ueno *et al.* (1992), who used a methanol:ethanol:water mixture — and the absence of this may result in the lower transition pressure. The measured volume change ($\Delta V/V_{trans}$) at 14 GPa was 18.6%. The NaCl phase was reported to be stable to at least 65 GPa and was found to persist on reducing the pressure to atmospheric pressure (Xia *et al.*, 1993). The equilibrium transition pressure was thus determined to be between 0 and 14 GPa.

References

Ueno, M., Onodera, A., Shimomura, O., and Takemura, K. (1992). *Phys. Rev. B* **45**, 10123.
Van Vechten, J. A. (1973). *Phys. Rev. B* **7**, 1479.
Vollstädt, H., Ito, E., Akaishi, M., Akimoto, S., and Fukunaga, O. (1990). *Proc. Jpn. Acad.* **B66**, 7.
Xia, Q., Xia, H., and Ruoff, A. L. (1993). *J. Appl. Phys.* **73**, 8198.

6. Aluminum Phosphide

AlP is the III–V analog of Si and has the zincblende structure with $a = 5.4635$ Å at ambient conditions. It is unstable in air, and this may be a reason for the relative paucity of experimental studies. Wanagel *et al.* (1976), using resistivity measurements in an opposed anvil device, first determined that AlP undergoes a sluggish transition to a conducting phase at a pressure slightly lower than that at which ZnS transforms to a metallic phase. Assuming a transition pressure of 15 GPa for ZnS, the transition pressure in AlP was reported as 14 GPa (Wanagel *et al.*, 1976). The diffraction study of Yu *et al.* (1978), using a sample loaded with a methanol:ethanol pressure-transmitting medium, determined the transition pressure as 17.0(5) GPa and ruled out the high-pressure phase as having either the β-tin or NaCl structures. Three observed diffraction lines could be indexed on a face-centered cubic structure, but the intensities were inconsistent with NaCl (Yu et al., 1978).

In the light of the suggestion of Froyen and Cohen (1983) that the behavior of AlP should be similar to that of AlAs, Greene et al. (1994) followed their EDX study of AlAs (see next section) with a combined EDX and optical reflectivity study of AlP. Taking care to avoid exposing their sample to air, and using no pressure-transmitting medium, they found a transition at 14.2 GPa on pressure increase. This is probably lower than the 17.0(5) GPa value obtained by Yu et al. (1978) because of the lack of any pressure medium. However, it is also the case that AlP samples degrade in time when loaded with methanol:ethanol. The correct value may be anywhere in the range 14–17 GPa. By assuming that the high-pressure phase would be sixfold coordinated, and that the volume change at the transition would be ~17% (as for AlAs — see next section), they found this phase to have a NiAs structure with a volume change of 17(1)% (Greene et al., 1994). It was not determined whether the structure is site-ordered or not. The fit to the observed d-spacings is good, but the observed reflection intensities do not agree at all well with those given by the NiAs structure. Greene et al. (1994) remark that the fit to the intensities is no worse than in the zincblende phase but, on the evidence presented, the fit is clearly poorer in the high-pressure phase. The NiAs identification of the structure should probably be regarded as likely to be correct but not yet certain. No definition of the 17(1)% volume change at the transition is given, but a comparison with AlAs suggests that it is $\Delta V/V_{trans}$. Zhang and Cohen (1987) calculated a value of 18.9% for $\Delta V/V_{trans}$, but this was for a transition from ZB to NaCl at 9.3 GPa. The reported lattice parameters for the NiAs phase at 19 GPa are $a = 3.466$ Å and $c = 5.571$ Å (Greene et al., 1994). No further transitions were found up to the maximum pressure reached, 43 GPa. The reverse transition pressure was determined as being between 8.4 and 4.8 GPa, giving an equilibrium transition pressure of 9.5(5.0) GPa. The optical reflectivity measurements showed that the NiAs phase exhibits metallic-like reflectivity in the near infrared (Greene et al., 1994).

References

Froyen, S. and Cohen, M. L. (1983). *Phys. Rev. B* **28,** 3258.
Greene, R. G., Luo, H., and Ruoff, A. L. (1994). *J. Appl. Phys.* **76,** 7296.
Wanagel, J., Arnold, V., and Ruoff, A. L. (1976). *J. Appl. Phys.* **47,** 2821.

Yu, S. C., Spain, I. L., and Skelton, E. F. (1978). *Solid State Commun.* **25,** 49.
Zhang, S. B. and Cohen, M. L. (1987). *Phys. Rev. B* **35,** 7604.

7. ALUMINUM ARSENIDE

The high-pressure behavior of AlAs has attracted relatively few experimental studies, perhaps because of its toxicity. At ambient conditions it has the zincblende structure and its almost perfect lattice match with GaAs (a = 5.660 Å for AlAs and a = 5.653 Å for GaAs) makes it a material of great technological interest. Weinstein *et al.* (1987) found a transition at 12.3(4) GPa on pressure increase by microscopic examination, while a subsequent Raman study on bulk-like thin films found the transition at 12.4(4) GPa on pressure increase (Venkateswaran *et al.*, 1992). In neither case, however, was the structure determined. An EDX study of AlAs by Greene *et al.* (1994) proposed that AlAs-II has the NiAs structure. To solve the structure the authors assumed that the high-pressure phase would be sixfold coordinated and that the volume change at the transition would be ~17%. Although the agreement between observed and calculated intensities is poor for the NiAs structure, the misfits were reported to be similar to those observed in the ZB phase. In fact, the fit is clearly poorer in the high-pressure phase. The authors described the fit between observed and calculated d-spacings of the NiAs structure as excellent, yet the position of one peak — (002) — is 0.019 Å from its calculated position, which is a large displacement. However, the peak is a weak shoulder (*Figure 1* of Greene *et al.*, 1994) and an accurate determination of its position is difficult. (The presence of the (002) reflection shows the structure to be site ordered.) As in the similar case of AlP, the identification of the structure as NiAs should probably be regarded as likely to be correct but not yet certain. The ZB-to-NiAs transition was found to start at 12 GPa on pressure increase and to be complete by 14 GPa. No further transitions were found up to the maximum pressure reached, 46 GPa. On pressure decrease, the reverse transition from NiAs to ZB occurred between 4.5 and 2 GPa. The equilibrium transition pressure was thus reported as 7(5) GPa, and the volume change ($\Delta V/V_{\mathrm{trans}}$) at 7 GPa is 17(1)% (Greene *et al.*, 1994).

REFERENCES

Greene, R. G., Luo, H., Li, T., and Ruoff, A. L. (1994). *Phys. Rev. Lett.* **72,** 2045.
Weinstein, B. A., Hark, S. K., Burnham, R. D., and Martin, R. M. (1987). *Phys. Rev. Lett.* **58,** 781.
Venkateswaran, U. D., Cui, L. C., Weinstein, B. A., and Chambers, F. A. (1992). *Phys. Rev. B* **45,** 9237.

8. ALUMINUM ANTIMONIDE

The number of high-pressure studies made on AlSb has been limited, probably because of its hygroscopicity and the consequent sample-handling problems. At ambient pressure AlSb has the zincblende structure with $a = 6.1355$ Å. The X-ray diffraction study of Jamieson (1963) reported a transition to a β-tin structure with a volume change ($\Delta V/V_{trans}$) of 16.5%. A subsequent diffraction study by Yu et al. (1978) found the transition at 8.0(3) GPa but reported that the high-pressure phase had the NaCl structure. The authors suggested that this different result may have been due to the more hydrostatic conditions achieved in their study. A further study by Baublitz and Ruoff (1983) using EDX techniques and pressure-transmitting fluids found the transition at 7.7(5) GPa, or slightly lower, but reported that the structure of the high-pressure phase was orthorhombic. They obtained a volume change (undefined) of 19.6(1.5)%. Based on the similarities of their data to data they had collected on GaAs-II (see Section III.11), the authors proposed that AlSb-II was an orthorhombic distortion of NaCl with spacegroup *Fmmm*. However, only reflections with h, k, and ℓ all even were observed, and there were large differences between observed and calculated intensities. An orthorhombic structure was also reported by Ves et al. (1986), who found the transition at 7.9(2) GPa.

An EDX study by Greene et al. (1995) attempted to resolve the uncertainty over the structure of AlSb-II by using a sample prepared and loaded under argon, and pressurized without pressure medium. Only four indexed lines of the high-pressure phase were reported, and these were found to be fitted by a β-tin structure. The volume change at the transition was 15(1)%; from the data presented this appears to be $\Delta V/V_{trans}$. However, poor agreement was found between the observed and calculated intensities in both the zincblende and high-pressure structures, probably because of the effects of preferred orientation (Greene et al., 1995). It was not possible to determine the atomic ordering of the high-pressure phase, despite the large difference in scattering power of Al ($Z = 13$) and Sb ($Z = 51$). On increasing the pressure further, a second phase transition was reported at 43 GPa (Greene et al., 1995). No information was reported on the structure of this third phase.

A subsequent diffraction study using ADX methods has clarified the structure of the high-pressure phase (Nelmes et al., 1997). The use of a methanol:ethanol pressure-transmitting medium was found to make no difference to the diffraction pattern of the high-pressure phase, except that sharper peaks were obtained with the medium. The diffraction patterns of the high-pressure phase were found not to be from a β-tin phase, which — among other things — cannot explain two of the lowest-angle diffraction lines (see *Figure 1* of Nelmes *et al.*, 1997). However, a fit to the patterns could be obtained with a site-ordered Cmcm structure (Fig. 10), with a very similar unit cell to that proposed by Baublitz and Ruoff (1983). The fit at 14.7 GPa gave $a = 5.353(1)$ Å, $b = 5.788(1)$ Å, $c = 5.086(1)$ Å, $y(\text{Al}) = 0.599(1)$, and $y(\text{Sb}) = 0.163(1)$ (Nelmes *et al.*, 1997). From a mixed-phase sample at 8.1 GPa, the volume change $(\Delta V/V_0)$ at the transition was found to be 19.5(2)%. There remain small misfits in a few (weak) peak positions, but an extensive search for an alternative unit cell suggests that the observed *d*-spacings do not exactly fit *any* lattice. This is a problem we have experienced in several other III–V and II–VI semiconductors and is likely to arise from stacking faults or deviatoric stress. Apart from these small misfits, the peak positions and intensities are well accounted for by the Cmcm structure, and we conclude that this structure — or a very similar one — is the structure of the high-pressure phase. Some calculations by Rodríguez-Hernández *et al.* (1996) did not consider Cmcm as a possible structure for AlSb-II, and instead proposed that it may have the NiAs structure. But this does not fit the data, and further calculations are now required.

We have extended our ADX studies to higher pressures, and, like Greene *et al.* (1995), find a further transition at 41(3) GPa. The diffraction pattern of the new phase is very broadened, but evidently quite different from Cmcm — as shown in Fig. 15. No attempt has yet been made to solve the structure in view of the data quality. Greene *et al.* (1995) found no evidence of any further transitions up to the maximum pressure reached in their study, 59 GPa.

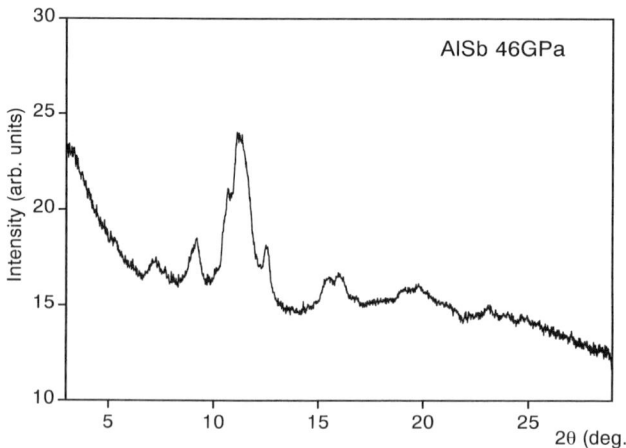

FIG. 15. Observed powder diffraction pattern from AlSb at 46 GPa. The structure is as yet unknown.

References

Baublitz, M. and Ruoff, A. L. (1983). *J. Appl. Phys.* **54,** 2109.
Green, R. G., Lou, H., Ghandehari, K., and Ruoff, A. L. (1995). *J. Phys. Chem. Solids* **56,** 517.
Jamieson, J. C. (1963). *Science* **139,** 845.
Nelmes, R. J., McMahon, M. I., and Belmonte, S. A. (1997). *Phys. Rev. Lett.* **79,** 3668.
Rodríguez-Hernández, P., Muñoz, A., and Mujica, A. (1996). *Phys. Stat. Sol.* (*b*) **198,** 455.
Ves, S., Strössner, K., and Cardona, M. (1986). *Solid State Commun.* **57,** 483.
Yu, S. C., Spain, I. L., and Skelton, E. F. (1978). *Solid State Commun.* **25,** 49.

9. Gallium Nitride

GaN is the most ionic of the III–V semiconductors and has the wurtzite structure at ambient pressure with $a = 3.190$ Å and $c = 5.189$ Å (Shultz and Thiemann, 1977). Samples are difficult to grow and this has limited the number of high-pressure studies.

A high-pressure transition in GaN was first identified by Perlin *et al.* (1992) at 47 and ~50 GPa using Raman and X-ray absorption techniques, respectively. The reverse transition started at 30 GPa and was complete around 20 GPa. The identity of the high-pressure structure could not be determined, but the sample was reported to blacken at the transition (Perlin *et al.*, 1992). The first *in situ* diffraction study of GaN was made by Xia *et al.* (1993) using EDX methods. No mention of the use of a pressure-transmitting fluid is made. The authors found their samples to be affected by strong texture effects in the wurtzite phase, although these effects were reported to disappear in the high-pressure phase. The phase transition was found to begin at 37 GPa, considerably lower than the 47 GPa found in the earlier Raman measurements, which were made on single-crystal samples in an argon or neon pressure-transmitting medium. Xia *et al.* (1993) reported the high-pressure phase as having a site-ordered NaCl structure and obtained a volume change ($\Delta V/V_{trans}$) of ~17.0% at 37 GPa. They obtained a lattice parameter of $a = 4.006(1)$ Å at 50 GPa. No further transitions were found to the maximum pressure reached, 70 GPa.

A more recent structural study to 60 GPa using ADX techniques, a laboratory-based X-ray source, and an alcohol:water pressure-transmitting fluid has been carried out by Ueno *et al.* (1994). The ordered NaCl phase was found to exist above 52.2(3.0) GPa and the volume decrease at the transition ($\Delta V/V_{trans}$) was determined as 17.9% (Ueno *et al.*, 1994). The lattice parameter of the NaCl phase at 60.6(5.9) GPa was measured as 3.985 Å.

References

Perlin, P., Jauberthie-Carillon, C., Itié, J.-P., San Miguel, A., Grzegory, I., and Polian, A. (1992). *Phys. Rev. B* **45**, 83.
Shultz, H. and Thiemann, K. H. (1977). *Solid State Commun.* **23**, 815.
Ueno, M., Yoshida, M., Onodera, A., Shimomura, O., and Takemura, K. (1994). *Phys. Rev. B* **49**, 14.
Xia, H., Xia, Q., and Ruoff, A. L. (1993). *Phys. Rev. B* **47**, 12925.

10. Gallium Phosphide

GaP has the zincblende structure under ambient conditions with $a = 5.4505$ Å. The high-pressure phase transition in GaP was first observed by Onodera *et al.* (1974) at a pressure later revised to 22(1) GPa (Piermarini and Block, 1975), a value supported by two later resistivity measurements (Bundy, 1975; Homan *et al.*, 1975). The first diffraction studies by Yu *et al.* (1978) located the transition at 22.0(5) GPa, but observed only four peaks from the high-pressure phase, which they reported were not consistent with either a cubic structure or the β-tin structure. A diffraction study by Baublitz and Ruoff (1982) using EDX methods found the transition at 21.5(8) GPa and reported the high-pressure phase as having the β-tin structure, an assignment confirmed by Hu *et al.* (1984). In both cases, however, the diffraction data contained extra reflections not accounted for by the β-tin structure. Also, in neither study was it possible to determine whether the structure was site-ordered or not, despite the very different atomic numbers of Ga ($Z = 31$) and P ($Z = 15$).

The reversibility of the high-pressure phase transition has been the subject of some disagreement. Pressure release from 32.5 GPa was reported to produce a mixture of crystalline and amorphous material (Jauberthie-Carillon and Guillemin, 1989). A further X-ray absorption study confirmed this, reporting that on rapid pressure release from 36 GPa the sample returned not to ZB but to a predominantly amorphous phase (Itié *et al.*, 1989). No evidence of amorphization was reported, however, in a sample recovered to ambient pressure from 29.6 GPa (Hu *et al.*, 1984). Further work is required on this.

The structure of GaP-II has now been studied using ADX diffraction techniques (Nelmes *et al.*, 1997). The high-pressure transition was located at 24(2) GPa and the diffraction profiles obtained above this pressure are very similar to the one reported by Baublitz and Ruoff (1982), including the "extra" reflections and the shoulders on the low-angle sides of the "β-tin" (200) and (220) reflections. The best (and very good) fit to the data is obtained with a Cmcm structure that is site-*dis*ordered (see Section II.1) with $a = 4.707(2)$ Å, $b = 4.949(1)$ Å, $c = 4.701(1)$ Å, $y(\text{Ga/P}) = 0.647(3)$, and $y(\text{Ga/P}) = 0.159(3)$ (Nelmes et al., 1997). The lack of long-range site ordering is made evident by the weakness of the (110) reflection and the absence of (111), as shown in *Figure 4* of Nelmes et al. (1997). (See Section II.1 concerning site disorder.) A two-phase refinement of a mixed ZB/Cmcm sample at 29.4 GPa gives a volume change ($\Delta V/V_0$) of 14.0(2)%. The Cmcm structure accounts for the extra peaks observed by previous authors, and for the shoulders on the "β-tin" (200) and (220) reflections observed by Baublitz and Ruoff (1982). We have observed no further transition up to the maximum pressure reached, 52 GPa.

A comprehensive pseudopotential calculation has been made which includes previously unconsidered phases, such as Cmcm, Imma, and SC16 (Mujica and Needs, 1997). The first high-pressure phase is predicted to be SC16, with a transition pressure of 14.7 GPa, considerably lower than the calculated transition pressures to Cmcm (17.7 GPa), diatomic-β-tin (17.8 GPa), and NaCl (18.3 GPa). SC16 is predicted to be unstable against the d-β-tin structure at 20.3 GPa and against Cmcm at 20.4 GPa. However, with further increases in pressure the d-β-tin structure is calculated to become increasingly more stable than Cmcm.

References

Baublitz, M., and Ruoff, A. L. (1982). *J. Appl. Phys.* **53,** 6179.
Bundy, F. P. (1975). *Rev. Sci. Instr.* **46,** 1318.
Homan, C. E., Kendall, D. P., Davidson, T. E., and Frankel, J. (1975). *Solid State Commun.* **17,** 831.
Hu, J. Z., Black, D. R., and Spain, I. L. (1984). *Solid State Commun.* **51,** 285.
Itié, J.-P., Polian, A., Jauberthie-Carillon, C., Dartyge, E., Fontaine, A., Tolentino, H., and Tourillon, G. (1989). *Phys. Rev. B* **40,** 9709.
Jauberthie-Carillon, C., and Guillemin, C. (1989). *J. Phys.: Condensed Matter* **1,** 6807.
Mujica, A., and Needs, R. J. (1997). *Phys. Rev. B* **55,** 9659; *Phys. Rev. B* **56,** 12653E (Erratum).
Nelmes, R. J., McMahon, M. I., and Belmonte, S. A. (1997). *Phys. Rev. Lett.* **79,** 3668.
Onodera, A., Kawai, N., Ishizaki, K., and Spain, I. L. (1974). *Solid State Commun.* **14,** 803.
Piermarini, G. J., and Block, S. (1975). *Rev. Sci. Inst.* **46,** 973.
Yu, S. C., Spain, I. L., and Skelton, E. F. (1978). *Solid State Commun.* **25,** 49.

11. Gallium Arsenide

GaAs is the III-V analog of Ge. Because of its technological importance and its apparently unusual high-pressure behavior, GaAs has attracted the most attention among the III–V systems. At ambient conditions GaAs has the zincblende structure with $a = 5.6532$ Å. A phase transition in GaAs was first found near 20 GPa by electrical measurements (Minomura and

Drickamer, 1962). Diffraction measurements by Yu *et al.* (1978a) reported a transition at 17.0(5) GPa to an unknown orthorhombic structure, while Shimomura *et al.* (1980) reported that the diffraction patterns from the high-pressure phase were similar to those from the supposed orthorhombic Pmm2 phase of InSb (Yu *et al.,* 1978b). Baublitz and Ruoff (1982), using EDX techniques, reported the transition at 17.2(7) GPa to an orthorhombic structure with spacegroup *Fmmm.* However, it was noted that the intensities calculated by such a structure were in disagreement with those observed. In addition, no reflections with h, k, and ℓ all odd were observed, and the observed reflections could therefore be equally well indexed on a unit cell halved along all three axes. The authors suggested that the effects of preferred orientation and the presence of planar defects would explain some of the intensity differences (Baublitz and Ruoff, 1982).

In a subsequent EDX study, Weir *et al.* (1989) reported the transition to GaAs-II as beginning at 16.6 GPa and being complete at 22.9 GPa. They reported the structure of GaAs-II as being orthorhombic, spacegroup *Pmm2* — as suggested by Shimomura *et al.* (1980) — with atoms at (0, 0, 0) and (0, 1/2, α), and $\alpha = 0.35$ at 22.9 GPa. This structure is shown in Fig. 16. The unit cell dimensions obtained at 22.9 GPa are $a = 2.482(6)$ Å, $b = 4.83(1)$ Å, and $c = 2.618(6)$ Å. At 24(1) GPa Weir *et al.* (1989) reported a second transition, to GaAs-III, with no volume change. They found this also to be orthorhombic, with spacegroup *Imm2,* and atoms at (0, 0, 0) and (0, 1/2, Δ) with $\Delta = 0.425$ at 28.1 GPa. The reported lattice parameters at the same pressure were $a = 4.92(1)$ Å, $b = 4.79(1)$ Å and $c = 2.635(6)$ Å.

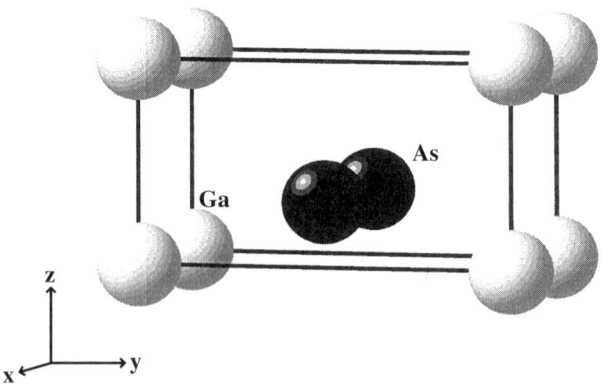

FIG. 16. Pmm2 structure obtained for GaAs-II by Weir *et al.* (1989), with Ga at (0, 0, 0) and As at (0, 1/2, $\alpha = 0.35$). The relative positions of atomic sites are very similar to the Cmcm structure (Fig. 10), but the arrangement of Ga and As on the sites is different.

Weir *et al.* (1989) appear to have considered the Pmm2 and Imm2 structures as both being site-ordered, but the difference reflections required to confirm this were not detected. The reported transition pressure to GaAs-III appeared consistent with a second resistance change observed by Minomura and Drickamer (1962). However, a more recent resistance study observed no such change (Zhang *et al.,* 1989). On further pressure increase, Weir *et al.* (1989) reported that the diffraction patterns of GaAs-III increasingly corresponded to that of a simple-hexagonal structure. They thus proposed a transition from GaAs-III to another phase, GaAs-IV, at 60–80 GPa, and GaAs-IV was found to be stable to 108 GPa, the highest pressure studied (Weir *et al.,* 1989). As pointed out in Section II.8, GaAs could not in fact have hexagonal symmetry if it were site-ordered (Fig. 8); it would have to be site-disordered, and hence be quasimonatomic.

A theoretical study of the proposed Pmm2 structure for GaAs-II by Zhang and Cohen (1989) found it to be energetically unfavorable, since each atom has six nearest neighbors of which four are like atoms. A larger $2a \times b \times 2c$ unit cell was found to have a lower energy — where a, b and c are the lattice parameters of the Pmm2 structure (Fig. 16) — with a different, NaCl-like ordering of the atoms in the x-z planes to maximize the number of unlike nearest neighbors. The spacegroup of this structure was quoted as *Amm2* by Vanderborgh *et al.* (1989) but, as discussed in Section II.10, it is in fact a special case of the Cmcm structure with $\Delta y = 0.5$. To relate the Pmm2 structure of Fig. 16 directly to Cmcm it should first be redescribed as Pm2m by interchanging the y and z axes. If the lattice parameters of Pm2m are a', b', and c' (respectively a, c, and b of Pmm2), then the Cmcm unit cell in Fig. 10 is $2a' \times 2b' \times c'$. In site-ordered Pm2m, the two atomic species would be at $(0, 0, 0)$ and $(0, \alpha, 1/2)$, respectively — with $\alpha \sim 0.35$ (see above). The NaCl-like x-y planes of Cmcm (see Fig. 10) are thus replaced in Pm2m by alternate x-y planes of one species (Ga) and the other (As) stacked along z — so that, as said, each atom has four like-atom nearest neighbors and only two unlike-atom nearest neighbors, compared with five or six unlike-atom nearest neighbors in Cmcm.

Because of the similar scattering power of Ga ($Z = 31$) and As ($Z = 33$), the diffraction profiles from the Pm2m and Cmcm structures are very similar. In particular, if $\Delta y = 0.5$ in the Cmcm structure, then the two differ only in their atomic ordering. (If the Pm2m and Cmcm structures (with $\Delta y = 0.5$) were both site-*dis*ordered, then they would be identical, with an $a' \times b' \times c'$ unit cell and spacegroup *Pmcm*.) Distinguishing the Pm2m, Cmcm ($\Delta y = 0.5$), Cmcm ($\Delta y \neq 0.5$), and Pmcm structures depends on the detection of very weak reflections, which are all absent in Pmcm and differ among the other three possibilities. This is made even more difficult because the diffraction profiles obtained from GaAs-II at room temperature are always characterized by broadened reflections and peak shifts. Despite considerable effort, initial ADX studies of GaAs were unable to observe any of the very weak additional reflections needed to obtain a decisive structure (McMahon and Nelmes, 1996). But it was concluded that the most probable structure was Cmcm without long-range site ordering, on the basis of (1) the fact that Pm2m and Pmcm both seem unlikely on energy grounds, and (2) the comparison with GaP (section III.10), which certainly *does* have a Cmcm structure lacking long-range site ordering (see Section II.1).

Since that work, we have used moderate heating of samples at high pressure to attempt to overcome the problems of peak broadening and peak displacement in GaAs-II. Figure 17 shows two diffraction profiles from the same sample of GaAs at ~18 GPa and 293 K before and after heating for 120 min at ~450 K. The diffraction profiles from the heated sample are found to be considerably sharper, and two very weak reflections are clearly observable at low angles. These reflections are in exactly the right place to be the (110) and (111) reflections from a site-ordered Cmcm structure, and neither are permitted in the Pm2m structure. Rietveld refinement of the sharpened profile in Fig. 17 yields a Cmcm structure with $a = 4.971(1)$ Å, $b = 5.272(1)$ Å, $c = 4.779(1)$ Å, $y(Ga) = 0.649(1)$, and

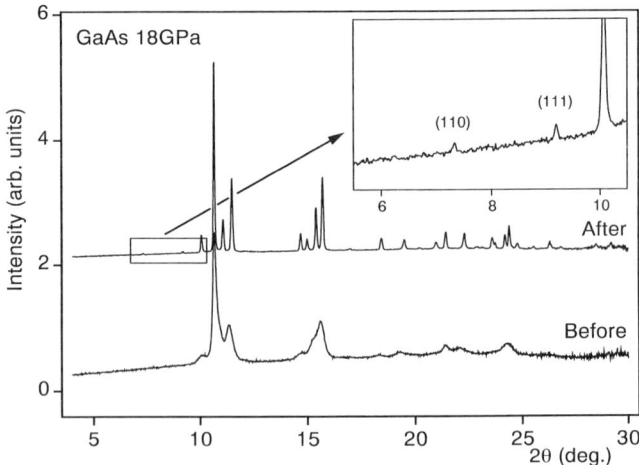

FIG. 17. Observed powder diffraction patterns obtained from GaAs at ~18 GPa, at room temperature, before and after heating for 120 min at ~450 K. The low-angle part of the pattern obtained after heating is enlarged in the inset to show the clearly visible (110) and (111) reflections. The pattern before heating is typical of those recorded in the literature for GaAs-II at room temperature.

$y(As) = 0.166(1)$. The coordinates correspond to an average value of 0.33 for α in the Pm2m subcell, close to the 0.35 obtained by Weir *et al.* (1989). We cannot yet show conclusively that the effect of heating is purely to sharpen the profile without any accompanying change in the underlying crystal structure; the GaAs-II phase obtained at room temperature may be randomly site-disordered and/or have $\Delta y = 0.5$. In that case, the structure obtained after heating would be a distinct phase. However, the most plausible interpretation of the structure before heating would then be an intermediate, nonequilibrium phase. It seems most probable that the true structure of GaAs-II is site-ordered Cmcm with $\Delta y \neq 0.5$, but that — without heating — it gives a broadened diffraction pattern and may also be only short-range ordered. Finally, we note that when pressure is decreased on sharpened GaAs-II (i.e., after heating), the transition to cinnabar discussed later in this section occurs at ~13 GPa instead of ~12 GPa.

The nature of the transition to GaAs-II has been the subject of a very detailed study by Besson *et al.* (1991) using a wide variety of experimental techniques. By comparing the results obtained from all these techniques on pressure increase and decrease, the authors were able to locate the true thermodynamic transition region of GaAs-I/GaAs-II at 12(1.5) GPa at 300 K. This may need to be reassessed following the discovery (see later discussion) that GaAs-II transforms first to a cinnabar phase on pressure decrease. However, Mujica and Needs (1996) have considered the Cmcm structure using pseudopotential methods and have reported the transition pressure for ZB-to-Cmcm as 12 GPa, in agreement with the value of Besson *et al.* (1991).

ADX studies to higher pressures show no discontinuities in diffraction profiles collected through the supposed transition to GaAs-III at 24(1) GPa (*Figure 6* of McMahon and Nelmes, 1996). In addition, reflections are observed above 24 GPa that cannot be explained by the proposed Imm2 structure of GaAs-III. The same reflections can also be seen in the 28.1 GPa data in *Figure 2* of Weir *et al.* (1989), and it is concluded, therefore, that there is no GaAs-II/GaAs-III transition at 24 GPa (McMahon and Nelmes, 1996). The sharp Pmm2-to-Imm2 structural change proposed by Weir *et al.* (1989) in any case seems improbable, as it requires a substantial structural rearrangement without any detectable volume change. However, although the Cmcm structure accounts for all the observed diffraction peaks at pressures above 24 GPa, there are some small but apparently reproducible mismatches in intensities of some weak reflections. In the profile at 35.2 GPa in *Figure 6* of McMahon and Nelmes (1996), the weak peak at $2\theta \sim 26°$ is too strong relative to the three reflections shown arrowed; also, the peak at 26° is a little displaced in *d*-spacing — by more than any other reflection. These misfits may all arise from the quite pronounced microstructural effects evident from the peak broadening, but more data are required in this pressure range to remove the residual uncertainties. If GaAs in fact remains site-ordered above 60–80 GPa, its symmetry remains orthorhombic — albeit quasi-hexagonal — and there is no transition to a phase IV at that pressure. In that case, the Cmcm phase is stable to at least 108 GPa, the highest pressure reached by Weir *et al.* (1989).

There has been equal interest more recently in the structural behavior of GaAs on pressure decrease. The generally accepted picture is that GaAs-II simply transforms back to the zincblende phase, at ~10 GPa (Besson *et al.*, 1991), except when the sample was recovered to ambient from 115 GPa, in which case completely amorphous GaAs has been obtained (Vohra *et al.*, 1990). However, Venkateswaran *et al.* (1991, 1992), Besson *et al.* (1990), and Tsuji *et al.* (1996) have all reported additional features on pressure decrease not accounted for by this simple picture.

ADX studies of GaAs on pressure decrease from GaAs-II have now revealed a previously uncharacterized phase with the cinnabar structure (McMahon and Nelmes, 1997). Cinnabar-GaAs first appears at 11.9(1) GPa, slightly above the reverse Cmcm-to-ZB transition at 11.2(3) GPa. At 8.3 GPa only peaks from the cinnabar and ZB phases are obtained. The refined cinnabar structure at 8.3 GPa has $a = 3.883(1)$ Å and $c = 8.657(2)$ Å, with $u(Ga) = 0.539(2)$ and $v(As) = 0.505(2)$. These coordinates are very similar to those reported for cinnabar ZnTe-II (Nelmes *et al.*, 1995), and, like ZnTe-II, cinnabar-GaAs is four-fold coordinated, with two nearest neighbors at 2.371(8) Å, two at 2.477(8) Å, and two like-atom next-nearest neighbors more than 1 Å further away at 3.488(6) Å for Ga and 3.478(6) Å for As (Fig. 9b). At 8.3 GPa, it is 7.4% denser than the ZB phase and 6.9% less dense than the Cmcm phase. As discussed by McMahon and Nelmes (1997), comparison with previous studies of GaAs on pressure decrease suggests that cinnabar-GaAs was probably observed, but not identified as such, by Vohra *et al.* (1990), Besson *et al.* (1990, 1991), and Tsuji *et al.* (1996). Furthermore, the optical reflectivity measurements of Vohra *et al.* (1990) suggest that the cinnabar phase is semiconducting with a bandgap of below 1.5 eV.

There is no evidence of the cinnabar phase of GaAs on pressure increase, indicating that it might not be a true equilibrium phase (McMahon and Nelmes, 1997). However, the phase transitions in GaAs exhibit large hysteresis, and a transition between two very different fourfold-coordinated structures may simply be too kinetically hindered at room temperature to occur before the transition to Cmcm. The close similarity of both the cinnabar and Cmcm structures of GaAs and ZnTe suggests that, like cinnabar-ZnTe, cinnabar-GaAs may indeed be an equilibrium phase. However, total energy calculations of Cmcm and cinnabar-GaAs show the two structures to be very similar in energy, but with that of cinnabar-GaAs slightly higher (Kelsey *et al.*, 1998; Mujica *et al.*, 1998). In both studies, the internal coordinates u

and v were found to be close to or exactly 0.5; in the latter case the symmetry of the structure becomes $P6_422$, or the enantiomorphic $P6_222$ (Mujica et al., 1998). It is not possible to rule out this higher-symmetry structure in GaAs completely from the observed (mixed-phase) diffraction data, though the 0.034(3) difference between u and v reported earlier does give a better fit than $u = v = 0.5$ (McMahon and Nelmes, 1997). However, $u = v = 0.5$ gives a *clearly* poorer fit to the single-phase diffraction data in the case of the similar cinnabar-ZnTe structure, the best-fitting refined values being $u = 0.540(2)$ and $v = 0.504(2)$ — and these are the same values within error as obtained for cinnabar-GaAs. This matter is discussed further in Section IV.

Yet further complexity in the behavior of GaAs has emerged in the observation by McMahon et al. (1998) of a transition to the cubic SC16 structure on heating GaAs-II to ~450 K just below 14 GPa (after making single-phase GaAs-II at ~24 GPa). This is the first observation of this structure in a III–V (or II–VI) system, although its existence has been postulated in GaAs previously in computational studies by Crain et al. (1994) and Mujica et al. (1995). The SC16 phase is stable on temperature decrease back to room temperature where all subsequent diffraction studies have been performed. At 18.9 GPa, the structure has $a = 6.594(1)$ Å with atoms at $(u\ (Ga) = 0.152(1), u, u)$ and $(v\ (As) = 0.640(1), v, v)^2$. The SC16 phase is recoverable to ambient pressure. On pressure increase it transforms to the Cmcm phase at 22.0(7) GPa, a pressure considerably higher than that at which the ZB phase starts to transform to Cmcm (~17 GPa). However, the true stability range of the SC16 phase is likely to be smaller than the 20 GPa suggested by these room-temperature results. High-temperature results show that the SC16 phase is not the equilibrium phase above 14.5 GPa, or below 13 GPa (McMahon et al., 1998), and the resulting range of 1.5 GPa is very similar to the 1.2 GPa calculated by Mujica et al. (1995). The large difference from the results at room temperature illustrates graphically the effects of kinetics in determining transition pressures in this system.

The cinnabar and SC16 phases make GaAs unique (so far) among the III–V and II–VI semiconductors in having two strongly distorted tetrahedral phases at high pressure. The bond lengths in the two phases are similar, although the cinnabar structure is the slightly more distorted: At 8.3 GPa, SC16 has nearest-neighbor (n-n) bond lengths of 2.369(19) Å and 2.426(19) Å (McMahon et al., 1998), while in cinnabar the bond lengths are 2.371(8) Å and 2.477(8) Å (McMahon and Nelmes, 1997). The cinnabar phase also has the greater deviations in the n-n bond angles from their

---

[2]The very similar structure with Ga and As interchanged can be excluded (McMahon et al., 1998).

ideal tetrahedral values: In SC16 the bond angles are ~100° and ~118°, while in cinnabar they vary from 91° to 141°. However, Mujica *et al.* (1998) have noted that the tetrahedra in cinnabar-GaAs are related to ideal tetrahedra by a simple rigid rotation of two of the bonds by 30° about the z-axis. In this sense, the tetrahedra in cinnabar-GaAs may be regarded as *less* distorted than those in the SC16 structure.

REFERENCES

Baublitz, M., and Ruoff, A. L. (1982). *J. Appl. Phys.* **53**, 6179.
Besson, J. M., Weill, G., Mansot, J. L., and Gonzalez, J. (1990). *High Pressure Research* **4**, 312.
Besson, J. M., Itié, J.-P., Polian, A., Weill, G., Mansot, J. L., and Gonzalez, J. (1991). *Phys. Rev. B* **44**, 4214.
Crain, J., Piltz, R. O., Ackland, G. J., Clark, S. J., Payne, M. C., Milman, V., Lin, J. S., Hatton, P. D., and Nam, Y. H. (1994). *Phys. Rev. B* **50**, 8389.
Kelsey, A. A., Ackland, G. J., and Clark, S. J. (1998). *Phys. Rev. B* **57**, R2029.
McMahon, M. I., and Nelmes, R. J. (1996). *Phys. Stat. Sol (b)* **198**, 389 (1996).
McMahon, M. I., and Nelmes, R. J. (1997). *Phys. Rev. Lett.* **78**, 3697.
McMahon, M. I., Nelmes, R. J., Allan, D. R., Belmonte, S. A., and Bovornratanaraks, T. (1998). *Phys. Rev. Lett.,* in press.
Minomura, S., and Drickamer, H. G. (1962). *J. Phys. Chem. Solids* **23**, 451.
Mujica, A., and Needs, R. J. (1996). *J. Phys.: Condens. Matter* **8**, L237.
Mujica, A., Needs, R. J., and Muñoz, A. (1995). *Phys. Rev. B* **52**, 8881.
Mujica, A., Muñoz, A., and Needs, R. J. (1998). *Phys. Rev. B* **57**, 1344.
Nelmes, R. J., McMahon, M. I., Wright, N. G., and Allan, D. R. (1995). *J. Phys. Chem. Solids* **56**, 545.
Shimomura, O., Kawamura, T., Fukamachi, T., Hosoya, S., Hunter, S., and Bienestock, A. (1980). *In* "High Pressure Science and Technology" (B. Vodar and P. Marteau, Eds.), Vol. 1, p. 534. Pergamon, Oxford.
Tsuji, K., Katayama, Y., Kanda, H., and Nosaka, H. (1996). *J. Non-Cryst. Solids* **198–200**, 24.
Vanderborgh, C. A., Vohra, Y. K., and Ruoff, A. L. (1989). *Phys. Rev. B* **40**, 12450.
Venkateswaran, U. D., Cui, L. J., Weinstein, B. A., and Chambers, F. A. (1991). *Phys. Rev. B* **43**, 1875.
Venkateswaran, U. D., Cui, L. J., Weinstein, B. A., and Chambers, F. A. (1992). *Phys. Rev. B* **45**, 9237.
Vohra, Y. K., Xia, H., and Ruoff, A. L. (1990). *Appl. Phys. Lett.* **57**, 2666.
Weir, S. T., Vohra, Y. K., Vanderborgh, C. A., and Ruoff, A. L. (1989). *Phys. Rev. B* **39**, 1280.
Yu, S. C., Spain, I. L., and Skelton, E. F. (1978a). *Solid State Commun.* **25**, 49.
Yu, S. C., Spain, I. L., and Skelton, E. F. (1978b). *J. Appl. Phys.* **49**, 4741.
Zhang, S. B., and Cohen, M. L. (1989). *Phys. Rev. B* **39**, 1450.
Zhang, S. B., Erskine, D., Cohen, M. L., and Yu, P. Y. (1989). *Solid State Commun.* **71**, 369.

12. GALLIUM ANTIMONIDE

GaSb is the least ionic of the III-V semiconductors and, as such, is the one expected to be the closest to silicon and germanium in its high-pressure behavior. At ambient pressure it has the zincblende structure with $a =$

6.0959 Å. A transition to a high-pressure metallic phase (GaSb-II) was first reported by Minomura and Drickamer (1962) at 8–10 GPa. Jamieson (1963) reported the same transition at 9 GPa and found GaSb-II to have the β-tin structure, a structure assignment that was subsequently supported by Yu *et al.* (1978) and Weir *et al.* (1987), who reported the transition at 6.2(3) GPa and 7.4(4) GPa, respectively, with a volume change ($\Delta V/V_{\text{trans}}$) of 17.1(1.2)% (Yu *et al.*, 1978). The transition pressure was determined as 7.65(10) GPa from a Raman Study (Aoki *et al.*, 1984). Despite the large difference in scattering power between Ga ($Z = 31$) and Sb ($Z = 51$), none of these diffraction studies was able to determine whether GaSb-II is site-ordered or not.

Weir *et al.* (1987) reported a second transition at 27.8(6) GPa to an SH-type structure from their EDX data, completing the diamond → β-tin → SH sequence reported for silicon and germanium (see Sections III.2 and III.3). However, as pointed out in Section II.8 and for GaAs earlier, GaSb-III cannot have the SH structure if it is site-ordered. Weir *et al.* (1987) reported a further transition to GaSb-IV at 61.0(7) GPa with a volume change ($\Delta V/V_0$) of ~5%. The structure of GaSb-IV — which was reported to be stable to at least 110 GPa — was tentatively indexed as the same Pmm2 orthorhombic structure initially proposed for GaAs-II (Fig. 16) and InSb-IV. As with GaSb-II and GaSb-III, however, the site ordering of GaSb-IV remained undetermined.

A more recent ADX study has revealed that the diffraction profiles from GaSb-II contain features that do not accord with the β-tin structure, and that these features become more apparent on pressure increase (McMahon *et al.*, 1994). In particular, a tetragonal lattice cannot account for the peak positions at low angles; nor can it explain three extra shoulders observed at higher angles (*Figure 1* of McMahon *et al.*, 1994). Inspection of the data of Weir *et al.* (1987) reveals that one of these shoulders is also observable in their published diffraction pattern at 23.3 GPa (their *Figure 1*). It is also apparent that GaSb-II is site-disordered, as the relatively strong (110) and (310) reflections required by an ordered structure are absent. (A diffraction pattern from GaSb-II is shown in Fig. 7, and a calculated profile for Ga and Sb in the d-β-tin structure is shown in the inset to Figure 3b. The effect of the orthorhombic distortion can be seen most readily in the insufficient splitting of the strong peak at $2\theta \sim 15°$ in Fig. 7; also, the complete absence of the quite strong (110) difference reflection is striking.) Comparison of the GaSb-II profiles with those obtained from Imma silicon revealed many similarities. An Imma structure, with each site assigned a 50:50 occupancy of Ga and Sb, was found to give a good fit to the data. At 18 GPa, the best-fitting Imma structure has $a = 5.276(1)$ Å, $b = 5.151(4)$ Å, $c = 2.886(1)$ Å, and $\Delta = 0.340(2)$ (McMahon *et al.*, 1994). The volume change ($\Delta V/V_{\text{trans}}$) at the transition is 18.3(1)%. At pressures near the transition from ZB the distortion of the Imma structure from β-tin ($a = b$ and $\Delta = 0.25$) is smaller. However, in a mixed-phase GaSb-I/GaSb-II profile collected at 7.6 GPa, the structure of GaSb-II is still clearly orthorhombic (McMahon *et al.*, 1994). Thus, GaSb transforms directly from an ordered ZB phase to a disordered Imma phase at 7 GPa.[3] No true β-tin phase is observed. The orthorhombic distortion of the unit cell and the displacement of Δ from 0.25 probably account for the anomalous effects reported in the EXAFS study of San-Miguel *et al.* (1992).

As pointed out in Section II.1, the fact that the average structure is site-disordered does not rule out the possibility of ordering over a short length scale. It may be that GaSb-II is site-ordered at nearest-neighbor distances, in which case its local symmetry (and in some senses true symmetry) would most probably be Imm2 — the structure shown in Fig. 6. It would be valuable if this could be distinguished from random site disorder by EXAFS measurements.

The ADX data (McMahon *et al.*, 1994) show no sign of the transition to GaSb-III at 27.8 GPa reported by Weir *et al.* (1987). Although the Imma structure does indeed become more "SH like" on pressure increase — that is, $b/c \rightarrow \sqrt{3}$ and $\Delta \rightarrow 0.5$ — the presence of the (312)/(132) doublet reflection, which would be absent in the SH structure, indicates that the structure is still Imma (see *Figure 4* of McMahon *et al.*, 1994), and the (312)/(132) peak is observable to at least 50 GPa. In further preliminary measurements at higher pressures, we have observed evidence of a transition between 63 and 71 GPa to an unknown phase. This is probably the transition reported by Weir *et al.* (1987) at 61.0(7) GPa. We have not yet identified the structure of this high-pressure phase fully, but it appears to be site-disordered body-centered cubic.

REFERENCES

Aoki, K., Anastassakis, E., and Cardona, M. (1984). *Phys. Rev. B* **30**, 681.
Jamieson, J. C. (1963). *Science* **139**, 845.
McMahon, M. I., Nelmes, R. J., Wright, N. G., and Allan, D. R. (1994). *Phys. Rev. B* **50**, 13047.
Minomura, S., and Drickamer, H. G. (1962). *J. Phys. Chem. Solids* **23**, 451.
San-Miguel, A., Polian, A., and Itié, J.-P. (1992). *High Pressure Research* **10**, 416.
Weir, S. T., Vohra, Y. K., and Ruoff, A. L. (1987). *Phys. Rev. B* **36**, 4543.
Yu, S. C., Spain, I. L., and Skelton, E. F. (1978). *Solid State Commun.* **25**, 49.

13. INDIUM NITRIDE

Because of the difficulty in sample preparation, there have been comparatively few diffraction measurements on bulk InN. At ambient pressure,

---
[3]See a discussion of some subsequent results in Section IV.

InN has the wurtzite structure with $a = 3.545$ Å and $c = 5.703$ Å (Pichugin and Tlachala, 1978). The first high-pressure transition was reported to start at 12.1(2) GPa by Ueno *et al.* (1993, 1994), using ADX techniques on a laboratory X-ray source. The transition is complete by ~15 GPa, and the high-pressure phase was reported as having a site-ordered NaCl structure with $a = 4.532$ Å at 18.2(5) GPa. The volume decrease ($\Delta V/V_{trans}$) at the transition is 17.6% (Ueno *et al.*, 1993). The identity of the high-pressure phase was confirmed as NaCl by Xia *et al.* (1994), who reported the transition pressure as 10 GPa with a volume change ($\Delta V/V_{trans}$) of 20(2)% using EDX techniques. However, a significantly higher transition pressure of ~23 GPa was reported by Perlin *et al.* (1993) from direct observation of the transition. They supported this observation with a calculated transition pressure of 25.4 GPa (Perlin *et al.*, 1993), but this is considerably higher than the value of 4.93 GPa calculated by Muñoz and Kunc (1993).

In addition to the observation of the NaCl phase, the diffraction study of Ueno *et al.* (1994) reported the appearance of two extra peaks at a pressure of 9.5 GPa from an unknown phase that appeared to be more compressible than the NaCl and wurtzite phase. They also observed unusual behavior in the pressure dependence of the axial ratio (c/a) of the wurtzite phase in the neighborhood of the transition to NaCl, and reported an extremely large value of 12.7(1.4) for B' (Ueno *et al.*, 1994).

Prompted by the nonlinear behavior of the axial ratio, Bellaiche and Besson (1996) and Besson *et al.* (1996) performed a total-energy calculation of the wurtzite phase of InN using the pseudopotential method. They found that their calculation reproduced the nonlinear behavior in the c/a ratio between 10 and 15.5 GPa, and concluded that InN would undergo a second-order isostructural phase transition at 16 GPa but for the fact that this pressure is (slightly) above that at which the wurtzite-to-NaCl transition is complete (Ueno *et al.*, 1994). However, it was postulated that the nonlinear pressure dependence of the wurtzite structure below 16 GPa might be the driving force behind the transition to the NaCl structure (Bellaiche *et al.*, 1996).

In their EDX study, Xia *et al.* (1994) reached a maximum of 35 GPa, and they found no further transitions to that pressure.

References

Bellaiche, L., Kunc, K., and Besson, J. M. (1996). *Phys. Rev. B* **54**, 8945.
Besson, J. M., Bellaiche, L., and Kunc, K. (1996). *Phys. Stat. Sol (b)* **198**, 469.
Muñoz, A., and Kunc, K. (1993). *J. Phys.: Condens. Matter* **5**, 6015.
Pichugin, I. G., and Tlachala, M. (1978). *Izv. Akad. Nauk SSSR. Neorg. Mater.* **14**, 175.
Perlin, P., Gorczyca, I., Porowski, S., Suski, T., Christensen, N. E., and Polian, A. (1993). *Jpn. J. Appl. Phys. Suppl.* **32-1**, 334.

Ueno, M., Yoshida, M., Onodera, A., Shimomura, O., and Takemura, K. (1993). *Jpn. J. Appl. Phys. Suppl.* **32-1,** 42.

Ueno, M., Yoshida, M., Onodera, A., Shimomura, O., and Takemura, K. (1994). *Phys. Rev. B* **49,** 14.

Xia, Q., Xia, H., and Ruoff, A. L. (1994). *Modern Physics Letters B* **8,** 345.

14. Indium Phosphide

InP has the zincblende structure at ambient conditions with $a = 5.8687$ Å. A high-pressure transition in InP was first observed by Minomura and Drickamer (1962), who found a transition to a metallic phase at ~13 GPa. Jamieson (1963) showed from diffraction measurements that this phase had the NaCl structure and reported a volume decrease ($\Delta V/V_{trans}$) of 19.6% at the transition. A subsequent diffraction study to 19 GPa using laboratory-source ADX techniques with a position-sensitive detector (Menoni and Spain, 1987) found the first transition at 10.80(5) GPa, and confirmed the high-pressure phase as having a site-ordered NaCl structure with a lattice parameter at 10.8 GPa quoted as 5.243 Å. However, *Table II* of Menoni and Spain (1987) appears to contain an error—the calculated *d*-spacing of the (200) reflection for $a = 5.243$ Å should be 2.622 Å rather than the 2.651 Å tabulated. The best least-squares fit to the tabulated *observed d*-spacings is 5.258 Å. The volume change ($\Delta V/V_{trans}$) was determined as 16.6(7)%, but this value reduces to 15.9% with the corrected lattice parameter. An optical absorption study reported the transition at 10.15(5)GPa (Müller *et al.*, 1980), while photoluminescence measurements locate it at 10.35(5) GPa (Kobayashi *et al.*, 1981). On further pressure increase, a second transition was reported at 18.90(5) GPa with a volume change ($\Delta V/V_{trans}$) of 9(2)% (Menoni and Spain, 1987). This was from a pattern obtained at the top pressure of the experiment just before gasket failure and the phase was tentatively indexed as β-tin-like; despite the large difference in scattering power between In ($Z = 49$) and P ($Z = 15$), no determination of the atomic ordering could be made because of the weakness of the diffraction pattern (Menoni and Spain, 1987). A further diffraction study analyzed the behavior of InP on pressure decrease (Patel *et al.*, 1989). The diffraction patterns at ambient pressure were reported to be different from those obtained before pressurizing the sample, although this difference may have been due to the presence of pressure-induced defects (Patel *et al.*, 1989).

Using ADX techniques, McMahon *et al.* (1993) located the phase transition from ZB to NaCl at 9.8(5) GPa, in good agreement with the values listed previously. The NaCl phase is clearly site-ordered, as shown by the nonzero intensity of the (111) reflection. The lattice parameter at 14.7 GPa is 5.200(2) Å, and the volume change ($\Delta V/V_0$) at the transition is 16.0(2)%. In a more recent ADX study, the NaCl phase was found to be stable to 28(1) GPa where there is a continuous transition to a site-ordered Cmcm structure (Nelmes et al., 1997). No change in the diffraction patterns was observed at 19 GPa, the pressure at which Menoni and Spain (1987) reported a (doubtful — see earlier discussion) transition to a β-tin-like phase. The fit to the Cmcm phase is very good apart from some difficulty in modeling the very broadened (111) reflection, and gives $a = 4.879(1)$ Å, $b = 5.088(2)$ Å, and $c = 4.923(3)$ Å, with $y(\text{In}) = 0.658(1)$ and $y(\text{P}) = 0.143(2)$ at 45.6 GPa, the maximum pressure reached.

A pseudopotential calculation, which considered Cmcm as a possible structure, reported a value of 5.6 GPa for the ZB-to-NaCl transition and 11–12 GPa for the NaCl-to-Cmcm transition (Mujica and Needs, 1997). The authors commented on the difficulty in locating the value of the NaCl-to-Cmcm transition accurately, but the calculated 11–12 GPa is very considerably lower than the experimental value of 28(1) GPa (Nelmes et al., 1997). The discrepancy in the observed and calculated values for the ZB-to-NaCl transition may arise because of hysteresis, but the continuous nature of the NaCl-to-Cmcm transition makes a large discrepancy between the pressure increase and true transition pressure unlikely in that case.

References

Jamieson, J. C. (1963). Science **139**, 845.
Kobayashi, T., Tei, T., Aoki, K., Yamamoto, K., and Abe, K. (1981). In "Physics of Solids under High Pressure" (J. S. Schilling and R. N. Shelton, Eds.), p. 141. North-Holland, Amsterdam.
McMahon, M. I., Nelmes, R. J., Wright, N. G., and Allan, D. R. (1993). *High Pressure Science and Technology, AIP Conf. Proc.* **309**, 629 (1993).
Menoni, C. S., and Spain, I. L. (1987). *Phys. Rev. B* **35**, 7520.
Minomura, S. and Drickamer, H. G. (1962). *J. Phys. Chem. Solids* **23**, 451.
Mujica, A., and Needs, R. J. (1997). *Phys. Rev. B* **55**, 9659; *Phys. Rev. B* **56**, 12653E (Erratum).
Müller, H., Trommer, R., Cardona, M., and Vogl P. (1980). *Phys. Rev. B* **21**, 4879.
Nelmes, R. J., McMahon, M. I., and Belmonte, S. A. (1997). *Phys. Rev. Lett.*, **79**, 3668.
Patel, D., Menoni, C. S., and Spain, I. L. (1989). *J. Appl. Phys.* **66**, 1658.

15. Indium Arsenide

InAs has the zincblende structure at ambient conditions with $a = 6.0583$ Å. The pioneering resistivity measurements of Minomura and Drickamer (1962) reported a transition in InAs at ~10 GPa. A subsequent resistivity study (Pitt and Vyas, 1973) found the transition at 6.9(2) GPa, whereas Raman measurements (Aoki et al., 1984) on single-crystal samples reported the transition at 7.60(15) GPa. Jamieson's (1963) diffraction study showed that the transition was to the NaCl structure with a volume change ($\Delta V/V_{\text{trans}}$) of 18.8%. Using EDX techniques, Vohra et al. (1985) confirmed a ZB-to-NaCl transition at 7.0(2) GPa with a volume change ($\Delta V/V_{\text{trans}}$) of 17.0(2)%. The reported lattice parameter of the NaCl phase at 8.46 GPa was 5.5005 Å. Vohra et al. (1985) reported a further transition to a β-tin type structure at 17.0(4) GPa with no measurable change in volume. The β-tin structure was found to be stable up to 27 GPa, the highest pressure studied. Despite the sizable difference in scattering power of In ($Z = 49$) and As ($Z = 33$), Vohra et al. (1985) were unable to determine whether or not the NaCl and β-tin structures were site-ordered.

ADX studies of InAs (Nelmes et al., 1995) have reported the transition to NaCl at 7 GPa, in agreement with other studies. Although the NaCl (111) and (311) difference reflections were found to be present — showing the NaCl phase to be site-ordered — their relative intensities are a factor of 2 weaker than they should be (Nelmes et al., 1995). The reasons for this are still unclear. In addition, the ADX studies show that the NaCl phase is stable to only 9 GPa, where additional reflections appear in the diffraction pattern, the intensities of which increase continuously with applied pressure — see *Figure 5* of Nelmes et al. (1995). At still higher pressures, the peak that was the (200) reflection in the NaCl phase develops a low-angle shoulder. These extra features, and their development as the pressure is increased, are remarkably similar to those observed in the continuous NaCl-to-Cmcm transitions found in CdTe and InP. This indicates that InAs transforms, not to β-tin at 17 GPa, but to a Cmcm, or Cmcm-like structure at 9 GPa (Nelmes et al., 1995). At 19.8 GPa, the structure has $a \sim 5.30$ Å, $b \sim 5.48$ Å, $c \sim 5.38$ Å, with $y(\text{In}) \sim 0.69$ and $y(\text{As}) \sim 0.18$. The structural change to Cmcm-like is probably what Vohra et al. (1985) detected at 17 GPa. However, the pressure in that study was determined by using an NaCl calibrant, the peaks of which would have overlapped and thus obscured the growth of the (021) and (221) reflections from the Cmcm phase at lower pressures — see *Figure 2* of Vohra et al. (1985).

The identification of a simple NaCl-to-Cmcm transition is still somewhat uncertain, however, as there is evidence for an intermediate phase between the NaCl and Cmcm phases (Nelmes et al., 1995). On pressure increase a (weak) extra reflection is observed at low angles which cannot be indexed on a Cmcm structure nor any plausible superlattice of it. On pressure decrease this reflection increases in intensity, while the (111) reflection from the NaCl phase does not reappear, and the strong peaks are displaced from their NaCl-phase positions. There is, as yet, no satisfactory interpretation of the intermediate phase, or the weakness of the NaCl difference reflections, and considerable further work on this system is required. We have observed no other transitions to the maximum pressure reached, 46 GPa.

The problems in understanding InAs have been underlined by the calculations of Mujica and Needs (1997). They find the ZB-to-NaCl transition at 3.9 GPa, and NaCl-to-Cmcm somewhere in the range 3.0–4.5 GPa, so that it is not clear if there is any region of stability for an NaCl phase at all. And, contrary to the experimental data, the calculations predict a further transition from Cmcm to Immm at 24(3) GPa.

References

Aoki, K., Anastassakis, E., and Cardona, M. (1984). *Phys. Rev. B* **30**, 681.
Jamieson, J. C. (1963). *Science* **139**, 845.
Minomura, S., and Drickamer, H. G. (1962). *J. Phys. Chem. Solids* **23**, 451.
Mujica, A., and Needs, R. J. (1997). *Phys. Rev. B* **55**, 9659; *Phys. Rev. B* **56**, 12653E (Erratum).
Nelmes, R. J., McMahon, M. I., Wright, N. G., Allan, D. R., Liu, H., and Loveday, J. S. (1995). *J. Phys. Chem. Solids* **56**, 539.
Pitt, G. D., and Vyas, M. K. R. (1973). *J. Phys. C: Solid State Phys.* **6**, 274.
Vohra, Y. K., Weir, S. T., and Ruoff, A. L. (1985). *Phys. Rev. B* **31**, 7344.

16. Indium Antimonide

InSb is the III-V analog of Sn, and under ambient conditions InSb has the zincblende structure with $a = 6.4794$ Å. Perhaps because of its apparently complex behavior at modest pressure, InSb has been the subject of detailed study for many years. The extensive work performed prior to 1992 — which resulted in the widely accepted phase diagram shown in Fig. 18a — has been reviewed critically by Nelmes et al. (1993), to which the reader is referred for a summary of previous work. (Since the structures and transitions have been so comprehensively revised by recent work, no summary of the many previous structural studies is given here.) Using ADX techniques, it has been shown that the complex behavior previously reported below 5 GPa can be explained by two distinct phase transition sequences involving three phases, labeled P2, P3, and P4. Either (1) the ZB phase (P1) transforms at ~2.1 GPa to a mixture of a site-disordered β-tin-like phase (P2) and an orthorhombic phase (P3), which then, over a period of hours, recrystallizes to a different orthorhombic phase (P4), or (2) P1 transforms directly to P4 at the higher pressure of ~3 GPa (Nelmes et al., 1993).

P2 has a site-disordered β-tin-like structure with $a = 5.697(1)$ Å and $c = 3.104(1)$ Å at 2.1 GPa (Nelmes et al., 1993). (See Section II.1 concerning site disorder.) Since P2 has been observed only as a component in mixed-phase samples, there is still the possibility that it has a slight orthorhombic distortion to Imma. But it seems likely to be the same phase as the true β-tin phase obtained previously on recovery to ambient pressure at low temperature from 2.5 GPa (Darnell and Libby, 1963) — or becomes this

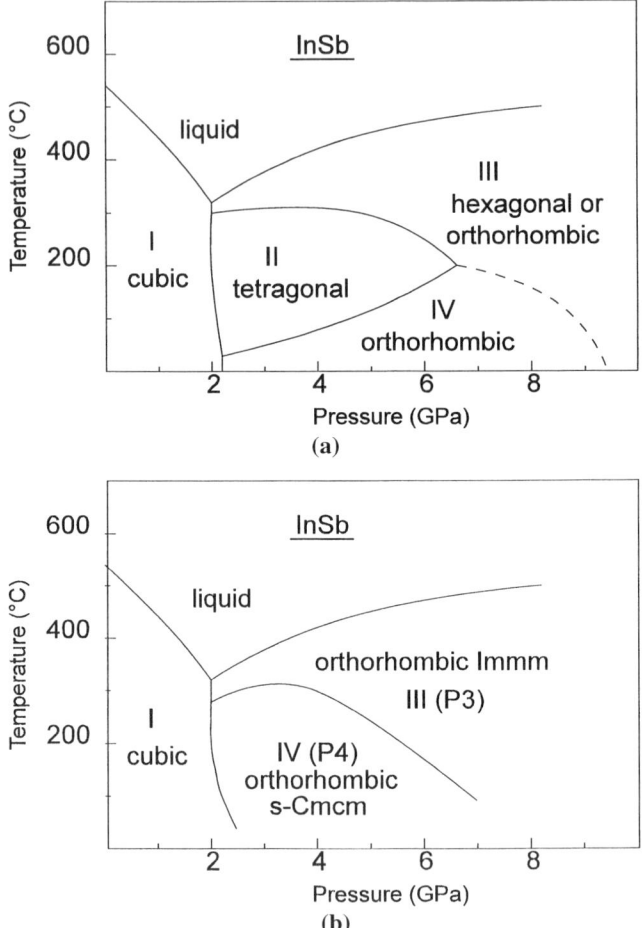

FIG. 18. (a) Previously accepted phase diagram of InSb, after Banus and Lavine (1967, 1969). Adapted with permission from *Figure 1* of Banus and Lavine (1969). Copyright 1969 American Institute of Physics. (b) Revised phase diagram of InSb, after Mezouar *et al.* (1996). Adapted with permission from *Figure 4* of Mezouar et al. (1996). Copyright 1996 WILEY-VCH Verlag Berlin GmbH.

on pressure decrease to ambient pressure. The P2 phase had not been detected at high pressure before this work and is *not* the InSb-II phase, which has a clearly orthorhombic structure (as detailed later). The volume change ($\Delta V/V_0$) at the ZB-to-P2 transition is 20.5(2)%.

P3 is the InSb-II phase but does *not* have the β-tin structure previously assigned to it. Although the structure of P3 was initially described as being orthorhombic with *Imm2* symmetry (Fig. 6), with $\Delta = 0.47$ (Guo *et al.*, 1993), the true structure has spacegroup *Immm* — that is, the Imm2 structure with

$\Delta = 0.50$ (Nelmes and McMahon, 1996). As described in Section II.7, a distinction between Imm2 with Δ close to 0.5 and Immm can be made through a comparison of pure difference reflections such as (110), (310), and (130) — whose intensities do not depend on the value of Δ — with other $k + \ell =$ odd reflections such as (101), (301), and (031), whose intensities vary with Δ but become pure difference reflections if $\Delta = 0.50$. Figure 19 shows two profiles we have obtained in subsequent ADX studies from the same sample of P3 InSb collected near to (n) and far from (f) the In K-absorption edge. The intensities of the (310), (130), and (301) reflections are all greatly enhanced near the edge because of the effects of anomalous dispersion, and the intensities of all three reflections reduce together away from the edge, strongly suggesting that $\Delta = 0.50$. In particular, the (301) peak — the only one of the three sensitive to Δ — has no residual intensity in the data collected far from the absorption edge, where the anomalous scattering is "switched off." We can certainly conclude that any deviation from 0.50 is <0.01. The structure of P3 is thus orthorhombic, spacegroup *Immm*, with In at (0,0,0) and Sb at (0,1/2,1/2). At ~2.3 GPa, $a = 5.847(1)$ Å, $b = 5.388(1)$ Å, and $c = 3.181(1)$ Å (Nelmes *et al.*, 1993). The volume change $(\Delta V/V_0)$ at the ZB-to-P3 transition at 2.1 GPa is 21.0(2)%, and P3 is thus 0.5(3)% denser than P2 at the same pressure (Nelmes *et al.*, 1993).

P4 is the equilibrium phase in the pressure range 2–9 GPa and can be identified as InSb-IV. However, rather than the Pmm2 structure proposed

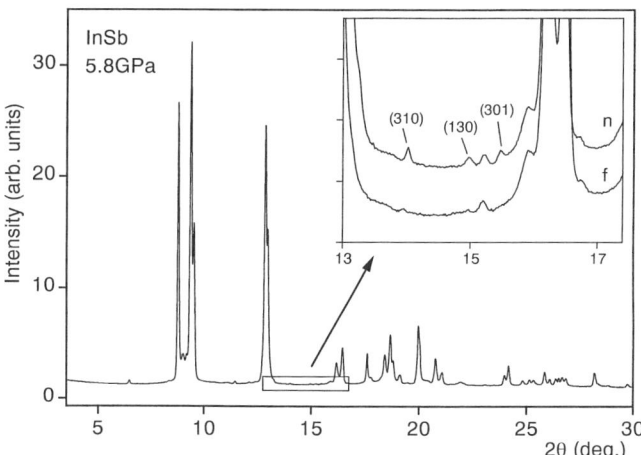

FIG. 19. Observed powder diffraction pattern from InSb at 5.8 GPa, collected near the In K-edge. Part of this pattern is enlarged in the inset (labeled "n"), where it is compared with the same part of a pattern collected far from (f) the edge. Extremely weak reflections enhanced by anomalous dispersion effects near the edge are indexed.

by Yu et al. (1978) — as shown in Fig. 16 for GaAs — the true structure has *Cmcm* symmetry, but is a site-ordered superstructure of the basic Cmcm structure (with a three times longer c axis), of which site-disordered Pmm2 can be regarded as a subcell (Nelmes and McMahon, 1995). (The relationship of Pmm2 to Cmcm is discussed in Section III.11 and the caption to Fig. 16). At 5.1 GPa, $a = 5.847(1)$ Å, $b = 6.140(1)$ Å, and $c = 16.791(1)$ Å, with In and Sb atoms in two different 4(c) positions at (0,0.120(1),1/4) and (0,0.620(1),1/4), respectively, and in two different 8(f) positions at (0,0.410(1),0.089(1)) and (0,0.910(1),0.081(1)), respectively (Nelmes and McMahon, 1995)[4]. In free refinements of all the variable parameters, the differences between y(In) and y(Sb) were found to be not significantly different from 0.5 for either site. This was therefore applied as a constraint in the final refinements. The volume change ($\Delta V/V_0$) at the ZB-to-P4 transition at 3.0 GPa is 19.5(1)%, and P3 is 0.5(1.0)% denser than P4 at the same pressure (Nelmes et al., 1993).

Whereas the basic Cmcm structure described in Section II.10 has eight atoms per unit cell, arranged in two NaCl-like planes, the InSb-IV superstructure has 24 atoms in six NaCl-like planes. The superstructure is shown in Fig. 20. In this case it is convenient to set the z axis, along which the NaCl-like planes are stacked, vertical; however, the basic Cmcm structure is normally drawn with this direction horizontal — see Fig. 10. The tripling of the unit cell along the z axis is revealed by the site ordering along z such as the sequence Sb, Sb, In, . . . labeled A, B, C in Fig. 20, and by small displacements in y and z from a $c/3$ repeat in site positions. In the basic Cmcm structure, all atoms lie in mirror planes and so the NaCl-like planes are necessarily flat. But in the superstructure this is true of only two of the planes — those given by the 4(c) positions and marked by arrows in Fig. 20. The other four planes (the 8(f) positions) are free to pucker, and do so a little — as shown in Fig. 20. Thus, although the $\pm(0.5 - \Delta y)/2$ zigzag along the x-axis chains of the basic Cmcm structure (Fig. 10) is absent, there are instead $\pm\Delta z/2$ zigzags along the x-axis and y-axis chains ($\Delta z = 0.089(1)-0.081(1)$). Because the unit cell is about three times larger along z than along y, this Δz corresponds in absolute magnitude to the $(0.5 - \Delta y)$ of ~0.02 found in many of the basic Cmcm structures (see Section II.10). The coordination is mixed: Atoms in the flat planes have six unlike-atom nearest neighbors at ~3 Å, while the atoms in the puckered planes have five such neighbors with a sixth one considerably farther away at ~3.5 Å (Nelmes and McMahon, 1995). The latter coordination is like that for all atoms in the basic Cmcm structure with $\Delta y \neq 0.5$. It is to be noted that the sequence of $c/3$ repeats labeled A, B, C in Fig. 20 (each corresponding to

---

[4]The 4(c) positions were incorrectly labeled 4(a) in this paper.

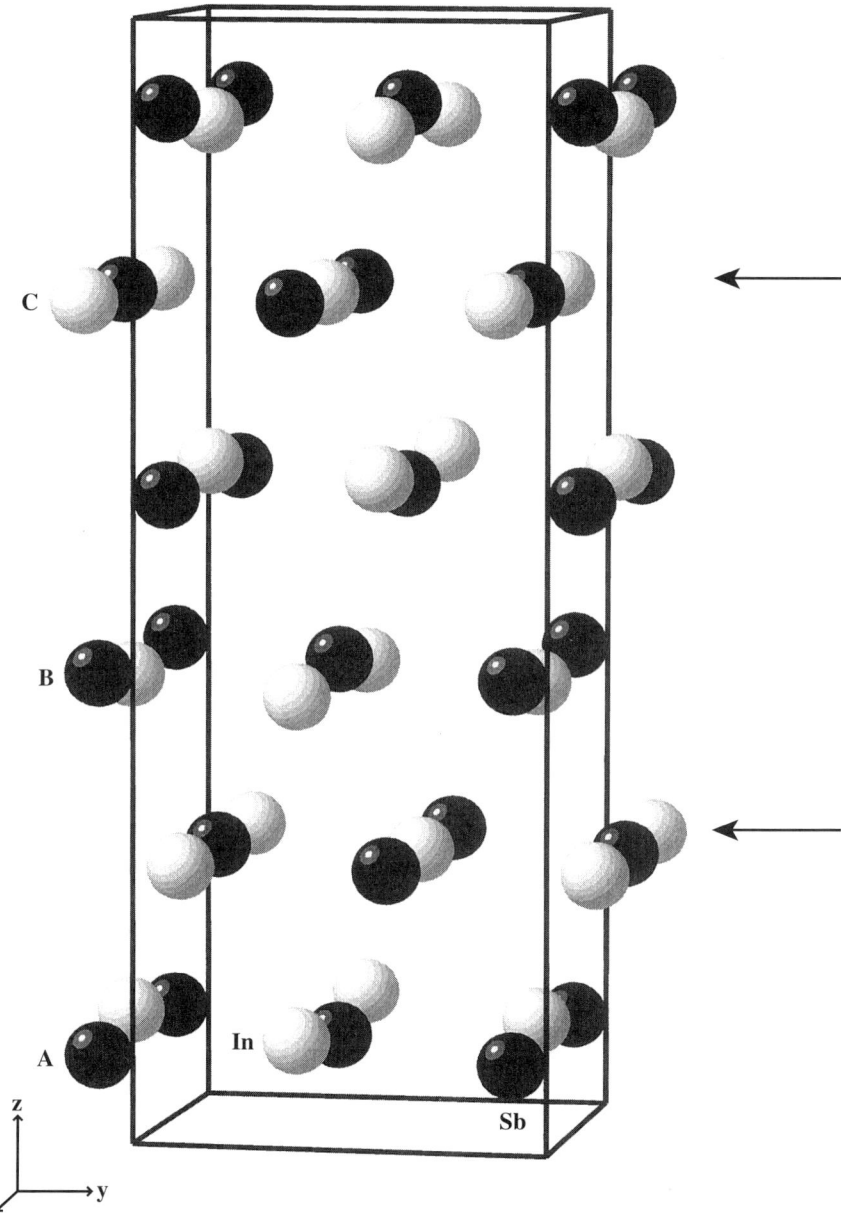

FIG. 20. The Cmcm superstructure of InSb-IV. The arrows mark the mirror planes at $z = 1/4$ and $3/4$. The sequence A, B, C is referred to in the text. Copyright 1995 American Physical Society.

the full c-axis repeat of the basic Cmcm structure) is ... puckered layer, puckered layer, flat layer ... and so on.

P4 is stable up to 9 GPa, where there is a transition to a phase previously reported as hexagonal (Banus and Lavine, 1969) or orthorhombic (Yu et al., 1978). Nelmes and McMahon (1996) showed from ADX data that the transition proceeds through a previously undetected site-disordered intermediate phase, P5, which is orthorhombic with spacegroup *Imma* — the same as the structure of GaSb-II. (See the discussion in Section III.12 about site disorder in GaSb-II.) The volume change ($\Delta V/V_0$) at 9.8 GPa is $-0.3(2)\%$, that is, the P5 phase is slightly *less* dense than the P4 phase. At 14.5 GPa, the lattice parameters are $a = 5.661(1)$ Å, $b = 5.399(1)$ Å, and $c = 3.003(1)$ Å, with $\Delta = 0.392(1)$. With time, or with moderate heating, P5 then transforms into InSb-III, which has the same Immm structure as P3 (Nelmes and McMahon, 1995). The volume reduction ($\Delta V/V_0$) at the P5-to-P3 transition is $1.8(2)\%$. Thus, the apparently distinct InSb-II and InSb-III phases of the phase diagram in Fig. 18a were shown to be one and the same phase.

If P3 is obtained at the ~2.1 GPa transition (sequence (1) above) and the pressure is increased above ~3 GPa on a time scale of about an hour or less, then the recrystallization to P4 is suppressed (Nelmes et al., 1993). Samples of P3 made this way show no transition at 9 GPa, and their patterns above that pressure are the same as for P3 obtained by transition from P4 at 9 GPa.

On further pressure increase, all P3 samples transform to another intermediate phase at ~17 GPa, the structure of which is as yet unknown, and then to a site-disordered bcc structure starting at 21(1)GPa (Nelmes and McMahon, 1996). The lattice parameter of the bcc phase at 36.7 GPa is 3.364(1) Å and the volume change ($\Delta V/V_0$) at the P3-to-bcc transition is $3.0(2)\%$. This last transition was previously reported at 28 GPa by Vanderborgh et al. (1989), who found the bcc phase to be stable to at least 66 GPa.

The ADX studies of InSb have thus revealed that the apparently well-established $P-T$ phase diagram of InSb, as shown in Fig. 18a, is substantially incorrect. Mezouar et al. (1996) have now made a redetermination of the InSb phase diagram to 8 GPa and 800 K with EDX methods. The new phase diagram to 8 GPa, which is shown in Fig. 18b, confirms and extends the ADX results and establishes P4 as the stable phase below ~8–9 GPa at room temperature, with P3 as the stable phase above that pressure. The prior appearance of the P3 phase (Immm) at ~2.1 GPa (at room temperature) is curious, as is the apparent involvement of an intermediate phase in each of the principal transitions, and these features all invite further investigation. The previous misidentification of a phase II field with a β-tin structure in the phase diagram of Fig. 18a appears to have arisen because this field was mapped by recovering samples to ambient pressure at low temperature. As noted earlier, this process does yield a sample with the β-tin structure, but it is a different phase (possibly P2).

REFERENCES

Banus, M. D., and Lavine, M. C. (1967). *J. Appl. Phys.* **38**, 2042.
Banus, M. D., and Lavine, M. C. (1969). *J. Appl. Phys.* **40**, 409.
Darnell, A. J., and Libby, W. F. (1963). *Science* **139**, 1301.
Guo, G. Y., Crain, J., Blaha, P., and Temmerman, W. M. (1993). *Phys. Rev. B* **47**, 4841.
Mezouar, M., Besson, J. M., Syfosse, G., Itié, J.-P., Häusermann, D, and Hanfland, M. (1996). *Phys. Stat. Sol. (b)* **198**, 403.
Nelmes, R. J., and McMahon, M. I. (1995). *Phys. Rev. Lett.* **74**, 106; *Phys. Rev. Lett.* **74**, 2618 (Erratum).
Nelmes, R. J., and McMahon, M. I. (1996). *Phys. Rev. Lett.* **77**, 663.
Nelmes, R. J., McMahon, M. I., Hatton, P. D., Piltz, R. O. and Crain, J. (1993). *Phys. Rev. B* **47**, 35; *Phys. Rev. B* **48**, 9949 (Corrigendum).
Vanderborgh, C. A., Vohra, Y. K., and Ruoff, A. L. (1989). *Phys. Rev. B* **40**, 12450.
Yu, S. C., Spain, I. L., and Skelton, E. F. (1978). *J. Appl. Phys.* **49**, 4741.

17. ZINC OXIDE

ZnO has the wurtzite structure at ambient conditions with $a = 3.2495$ Å and $c = 5.2069$ Å. The existence of a transition from the wurtzite phase to an NaCl phase at high pressure was first established by Bates *et al.* (1962), who found that NaCl could be obtained at ambient pressure by quenching from 10 GPa. Using *in situ* diffraction methods, Jamieson (1970) confirmed site-ordered NaCl as the high-pressure phase, but found the wurtzite-to-NaCl transition to be reversible. Yu *et al.* (1978) reported the transition to the NaCl phase at 8.0(3) GPa with a volume change ($\Delta V/V_{trans}$) of 16.6(1.2)%. Karzel *et al.* (1996) reported the transition to occur at 8.7(5) GPa with a volume change ($\Delta V/V_0$) of 16.7(3)%, and found the reverse transition at 2.0(5) GPa. (They determined these transition pressures as the pressure at which half the sample has transformed.) Gerward and Staun Olsen (1995) determined the equation of state of both the wurtzite and NaCl phases to ~29 GPa — the highest pressure at which ZnO has been measured so far — using EDX techniques. They reported that a "large fraction" of the NaCl phase was retained at ambient pressure on pressure release. The reported lattice parameter of the NaCl phase at ambient pressure was 4.280(4) Å, in agreement with the value reported by Bates *et al.* (1962), while at 10 GPa the reported lattice parameter was 4.211(3) Å (Gerward and Staun Olsen, 1995). In both the wurtzite and NaCl phases, Gerward and Staun Olsen (1995) determined the value of B' as ~9, considerably larger than the value of ~4 typical for a material like ZnO. The authors recognized this problem and stated that their result might be at least partly due to experimental difficulties in measuring the pressure in the transition range (Gerward and Staun Olsen, 1995).

References

Bates, C. H., White, W. B., and Roy, R. (1962). *Science* **137**, 993.
Gerward, L., and Staun Olsen, J. (1995). *J. Synchrotron Rad.* **2**, 233.
Jamieson, J. C. (1970). *Phys. Earth Plan. Inter.* **3**, 201.
Karzel, H., Potzel, W., Köfferlein, M., Schiessl, W., Steiner, M., Hiller, U., Kalvius, G. M., Mitchell, D. W., Das, T. P., Blaha, P., Schwarz, K., and Pasternak, M. P. (1996). *Phys. Rev. B* **53**, 11425.
Yu, S. C., Spain, I. L., and Skelton, E. F. (1978). *Solid State Commun.* **25**, 49.

18. Zinc Sulfide

ZnS has the prototype zincblende structure at ambient conditions with $a = 5.4102$Å, and also the wurtzite structure with $a = 3.8226$ Å and $c = 6.2605$ Å. Only the ZB form has been studied extensively under pressure. A high-pressure transition was first reported by Minomura and Drickamer (1962). The high-pressure phase was subsequently identified as NaCl by Smith and Martin (1965), who reported the transition pressure as 11.7 GPa with a volume change ($\Delta V/V_0$) of 21%. The transition pressure was located at 15.0(5) GPa by Piermarini and Block (1975) by visual inspection, at 15.0(5) GPa by Yu *et al.* (1978) using diffraction techniques, and at 14.7(7) and 15.4(5) GPa by Ves *et al.* (1990) using diffraction and optical techniques, respectively. In both diffraction studies the high-pressure structure was confirmed as NaCl with volume changes ($\Delta V/V_{trans}$) of ~15% (Yu *et al.*, 1978) and 15.7% (Ves *et al.*, 1990). The lattice parameter of the NaCl phase at 17.1(3.5) GPa is 4.839(3) Å (Zhou *et al.*, 1991)[5].

In contrast to previous studies which reported that the high-pressure phase was metallic (Minomura and Drickamer, 1962) and opaque (Piermarini and Block, 1975; Weinstein, 1977), Ves *et al.* (1990) found that the NaCl phase is an indirect-gap semiconductor with a gap of ~2eV. The presence of a gap was confirmed by Zhou *et al.* (1991), who performed diffraction and absorption measurements to 45 and 36 GPa, respectively, and determined its value to be ~2.1eV "a few GPa" past the transition pressure. A pseudopotential study by Nazzal and Qteish (1996) has suggested that the cinnabar structure is energetically stable between the ZB and NaCl phases. The calculated transition pressures are 11.4 GPa for the transition from ZB to cinnabar, and 14.5 GPa for the transition from cinna-

---

[5]The authors report a (110) reflection, which should be absent for NaCl.

3 GROUP IV, III–V, AND II–VI SEMICONDUCTORS

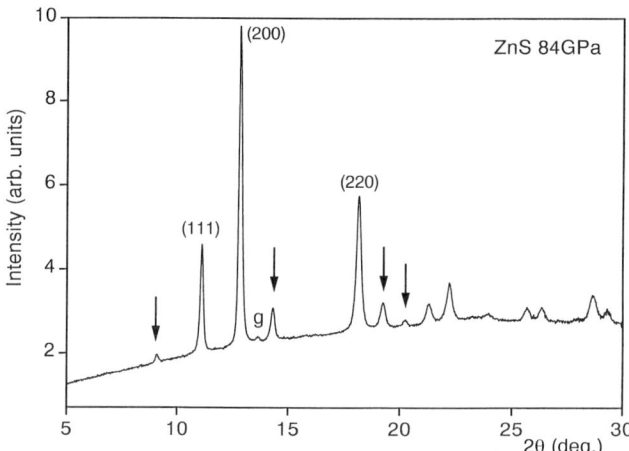

FIG. 21. Observed powder diffraction pattern obtained from ZnS at 84 GPa. The indices label the main peaks retained from the NaCl phase. The arrows mark non-NaCl reflections, and g marks a weak gasket line.

bar to NaCl. We have carried out ADX studies to look for evidence of this phase, but have not found it either on pressure increase or decrease at room temperature.

Using ADX techniques, we have observed the ZB-to-NaCl transition at 14.5(5) GPa and can confirm that the NaCl phase is site-ordered. In a preliminary study to much higher pressures, we have found a transition at 69(3) GPa to a new phase with a structure that appears to be a Cmcm-like distortion of NaCl. Data collected at 84 GPa — the maximum pressure reached so far — are shown in Fig. 21, and arrows mark non-NaCl peaks. More data are required on this new phase — in particular, to determine whether its structure *is* Cmcm.

REFERENCES

Minomura, S., and Drickamer, H. G. (1962). *J. Phys. Chem. Solids* **23**, 451.
Nazzal, A. and Qteish, A. (1996). *Phys. Rev. B* **53**, 8262.
Piermarini, G. J., and Block, S. (1975). *Rev. Sci. Inst.* **46**, 973.
Smith, P. L., and Martin, J. E. (1965). *Phys. Lett.* **19**, 541.
Ves, S., Schwarz, U., Christensen, N. E., Syassen, K., and Cardona, M. (1990). *Phys. Rev. B* **42**, 9113.
Weinstein, B. A. (1977). *Solid State Commun.* **24**, 595.
Yu, S. C., Spain, I. L., and Skelton, E. F. (1978). *Solid State Commun.* **25**, 49.
Zhou, Y., Campbell, A. J. and Heinz, D. L. (1991). *J. Phys. Chem. Solids* **52**, 821.

19. ZINC SELENIDE

ZnSe is isoelectronic with GaAs and Ge and has the zincblende structure under ambient conditions with $a = 5.6676$ Å. A high-pressure transition was first reported by Samara and Drickamer (1962) at 16.5 GPa. But subsequent resistivity measurements, using the modern pressure scale, have located the same transition at 13 GPa (Tiong et al., 1989) and 13.5 GPa (Itkin et al., 1995), while optical studies have located the transition at 13.7(3) GPa (Piermarini and Block, 1975) and 13.5(2) GPa (Ves et al., 1985). The diffraction studies of Smith and Martin (1965), Karzel et al. (1996), and Nelmes and McMahon (1998) have reported that the high-pressure phase has the NaCl structure with a volume change ($\Delta V/V_0$) at the transition of 13.3% (Karzel et al., 1996) or 13.0(1)% (Nelmes and McMahon, 1998). The lattice parameter of the NaCl phase at 13.6 GPa is 5.110(1) Å and the structure is site-ordered (Nelmes and McMahon, 1998).

The Raman study of Lin et al. (1997) reported two transitions at 4.7 and 9.1 GPa, in contrast to other Raman studies of Arora and Sakuntala (1995) and Arora et al. (1988) which showed no such transitions. However, Greene et al. (1995) also observed anomalous features in both Raman and EDX diffraction data at 5 GPa. In addition, they reported that the NaCl phase has a small rhombohedral distortion, perhaps due to nonhydrostatic conditions, and starts to transform to an eight-fold coordinated simple hexagonal (SH) phase, ZnSe-III, at ~48 GPa (Greene et al., 1995). On this interpretation, the powder patterns have to be indexed as an approximately equal mixture of NaCl and SH, without any further transformation from NaCl to SH, over a large pressure range up to 120 GPa, the highest pressure reached; the SH phase would also have to be site-disordered (see Section II.8). There are also some weak reflections which are not explained by the NaCl or SH phases, but are ascribed to vacancy defects.

ADX studies of ZnSe have been made to 60 GPa (McMahon and Nelmes, 1996). We note that the data show no apparent structural discontinuity at 5 GPa. Also, no evidence of the cinnabar phase was observed on either pressure increase or decrease, in disagreement with a pseudopotential calculation (Côté et al., 1997) which suggests that cinnabar should be stable from 10.2 to 13.4 GPa. Above 20 GPa an increasingly poor fit to the NaCl structure is observed (Nelmes and McMahon, 1998). However, the relative displacements of the NaCl peaks is characteristic of the effects of deviatoric stress (Singh, 1993) rather than the rhombohedral distortion proposed by Greene et al. (1995). Then a continuous transition to a site-ordered Cmcm structure is observed with the first appearance of new reflections at 30.0(5) GPa (McMahon and Nelmes, 1996). The rate of increase of the Cmcm distortion with pressure is much less than that observed in — for example — CdTe, as shown in *Figure 4* of McMahon and Nelmes (1996). However, the additional reflections of Cmcm become clearly visible by 48 GPa (see *Figure 3* of McMahon and Nelmes, 1996), and it is probably this that Greene *et al.* (1995) interpreted as a partial transition to SH. At 60 GPa the refined structure of ZnSe-III is $a = 4.728(1)$ Å, $b = 4.800(1)$ Å, and $c = 4.703(2)$ Å, with $y(\text{Zn}) = 0.704(1)$ and $y(\text{Se}) = 0.196(1)$ (McMahon and Nelmes, 1996). The pseudopotential calculation by Côté *et al.* (1997) has suggested that NaCl should transform to Cmcm at 36.5 GPa.

A little above 48 GPa in the ADX data, there are some very weak indications of the onset of the same type of distortions of Cmcm as are seen clearly in CdSe above 36 GPa (see Fig. 24). We simply note this as a matter for further future work. Even at 60 GPa, the distortions remain too small to degrade the Cmcm fit (see earlier discussion).

References

Arora, A. K., and Sakuntala, T. (1995). *Phys. Rev. B* **52**, 11052.
Arora, A. K., Suh, E.-K., Debska, U., and Ramdas, A. K. (1988). *Phys. Rev. B* **37**, 2927.
Côté, M., Zakharov, O., Rubio, A., and Cohen, M. L. (1997). *Phys. Rev. B* **55**, 13025.
Greene, R. G., Luo H., and Ruoff, A. L. (1995). *J. Phys. Chem. Solids* **56**, 521.
Itkin, G., Hearne G. R., Sterer, E., Pasternak, M. P., and Potzel W. (1995). *Phys. Rev. B* **51**, 3195.
Karzel, H., Potzel, W., Köfferlein, M., Schiessl, W., Steiner, M., Hiller, U., Kalvius, G. M., Mitchell, D. W., Das, T. P., Blaha, P., Schwarz, K., and Pasternak, M. P. (1996). *Phys. Rev. B* **53**, 11425.
Lin, C., Chuu, D., Yang, T., Chou, W., Xu, J., and Huang, E. (1997). *Phys. Rev. B* **55**, 13641.
McMahon, M. I. and Nelmes, R. J. (1996). *Phys. Stat. Sol. (b)* **198**, 389.
Nelmes, R. J., and McMahon, M. I. (1998). To be submitted to *Phys. Rev. B*.
Piermarini, G. J., and Block, S. (1975). *Rev. Sci. Inst.* **46**, 973.
Samara, G. A., and Drickamer, H. G. (1962). *J. Phys. Chem. Solids* **23**, 457.
Singh, A. K. (1993) *J. Appl. Phys.* **73**, 4278.
Smith, P. L., and Martin, J. E. (1965). *Phys. Lett.* **19**, 541.
Tiong, S. R., Hiramatsu, M., Matsushima, Y., and Ito E. (1989). *Jpn. J. Appl. Phys.* **28**, 291.
Ves, S., Strössner, K., Christensen, N. E., Kim, C. K., and Cardona M. (1985). *Solid State Commun.* **56**, 479.

20. Zinc Telluride

At ambient conditions ZnTe has the zincblende structure with $a = 6.1037$ Å. The resistivity measurements of Samara and Drickamer (1962) revealed three sharp discontinuities in the conductivity of ZnTe below 15 GPa, but subsequent unpublished diffraction measurements by Smith and Martin (1965) revealed that only two of these corresponded to structural phase transitions — to ZnTe-II at about 9 GPa and then to ZnTe-III at about 11 GPa.

The first transition was characterized in resistivity (Ohtani et al., 1980) and optical (Strössner et al., 1987a) measurements which revealed ZnTe-II to be a transparent semiconductor with a bandgap of 2.5eV. The transition pressure was reported as 8.5 GPa (Ohtani et al., 1980), 9.4(3) GPa (Strössner et al., 1987a), and 9.3 GPa (Strössner et al., 1987b). The structure was not solved from the initial diffraction measurements, but the data excluded both NaCl and β-tin as possible solutions (Strössner et al., 1987b). ZnTe-II has since been solved as having a cinnabar structure by several groups using ADX (McMahon et al., 1993; Nelmes et al., 1995), EDX (Kusaba and Weidner, 1993; Qadri et al., 1993), and combined EDX/EXAFS (San-Miguel et al., 1993) techniques. At 8.9 GPa, detection of the (003) reflection confirms site-ordering (Section II.9), and the refined cinnabar structure has $a = 4.105(1)$ Å and $c = 9.397(1)$ Å, with $u(Zn) = 0.540(2)$ and $v(Te) = 0.504(2)$ (Nelmes et al., 1995). The volume change ($\Delta V/V_0$) at the ZB-to-cinnabar transition is $\sim$9% at 9.6 GPa (Kusaba and Weidner, 1993). The structure is very close to being four-fold coordinated, with two nearest-neighbor atoms at 2.528(4) Å, two at 2.646(5) Å, and the next two more than 1 Å farther away at 3.743(9) Å. This is similar to the exact four-fold coordination proposed in the EXAFS study of San-Miguel et al. (1992), but the ADX diffraction results are able to reveal the 0.12 Å difference in the nearest-neighbor distances.

The transition to ZnTe-III — which is metallic (Ohtani et al., 1980; Strössner et al., 1987a) — has been reported to occur at 13 GPa (Ohtani et al., 1980), 11.9(3) GPa (Strössner et al., 1987a), and 11.9 GPa (Strössner et al., 1987b). The structure of ZnTe-III resisted repeated attempts at solution — diffraction studies reported a monoclinic (Strössner et al., 1987b) or β-tin (Qadri et al., 1993) structure, while an EXAFS study reported the coordination to be NaCl-like (San Miguel et al., 1992) — until ADX techniques showed it to have a previously unobserved site-ordered distorted-NaCl structure with spacegroup *Cmcm* (Nelmes et al., 1994). This was the first report of the Cmcm structure (Section II.10), which has since been observed in a large number of the other II–VI and III–V systems. At 15.7 GPa, the structure has $a = 5.379(1)$ Å, $b = 5.971(1)$ Å, and $c = 5.010(2)$ Å, with $y(Zn) = 0.640$ (1) and $y(Te) = 0.190(1)$ (Nelmes et al., 1994). From a mixed-phase sample at 11.5 GPa, the volume change ($\Delta V/V_0$) at the cinnabar-to-Cmcm transition is 5.7(2)%. The structure has an unexpected five-fold coordination of closest near neighbors, with five unlike atoms at $\sim$2.7 Å, and a further three atoms (one unlike and two like) at 3.0–3.4 Å.

We have now extended our studies of ZnTe to higher pressures. The fit of Cmcm to the data gets gradually poorer because of deviatoric stress (we believe), but there is no evidence of a further transition until $\sim$85 GPa, where the appearance of new diffraction peaks reveal a change in structure

3 GROUP IV, III–V, AND II–VI SEMICONDUCTORS

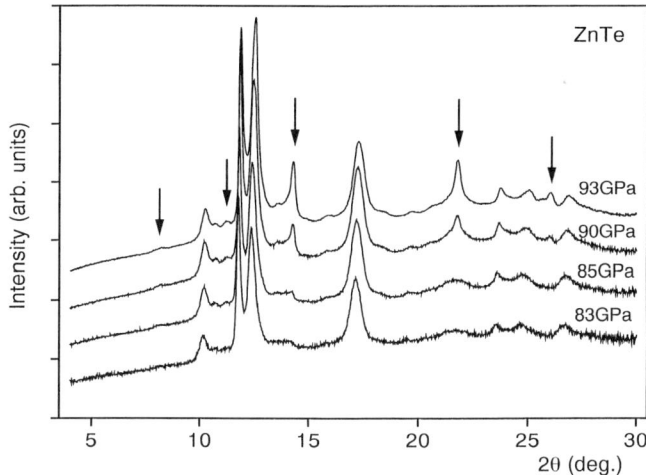

FIG. 22. Observed powder diffraction patterns from ZnTe through the transition at ~85 GPa from the Cmcm phase to a new phase characterized by the appearance of the marked reflections.

(see Fig. 22). Further work is needed to characterize this new phase in more detail. We find no evidence of any further transition up to the highest pressure reached in our studies, 93 GPa (Fig. 22).

The cinnabar and Cmcm structures have been the subject of two theoretical studies (Lee and Ihm, 1996; Côté et al., 1997). These studies confirm a zincblende-to-cinnabar transition, with calculated transition pressures of 8.06 GPa (Lee and Ihm, 1996) and 5.9 GPa (Côté et al., 1997), and a cinnabar-to-Cmcm transition with calculated transition pressures of 10.24 GPa (Lee and Ihm, 1996) and 11.1 GPa (Côté et al., 1997). The calculations also suggest that the Cmcm structure should be stable to ~100 GPa (Lee and Ihm, 1996), in disagreement with the latest diffraction results.

ZnTe is peculiar among the II–VI systems in having no NaCl phase (at room temperature). However, *in situ* high-temperature high-pressure diffraction measurements using EDX techniques have been reported that find a transition to an NaCl phase (ZnTe-IV) at high pressure and temperature. The triple point between the cinnabar, Cmcm, and NaCl phases has been located at 13 GPa and 453 K (Shimomura et al., 1997).

REFERENCES

Côté, M., Zakharov, O., Rubio, A., and Cohen, M. L. (1997). *Phys. Rev. B* **55**, 13025.
Kusaba, K., and Weidner, D. J. (1993). "High Pressure Science and Technology," *AIP Conf. Proc.* **309**, 553.
Lee, G-D., and Ihm, J. (1996). *Phys. Rev. B* **53**, R7622.

McMahon, M. I., Nelmes, R. J., Wright, N. G., and Allan, D. R. (1993). "High Pressure Science and Technology," *AIP Conf. Proc.* **309**, 633.
Nelmes, R. J., McMahon, M. I., Wright, N. G., and Allan, D. R. (1994). *Phys. Rev. Lett.* **73**, 1805.
Nelmes, R. J., McMahon, M. I., Wright, N. G., and Allan, D. R. (1995). *J. Phys. Chem. Sol.* **56**, 545.
Ohtani, A., Motobayashi, M., and Onodera, A. (1980). *Phys. Lett.* **75A**, 435.
Qadri, S. B., Skelton, E. F., Webb, A. W., and Hu, J. Z. (1993). "High Pressure Science and Technology," *AIP Conf. Proc.* **309**, 319.
Samara, G. A., and Drickamer, H. G. (1962). *J. Phys. Chem. Solids* **23**, 457.
San-Miguel, A., Polian, A., Itié, J.-P., Marbeuf, A., and Triboulet, R. (1992). *High Pressure Research* **10**, 412.
San-Miguel, A., Polian, A., Gauthier, M., and Itié, J.-P. (1993). *Phys. Rev. B* **48**, 8683.
Shimomura, O., Utsumi, W., Urakawa, T., Kikegawa, T., Kusaba, K., and Onodera, A. (1997). Special issue of *The Review of High Pressure Science and Technology*, Vol. 6, p. 207 (ISSN-0917-6373).
Smith, P. L., and Martin, J. E. (1965). *Phys. Lett.* **19**, 541.
Strössner, K., Ves, S., Kim, C. K., and Cardona, M. (1987a). *Solid State Commun.* **61**, 275.
Strössner, K., Ves, S., Hönle, W., Gebhardt, W., and Cardona, M. (1987b). *In* "Proceedings of the 18th International Conference on the Physics of Semiconductors" (O. Engstrom, Ed.), p. 1717. World Scientific, Singapore.

21. Cadmium Oxide

CdO has the NaCl structure at ambient conditions with $a = 4.6942$ Å. It retains this structure "to at least 30 GPa" (Liu and Bassett, 1986). No further work appears to have been done on this system.

Reference

Liu, L., and Bassett, W. A. (1986). In "Elements, Oxides and Silicates," p. 96. Oxford University Press, New York.

22. Cadmium Sulfide

At ambient conditions CdS crystallizes in both the wurtzite and zincblende structures. For ZB, $a = 5.818$ Å, and for wurtzite, $a = 4.1362$ Å and $c = 6.714$ Å. Only the wurtzite form has been extensively studied under pressure. A phase transition in the wurtzite phase at 2.75 GPa was first identified by optical means (Edwards and Drickamer, 1961). Subsequent studies showed the transition is accompanied by an abrupt decrease in the electrical resistivity, at 2.3(1) GPa (Samara and Drickamer, 1962; Samara and Giardini, 1965) and is to a phase with the NaCl structure (Rooymans, 1963; Owen *et al.*, 1963; Kabalkina and Troitskaya, 1964; Mariano and Warekois, 1963) which is semiconducting, with an indirect gap of 1.6–1.7 eV (Batlogg *et al.*, 1983; Savić and Urošević, 1987). The volume decrease ($\Delta V/V_0$) at the transition was found to be 16.0% by Cline and Stephens (1965), who determined a transition pressure of 2.34(6) GPa.

On further compression, the resistivity of CdS starts to rise again, passing through a maximum at 20–30 GPa before decreasing once again (Samara and Drickamer, 1962; Bundy, 1975). Using a combination of EDX and ADX on a rotating-anode source, Suzuki *et al.* (1983) reported that the NaCl phase of CdS transforms to a new phase, CdS-III, at ~50 GPa, and that single-phase CdS-III samples are obtained above 58 GPa. CdS-III was found to be stable up to at least 68 GPa and was reported to have an orthorhombic structure with spacegroup *Pmmn*. Suzuki *et al.* (1983) reported lattice parameters of $a = 3.471(8)$ Å, $b = 4.873(7)$ Å and $c = 3.399(7)$ Å at 61 GPa. The volume change (not defined) at the NaCl/III transition was found to be ~0.8%. Although calculated intensities were tabulated, no atomic coordinates were reported. The Pmmn structure is an orthorhombic distortion of NaCl. It is exactly NaCl if $a = c = b/\sqrt{2}$ and if the atoms are at (0, 1/2, 1/4) and (0, 0, 1/4). The NaCl-to-Pmmn transition is equivalent to the transition observed in CrN on cooling (Nasr-Eddine and Bertaut, 1971).

We have recently embarked on our own ADX studies of CdS to 60 GPa. The transition to the NaCl phase is observed at 2.3(3) GPa and the structure is confirmed as site-ordered by the appearance of reflections such as (111). The transition to CdS-III is observed at 51(9) GPa. Refinements of a CdS-III profile at 60 GPa using the site-ordered Pmmn structure give $a = 3.493(1)$ Å, $b = 4.877(2)$ Å, and $c = 3.412(1)$ Å, with Cd and S at (0, 0.5, 0.319(1)) and (0, 0, 0.148(3)), respectively. However, the Pmmn structure requires peak splittings — such as the NaCl (111) reflection splitting into the (110) and (011) reflections of Pmmn, as shown in Fig. 23 — which are

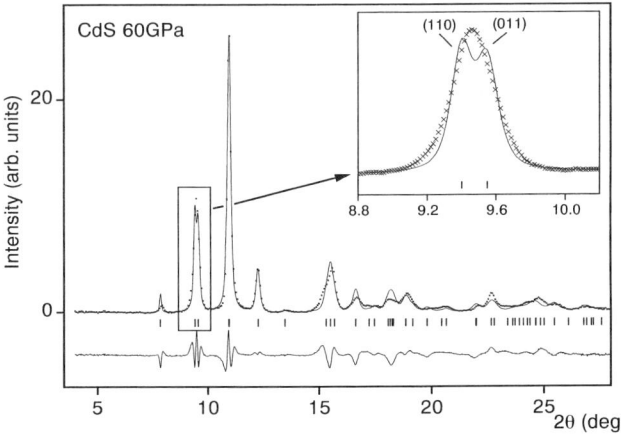

FIG. 23. Observed powder diffraction patterns from CdS at 60 GPa (points), and the best fitting calculated profile with a Pmmn structure. Enlargement of the low-angle peak in the inset shows that this structure predicts a splitting not found in the data.

observed neither by Suzuki et al. (1983) nor by us (Fig. 23). The transition to CdS-III looks superficially very similar to the continuous NaCl-to-Cmcm transition observed in other systems such as CdTe and ZnSe (*Figure 3* of McMahon and Nelmes, 1996). Also, the NaCl-to-Cmcm transition requires no splitting of the (111) reflection. A fit to the same CdS-III data at 60 GPa using a site-ordered Cmcm structure gives $a = 4.883(3)$ Å, $b = 4.881(4)$ Å, and $c = 4.875(7)$ Å with $y(Cd) = 0.699(1)$ and $y(S) = 0.174(3)$, but, though this solution removes the problem of splitting (111), the fit to the peak positions in the diffraction profile is otherwise somewhat poorer than that obtained using the Pmmn structure. However, the peak positions in the NaCl phase immediately prior to the transition to CdS-III are also displaced from their calculated positions, probably because of deviatoric stress. The exact structure of CdS-III thus still remains to some extent uncertain: Pmmn gives the better fit to peak positions but requires peak splittings that are clearly not observed, such as for (111), whereas Cmcm gives a poorer fit to peak positions but requires no (111) splitting. We can find no other possible unit cell to fit the data. In either case, the structure would appear to be site-ordered distorted-NaCl, and the NaCl to distorted-NaCl transition then occurs at ~50 GPa without the mixed-phase region reported by Suzuki et al. (1983).

References

Batlogg, B, Jayaraman, A., Van Cleve, J. E., and Maines, R. G. (1983). *Phys. Rev. B* **27**, 3920.
Bundy, F. P. (1975). *Rev. Sci. Inst.* **46**, 1318.
Cline, C. F., and Stephens, D. R. (1965). *J. Appl. Phys.* **36**, 2869.
Edwards, A. L. and Drickamer, H. G. (1961). *Phys. Rev.* **122**, 1149.
Kabalkina, S. S., and Troitskaya, Z. V. (1964). *Soviet Phys. — Doklady* **8**, 800.
Mariano, A. N., and Warekois, E. P. (1963). *Science* **142**, 672.
McMahon, M. I., and Nelmes, R. J. (1996). *Phys. Stat. Sol. (b)* **198**, 389.
Nasr-Eddine, M., and Bertaut, E. F. (1971). *Solid State Commun.* **9**, 717.
Owen, N. B., Smith, P. L., Martin, J. E., and Wright, A. J. (1963). *J. Phys. Chem. Solids* **24**, 1519.
Rooymans, C. J. M. (1963). *Phys. Lett.* **4**, 186.
Samara, G. A., and Giardini, A. A. (1965). *Phys. Rev.* **140**, A388.
Samara, G. A., and Drickamer, H. G. (1962). *J. Phys. Chem. Solids* **23**, 457.
Savić, P., and Urošević, V. (1987). *Chem. Phys. Lett.* **135**, 393.
Suzuki, T., Yagi, T., Akimoto, S., Kawamura, T., Toyoda, S., and Endo, S. (1983). *J. Appl. Phys.* **54**, 748.

23. Cadmium Selenide

Under ambient conditions CdSe exists in both the (metastable) zinc-blende and the wurtzite structures with lattice parameters of $a = 6.052$ Å, and $a = 4.2999$ Å and $c = 7.0109$ Å, respectively. Only the wurtzite phase appears to have been studied under pressure. The transition in wurtzite-CdSe at 2.7 GPa was first observed in optical studies (Edwards and Drickamer, 1961). Subsequent diffraction studies found the high-pressure phase to have the NaCl structure (Rooymans, 1963; Mariano and Warekois, 1963) accompanied by an abrupt decrease in the electrical resistivity (Ignatchenko and Babushkin, 1993), and a volume change ($\Delta V/V_0$) of 16.4% (Cline and Stephens, 1965). The NaCl phase is not metallic and has an indirect gap of ~0.65 eV at 3 GPa (Cervantes et al., 1996). There have been only a few studies of CdSe at higher pressures. Absorption (Cervantes et al., 1996) and conductivity (Ignatchenko and Babushkin, 1993) measurements have reported that CdSe remains nonmetallic up to 50 GPa, with the bandgap falling to 0.5 eV at 45 GPa (Cervantes et al., 1996). There is also some evidence of a transition, possibly electronic, at ~40 GPa (Ignatchenko and Babushkin, 1993). In a pseudopotential calculation, a transition from NaCl to the CsCl structure was predicted at 94 GPa (Zakharov et al., 1995) with metallization occurring near 180 GPa (Cervantes et al., 1996). However, the existence of transitions to other intermediate phases between NaCl and CsCl at lower pressures could not be excluded (Zakharov et al., 1995). A more recent calculation has considered the cinnabar and Cmcm structures as possible intermediate phases (Côté et al., 1997). Although the cinnabar phase was found not to be stable, a transition from NaCl to Cmcm was predicted at 29 GPa (Côté et al., 1997).

We have carried out ADX studies of CdSe to 85 GPa. We confirm the NaCl phase to be site-ordered and find a continuous transition to a site-ordered Cmcm structure at 27.0(5) GPa (Fig. 24). At 34.4 GPa, the refined

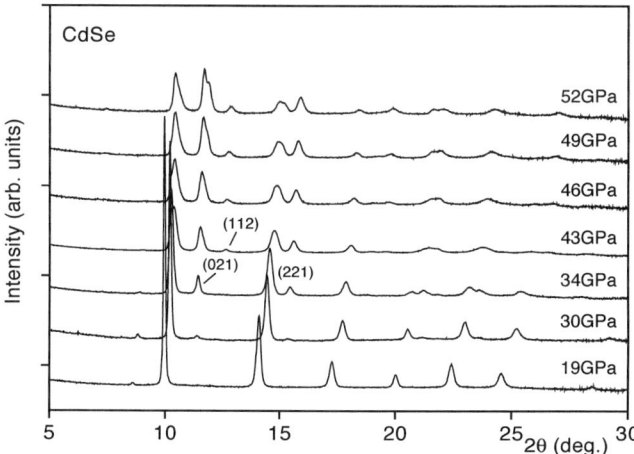

FIG. 24. Observed powder diffraction patterns from CdSe through the NaCl-to-Cmcm transition at 27 GPa, and the Cmcm-to-distorted-Cmcm transition at 36 GPa.

structure has $a = 5.200(1)$ Å, $b = 5.222(1)$ Å and $c = 5.159(1)$ Å, with $y(\text{Cd}) = 0.703(1)$ and $y(\text{Se}) = 0.214(1)$. At 36(1) GPa, there is evidence of a further transition to a distorted-Cmcm structure. This transition is evident from the emergence of the (110) and (112) reflections, and the splitting of the (021) reflection. The growth of the (021) and (221) reflections of Cmcm above 27 GPa can be seen in the 30 and 34 GPa patterns of Fig. 24, followed by the progressive growth of (112) above 36 GPa, and the splitting of (021) which becomes clearly defined in the data at 52 GPa. This distorted-Cmcm structure has so far remained unsolved despite considerable effort. (There is evidence of a similar but much weaker distortion of the Cmcm phase of ZnSe above ~50 GPa — see section III.19.) We find no evidence of any further transition up to the maximum pressure reached in our studies, 85 GPa.

References

Cervantes, P., Williams, Q., Côté, M., Zakharov, O., and Cohen, M. L. (1996). *Phys. Rev. B* **54,** 17585.
Cline, C. F., and Stephens, D. R. (1965). *J. Appl. Phys.* **36,** 2869.
Côté, M., Zakharov, O., Rubio, A., and Cohen, M. L. (1997). *Phys. Rev. B* **55,** 13025.
Edwards, A. L., and Drickamer, H. G. (1961). *Phys. Rev.* **122,** 1149.
Ignatchenko, O. A., and Babushkin, A. N. (1993). *Phys. Solid State* **35,** 1109.
Mariano, A. N., and Warekois, E. P. (1963). *Science* **142,** 672.
Rooymans, C. J. M. (1963). *Phys. Lett.* **4,** 186.
Zakharov, O., Rubio, A., and Cohen, M. L. (1995). *Phys. Rev. B* **51,** 4926.

24. Cadmium Telluride

CdTe is isoelectronic with InSb and Sn and has the zincblende structure at ambient pressure with $a = 6.482$ Å. A phase transition at about 3.5 GPa was first found optically (Edwards and Drickamer, 1961), and then confirmed in conductivity measurements which revealed a further transition at ~10 GPa (Samara and Drickamer, 1962). Diffraction measurements subsequently identified the transition at 3.5 GPa as being to a phase with the NaCl structure (Mariano and Warekois, 1963; Owen *et al.*, 1963; Borg and Smith, 1967), with a volume change ($\Delta V/V_0$) of 16.4% (Borg and Smith, 1963; Cline and Stephens, 1965). The NaCl phase is an indirect-gap semiconductor with a gap of ~0.08 eV at 6.8 GPa (Gonzalez *et al.*, 1995). ADX studies have revealed, however, that the ZB-to-NaCl transition at 3.5 GPa is actually two closely spaced transitions: ZB to site-ordered cinnabar at 3.4(2) GPa and then cinnabar to NaCl at 3.9(2) GPa (Nelmes *et al.*, 1993). This was the first report of a cinnabar phase outside the mercury chalcogenides. Although a single-phase cinnabar sample could not be obtained on pressure increase (Nelmes et al., 1993), such samples were obtained on pressure decrease from the NaCl phase (McMahon et al., 1993). The cinnabar phase was found to be stable with decreasing pressure from 3.6 to 2.7 GPa, and the structural P-dependence was determined over this range. At 3.2 GPa the structure refinement gave $a = 4.319(1)$ Å and $c = 10.265(2)$ Å, with Cd and Te at $(0.622(2), 0, 1/3)$ and $(0.565(2), 0, 5/6)$, respectively. The volume changes ($\Delta V/V_0$) at the NaCl-to-cinnabar and cinnabar-to-ZB transitions are $3.4(2)\%$ and $12.2(1)\%$, respectively. The cinnabar structure of CdTe has $4 + 2$ coordination — like cinnabar-HgTe — and, on compression, moves closer to the sixfold coordination of NaCl (McMahon et al., 1993), although the transition to NaCl still involves large discontinuities in structural geometry. The study of the structural P dependence shows that the variation of the shortest nearest-neighbor bond (the "a" bond) is continuous with that of the Cd–Te bonds in the ZB phase, whereas the second shortest "b" bonds have a variation that is continuous with that of the Cd–Te distance in the NaCl phase (McMahon et al., 1993). Similar behavior has been found in the cinnabar phase of HgTe (see Section III.28). The energy gap in cinnabar-CdTe is similar to that of the ZB phase (González et al., 1995).

In some more recent high-temperature EDX studies at LURE, Martinez-Garcia et al. (1998) have mapped the P–T phase diagram up to 1173 K and 5 GPa. The results show that the pressure range of the cinnabar phase decreases with temperature until a triple point is reached at 2.6(1) GPa and 735(20) K. Above that temperature there is a direct transition between the ZB and NaCl phases.

The transition in CdTe at ~10 GPa was initially reported to be to the β-tin structure (Owen et al., 1963; Borg and Smith, 1967). Further diffraction work reproduced these results and apparently revealed yet another transition, to an orthorhombic structure with spacegroup Pmm2, above 12 GPa (Hu, 1987). However, a detailed ADX study found the transition at 10.1(2) GPa to be a continuous transition to a site-ordered Cmcm structure (Nelmes et al., 1995). The transition is made evident at first only through the appearance of extra reflections due to the shearing of the NaCl planes; splitting of the "NaCl" reflections due to the orthorhombic distortion of the Cmcm unit cell becomes evident only at ~12 GPa. This probably accounts for the apparent further transition at 12 GPa reported by Hu (1987). Thus, the phases previously reported as β-tin and Pmm2 are one and the same, and this phase has the Cmcm structure (Nelmes et al., 1995).

The Cmcm structure is site-ordered. At 18.6 GPa it has $a = 5.573(1)$ Å, $b = 5.960(2)$ Å, and $c = 5.284(4)$ Å, with $y(\text{Cd}) = 0.650(3)$ and $y(\text{Te}) = 0.180(3)$ (Nelmes et al., 1995). The pressure dependence of the structural

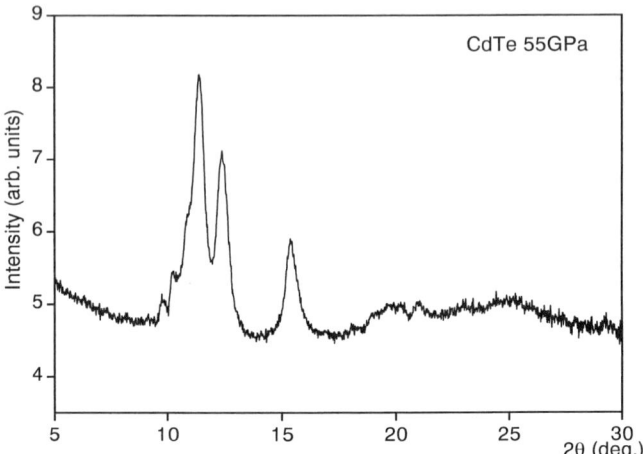

FIG. 25. Observed powder diffraction pattern obtained from CdTe at 55 GPa.

coordinates of the Cmcm phase has been determined from 10 to 28 GPa. As shown in *Figure 12* of Nelmes *et al.* (1995), the results reveal in detail how the sixfold coordination of the NaCl structure evolves into the 5 + 3 coordination of the Cmcm structure (see the discussion in Section II.10).

In further work, we have found the Cmcm structure to be stable to 40 GPa, although at these pressures the fit is degraded, probably because of deviatoric stress. At 42(2) GPa, we observe a transition to a phase giving the broadened complex pattern shown in Fig. 25. This is (probably) the same transition reported by Hu (1987) at 34 GPa. The structure of the new phase remains unsolved. No evidence of any further transition has been found to 55 GPa.

Calculations by Ahuja *et al.* (1997) support a ZB-to-cinnabar transition at 3.3 GPa, but disagree with experiment in giving pressures of 10.0 and 15.0 GPa for cinnabar-to-NaCl and NaCl-to-Cmcm, and in predicting a further transition from Cmcm to CsCl at 28.0 GPa.

REFERENCES

Ahuja, R., James, P., Eriksson, O., Wills, J. M., and Johansson, B. (1997). *Phys. Stat. Sol. (b)* **199,** 75.
Borg, I. Y., and Smith, D. K. Jr. (1967). *J. Phys. Chem. Solids* **28,** 49.
Cline, C. F., and Stephens, D. R. (1965). *J. Appl. Phys.* **36,** 2869.
Edwards, A. L., and Drickamer, H. G. (1961). *Phys. Rev.* **122,** 1149.
González, J., Pérez, F. V., Moya, E., and Chervin, J. C. (1995). *J. Phys. Chem. Solids* **56,** 335.
Hu, J. Z. (1987). *Solid State Commun.* **63,** 471.
Mariano, A. N., and Warekois, E. P. (1963). *Science* **142,** 672.

Martinez-Garcia, D., Le Godec, Y., Mezouar, M., Syfosse, G., Itié, J.-P., and Besson, J. M. (1998). To be published.
McMahon, M. I., Nelmes, R. J., Wright, N. G., and Allan, D. R. (1993). *Phys. Rev. B* **48,** 16246.
Nelmes, R. J., McMahon, M. I., Wright, N. G., and Allan, D. R. (1993). *Phys. Rev. B* **48,** 1314.
Nelmes, R. J., McMahon, M. I., Wright, N. G., and Allan, D. R. (1995). *Phys. Rev. B* **51,** 15723.
Owen, N. B., Smith, P. L., Martin, J. E., and Wright, A. J. (1963). *J. Phys. Chem. Solids* **24,** 1519.
Samara, G. A., and Drickamer, H. G. (1962). *J. Phys. Chem. Solids* **23,** 457.

25. MERCURY OXIDE

At atmospheric pressure HgO exists in both an orthorhombic form and, less commonly, in an HgS-like cinnabar form (Aurivillius, 1956; Aurivillius and Carlsson, 1958). Other, more complex, forms have also been reported (Benjamin, 1982). The orthorhombic form has spacegroup *Pnma* with $a = 6.612$ Å, $b = 5.520$ Å, and $c = 3.521$ Å and Hg and O atoms on the 4(c) sites at (0.115, 1/4, 0.245) and (0.36, 1/4, 0.58), respectively (Aurivillius, 1956). The cinnabar structure has $a = 3.577$ Å and $c = 8.681$ Å, with $u(Hg) = 0.745$ and $v(O) = 0.46$ (Aurivillius and Carlsson, 1958). The orthorhombic structure has planar -O–Hg–O–Hg- zigzag chains running parallel to the x axis (Aurivillius, 1956), while, as in HgS, cinnabar-HgO is 2 + 4 coordinated and has -O–Hg–O–Hg- spiral chains running parallel to the z axis (Aurivillius and Carlsson, 1958). Resistivity studies on a cinnabar-HgO sample to 35 GPa by Tsidil'kovskii *et al.* (1985) revealed a transition to a metallic state at pressures above 10 GPa, and the authors speculated that this might be accompanied by a structural transition to an NaCl phase. Subsequent studies on samples recovered to ambient conditions reported that a tetragonal phase was obtained if an orthorhombic-HgO sample was pressurized to above 10 GPa (Voronin and Shchennikov, 1989).

A recent combined EXAFS and EDX study of orthorhombic HgO by San Miguel *et al.* (1995) apparently showed two transitions well below 10 GPa, at 2 and 5 GPa. But a subsequent combined Raman/ADX study of orthorhombic-HgO by Zhou *et al.* (1998) found no transitions in orthorhombic HgO below 10 GPa. Rather, they attributed the transitions reported by San-Miguel *et al.* (1995) to partial decomposition of HgO in the synchrotron beam, leading first to the freezing of liquid mercury in the sample at 2 GPa, and then to a transition between the α and β forms of mercury at 5 GPa (Zhou *et al.,* 1998). (We have found this behavior in our own synchrotron ADX studies, and the fit to elemental mercury is good.) At higher pressure, Zhou *et al.* (1998) found a first-order transition at 14(1) GPa to a site-ordered tetragonal phase with spacegroup *I4/mmm*. At 19.3 GPa, the structure has $a = 3.370(1)$ Å and $c = 4.651(2)$ Å, with Hg at (0, 0, 0) and O at (0, 0, 1/2) (Zhou *et al.,* 1998). This structure is only slightly distorted from that of NaCl and would be exactly NaCl if $c/a = $

$\sqrt{2} = 1.414$. At 19.3 GPa, $c/a = 1.380$. However, Zhou et al. (1998) remark that the fit to their diffraction data using the I4/mmm structure contains systematic deviations in peak positions, and that, indeed, the I4/mmm structure is incompatible with the Raman spectra they recorded from HgO-II. Attempts at fitting a lower symmetry structure were unsuccessful, however, and the I4/mmm structure was regarded as a proper description of the "average" structure.

We have also been carrying out ADX studies of orthorhombic HgO. We locate the transition to HgO-II at 11.6(7) GPa and support the site-ordered I4/mmm interpretation of the structure of HgO-II. As in the data of Zhou et al. (1998), we observe significant displacements between observed and calculated peak positions — in particular the (002) reflection. However, we observe no weak reflections or peak splittings that would indicate a structure of lower symmetry.

At higher pressures, the Raman measurements of Zhou et al. (1998) reveal a new mode appearing at 14 GPa, while above 28 GPa some of the modes overlap. Reflectivity measurements showed a change at 14 GPa and a transition near 28 GPa — which the authors suggested was electronic in nature — to a metallic state. Our ADX studies to 34 GPa show evidence of a transition to a true NaCl phase beginning at 26(3) GPa. Single-phase samples were obtained above 30(1) GPa. The fit of the NaCl structure to the profiles above 30 GPa contains significant displacements between observed and calculated peak positions, as noted earlier for the tetragonal phase. However, the simplicity of the diffraction pattern, and the complete absence of additional weak peaks or peak splittings, suggests that NaCl is the correct structure.

REFERENCES

Aurivillius, K. (1956). *Acta Chem. Scand.* **10,** 852.
Aurivillius, K., and Carlsson, I.-B. (1958). *Acta Chem. Scand.* **12,** 1297.
Benjamin, D. J. (1982). *Mat. Res. Bull.* **17,** 179.
San-Miguel, A., Gonzalez Penedo, A., Itié, J.-P., Polian, A., and Bordet, P. (1995). In "Proceedings of the Joint XV AIRAPT and XXXIII EHPRG International Conference" (Trzeciakowski, W. A., Ed.), p. 438. World Scientific, Singapore.
Tsidil'kovskii, I. M., Shchennikov, V. V., and Gluzman, N. G. (1985). *Sov. Phys. Semicond.* **19,** 901.
Voronin, V. I., and Shchennikov, V. V. (1989). *Sov. Phys. Crystallogr.* **34,** 293.
Zhou, T., Schwarz, U., Hanfland, M., Liu, Z. X., Syassen, K., and Cardona, M. (1998). *Phys. Rev. B* **57,** 153.

26. MERCURY SULFIDE

Under ambient conditions HgS exists in both a (red) cinnabar form, with $a = 4.145$ Å, $c = 9.496$ Å, $u(\text{Hg}) = 0.7198$, and $v(\text{S}) = 0.4889$ (Auvray and

Genet, 1973), and a metastable (black) zincblende form, with $a = 5.851$ Å. Under pressure, the ZB form transforms to the cinnabar form at ~5 GPa (Bridgman, 1940). The cinnabar form has a 2 + 4 coordinated structure, with -S–Hg–S- spiral chains running parallel to the z axis (Fig. 9a). This structure is thus very similar to that observed in cinnabar-HgO, but is different from the 2 + 2 + 2 and 4 + 2 coordinated structures observed in the cinnabar phases of HgSe, CdTe, and HgTe. The concurrent EDX studies of Huang and Ruoff (1983) and Werner *et al.* (1983) both reported a transition from the cinnabar form to an NaCl structure, and Huang and Ruoff (1983) obtained a lattice parameter of 5.070(5) Å at 30 GPa. However, the transition pressure of 13 GPa reported by Huang and Ruoff (1983) was questioned by Werner *et al.* (1983), who were unable to detect any transitions up to 24 GPa using diffraction techniques, and who observed Raman peaks from the cinnabar phase at 20 GPa. They concluded, therefore, that there was no transition to NaCl below this pressure (Werner *et al.*, 1983). But, in a subsequent study, Huang and Ruoff (1985) confirmed their view of a cinnabar-to-NaCl transition at 13 GPa and attributed the existence of the cinnabar Raman lines at 20 GPa to a residual component of the cinnabar phase.

We have studied cinnabar-HgS to 55 GPa using ADX techniques. The superior resolution afforded by ADX techniques and the sensitivity to weak reflections has enabled us to distinguish the cinnabar and NaCl phases — through both peak splittings and the additional weak reflections in the cinnabar profiles — to much higher pressures than was previously possible. We observe clear evidence of reflections from the cinnabar structure to 20.5 GPa, above which we observe peaks only from a site-ordered NaCl phase. The cinnabar-to-NaCl transition occurs at 20.5(7) GPa, a value in complete accord with the Raman measurements of Werner *et al.* (1983). We have found that HgS samples develop very strong preferred orientation that onsets over a small pressure range at ~12 GPa in the cinnabar phase. This can give rise to quite large changes in peak intensities, and we suggest this as a possible cause for the apparent transition reported by Huang and Ruoff (1983, 1985) at 13 GPa.

We have studied the NaCl phase of HgS to 55 GPa and have found evidence of a continuous transition to HgS-III at 52(3) GPa. We have, as yet, only limited diffraction information on this phase, but the diffraction profiles suggest that the structure is a distortion of the NaCl structure, possibly Cmcm.

REFERENCES

Auvray, P., and Genet, F. (1973). *Bull. Soc. Minéral Cristallogr.* **96**, 218.
Bridgman, P. W. (1940). *Proc. Am. Acad. Sci.* **74**, 21.

Huang, T., and Ruoff, A. L. (1983). *J. Appl. Phys.* **54**, 5459.
Huang, T., and Ruoff, A. L. (1985). *Phys. Rev. B* **31**, 5976.
Werner, A., Hochheimer, H. D., Strössner, K., and Jayaraman, A. (1983). *Phys. Rev. B* **28**, 3330.

27. MERCURY SELENIDE

Under ambient conditions HgSe has the zincblende structure with $a = 6.084$ Å and is semimetallic (Kafalas *et al.*, 1962). A transition pressure of 0.74 GPa first reported by Bridgman (1940) was subsequently confirmed by Kafalas *et al.* (1962), who found the transition to be accompanied by a large increase in resistivity, signaling a change from a semimetal to a semiconductor. The transition has also been located by Mariano and Warekois (1963) at 1.5 GPa, by Jayaraman *et al.* (1963) at 0.8 GPa, and by Ford *et al.* (1982) at 0.95 GPa.

The high-pressure phase has a cinnabar structure (Kafalas *et al.*, 1962; Mariano and Warekois, 1963; Onodera *et al.*, 1982) and is a semiconductor with a gap reported to be 0.78 eV (Onodera *et al.*, 1982; Ohtani *et al.*, 1982) or 0.5–0.9 eV (Gluzman and Shchennikov, 1979). The data of Kafalas *et al.* (1962) show the (003) difference reflection to have been detected; the structure is thus site-ordered (Section II.9). The reported volume change ($\Delta V/V_0$) at the ZB-to-cinnabar transition is 9.0% (Jayaraman *et al.*, 1963). We have determined the detailed structure of cinnabar-HgSe using ADX techniques. At 4.0 GPa, Rietveld refinements give $a = 4.120(1)$ Å and $c = 9.560(1)$ Å, with $u(\text{Hg}) = 0.662(1)$ and $v(\text{Se}) = 0.550(1)$. The ZB-to-cinnabar transition is observed to start at 1.15(5) GPa (McMahon *et al.*, 1997), and the measured volume change ($\Delta V/V_0$) at 1.41 GPa is 9.9(1)%. In contrast to earlier assumptions (Kafalas *et al.*, 1962; Gluzman and Shchennikov, 1979) that the coordination of cinnabar-HgSe is 2 + 4 coordinated as in cinnabar-HgS, each atom has pairs of unlike neighbors at 2.541(4) Å, 2.891(4) Å, and 3.240(5) Å at 4.0 GPa. The structure thus has 2 + 2 + 2 coordination. Observation of the (003) reflection confirms that the phase is site-ordered.

ADX techniques have also revealed a second phase that can be formed from the ZB phase of HgSe on pressure increase (McMahon *et al.*, 1996). It appears when — in the very sluggish ZB-to-cinnabar transition at 1.15 GPa — the ZB phase is taken well above 1.15 GPa to about 2.1 GPa; that is, it is a transition from ZB occurring well outside the range in which ZB is the equilibrium phase with respect to cinnabar. Since the new phase itself transforms to cinnabar at slightly above 2.25 GPa, it can be regarded as a "hidden" intermediate phase (McMahon *et al.*, 1996). Although single-phase samples of this phase have not been obtained, the structure has been successfully solved as having an orthorhombic distorted-ZB structure, spacegroup $C222_1$ (McMahon et al., 1996). Observation of the (002) reflection confirms the structure is site-ordered. The refined structure at 2.25 GPa has $a = 5.992(1)$ Å, $b = 5.879(1)$ Å, and $c = 6.045(2)$ Å and is site-ordered with Hg and Se at (0.302(1), 0, 0) and (0, 0.207(2), 1/4), respectively. The volume change ($\Delta V/V_0$) at the ZB-to-$C222_1$ transition is 1.2(1)% (McMahon et al., 1997), whereas from a mixed-phase $C222_1$/cinnabar profile at 2.25 GPa, the volume change at the $C222_1$-to-cinnabar transition is 8.6(1)%. The structure is discussed in Section II.11 and shown in Fig. 11. As discussed in more detail there, the atomic coordinates of the $C222_1$ structure are remarkably similar to those of Al/Ga and P in the low-cristobalite phase of $AlPO_4$ and $GaPO_4$, and the ZB-to-$C222_1$ transition is thus analogous to the high–low cristobalite transition in these materials (McMahon et al., 1996). The existence of the $C222_1$ phase may prove useful in understanding the unusual behavior reported in the pressure dependence of the elastic constants of the ZB phase of HgSe (Ford et al., 1982).

At higher pressures, a combined resistivity and diffraction study to 20 GPa reported a transition at 15.5 GPa to what was tentatively indexed as an NaCl phase. (Onodera et al., 1982; Ohtani et al., 1982). The NaCl structure was confirmed by Huang and Ruoff (1983), who found $a = 5.360$ Å at 21 GPa. Our own ADX measurements have located the onset of the transition to the site-ordered NaCl phase at 14.6(6) GPa with a volume change ($\Delta V/V_0$) at 0.2(4)% measured at 15.7 GPa. However, the similarity of the diffraction patterns from the cinnabar and NaCl phases and the existence of microstructural effects in the NaCl-phase diffraction profiles (see later discussion) make an accurate determination of the volume change difficult. There is some evidence to suggest the existence of another semiconducting phase obtainable only at high temperature in the range 15–17 GPa (Onodera et al., 1982; Ohtani et al., 1982).

A further transition, to HgSe-IV, was found by Huang and Ruoff (1983) at 28 GPa. They initially reported it as having an orthorhombic structure, but later retracted this suggestion, in favor of a body-centered structure, closely related to β-tin (Huang and Ruoff, 1985). However, the β-tin-like structure could not account for five additional diffraction peaks, and an extra hexagonal phase was included to explain two of them (Huang and Ruoff, 1985). HgSe-IV has since been determined to have the orthorhombic Cmcm structure using ADX techniques (Nelmes et al., 1997). The site-ordered Cmcm structure accounts for the five additional lines observed by Huang and Ruoff (1985), and a fit at 35.6 GPa gives $a = 5.153(1)$ Å, $b = 5.559(1)$ Å, $c = 4.972(2)$ Å, $y(Hg) = 0.644(1)$, and $y(Se) = 0.141(2)$. The transition from the NaCl phase is first-order with a small volume change ($\Delta V/V_0$) of 0.9(5)% (Nelmes et al., 1997). The fit of the Cmcm structure to the diffraction data contains some small misfits, and the fit is clearly not as good as that obtained with Cmcm for HgTe-IV (see Section III.28). In HgSe, we attribute these misfits to microstructural effects, for which there is direct evidence in the NaCl phase of the same sample (Nelmes et al., 1997). No further transitions have been found up to the maximum pressure reached, 50 GPa.

REFERENCES

Bridgman, P. W. (1940). *Proc. Amer. Acad. Sci.* **74,** 21.
Ford, P. J., Miller, A. J., Saunders, G. A., Yoğurtçu, Y. K., Furdyana, J. K., and Jaczynski, M. (1982). *J. Phys. C: Solid State Phys.* **15,** 657.
Gluzman, N. G., and Shchennikov, V. V. (1979). *Sov. Phys. Solid State* **21,** 1844.
Huang, T., and Ruoff, A. L. (1983). *Phys. Rev. B* **27,** 7811.
Huang, T., and Ruoff, A. L. (1985). *Phys. Rev. B* **31,** 5976.
Jayaraman, A., Klement, W. Jr., and Kennedy, G. C. (1963). *Phys. Rev.* **130,** 2277.
Kafalas, J. A., Gatos, H. C., Levine, M. C., and Banus, M. D. (1962). *J. Phys. Chem. Solids* **23,** 1541.
Mariano, A. N., and Warekois, E. P. (1963). *Science* **142,** 672.
McMahon, M. I., Nelmes, R. J., Liu, H., and Belmonte, S. A. (1996). *Phys. Rev. Lett.* **77,** 1781.
Nelmes, R. J., McMahon, M. I., and Belmonte, S. A. (1997). *Phys. Rev. Lett.* **79,** 3668.
Ohtani, A., Seike, T., Motobayashi, M., and Onodera, A. (1982). *J. Phys. Chem. Solids* **43,** 627.
Onodera, A., Ohtani, A., Motobayashi, M., Seike, T., Shimomura, O., and Fukunaga, O. (1982). *In* "Proceedings of the 8th AIRAPT Conference," Vol. 1 (C. M. Backman, T. Johannisson, and L. Tengner, Eds.), p. 321. Arkitektopia, Uppsala.

28. MERCURY TELLURIDE

Under ambient conditions HgTe has the zincblende structure, with $a = 6.4603$ Å, and is semimetallic (Kafalas et al., 1962). A transition at 1.3 GPa first reported by Bridgman (1940) was subsequently found to be accompanied by a large increase in resistivity (Blair and Smith, 1961), signaling a change from a semimetal to a semiconductor. The high-pressure phase has the cinnabar structure (Mariano and Warekois, 1963; Onodera et al., 1982) and is a semiconductor with a gap of 0.7 eV at 2 GPa falling to zero at ~6.5 GPa (Onodera et al., 1982; Ohtani et al., 1982). The volume change ($\Delta V/V_0$) at the ZB-to-cinnabar transition is 11(2)% (San-Miguel et al., 1995). Cinnabar-HgTe was the first cinnabar structure to be refined *in situ* at high pressure (Wright et al., 1993). The results were obtained from ADX data and revealed that rather than having the 2 + 4 coordination observed in cinnabar-HgS and cinnabar-HgO, cinnabar-HgTe has 4 + 2 coordination. The refined structure at 3.6 GPa is site-ordered and has $a = 4.383(1)$ Å and $c = 10.022(1)$ Å, with $u(\text{Hg}) = 0.641(1)$ and $v(\text{Te}) = 0.562(1)$; each atom has three pairs of unlike neighbors at 2.732(4) Å, 2.995(4) Å, and 3.460(5) Å. The refined c/a ratio at 3.6 GPa is 2.287(2), similar to that reported by Qadri *et al.* (1986), but significantly different from that reported by Werner *et al.* (1983), who appear to have misindexed one of their four observable reflections — as discussed by Wright *et al.* (1993). The pressure evolution of cinnabar-HgTe has been determined by San-Miguel *et al.* (1995) in a combined EXAFS/ADX–diffraction study. As in cinnabar-CdTe, the structure tends toward that of NaCl on pressure increase, although the transition to NaCl is strongly first-order with large discontinuities in structural parameters. The study of the structural P dependence shows that the variation of the shortest nearest-neighbor bond (the "a" bond) is continuous with that of Hg–Te bonds in the ZB phase, whereas the second shortest "b" bonds have a variation that is continuous with that of the Hg–Te distance in the NaCl phase (San-Miguel *et al.*, 1995). Similar behavior has been found in the cinnabar phase of CdTe (see Section III.24).

The orthorhombic $C222_1$ structure observed in HgSe as a "hidden" intermediate phase between the ZB and cinnabar phases (see Section III.27, on HgSe, for a more detailed discussion) is also found in HgTe (McMahon *et al.*, 1996b). The $C222_1$ phase is first observed at 2.25 GPa and the volume change $(\Delta V/V_0)$ at the transition is 1.2(1)%. At 2.55 GPa, it has $a = 6.295(2)$ Å, $b = 6.241(2)$ Å, $c = 6.364(2)$ Å, x(Hg) $= 0.315(1)$ and y(Te) $= 0.205(2)$ (McMahon *et al.*, 1996b). Because the distortion from ZB is less than in HgSe, and because of the more similar scattering power of Hg and Te, it is difficult to determine with certainty that the $C222_1$ phase is site-ordered. However, there is no reason or evidence to expect it to differ from that of HgSe. $C222_1$ transforms to cinnabar at 2.6 GPa with a volume change of 8.7(1)%. Unusual behavior has been noted in the pressure dependence of the elastic constants (Miller *et al.*, 1981) and thermal expansion (Besson *et al.*, 1996) of the ZB phase of HgTe. The existence of the previously unknown $C222_1$ phase may be relevant to understanding this behavior.

At higher pressures, a combined resistivity and diffraction study to 20 GPa (Onodera *et al.*, 1982; Ohtani *et al.*, 1982) reported a transition from cinnabar to a metallic NaCl phase at 8 GPa and a further transition at 12 GPa to an unidentified phase, HgTe-IV. Subsequent EDX studies to 20 GPa (Werner *et al.*, 1983) and 15 GPa (Huang and Ruoff, 1983) confirmed the identity of the site-ordered NaCl phase and reported $a = 5.843$ Å at 8.9 GPa (Huang and Ruoff, 1983). The reported volume change $(\Delta V/V_0)$ at the cinnabar-to-NaCl transition is 3(1)% (San-Miguel *et al.*, 1995). However, Werner *et al.* (1983) and Huang and Ruoff (1983) disagreed on the identity of HgTe-IV, reporting it as having the β-tin structure (Werner *et al.*, 1983) or an orthorhombic structure (Huang and

Ruoff, 1983). Subsequently, Huang and Ruoff (1985) revised their interpretation of the structure to β-tin-like. McMahon et al. (1996a), using ADX techniques, found that the transition to HgTe-IV occurs at 10.2(3) GPa and that the diffraction pattern could not be indexed on the β-tin structure. However, a site-ordered Cmcm structure accounts for all the observed reflections. At 18.5 GPa, the refined structure has a = 5.5626(2) Å, b = 6.1516(5) Å, and c = 5.1050(8) Å, with y(Hg) = 0.624(1) and y(Te) = 0.152(1). The closest-neighbor coordination is fivefold as found in ZnTe and CdTe, but less clearly so — especially for the Hg atom. The volume change ($\Delta V/V_0$) at the NaCl-to-Cmcm transition is 1.2(1)% at 11 GPa. McMahon et al. (1996a) reported a second, different Cmcm structure in which the Hg and Te sites are interchanged and have coordinates that are slightly altered. This structure has a less plausible coordination (two like nearest neighbors) but could not be excluded — indeed, it gave a slightly better fit to the data. However, this alternative configuration of the structure has been ruled out in a subsequent X-ray absorption (XANES) study by Briois et al. (1997).

At still higher pressures Huang and Ruoff (1985) reported a transition to HgTe-V at 38.1 GPa with a structure thought to be distorted-CsCl. Nelmes et al. (1995) observed the onset of this transition at the significantly lower pressure at 28 GPa, although a pressure of ~50 GPa was required to transform all of the Cmcm phase to a single-phase profile of HgTe-V. The structure of HgTe-V is disordered body-centered cubic with a = 3.299(1) Å at 51 GPa (Nelmes et al., 1995). (See Section II.1 concerning site disorder.) The volume change ($\Delta V/V_0$) at the Cmcm-to-bcc transition is 3.0(3)%. The relationship of this result to the observations of Huang and Ruoff (1985) is not clear. Although the d = 2.361 Å peak they observed at 41 GPa can be indexed as the (110) peak of the bcc structure with a = 3.339 Å, the other two reflections they reported at 2.085 Å and 1.469 Å cannot be explained by such a structure.

References

Besson, J. M., Grima, P., Gauthier, M., Itié, J.-P., Mézouar, M., Häusermann, D., and Hanfland, M. (1996). *Phys. Stat. Sol. (b)* **198,** 419.
Blair, J., and Smith, A. C. (1961). *Phys. Rev. Lett.* **7,** 124.
Bridgman, P. W. (1940). *Proc. Amer. Acad. Sci.* **74,** 21.
Briois, V., Brouder, Ch., Sainctavit, Ph., San Miguel, A., Itié, J.-P., and Polian, A. (1997). *Phys. Rev. B* **56,** 5866.
Huang, T., and Ruoff, A. L. (1983). *Phys. Stat. Sol. (a)* **77,** K193.
Huang, T., and Ruoff, A. L. (1985). *Phys. Rev. B* **31,** 5976.
Kafalas, J. A., Gatos, H. C., Levine, M. C., and Banus, M. D. (1962). *J. Phys. Chem. Solids* **23,**1541.

Mariano, A. N., and Warekois, E. P. (1963). *Science* **142**, 672.
McMahon, M. I., Wright, N. G., Allan, D. R., and Nelmes, R. J. (1996a). *Phys. Rev. B* **53**, 2163.
McMahon, M. I., Nelmes, R. J., Liu, H., and Belmonte, S. A. (1996b). *Phys. Rev. Lett.* **77**, 1781.
Miller, A. J., Saunders, G. A., Yoğurtçu, Y. K., and Abey, A. E. (1981). *Phil. Mag.* **43**, 1447.
Nelmes, R. J., McMahon, M. I., Wright, N. G., and Allan, D. R. (1995). *J. Phys. Chem. Solids* **56**, 545.
Ohtani, A., Seike, T., Motobayashi, M., and Onodera, A. (1982). *J. Phys. Chem. Solids* **43**, 627.
Onodera, A., Ohtani, A., Motobayashi, M., Seike, T., Shimomura, O., and Fukunaga, O. (1982). *In* "Proceedings of the 8th AIRAPT Conference," Vol. 1 (C. M. Backman, T. Johannison, and L. Tengner, Eds.), p. 321. Arkitektopia, Uppsala.
Qadri, S. B., Skelton, E. F., Webb, A. W., and Dinan, J. (1986). *J. Vac. Sci. Technol. A* **4**, 1974.
San-Miguel, A., Wright, N. G., McMahon, M. I., and Nelmes, R. J. (1995). *Phys. Rev. B* **51**, 8731.
Werner, A., Hochheimer, H. D., Strössner, K., and Jayaraman, A. (1983). *Phys. Rev. B* **28**, 3330.
Wright, N. G., McMahon, M. I., Nelmes, R. J., and San-Miguel, A. (1993). *Phys. Rev. B* **48**, 13111.

IV. Discussion

The transitions and structures that emerge from Section III as the best present interpretations (in our view) of the behavior of the group IV, III–V, and II–VI semiconductors at room temperature are summarized in Table I. First and foremost, it is clear that the structural systematics have been radically transformed in the past few years. Among the high-pressure phases of the III–V and II–VI systems, the *only* structural interpretations that are the same as 5 years ago are NaCl, cinnabar in HgSe and HgTe, bcc in InSb, and possibly Pmmn in CdS. The structures of all the others have been found to be quite different from what had been supposed; many further new phases have been discovered and identified; conversely, several supposed transitions have been shown to be mistaken interpretations of previous data; and some transition pressures have been very significantly revised. There is considerably more complexity in the many lower symmetry structures revealed, and phenomena such as intermediate phases. Yet there is also a certain simplification in the discovery of an apparent underlying sequence common to many of the III–V and II–VI systems.

Of all the changes and new discoveries one of the most striking is the complete disappearance of the diatomic β-tin (d-β-tin) structure that played such a prominent part in previous interpretations of the systematics. In all cases, the true structure of the phases concerned has been shown to be the orthorhombic Cmcm structure, which is quite significantly different. In particular, Cmcm is made up of a stacking of NaCl-like planes (Fig. 10), which d-β-tin is not (Fig. 3a). The absence of d-β-tin is made the more surprising by the fact that *ab initio* calculations show that this structure has

TABLE I

KNOWN TRANSITIONS, PHASES, AND STRUCTURES OF THE GROUP IV, III–V, AND II–VI SEMICONDUCTORS AT ROOM TEMPERATURE,
EXCEPT THE BORON-V GROUP
(SEE SECTION III.4)

Si	Diam	11 GPa	β-tin	13.2 GPa	Imma	15.6 GPa	SH	37.6 GPa	mono	41.8 GPa	hcp	78 GPa	fcc → 248 GPa
	BC8	2 GPa	R8	10.1 GPa	β-tin								
Ge	Diam	10.6 GPa	β-tin	~75 GPa	Imma	~85 GPa	SH	102 GPa	dhcp → 125 GPa				
	BC8†/ST12	7.6 GPa	β-tin										

AlN	Wurtz	22.9 GPa	NaCl → 65 GPa		
AlP	ZB	14–17 GPa	NiAs → 43 GPa		
AlAs	ZB	13 GPa	NiAs → 46 GPa		
AlSb	ZB	7.8 GPa	Cmcm	42 GPa	unknown → 59 GPa
GaN	Wurtz	~50 GPa	NaCl → 70 GPa		
GaP	ZB	23 GPa	*Cmcm → 52 GPa		
GaAs	ZB	[SC16?]§	17 GPa	Cmcm	60–80 GPa	GaAs-IV? → 108 GPa
	ZB	8.1 GPa	Cinn	11.7 GPa	Cmcm?	
	ZB		11.2 GPa		Cmcm?	
GaSb	ZB	7.6 GPa	*Imma	61 GPa	*bcc? → 110 GPa
InN	Wurtz	10–12 GPa	NaCl → 35 GPa		
InP	ZB	~10 GPa	NaCl	28 GPa	Cmcm → 46 GPa
InAs	ZB	7 GPa	NaCl?	9 GPa	Cmcm? → 46 GPa
InSb	ZB	~3 GPa	s-Cmcm	9 GPa	(*Imma) 9 GPa	Immm	~17 GPa	(unknown)	21 GPa	*bcc → 66 GPa
	{ZB	~2.1 GPa	(*β-tin-like)	(Immm)	~2.1 GPa	s-Cmcm}

ZnO	Wurtz	8 GPa	NaCl → 29 GPa		
ZnS	ZB	15 GPa	NaCl	69 GPa	d-NaCl → 84 GPa
ZnSe	ZB	13.5 GPa	NaCl	30 GPa	Cmcm	~50 GPa	d-Cmcm? → 120 GPa
ZnTe	ZB	9.3 GPa	Cinn	11 GPa	Cmcm	85 GPa	unknown → 93 GPa

TABLE I (*continued*)

CdO	NaCl → 30 GPa						
CdS	Wurtz	2.3 GPa	NaCl	~50 GPa	Pmmm?/Cmcm? → 68 GPa		
CdSe	Wurtz	2.7 GPa	NaCl	27 GPa	Cmcm	36 GPa	d-Cmcm? → 85 GPa
CdTe	ZB	3.4 GPa	Cinn	3.9 GPa	NaCl	10.1 GPa	Cmcm 42 GPa unknown → 55 GPa
HgO	Pnma	12 GPa	I4/mmm?	26 GPa	NaCl → 34 GPa		
HgS	Cinn	20 GPa	NaCl	52 GPa	d-NaCl → 55 GPa		
HgSe	ZB	1.2 GPa	Cinn	15 GPa	NaCl	25 GPa	Cmcm → 50 GPa
	{ZB	2.1 GPa	d-ZB	2.4 GPa	Cinn}		
HgTe	ZB	1.3 GPa	Cinn	8 GPa	NaCl	10 GPa	Cmcm 28 GPa *bcc → 51 GPa
	{ZB	2.25 GPa	d-ZB	2.6 GPa			

All the results shown are those obtained at room temperature. Phases obtained by heating — such as Si-IV — are not included but are discussed in the text. Phases are identified by the best present assignment of their structures, as discussed in Section III. Structures are referred to by abbreviations (such as "ZB" for zincblende), standard names (such as "β-tin" or "BC8"), symmetry class (such as "mono" for monoclinic), or space group symmetry (such as "Imma"). The prefixes "d-" and "s-" denote "distorted" and "superstructure," respectively. All the names and descriptions used are identified in the introduction to Section II. Where a transition has been shown to occur but the structure obtained has yet to be characterized, it is tabulated as "unknown." Structural assignments about which significant uncertainty is expressed in Section III are marked "?". Phases which lack long-range site ordering (see Section II.1) are marked with an asterisk. The phases shown in parentheses for InSb have not yet been obtained as single-phase samples and appear to be intermediate phases. Sequences shown in {} brackets for InSb, HgSe, and HgTe are reproducibly observable alternative behaviors. In most cases, the phases shown and the quoted transition pressures, P_{tr}, are those obtained on pressure increase. Results obtained on pressure decrease are on a shaded background. The values given for P_{tr} should be regarded as best estimates derived from the range of experimental values in the literature, taking account of the probable reliability of the different techniques used. Full details of reported values are given in Section III. The range of values is large in some cases, and the estimates tabulated are correspondingly approximate. In many cases, transition pressures have also been measured on pressure decrease (or on pressure increase for the sequences shown here on pressure decrease), and this information is given in Section III. The quoted P_{tr} should thus *not* generally be taken to be equilibrium values; there is often a considerable difference between the pressures at which the transitions in these systems are observed on upstroke and downstroke at room temperature. At the end of each pressure-increase sequence is given the highest pressure to which the system has been studied so far.

† The BC8 phase of Ge transforms to a hexagonal-diamond-like phase at room temperature; the same transformation occurs in Si, but at high temperature.

§ The SC16 phase of GaAs has been tentatively included because, although heating to ~450 K is required to produce it, calculations strongly suggest it is an equilibrium phase at room temperature.

an energy even lower than that of Cmcm in some cases — for example, in GaAs and GaP (Mujica et al. 1995; Mujica and Needs, 1997). Nelmes et al. (1997) point out that the competing structures — NaCl, Cmcm, and Immm — are all made up of flat NaCl-like layers, differing only in the way they are stacked, and suggest that intrinsically different bonding in the layers from that between the layers, as found in calculations of a Cmcm-like GaAs structure by Zhang et al. (1989), could perhaps favor the formation of these structures. It is interesting to note that there is as yet no example of a site-ordered tetragonal structure in any of the III–V or II–VI systems, with the one probable exception of the I4/mmm phase of HgO. There is the β-tin-like (probably true β-tin) P2 phase of InSb, and the Imma phases of GaSb and InSb that are only a little distorted from β-tin, but these are all quasimonatomic site-disordered phases in which truly NaCl-like layers cannot be defined — except possibly on a very short length scale, as discussed further later.

Another change to have affected the overall picture of the systematics has been the occurrence of the cinnabar structure in systems other than the mercury chalcogenides — to which it was previously believed to be confined. New phases with this structure have so far been found experimentally in CdTe, ZnTe, and GaAs and have been proposed on the basis of total-energy calculations as possible phases in ZnS and ZnSe (Sections III.18 and III.19) — though there is no evidence yet for the latter two in room-temperature studies. And a transition to a metastable cinnabar phase on pressure decrease has been predicted in GaP from calculations by Mujica et al. (1998). Determinations of structural pressure dependence in CdTe and HgTe (McMahon et al., 1993; San-Miguel et al., 1995) have revealed that the magnitude and variation of the shortest bond length is continuous with the P-dependence of the Cd–Te (or Hg–Te) distance in the zincblende phase, and that there is a similar relationship between the second shortest cinnabar bond length and the NaCl phase. Cinnabar — which can be understood as a distorted NaCl structure (Section II.9) — thus appears to be an intriguingly specific intermediate step between the zincblende and NaCl structures. The almost exactly fourfold-coordinated cinnabar phases in GaAs and ZnTe can now also be seen in another way as members of the growing family of distorted tetrahedral structures in the III–V and II–VI systems. Indeed, as remarked in Sections II.9 and III.11, it has been questioned whether these fourfold-coordinated structures are correctly named as cinnabar. The present balance of evidence (Section III.11) is that the cinnabar structures of GaAs and ZnTe both show a small but significant distortion of the Ga–As and Zn–Te bond lengths such that two are ~0.1 Å longer than the other two in each tetrahedron, and that their spacegroup symmetry is $P3_121$, as for all other cinnabar phases. If further work showed that there were in fact *no* differences among the tetrahedral bond lengths, in these or other phases not yet found, the spacegroup would change to $P6_222$ (or enantiomorphic $P6_422$) — with a halved unit cell volume if the structure were also site-disordered. Quite apart from this possibility, Mujica *et al.* (1998) point out that these structures are made up of twisted but otherwise symmetric tetrahedra, very unlike the 2-coordinated chains of HgS cinnabar. Of course, there are also the intermediately coordinated cinnabar structures of HgSe, HgTe, and CdTe (Section II.9). In view of the wide range of variations from the prototype cinnabar, HgS, Mujica *et al.* (1998) suggest that it would be preferable to refer to all of these as the $P3_121$ structure — or $P6_222$ in the higher symmetry case. There are advantages and disadvantages to this, and we leave the matter to further debate, which — as already remarked — needs to consider the whole matter of a more systematic approach to the naming and description of structures (Section II.1). Whatever the eventual conclusions, the known cinnabar types of structure have become considerably more various and central to the overall systematics.

It had appeared until recently that not only tetragonal but also hexagonal symmetry would prove to be rare in the site-ordered III–V and II–VI systems, with the recognition that proposed simple hexagonal structures must in fact have orthorhombic symmetry (Section II.8). However, the discovery of the NiAs structure in AlP and AlAs, and the growing number of cinnabar phases, has had the effect of *increasing* the representation of hexagonal structures in the overall systematics. The discovery of the NiAs phases is also interesting and significant in introducing a different behavior into systematics that otherwise appear quite uniform across all the III–V's and II–VI's once the NaCl phase is reached. It would now be of great interest to follow the transitions of AlP and AlAs to higher pressures.

The reinterpretation of all the supposed β-tin phases as Cmcm, the discovery of this structure in many other systems, and the wider occurrence of cinnabar have, as said, radically altered the nature of the systematics — from a picture of (mostly) a simple progression through high-symmetry structures of increasing coordination, to one in which *all* systems transform into complex, lower-symmetry structures. It is to be expected that there will be a return to simple high-symmetry structures in all cases at sufficiently high pressure (as already established in GaSb, InSb, and HgTe), but it is clear that there is much more complexity than ever previously suspected in the intermediate range. The lower-symmetry structures all have variable atomic coordinates, and hence the nature of the structure can vary with pressure. This is a large qualitative change that makes the structural study of the semiconductors under pressure considerably more interesting and challenging. The solution of lower-symmetry structures, and the accurate determination of variable atomic coordinates, is much more demanding of diffraction studies than is the identification of a high-symmetry structure — as discussed further later. And it is a challenge for computational work to achieve the precision to predict atomic coordinates and discriminate between different lower-symmetry structures.

Another aspect of increased complexity has come from the discovery of intermediate phases in InSb (Section III.16), some phases found only on pressure decrease in GaAs (Section III.11), and "hidden" phases in HgSe and HgTe (Sections III.27 and III.28). The role of the intermediate phases in the transitions of InSb remains a mystery as yet. No completely clear evidence has been found so far of similar behavior in other systems, except — perhaps — InAs, as discussed in Section III.15. Only a few systems have yet been studied in detail on pressure decrease, and other behavior like that of GaAs may well be found (such as the cinnabar phase predicted in GaP by Mujica *et al.* (1998)). The "hidden" distorted zincblende phases in HgSe and HgTe are an intriguing addition to the known structures, suggesting a possible underlying sequence of steps from zincblende, first through distorted zincblende and then distorted NaCl (cinnabar) to NaCl, and then on to another distorted NaCl form (Cmcm). It may possibly be fruitful to see the similarities and differences among all the individual systems in terms of varying degrees of expression of such a sequence. In any case, the wide occurrence of the Cmcm structure, and the cinnabar phase in GaAs, lead to a far greater similarity of systematics between the III–V's and II–VI's than was previously suspected.

The work of the past few years has clarified the long-uncertain matter of site ordering in the III–V and II–VI systems. Most phases have been shown to be ordered, and the exceptions to this are interesting. The known examples are the Cmcm phase of GaP (and possibly of GaAs before heating), the Imma phases of GaSb and InSb, the β-tin-like P2 phase of InSb, and the bcc phases of GaSb, InSb, and HgTe. First, it is clear that site disordering is found more in the III–V's, and this might be related to the observation that the III–V's generally form poorer powder patterns in their high-pressure phases — with more peak broadening — than the II–VI's. As discussed in Section II.1, diffraction studies cannot determine the length scale of site ordering when difference reflections are absent, except to establish an upper bound relative to the crystallite size. It may be that some of these structures are uniformly site-ordered on the length scale of several unit cells (Crain *et al.* (1993) estimate ~10 in the P2 phase of InSb — see Section II.1), and that the smaller crystallite size and tendency to disorder simply reflects an intrinsically greater difficulty in making structural transformations coherently over macroscopic length scales in the more covalent systems (Crain *et al.*, 1994), rather than random disordering. On the other hand, there are also some indications that any ordering may be intrinsically very short range in all cases. One comes from the Imma phase of GaSb. It can be seen in the inset to Fig. 3b that the (110) reflection is quite strong for GaSb in a β-tin-like structure (such as Imma) — simply because Ga and Sb are quite different in atomic number. But in the diffraction pattern of Fig. 7 the (110) reflection is *completely* absent; it is at least a factor of 15 weaker than it would be in a long-range site-ordered structure, and an effect of this magnitude implies that any ordering can be only on the length scale of about two unit cells (or less). If the length scale of any ordering is not intrinsically very short, why is it that cases are never seen — in Imma GaSb or any of the other disordered phases — in which domain sizes happen to be large enough to give detectable, if broadened, difference reflections? These reflections are always completely absent in data from all of the many samples studied to date. Another, different indication of intrinsically short range in any ordering comes from the evidence that site-ordered binary systems do not form the β-tin structure. As rehearsed earlier, it appears so far that β-tin may be a preserve of monatomic systems for some reason, and the occurrence of the β-tin-like Imma phase of Si and Ge in GaSb and InSb (and the β-tin-like P2 phase of InSb) would then suggest that they are disordered on a short enough length scale to make them behave as quasi-monatomic systems. If this is the case, it is of course of considerable interest show it to be so and to establish what the required length scale for such behavior is. EXAFS can in principle distinguish site ordering from site disordering at nearest-neighbor distances, and it would be a particularly interesting experimental challenge to attempt that for, say, Imma GaSb. And there is also a challenge for computational studies in modeling a randomly site-disordered system. The outcome would illuminate further the question as to why site disordering is found more frequently in the III–V systems. And if the β-tin-like phases were shown to be site-ordered on *any* length scale, that would add a new twist to the puzzle of the absence of long-range site-ordered diatomic β-tin.

A general effect of site ordering is to be seen in the many differences now apparent between the group IV systems, and the III–V's and II–VI's. Apart from the diamond and zincblende structures, there are as yet *no other* structural arrangements common to the group IV systems and the site-ordered III–V's and II–VI's. That this divergence of behavior arises from the site ordering of the binary systems appears clear from the converse *similarities* shown — uniquely — by the site-disordered, quasi-monatomic Imma and β-tin-like phases in GaSb and InSb. Some similarities of a wholly different and unexpected kind have emerged in the observation of a high–low cristobalite-type transition in HgSe and HgTe (Section II.11), the close similarity of the Si positions in the keatite polymorph of SiO_2 to the ST12 structure of Ge (Section II.13), and the observation that the structural coordinates in the cinnabar phases of ZnTe and GaAs are similar to those of the Al and P atoms in berlinite — the binary equivalent of quartz. These results suggest that the structures of SiO_2 and its polymorphs, and indeed other tetrahedrally bonded networks such as are found in the polymorphs of H_2O ice, may provide a fruitful source of structural models for other high-pressure or metastable phases of the tetrahedrally coordinated semiconductors.

The changes to all previous pictures of the structural systematics are so substantial that the questions naturally arise as to why, and whether what has been discovered in the more recent work is even the same behavior. The answer to the latter question is emphatically that it *is* the same. In nearly all cases, the evidence for the revised interpretations can be seen in previously published data—a host of neglected weak peaks, shoulders and other asymmetries, and misfits. In fact, the room-temperature behavior of the semiconductors appears to be remarkably robust: For example, the powder patterns described for β-tin Si by Jamieson (1963), with a dry sample held in an amorphous-boron annulus between tungsten-carbide pistons, show precisely the same very strong preferred-orientation effects as are obtained today when loaded with methanol:ethanol in a diamond-anvil cell. Present work is yielding the same behavior and diffraction patterns, but often (not always) with much better signal-to-noise. And analysis is much more stringent. Many of the d-spacing misfits that were previously taken to be acceptable would now be worse than a poor fit; *they would prevent a Rietveld analysis ever being started.* A good example of all this is the study of InSb by Nelmes *et al.* (1993) in which the results are (i) very different from all previous interpretations and (ii) entirely consistent with all previous data — as the authors show in detail.

The results summarized in Table I are thus believed to represent the true (as so far best determined), reproducible behavior of the group IV, III–V, and II–VI systems, as it always has been when samples are compressed in a diamond-anvil pressure cell at room temperature with a conventional pressure-transmitting medium (methanol:ethanol, water, etc.) or without a medium — except that the absence of a medium alters the transition pressure in some cases. Very recent work in which samples have been heated to anneal them, or have been compressed under highly hydrostatic conditions with a gas medium, have indicated some different behaviors. As discussed in Section III.11, moderate heating of Cmcm-GaAs sharpens the powder pattern considerably and reveals reflections that show the structure to be site-ordered. The as yet unanswered question is whether Cmcm-GaAs as formed at room temperature without any heating is the same phase — but gives too broadened a diffraction pattern to allow those reflections to be detected — or is in fact a different, site-disordered phase. In the case of GaSb, we have found that similar modest heating (to ~470 K) of the Imma phase clearly *does* bring about a transition, to a site-disordered β-tin phase. In some other new work, Mezouar *et al.* (1997) have made a careful comparative study of GaSb samples compressed with and without nitrogen gas as a medium and heated under pressure in a large-volume pressure cell, and they have concluded that the Imma phase is stabilized by strain. It will be of considerable interest and importance to see whether other such cases emerge. There is significant evidence that strain more commonly alters transition pressures than the structures formed, but further work clearly needs to be done. (It is of considerable interest in relation to the above discussion of the absence of site-ordered β-tin structures in the III–V and II–VI systems to note that, even when produced under conditions favorable to achieving complete equilibrium, β-tin GaSb remains site-disordered. Both in our experiments and in those of Mezouar *et al.* (1997), the β-tin powder patterns are considerably sharper than those of the Imma phase (Fig. 7), and the fact that the (110) difference reflection is still completely absent strengthens the conclusion that any site ordering cannot extend over a length scale of much more than about two unit cells, or less. What, then, is the true equilibrium structure — which is presumably long-range site-ordered — β-tin, or a different structure altogether?)

The established systematics, to which the various foregoing revisions and discussions apply, mostly concern the transformation from fourfold coordination to approximately sixfold coordination in the III–V's and II–VI's. Little has been determined before of their evolution to eightfold and higher coordination — unlike the group IV systems. One of the significant outcomes of the new work has been some first glimpses of this next major step. In some cases it is apparently simple: GaSb, InSb, and HgTe all transform from their NaCl-like (Cmcm and Immm) or β-tin-like (Imma) phases directly to site-disordered bcc. But AlSb, ZnSe, ZnTe, CdSe, and CdTe all show clear evidence of transforming from Cmcm to even more complex structures (Sections III.8, III.19, III.20, III.23, and III.24). Whether this will prove to be simply a more complex NaCl-like phase prior to an abrupt transition back to higher symmetry, or — following the sequence zincblende, distorted zincblende, distorted NaCl, NaCl, distorted NaCl — something closer to distorted CsCl, remains to be discovered. Either way, it will open up an entirely new chapter in the structural systematics and is an exciting prospect. However, it is already clear that there are sample problems to overcome before the structures are likely to be solved. In some cases there appear to be significant *d*-spacing displacements, and in others the present data quality is probably insufficient in relation to the apparent structural complexity (Figs. 15 and 25). In the meantime, the only other clue to possible higher pressure structures are calculations showing that Cmcm is predicted to transform to Immm in InP and InAs (Mujica and Needs, 1997). It is interesting to note that in the only case where the structure of a non-bcc post-Cmcm phase is known — in InSb — it is indeed Immm. However, Immm will not account for the patterns obtained for the post-Cmcm phases in AlSb, ZnSe, ZnTe, CdSe, and CdTe. Also, the behavior of InSb is unusual in so many respects, and may also be atypical in this; in any case, Immm does not take the story beyond the NaCl-like stage.

The problems that presently stand in the way of progress on the post-Cmcm structures are representative of a general difficulty for structure solution and refinement from high-pressure powder diffraction data. The results obtained with the more sensitive techniques of recent years have revealed how common it is that the effects of pressure are to produce relative structural complexity in the range of the first few tens of gigapascals. At the same time, it is difficult or impossible with existing techniques to avoid microstructural effects in the samples that broaden reflections and distort d-spacing relationships, thereby reducing the information content of the data — often below the level required to solve the more complex structures. New approaches will be needed to overcome these problems. Heating of samples to anneal them, and possibly the use of gas media, will improve data quality in many cases, as reported earlier for GaAs and GaSb. Even then, it is almost certain that strong microstructural effects will remain in many cases, particularly in the very high pressure range that now looks so interesting. The challenge is to couple improvements to sample quality with a full range of procedures to characterize and model the microstructure, and also to find ways to supplement reduced information content with other sources of information such as $hk\ell$-dependent effects and anomalous dispersion. And it will surely become increasingly desirable and usual for those pursuing structure analysis to use total-energy calculations to guide choices between different possible solutions. We believe the structures given without any question mark in Table I, or other doubts expressed in the text, are correct as far as it is possible to be certain about any experimental result. But the lesson of hindsight is that even quite small misfits may signal a significantly mistaken interpretation, and we will not be surprised if the few we are aware of will lead in due course to some further revisions! However, it is important to stress that the standards of certainty now being applied are *very* high — comparable to those of high-quality ambient-pressure work.

A recurring issue in much of the preceding discussion is the problem of establishing the equilibrium phase — particularly in the conditions just discussed which so evidently strain the sample. For example, why does the diatomic β-tin phase not form, even where it is reproducibly calculated to have a lower energy than the Cmcm phase? What *is* the true site-ordered equilibrium structure of GaSb-II? Why is the SC16 phase of GaAs not formed at room temperature — though its calculated energy is clearly below that of Cmcm and cinnabar? Where else may these phases be the equilibrium form but be unknown because they do not form? And what of possible equilibrium phases that are not known because they have not even been considered for calculation? The answers to some of these questions will depend on kinetics; it is clear from the reconstructive nature of the major transitions, and the large hysteresis that often accompanies them, that there are considerable energy barriers in many cases at room temperature. The transitions and structures observed may then be determined as much by the energetics of transition pathways as by energy minimization of phases. The "hidden" phases of HgSe and HgTe, the several different phases found on pressure decrease in Si, Ge, and GaAs, and the two different sequences observed in InSb, all illustrate this. We simply do not yet know how many of the phases in Table I represent the true equilibrium systematics, and it may well never be possible to discover that fully from room-temperature experiments. As already suggested, accurate energy calculations may assist once they can be dependably related to the actual experimental conditions (e.g., finite temperature). At present it appears that these factors affect the III–V's more than the group IV's and II–IV's, perhaps through the influence of covalency on the degree of structural connectivity expected to be retained through the transformations, coupled with the constraints of avoiding like-atom nearest neighbors in the binary systems (Crain *et al.*, 1994b). To take the role and influence of transition pathways fully into account, much more yet needs to be learned about the *mechanisms* of reconstructive transformations in these systems. It is possible that new information may come in due course from the modeling of microstructural characteristics that is now being developed to assist structure solving and refinement. For example, measurements of $hk\ell$ dependence in powder pattern peak widths give information about anisotropic particle size, and this coupled with modeling of preferred orientation could yield presently unknown information about the absolute crystallographic relationship between low and high-pressure phases.

The use of sample heating for the purposes of annealing is but a small part of a large new area of work now needed and starting, namely the study of the full $P-T$ dependence of the structural systematics. It cannot be stressed too much how very little is yet known away from room temperature. There is almost no information at all at low temperature — but, given the problems with kinetics at room temperature, the general value of low-temperature studies perhaps seems doubtful. Only for GaAs, InSb, ZnTe, CdTe, and HgTe are there any extensive high-temperature results and a reasonably complete $P-T$ phase diagram. The small amount yet known suggests that systematic high-temperature studies under pressure will prove richly rewarding and informative.

REFERENCES

Crain, J., Ackland, G. J., Piltz, R. O., and Hatton, P. D. (1993). *Phys. Rev. Lett.* **70**, 814.
Crain, J., Piltz, R. O., Ackland, G. J., Clark, S. J., Payne, M. C., Milman, V., Lin, J. S., Hatton, P. D., and Nam, Y. H. (1994). *Phys. Rev. B* **50**, 8389.
Jamieson, J. C. (1963). *Science* **139**, 762.
McMahon, M. I., Nelmes, R. J., Wright, N. G., and Allan, D. R. (1993). *Phys. Rev. B* **48**, 16246.
Mezouar, M., Libotte, H., Deputier, S., Le Bihan, T., and Häusermann D. (1997). In preparation.
Mujica, A., and Needs, R. J. (1997). *Phys. Rev. B* **55**, 9659; *Phys. Rev. B* **56**, 12653E (Erratum).
Mujica, A., Needs, R. J., and Muñoz, A. (1995). *Phys. Rev. B.* **52**, 8881.
Mujica, A., Muñoz, A., and Needs, R. J. (1998). *Phys. Rev. B* **57**, 1344.
Nelmes, R. J., McMahon, M. I., Hatton, P. D., Piltz, R. O., and Crain, J. (1993). *Phys. Rev. B* **47**, 35; *Phys. Rev. B* **48**, 9949 (Corrigendum).
Nelmes, R. J., McMahon, M. I., and Belmonte, S. A. (1997). *Phys. Rev. Lett.* **79**, 3668.
San-Miguel, A., Wright, N. G., McMahon, M. I., and Nelmes, R. J. (1995). *Phys. Rev. B* **51**, 8731.
Zhang, S. B., Erskine, D., Cohen, M. L., and Yu, P. Y. (1989). *Solid State Commun.* **71**, 369.

V. Concluding Remarks

After 30 years of study to the early 1990s, it had seemed that the structural systematics of the group IV, III–V, and II–VI semiconductors were largely known and understood. We have tried to show in this review how greatly the results of improved experimental techniques have altered this picture in the past few years, stimulating a new surge of theoretical interest and revealing a great deal of exciting and informative experimental work yet to be done.

We suggest the following as some of the more significant questions and challenges identified in Section IV. Why is the diatomic β-tin structure not found? Are there other cinnabar phases, such as those proposed from calculations in ZnS, ZnSe and GaP — perhaps at high temperature? What do the NiAs phases of AlP and AlAs transform to at higher pressures? Are there any other examples of the distorted-zincblende "hidden" phase of HgSe and HgTe? A theoretical study of this behavior would seem worthwhile. Can EXAFS studies distinguish site disordering from site ordering at nearest-neighbor distances in the Imma phase of GaSb? Why are site-disordered phases formed in the particular systems that they are; what is the length scale of any ordering; what is the significance of quasi-monatomic behavior; and can calculations be carried out to illuminate the answers in these questions? Why do the III–V systems exhibit much more pronounced microstructural effects — such as peak broadening — than the II–VI systems on passing through high-pressure transitions? What is the nature of the post-Cmcm structures and transitions? Among other possibilities, are transitions to Immm seen in systems other than InSb? Can more details be determined of the transition pathways? It now seems clear that experimental studies and theoretical calculations need to be brought to a meeting point, preferably at high temperature. In this way the issue of true equilibrium may be more reliably and completely addressed. And more extensive high-temperature studies are now required to improve sample quality and overcome kinetic effects. They are also needed to extend knowledge of the systematics to high temperature and map out $P-T$ phase diagrams for more of the systems.

Finally, we note that the review contains a large amount of factual and numerical information. Despite our best efforts, there are likely to be a few errors. We would welcome corrections.

Acknowledgments

We gratefully acknowledge the substantial contribution to the work at SRS Daresbury by all our colleagues and co-workers especially D. R. Allan, N. G. Wright, H. Liu, S. A. Belmonte, J. S. Loveday, and A. San-Miguel. For access to their work prior to publication, we thank D. Martinez-Garcia and colleagues; A. A. Kelsey, G. J. Ackland, and S. J. Clark; U. Schwarz, K. Syassen, and colleagues; A. Mujica, A. Muñoz, and R. J. Needs; and M. Mezouar and colleagues. The SRS program is supported by research grants from the Engineering and Physical Sciences Research Council (EPSRC) and also the Royal Society, by funding from the Council for the Central Laboratory of the Research Councils (CCLRC), and by facilities made available by Daresbury Laboratory. We thank G. Bushnell-Wye, A. A. Neild, and R. Proc for valued support in constructing and maintaining beamline equipment, D. M. Adams of Diacell Products for much assistance in the development of pressure cells for this program, and A. Polian, J.-P. Itié, and J. M. Besson for assistance with samples and pressure cell loading and for much helpful discussion of experimental results. Our special thanks are due to H. Iwasaki, Y. Fujii, O. Shimomura, K. Takemura, T. Kikegawa, H. Fujihisa, and their colleagues, for generous sharing of their image-plate techniques on which the SRS facilities are based. The initial and continuing development of these facilities and their subsequent use has been greatly aided by an EPSRC Senior Fellowship and a Senior Visiting Fellowship at CLRC (RJN), and by a Royal Society Research Fellowship (MIM). We have benefited greatly from stimulating discussions with those carrying out associated computational studies, particularly A. Mujica, R. J. Needs, and G. J. Ackland. Finally, we wish to acknowledge our indebtedness to all those who charted the territory, particularly A. L. Ruoff and his colleagues.

CHAPTER 4

Optical Properties of Semiconductors under Pressure

A. R. Goñi

INSTITUT FÜR FESTKÖRPERPHYSIK
TECHNISCHE UNIVERSITÄT BERLIN
BERLIN, GERMANY

K. Syassen

MAX PLANCK-INSTITUT FÜR FESTKÖRPERFORSCHUNG
STUTTGART, GERMANY

I. INTRODUCTION	248
II. EXPERIMENTAL ASPECTS	251
1. Diamond Anvil Cells	251
2. Pressure Medium	253
3. Pressure Measurement	254
4. Optical Spectroscopy at High Pressure	256
III. ELECTRONIC BAND STRUCTURE UNDER PRESSURE	259
1. Linear Combinations of Atomic Orbitals and Tight-Binding Method	260
2. Penn Model and the Dielectric Theory of the Covalent Bond	264
3. The $\vec{k}\cdot\vec{p}$ Method	267
4. Uniaxial Stress Effects	269
5. Quantum Well Structures	271
IV. REFRACTIVE INDEX DISPERSION	273
1. Refractive Index and Optical Dispersion	274
2. Volume Dependence of the Low-Frequency Dielectric Function	277
3. Method of Optical Interferences	279
4. Experimental Results	282
V. OPTICAL ABSORPTION NEAR THE FUNDAMENTAL GAP	285
1. Absorption Spectral Functions	285
2. Results and Discussion	289
VI. EFFECT OF PRESSURE ON EXCITON ABSORPTION	297
1. Energy Spectrum of Excitons	298
2. Exciton Absorption	303
3. Pressure Effects on Excitons	307
VII. PHOTOLUMINESCENCE STUDIES UNDER PRESSURE	321
1. Optical Emission in Semiconductors	321

 2. Pressure Effects on PL Emission in Bulk Materials 327
 3. Low-dimensional Structures under Pressure: PL Studies 333
VIII. PHOTOMODULATED REFLECTANCE.. 342
 1. Photomodulated Optical Response 342
 2. Results for Bulk Materials... 345
 3. Quantum Wells and Epilayers 346
IX. OPTICAL PROPERTIES OF ELECTRON GASES UNDER PRESSURE............. 351
 1. Elementary Excitations of the Electron Gas.................................. 352
 2. Inelastic Light Scattering by Elementary Excitations of the Electron Gas..... 358
 3. LO-Phonon–Plasmon Coupled Modes in Bulk Semiconductors.... 363
 4. Elementary Excitations of the 2-D Electron Gas Under Pressure.... 370
 5. Electron–Electron Interactions in Double-Layer 2-D Electron Gases 377
X. HIGH-PRESSURE PHASES.. 385
 1. General Remarks... 385
 2. Results for Polar Compounds... 386
 3. Results for the Elemental Semiconductors 391
A. APPENDIX: LINEAR OPTICAL RESPONSE OF SOLIDS 392
 A1. Electromagnetic Waves in a Medium 394
 A2. Interband Absorption... 397
 A3. Reflectance and Transmittance.. 400
 A4. Lorentz–Drude Oscillators 402
 A5. Dispersion Relation and Sum Rules 405
 A6. Analysis of High-Pressure Reflectance Spectra 407
 Acknowledgments 409
 References 410

I. Introduction

The effect of a large hydrostatic pressure on semiconductor optical properties can be best illustrated with a simple experiment, namely, the optical inspection with a microscope of light transmitted through a piece of GaAs as it is being pressurized in a diamond anvil cell (DAC). Figure 1 shows microphotographs taken at different pressures. As the pressure increases a striking change in color of the GaAs is observed,[1] from opaque to red and finally orange-yellow. At around 17 GPa the sample becomes opaque again because of the occurrence of a structural phase transition to a metallic high-pressure phase. In the stability range of the semiconducting phase all changes are reversible and are just a result of a large increase of the lowest direct bandgap with pressure from 1.43 eV at ambient to 2.8 eV at 17 GPa.

The advantages of pressure for studying semiconducting materials have been realized shortly after Bridgman's electrical transport studies of Si and Ge [1]. High-pressure optical investigations of tetrahedrally coordinated semiconductors were pioneered 40 years ago by Paul and co-workers [2–4]

---

[1] Due to cost limitations, this figure appears as black/white. For color prints, please write to syassen@servix.mpi-stuttgart.mpg.de.

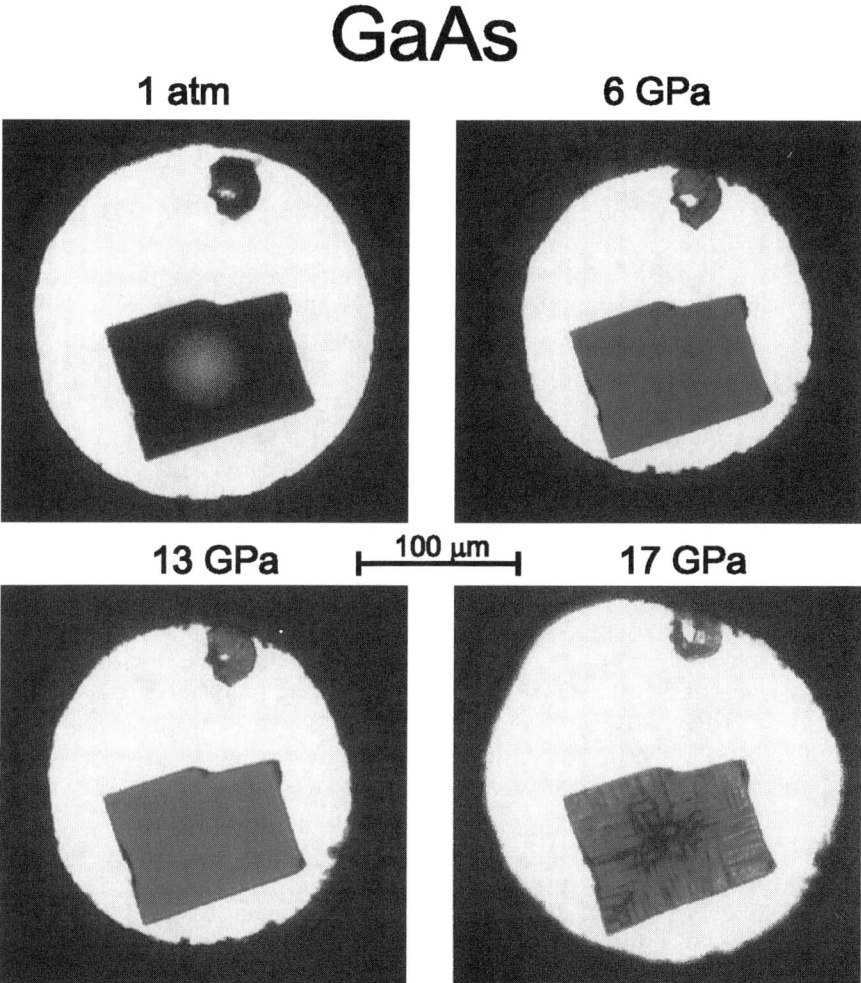

FIG. 1. Microphotographs of a GaAs sample in the gasketed diamond anvil cell taken at different pressures. The round spot in the upper left photograph indicates the typical size of a laser beam focused onto the sample. At 17 GPa a small fraction of the sample (opaque regions) has started to transform to a high-pressure metallic phase.

and by Drickamer's group [5–7]. In the period up to 1970 the systematic studies of optical properties under hydrostatic pressure (mostly below 1 GPa) [8, 9], under uniaxial stress [10, 11], and by modulation spectroscopy [12] had led to a fairly complete picture of the electronic band structure of semiconductors and its dependence on homogeneous strain. The appearance of the dielectric theory [13–16] marks the point where systematic trends in the experimental observations were understood, at least at a semiempirical level. In parallel, the basic effects of pressure on the lattice dynamics of semiconductors had also been explored [17, 18]. The development of the gasketed DAC into a tool for optical measurements at hydrostatic pressures well above 10 GPa [19–24] has greatly reduced the experimental complexity and, through the enlarged pressure range and easy operation at cryogenic temperatures, has opened many new possibilities for studying the physics of semiconductors under pressure.

Basically, high pressure allows for a controlled reduction of interatomic distances ($\approx$3% in GaAs at 10 GPa) and a continuous modification of the chemical bonding. On the other hand, the interatomic distance is also the basic input parameter in most theoretical models of solid-state properties. For this reason there exists a close link between theory and high-pressure experiments in general. Moreover, many of the conventional semiconductors belong to those materials for which the one-electron theories can provide an appropriate description. Thus, the optical properties of semiconductors under pressure are a very popular testing ground for theoretical models of the electronic excitation spectrum.

The application of pressure also plays a major role in optical investigations of electronic states in artificial semiconductor structures such as quantum wells and superlattices. This is partly because the relative ordering of energy levels at the interface between the different materials of a semiconductor heterostructure can be tuned continuously for a given sample. This significantly reduces the need for a systematic modification of the electronic structure through chemical composition and geometrical design parameters.

The present review mainly deals with tetrahedral semiconductors and their layered heterostructures. We begin by giving a brief account of experimental aspects (Section II), followed by a summary of the basic band-structure background needed for the discussion of pressure effects in tetrahedral semiconductors (Section III). We then treat the effect of pressure on the low-frequency electronic dielectric constant (Section IV) and the optical absorption near the fundamental bandgap (Section V). There follows a discussion of excitonic absorption in bulk materials and two-dimensional systems (Section VI). Typical results obtained by photoluminescence spectroscopy and photoreflectance measurements are presented in Sections VII and VIII. Inelastic light-scattering studies on electron gases in 3-D and 2-D systems are discussed in some detail in Section IX. We conclude with a brief look at the optical properties of high-pressure phases of tetrahedral semiconductors (Section X). Throughout the text we refer to basic relations on the optical response of solids. The Appendix presents a condensed summary of such relations.

We have limited the discussion to hydrostatic pressure effects. The tuning of biaxial strain through the application of hydrostatic pressure comes into play here and we have included a very brief account of nonhydrostatic stress effects on electronic states in semiconductors. For more extensive reviews of uniaxial stress effects in semiconductors we refer to Refs. [25, 26]. There has been a growing activity in the field over the past decade or so. Throughout we have tried to include a few key references concerning topics which are not covered in this article but are part of the published literature. A source of additional information is provided by several collections of papers devoted to high-pressure semiconductor research [27–31].

II. Experimental Aspects

A large body of experimental information on electronic and vibrational properties of semiconductors under hydrostatic pressure was obtained by using "large-volume" cells with optical windows operating at pressures up to about 1 GPa. The term "large volume' refers to pressure chambers accommodating crystals several millimeters in size. These techniques continue to play an important role in high-pressure semiconductor research, but they are nowadays used mostly in studies of electrical transport phenomena and electroluminescence devices. The development of the gasketed diamond anvil cell [19, 20] in combination with the ruby pressure sensor [21, 23] and hydrostatic pressure media [22] including condensed rare gases [32, 33] has revolutionized high-pressure research in general and has opened new possibilities for optical semiconductor research under pressure. The principle of the diamond anvil cell and various DAC constructions have been reviewed extensively (see, e.g., Refs. [34–37]). Here, we address a few points relevant in optical spectroscopies of semiconducting materials.

1. DIAMOND ANVIL CELLS

The heart of a DAC consists of a pair of diamonds and a metal gasket as shown in Fig. 2. A hole in the gasket forms the pressure chamber into which the sample is placed together with a small chip of ruby for pressure measurement. The gasket hole is filled with a suitable hydrostatic medium, for instance, a 4:1 mixture of methanol/ethanol for room-temperature experiments or condensed helium for low-temperature studies. The pressure range of interest in the experimental investigation of semiconducting properties is determined by the phase stability region of the ambient pres-

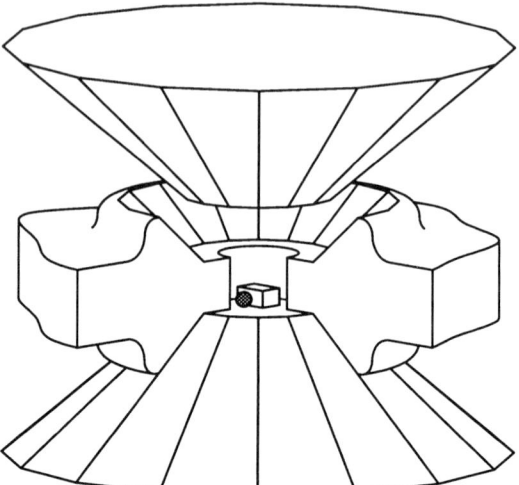

FIG. 2. The center parts of a diamond anvil high-pressure cell: diamond anvil tips and preindented gasket with sample and ruby chip placed in the cylindrical hole.

sure modification. Most semiconductors transform to denser, inevitably polycrystalline phases at pressures below 20 GPa. For experiments in this pressure range the typical dimensions are 500 to 600 μm diameter for the diamond tip face, a hole diameter of 200 μm, and a starting gasket thickness around 80 to 100 μm. At pressures near 10 GPa the gasket thickness becomes reduced to 40–50 μm, the exact value depending on the gasket material.

The diamonds are often selected as type IIa quality in order to minimize absorption in the visible and near UV spectral range up to the intrinsic diamond optical absorption edge, which is at 5.3 eV at zero pressure. The red shift of the optical absorption edge of diamond anvils under compressive stress is small [38–40] and is irrelevant at pressures below 20 GPa. For polarized optical measurements the diamonds are selected for low birefringence. A situation encountered in light-scattering experiments using laser excitation is that the scattered light intensity of the sample is quite weak and the extrinsic luminescence emitted by the diamond windows becomes a disturbing factor. In this case, the diamonds need to be selected for very low luminescence from a batch of stones preclassified as low-luminescence-grade. An overview of extrinsic luminescence properties of diamond is given in Refs. [41, 42].

Samples for high-pressure experiments in the DAC are prepared by mechanical polishing into the shape of a plane-parallel plate with thickness

4 OPTICAL PROPERTIES OF SEMICONDUCTORS UNDER PRESSURE

ranging from 5 to 30 μm. Defect formation induced by the polishing procedure may lead to a degradation of sharp spectral features in optical emission or absorption as well as broadening of Raman lines. In order to avoid defect formation the final polishing is done with the smallest possible grain size, followed by a Siton etching if suitable for the particular material of interest.

A large variety of DAC constructions have been reported which meet the needs of different optical spectroscopy methods [36]. The DAC version shown in Fig. 3 is similar to that developed by Syassen and Holzapfel [43], but reduced in size in order to fit into a cryostat with 50-mm diameter central tube. The force acting on the diamonds is generated via a symmetrical lever-screw mechanism, which can be operated from outside the cryostat, so that pressure can be charged without removing the DAC from the cryostat.

2. PRESSURE MEDIUM

Hydrostatic pressure conditions are essential for low-temperature optical studies because electronic levels of semiconductors are quite sensitive to uniaxial stresses. On the other hand, all pressure media are solid at 10 K, and truly hydrostatic conditions are not possible in a solid medium. In fact, with the solid rare gases Ar and Xe, pronounced nonhydrostatic effects

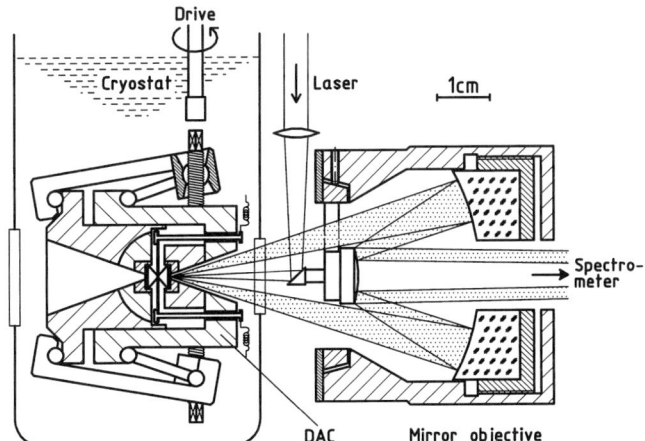

FIG. 3. A diamond anvil high-pressure cell (DAC) mounted inside a cryostat. In this version of the DAC, the force acting on the diamonds is generated via a lever-screw mechanism, which can be operated from outside the cryostat. In Raman and photoluminescence studies, a Schwarzschild-type microscope objective is used to collect light emitted by the sample. See text for details.

are observed at pressures less than 3 GPa [44, 45]. In connection with PL measurements of the exciton transitions in Cu_2O under pressure [46, 47], it was demonstrated [45] that the uniaxial stress generated in He at 10 K and at pressures around 1 GPa is less than 0.01 GPa. However, with He as medium a small pressure change performed at low temperature at about 6 GPa can result in the generation of uniaxial stresses on the order of 0.1 GPa [45]. Thus, even if He is used as a pressure medium, it may be necessary in high-resolution studies, for instance of exciton transitions or phonon linewidths, to warm up to temperatures above the melting line of He [48, 49] before changing the pressure. The melting temperature T_m is approximately given by the empirical relation [49]

$$T_m = 61 P^{0.639}, \qquad (2.1)$$

where temperature is in kelvins and pressure in gigapascals. It should be noted that free-standing semiconductor layers of thickness less than 1 μm always appear to be subjected to some uniaxial stress right at the solidification point of He. This effect has, for instance, been observed in exciton absorption studies of thin ZnTe layers [50, 51]. A strain-induced splitting of exciton lines is not observed if the sample thickness is larger than 1 μm [50].

The DAC can be filled with He using a high-pressure gas compression system at room temperature or the cryogenic loading procedure. In the latter case the sample and the ruby are placed in the DAC, leaving a small gap between gasket and diamond anvil. The DAC is inserted in the cryostat, flooded with superfluid He, and then the cell is closed, thereby trapping the He inside the gasket hole at an initial pressure of less than 1 GPa.

The temperature cycle involved in the pressure change is time consuming if the whole DAC is heated. For this purpose a local heating is advantageous. This is realized in the DAC shown in Fig. 3. The gasket is formed into a strip a few millimeters in width. With a current of about 10 A passed through the gasket, a local temperature increase of about 150° is obtained, provided the diamonds are mounted on ceramic backing plates in order to reduce heat losses.

3. Pressure Measurement

Pressure in a DAC is commonly measured by the ruby luminescence method. The initial calibration [23] of the red shift of the R_1 and R_2 lines as a function of pressure has been modified slightly in later studies [52–54]. According to Ref. [53] the empirical relation between pressure in GPa and wave number ν is

$$P = 248.4[(\nu_0/\nu)^{7.665} - 1]. \qquad (2.2)$$

The zero-pressure wave number ν_0 varies slightly (± 1 cm^{-1}) for different samples, depending on Cr concentration and internal strain.

It is usually assumed that pressure- and temperature-induced R-line shifts are independent of each other. There is also experimental evidence for the validity of this assumption [55–58]. Thus, given the temperature dependence of the R-line frequency at zero pressure, the preceding pressure calibration [Eq. (2.2)] can be used for pressure measurement at low temperatures. The frequency shift $\nu(T)$ of the R-lines with temperature was first measured by McCumber and Sturge [59] and has been studied repeatedly thereafter (see [60] and references therein). Figure 4 shows $\nu(T)$ data for the two R lines measured from 4 K to about 300 K [61]. The temperature-induced shift has been attributed to a coupling of the Cr^{3+} crystal field levels to acoustic phonons [59]. Empirical $\nu(T)$ relations have been proposed (see

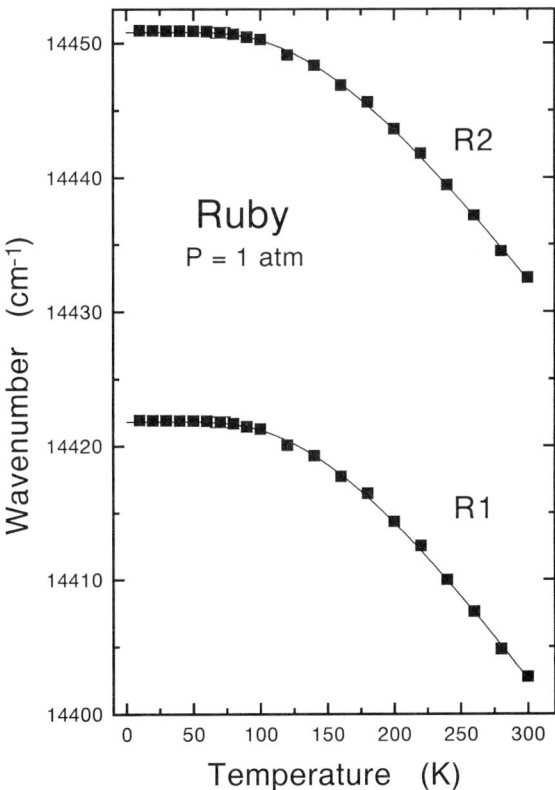

FIG. 4. Energies of the ruby R lines as a function of temperature. The solid lines correspond to the result of a least-squares fit using Eq. (2.3). After Ref. [61].

again [60] and references therein). We like to point out here that the temperature dependence of the R-lines can be well represented by a semi-empirical relation proposed by Viña et al. [62] to describe the temperature-dependent shift of semiconductor bandgaps:

$$\nu(T) = \nu_0(T = 0 \text{ K}) - \frac{\alpha}{\exp(\Theta/T) - 1}. \quad (2.3)$$

The temperature Θ characterizes the phonon dispersion and α is an electron–phonon coupling parameter. The parameters of the fitted curves in Fig. 4 are as follows:

	ν_0 [cm^{-1}]	α [cm^{-1}]	Θ [K]
R1	14421.8 ± 0.4	76.6 ± 6.9	482 ± 20
R2	14450.8 ± 0.4	72.4 ± 6.5	478 ± 20

The effect of temperature on the R-line frequencies near room temperature is about -0.14 cm^{-1}/K. A temperature increase of 5° has about the same effect as an increase in hydrostatic pressure by 0.1 GPa (-0.76 cm^{-1}). Therefore, the local temperature needs to be controlled in pressure experiments, if very accurate pressure determination by the ruby luminescence method is important.

We note that pressure sensors based on Sm-doped materials have been proposed, which show a larger shift with pressure of $4f$ emission lines as compared to the R lines of ruby [63]. Even higher resolution in pressure may be obtained by using the established bandgap shift of a semiconductor material such as GaAs to measure pressure.

4. Optical Spectroscopy at High Pressure

The experimental configuration depicted in Fig. 3 is typical for low-temperature photoluminescence (PL) and inelastic light-scattering experiments. A Schwarzchild-type microscope objective is shown as the light-collecting optical element in Fig. 3. The main advantage compared to lens optics is that the working distance of the mirror objective is about a factor of 2 larger than its focal length. For a given focal length this leaves extra space for cryostat windows, mirrors, etc., between sample and collecting optics. The absence of chromatic aberrations in mirror objectives is not an essential feature in most light-scattering experiments in the visible, because a relatively narrow spectral range is covered. It can, however, be advantageous in situations where alignment procedures are performed in the visible range and the actual measurements need to be done in the infrared or UV range.

In PL and Raman measurements the laser spot on the sample is typically 30 μm in diameter. The developments in recent years of multichannel optical detection systems has resulted in an enhanced sensitivity compared to conventional scanning spectrometers. As a result, it is often sufficient to use laser powers of 1 to 10 mW falling on the sample inside the DAC. Resonant Raman spectroscopy requires some flexibility in excitation wavelength because of the large shift of bandgaps under pressure, or, alternatively, pressure can be used to tune into resonance conditions.

Absorption (transmittance) measurements often require high resolution *and* the coverage of a wide spectral range. These measurements are therefore frequently done with a conventional scanning-type spectrometer. A sketch of a typical experimental setup for high-pressure optical absorption measurements is shown in Fig. 5. White light from a xenon arc or tungsten halogen lamp illuminates an entrance pinhole. A demagnified image of the pinhole is produced on the sample inside the DAC by an achromatic lens. The spot diameter (~30 μm) has to be chosen considerably smaller than the lateral dimensions of the sample (typically 100 μm) in order to avoid scattered light due to multiple reflections in the pressure cell. The transmitted light is spatially filtered by the field stop placed in the image plane of a collecting objective. The field stop helps to further suppress stray light. Finally, the light is passed through a grating spectrometer and then detected by either a liquid-nitrogen cooled Ge detector (in the wavelength range 0.8–1.7 μm) or a cooled GaAs photomultiplier (230–870 nm). In quantitative measurements of very low transmittances it is important to optimize the stray-light suppression at false wavelengths and to possibly correct for the detector dark signal. For a DAC with dimensions given earlier, the sample thickness can vary from ~1 μm to a maximum of about 50 μm (at 10 GPa). This enables the measurement of absorption coefficients of bulk materials ranging from 5×10^4 cm^{-1} down to about 100 cm^{-1}.

If information on electronic excitations at energies above the lowest

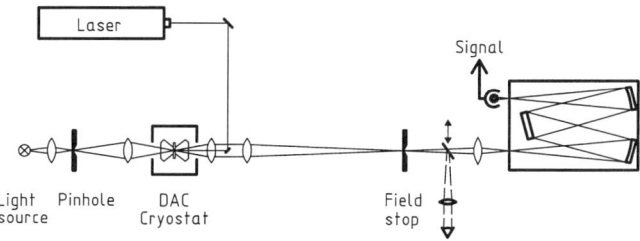

FIG. 5. Block diagram of a typical experimental setup for high-pressure optical absorption measurements. The laser is for determining the pressure *in situ* by the ruby luminescence method.

direct gap cannot be obtained by absorption measurements, photomodulated reflectance (PR) spectroscopy [12, 64] provides an alternative. The method can be considered as the contactless variant of electromodulation spectroscopy. It is particularly useful in investigations of electronic transitions in semiconductor quantum well systems under pressure [65]. Some physical background is given in Section VIII. Here we discuss briefly the experimental aspect.

Shown in Fig. 6 is a schematic drawing of an apparatus for PR measurements in combination with a DAC [61]. A monochromatic probe beam $I_0(\hbar\omega_1)$ of variable energy $\hbar\omega_1$, modulated at frequency f_1, is focused on the sample inside the DAC using a mirror-type objective. The reflected probe light is detected by a photomultiplier or other suitable detector. With the lock-in amplifier tuned to frequency f_1, one would measure a signal $R_1(\hbar\omega_1)$ which is proportional to the sample reflectance. A modulation of the reflectance is achieved by a second light beam (laser at energy $\hbar\omega_2$, modulation frequency f_2) which is focused onto the same spot of the sample. Suitable filtering blocks the laser light from reaching the detector. The detector thus also sees a weak modulated intensity $R_2(\hbar\omega_1) \ll R_1(\hbar\omega_1)$ at frequency f_2 corresponding to the modulation of the sample reflectance. This signal can in principle be measured by tuning the lock-in to the frequency f_2. In practice the lock-in is tuned to the sum frequency $f_1 + f_2$. In this way only the true PR signal is measured, even in the presence of

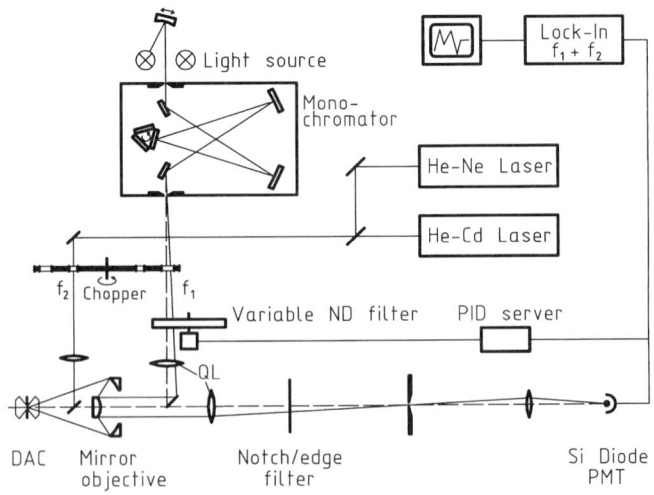

FIG. 6. Schematic drawing of an apparatus for reflectance and photomodulated reflectance spectroscopy under pressure. See text for explanations. After Ref. [61].

laser-excited bandgap luminescence. An automatic normalization has been suggested by Shen *et al.* [66, 67]. For this purpose the dc component of the detector is fed into a servo unit connected to a variable neutral-density filter, which keeps the dc signal constant independent of $\hbar\omega_1$.

Normal reflectance spectroscopy at pressures below 1 GPa has been used in the early period of high-pressure semiconductor research in order to investigate optical transitions at energies above the fundamental gap [8, 9]. Nowadays, it is not one of the standard methods in high-pressure studies of semiconductors in their ambient pressure modifications, but is used in investigations of their high-pressure phases. In this case one is mostly interested in a crude overall picture of the electronic properties, in particular with respect to possible metallic behavior. The optical system shown in Fig. 6 can be used for quantitative absolute reflectance measurements in a DAC. For this purpose all filters are removed and the lock-in is tuned to frequency f_1. The reflectance often is measured under quasi-hydrostatic conditions, where the sample is in direct optical contact with the diamond window and otherwise embedded in a solid medium of low shear strength. In order to determine the absolute reflectance, the intensity reflected at the diamond–sample interface needs to be normalized in such a way that also the weak absorption in the diamond window is taken into account. For this purpose a reference spectrum is measured separately by focusing onto the inner diamond interface of the empty DAC. The absolute reflectance of this diamond–air interface can be calculated from the refractive index dispersion of diamond [68]. The variation of the refractive index of diamond with pressure [40, 69, 70] can be neglected. The optical system in Fig. 6 is designed to provide a narrow field of depth at the sample position (confocal optics). In this way the reflection at the outer diamond surface is not reaching the detector. Because of the narrow field of depth one has to correct for chromatic aberrations in the diamond window, if spectra are measured over a wide spectral range. This is achieved by introducing the two quartz lenses labelled QL in Fig. 6.

Results of nonlinear optical spectroscopies are not covered in this article. For a brief description of experimental configurations we refer to the review by Reimann [45]. For time-resolved DAC experiments on the subpicosecond time scale, see Refs. [71–74].

III. Electronic Band Structure under Pressure

A short account of the effects of high pressure on fundamental semiconductor band structure parameters is almost imperative for the discussion of the pressure dependence of the optical properties, for they are largely determined by the magnitude of bandgaps and the characteristics of the electronic states at valence- and conduction-band extrema. Here we mainly focus on empirical models which provide a simple description of the electronic structure, which are closely related to magnitudes measured by means of optical spectroscopies under hydrostatic pressure.

1. LINEAR COMBINATIONS OF ATOMIC ORBITALS AND TIGHT-BINDING METHOD

A common characteristic of tetrahedrally coordinated semiconductors is a strongly covalent bonding involving hybridization of valence s and p states. The semiconductor band structure is then naturally obtained in a tight-binding approach by linear combinations of such atomic orbitals. Typical crystal structures for the group IV elementary semiconductors and $A^N B^{8-N}$ binary compounds are of the diamond and zincblende or wurtzite type, respectively. The four sp^3 outermost orbitals of an atom form four hybrids oriented along the diagonals of a cube, which mix with the hybrids of the nearest neighbors. This leads, as in the case of a diatomic molecule, to formation of bonding and antibonding orbitals. In the solid these orbitals become broadened into bands, the four of them lowest in energy being filled up with electrons.

The bonding–antibonding energy splittings are at the origin of bandgaps which open up in the band structure of crystals. The *covalent energy* is defined as the expectation value of the interaction between two hybrid orbitals $|h^{c,a}\rangle$ (c ≡ cation, a ≡ anion) of adjacent atoms [75],

$$V_2 = - \langle h^c | \mathcal{H} | h^a \rangle \tag{3.1}$$

$$= \frac{1}{4}\left(-V_{ss\sigma} + \frac{2}{\sqrt{3}} V_{sp\sigma} + 3 V_{pp\sigma}\right), \tag{3.2}$$

which is written in terms of the Coulomb interaction parameters $V_{ll'm}$ between orbitals with angular momentum l, l' and molecular quantum number m. For a polar lattice with two different kind of atoms, the energies of the corresponding hybrids are different, and this difference is called the *polar energy*,

$$V_3 = \frac{\epsilon_h^c - \epsilon_h^a}{2}, \tag{3.3}$$

with

$$\epsilon_h^{c,a} = \frac{\epsilon_s^{c,a} + 3\epsilon_p^{c,a}}{4}, \tag{3.4}$$

where $\epsilon_s^{c,a}$, $\epsilon_p^{c,a}$ are the atomic energy eigenvalues.

In crystals the energies of the electron states are mainly determined by interactions of first nearest neighbors. Thus, we shall describe band states in terms of molecular orbitals constructed using linear combinations of hybrids from adjacent atoms. The symmetric (antisymmetric) combinations lead to bonding (antibonding) molecular orbitals. The dependence of the bond orbital energies on the interatomic distance is sketched in Fig. 7. The equilibrium lattice constant corresponds to the value of d which minimizes the total energy of the crystal. The energy splitting between bonding (B) and antibonding (AB) states is given by

$$\epsilon_{AB} - \epsilon_B = 2\sqrt{V_2^2 + V_3^2}. \tag{3.5}$$

If d decreases with respect to d_0, the energy of the bonding state changes relatively little, being close to its minimum, whereas the antibonding state moves up appreciably in energy. Thus, the B–AB splitting is expected to increase with pressure. The main contribution to the pressure dependence

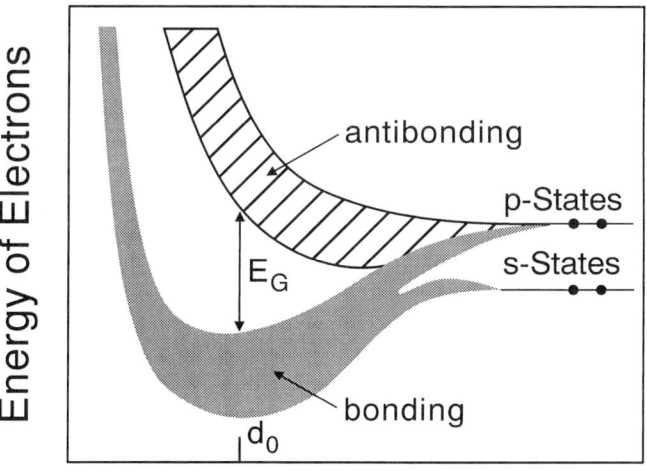

FIG. 7. Development of the atomic energy levels of a group IV element as the interatomic distance decreases. The distance d_0 indicates the equilibrium distance in the solid.

arises from the covalent energy V_2, which is a function of the overlap integrals $V_{ll'm}$ between nearest-neighbor orbitals [see Eq. (3.2)]. The polar energy V_3, in contrast, is essentially independent of pressure, because it is given by the energy difference of atomic levels [14–16].

Within this simple picture we can understand, for instance, the increasing trend of the lowest direct bandgap (E_0) at the Brillouin-zone center for the group IV elements upon reduction of interatomic distance. The top of the valence band for all these materials is formed by Γ_{8v} bonding p states. The E_0 gap is the energy separation between these and the Γ_{6c} conduction band edge states, formed from antibonding s orbitals. As the lattice constant increases in going to the heavier elements, E_0 decreases from 7.1 eV for diamond, an insulator, to ≈ 3.5 and 0.8 eV for Si and Ge, respectively, to collapse for Sn, which is a semimetal [76]. Exactly on the same lines we can explain the positive pressure coefficient of E_0 observed in most tetrahedral semiconductors [16, 77, 78].

A feature of the tight-binding method is that it provides a real space picture of the electronic interactions which give rise to the band structure. Moreover, using only a few interaction parameters it is possible to obtain energy bands and electronic wave functions in the entire Brillouin zone [79–81]. Spin–orbit interactions can also be incorporated into the calculations by including relativistic terms in the Hamiltonian and by an appropriate choice of the parameters [82]. The effect of hydrostatic pressure is introduced in the theory by accounting for the dependence of the overlap parameters on interatomic distance. The scaling rule is of the form [75]

$$V_{ll'm} = \eta_{ll'm} \frac{\hbar^2}{m_0 d^{\alpha_{ll'm}}}, l, l' = s, p, d, \ldots, m = \sigma, \pi, \ldots, \quad (3.6)$$

where $\eta_{ll'm}$ is a geometrical factor which depends on crystal symmetry [83, 75], m_0 is the electron mass, and $\alpha_{ll'm}$ is the exponent which determines the pressure dependence of the matrix element. Harrison [75] has deduced a d^{-2} scaling law for overlap parameters between s and p orbitals by comparing the bandwidths obtained in tight binding with those given within the free-electron model. More sophisticated calculations, checked against determination of deformation potentials under hydrostatic [84–87] and uniaxial stress [87–89] for many different bulk semiconductors, indicate a clear deviation from the d^{-2} scaling. In empirical TB calculations the exponents $\alpha_{ll'm}$ are usually adjusted such that the pressure dependence of the direct gap at the Γ point and both indirect Γ–X and Γ–L gaps are well reproduced [90]. Because strain effects and changes in chemical composition are readily accounted for within the TB approach, this method has been applied to low-dimensional structures such as superlattices [91], monolayers [92, 90], and quantum wires [93].

A general result is that the d dependence is much more pronounced for s than for p states. As an example, the values of the exponents for GaAs [90] are $\alpha_{ss} = 4.2$ and $\alpha_{pp} = 1.5$. Martinez has explained this in terms of the dependence on volume of the band dispersion calculated using pseudopotentials [94]. He showed that the contributions to the band deformation potentials arising from kinetic and potential energies largely cancel each other for antibonding p states, for which the electronic charge concentrates in the center of the bond, rather than for s states, which are localized around the atoms.

This difference in the interatomic distance scaling law for s and p-type molecular orbitals is also at the origin of the distinct behavior with hydrostatic pressure of the three lowest band gaps in cubic materials, E_0, $E_{\Gamma-L}$, and $E_{\Gamma-X}$, which is well established through a large body of experimental [16, 77, 78] and theoretical investigations [16, 95–97]. All three gap energies are referred to the top of the valence band at the Γ point, which is fairly insensitive to pressure, being of p-bonding character. The E_0 gap exhibits the larger pressure coefficient of ~0.1 eV/GPa because the corresponding conduction band minimum at Γ is pure antibonding s-like. The L valleys present almost 50% admixture with antibonding p states and they shift up in energy with pressure at a rate which is roughly one-half of that of E_0. Finally, the pressure coefficient of the Γ–X indirect gap is negative in sign and $\sim\frac{1}{10}$ that of E_0 in magnitude. This is, in part, a consequence of the fact that although the X_{1c} is a conduction-band state, it has bonding character [97]. Moreover, there is an additional contribution to the reduction of this indirect gap with pressure from excited d states resonant with the conduction band [97].

Figure 8 illustrates the systematics in the pressure dependence of semiconductor band states by showing the calculated valence and conduction band structure of GaAs at ambient pressure and at 16 GPa [98]. These LMTO-ASA calculations include spin–orbit interaction and have been corrected for the so-called local-density-gap error (see later discussion), so they reproduce both the experimental gap energies and their change with pressure very accurately. In this way, the use of hydrostatic pressure allows a controlled tuning of the band structure of several direct-gap semiconductors toward a degeneracy and finally a crossover of zone center (Γ) and zone boundary (X) conduction-band states. The prototype example in GaAs [99–102] (see Fig. 8) with a Γ–X crossover at about 4.2 ± 0.2 GPa [102]. This band crossing has profound consequences on the optical properties and leads to interesting phenomena such as Γ–X state mixing effects in heterostructures.

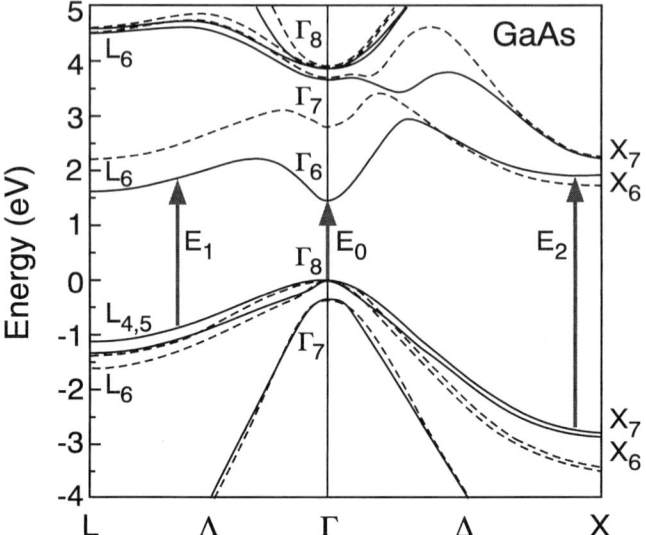

FIG. 8. Calculated electronic band structure of GaAs at ambient pressure (solid lines) and at a pressure of 16 GPa (dashed lines). Spin–orbit splitting is included. Vertical arrows indicate the main direct interband transitions at the Γ point (E_0), along the Γ–L direction (E_1), and near the X-point (E_2). After Ref. [98].

2. PENN MODEL AND THE DIELECTRIC THEORY OF THE COVALENT BOND

Phillips and Van Vechten [14, 15] have developed an empirical theory for covalent semiconductors of the type $A^N B^{8-N}$ based on a simplified model band structure with nearly free-electron dispersion and a single, isotropic gap, the Penn gap [103]. The few parameters of the theory are obtained from data of the low-frequency electronic dielectric constant of the semiconductor using a generalized expression of Eq. (A.39) for the dielectric function of a harmonic oscillator:

$$\epsilon_\infty = 1 + D \cdot A \cdot \frac{(\hbar \omega_p)^2}{E_g^2}, \tag{3.7}$$

with

$$A = 1 - B + \frac{B^2}{3}; \quad B = \frac{E_g}{4E_F}, \tag{3.8}$$

where D is a phenomenological factor, ω_p is the effective plasma frequency, E_F is the Fermi energy, and E_g is the Penn gap.

Within the nearly-free electron model of Penn, the Brillouin zone (BZ) has spherical shape and the Penn gap opens up at the Fermi wave vector $\vec{G}_F$. It is given by twice the Fourier component of the periodic potential $E_g = 2|U(\vec{G}_F)|$. If there are two different atoms per unit cell, the Penn gap can be written as

$$E_g = |E_h + iC| = \sqrt{E_h^2 + C^2}, \tag{3.9}$$

where E_h is the pure covalent contribution, which is determined by the average electrostatic potential, and C is the polar contribution, which is proportional to the potential difference between anion and cation for polar lattices. Here we point out the close relation of E_g with the bonding–antibonding splitting of Eq. (3.5) obtained within tight binding. A key parameter of the theory is the Phillips ionicity, which is a measure of the polarity of the bond:

$$f_i = \frac{C^2}{E_h^2 + C^2} = \frac{C^2}{E_g^2}. \tag{3.10}$$

The fact that such a simple model enables the interpretation of observed systematics in several physical properties of semiconductors such as crystal structure, bandgaps, and ionization energies is not accidental [104]. Compounds of the $A^N B^{8-N}$ type have four valence electrons per atom, which fill up the first four valence bands. The Jones zone corresponds to the fourth filled Brillouin zone in the extended scheme [104], which is nearly spherical. The average gap at the Jones zone (~4 eV) falls in the visible range of the optical spectrum; thus, the dielectric function is well described by an expression such as (3.7).

The factor D in Eq. (3.7) is introduced to account for the effect of filled $3d$ states on the dielectric response of semiconductors. These d levels can be resonant with the valence band and then mix with the valence p states. They contribute to the overall oscillator strength for optical transitions by recombining with holes in the p bands, as indicated schematically in Fig. 9. This contribution enters through the f-sum rule [Eq. (A.51)] as a factor in the plasma frequency, which takes into account the enhancement of the effective number of electrons participating in interband optical transitions: $D = n_{\text{eff}}/4$ [105].

One important result concerns the extension of the Penn gap equation (3.9) to several direct bandgaps in the Brillouin zone — $E_0(\Gamma_c - \Gamma_v)$,

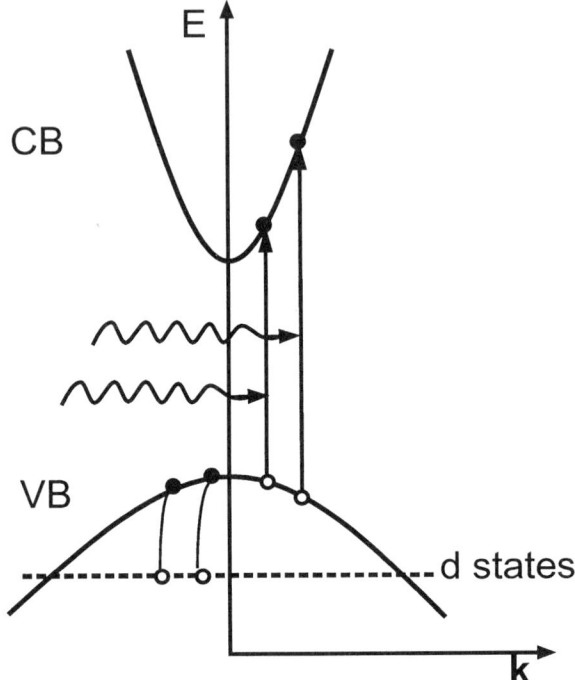

FIG. 9. Schematic representation of the way in which filled $3d$ states contribute to enhance the effective number of valence electrons participating in interband optical transitions.

$E_1(\Lambda_c - \Lambda_v)$ and $E_2(\Delta_c - \Delta_v)$ as indicated by arrows in Fig. 8 — which largely determine the optical properties of the semiconductor [14, 15],

$$E_{g,i} = (E_{h,i} - (D-1) \cdot \Delta E_i) \cdot \sqrt{1 + \frac{C^2}{E_{h,i}^2}}, \qquad i = 0, 1, 2, \qquad (3.11)$$

where the term $(D - 1) \cdot \Delta E_i$ in the bracket represents the shrinkage of the gap due to the effect of the filled d states on the valence bands.

Within this simple model it is now straightforward to discuss the pressure dependence of direct bandgaps in tetrahedral semiconductors [16]. It is mainly governed by the variation with pressure of the covalent energy E_h, whereas the polar contribution C is taken as pressure independent. By comparing with the gap change with lattice constant for the group IV elements, Camphausen et al. [16] have found a potential law for the covalent part:

$$E_{h,i} \propto d^{-s_i}, \qquad \bar{s} = 2.48. \qquad (3.12)$$

We point out that the average exponent $\bar{s}$ is close to 2 as expected from the distance dependence of the tight-binding overlap integrals. Further contributions arise from the d-state correction since the parameter ΔE_i also depends on interatomic distance [16]. Equation (3.11) implies that the stronger the ionicity of the bond, the smaller the variation of the gaps with pressure. This can be illustrated by comparing the linear pressure coefficients of the E_0 gap for the compound family of the Ge row. The values of the linear coefficient are 121 meV/GPa for Ge [106], 108 meV/GPa for GaAs [100, 102], and 70 meV/GPa for ZnSe [107], which decrease with increasing ionicity.

3. The $\vec{k} \cdot \vec{p}$ Method

The $\vec{k} \cdot \vec{p}$ method is particularly useful in obtaining analytic expressions for band dispersions and effective masses around high-symmetry points in terms of the corresponding bandgaps and optical matrix elements [108–112]. For this reason it is also suitable for the interpretation of optical spectra. The method is named after the form of the interaction which couples different band states at given wavevector $\vec{k}$. Basically, the method consists in finding the perturbative solution of a one-electron Schrödinger equation for the lattice-periodic part of the Bloch function:

$$\varphi_{n,\vec{k}}(\vec{r}) = \frac{1}{\sqrt{V}} u_{n,\vec{k}}(\vec{r}) \cdot e^{i\vec{k}\cdot\vec{r}}. \tag{3.13}$$

Around the BZ center ($\vec{k} \approx 0$) one has that

$$\{\mathcal{H}_0 + W(\vec{k})\} u_{n,\vec{k}} = E_{n,\vec{k}} u_{n,\vec{k}}, \tag{3.14}$$

where

$$\mathcal{H}_0 = \frac{p^2}{2m_0} + U(\vec{r}) + \frac{\hbar}{4m_0^2 c^2}(\vec{\nabla} U \times \vec{p}) \cdot \vec{\sigma} \tag{3.15}$$

is the unperturbed Hamiltonian which includes the spin–orbit (SO) interactions, and

$$W(\vec{k}) = \frac{\hbar^2 k^2}{2m_0} + \frac{\hbar \vec{k} \cdot \vec{p}}{m_0} \tag{3.16}$$

is a perturbation for small values of $\vec{k}$. In Eq. (3.15) $\vec{\nabla} U$ represents the gradient of the crystal potential and $\vec{\sigma}$ is the electron spin operator, whose components are the Pauli spin matrices.

We now briefly describe the Kane model [109] for a tetrahedral semiconductor, which gives the $\vec{k} \cdot \vec{p}$ solutions around the Γ point but considers only four spin-degenerate bands: the lowest conduction and the threefold degenerate upper valence bands. Because of spin–orbit interaction the valence band splits into the heavy/light hole doublet and the split-off band. The eigenfunctions of $\mathcal{H}_0$ from Eq. (3.15) are classified according to the total angular momentum $\vec{j} = \vec{l} + \vec{s}$, which is a good quantum number. As basis functions we use two s-like states $|s \uparrow, \downarrow \rangle$ with up ($\uparrow$) and down ($\downarrow$) spin components and six p-type orbitals $|x \uparrow, \downarrow \rangle$, $|y \uparrow, \downarrow \rangle$, and $|z \uparrow, \downarrow \rangle$, which combine to give the eight Γ-point wavefunctions listed in Table I. The energy eigenvalues referred to the top of the valence band are also indicated, where E_0 is the bandgap and Δ_0 is the spin–orbit splitting (see later discussion).

At finite wave vector $\vec{k}$ the bands couple via the $\vec{k} \cdot \vec{p}$ interaction $W(\vec{k})$. By using the wave functions of Table I and the matrix element of the momentum operator, defined as

$$p = p_x = p_y = p_z = -i \langle s | p_x | x \rangle, \quad (3.17)$$

the $\vec{k} \cdot \vec{p}$ Hamiltonian can be readily diagonalized, leading to parabolic band dispersions of the type

$$E_i(k) = E_i(0) + \frac{\hbar^2 k^2}{2 m_i^*} \quad (3.18)$$

TABLE I

Γ Point Wave Functions and Energy Eigenvalues of The Kane Model

	Bands	j,	j_z	Wave Functions	Energies
CB:	electrons	$\frac{1}{2}$,	$\pm \frac{1}{2}$	$u_e = |s \uparrow, \downarrow >$	E_0
VB:	heavy holes	$\frac{3}{2}$,	$\pm \frac{3}{2}$	$u_{hh} = \frac{1}{\sqrt{2}} |(x \pm iy) \uparrow, \downarrow >$	0
	light holes	$\frac{3}{2}$,	$\pm \frac{1}{2}$	$u_{lh} = \frac{1}{\sqrt{6}} |(x \pm iy) \downarrow, \uparrow > \mp \sqrt{\frac{2}{3}} |z \uparrow, \downarrow >$	0
	split-off	$\frac{1}{2}$,	$\pm \frac{1}{2}$	$u_{so} = \pm \frac{1}{\sqrt{3}} |(x \pm iy) \downarrow, \uparrow > + \frac{1}{\sqrt{3}} |z \uparrow, \downarrow >$	$-\Delta_0$

with isotropic effective masses

$$\frac{m_0}{m_e^*} = 1 + \frac{2p^2}{3m_0}\left(\frac{2}{E_0} + \frac{1}{E_0 + \Delta_0}\right) \qquad (3.19)$$

$$\frac{m_0}{m_{hh}^*} = 1 \qquad (3.20)$$

$$\frac{m_0}{m_{lh}^*} = 1 - \frac{4p^2}{3m_0 E_0} \qquad (3.21)$$

$$\frac{m_0}{m_{SO}^*} = 1 - \frac{2p^2}{3m_0(E_0 + \Delta_0)}. \qquad (3.22)$$

In this way we obtain a very simple picture for the interpretation of the effects of pressure on the energy bands near the zone center. The main contribution to the pressure dependence of the effective masses comes from the change in bandgap E_0. The spin–orbit coupling Δ_0 is fairly insensitive to pressure since it is given by the derivative of the crystal potential in region of the atomic cores, where the electronic charge remains unaltered because of the Pauli exclusion principle [106, 96]. The matrix element of the momentum operator changes also little with pressure because it is roughly inversely proportional to the lattice parameter a_0 [113]. Thus, with increasing pressure the gap becomes wider and consequently the light particles, that is, electrons, light, and split-off holes, turn heavier and the bands become flatter.

4. Uniaxial Stress Effects

The appearance of uniaxial stress components in high-pressure experiments are unavoidable. In many important cases of quantum well heterostructures made of materials with a lattice mismatch, large biaxial stresses build up during isomorphic growth. Furthermore, even for lattice-matched heterostructures a biaxial strain would be induced by compression under hydrostatic conditions if the constituent materials have different bulk moduli [114, 115]. In both cases the biaxial strain in the x, y-plane of the quantum wells for growth in the (001) direction (z) is determined by the lattice parameter mismatch:

$$\epsilon_{xx} = \epsilon_{yy} = \frac{a^B - a^A}{a_0}, \qquad \epsilon_{zz} = -\frac{2C_{12}}{C_{11}}\epsilon_{xx}, \qquad (3.23)$$

where C_{ij} are the elastic constants [116]. In addition to the energy shifts of the bands due to the hydrostatic component, a biaxial stress causes splittings of degenerate states by lowering the crystal symmetry.

Within the framework of $\vec{k} \cdot \vec{p}$ theory uniaxial stress effects are taken into account using the orbital-strain Hamiltonian derived by Pikus and Bir [117]. For a given band (i) at $\vec{k} = 0$ it is written as

$$\mathcal{H}_\epsilon^{(i)} = -a^{(i)}(\epsilon_{xx} + \epsilon_{yy} + \epsilon_{zz}) - 3b^{(i)}\left[\left(j_x^2 - \frac{j^2}{3}\right)\epsilon_{xx} + \text{c.p.}\right]$$
$$- \frac{6d^{(i)}}{\sqrt{3}}\left[\frac{1}{2}(j_x j_y + j_y j_x)\epsilon_{xy} + \text{c.p.}\right] \qquad (3.24)$$

(c.p. stands for cyclic permutations). The first term in Eq. (3.24) represents the energy shift of the band due to the hydrostatic part of the deformation determined by the hydrostatic deformation potential $a^{(i)}$. Absolute values for $a^{(i)}$ are difficult to obtain experimentally, because one usually measures the pressure dependence of gaps, which yields just the *relative* deformation potential between conduction and valence band. The parameters $b^{(i)}$ and $d^{(i)}$ are uniaxial deformation potentials for tetragonal and rhombohedral-type distortions, respectively. Optical measurements have been used to determine the strain shifts and splittings of band extrema [11, 118–123]. Simple theoretical schemes to estimate deformation potentials can be found in Ref. [124]. For a review of uniaxial stress effects in semiconductors we refer to [25].

In strained superlattices, the tetragonal distortion of the lattice given in Eq. (3.23) lifts the degeneracy of heavy and light holes at Γ. The strain Hamiltonian takes the simple form

$$\mathcal{H}_\epsilon^{(i)} = -\delta E_\text{H} - \frac{3}{2}\delta E_{001}\left(j_z^2 - \frac{j^2}{3}\right) \qquad (3.25)$$

$$\delta E_\text{H} = 2a^{(i)}\frac{C_{11} - C_{12}}{C_{11}}\epsilon_{xx} \qquad (3.26)$$

$$\delta E_{001} = -2b^{(i)}\frac{C_{11} + 2C_{12}}{C_{11}}\epsilon_{xx} \qquad (3.27)$$

Here we neglected the coupling with the split-off band, which leads to higher order terms in the perturbation δE_{001} [11]. Using the wave functions of Table I we obtain for the strain-induced splittings of the upper hole bands

$$\Delta E_{hh} = -\delta E_H - \frac{1}{2}\delta E_{001} \tag{3.28}$$

$$\Delta E_{lh} = -\delta E_H + \frac{1}{2}\delta E_{001}. \tag{3.29}$$

For most of the group IV and III–V semiconductors the sign of the valence band deformation potentials are $a > 0$ and $b < 0$. Thus, a tensile strain in the x, y-plane ($\epsilon_{xx} > 0$) results in a downward energy shift of the heavy hole band, whereas the light holes are almost unaffected because of cancelation of the hydrostatic and the uniaxial contributions.

Another interesting problem concerns the splitting of the conduction band X valleys in strained superlattices. Most of the quantum well structures are epitaxially grown on (001) oriented substrates. The built-in biaxial strain splits the sixfold degenerate X valleys in on X_{xy} quadruplet and an X_z doublet. The energy shifts of these states are given by [119]

$$\Delta E(X_z) = -\delta E_H^X + \frac{2}{3}\delta E_{001}^X \tag{3.30}$$

$$\Delta E(X_{xy}) = -\delta E_H^X - \frac{1}{3}\delta E_{001}^X, \tag{3.31}$$

where the hydrostatic and tetragonal deformation energies δE_H and δE_{001} are defined as in Eqs. (3.26) and (3.27) but replacing the deformation potential parameters by

$$a_X = \Xi_d + \frac{1}{3}\Xi_u, \qquad b_X = \frac{1}{2}\Xi_u. \tag{3.32}$$

For the X valleys the signs of a_X and b_X are reversed with respect to the valence band parameters so that the X_z states become lower in energy for a biaxial expansion in the x, y-plane.

5. QUANTUM WELL STRUCTURES

The restriction of the free carrier motion in the growth direction (z hereafter), which occurs at the interfaces in quantum well (QW) and superlattice (SL) structures, leads to the formation of discrete energy levels and minibands, respectively. The confinement energies of electrons and holes are usually obtained from $\vec{k} \cdot \vec{p}$ calculations within the envelope function approximation (EFA) [125, 126]. In the case of strained heterostructures, the effects of uniaxial stress are incorporated via the Pikus–Bir Hamiltonian (3.24). Quantum well states are well described as a superposition of Bloch wavefunctions, for which the Fourier transform of the expansion coefficients is just the so-called envelope function $F(z)$. In the effective mass approximation the envelope function is obtained as the solution of a Schrödinger equation, in which only the superlattice potential $U_{SL}(z)$ appears explicitly,

$$\left[-\frac{\hbar^2}{2\tilde{m}^*} \frac{\partial^2}{\partial z^2} + U_{SL}(z) \right] \cdot F_n(z) = \epsilon_n \cdot F_n(z), \tag{3.33}$$

where $\tilde{m}^*$ is the effective mass tensor. The Hamiltonian of Eq. (3.33) is a matrix whose dimensions are determined by the number of bands being considered. For the description of near-bandgap properties at the Γ point, it is frequently enough to work with the lowest conduction band and the three uppermost hole bands of Table I (8×8 Hamiltonian).

For a single band and in the case of an infinitely deep well of width L_z, the energy eigenvalues and wave functions of Eq. (3.33) are readily obtained as

$$\epsilon_n = \frac{\hbar^2}{2m^*} \cdot \left(\frac{n\pi}{L_z}\right)^2, \qquad n = 1, 2, 3, \ldots \tag{3.34}$$

$$F_n(z) = A \cdot \sin\left(\frac{n\pi z}{L_z}\right). \tag{3.35}$$

Equation (3.34) implies that the confinement energy grows quadratically with the magnitude of the effective confinement wave vector $k_n = n\pi/L_z$ given by the subband quantum number n and the well width L_z, but it decreases inversely proportional to the effective mass of the confined particle.

Since U_{SL} is a piecewise constant potential which depends on the band alignments, Eq. (3.33) can be separated into four Kronig–Penney type equations. This strictly holds only for the heavy holes, for which the effective mass of the confined states is that of the bulk. The light particles, however, are coupled by $\vec{k} \cdot \vec{p}$ interaction. Because of confinement, the bandgap increases, resulting in an additional nonparabolicity, which can be taken into account by assuming that the effective masses of light particles in the wells are not constants but a function of energy. The electron and light-hole masses are thus calculated using the $\vec{k} \cdot \vec{p}$ expressions (3.19) and (3.21), which have been generalized for the QW case by renormalizing the gaps by adding the confinement energies ϵ_n [114]:

$$\frac{m_0}{m_e^*(\epsilon_n)} = 1 + \frac{2p^2}{3m_0}\left(\frac{2}{E_0 + \epsilon_n} + \frac{1}{E_0 + \Delta_0 + \epsilon_n}\right) \quad (3.36)$$

$$\frac{m_0}{m_{lh}^*(\epsilon_n)} = 1 - \frac{4p^2}{3m_0}\frac{1}{E_0 + \epsilon_n}\cdot\left[1 - \frac{\epsilon_n}{2(\Delta_0 - \epsilon_n)}\right]. \quad (3.37)$$

The second term in square brackets of Eq. (3.37) is particularly important when confinement energies of light holes become comparable to the valence band spin–orbit splitting. In this way, the electron and light-hole confinement energies are obtained from Kronig–Penney-type equations such as (3.33) by iteration using the corresponding effective masses (3.36) and (3.37).

One fundamental parameter for the determination of the subband structure in the EFA is the bandgap discontinuity ΔE_0. For these calculations the band offsets, defined as fractions of the gap energy difference between well and barrier materials ($\Delta U_c = Q_c \cdot \Delta E_0$, $\Delta U_v = Q_v \cdot \Delta E_0$, $Q_c + Q_v = 1$), need to be known *a priori*. Their determination is thus an important task, for which optical methods combined with high pressure often prove to be helpful [127–129].

The effect of pressure on QW states can now be easily established by inspection of Eqs. (3.36) and (3.37). With increasing pressure the light particles becomes heavier because of the increase of the bandgap. This, in turn, tends to reduce the confinement energy of electrons and light holes. On the other hand, the dependence on pressure of the band offsets would change the height of the potential barriers for confinement. In first approximation, the band offsets vary in proportion to the difference in the pressure coefficients of the gaps for the two materials,

$$\frac{\partial \Delta U_{c,v}}{\partial P} \approx Q_{c,v} \cdot \left(\frac{\partial E_0^B}{\partial P} - \frac{\partial E_0^A}{\partial P}\right), \quad (3.38)$$

assuming pressure-independent $Q_{c,v}$ ratios. For the GaAs/AlGaAs system the change of the offsets with pressure is negligible [127, 130], whereas for GaSb/InAs QWs, it amounts to ~60 meV/GPa [130–132].

IV. Refractive Index Dispersion

The complex dielectric function $\epsilon(\omega)$ is a fundamental property in semiconductor optics (see Appendix). The light dispersion in the material is governed by its real part $\epsilon_1(\omega)$, whereas the magnitude of the imaginary part $\epsilon_2(\omega)$ determines the amount of radiation that is absorbed by the medium. In particular, the static dielectric constant $\epsilon(0)$ appears as an important parameter in related fields such as charge screening and other collective phenomena (excitons, plasmons, etc.), and in lattice-dynamical properties (optical phonon splittings, polarons, and phonon polaritons). Many of these phenemona are also studied under external hydrostatic pressures, for the analysis of which the knowledge of the dependence on pressure of the low-frequency dielectric constant is needed [133, 36]. In the following discussion of the low-frequency limit we will actually refer to ϵ_∞, that is, the so-called infrared or low-frequency electronic dielectric constant, rather than $\epsilon(0)$. The difference between these dielectric constants resides in that ϵ_∞ is of purely electronic origin arising from optical interband transitions and it does not contain any contribution from the optical phonons of the lattice. In homopolar materials such as Ge and Si, both dielectric constants are equal; in polar semiconductors, ϵ_∞ is smaller than $\epsilon(0)$ typically by 10–20%. The volume dependence of ϵ_∞ plays a principal role in semi-empirical models of semiconductor properties under pressure [14–16, 134–136]. In this section we will present the pressure-induced variation of the low-frequency dielectric constant ϵ_∞ of several representative semiconductors as experimentally obtained from measurements of the refractive index at photon energies below the lowest direct gap. The experimental data will be also compared to the results of recent self-consistent band-structure calculations within the local-density approximation (LDA) for the pressure dependence of the dielectric function [96, 137–142].

1. REFRACTIVE INDEX AND OPTICAL DISPERSION

In the spectral region of vanishing absorption (absorption coefficient $\alpha(\omega) = 0$) the refractive index of a material is simply given by the square root of the real part of the frequency-dependent dielectric function [see Eq. (A.11)]. In this case, the refractive index can be measured under pressure by the interference method as described later, which allows us to determine the pressure dependence of the static dielectric constant by extrapolation of a simple model dielectric function used to fit the experimental data. Empirical models for $\epsilon_2(\omega)$ and its Kramers–Kronig counterpart $\epsilon_1(\omega)$, which essentially consider the different contributions to the dielectric function arising from optical transitions related to Van Hove singularities at critical points (CPs) in the joint density of states [113, 143], have been utilized for the analysis of the refractive index data in pressure experiments carried out for a variety of semiconductors: diamond [69, 144], Si [4, 145],

Ge [4, 146, 147], SiC and BN [144], GaN [148], GaP [149], GaAs [150, 151, 147], InSb [152], ZnSe and ZnTe [51, 153]. GaSe [154], InSe [155], and several alkali halides [156]. A detailed description of the oscillator model which accounts for the contribution of the main CPs at the different optical gaps of III–V semiconductors, including broadening effects in order to achieve considerably good agreement with the ellipsometry data of Ref. [157], is given elsewhere [158]. Here we briefly present a simplified two-oscillator model used for the analysis of the index of refraction data of Ge and GaAs under pressure.

The main contribution to $\epsilon_2(\omega)$ in tetrahedral semiconductors arises from transitions at critical points along the Γ–L direction (E_1 gap) and near the X point (E_2 gap) of the Brillouin zone [157]. In this model both contributions are combined into a single oscillator at energy E_1. This critical point can be treated as a two-dimensional one of the P_0 type (see Table III), because of the large difference between the longitudinal and transverse effective masses [143]. Hence, the contribution to the imaginary part of the dielectric constant is

$$\Delta\epsilon_2(\omega) = \begin{cases} \dfrac{\pi B}{X_1^2} & \text{for } E_1 < \hbar\omega < E_c \\ 0 & \text{otherwise,} \end{cases} \quad (4.1)$$

with $X_1 = \hbar\omega/E_1$. The cutoff energy E_c is introduced to avoid the unphysical extension of the parabolic bands to high energies and to fulfill optical sum rules. The Kramers–Kronig transformation of Eq. (4.1) can be obtained analytically [113], leading to

$$\Delta\epsilon_1(\omega) - 1 = -\dfrac{B}{X_1^2} \ln\left[\dfrac{1 - X_1^2}{1 - X_c^2}\right], \quad X_c = \dfrac{\hbar\omega}{E_c}. \quad (4.2)$$

Transitions at the fundamental direct gap E_0 produce strong dispersion of $n(\omega)$ in its vicinity. This CP corresponds to a 3-D interband minimum (M_0 type) and, assuming parabolic bands, $\epsilon_2(\omega)$ behaves as a square-root-like edge,

$$\Delta\epsilon_2(\omega) = \begin{cases} \dfrac{C_0''}{X_0^2}\sqrt{X_0 - 1} & \text{for } \hbar\omega > E_0 \\ 0 & \text{otherwise,} \end{cases} \quad (4.3)$$

with $X_0 = \hbar\omega/E_0$. A similar contribution is usually added for the spin–orbit split-off gap $E_0 + \Delta_0$. For the real part of the dielectric constant one gets

$$\Delta\epsilon_1(\omega) = \frac{C_0''}{X_0^2}[2 - \sqrt{1 + X_0} - \sqrt{1 - X_0}\,\Theta(1 - X_0)] \qquad (4.4)$$

(Θ is the unit-step function). Excitonic effects are accounted for separately by adding a δ-function-like Lorentz oscillator (see Refs. [158, 147] for details).

The oscillator model for $\epsilon(\omega)$ is illustrated in Fig. 10 for Ge at ambient pressure. The function $\epsilon_2(\omega)$ (solid curve) is given by the combination of Eqs. (4.1) and (4.3) and a δ-function to account for discrete excitons at E_0, while $\epsilon_1(\omega)$ is obtained using Eqs. (4.2) and (4.4). The model has only two adjustable parameters C_0'' and B, corresponding to the strength of the oscillators at the E_0 and E_1 gaps, respectively. The other parameters, for Ge and GaAs are listed in Table 1 of Ref. [147]. We emphasize the fact

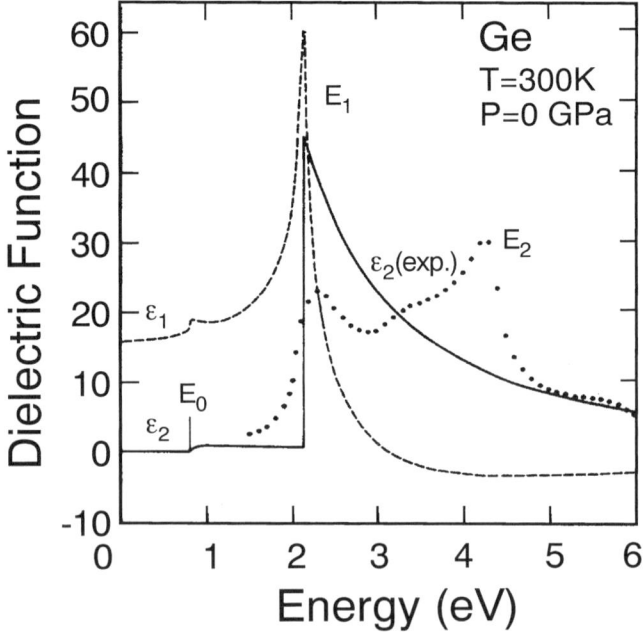

FIG. 10. Real and imaginary part of the model dielectric function used to describe the refractive-index dispersion in Ge at ambient pressure. The dotted curve represents experimental data from Ref. [157]. After Ref. [147].

that this model fulfills both the f-sum rule for the total effective number of electrons and the static sum rule [159] (see Appendix). The former sum rule allows us to relate the cutoff frequency E_c, which is of the order of the valence-electron plasma frequency (~15 eV), with the oscillator strength parameters by imposing the constancy of the effective number N_{eff} of valence electrons of Eq. (A.53) contributing to the optical response [159]. Further, the cutoff E_c is determined uniquely only after fulfillment of the static sum rule of Eq. (A.50). For comparison, we also show in Fig. 10 the experimental data for $\epsilon_2(\omega)$ [157]. The strong CP at E_2 (corresponding to interband transitions along the Γ–X direction) does not appear explicitly in the two-oscillator model, but its contribution to $\epsilon_1(\omega)$ in the region below the direct gap is retained due to the sum-rules constraints.

For the analysis of high-pressure data we take into account the experimental dependence of the E_0 and E_1 bandgaps on pressure. Assuming that the effective number of valence electrons N_{eff} does not change with pressure, the pressure coefficient of the cutoff frequency E_c turns out to be the same as for the E_1 gap. We show later that in spite of the crudeness of this model dielectric function, it describes very well the frequency dependence of the refractive index at photon energies below the direct gap E_0 even at different pressures.

2. VOLUME DEPENDENCE OF THE LOW-FREQUENCY DIELECTRIC FUNCTION

In the presence of an external static electric field, the polarization response of the semiconducting material is represented by the dielectric constant ϵ_∞. It actually describes the most significant part of the polarization, arising from electronic interband transitions as compared with the much smaller contribution of the lattice phonons. Its value can be determined by extrapolation to zero frequency of the dielectric function measured by optical means such as ellipsometry or from the interference pattern in transmission. The macroscopic polarization vector $\vec{P}$ induced by the electric field $\vec{E}$ results from the superposition of the electric dipole moments of N polarizable entities or dipoles per unit volume characterized by a polarizability α_p per primitive unit cell,

$$\vec{P} = \alpha_p \cdot N \cdot \vec{E}_{\text{loc}}, \qquad (4.5)$$

where $\vec{E}_{\text{loc}}$ is the *local* electric field at the site of each polarizable center. Although the dielectric constant is a macroscopic quantity, its connection to the microscopic polarizability α_p is attained through the *local-field equation*

$$\vec{E}_{loc}(\vec{r}) = \vec{E}(\vec{r}) + 4\pi\gamma\vec{P}(\vec{r}) \qquad (4.6)$$

The factor γ accounts for the local-field effects and depends on the nature of the dipoles. From Eqs. (A.5), (4.5), and (4.6) one finally obtains the Clausius–Mossotti equation for relating ϵ_∞ with α_p:

$$4\pi\gamma N\alpha_p = \frac{\epsilon_\infty - 1}{\epsilon_\infty + (\gamma^{-1} - 1)}. \qquad (4.7)$$

Here we distinguish between two special cases. The one is for strongly localized polarizable centers in a cubic arrangement, for which the factor γ is equal to $\frac{1}{3}$. In this case, Eq. (4.7) leads to the well-known Lorentz–Lorenz formula for the relation between atomic polarizability and static dielectric constant, which applies for most ionic crystals such as the alkali halides. The opposite case occurs for delocalized electrons, for which the local-field effects vanish ($\gamma = 0$). This holds also for band electrons in tetrahedral semiconductors because of the large overlap between orbitals forming covalent bonds. Equation (4.7) thus simplifies to the Sellmeyer formula: $\epsilon_\infty = 1 + 4\pi N\alpha_p$ (see Appendix).

Within this macroscopic treatment the pressure dependence of ϵ_∞ is simply obtained by taking the volume derivative of Eq. (4.7), which gives [133]

$$\frac{\epsilon_\infty}{\epsilon_\infty - 1} \cdot \frac{1}{\gamma(\epsilon_\infty - 1) + 1} \cdot \frac{\partial \ln \epsilon_\infty}{\partial P} = \kappa - \kappa \cdot \frac{\partial \ln \alpha_p}{\partial \ln V}. \qquad (4.8)$$

where $\kappa = -\partial \ln V/\partial P$ is the compressibility of the material. With this definition, the volume derivative of ϵ_∞ is

$$r = \frac{\partial \ln \epsilon_\infty}{\partial \ln V} \sim -\left(1 - \frac{\partial \ln \alpha_p}{\partial \ln V}\right). \qquad (4.9)$$

In large-bandgap ionic materials such as alkali halides, and particularly for solid noble gases, α_p changes very little with pressure. Consequently, the changes of the macroscopic dielectric constant with pressure are governed by the increase in the density of polarizable centers giving *negative* values for r. In contrast, in covalent semiconductors the bond polarizability depends strongly on bond length. In this case, large changes in α_p cause the decrease of the dielectric constant ϵ_∞ with increasing pressure. This effect is expected to be less pronounced for materials with partly ionic character of the chemical bond.

The issue of the volume dependence of the dielectric constant can be approached from a microscopic point of view, namely from that of the changes in the electronic band structure induced by the application of an external hydrostatic pressure. The close relation between ϵ_∞ and semiconductor band structure has been already pointed out by the discussion of the Penn gap model [103] and the dielectric theory of the covalent bond [14, 15] in a previous section. The low-frequency dielectric constant ϵ_∞ can be expressed in terms of the average optical gap (or Penn gap) $E_g^2 = E_h^2 + C^2$ using Eq. (3.7). The factor A in Eq. (3.7) is constant ($A \sim 1$) and the plasma frequency varies as the square root of the density ($\omega_p \sim V^{-1/2}$). The volume dependence of homopolar contribution E_h and the factor $D = n_{\text{eff}}/4$ which accounts for the effects of occupied d states on interband transition probability is estimated to be $E_h \sim V^{0.83}$ (Ref. [14, 15]) and $D - 1 \sim V^{4.3}$ (Ref. [16]). The ionic part C of the Penn gap is, as usual, taken as pressure independent, $dC/dP \approx 0$. Thereafter, the volume derivative of ϵ_∞ can be written in terms of the Phillips–Van Vechten ionicity $f_i = C^2/E_g^2$ as

$$\frac{\epsilon_\infty}{\epsilon_\infty - 1} \cdot \frac{\partial \ln \epsilon_\infty}{\partial \ln V} = -1 - 2(1 - f_i) \frac{\partial \ln E_h}{\partial \ln V} + \frac{\partial \ln D}{\partial \ln V} \tag{4.10}$$

$$r = \frac{\partial \ln \epsilon_\infty}{\partial \ln V} \approx \frac{5(\epsilon_\infty - 1)}{3\epsilon_\infty} \cdot (0.9 - f_i). \tag{4.11}$$

According to Eq. (4.11) the borderline between *covalent* and *ionic* pressure behavior of ϵ_∞ (corresponding to a change in the sign of r) is expected for $f_i \approx 0.9$.

Along the same lines follows the interpretation of the pressure dependence of the dielectric constant when described by empirical oscillator models for the different critical points. In this case the decrease of ϵ_∞ with pressure is attributed to a blueshift of the oscillator energies as pressure increases. This effect is larger with lower energy of the main oscillator contributing to the optical response of the particular semiconductor.

3. METHOD OF OPTICAL INTERFERENCES

The intensity of light transmitted through a plane-parallel sample of thickness d and refractive index $n(\lambda)$ presents a series of maxima (minima) as a function of wavelength due to the constructive (destructive) interference of multiple-reflected beams between both sample–medium interfaces, as given by Eq. (A.32) for a transparent medium ($\alpha = 0$) at normal incidence. The conditions for occurrence of the intensity maxima, as obtained from the phase change of Eq. (A.33), are

$$2dn(\lambda_m) = m\lambda_m, \qquad m = 1, 2, 3, \ldots. \qquad (4.12)$$

The integer m refers to the order of interference and λ_m is the wavelength of the corresponding maximum.

Figure 11 shows representative transmission spectra of Ge measured at two different pressures. The interference pattern is clearly observed below the absorption edge at the energy of the direct gap E_0. The assignment of the order m for each maximum was performed at ambient pressure from the wavelength separation $\lambda_m - \lambda_{m+1}$ between two consecutive maxima and by using literature data of $n(\lambda)$ tabulated, for example, in Refs. [157, 160]. The sample thickness at ambient pressure d_0 can now be determined from the wavelengths of the interference maxima in the zero-pressure spectrum. One needs to keep track of the indexing of each maximum during

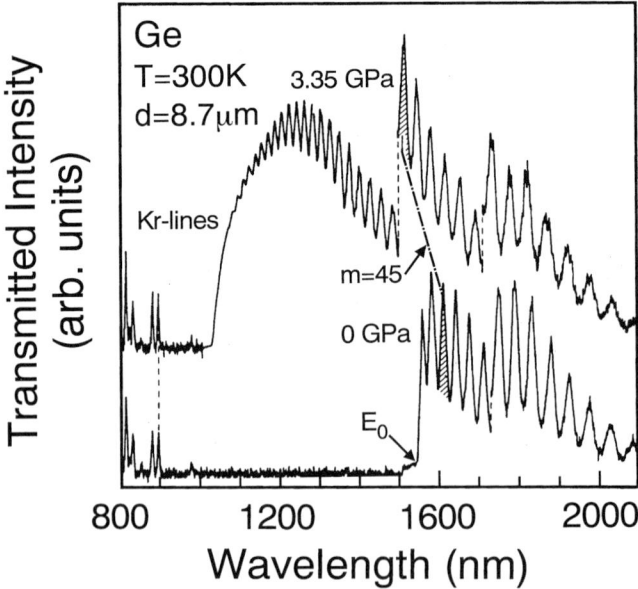

FIG. 11. Near-infrared transmission spectra of an 8.7-μm-thick Ge sample at two different pressures. The shift to lower wavelengths with increasing pressure is indicated for the interference maximum of order $m = 45$. After Ref. [147].

a whole pressure cycle by recording the number of maxima passing by at a fixed wavelength, while pressure is changed. The missing of interference orders is the most important source of systematic errors for the determination of the pressure dependence of the refractive index. For that purpose an electric motor drive has been used to change the pressure in a continuous manner and at low speed (~0.05 GPa/min). In Fig. 11 the pressure-induced shift of the interference maximum of order 45 is indicated. For the evaluation of $n(\omega)$ from the transmission spectra measured at different pressures using Eq. (4.12), it is also necessary to take into account the thickness reduction by scaling d according to the Murnaghan's equation of state [161],

$$\frac{d(P)}{d_0} = \left(1 + \frac{B'_0 P}{B_0}\right)^{-1/3B'_0}, \qquad (4.13)$$

with B_0 and B'_0 being the isothermal bulk modulus and its pressure derivative, respectively. One has to take special care in always illuminating the same portion of the sample to avoid possible errors due to changes in the thickness because of nonparallelism of the sample surfaces. The effect of a small tilting of the sample inside the pressure cell is, in contrast, negligible, because the effective sample thickness when tilted by an angle θ changes as $\cos\theta$, and θ is at most 0.1 rad. In this way it is possible to determine the refractive index at high pressures with an estimated error of less than 0.3%.

As an example, Fig. 12 shows refractive index data of Ge measured at several pressures and at 300 K. The scatter of the data points is mainly caused by uncertainties in the determination of the maximum positions λ_m due to spurious interferences coming from multiple reflections inside the pressure cell. For Ge $n(\omega)$ decreases monotonically with pressure. Such behavior will be discussed later in comparison to other compounds with different bonding characteristics.

An alternative method for measuring the refractive index of transparent substances under high hydrostatic pressure in connection with the DAC has been proposed [162] and demonstrated for 11 alkali halides [163]. By this technique the transmission spectrum is acquired with a Fourier-transform spectrometer, and the frequency position of the characteristic satellite peak in the interferogram corresponding to the period of the Fabry–Perot oscillations of the transmitted light is compared to that of a reference sample placed *side by side* in the gasket hole. The ratio of the satellite peak positions in the Fourier-transform interferogram is equal to the ratio of the respective refractive indices [162]. This method has the advantage that neither the sample thickness nor the order of the fringe is necessary for the determination of the refractive index under pressure. On

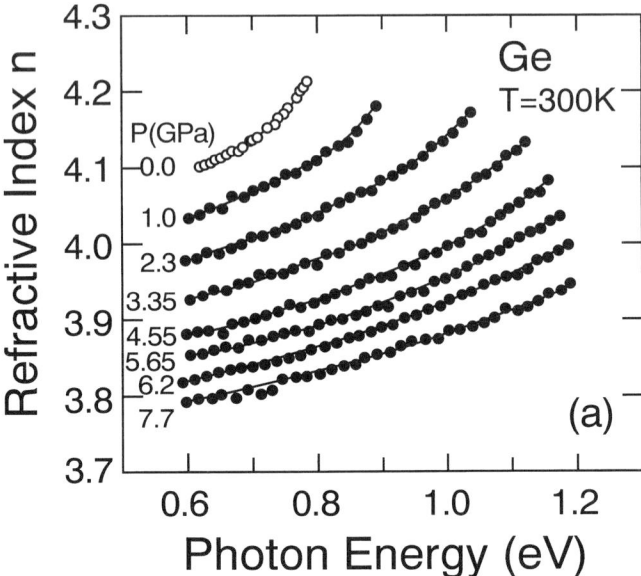

FIG. 12. The refractive index of Ge as a function of photon energy at several pressures (solid circles). Open circles are literature data [157, 160] at normal pressure. The solid curves represent the results of fits to the experimental data using an empirical two-oscillator model. After Ref. [147].

the other hand, the sample preparation and loading of the pressure cell become somewhat more complicated.

4. Experimental Results

Figure 13 illustrates the systematic trends concerning the dependence on pressure of the low-frequency dielectric constant ϵ_∞ observed for several materials depending upon their bonding character. The selected materials are representative of the different bond types: He and Xe are molecular solids with very weakly bonded closed-shell atoms, the bonding in Ge is purely covalent whereas in CsI it is totally ionic, and the covalently bonded GaAs and ZnSe pertain to the Ge row of the periodic table but exhibit an increasingly ionic bond character. The infrared dielectric constant increases with pressure for He [164], Xe [165], and CsI [156], but ϵ_∞ diminishes for the other semiconductors with covalent bond character. In the former case the pressure behavior of ϵ_∞ is exclusively dominated by the increase of the density of polarizable centers while the volume reduces, as indicated by

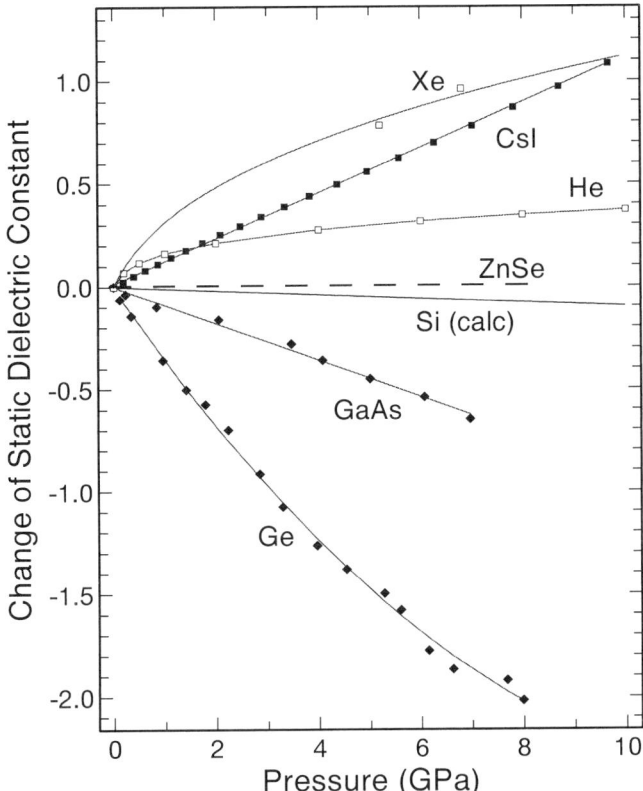

FIG. 13. Change of the low-frequency electronic dielectric constant of Ge [147], GaAs [147], ZnSe [153, 186], CsI [156], He [164], and Xe [165] as a function of pressure derived from measurements of the refractive index dispersion. The solid line for Si represents the results of all-electron full-potential LMTO calculations [141].

Eq. (4.8). The volume derivative of ϵ_∞ is just given by the volume compressibility [see Eq. (4.9)] and ϵ_∞ should scale linearly with density. This is true for He and Xe but does not apply to CsI, which displays a strongly supralinear behavior when plotted as a function of density. This has been explained [156] as due to the closure of the bandgap of CsI with increasing pressure, which leads to a pressure-induced enhancement of the microscopic polarizability α_p giving an additional contribution to r [see Eq. (4.9)].

For materials with covalent bonds the most significant effect which determines the negative sign of $\partial \ln \epsilon_\infty / \partial P$ is the *stiffening* of the bonds, hence reducing their polarizability, when high hydrostatic pressure is applied. From the point of view of the electronic band structure Eq. (3.7) tells us that the actual value of the dielectric constant ϵ_∞ is inversely proportional to the magnitude of the average optical gap E_g. Generally speaking, the contribution to the electronic polarizability associated with a certain interband optical transition is determined by its probability, which decreases in proportion to the transition energy. The effect of pressure is to shift the direct gaps to higher energies and thus to reduce the electronic contribution to the dielectric constant.

Obviously, this reduction of α_p with pressure diminishes in its importance the more ionic the bonds are, as indicated by the striking decrease in slope of the ϵ_∞-versus-pressure curves in Fig. 13 for the sequence of covalently bonded semiconductors with increasing ionicity f_i of Ge ($f_i = 0$), GaAs ($f_i = 0.31$), and ZnSe ($f_i = 0.74$). Indeed, ZnSe appears to be close to the crossover point between covalent (ϵ_∞ decreases) and ionic behavior (ϵ_∞ increases with pressure). This can be interpreted within the framework of the Phillips–Van Vechten theory [14–16] as due to the larger ionic contribution to the average optical gap, C, which in first approximation is pressure independent. Therefore, for materials with strong ionic character of the covalent bond, E_g increases more slowly with increasing pressure, resulting in a less pronounced variation of ϵ_∞. This method, however, is too simple to yield quantitative agreement for the pressure dependence of ϵ_∞ in many cases of even elementary semiconductors such as Si. Moreover, some of its assumptions, for instance that of $dC/dP \equiv 0$, could not be justified by later *ab initio* band-structure calculations. An analysis of the possibilities and the limits of such an empirical method can be found in Ref. [142].

In recent years total-energy band structure calculations have improved their accuracy so far that they are able to reproduce well the complex dielectric function $\epsilon(\omega)$ of semiconductors and also to predict its pressure dependence [96, 137, 138, 141]. Although these calculations are based on the LDA approximation, which is known to be incapable of producing the correct static dielectric function because of the so-called gap problem (LDA underestimates bandgaps by about 50% because of the poor treatment of exchange and correlation), they lead to very satisfactory results when the many-body effect of a quasiparticle energy shift in the optical response is included in the theory [166, 167]. This is achieved by introducing a $\vec{k}$-dependent shift in the conduction bands so as to fit some of the bandgaps to their experimental values. The theoretically predicted pressure dependence of the low-frequency dielectric constant ϵ_∞ of Ge and GaAs obtained by this method [141] is in remarkably good agreement with the experimental results of Fig. 13.

An interesting result of these total-energy calculations [137, 141] concerns the variation with pressure of ϵ_∞ for Si, which, in contrast to that of Ge, is predicted to be very small. This becomes less surprising after the spectral form of $\epsilon_2(\omega)$ for these materials is examined (see Figs. 3 and 4 of Ref. [141]). The average optical gap is for all three semiconductors about 5 eV. Although for Si the optical oscillator strength is concentrated in a narrow band ~2 eV in width around the Penn gap, this distribution is much wider (about 5 eV) for Ge and GaAs, leading to an appreciable optical oscillator strength down to the near-infrared region. This makes ϵ_∞ of Si much more insensitive to pressure compared to that of Ge and GaAs, as is also evident from the only available "low-pressure" refractive-index measurements for Si [4]. The interpretation of such behavior is supported by refractive-index measurements on diamond under pressure [69, 144]. For this very wide-gap material one also finds a comparably small pressure coefficient of ϵ_∞.

V. Optical Absorption Near the Fundamental Gap

Since the pioneering work of Welber *et al.* [99] on the pressure dependence of the direct bandgap of GaAs, optical absorption measurements with the diamond anvil cell have been continuously pursued for almost every semiconductor in order to determine the variation with pressure of the one of the fundamental properties of semiconductors, namely the bandgap. This technique is characterized by its *relative* simplicity and a straightforward interpretation of the measured spectra in terms of the dielectric function of the semiconductor. The measurement of the reflectivity at normal incidence in the DAC is the complementary technique to absorption, because by these means it is possible to obtain information about higher-energy interband optical transitions, for which the absorption coefficient is so high that it would require the use of unrealistically thin samples. To date, many results on the pressure dependence of the E_1 or E_2 optical gaps in several semiconducting materials still must be traced back to the reflectivity work of Zallen and Paul (see, e.g., Refs. [8, 9]). A drawback of this technique resides in its limited sensitivity and spectral resolution. This, however, can be overcome by modulating an external or internal sample parameter such as temperature or electric field, in such a way that one measures the derivative of the reflectivity. These experiments conform to the so-called modulation spectroscopy, which is the subject of another section.

1. Absorption Spectral Functions

The frequency-dependent absorption coefficient $\alpha(\omega)$ is directly related through Eq. (A.15) to the imaginary part of the dielectric function $\epsilon_2(\omega)$, which is a measure of the power dissipated in a material medium during the passage of a light beam. From a microscopic point of view the main contribution to $\epsilon_2(\omega)$ and thus $\alpha(\omega)$ is given by the probability for optical interband *direct* transitions between valence and conduction band states [see Eq. (A.23)], which in the dipole approximation is proportional to the square of the optical matrix element $M_{v,c}$ and the joint density of states $J_{v,c}(\omega)$. The main contributions to the absorption coefficient arise from the singularities of $J_{v,c}$ that is, the critical points (CPS), as given by Eq. (A.25). Its magnitude and lineshape depends on the CP type and dimensionality. The selection rules for direct optical transitions are determined by the dipole matrix element (A.20). In the vicinity of a CP at $\vec{k}_0$ the matrix element $M_{v,c}(\vec{k})$ is a slowly varying function and it can be expanded in powers of k [168]:

$$M_{v,c}(\vec{k}) = M_{v,c}(\vec{k}_0) + (\vec{k} - \vec{k}_0) \cdot \vec{\nabla}_k M_{v,c}|_{\vec{k}_0} + \ldots . \qquad (5.1)$$

The first term corresponds to dipole *allowed* transitions. In this case the dipole matrix element is nothing else but the expectation value of the momentum operator $\vec{p}$ which transforms like a vector under inversion, therefore connecting states of different parity. A typical example is the fundamental bandgap E_0 at the Γ point of BZ in tetrahedral semiconductors between the top of the valence band which has p character and the bottom of the conduction band with s character. This is a 3-D CP of the M_0 type for which the joint density of states exhibits a square-root dependence on energy, if excitonic effects are completely neglected. Thus, the absorption edge is also square-root-like:

$$\alpha(\omega) \sim \frac{|M_{v,c}|^2}{\hbar \omega} (\hbar \omega - E_0)^{1/2}. \qquad (5.2)$$

The matrix element for dipole *forbidden* transitions, such as between states of the same parity, is given by the second term of Eq. (5.1). For 3-D M_0-type CP one obtains

$$\alpha(\omega) \sim \frac{|\vec{\nabla}_k M_{v,c}|^2}{\hbar \omega} (\hbar \omega - E_0)^{3/2}. \qquad (5.3)$$

To this point we have considered only the case of direct optical transitions, that is, k-conserving vertical transitions in $\vec{k}$ space. Many semiconductors are indirect-bandgap materials at ambient pressure or become indirect after the pressure-induced Γ–X crossover. In this case the valence and conduction band minima occur at different points of the BZ and the wave vector difference between them is typically 1000 times larger than the photon wave vector. Therefore, indirect optical transitions by photon absorption can take place only if assisted by phonons with the necessary wave vector $\vec{Q}$ for momentum conservation. The transition probability between an initial valence band state $|i\rangle$ with wave vector $\vec{k}_i$ and a final conduction band state $|f\rangle$ at $\vec{k}_f$ is given in second-order perturbation theory by [168]

$$W_{fi} = 2\pi\hbar \sum_{\vec{k}_i,\vec{k}_f,m} \frac{|\langle f|\mathcal{H}'_{\text{e-ph}}|m\rangle|^2 \cdot |\langle m|\mathcal{H}'_R|i\rangle|^2}{(E_c(\vec{k}_i) - E_v(\vec{k}_i) - \hbar\omega)^2} \quad (5.4)$$
$$\times \delta(E_c(\vec{k}_f) - E_v(\vec{k}_i) - \hbar\omega \pm \hbar\Omega_Q),$$

where m denotes virtual intermediate states, and $\mathcal{H}'_{\text{e-ph}}$ and $\mathcal{H}'_R$ are the electron–phonon and radiation interaction Hamiltonians, respectively. The $+(-)$ sign corresponds to a process where a phonon with energy $\hbar\Omega_Q$ is created (destroyed), and the delta function accounts for energy conservation between final and initial states. An example of such a two-step absorption process is schematically shown in Fig. 14, together with the corresponding Feynman diagram. In this case a photon with energy $\hbar\omega$ is absorbed by creating a virtual electron–hole pair, from which the electron is subsequently scattered by a phonon and changes its wave vector from $\vec{k}_i$ to $\vec{k}_f$. The other possible process, which involves the scattering of the hole, has been neglected because it is much less probable [168].

The indirect absorption coefficient is obtained from the transition probability (5.4) by integration assuming parabolic band extrema with scalar effective masses [168],

$$\alpha(\omega) = \frac{C(T)|M_R|^2|M_{\text{e-ph}}|^2}{n(\omega)\hbar\omega} \int dE_v(\vec{k}_i) \int dE_c(\vec{k}_f) \frac{\sqrt{E_v(\vec{k}_i)}\sqrt{E_c(\vec{k}_f) - E_g}}{(E_c(\vec{k}_i) - E_v(\vec{k}_i) - \hbar\omega)^2} \quad (5.5)$$
$$\times \delta(E_c(\vec{k}_f) - E_v(\vec{k}_i) - \hbar\omega \pm \hbar\Omega_Q),$$

where E_g is the indirect gap energy, $n(\omega)$ is the refractive index of the material, and $C(T)$ is a factor which contains the temperature-dependent Bose–Einstein occupation numbers of the phonons. The integration over the delta function is straightforward and with the substitutions $E = E_v(\vec{k}_i)$, $E_{\text{ind}} = E_g \pm \hbar\Omega_Q$, and $E_c(\vec{k}_f) - E_v(\vec{k}_i) = E_0 + E(1 + m_h^*/m_e^*)$;

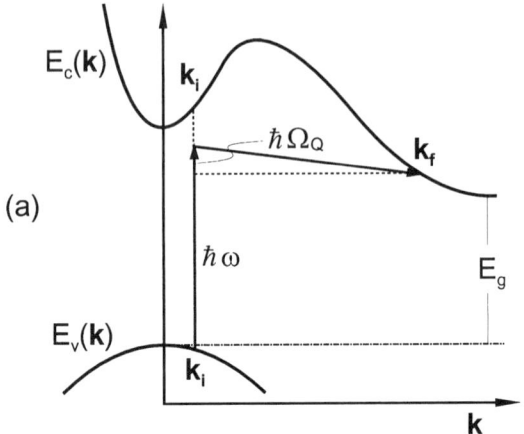

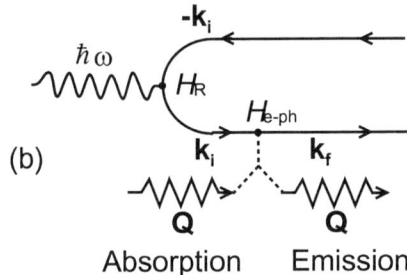

FIG. 14. (a) Schematic representation in $\vec{k}$-space of an indirect absorption process, where a photon is absorbed by creating an electron–hole pair with simultaneous absorption (emission) of a near-zone-boundary phonon. (b) The corresponding Feynman diagram.

we obtain

$$\alpha(\omega) = \frac{C(T)|M_R|^2|M_{e-ph}|^2}{n(\omega)\hbar\omega} \mathscr{A}^2(\omega) \qquad (5.6)$$

$$\mathscr{A}^2(\omega) = \int_0^{\hbar\omega - E_{ind}} \frac{dE\sqrt{E}\sqrt{\hbar\omega - E_{ind} - E}}{(E_0 - \hbar\omega + E(1 + m_h^*/m_e^*))^2}, \qquad (5.7)$$

where E_{ind} corresponds to the onset energy for indirect absorption.

In Eq. (5.7) we have intentionally introduced the direct gap energy E_0 in the energy denominator to emphasize the fact that when the photon energy approaches E_0 this expression diverges and the absorption becomes direct in nature. We thus have to distinguish between the resonant and nonresonant cases. In the nonresonant case, that is, for $\hbar\omega \ll E_0$, the energy denominator in (5.7) can be taken as constant for integration leading to a square law for the indirect absorption edge (without excitonic corrections):

$$\alpha(\omega) \sim \frac{1}{\hbar\omega} \int_0^{\hbar\omega - E_{\text{ind}}} dE \sqrt{E} \sqrt{\hbar\omega - E_{\text{ind}} - E}$$
$$\sim \frac{1}{\hbar\omega} (\hbar\omega - E_{\text{ind}})^2. \tag{5.8}$$

The variation of the energy denominator with photon energy for the resonant case has been calculated by Hartman [169] and extended by Dumke et al. [170] to account for heavy–light hole degeneracy:

$$\mathscr{A}^2 = \frac{m_{\text{lh}}^{*3/2} \mathscr{A}_l^2 + m_{\text{hh}}^{*3/2} \mathscr{A}_h^2}{m_{\text{lh}}^{*3/2} + m_{\text{hh}}^{*3/2}}$$

$$\mathscr{A}_{l,h}^2 = \frac{1}{c_{l,h}^2} \left[\frac{2 + (c_{l,h} - 2)x}{2\sqrt{1 + (c_{l,h} - 2)x + (1 - c_{l,h})x^2}} - 1 \right]. \tag{5.9}$$

Here, $c_{l,h} = 1 + m^*_{lh,hh}/m^*_e$ and the reduced frequency $x = (\hbar\omega - E_{\text{ind}})/(E_0 - E_{\text{ind}})$. If the photon energy $\hbar\omega$ approaches the direct gap E_0 ($x \to 1$), Eq. (5.9) diverges because of the contribution of intermediate states with the resonant energy denominator.

As usual, the effect of pressure on the optical absorption manifests itself in the variation of the energy gaps and of the electron and light hole effective masses. The latter can be calculated within $\vec{k} \cdot \vec{p}$ theory [see Eqs. (3.19) and (3.21)] by assuming a linear dependence of the two effective-mass parameters on direct bandgap energy:

$$m_e^*(P) = m_e^*(0) \frac{E_0(P)}{E_0(0)} \tag{5.10}$$

$$m_{\text{lh}}^*(P) = m_{\text{lh}}^*(0) \frac{E_0(P)}{E_0(0)}. \tag{5.11}$$

2. Results and Discussion

Optical absorption experiments with the DAC have contributed substantially to establishing the high-pressure behavior of energy gaps (direct as well as indirect) in tetrahedrally bonded semiconductors of the group IV [106, 171–173], III–V [99, 102, 174–179], II–VI [153, 180–188], I–VII [189], layered compounds [190–192, 154], and chalcopyrites [193–195]. Works where the main emphasis is on excitonic properties under pressure are discussed later. An important result is that the measured dependence on lattice constant of the bandgaps has become a stringent test for the accuracy of recently developed parameter-free *ab initio* methods for band-structure calculations [95–97, 196, 197, 141]. In this section we will present a few examples which are representative of the effect of pressure on the absorption edges and their lineshapes for direct and indirect gap materials.

The absorption coefficient α is obtained from Eq. (A.35) according to

$$\alpha(\omega) = \frac{1}{d} \cdot \ln\left[\frac{I_0(\omega)}{I_s(\omega)}\right] - c_0. \qquad (5.12)$$

where I_s is the intensity of light transmitted through the sample and I_0 is the reference intensity measured with light passing through the pressure cell next to the sample. The constant c_0 describes spurious background signal and it is adjusted for every spectrum to yield zero absorption at photon energies for which the sample should be totally transparent. If the thickness of the absorbing medium is not well known, it is more convenient to analyze the spectra in terms of the optical density OD defined in Eq. (A.36).

One common problem in the analysis of absorption spectra lies in the methods by which different authors extract the energy position of the bandgap from the measured absorption edge. For direct optical transitions, due to excitonic effects which are important even at room temperature, the absorption edge is not a square root as predicted by Eq. (5.2) but a steplike function. The gap energy is often determined from the absorption saturation point. This, however, could lead to a large underestimation of the gap energy if the spectrum is not representative of the full direct absorption edge profile, that is, when the maximum measurable absorption is too low because of stray-light problems. In the case that the spectra show a well-developed edge, the position of the gap is at best assigned to the zeros of the second derivative obtained by numerical differentiation of the spectra.

Figures 15 and 16 show absorption spectra of a 3-μm-thick Ge and a 1.7-μm-thick GaAs sample, respectively, at different pressures. The low-energy edge in the spectra, which is strongly steepened by excitonic effects, corresponds to absorption across the direct gap E_0, whereas the second feature about 300 meV higher in energy is attributed to optical transitions at the $E_0 + \Delta_0$ gap, with Δ_0 the valence-band spin–orbit splitting at Γ. The

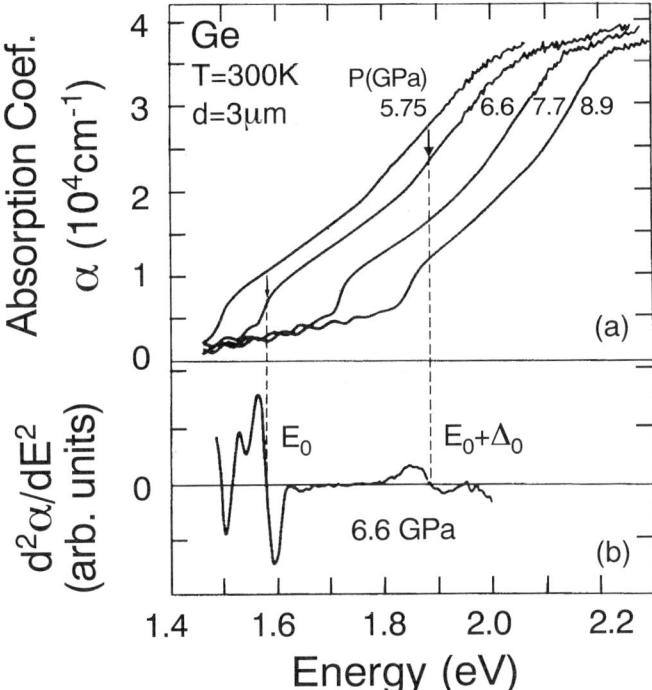

FIG. 15. (a) Absorption spectra of a 3-μm-thick Ge sample at different pressures and at 300 K. (b) Second derivative with respect to photon energy of the spectrum at 6.6 GPa. Arrows indicate the position of the gaps as obtained from the zeros of second derivative spectra. After Ref. [106].

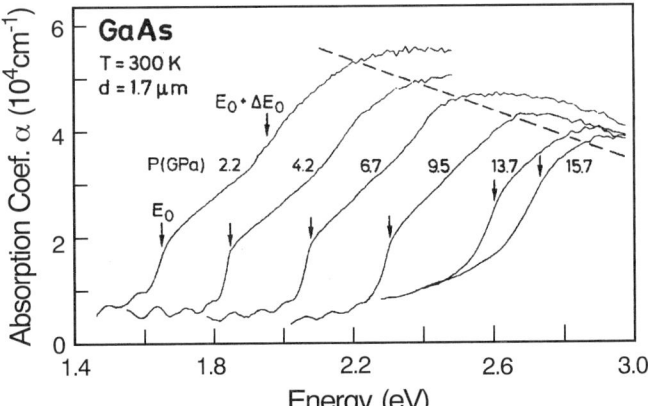

FIG. 16. Absorption spectra of a 1.7-μm-thick GaAs sample at different pressures and at 300 K. Arrows indicate the position of the direct absorption edge. After Ref. [102].

arrows in Figs. 15 and 16 correspond to the energy of the direct absorption edges as defined by the zeros of the second derivative spectra (see an example in Fig. 15b). We emphasize here that because of the use of extremely thin samples, it was possible to measure the absorption profile for photon energies much higher than the fundamental bandgap up to coefficients of $\sim 4 \times 10^4$ cm^{-1}. The dashed line in Fig. 16 indicates, for instance, the absorption level at which stray light becomes dominant.

The results for the E_0 and $E_0 + \Delta_0$ gap energies of Ge, GaAs, and ZnSe [153] are plotted in Fig. 17 as a function of pressure. Both gaps shift to higher energies with pressure, but while for Ge the dependence is essentially linear, for GaAs and ZnSe one finds a sublinear blueshift. A common practice among researchers in the field is to fit the pressure dependence of the direct gaps using second-order polynomials. Such a dependence, however, has no physical meaning and it turns out that the second-order

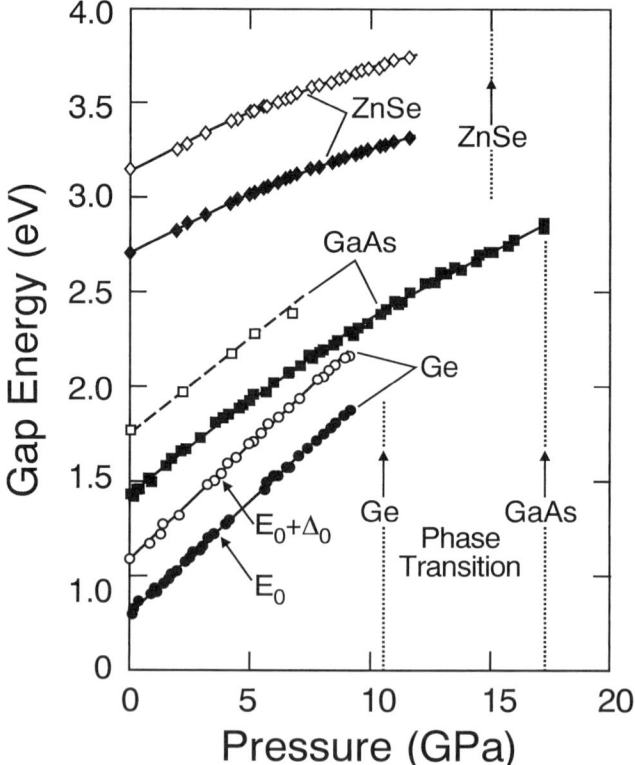

FIG. 17. Direct bandgap energies at 300 K of Ge [205], GaAs [205], and ZnSe [153] as a function of pressure as determined from optical absorption measurements.

coefficient depends on the particular experimental conditions such as the pressure range in which measurements are carried out. This makes difficult a direct comparison of the large body of experimental data available in the literature (see Ref. [198] for a discussion of this problem). The linear pressure coefficients for Ge, GaAs, and ZnSe are $dE_0/dP = 0.121(2)$ eV/GPa [106], 0.108(2) eV/GPa [102], and 0.079(2) eV/GPa [153], respectively. These values agree within 5% with results of other investigations.

Theoretical results based on *ab initio* pseudopotential and LMTO band-structure calculations [95, 197, 96] agree with these experimental pressure coefficients to within 5%. Although all these calculations were performed within the local-density approximation (LDA), which is known to yield absolute values for the bandgaps that are too small because of an overestimation of exchange effects, the gap variation with pressure (or volume) is well accounted for. This is because the quasiparticle self-energy correction to the gap energy is a slowly varying function of density ($\rho^{1/3}$), its contribution to the total gap change being only 1–2% [199].

We notice the difference in the pressure behavior of the E_0 gap of Ge and isoelectronic GaAs and ZnSe, all of them having very similar zero-pressure lattice parameter and pressure–volume relations [200–202]. A qualitative explanation for this difference can be given in terms of the Phillips–Van Vechten theory [14–16]. As expressed by Eq. (3.9) the energy gap in covalent materials is composed of a covalent or homopolar (E_h) and an ionic part (C). Although the covalent part E_h increases strongly with pressure for states with different bonding and antiboding character, as is the case for the E_0 gap, the ionic part C is hardly affected by volume changes. The quantity E_h is the same for all three materials, whereas $C = 0$ for Ge but $C = 2.9$ eV and 5.6 eV for GaAs and ZnSe [14, 15], respectively. Consequently, the E_0 gap should be less pressure dependent the greater the ionicity of the compound, as is experimentally observed.

Another result concerns the pressure dependence of the spin–orbit splitting parameter Δ_0 in Ge. This dependence is obtained from the difference of the E_0 and $E_0 + \Delta_0$ data measured at the same pressure and at room temperature or also at 10 K [203]. The low-temperature results are shown in Fig. 18. We find that the spin–orbit splitting of Ge increases slightly with pressure according to

$$\Delta_0[\text{eV}] = 0.294(5) + 0.0016(5) \cdot P[\text{GPa}]. \tag{5.13}$$

This result, corresponding to $d \ln \Delta_0/d \ln V = -0.40(12)$, agrees quite well with the value of 2 meV/GPa predicted by relativistic band-structure calculations [96, 204]. The small volume derivative of the spin–orbit splitting indicates a relative rigidity of the valence charge distribution in the atomic

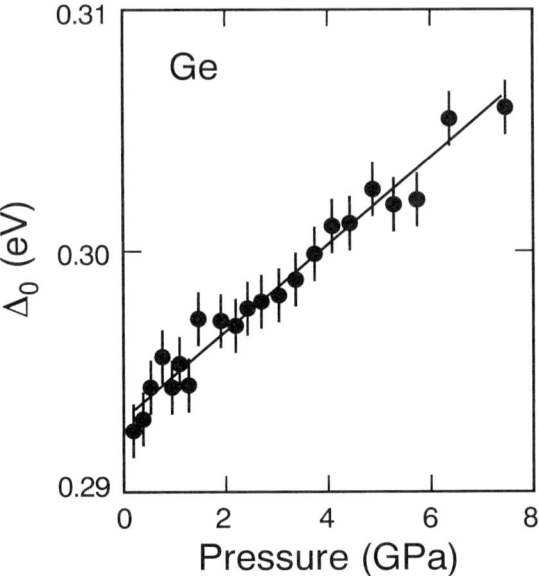

Fig. 18. Pressure dependence of the valence band spin–orbit splitting Δ_0 of Ge. After Ref. [203].

core region due to Pauli's exclusion principle. This is a quite general characteristic of the charge density in crystals; thus, this result is expected to be valid for other semiconductors also (see, for instance, Ref. [205] for GaAs and Ref. [153] for ZnSe).

The gradual broadening with pressure of the direct absorption edge at E_0 in GaAs (see spectra in Fig. 16) is attributed to the fact that above the pressure of the conduction band crossover, the indirect gap Γ_v–X_c is the lowest in energy. Using thick samples one can enhance the weak indirect absorption in order to study the pressure dependence of the indirect gap. Figure 19 shows some selected absorption spectra of GaAs measured with a 29-μm-thick sample for different pressures and at 300 K. For $P > 4$ GPa, the absorption exhibits a smooth onset followed by a quadratic dependence on photon energy as expected for an indirect bandgap without excitonic effects [see Eq. (5.8)]. The energy E_{ind} of the onset for indirect optical transitions is taken as the point where the experimental data show absorption above a linear background level.

The pressure dependence of the indirect gap $E_g^{\Gamma-X}$ is obtained from the absorption onset energy E_{ind} by taking into account that indirect transitions are assisted by emission or absorption of zone-edge phonons, so as to conserve crystal momentum. By analogy with Si [206, 207] one infers that at room temperature the indirect absorption onset corresponds to a process

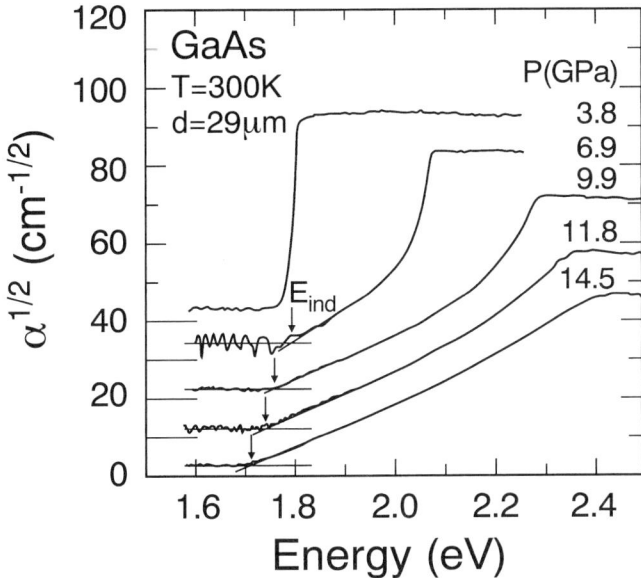

FIG. 19. Absorption spectra in the energy region below the direct gap obtained with a 30-μm-thick GaAs sample at different pressures above the Γ–X crossover. For clarity the curves are displaced vertically. After Ref. [208].

involving X-point LA phonons, which for GaAs have an energy of 28 meV [77, 78]. By subtracting the small pressure-dependent phonon frequency from the experimental results for E_{ind}, one obtains the values of the Γ–X indirect gap shown in Fig. 20 together with data for other semiconductors. The solid line through the GaAs data points corresponds to a linear relation with a negative pressure coefficient of −13.5(13) meV/GPa. The Γ–X indirect gap of GaAs thus decreases with pressure at a rate which is roughly one-tenth that of the direct gap E_0. We see that the Γ–X conduction band crossover for GaAs occurs at 4.2 ± 0.3 GPa (see Fig. 20) in agreement with luminescence measurements [100, 101]. As illustrated by the data of Fig. 20, a similar variation with pressure of the indirect Γ–X gap is found for many group IV and III–V compounds; the corresponding pressure coefficients are in the range from −10 to −20 meV/GPa.

We briefly turn to a more detailed analysis of the experimental absorption spectra in the energy range of the indirect bandgap. The spectra of Fig. 19 clearly indicate a departure of the indirect absorption from the expected square law for photon energies close to the direct gap E_0. As already discussed, this effect is attributed to a divergent energy denominator in the transition probability as the photon energy approaches the direct gap and

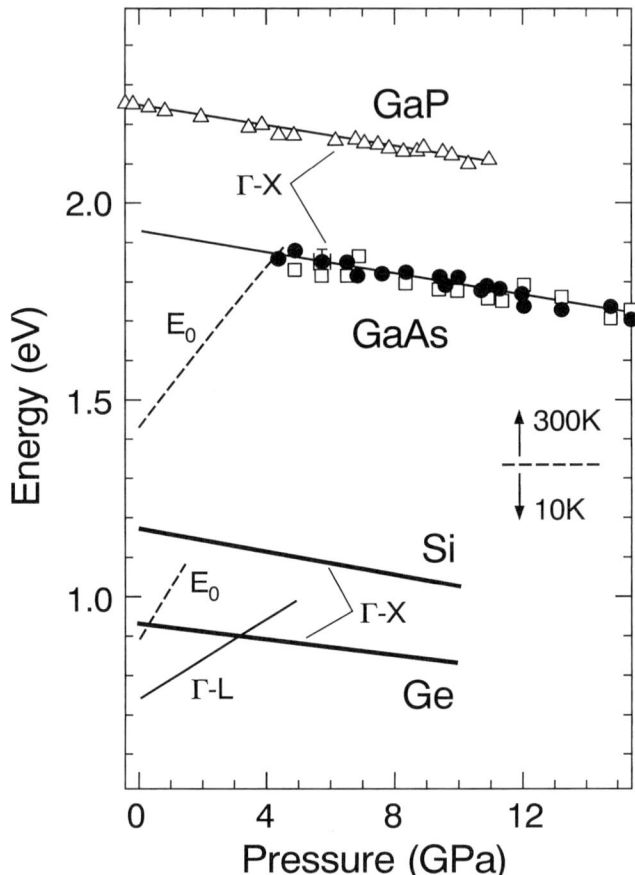

FIG. 20. Indirect $\Gamma-X$ gap energies of GaAs [102] and GaP [179] as a function of pressure. The open squares represent the energy of the indirect gap of GaAs obtained from fitting the model of Hartman [Eq. (5.9)] to optical absorption spectra. Pressure shifts of the $\Gamma-X$ gap of Si [342] and Ge [203] are also shown.

the virtual intermediate transitions for indirect absorption become resonant. Such energy dependence of the energy denominator is accounted for in the indirect absorption model due to Hartman [169] which is expressed by Eq. (5.9). Figure 21 shows absorption spectra of GaP [208], which is an indirect-gap material even at ambient pressure. The solid curves represent the results of least-squares fits to the measured spectra using the model Eq. (5.9). Three adjustable parameters were used: the indirect-gap energy $E_g^{\Gamma-X}$, a prefactor, and a constant background. Figure 21 suggests that the Hartman model describes the indirect absorption spectra very well even

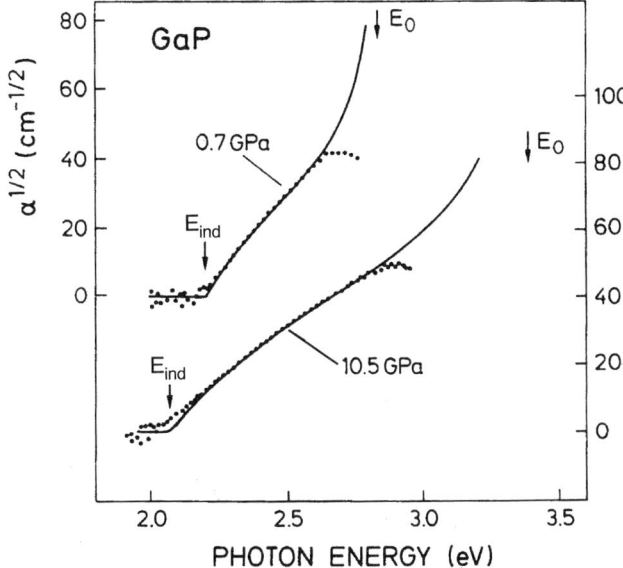

FIG. 21. Square root of the absorption coefficient of GaP at two different pressures (dots) as a function of photon energy. The solid curves represent the spectral dependence of the indirect absorption according to the Hartman model [Eq. (5.9)]. After Ref. [208].

up to the energy where absorption saturation due to stray light occurs. By using the pressure-dependent gap energies and effective masses the model also accounts for the effect of pressure on the indirect absorption (see Fig. 21).

VI. Effect of Pressure on Exciton Absorption

The optical properties of undoped semiconductors and, in particular, the near-bandgap optical absorption are strongly influenced by the effect of exciton formation, that is, of a photoexcited electron–hole pair coupled by Coulomb interaction. The energy spectrum of excitons resembles that of the hydrogen atom obtained by using the electron and hole effective masses and the static dielectric constant of the host material to account for the screening of the Coulomb potential [209–211]. Direct optical transitions into discrete exciton states give rise to a hydrogen-like series of absorption lines for photon energies just below the bandgap [212, 213], and transitions to the exciton continuum cause an enhancement of the absorption above the gap energy [209, 210]. Provided that the fine structure of discrete states is resolved in absorption spectra, valuable information about the dependence on lattice constant of fundamental exciton parameters such as the binding energy, optical matrix elements and lifetimes is obtained from high-pressure experiments. A particularly interesting problem concerns the pressure-induced broadening of exciton absorption lines, which is observed in direct-gap materials such as GaAs [214] and Ge [203] above the pressure of the Γ–X crossover. Because in high-quality crystals the exciton linewidth is primarily determined by electron–phonon scattering, one is able to obtain from the broadening of the exciton line above the conduction-band crossover accurate values of phonon deformation potentials for phonon-assisted Γ–X intervalley scattering processes.

1. ENERGY SPECTRUM OF EXCITONS

We begin this section with a brief account of the exciton parameters and their relation with the crystal band structure. Since the exciton energy spectrum depends on the dimensionality of the system, we discuss the properties of excitons in three and two dimensions and the intermediate case of a strongly anisotropic bulk material. The latter case holds not only for a layered semiconductor such as InSe, but sometimes also for quantum well heterostructures.

a. Excitons in the Bulk

When a photon is absorbed by promoting an electron into the empty conduction band of a semiconductor, the photoexcited electron and the hole left behind in the valence band are mutually attracted via Coulomb interaction, forming a two-particle complex. Although we are dealing here with a many-body problem it turns out that the two-particle picture is a very good approximation for the description of band excitons. The reason is that the Hartree part of the Coulomb interaction of the conduction-band electron with the almost filled valence band cancels out with the contribution of the positive background. There remains only the contribution from the exchange interaction of the electron in its initial valence band state, which is missing in the final state because of the orthogonality of conduction- and valence-band wave functions [168]. This exchange Coulomb interaction is negative, so it can be reviewed as an attractive electrostatic potential between electron and hole. The Hamiltonian for an electron (hole) with coordinate $\vec{r}_e$ ($\vec{r}_h$) and wave vector $\vec{k}_e$ ($\vec{k}_h$) simply reads

$$\mathcal{H} = \mathcal{H}_0 + \mathcal{H}' \tag{6.1}$$
$$= \sum_{i=e,h} \left[\frac{\hbar^2 k_i^2}{2m_i} + U(\vec{r}_i) \right] - \frac{e^2}{\epsilon(0)|\vec{r}_e - \vec{r}_h|}.$$

The exciton wave function $\Phi_{\vec{K}}$ ($\vec{K} = \vec{k}_e - \vec{k}_h$) is a superposition for products of electron and hole Bloch functions:

$$\Phi_{\vec{K}}(\vec{r}_e, \vec{r}_h) = \sum_{\vec{k}} A_{v,c}(\vec{k}) \cdot \varphi_{v,\vec{k}}(\vec{r}_h) \cdot \varphi_{c,\vec{k}+\vec{K}}(\vec{r}_e). \tag{6.2}$$

Excitonic effects are most significant in the optical properties of semiconductors at the critical points. This means that for constructing the exciton wave function one needs to consider only Bloch functions with wave vectors in a small region around the critical point at $\vec{k}_0$. In this case the Bloch factors are taken as wave-vector independent so that the exciton wave function simplifies to

$$\Phi_{\vec{K}}(\vec{r}_e, \vec{r}_h) = u_{v,0}(\vec{r}_h) \cdot u_{c,0}(\vec{r}_e) \sum_{\vec{k}} A_{v,c}(\vec{k}) e^{-i\vec{k}\cdot\vec{r}_h} e^{i(\vec{k}+\vec{K})\cdot\vec{r}_e}$$
$$= u_{v,0}(\vec{r}_h) \cdot u_{c,0}(\vec{r}_e) \cdot F_{v,c,\vec{K}}(\vec{r}_e, \vec{r}_h), \tag{6.3}$$

where we have set $\vec{k}_0 = 0$ without loss of generality. The function $F_{v,c,\vec{K}}$ defines the exciton envelope. In the effective mass approximation one obtains, by using the two-particle Hamiltonian (6.1) and the exciton wave function (6.3), a Schrödinger equation for the exciton envelope function [168]:

$$\left[-\frac{\hbar^2 \nabla_{r_e}^2}{2m_e^*} - \frac{\hbar^2 \nabla_{r_h}^2}{2m_h^*} - \frac{e^2}{\epsilon(0)|\vec{r}_e - \vec{r}_h|} \right] F_{v,c,\vec{K}}(\vec{r}_e, \vec{r}_h)$$
$$= (E - E_g) F_{v,c,\vec{K}}(\vec{r}_e, \vec{r}_h). \tag{6.4}$$

The energies are now measured relative to the bandgap energy E_g.

Equation (6.4) can be separated into two decoupled equations for the center of mass (CM) and the relative motion by introducing the corresponding coordinates defined by

$$\vec{r} = \vec{r}_e - \vec{r}_h, \quad \vec{R} = \frac{m_e^* \vec{r}_e + m_h^* \vec{r}_h}{M}, \tag{6.5}$$

respectively, with $M = m_e^* + m_h^*$ and the reduced mass $1/\mu = 1/m_e^* + 1/m_h^*$. The exciton envelope function also splits into two components: a plane wave for describing the uniform CM motion and an envelope for the relative motion, which gives the probability of finding electron and hole separated by a distance r:

$$F_{v,c,\vec{K}} = e^{i\vec{K}\cdot\vec{R}} \sum_{\vec{k}} A_{v,c}(\vec{k}) e^{i\vec{k}\cdot\vec{r}} = e^{i\vec{K}\cdot\vec{R}} F_{v,c}(\vec{r}). \tag{6.6}$$

For the relative envelope wave function one obtains a Schrödinger equation,

$$\left[-\frac{\hbar^2 \nabla_r^2}{2\mu} - \frac{e^2}{\epsilon(0)r} \right] F_{v,c}(\vec{r}) = \left(E - E_g - \frac{\hbar^2 K^2}{2M} \right) F_{v,c}(\vec{r}), \tag{6.7}$$

which is formally identical to that of the hydrogen atom. The solutions are thus a Rydberg series of discrete exciton states with envelope functions which are the product of a Hermite polynomial for the radial part and a spherical harmonic for the angular part:

$$F_{nlm}(\vec{r}) = H_{nl}(r) \cdot Y_l^m(\theta, \varphi). \tag{6.8}$$

The energy eigenvalues are given by

$$E = E_g + \frac{\hbar^2 K^2}{2M} - \frac{\mathcal{R}}{n^2}, \quad n = 1, 2, 3\ldots \tag{6.9}$$

$$\mathcal{R} = \frac{\mu e^4}{2\hbar^2 \epsilon(0)^2}, \quad a_B = \frac{\hbar^2 \epsilon(0)}{\mu e^2}, \tag{6.10}$$

where $\mathcal{R}$ and a_B are the exciton Rydberg (or binding energy) and the effective exciton Bohr radius, respectively. The discrete spectrum is followed by a continuum of states with kinetic energies $E = \hbar^2 k^2/2\mu$ relative to E_g. Using the parameters of GaAs, for example, we obtain a binding energy of ~5 meV and a Bohr radius of about 100 Å.

For a strongly anisotropic material with axial symmetry, such that either the dielectric constant or the effective masses or both differ much in the direction parallel ($\parallel$) and perpendicular ($\perp$) to the symmetry axis, the hydrogen-atom-like effective Hamiltonian is also anisotropic, and analytical solutions of the Schrödinger equation for the exciton envelope function can be found only in very special cases. In a perturbative approach Gerlach and Pollmann [215] have obtained approximate solutions for different bound states with quantum numbers n, l, m in terms of the anisotropy parameter $\alpha = 1 - (\mu_\perp \epsilon(0)_\perp)/(\mu_\parallel \epsilon(0)_\parallel)$:

$$\mathcal{R} = \frac{e^4 \mu_\perp}{2\hbar^2 \epsilon_\perp \epsilon_\parallel n^2} \cdot z_{lm}^2(\alpha). \tag{6.11}$$

For the 1s exciton ground state the correction factor z_{lm} takes the form

$$z_{00}(\alpha) = \begin{cases} \frac{1}{\sqrt{\alpha}} \arcsin(\sqrt{\alpha}) & \alpha > 0 \\ 1 & \alpha = 0 \\ \frac{1}{\sqrt{|\alpha|}} \operatorname{arcsh}(\sqrt{|\alpha|}) & \alpha < 0. \end{cases} \tag{6.12}$$

If the carriers are free to move only in a plane, that is, $\mu_\parallel \to \infty$, the anisotropy parameter is equal to 1 and $z_{00} = \pi/2$, giving an enhancement factor of about 2.5 for the exciton Rydberg.

b. 2-D Excitons

In two dimensions the hydrogenic Schrödinger equation for the relative motion is solved in polar coordinates, yielding for the envelope function the general form [216]

$$F(\vec{r}) = \frac{1}{\sqrt{2\pi}} R(r) e^{im\theta}. \tag{6.13}$$

The solutions with energies lower than the bandgap correspond again to the discrete spectrum, but the radial part of the envelope function is now given in terms of Laguerre polynomials,

$$R(r) = e^{-\rho/2} \rho^{|m|} L_{n+|m|}^{2|m|}(\rho), \rho = \frac{r}{a_B \lambda}, \frac{1}{\lambda^2} = -\frac{4E}{\mathcal{R}}, \tag{6.14}$$

and the energies (referred to E_g and with $K \equiv 0$) are

$$E_n = -\frac{\mathcal{R}}{\left(n+\frac{1}{2}\right)^2}, n = 0, 1, 2, \ldots, \tag{6.15}$$

where the exciton Rydberg and effective Bohr radius are defined as in (6.10). For the continuum the eigenvalues have the same formal expressions as in the 3-D case. We note that in 2-D the ground-state energy is $E_0 = -4\mathcal{R}$.

Quantum well systems correspond to an intermediate case between the ideal 3-D and 2-D limits. On the one hand, the CM motion is strictly constrained to the plane of the quantum well, whereas the spatial extension of the electron and hole wave functions in the growth direction cannot be neglected. Nevertheless, the most important consequence of the confinement of the carriers forming an exciton in quantum wells concerns the enhancement of the electrostatic attraction which reflects itself in the increase of the exciton binding energy. The usual approach for calculating exciton binding energies in QWs consists in finding the variational parameters of a trial envelope function $\Psi(\vec{r}, z_e, z_h)$ for an effective-mass two-band Hamiltonian [217, 218]

$$\mathcal{H}_{exc}^{QW} = \sum_{i=e,h} \left[-\frac{\hbar^2 \partial_{z_i}^2}{2m_i^*} + U_i \cdot \theta(|z_i| - L/2) \right] \\ - \frac{\hbar^2 \nabla_{x,y}^2}{2\mu_\perp} - \frac{e^2}{\epsilon(0)\sqrt{r^2 + (z_e - z_h)^2}}, \tag{6.16}$$

where $\theta(z)$ is the unit step function, U_i is the well depth, and L is the well width. The trial function for the exciton ground state is the product of the electron and hole envelopes $\zeta_e(z_e)$, $\zeta_h(z_h)$ of the QW confined states times a function ϕ_{e-h} which depends on the electron–hole separation:

$$\psi(\vec{r}, z_e, z_h) = \zeta_e(z_e) \cdot \zeta_h(z_h) \cdot \phi_{e-h}(\vec{r}, z_e - z_h). \tag{6.17}$$

The best choice for ϕ_{e-h} is a 1s-like function of the hydrogen atom,

$$\phi_{e-h}(r, z) = c e^{-\frac{\sqrt{r^2+z^2}}{\lambda}}, \tag{6.18}$$

with two variational parameters c and λ, which are to be determined by minimization of the energy for different well widths L. The result is a monotonic increase of the exciton binding energy by more than a factor of 2 for decreasing well widths down to $L \sim 50$ Å. Below this value the binding energy starts to decrease, because for very thin wells the confinement effects are reduced by substantial penetration of the wave functions into the barriers [217, 218].

2. Exciton Absorption

Among the diverse optical properties of semiconductors, the most striking excitonic effects are observed in the near-bandgap absorption. In high-quality crystals, different absorption lines corresponding to optical transitions to discrete exciton states can be resolved at low temperatures [219–221]. This enables the determination of several exciton parameters from a lineshape analysis of the absorption spectra, which can be compared to results of band-structure calculations in order to obtain details of the electronic structure. In this subsection we summarize the results concerning the spectral lineshape of exciton absorption.

a. *Absorption Coefficient for Direct Excitonic Transitions*

In order to account for excitonic effects on the optical absorption, we proceed in an analogous way as for uncorrelated electron–hole pairs by calculating the imaginary part of the dielectric function $\epsilon_2(\omega)$ for direct optical interband transitions. As we will see, exciton formation affects mainly the optical matrix element enhancing the oscillator strength for optical transitions, whereas the combined density of states remains unaltered (besides for the additional delta-like peaks corresponding to the discrete exciton states). For vertical transitions in $\vec{k}$ space the wave vectors of the photoexcited electron and hole are the same. This leads to the formation of excitons with vanishing CM momentum ($\vec{K} \equiv 0$).

Essentially, we need to calculate the dipole matrix element for optical transitions between the initial state Ψ, which corresponds to a filled valence band with N electrons and an empty conduction band, and a final exciton state with wave function $\Phi_0(\vec{r}_e, \vec{r}_h)$ given by (6.2). The optical matrix element is [168]

$$\int \Phi_0 \hat{e}_L \cdot \vec{p}_i \Psi \, d\tau_1 \cdots d\tau_N = \sum_{\vec{k}} A_{v,c}(\vec{k}) \langle c\vec{k} | \hat{e}_L \cdot \vec{p} | v\vec{k} \rangle, \qquad (6.19)$$

TABLE II

Various Quantities Entering Into Eq. (6.22) for Calculating, Within the Effective Mass Approximation, the Absorption Coefficient for Dipole-Allowed Direct Excitonic Transitions in Three and Two Dimensions. The Quantity $\mathscr{V}(\mathscr{S})$ is the Quantization Volume (Surface) for The Normalization of The Bloch Wavefunction in 3D(2D).

	3-D		2-D
Discrete states	$\|F_0(0)\|^2 = \dfrac{1}{\pi a_B^3 n^3}$ $n = 1, 2, \ldots$		$\|F_0(0)\|^2 = \dfrac{1}{\pi a_B^2 (n + 1/2)^3}$ $n = 0, 1, \ldots$
		$\dfrac{1}{a_B^3} = \left(\dfrac{2\mu}{\hbar^2}\right)^{3/2} \mathscr{R}^{3/2}$ $\mathscr{R} = \dfrac{\mu e^4}{2\hbar^2 \varepsilon_0^2}$ $J_{v,c} = \delta(E - E_n)$	
	$E_n = E_g - \dfrac{\mathscr{R}}{n^2}$		$E_n = E_g - \dfrac{\mathscr{R}}{(n + 1/2)^2}$
Continuum spectrum	$\|F_0(0)\|^2 = \dfrac{1}{\mathscr{V}} \dfrac{\lambda e^\lambda}{sh\lambda}$		$\|F_0(0)\|^2 = \dfrac{1}{\mathscr{S}} \dfrac{e^\lambda}{ch\lambda}$
		$\lambda = \sqrt{\dfrac{\pi^2 \mathscr{R}}{E - E_g}}$	
	$J_{v,c} = \mathscr{V} \dfrac{\sqrt{2}\,\mu^{3/2}}{\pi^2 \hbar^3} \sqrt{E - E_g}$		$J_{v,c} = \mathscr{S} \dfrac{\mu}{\pi \hbar^2}$

where $M_{v,c} \equiv \langle c\vec{k}|\hat{e}_L \cdot \vec{p}|v\vec{k}\rangle$ is nothing but the dipole matrix element (5.1) for uncorrelated electrons. In first approximation it is wave-vector independent so it can be pulled out of the summation in (6.19), giving

$$\int \Phi_0 \hat{e}_L \cdot \vec{p}_i \Psi \, d\tau_1 \cdots d\tau_N = M_{v,c} \cdot F_0(0), \tag{6.20}$$

where $F_0(0) = \sum_{\vec{k}} A_{v,c}(\vec{k})$ is after Eq. (6.3) the exciton envelope function at the origin ($\vec{r} = 0$). Finally, we just have to replace the dipole matrix element (6.20) in Eq. (A.23) for the absorption coefficient to obtain

$$\alpha_{exc}(\omega) = \alpha_0(\omega) \cdot |F_0(0)|^2, \tag{6.21}$$

where $\alpha_0(\omega)$ represents the direct absorption coefficient calculated without taking into account the Coulomb interaction in the final state between electron and hole. This is an important result, which states that the enhancement of the optical matrix element for dipole allowed transitions depends on the probability of finding electron and hole at the same location ($r = 0$). This implies, for instance, that only optical transitions to s-like exciton states are allowed, since these are the only ones which have nonvanishing

4 OPTICAL PROPERTIES OF SEMICONDUCTORS UNDER PRESSURE

envelopes at the origin. Equation (6.21) is very general and holds in any dimension. In Table II we summarize the results for the exciton absorption coefficient for dipole allowed direct transitions in 3-D [209, 210] and 2-D [216] calculated using

$$\alpha_{\text{exc}}(\omega) = \frac{16\pi^2 e^2}{m_0^2 c n(\omega) \omega} \cdot |M_{v,c}|^2 \cdot J_{v,c}(\omega) \cdot |F_0(0)|^2. \tag{6.22}$$

b. Finite Linewidth and Absorption Lineshape

Excitonic states are broadened through exciton–phonon interaction, giving rise, in the weak exciton–phonon coupling limit, to a Lorentzian lineshape [211]. This is a good *Ansatz* for high-quality bulk crystals. In quantum well structures, however, compositional disorder and well-width fluctuations are responsible for the inhomogeneous broadening of the excitons with Gaussian lineshape [222]. A finite linewidth of the exciton states can be introduced in the expressions for the 3-D and 2-D absorption coefficients of Table II by means of a Lorentzian [223] or a Gaussian [224] convolution function, respectively. The absorption coefficient as a function of incident photon energy $\hbar\omega$ then takes the forms

$$\alpha^{3D}(\omega) = \frac{C_0 \mathcal{R}^{1/2}}{\hbar\omega} \left\{ \sum_{m=1}^{\infty} \frac{2\mathcal{R}}{m^3} \frac{\Gamma_m}{(\hbar\omega - E_m)^2 + \Gamma_m^2} \right.$$
$$\left. + \int_{E_g}^{\infty} dE \frac{1}{1 - e^{-2\pi z}} \frac{\Gamma_c}{(\hbar\omega - E)^2 + \Gamma_c^2} \right\}, \tag{6.23}$$

$$C_0 = \frac{8(2\mu)^{3/2} e^2 |M_{v,c}|^2}{m_0^2 c \hbar^2 n(\omega)}, \quad z^2 = \frac{\mathcal{R}}{E - E_g} \tag{6.24}$$

and

$$\alpha^{2D}(\omega) = \frac{C_0}{\hbar\omega} \left\{ \sum_{m=0}^{\infty} \frac{2\mathcal{R}}{\left(m + \frac{1}{2}\right)^3} \frac{1}{\Gamma_m} \exp\left(-\frac{(\hbar\omega - E_m)^2}{2\Gamma_m^2}\right) \right.$$
$$\left. + \int_{E_g}^{\infty} \frac{dE}{\Gamma_c} \frac{1}{1 + e^{-2\pi z}} \exp\left(-\frac{(\hbar\omega - E)^2}{2\Gamma_c^2}\right) \right\}, \tag{6.25}$$

$$C_0 = \frac{8\sqrt{2\pi} e^2 \mu |M_{v,c}|^2}{m_0^2 c \hbar n(\omega)}. \tag{6.26}$$

The first and second term in Eqs. (6.23) and (6.25), denoted as α_1 and α_2 (see Fig. 22), represent the contribution of the discrete states and the exciton continuum, respectively. An additional constant term α_0 is introduced to include the contribution of residual absorption from indirect transitions. The integrals can be performed analytically [147, 225]; thus, Eqs. (6.23) and (6.25) can be evaluated directly.

Experimental spectra near the exciton absorption edge can be now fitted by using Eqs. (6.23) and (6.25) with $C_0, E_g, \mathscr{R}, \Gamma_1$ as adjustable parameters. If the individual exciton lines are not well resolved we use the same linewidth $\Gamma_m = \Gamma_1$ for all discrete states and the empirical relation $\Gamma_c \approx 2\Gamma_1$ for the exciton continuum [226]. Figure 22 illustrates the result of a fit to the absorption data of Ge at low temperatures. For quantum well structures, several exciton lines are observed in the absorption spectra corresponding to different heavy and light-hole excitons because of the effects of confinement, which splits the conduction and valence band states into 2-D subbands. A description of the spectral dependence of the absorption edge in QW's is obtained by adding expressions such as Eq. (6.25) for each allowed optical transition. An example of spectra obtained for a GaInAs/AlInAs MQW structure is displayed in Fig. 23, where the features in absorption correspond to excitonic transitions between the first confined electron and

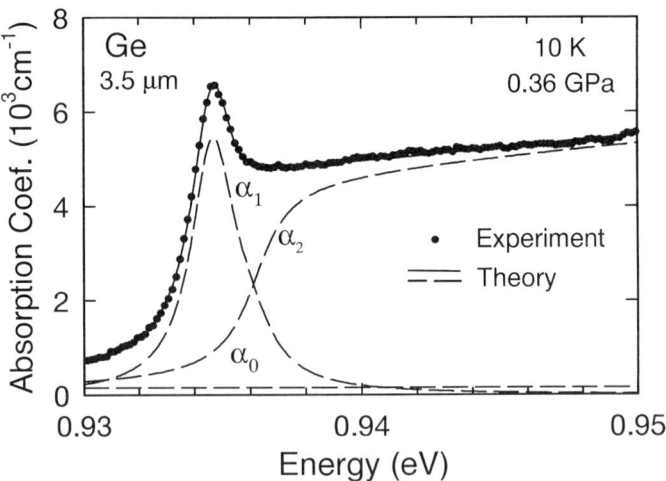

FIG. 22. Direct exciton absorption spectrum of Ge at 0.36 GPa and at 10 K. The solid dots are experimental data. The solid line corresponds to the fitted exciton absorption model, Eq. (6.23). The dashed lines represent the contributions of the bound exciton states (α_1), the exciton continuum (α_2), and the indirect absorption (α_0), respectively. After Ref. [203].

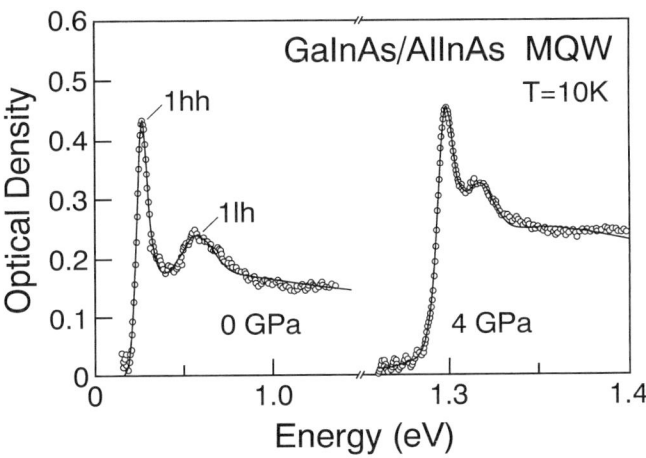

FIG. 23. Exciton absorption spectra of a $Ga_{0.47}In_{0.53}As/Al_{0.48}In_{0.52}As$ MQW system for two different pressures at 10 K. The solid lines represent the results of fits to the experimental data using Eq. (6.25). After Ref. [114].

the first heavy- (1hh) and light-hole subbands (1lh). Obviously, the excitonic features of the absorption edge in bulk materials as well as in QW structures are very well accounted for within the preceding model.

3. PRESSURE EFFECTS ON EXCITONS

The effect of pressure on excitonic linear and two-photon optical absorption has been investigated for a number of bulk semiconductors for which the exciton binding energy varies by two orders of magnitude ranging from a few milli-electron volts in group IV [203] and III–V [214] compounds, to tenths of a milli-electron volt in II–VI [51, 153, 227–229], I–VII [230], and layered materials [192, 154, 226], to about 139 meV for CuCl. The high-pressure behavior of excitons in systems of reduced dimensions such as a ZnSe/ZnMgSe single quantum well [231], GaInAs/AlInAs [114] and InGaAs/GaAs multiple quantum wells [232], and CuCl nanocrystals in a LiCl matrix [233] has been also studied by means of optical absorption. All these cases are concerned with excitons of the Wannier type (the exciton wave function is strongly localized in $\vec{k}$-space), which means that their energy spectrum is adequately described by a hydrogen-like model. Here we focus on the pressure dependence of exciton parameters such as binding energy, oscillator strength and linewidth, with particular emphasis in the last case on the pressure-induced broadening of exciton absorption lines due to the reduction of the exciton lifetime by phonon-assisted intervalley scattering processes above the conduction-band crossover. We close this section by demonstrating how useful high-pressure exciton absorption experiments can be in determining band structure parameters of bulk and QW structures from the presence dependence of the exciton absorption.

To illustrate the effect of pressure on the exciton absorption, we show in Fig. 24 several high-resolution spectra of Ge measured in the spectral region of the direct gap E_0 at different pressures and at 10 K. At low pressures a sharp peak is observed at the absorption edge, which is assigned to optical transitions into discrete exciton states. The absorption lines corresponding to different quantum numbers are not resolved, but the observed peak is dominated by the $n = 1$ ground state of the exciton. With increasing pressure the absorption edge shifts to higher energy (please note that for clarity the spectra in Fig. 24 are plotted relative to the peak energy E_1), while the exciton peak broadens monotonically. At about 3 GPa the peak feature has almost disappeared and the absorption profile becomes edge-like. These spectra changes, which are fully reversible for decreasing pressure, are very well reproduced by the model function (6.23) for the 3-D

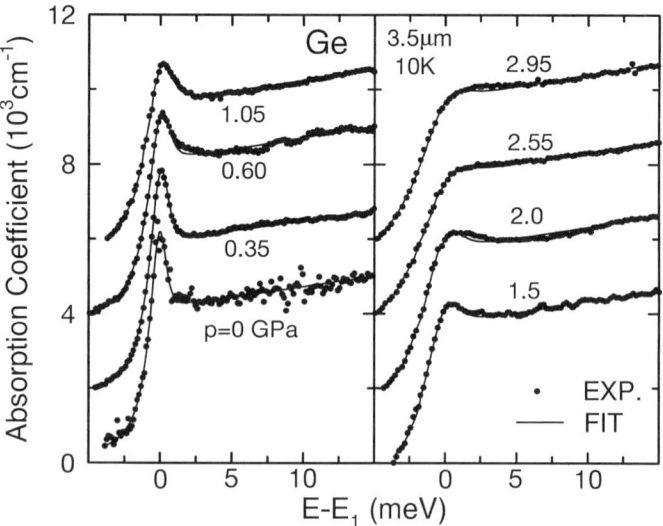

FIG. 24. Variation of the exciton absorption spectrum of Ge with pressure. Spectra are plotted relative to the energy E_1 of the lowest discrete exciton state. The solid dots are the experimental data and the solid curves correspond to the lineshape model function. After Ref. [203].

exciton absorption including a Lorentzian broadening (solid curves in Fig. 24).

a. Exciton Absorption Strength and Binding Energy under Pressure

An obvious but important point is that optical absorption measurements enable a very accurate determination of the bandgap variation with pressure. The gap energy is obtained from lineshape fits of Eq. (6.23) to the exciton absorption spectra according to $E_g = E_1 + \mathcal{R}$. Gap energies determined by other optical means are systematically underestimated by at least the exciton binding energy. The method of measuring the linear absorption, however, fails in the case of materials characterized by very large values of the absorption coefficient at the fundamental band gap, since this technique is limited by the thickness of the sample. For that purpose two-photon spectroscopy is particularly suitable because it allows the detection of the change in transmission of a weak probe beam induced by a pulsed, high-power laser [233].

The parameter C_0 of Eqs. (6.24) and (6.26) determines the strength of the exciton absorption as predicted by Elliott's theory [209, 210]. For excitons formed out of states near the conduction- and valence-band edges at the Γ point of the Brillouin zone, as is for example the case with Ge and GaAs, the reduced effective mass μ is, within $\vec{k} \cdot \vec{p}$ theory, proportional to the direct bandgap E_0 [see Eqs. (3.19) to (3.22)]. One thus expects that $C_0 \sim E_0^{3/2}$. The values of C_0 as obtained from fits to the exciton absorption spectra of Ge and GaAs for different pressures are plotted together versus $E_0^{3/2}$ in Fig. 25. The data points indeed lie on a straight line. The same behavior is found for the C_0 data of GaAs at 100 and 20 K. This, in fact, implies that the matrix element $M_{v,c}$ of the electron–photon interaction for dipole-allowed transitions, which is given by the interband momentum operator $p \approx 2\pi/a_0$ [113], is essentially independent of pressure. One can obviously neglect the small variation with pressure of the lattice parameter a_0 compared to the change of E_0. These results, which were obtained for different materials and at different temperatures, clearly indicate that it is possible to determine *absolute* values of the exciton absorption strength from the lineshape analysis of the measured spectra.

Figure 26 shows the results for the exciton binding energy in Ge (10 K) and GaAs (200 K) as a function of the direct gap energy E_0. The solid lines in Fig. 26 represent fits to the data points using a linear function. Again this behavior agrees with that expected from $\vec{k} \cdot \vec{p}$ theory, since after Eq. (6.10) the exciton Rydberg $\mathcal{R}$ varies in proportion to the reduced effective mass μ, that is, to the gap E_0, provided that the change with

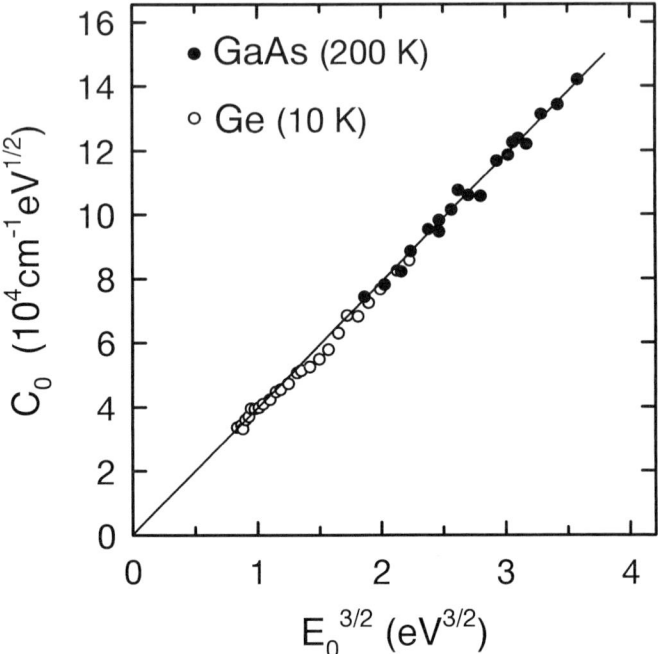

FIG. 25. The absorption strength parameter C_0 as defined in Eq. (6.24) versus energy gap $E_0^{3/2}$ for Ge and GaAs. The solid line is a fit to the data with a linear function.

pressure of the static dielectric constant $\epsilon(0)$ can be neglected [147]. The different slope of $\mathscr{R}$ versus E_0 for Ge and GaAs is due to the difference in the values of $\epsilon(0)$, which enters as the square in the expression (6.10) for the Rydberg. We point out that this behavior of the exciton Rydberg is very general and is found in several II–VI compounds as well [233].

The validity of the description of the E_0 exciton of diamond- and zinc-blende-type semiconductors in terms of a hydrogenic model is not obvious because of the degeneracy of the light- and heavy-hole bands. Equation (6.10) can nevertheless be used to estimate the Rydberg energy, if an average value $\bar{\mu}$ for the reduced mass is introduced:

$$\frac{1}{\bar{\mu}} = \frac{1}{m_e^*} + \frac{1}{2}\left(\frac{1}{m_{lh}^*} + \frac{1}{m_{hh}^*}\right). \tag{6.27}$$

By using the literature data [77, 78] for Ge ($m_e^* = 0.038$, $m_{lh}^* = 0.043$, $m_{hh}^* = 0.32$, $\epsilon(0) = 16.0$) and GaAs ($m_e^* = 0.068$, $m_{lh}^* = 0.076$, $m_{hh}^* = 0.36$, $\epsilon(0) = 12.5$) one finds from Eq. (6.10) values for $\mathscr{R}$ that are only 10% smaller than the experimental ones at ambient pressure. Actually, the binding energy in GaAs has been calculated more accurately [234] by a perturba-

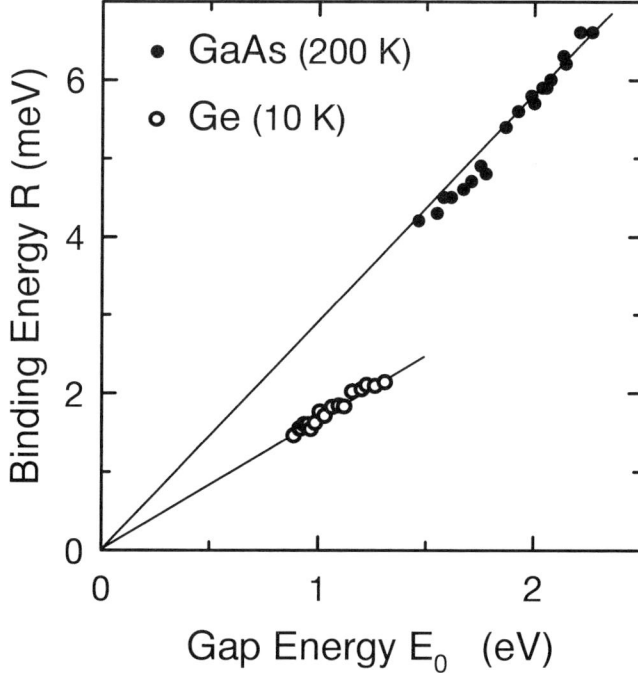

FIG. 26. The exciton Rydberg $\mathscr{R}$ given by Eq. (6.10) as a function of direct gap energy E_0 for Ge and GaAs. The solid lines represent fitted linear relations.

tion method from the theoretical energy position of the 1s and 2s levels of the fourfold-degenerate E_0 exciton, giving excellent agreement within 1% with the measured value.

A different pressure dependence of the Rydberg energy is found in layered semiconductors such as GaSe [154] and InSe [226] for the lowest-gap exciton: $\mathscr{R}$ decreases strongly with pressure. Figure 27 illustrates this peculiar dependence on pressure of the exciton binding energy in the case of InSe. The Rydberg is now given by Eq. (6.11), which takes into account the anisotropy of effective masses and dielectric constant in layered crystals. The symbols $\parallel$ and $\perp$ denote the corresponding components parallel and perpendicular to the c axis, the one perpendicular to the covalently bonded layers. By using the reduced effective masses of InSe from Ref. [235] ($\mu_\perp = 0.045$, $\mu_\parallel = 0.131$) and the dielectric constants at ambient conditions [236] ($\epsilon(0)_\perp = 9.7$, $\epsilon(0)_\parallel = 6.9$) one obtains a value of $z_{00} \approx 0.8$ and $\mathscr{R} = 14.3$ meV, in good agreement with the measured value (see Fig. 27).

The pressure dependence of the parameters in Eq. (6.11) is reasonably well known except for $\epsilon(0)_\parallel$. Within $\vec{k} \cdot \vec{p}$ theory we have $\mu_\perp \propto E_g$. The lowest direct gap of InSe, however, exhibits a strongly nonlinear variation with pressure, having a minimum at about 0.5 GPa, which is due to the

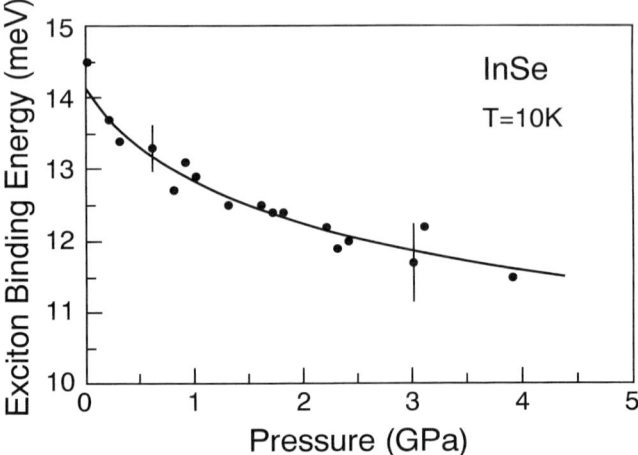

FIG. 27. Exciton binding energy $\mathscr{R}$ of InSe at 10 K as a function of pressure. The solid line corresponds to a fit using a second-order polynomial. After Ref. [226].

large anisotropy of intralayer and interlayer bonding [226]. The component $\epsilon(0)_\perp$ is, on the contrary, only weakly dependent on pressure [155]. Thus, to account for the observed dependence of $\mathscr{R}$, one finds from Eq. (6.11) that $\epsilon(0)_\parallel$ must increase linearly in the pressure range up to 4 GPa at the large rate of 1.35 GPa^{-1}. A similar pressure behavior of the exciton Rydberg has been observed for GaSe as well [154]. In this work it was argued that a strong increase of $\epsilon(0)_\parallel$ due to a pressure-induced charge redistribution from intralayer to interlayer space might be responsible for it. With increasing pressure GaSe tends to become a more isotropic material, at least for some macroscopic properties such as the static dielectric constant. In that case the variation of $\mathscr{R}$ with pressure can be accounted for by assuming that $\epsilon(0)_\parallel \approx \epsilon(0)_\perp$ at about 5 GPa [154]. For InSe the decrease of $\mathscr{R}$ between ambient pressure and 4 GPa indicates an even stronger increase of $\epsilon(0)_\parallel$ as compared to GaSe.

b. Pressure-Induced Exciton Line Broadening

We have pointed out earlier that pressure induces a clear broadening of the discrete exciton line, which has been observed in absorption spectra of Ge [203], GaAs [214], InSe [226], and In$_{0.2}$Ga$_{0.8}$As/GaAs MQW structures [232]. In the case of GaAs, this broadening starts at the Γ–X crossover

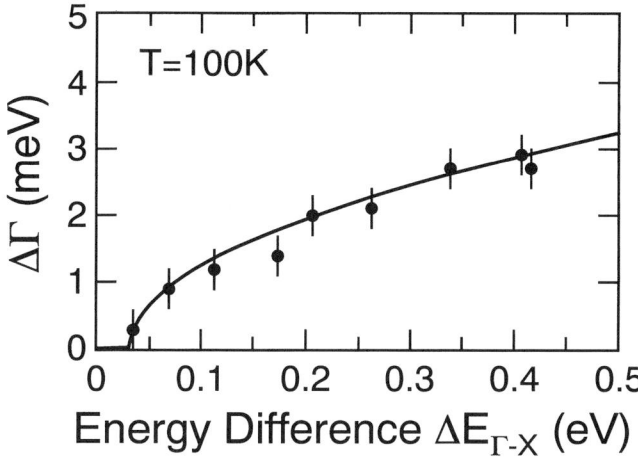

FIG. 28. Broadening $\Delta\Gamma$ of the direct exciton of GaAs for pressures above the Γ–X conduction band crossover at 100 K as a function of the conduction band energy difference $\Delta E_{\Gamma X}$. Solid curves are the results of fitting Eq. (6.29) to the experimental data. After Ref. [214].

pressure P_c. In Fig. 28 we have plotted the line broadening $\Delta\Gamma$, that is, the difference in linewidth referred to ambient pressure, at 100 K and as a function of the energy difference $\Delta E_{\Gamma X}$ between the Γ and X conduction-band valleys. For $P > P_c$, such broadening is interpreted in terms of phonon-assisted intervalley scattering of electrons from Γ to X conduction-band states, which gives rise to a decrease of the exciton lifetime. A similar behavior of the exciton linewidth is observed for Ge at low temperatures, as illustrated in Fig. 29. For pressures below ~0.6 GPa the half-width increases only slightly. In this range the line broadening is determined by phonon-assisted intervalley scattering from Γ to the lower-lying L minimum. There is also a pressure-independent contribution due to inhomogeneous broadening which depends on sample quality. Above 0.6 GPa the X valley becomes the absolute conduction-band minimum, and Γ–X intervalley scattering becomes the dominant mechanism for exciton line broadening. An important result is that we now can make use of this effect in order to determine the phonon deformation potential constant for intervalley scattering, which otherwise has to be derived indirectly from high-field transport experiments [237, 238] or by measuring relaxation times of photoexcited carriers [239–243].

To interpret the exciton-broadening data, one can use a simple model for intervalley electron scattering within the effective-mass approximation by considering the phonon-induced deformation potential interaction as

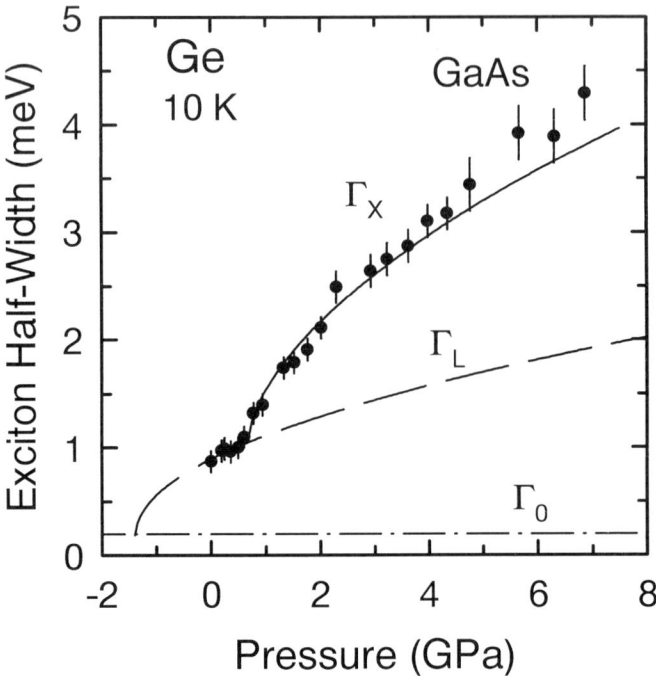

FIG. 29. Pressure dependence of the half-width of the direct exciton line of Ge. The solid (dashed) line corresponds to the fitted model [Eq. (6.29)] including both the Γ–L and Γ–X intervalley scattering (only Γ–L scattering). The constant Γ_0 accounts for other scattering processes such as impurity and defect scattering. After Ref. [203].

scattering mechanism [244]. The homogeneous half-width of the direct exciton can thus be expressed as a sum of different contributions

$$\Gamma_c = \frac{\hbar}{2}(P_{\Gamma L} + P_{\Gamma X} + P_0), \tag{6.28}$$

where $P_{\Gamma L}$ and $P_{\Gamma X}$ are the intervalley scattering probabilities from the Γ valley to the L and X valleys, respectively. P_0 includes additional contributions from inhomogeneous broadening and defect scattering. According to Conwell [244], the intervalley scattering probability is given by

$$P_{\Gamma i} = \frac{N_i m_i^{3/2} D_{\Gamma i}^2}{\sqrt{2\pi}\hbar^2 \rho E_Q}[(n_Q + 1)(\Delta E_{\Gamma i} - E_Q)^{1/2} + n_Q(\Delta E_{\Gamma i} + E_Q)^{1/2}], \tag{6.29}$$

where N_i ($i = L, X$) is the number of equivalent valleys, m_i the corresponding density-of-states effective mass, and ρ the material density. E_Q and n_Q

are energy and occupation number of the involved phonons, $\Delta E_{\Gamma i}$ is the energy difference between the Γ and L or X valleys, and $D_{\Gamma i}$ the phonon deformation potential constant for Γ–L or Γ–X intervalley scattering. The first and second term in the square brackets of Eq. (6.29) correspond to phonon emission and absorption, respectively. At low temperatures, however, only phonon emission processes need to be considered.

By symmetry, only zone-edge LA phonons can participate in Γ–L intervalley scattering processes in Ge and GaAs, whereas for Γ–X processes only zone-edge LO phonons are allowed [245, 246]. However, experimentally one is dealing with energy-conserving scattering processes between electron states which are spread over a certain wave-vector range around high-symmetry points of the BZ such as Γ, X, or L. Therefore, for a comparison with theoretical results one should keep in mind that measured deformation potentials are to be regarded as effective values averaged over wave-vector regions around conduction valleys, which pick up different contributions from several phonon branches because of relaxation of the symmetry selection rules.

The solid and dashed curves in Figs. 28 and 29 are obtained by fitting Eq. (6.29) to the measured exciton half-widths as a function of pressure with only the deformation potential constants for the different intervalley scattering processes as adjustable parameters. More details about the energies and other parameters of the calculations are given elsewhere [214, 203]. For GaAs the agreement between theory and experiment is excellent, yielding a value for the Γ–X deformation potential of $D_{\Gamma X} = 4.8(3)$ eV/Å. Values for $D_{\Gamma X}$ obtained with high-field transport [237], standard PL [247], and hot-electron luminescence [242, 243] or Raman [239, 240] methods lie in the range 6–10 eV/Å, except for the measurements on AlGaAs alloys [241], from which a value of 3.4 eV/Å is deduced. Pseudopotential calculations [245, 246] give values of $D_{\Gamma X} = 3.0$–4.1 eV/Å, but when the lowered symmetry of the real intervalley scattering process is taken into account, the calculated value increases significantly to 6.5 eV/Å.

In evaluating the deformation potential $D_{\Gamma L}$ of Ge, only data points for $P < 0.6$ GPa have been considered. Because it is not possible to separate intervalley scattering contributions from that of the background scattering rate Γ_0, only an upper bound for $D_{\Gamma L}$ of 4.5(3) eV/Å was given. This result is in very good agreement with the value of 4.2(2) eV/Å obtained from time-resolved hot electron luminescence [248]. The Γ–X scattering plays the main role in the broadening of the exciton as soon as the X valley becomes the lowest conduction-band minimum. The intervalley deformation potential $D_{\Gamma X}$ of Ge can be determined with a high degree of accuracy from a fit of Eq. (6.29) to the data points of Fig. 29, yielding the value of 2.2(3) eV/Å. To our knowledge, there are no reliable experimental values for the intervalley deformation potentials of Ge in the literature apart from that of Ref. [248]. More recent self-consistent calculations using *ab initio* tight-binding methods [249] yield the values of $D_{\Gamma L} = 3.89$ eV/Å and $D_{\Gamma X} = 2.27$ eV/Å, which are in excellent agreement with the experimental results.

There is a striking difference between the intervalley deformation potentials of Ge and those of isoelectronic GaAs. The value of $D_{\Gamma X}$ determined here for Ge is about half of that measured for GaAs by the same method. A similar observation has been reported regarding the much longer time constant for Γ–L scattering in Ge compared to GaAs, as deduced from subpicosecond time-resolved Raman spectroscopy [250]. Furthermore, the self-consistent calculations [249] for Ge and GaAs strongly support this result. The difference between the deformation potentials of Ge and GaAs can be easily understood in a tight-binding picture. The deformation potential constant for Γ–X intervalley scattering is proportional to the matrix element $M_{\Gamma X} = |\langle X|\Delta V|\Gamma\rangle|$, where ΔV stands for the change in the potential due to the atomic displacements obtained from the phonon eigenvector. The main contribution arises from the $V_{sd\sigma}$ interatomic potential [75] between the cation s-orbital at Γ and the anion d_{3z^2-1}-state at the X point. The increase in the cation s-component due to the antisymmetric part of the potential in polar GaAs together with the much larger d-character of the wave function at the X-valley [251, 97] result in a strong enhancement of the matrix element of electron–phonon interaction and hence of the intervalley deformation potential of GaAs.

Finally, we point out that such a pressure-induced broadening of the direct exciton has been also observed for InSe [226], possibly indicating the existence of an indirect gap with negative pressure coefficient at an energy close to that of the direct fundamental gap. For the γ-InSe polytype with a rhombohedral crystal structure, the direct gap occurs at the Z point of the BZ, whereas the next conduction-band minimum lying close in energy corresponds to the M point. Other indications of a conduction-band crossover at about 0.8 GPa come from room-temperature resistivity measurements under pressure [252]. The pressure coefficient of the Z–M gap of γ-InSe is estimated to be about -100 meV/GPa, which is in accordance with the pressure behavior of the exciton linewidth. Furthermore, tight-binding LMTO calculations performed at reduced crystal volume [253] predict a similar pressure dependence for the Z–M energy difference as experimentally obtained, but the calculated absolute value of this gap is too large.

c. Band-Structure Parameters from Exciton Binding Energies

The method of the exciton absorption under pressure not only is applicable for the accurate determination of the variation with pressure of bandgaps, but it can yield further information on other band-structure parameters of semiconductors such as effective masses and exchange interactions. In this sense, nonlinear optical absorption techniques such as two-photon spectroscopy have proved to be a powerful tool for such studies because of their high sensitivity. When the magnitude of the two-photon absorption is measured by the intensity of a related luminescence emission as a function of twice the laser photon energy rather than by conventional direct means [45], the sensitivity enhancement is such that it allows the detection of at least one photon per laser pulse! For a comprehensive description of this method and results we refer the reader to the review articles of Reimann et al. [233, 45] and references therein. Here we briefly discuss two examples in which two-photon spectroscopy appears to be particularly useful. We close this section going back to linear absorption measurements but in MQW samples, where we show how the change with pressure of the effective mass character of the first confined light-hole subband can be derived from the pressure dependence of the related exciton binding energy [114].

Next we discuss the pressure dependence of the electron–hole exchange interaction in materials with large overlap of electron and hole wave functions. Although the exchange interaction between the conduction-band electron and the valence-band hole that form the exciton is usually neglected, it can amount to a few milli-electron volts in cases like Cu_2O, ZnTe, and $AgGaS_2$. This interaction leads to the splitting of the 1s exciton ground state into *para* (total spin $S = 1$) and *ortho* ($S = 0$) excitons [254]. The paraexciton, which is optically forbidden because the electron–photon Hamiltonian does not depend on spin, becomes allowed because of spin–orbit interaction and is observed in two-photon absorption spectra of ZnTe [255] and $AgGaS_2$ [256] as a weak peak on the low-energy side of the 1s exciton. From the change with pressure of the *ortho–para* energy difference, one obtains directly the pressure dependence of the exchange energy E_e, which is three times larger than that of the exciton Rydberg, as expected from the change of the wave-function overlap [255, 256].

Furthermore, if the fine structure in the spectral region of the 2p exciton absorption is resolved as for ZnTe [255], ZnSe [228], and ZnS [229], then it is possible to determine the spherical valence band parameter μ as a function of pressure. This fine structure arises from the coupling between the $J = \frac{3}{2}$ angular momentum of the hole and the orbital momentum $L = 1$ of the exciton envelope function [257, 258]. The splittings of the exciton 2p states thus depend on the actual form of the fourfold degenerate Γ_8 valence band, which in the spherical approximation is described using a single parameter μ (~0.2 for the cited semiconductors). Within experimental error this parameter turns out to be independent of pressure for all three materials in the range up to 4 GPa [233].

An example of well-resolved excitonic features in the low-temperature absorption spectra of a $Ga_{0.47}In_{0.53}As/Al_{0.48}In_{0.52}As$ MQW structure is shown in Fig. 23. From these data it is possible to determine the pressure dependence of exciton binding energies (E_b) among other exciton parameters. The values of E_b^{1hh} and E_b^{1lh} for the binding energies of the heavy- and light-hole excitons, respectively, corresponding to the first confined subbands of the quantum wells are shown in Fig. 30 as a function of pressure. The binding energy of the heavy-hole exciton exhibits a linear increase with increasing pressure, whereas the light-hole binding energy diminishes abruptly by a factor of 3 in the pressure range below 1.5 GPa. Above this pressure E_b^{1lh} decreases slowly. We will demonstrate that this unusual behavior is simply due to the change with pressure of the in-plane reduced effective mass $\mu_\perp$.

In order to gain further insight into the problem of the pressure dependence of exciton binding energies of QWs, the in-plane subband dispersion has been studied theoretically within the envelope-function approximation (EFA) [125]. For a finite in-plane wave vector $\vec{k}_\perp = (k_x, k_y) \neq 0$, heavy-

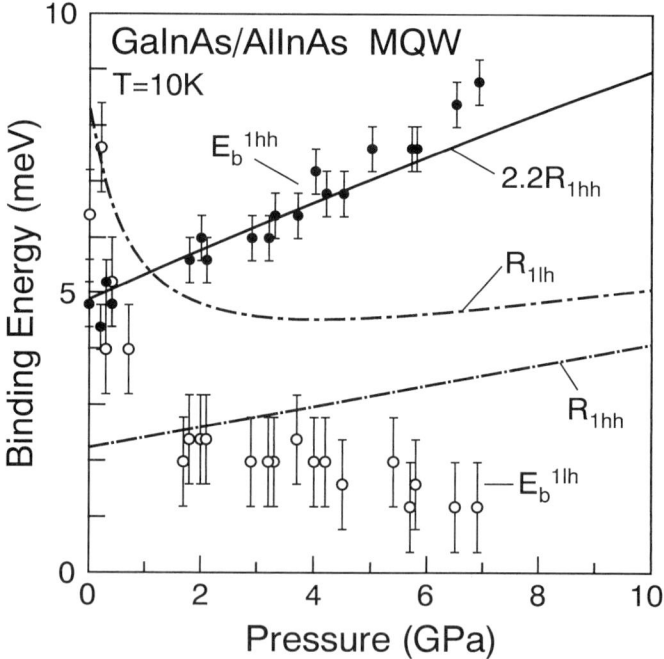

FIG. 30. Binding energies of the first heavy-hole and light-hole excitons in GaInAs/AlInAs MQWs at 10 K and as a function of pressure. The curves represent calculations of the exciton binding energies using in-plane reduced masses as obtained from $\vec{k} \cdot \vec{p}$ theory. After Ref. [114].

hole and light-hole subband states are mixed through $\vec{k}\cdot\vec{p}$ interaction. Their envelope functions are taken as a linear combination of the $k = 0$ eigenfunctions, and the dispersion relations are obtained by numerical diagonalization of the 4×4 Luttinger Hamiltonian (3.33). The effect of pressure is again introduced into the theory by considering the enhancement of the effective masses in proportion to the direct bandgap and, which is of crucial importance in this MQW sample [114], by accounting for the additional splitting of the hole subbands due to the pressure-induced biaxial strain of the wells according to Eqs. (3.28) and (3.29).

Results of the in-plane wave-vector dispersion along the (01) direction for the first three confined hole levels are shown in Figs. 31a–c for three different pressures. We call the attention to a striking result: At ambient pressure the first light-hole subband (1lh) exhibits, as a consequence of the coupling, an *electron-like* dispersion at the zone center. The positive curvature leads to a negative hole mass. With increasing pressure all three subbands become flatter, that is, the effective masses increase with pressure. The reduction of the energy separation between the 1hh and the 1lh subbands enhances the coupling between them. At a pressure of ≈ 3.7 GPa, the calculation predicts a crossover from electron-like to hole-like mass character of the 1lh subband.

To illustrate this effect, the effective hole masses have been calculated from the dispersion curves by taking the second derivative of the energy eigenvalues with respect to $k_\perp$ at the zone center. The resulting masses are shown in Fig. 32 as a function of pressure. The first electron and heavy-hole mass increase linearly with pressure. The light-hole mass, in contrast, clearly shows a change in sign at 3.7 GPa corresponding to the crossover from electron-like to hole-like mass character. At 10.5 GPa, when the 1hh

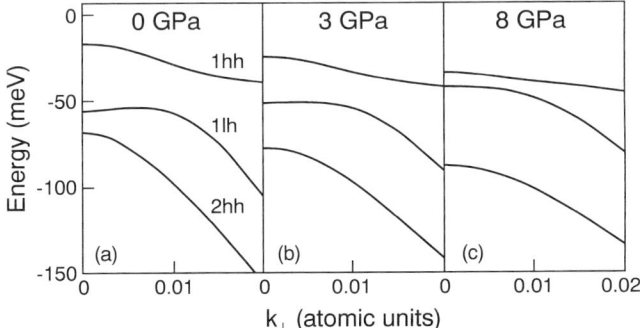

FIG. 31. Calculated in-plane subband dispersion curves of the first three valence-band confined states of a GaInAs/AlInAs MQW structure for three different pressures. After Ref. [114].

and the 1lh subbands become degenerate, the light-hole mass is twice the heavy-hole one.

One can now use the calculated effective masses in order to explain the pressure dependence of the exciton binding energies. We assume that the factor z_{00}^2, which appears in Eq. (6.11) for the exciton Rydberg in the anisotropic case, does not change appreciably with pressure. Thus, according to Eq. (6.11) the pressure dependence of $E_b = \mathcal{R}$ is given by that of the in-plane reduced effective mass $\mu_\perp$. From the results of Fig. 32 the reduced heavy-hole mass and thus the binding energy are expected to increase linearly with pressure. The dashed–dotted line in Fig. 30 corresponds to the calculated values of the heavy-hole Rydberg, which when multiplied by a factor of 2.2 (solid line in Fig. 30) fall on top of the experimental data points of E_b^{1hh} versus pressure. We notice that this number is very close to the value of $z_{00}^2 = 2.5$ expected for a system with an infinite reduced mass parallel to the anisotropy axis, as is the case for excitons confined in semiconductor QWs.

In a similar way one can explain the unusual pressure dependence of the light-hole binding energy E_b^{1lh}. The increasingly negative light-hole mass causes the strong decrease of $\mu_\perp$ and therefore of E_b up to the point at

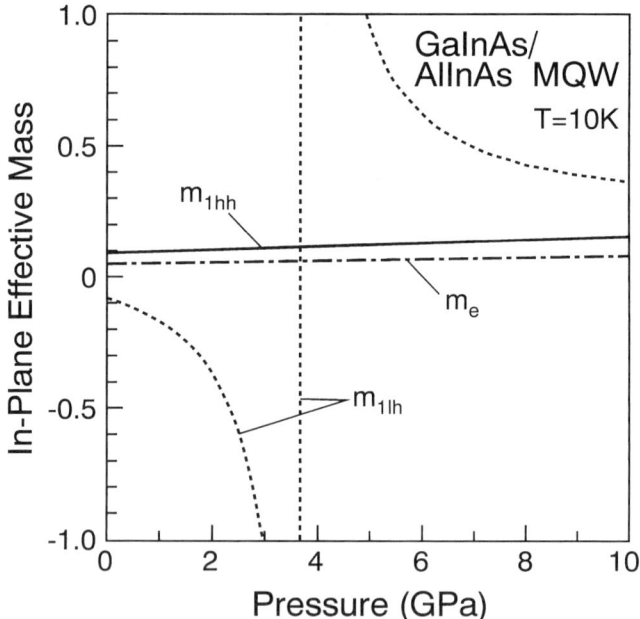

FIG. 32. In-plane effective electron and hole masses (in atomic units) as obtained from dispersion relations like the ones shown in Fig. 31. After Ref. [114].

which flat-band conditions ($m^*_{\text{lh},\perp} \to \infty$) are reached. Above the crossover pressure $m^*_{\text{lh},\perp}$ is positive and diminishes. This compensates in part for the increase of $m^*_{\text{e},\perp}$ with pressure. As a result of that, E_b^{1lh} is a slowly varying function of pressure in this region. The calculated pressure dependence of E_b^{1lh} ($z_{00}^2 = 1$) is represented by the dot–dashed curve in Fig. 30. The qualitative agreement with the experiment is very good. This appears to be the first optical demonstration of the change in character of hole mass in semiconductor QWs due to the application of external hydrostatic pressure.

VII. Photoluminescence Studies under Pressure

The study of the photoluminescence (PL), that is the optical emission due to radiative recombination of photogenerated minority carriers by laser excitation at energies above that of the bandgap, has found extensive application as a characterization tool for near-band-edge states in semiconductor bulk materials as well as various low-dimensional structures. This technique is therefore particularly suitable for tracking the evolution of the fundamental bandgap with pressure, the PL intensity being very sensitive to the occurrence of a pressure-induced Γ–X conduction-band crossover. The irreversible quenching of PL emission is also a clear indication for a structural phase transition at high pressures. Since at low temperatures the optical emission is dominated by impurity-related recombination processes, PL measurements under pressure have been widely used for studies of the electronic properties of all sorts of defects in semiconductors. Such experiments have often led to the discovery of new defect states which at ambient pressure conditions are resonant within the conduction band but become optically *active* when they are pushed into the bandgap at sufficiently high pressures. In this section we will give a general overview of the main issues just mentioned for high-pressure PL, with emphasis on the band-structure properties, but without going into the details of defect states themselves. Here we are also concerned exclusively with spontaneous radiation processes at low excitation powers, leaving out of the discussion stimulated emission phenomena. For a thorough account of the effects of hydrostatic pressure as well as uniaxial stress components on gain and threshold current in laser diodes and quantum well lasing devices, we refer the reader to Refs. [259–264].

1. Optical Emission in Semiconductors

In photoluminescence one is dealing with the spontaneous emission of light by radiative recombination of electron–hole pairs characterized by non-thermal-equilibrium distributions. Electrons and holes are created in pairs by absorption of photons with energies well above the bandgap, and then they relax toward lower energy states, reaching quasi-thermal equilibrium before recombining radiatively. The relaxation process, which occurs mainly via emission and absorption of phonons, runs independently for electrons and holes, often leading to distribution functions with quasi-Fermi levels and effective temperatures, which are different for both types of carriers. The emission rate $R_{v,c}$ for optical transitions between conduction- and valence-band states is simply given by [76]

$$R_{v,c} = A_{v,c} \cdot f_c (1 - f_v), \tag{7.1}$$

where $f_c(f_v)$ are the electron (hole) quasi-thermal distribution functions and $A_{v,c}$ is the Einstein coefficient for spontaneous emission between valence and conduction band. The coefficient $A_{v,c}$ is a measure of the transition probability by photon emission; thus, it is determined by the magnitude of the optical dipole-matrix element $M_{v,c}$ [Eq. (A.20)] and the joint density of states $J_{v,c}$ [Eq. (A.24)]. In the following we briefly discuss PL emission processes commonly found in semiconductors, which are classified by the type of state involved, such as band states, excitons, or impurities. A general review on the theory of PL emission is given in Ref. [265].

In very pure semiconductor materials the radiative recombination occurs preferentially between conduction and valence-band extrema. As for optical absorption we distinguish here between direct and indirect (in $\vec{k}$-space) optical transitions. The PL lineshape for direct *band-to-band* transitions can be easily calculated by assuming a constant dipole matrix element and using the square-root-like density of states of the bulk material. Further, for low excitation powers and nonintentionally doped semiconductors, the quasi-Fermi functions are well approximated by Boltzmann distributions. By substituting into Eq. (7.1) one obtains for the PL intensity

$$I_{PL}(\hbar\omega) \propto \begin{cases} \sqrt{\hbar\omega - E_g} \cdot \exp\left(-\dfrac{\hbar\omega - E_g}{k_B T}\right) & \text{for } \hbar\omega > E_g \\ 0 & \text{otherwise.} \end{cases} \tag{7.2}$$

To account for the experimental finite linewidth one just needs to convolute Eq. (7.2) with a Lorentzian or Gaussian function in case of homogeneous or inhomogeneous broadening, respectively.

Indirect optical transitions take place only if assisted by phonons, being orders of magnitude less intense than dipole-allowed direct transitions. Since energy is conserved in this two-step process, the PL emission line appears down-(up-)shifted with respect to the indirect bandgap by the energy of the emitted (absorbed) phonon. This is the so-called *phonon replica* emission line. At low temperatures only phonon emission is important. The zero-phonon line (ZPL) can be also observed if there is substantial elastic scattering of carriers due to defects or, as in quantum well structures, due to interface roughness. In this case, the necessary change in crystal momentum of the carrier is taken up by the lattice without loss of energy. The large difference in intensity between direct and indirect PL emission serves in high-pressure experiments, for example, as indicator of a Γ–X crossover.

Since semiconductor materials contain always a certain background concentration of shallow impurities, at low temperatures free carriers freeze out mostly on the donors and acceptors. The PL emission is then dominated by *free-to-bound* transitions, that is, in an *n*-type material electrons trapped on the donors will recombine radiatively with valence-band holes, whereas for a *p*-type semiconductor the recombination process involves a conduction-band electron and a hole attached to an acceptor. The emitted photon energy is given by $E_g - E_D(E_A)$, where $E_D(E_A)$ is the shallow donor (acceptor) binding energy. As already mentioned, the free-to-bound PL intensity depends on the number of nonionized impurities, which varies exponentially with temperature as $1 - \exp(-E_{A,D}/(k_B T))$. A similar reduction of PL intensity is obtained by increasing the laser excitation power. Thus, the binding energy of the shallow impurity can be deduced from the corresponding Arrhenius plot of $\ln(I_{PL})$ versus T^{-1}. By measuring the temperature dependence of the PL intensity at different pressures, it is possible to study the pressure dependence of the donor (acceptor) binding energy.

When both impurity species are present in comparable concentrations (GaN is a good example), *donor–acceptor pair* (DAP) transitions become apparent in low-temperature PL spectra. This transition can be represented by the reaction

$$D^0 + A^0 \rightarrow \hbar\omega + D^+ + A^-, \tag{7.3}$$

indicating that in the final state both impurities become ionized. This means that there is a Coulomb energy term, which depends on the donor–acceptor separation R and which tends to increase the energy of the emitted photon, since the final state energy is lowered by the electrostatic attraction between the ionized impurities: $\hbar\omega = E_g - E_A - E_D + e^2/(\epsilon(0)R)$. The DAP emission line is usually accompanied by phonon sidebands which result from the simultaneous emission of phonons in the recombination process due to the enhancement of the electron–phonon interaction for the bound charges (this can be viewed as the recoil of the ionized impurities).

Exciton formation plays an important role in the emission properties of high-quality semiconductors at low temperatures as well. This gives rise to a sharp exciton peak in the PL spectra at the energy of the ground-state exciton. In this case we are dealing with so-called *free-exciton* recombination. The PL peak should be of Lorentzian type, but because of polariton effects it very often exhibits a quite asymmetric lineshape with a high energy shoulder [76]. Excitons, however, have the tendency to bind to shallow impurities (neutral or ionized) and thus gain additional binding energy. Excitons bound to neutral donors (acceptors) are usually denoted by $D^0X(A^0X)$ and they can be regarded as analogs of the hydrogen molecule H_2, except for the different binding energies. Excitons bound to ionized impurities are denoted as D^+X and A^-X, the former being analogous to an H_2^+ molecule. Since the extra binding energy of the exciton bound to the impurity is typically smaller than the binding energy of the exciton itself, when temperature is raised one observes a transition from bound to free excitonic emission with a consequent upward shift of the PL peak. If the temperature is increased further, the exciton will dissociate, leading to band-to-band luminescence. Bound-exciton emission also shows phonon replicas in the PL spectrum, in particular for strongly polar semiconductors.

In heavily doped semiconductors (with doping levels well above that of the Mott transition, for which the carriers become delocalized and fill up the conduction or valence band up to the Fermi energy E_F), the PL lineshapes are strongly affected by band-filling and many-body effects. The latter, for instance, are responsible for the reduction of the bandgap depending on the free carrier concentration [226]. A rule of thumb states that in n-type materials the reduction of the gap is of the order of the Fermi energy [267]. Because of the filling of the conduction (valence) band with electrons (holes) up to E_F, the PL linewidth becomes proportional to the Fermi energy and, thus, to the free carrier density. For describing the PL lineshape in doped bulk materials one has to take into account the breakdown of wave-vector conservation for interband optical transitions due to the strong carrier–carrier and carrier–ionized-impurity scattering [268, 269]. By integration of Eq. (7.2) over the energy difference E' of the emitted photon with respect to that of the gap (here direct as well as indirect transitions in $\vec{k}$-space contribute to the PL intensity at given energy), one obtains

$$I_{\text{PL}}(\hbar\omega) \sim \int_0^{\hbar\omega} \sqrt{E'} \sqrt{\hbar\omega - E_g - E'} f_e f_h \, dE'. \tag{7.4}$$

In the case of n-doping, for example, the only holes present in the sample for recombination are the photogenerated ones, whose concentration is in the range of 10^{15} per cm^3 for a typical excitation power of 5 W/cm^2. Thus, the hole Fermi energy is many orders of magnitude smaller than that of electrons and the product $\sqrt{E'} f_h$ can be approximated by a delta function. Now, because of degeneracy of band-edge and donor states, the square root density of states needs to be modified by introducing a Gaussian broadening. The PL lineshape function is then written as [270]

$$I_{\rm PL}(\hbar\omega) \sim f_e(\hbar\omega - E_g, E_F) \times \int_0^{\hbar\omega - E_g} \sqrt{\hbar\omega - E_g - E''} \exp(-(E'')^2) dE''. \quad (7.5)$$

As an example, Fig. 33 shows a PL spectrum (points) of heavily n-doped GaAs at 10 K and ambient pressure. The solid curve represents the result of fitting Eq. (7.5) to the data points using the renormalized gap E_g and the Fermi energy E_F as adjustable parameters. The agreement between model lineshape and experiment is excellent. In this way it is possible to

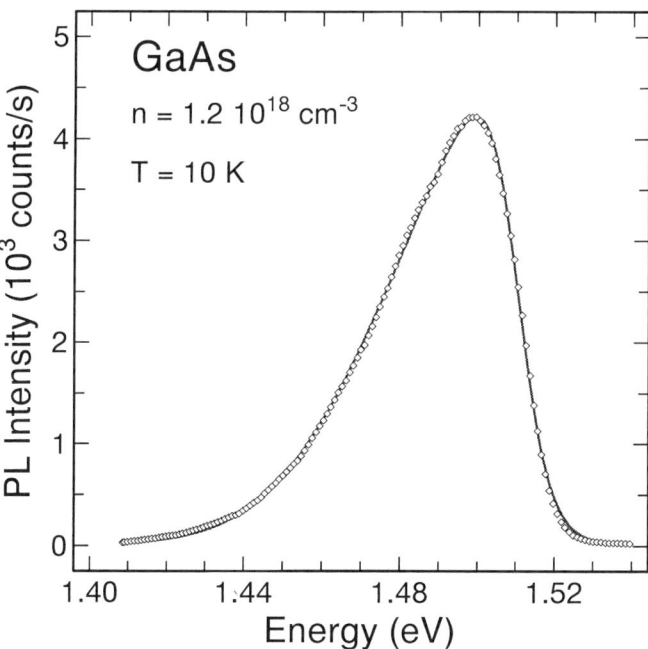

FIG. 33. Photoluminescence spectrum of heavily n-doped GaAs ($n = 1.2 \times 10^{18}$ cm^{-3}) at ambient pressure and 10 K. The solid curve represents the result of a least-squares fit using the lineshape function of Eq. (7.5).

determine the Fermi energy under pressure. This will be discussed in more detail in Section IX, on the properties of the electron gas, where the luminescence have been used for monitoring the abrupt decrease of the carrier density in n-GaAs and n-InP at the $\Gamma-X$ conduction-band crossover due to the transfer of electrons between valleys [271, 270]. A marked reduction in the PL linewidth with pressure was previously observed for Te-doped GaAs by Olego *et al.* [272] at about 3 GPa and interpreted in similar terms. This behavior is obviously absent for p-doping [272].

In high-quality quantum well structures the luminescence is mainly of excitonic origin. At low temperatures PL peaks are very sharp (FWHM ~ 1 meV), but their width is now determined by the amount of inhomogeneous broadening present due to interface roughness and well-width fluctuations. In modulation-doped heterostructures the description of PL lineshapes becomes a more difficult task because of the enhancement of the exciton binding as well as of many-body effects due to confinement of the carriers. A phenomenological formula, which includes these effects, is [273, 274]

$$I_{PL}(\hbar\omega) \sim \text{er f}(\hbar\omega - E_g) f_e f_h + I_{exc}(\hbar\omega), \qquad (7.6)$$

where the error function stands for the steplike 2-D density of states in quantum wells. In the case of n-type doping, f_e is the Fermi distribution and f_h is again approximated by a Gaussian. The last term in Eq. (7.6) takes the form of a Gaussian function at the gap and accounts for many-body effects in the Fermi sea [274]. In Section IX we demonstrate how, through optical emission, it is possible to determine the pressure dependence of 2-D electron densities in modulation-doped GaAs single and double quantum wells [273, 275–277]. As reported before for GaAs/AlGaAs MQW [278, 279], a reduction of the free carrier density with pressure is also observed in these structures, which is directly inferred from the pressure-induced narrowing of the main PL emission peak.

We mention another spectroscopic technique, photoluminescence excitation (PLE), which is capable of giving information similar to that provided by optical absorption, but especially in studies of epilayers or QW structures, it is applicable without the need to remove the substrate (if it is opaque in the spectral region of interest). In PLE one measures the emission of the sample with the spectrometer set to a fixed position corresponding to a particular PL peak, while the photon energy of the excitation (typically a continuously tunable dye or Ti:sapphire laser) is varied within a certain spectral range. Since the carriers have to relax from the energy at which they have been photogenerated to the final state from which they recombine radiatively, the recorded PLE signal is not simply proportional to the absorption coefficient at the energy of excitation. Interpretation of PLE spectra is not always straightforward, because the relaxation process of the carriers is not well known.

2. Pressure Effects on PL Emission in Bulk Materials

In high-quality undoped bulk semiconductors the main luminescence signal is, at low temperatures, usually linked to recombination processes of free or shallow-impurity-bound excitons. The PL linewidths are in these cases among the narrowest ones obtained with various optical spectroscopies. This enables a very accurate determination of the pressure dependence of the fundamental bandgap, provided the changes of the binding energies with pressure can be neglected. Since this is generally valid, high-pressure low-temperature PL experiments have been pursued systematically for several bulk compounds, including ternary and quaternary alloys: for instance, SiC [280], GaAs [281, 100, 101, 247], GaP [282], GaN [283–285], InP [174, 286–288], GaInAs [289, 290], GaInP [291], GaInAsP [292], AlInAs [293], AlGaAs [294, 295, 241], ZnSe [296], ZnMnSe [297], ZnMnS [298], CdS [299], CdTe [300, 301], CdSeTe [302], CdMnTe [303], CdMnSe [304], CdCoSe [304], and Cu_2O [47]. Because of the high spectral resolution, the degree of hydrostaticity of the pressure-transmitting medium plays an important role in low-temperature PL measurements. Splittings and shifts of the PL lines due to uniaxial stress components are readily detected. For that reason the use of condensed He as pressure medium is highly recommended for cryogenic-pressure experiments.

The pressure dependence of the lowest bandgap of GaAs constitutes a classic example which is representative of the results obtained by this method. Figure 34 shows several PL spectra of high-purity n-type GaAs measured at different pressures and at 5 K, as reported in Ref. [100]. At low pressures the PL spectra are dominated by the strong bound-exciton line D_Γ^+ (exciton bound to an ionized shallow donor related to the Γ conduction band minimum). At the high-energy side of this line the emission from both the ground ($n = 1$) and first excited state ($n = 2$) of the free exciton (F_Γ) is also observed. The broad and much weaker PL band ~ 29 meV below D_Γ^+ corresponds to band-to-Si-acceptor recombination. The principal direct-edge transitions shift to higher energies with increasing pressure without changing appreciably in intensity up to about 4 GPa. At this pressure the conduction band crossing of Γ and X states occurs and the direct luminescence fades abruptly, while two X-related bound-exciton peaks D_X^0 become apparent in the spectra. The characteristic phonon replica of the bound exciton corresponding to different zone-edge and zone-center phonons together with the phonon replica of the free exciton at X are also

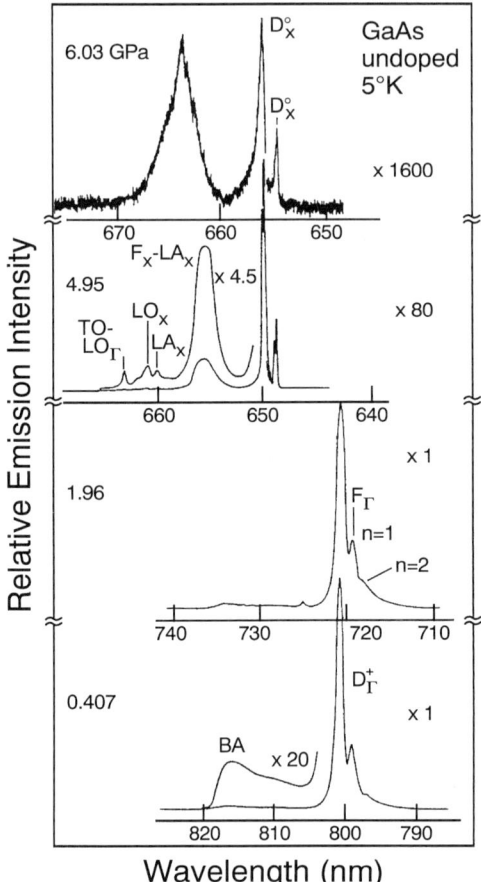

FIG. 34. Representative photoluminescence spectra of undoped n-type GaAs for different pressures and at 5 K. Spectra are aligned to the donor-bound-exciton lines. See the text for the assignment of the observed features. After Ref. [100].

resolved at the low-energy side of D_X^0. As indicated by the scale factors in Fig. 34, the indirect X emission is at least 100 times weaker than the direct one. The broad band in the 6.03-GPa spectrum is assigned to the phonon-replica transition F_X–LA_X of the X-related free exciton, which is broadened by sample heating effects due to the larger laser powers used for excitation at high pressures.

The results for the pressure dependence of the observed PL transitions are summarized in Fig. 35 [100]. Since the optical transitions involve elec-

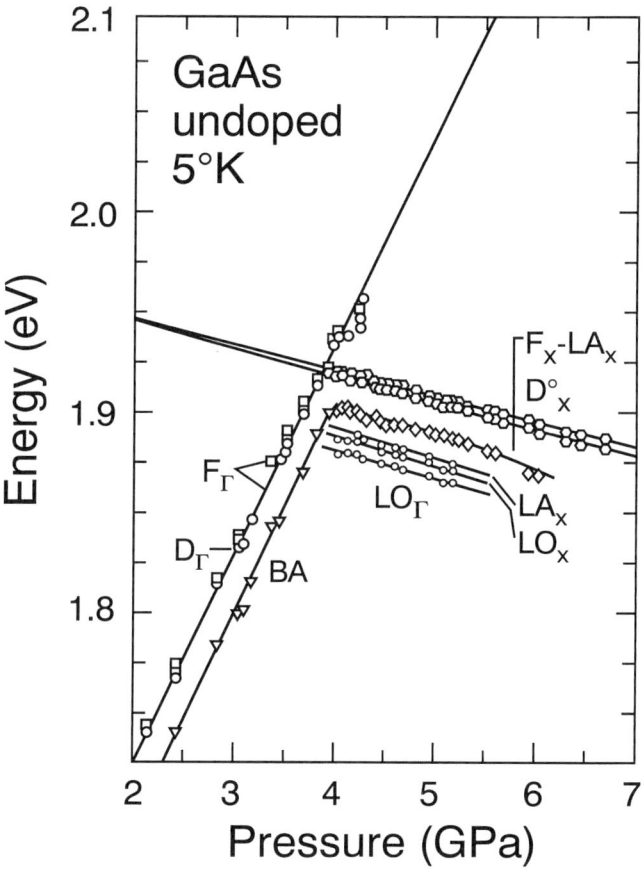

FIG. 35. Observed photoluminescence transitions versus pressure near crossover in undoped *n*-type GaAs at 5 K. After Ref. [100].

tronic states attached to the different conduction-band minima, they yield an accurate determination of pressure coefficients of the direct Γ_v–Γ_c and indirect Γ_v–X_c gaps. Furthermore, the PL data in the indirect-gap regime also permit the observation of near-zone-boundary phonons, which otherwise can be optically determined by second-order Raman techniques.

The signature of the pressure-induced Γ–X crossover is the strong decrease of the PL intensity and the appearance of weaker X lines, which redshift with pressure at a much lower rate than that at which the direct gap increases. The determination of the crossover pressure, however, based only on the quenching of the Γ luminescence, could lead to erroneous results if the band crossing takes place close to a phase-transition pressure. A typical example is InP, for which the crossover occurs right above the structural phase transition at 10.4 GPa (as we show later from Raman scattering from coupled plasmon–LO-phonon modes in n-doped InP), whereas the reported crossover pressure from PL data varies from 8 GPa [286, 287] to 10 GPa [291, 288].

Another interesting problem, which in recent years has attracted much attention in the high-pressure community, is that of the band-structure effects of spontaneous atomic ordering in semiconductor alloys such as InGaP [305–310] and ZnCdSe [311], when grown under certain conditions by vapor phase epitaxial techniques. For example, $In_{0.5}Ga_{0.5}P$ alloys will form a CuPt-like structure by ordering of group III atoms in alternating planes with doubling of the periodicity in the [111] direction. For the ordered material the bandgap is reduced by as much as 100 meV, as determined by photoluminescence [308, 309]. This effect has been interpreted as the result of band folding of the L into the Γ point and the consequent repulsion (anticrossing) of these states due to the new periodicity in the Γ–L direction [312]. In this respect high-pressure PL experiments provide a means of testing that model. Because of the fact that the pressure coefficient of the Γ–L indirect gap is roughly one-half of that of the direct gap at Γ, with increasing pressure the state repulsion is enhanced, leading to an appreciable sublinearity in the pressure variation of the direct gap in the ordered alloy [305, 308, 309]. From the behavior of the PL-peak maximum with pressure it is possible to infer the strength of the coupling potential.

Certainly, high-pressure cryogenic-PL experiments have proved to be a powerful tool for studying the interplay between localizing potential and band structure in impurity and defect-state formation in many materials: SiC [172, 313], GaAs [314–323], AlGaAs [324, 325], InGaAs [326], GaP [327], GaN [285, 284, 328–330], InP [331, 332], ZnSe [333], ZnS [334], and chalcopyrites [194, 335] and references therein. A general statement can be made about the pressure behavior of shallow and deep levels by using the integral form of the Schrödinger equation for the description of electronic states in semiconductors [336, 316],

$$(E_n(\vec{k}) - E) \cdot \psi(\vec{k}) + \int \psi(\vec{k}) U_b(\vec{k}, \vec{k}') \psi(\vec{k}') d\vec{k}' = 0 \quad (7.7)$$

$$\psi(\vec{r}) = \int \psi(\vec{k}) \varphi_{n,\vec{k}}(\vec{r}) d\vec{k}, \quad (7.8)$$

where E and $\psi(\vec{k})$ are the bound-state energy and wavefunction, respectively, U_b is the impurity potential, $E_n(\vec{k})$ is the effective-mass band-structure Hamiltonian, and $\varphi_{n,\vec{k}}(\vec{r})$ are Bloch functions. The wave function $\psi(\vec{r})$ and its Fourier transform $\psi(\vec{k})$ represent the bound-state *envelopes*, which take the form of a Gaussian. The spatial extent $\Delta \vec{r}$ of $\psi(\vec{r})$ and the spread in momentum $\Delta \vec{k}$ of $\psi(\vec{k})$ are thus inversely proportional to each other, and it follows that the deeper the defect level (wave function strongly localized in real space), the larger the wave vector region of the band structure from which the bound-state envelope picks up a contribution.

From the preceding discussion it is clear why, when hydrostatic pressure is applied, shallow donor and acceptor states follow the corresponding band extremum which is closest in energy, as shown for GaAs in Fig. 35. After the Γ–X crossover the shallow donors become tied to the X minima and their binding energy changes due to the different effective masses of electrons at the Γ and the X point. There is no criterion, however, to state *a priori* the validity of the shallow- or deep-level model. Hexagonal GaN with a bandgap of 3.5 eV at 10 K, for instance, exhibits in low-temperature PL spectra a strong peak I_2 (3.472 eV at 1 bar), which corresponds to recombination of an exciton bound to a neutral donor, and a series of donor–acceptor-pair emission lines (zero-phonon line at 3.271 eV at 1 bar), as shown in Fig. 36 [284, 329, 285]. In spite of the large binding energies of the bound states involved in the optical transitions, with increasing pressure these lines shift in parallel to higher energies (see Fig. 36) at the same rate as the bandgap, as obtained from absorption measurements [148]. Incidentally, notice that the broad PL band at 2.2 eV (1 bar, 9 K), the so-called *yellow* luminescence, whose origin remains unrevealed to date, displays a weaker pressure dependence (~30%) than the DAP and I_2 lines [284, 285, 337]. This speaks against assignment of the yellow band in hexagonal GaN to optical transitions between band-edge-related impurity states.

In contrast to shallow states, the pressure dependence of deep levels is characterized by not following any particular band edge. For example, Fig. 37 displays the variation with pressure of the PL-peak position of emission lines related to N isoelectronic traps in GaAs [316, 314]. Their nonlinear pressure behavior cannot be ascribed to a single conduction band minimum; on the contrary, there is a large amount of interband mixing as predicted by tight-binding calculations [338]. We also point out the interesting fact that the N-deep level in GaAs is silent in PL at ambient pressure. Above 2.2 GPa (see Fig. 37) it emerges from the conduction band forming an excitonic bound state which then can recombine radiatively, also assisted by phonons. A similar behavior under pressure is observed for N atoms and pairs NN_i in GaP [339, 327]. For GaP being an indirect-gap semiconductor already at ambient pressure, the contribution from the X valley becomes appreciable much earlier in pressure (at about 4 GPa) leading to a turnover in the pressure derivative of the deep-level binding energies from ~ 100 meV/GPa to −14 meV/GPa [327].

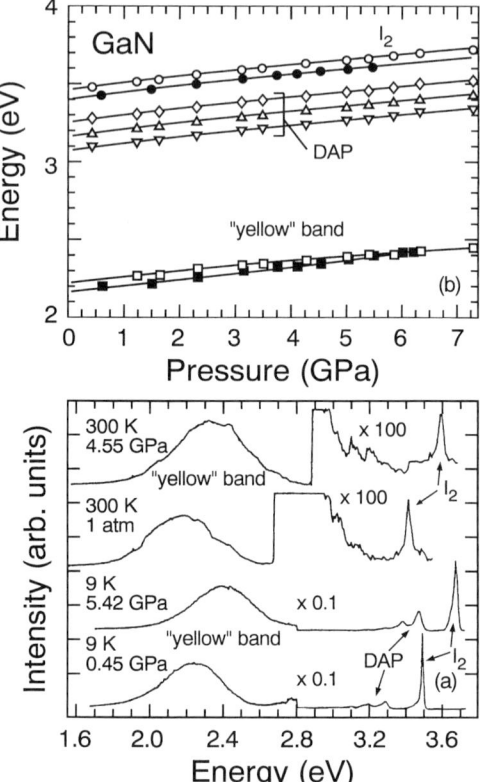

FIG. 36. (a) Photoluminescence spectra of non-intentionally doped GaN epitaxial layers at selected pressures at 9 and 300 K [284]. The vertical scale is the same for the four spectra. (b) Photoluminescence-peak energies of the I_2 line, the DAP lines, and the "yellow" band of GaN as a function of pressure at 9 K (open symbols) and 300 K (solid symbols). Solid curves are fits to the data points using second-order polynomials. After Ref. [284].

Another interesting effect of applied pressure on deep defects concerns a shallow–deep competition observed in II–VI semiconductors. A review of this doping-related problem in ZnSe is found in Ref. [333]. The binding energies of deep acceptor states tend to decrease with pressure, suggesting that compression acts to destabilize these states against competing shallow hole levels. Furthermore, a previously unknown deep donor level in n-type Ga and Cl-doped ZnSe is found to emerge into the gap at about 2.5 GPa

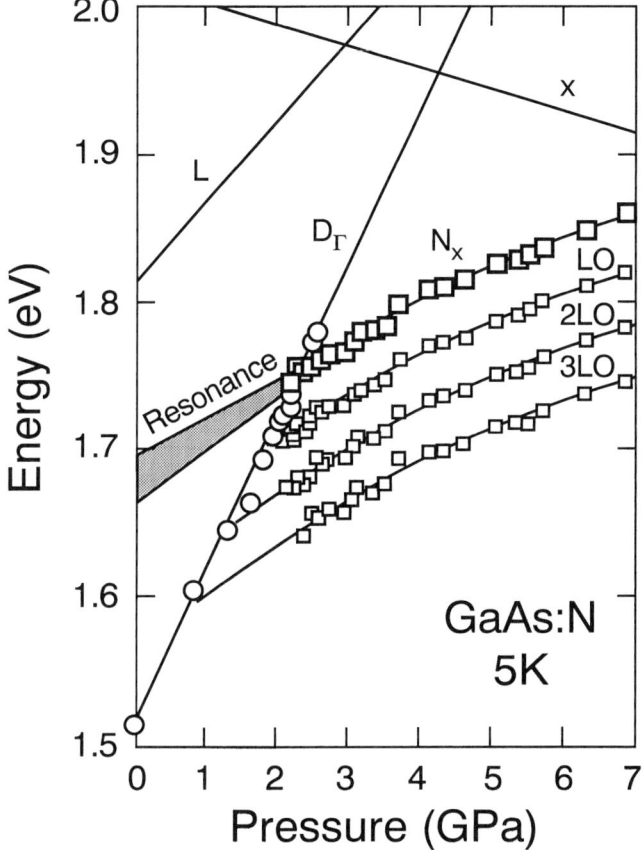

FIG. 37. Photoluminescence energies of nitrogen deep traps and resonances in GaAs:N versus pressure at 5 K. After Ref. [316].

[333]. In this way, valuable information on the complex issue of doping in II–VI materials comes to light from high-pressure experiments.

3. Low-Dimensional Structures under Pressure: PL Studies

Epitaxially grown semiconductor heterostructures such as quantum wells and superlattices are an important part of state-of-the-art electronics, particularly for optoelectronic applications such as light-emitting diodes and solid-state lasers. To a large extent this is a consequent of the fact that a device can be designed to have specific properties by tailoring the electronic band structure through the control of heterostructure parameters such as material composition, layer widths, and built-in stress. For this purpose the knowledge of the band alignments for the different constituents of the structure, that is, the valence- and conduction-band offsets, is absolutely necessary. As shown first by Venkateswaran et al. [340] and Wolford et al. [127], PL measurements under hydrostatic pressure provide a simple but accurate means of determining this parameter, among other band-structure properties. What follows is a list of references covering recent high-pressure PL work on heterostructure systems: Si/SiGe [341], SiC/Si [342], GaAs/AlGaAs [74, 127, 279, 340, 343–354], GaAs/AlAs [295, 349, 355–368], GaAs/InGaAs [232, 369–372], InAs/GaAs [90, 373], InAs/InGaAs [132], InGaAs/AlGaAs [374–377], GaAs/InGaP [128, 378, 376], GaAs/GaAsP [379], InP/InGaAs [380], InP/InAlAs [381], InGaP/InAlP [382–384], InP/InGaP [385], GaP/AlP [386], GaAs/GaSb [387], GaInSb/GaSb [388], ZnSe/ZnS [389], ZnSe/ZnSSe [390], CdTe/ZnTe [391, 392], CdTe/CdZnTe [393], CdTe/CdMnTe [394–396], CdSeTe/ZnSeTe [397], HgTe/CdTe [398], and CdS nanocrystals in a glass matrix [399, 400]. For lack of space we are not able to cover here the topic of magnetoluminescence investigations in quantum wells under pressure. Nevertheless, we would like to refer to work done in III–V heterostructure systems [279, 386] and diluted magnetic II–VI semiconductors [394, 396, 304], and references therein.

Band offsets can be determined by high-pressure spectroscopy in a simple way if the X minima of a heterostructure are type II, that is, if electrons are confined in the barriers, then holes reside in the wells, or vice versa, so that the associated optical transition is indirect in real space, taking place across the heterointerface. For example, in the case of GaAs/AlGaAs [127, 345] and GaAs/AlAs [352, 360] QW structures, high pressure can cause the Γ states in the wells and barriers to cross in energy with the X states of the barriers, for which the type-II PL emission with energy E^X might be observed, as indicated schematically in Fig. 38. The absolute valence-band offset ΔU_v is given by

$$\Delta U_v = E^X_{\text{barrier}} - E^X + \epsilon_{hh} + \epsilon_X, \qquad (7.9)$$

where E^X_{barrier} is the type-I transition energy for Γ–X recombination in the barrier material, and ϵ_{hh} and ϵ_X are the confinement energies of heavy holes in the wells and electrons in the barrier X valleys, respectively. Here we neglect the difference in exciton binding energies for type-I and type-II X-excitons. For an absolute determination of ΔU_v we need to estimate the confinement of the carriers. Because of the typically very large effective masses of the conduction-band X-valleys ($m^*_{X,l} = 1.98$ and $m^*_{X,t} = 0.27$ for GaAs [77]) the confinement energy ϵ_X is negligibly small. The energy of

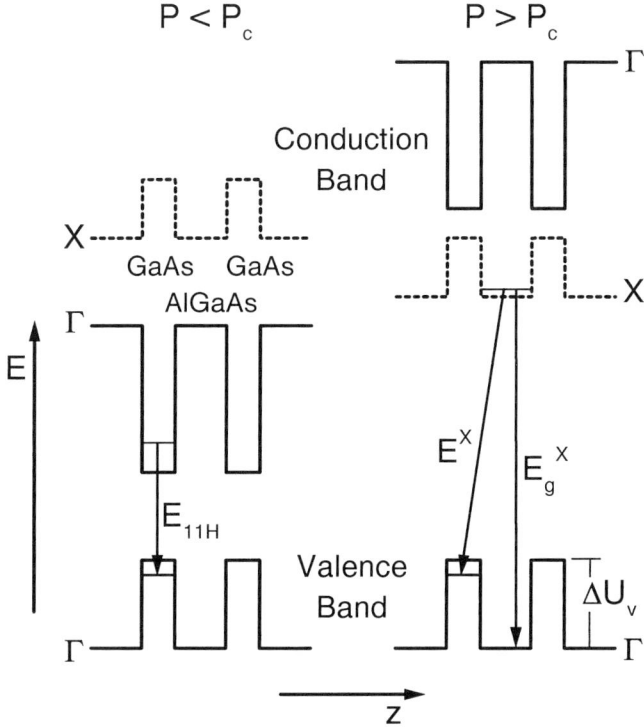

FIG. 38. Schematic alignment of the conduction- and valence-band edges in GaAs/AlGaAs heterostructures before (type I) and after (type II) of Γ–X crossover. The valence-band offset ΔU_v is obtained directly from the energy difference $E_g^X - E^X$ [345].

the QW heavy-hole states is readily obtained from $\vec{k} \cdot \vec{p}$ theory (see Section III).

As can be seen from Eq. (3.38) the band offsets are expected to vary with pressure in proportion to the difference in the pressure coefficients of the well and barrier gaps. The preceding method has the difficulty that if the type-I X-emission from the barrier is not observed, then the determination of the pressure coefficient of ΔU_v becomes fairly inaccurate without exact knowledge of the pressure derivative of E_{barrier}^X. This has been pointed out for the case of short-period GaAs/AlAs SL, where the erroneous assumption of equal pressure coefficients for the indirect gaps of GaAs and AlAs led to discrepancies by more than a factor of 2 for $d\Delta U_v/dP$ [295]. In that work it is also shown not only that the valence-band offset is pressure dependent, but that its pressure coefficient also depends on the SL period, that is, on the well width [295]. The pressure shift increases with increasing period, tending toward a value of about 5 meV/GPa for long periods [356, 349]. For very short-period SL (less than 5 × 5) ΔU_v even decreases with pressure. The reason is still not known, but one may speculate that the band offsets and their pressure coefficient are sensitive to the amount of interdiffusion of cation species present at the interface. Because of growth conditions, surface segregation is larger for very short periods [401].

Many heterostructure systems do not have, however, a type-II X-minimum alignment of the band edges like the InGaAs/GaAs and InGaP/GaAs systems, and then the method described before for determination of band discontinuities does not work. Nevertheless, as proposed by Whitaker et al. [376], it is still possible to obtain the band offsets of the InGaAs and GaAsP alloys against GaAs by comparison of the PL-emission energies below and above the Γ–X crossover pressure P_c measured for heterostructures containing Al in the barriers (instead of pure GaAs) and by using a GaAs/AlGaAs QW as reference. The point is that the systems with AlGaAs barriers indeed possess the type-II alignment of the X minima; thus above P_c recombination takes place from the *same* X state in the AlGaAs barrier to heavy-hole states confined in the alloy and the GaAs well for the sample under study and the reference, respectively. To illustrate how this method works, we reproduce in Fig. 39 the PL results of Ref. [376] for two $In_xGa_{1-x}As$/AlGaAs MQW samples with $x = 0.05$ and 0.1 and a similar GaAs/AlGaAs reference structure. The difference in the direct luminescence energy ΔE_g is the change in the bandgap between GaAs and InGaAs, whereas in the indirect case the energy separation ΔE_v between the X-emission lines is a measure of the valence-band offset. The valence-band offset ratio $Q_v = 1 - Q_c$ is then obtained as $Q_v = \Delta E_v/\Delta E_g$ to be $Q_v : Q_c = 0.38 : 0.62$ for the InGaAs/GaAs system [376, 369].

A high-pressure method for band-offset determination, which does not make use of type-II X-PL emission, consists in measuring the pressure dependence of the harmonic-oscillator-like electron-energy-level spacings in $Al_xGa_{1-x}As$ *parabolic* quantum wells by using PLE spectroscopy, as reported by Cheong et al. [353]. For parabolic wells the level spacing is directly determined by the curvature of the well potential, which for an infinitely deep wells turns out to be proportional to the conduction-band offset [353],

$$\hbar\omega_e = \hbar \left(\frac{8Q_c \Delta E_g}{L_z^2 m_e^*}\right)^{1/2}, \qquad (7.10)$$

where L_z is the width of the parabolic well. In this way, the corrected Q_c value at atmospheric pressure for the $GaAs/Al_xGa_{1-x}As$ system ($x \approx 0.3$)

4 OPTICAL PROPERTIES OF SEMICONDUCTORS UNDER PRESSURE 337

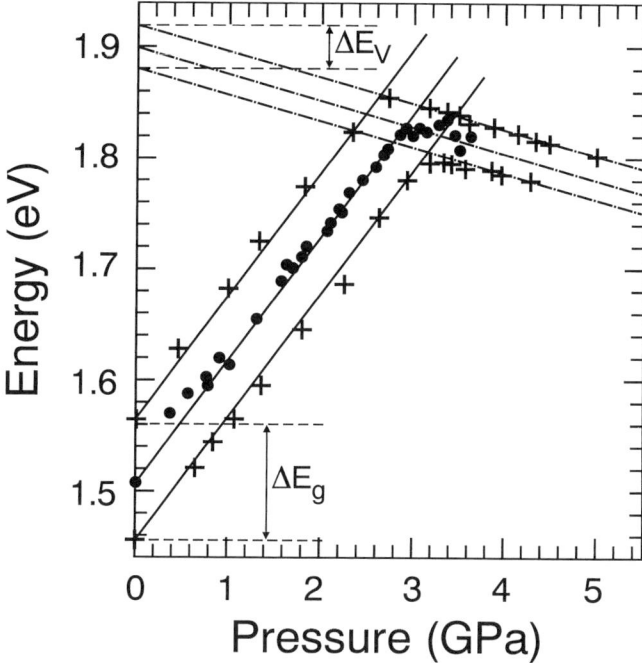

FIG. 39. Photoluminescence peak energies of the GaAs/AlGaAs reference sample (upper crosses) and the two $In_xGa_{1-x}As$/AlGaAs samples (solid circles, $x = 0.05$ and lower crosses, $x = 0.1$) as a function of pressure. The quantities ΔE_g and ΔE_v, which give the band offset ratio, are marked. After Ref. [376].

reads 0.63 ± 0.04 instead of 0.70 ± 0.04 [340] and 0.69 ± 0.03 [127], and the pressure coefficient of ΔU_c is 7.3 ± 2.5 meV/GPa. We remark that the assumption of a pressure-independent relative valence-band offset, that is, a constant $Q_v : Q_c$ ratio, is still valid for the discussed heterostructures.

The breakthrough of the past years in epitaxial-growth techniques makes possible the fabrication of high-quality heterostructures of group IV elements and III–V and II–VI compounds with large lattice mismatch. In this case, the details of the subband structure are complicated by the effect of the built-in stress in the QW structure. The reduction of symmetry associated with the biaxial strain gives rise to additional energy splittings of hole subbands and L or X conduction-band valleys. Very unusual band alignments might be obtained in strained-layer QWs. A good example of such a situation are CdTe/ZnTe superlattices. As established by high-pressure PL experiments [391, 392], these SLs are characterized by having simultaneously type-I alignment for heavy holes and type-II for light holes, such that the first light-hole subband confined in ZnTe is the valence-band ground state.

The InGaAs/GaAs system is also interesting because of the extremely large lattice mismatch in excess of 7% between InAs and GaAs. Figure 40 shows PL and absorption results obtained for an $In_{0.2}Ga_{0.8}As/GaAs$ MQW sample at 10 K and up to 8 GPa [232]. A controversial point here is concerned with the confinement of the light holes, in the sense of to what extent the large compressive stress of the InGaAs layers suffices for expelling the light holes out of the wells. As illustrated in Fig. 40b, the valence-band offset is obtained directly from the energy difference between the emission lines X and X_1, being about 80 meV [232], although from absorption the energy separation between first heavy- and light-hole subbands is only 60 meV, thus indicating that the light holes are confined to the InGaAs wells, in contrast with a previous report [371]. Furthermore, above the Γ–X crossover the pseudodirect optical transition between X_Z states of the strained InGaAs layers and the heavy holes was observed for the first time in absorption [232]. This result together with the observation of the zero-phonon emission line from X_{XY} well states enables the determination of the shear deformation potential Ξ_u [see Eqs. (3.30) and (3.31)] for the X valleys of $In_{0.2}Ga_{0.8}As$. One obtains $\Xi_u = 5.5 \pm 0.4$ eV, which is in excellent

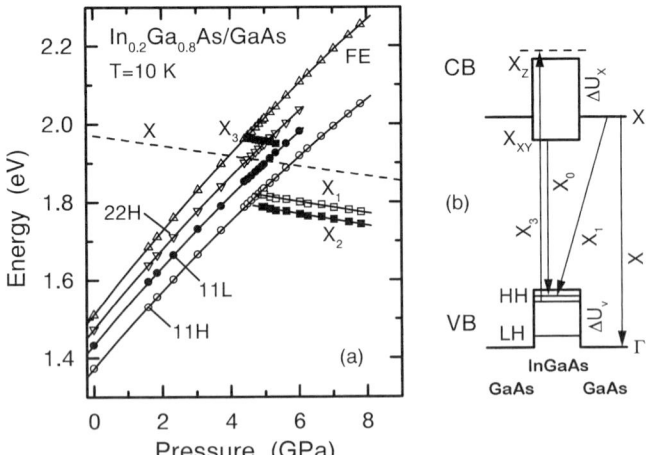

FIG. 40. (a) Energies of photoluminescence and absorption bands of an $In_{0.2}Ga_{0.8}As/GaAs$ MQW sample as a function of pressure [232]. The solid lines represent fitted linear or quadratic relations. The dashed line shows the variation with pressure of the Γ–X indirect gap in bulk GaAs. (b) Sketch of the X-point conduction-band and Γ-point valence-band profile of the InGaAs/GaAs MQW structure above the crossover pressure. Arrows indicate observed optical transitions.

agreement with the interpolated value between that of GaAs (6.0 eV) and that of InAs (3.7 eV) calculated within a tight-binding approach [88]. For comparison, we also give the shear deformation potential of the X point of AlAs, $\Xi_u = 6.9 \pm 0.6$ eV, obtained from PL measurements on GaAs/AlAs SL under (100) uniaxial stress [368].

One issue which has not been discussed so far is that of carrier dynamics. In this respect, the study of the time evolution of PL intensities in bulk [101, 241] and QW structures [348, 359, 350, 74] under pressure can yield important information about intervalley scattering and tunneling rates. For bulk GaAs, Leroux *et al.* have observed a threefold enhancement of electron radiative lifetime above the Γ–X crossover [101], because of the much lower efficiency of indirect recombination processes. At room temperature and for a p-type sample with hole concentration of 1.5×10^{18}cm^{-3}, the PL decay time varies from about 4 to 12 nsec between ambient and 5 GPa. Here one would raise the question how Γ–X intervalley scattering times for electrons compare with these results. Deformation potentials for phonon-assisted intervalley scattering can be measured by looking at the pressure-induced broadening of the exciton linewidth in absorption (see Section VI) or luminescence above the crossover. The latter experiment has been performed in GaAs/AlGaAs MQWs at 80 K [247], from which a deformation potential value of $D_{\Gamma X} = 10.7 \pm 0.7$ eV/Å is obtained for GaAs. This value is twice as large as the one determined in bulk GaAs by optical absorption [214], which suggests that interface roughness might play a role as an additional scattering channel in quantum wells.

In fact, the type-I PL emission in GaAs/AlAs short-period SLs also exhibits almost a doubling of its linewidth in a small pressure range of 0.5 GPa above the crossover [360]. Since very-short-period superlattices are expected to behave similarly to an alloy in terms of scattering potential, it is interesting to compare results obtained for both systems. For instance, time-resolved PL measurements under high pressure in $Al_xGa_{1-x}As$ alloys clearly indicate the importance of alloy-disorder-activated processes for the interpretation of the observed decay times in terms of Γ–X intervalley scattering [294, 241]. The authors were able to separate the contributions from phonon-assisted and alloy-disorder scattering, the latter having a strength of about 25% of the total phonon-assisted one, from which they deduce a Γ–X deformation potential of 3.4 eV/Å for $Al_{0.38}Ga_{0.62}As$ [241]. Experiments on very thin SL structures under uniaxial stress [402], by which means it is possible to distinguish between zero-phonon and phonon-assisted recombination, imply that the interface disorder has an efficiency for intervalley coupling comparable to that for alloy disorder. For further assessment of intervalley scattering efficiencies in QW structures, also concerning carrier-transfer processes between well and barrier states, we refer to the high-pressure time-resolved PL measurements of Ref. [348] and pump-and-probe experiments of Ref. [71].

An interesting situation is attained in QW structures when hydrostatic pressure is used to drive the Γ conduction band states of the wells, for example, toward degeneracy with X valleys in the barriers. In this case, state mixing occurs because of breakdown of translational symmetry in the growth direction of the structure, which leads to folding of X_Z states into the Brillouin-zone center. In fact, the observation of type-II recombination across the heterointerface is a direct consequence of level anticrossing effects between well and barrier states. Pressure-induced Γ–X mixing has been studied theoretically [365, 351] and experimentally in GaAs/AlGaAs MQWs [346, 354] and short-period GaAs/AlAs superlattices [358, 362, 367]. In particular, Jaros et al. [365] have addressed the issue of the important role played by interface imperfections in the enhancement of the optical matrix element for type-II X-transitions due to the mixing with Γ states.

Evidence for mixing between Γ and X_z conduction-band states at the pressure-induced Γ–X crossover has been obtained from the pressure behavior of the emission peaks in two particular InAs/GaAs heterostructures. One of the samples consists of a single highly strained InAs monolayer (ML) embedded in bulklike GaAs (0.8 ML effective thickness) [90]. The second one contains InAs dots in GaAs formed at the step edges provided by the terrace configuration induced by the surface tilt of the misoriented GaAs substrate [373]. The dots are 0.3 ML thick with a lateral extent of about 7 nm. For both samples the PL emission is dominated below the crossover pressure P_c by direct recombination (labeled as M and D for the monolayer and the dots, respectively) between Brillouin-zone-center electron and heavy-hole states bound to the InAs. Above the crossover, the PL peaks denoted as M_1 (D_1) corresponding to the type-II transition from the X valley in GaAs to the heavy-hole state in the InAs monolayer (dots) are apparent in the spectra. The measured peak positions of these PL lines are plotted in Fig. 41 (symbols) as a function of pressure for a narrow range below and above P_c.

In first-order perturbation theory the energy position of the two interacting InAs-Γ and GaAs-X conduction-band states can be expressed as [346]

$$E_{\pm} = \frac{1}{2}\{(E_\Gamma + E_X) \pm [(E_\Gamma - E_X)^2 + 4V^2]^{1/2}\}, \qquad (7.11)$$

where E_Γ and E_X are the noninteracting state energies, which vary linearly with pressure in the short pressure range considered here, and V is the

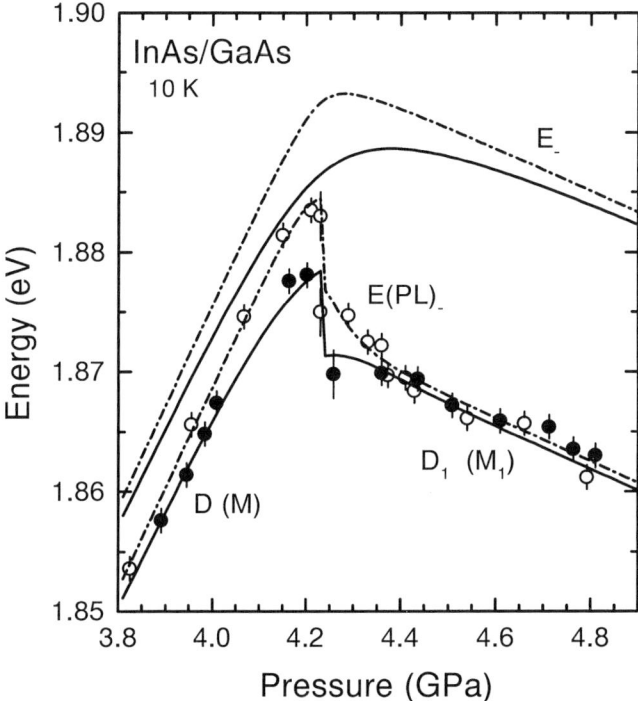

FIG. 41. Calculated results for the anticrossing behavior of the Γ and X_Z states of InAs/GaAs monolayer and dot samples under pressure [373]. The solid and dot–dashed curves as well as the solid and open circles refer to the dot and monolayer sample, respectively. The curves through the data points for the photoluminescence peaks D(M) and $D_1(M_1)$ are obtained using Eq. (7.12) by taking into account the change in exciton-binding energy at the crossover pressure.

interaction potential. Since the observed PL peaks correspond to the recombination of excitons, the PL peak position is

$$E_{PL}(P) = E_-(P) - \mathcal{R}(P), \tag{7.12}$$

where $\mathcal{R}$ is the exciton binding energy given by Eqs. (6.11) and (6.12), which takes into account the anisotropy of effective masses. The effect of state mixing is accounted for in the calculation of the binding energy by using an approximate expression [373]

$$\mu\| = A_\Gamma \cdot \mu\|(\Gamma) + A_X \cdot \mu\|(X), \tag{7.13}$$

where

$$A_\Gamma = \frac{V^2}{(E_\Gamma - E_\pm)^2 + V^2}$$
$$A_X = 1 - A_\Gamma$$

are the squared coefficients of Γ and X_z state wave functions forming the mixed states. Equation (7.13) implies that a mixing takes place only between the z components of the wave functions.

Calculated energies E_- and E_{PL} are presented in Fig. 41 together with the experimental peak energies. The solid and dot-dashed curves correspond to the dot and monolayer sample, respectively. The parameters used in the calculation, which are the same for both samples with the exception of the interaction potential V, are listed in Table II of Ref. [373]. The values for P_c and interaction potential were obtained from a least-squares fit. The agreement between experimental data and calculated results is within experimental uncertainty. State mixing due to the potential step at the InAs/GaAs interfaces is revealed by the anticrossing behavior of the emission peaks, that is, the departure of E_{PL} from a linear pressure dependence while approaching P_c from below and above. The observed jump in the energy E_{PL} at the crossover pressure originates from the abrupt change in the excitonic properties as a consequence of the large difference in binding energy between the Γ and X excitons.

The interaction potential for the monolayer (3 ± 1 meV) is comparable in magnitude to the result of Meynadier *et al.* [403]. They obtained an interaction potential of 1 meV for a GaAs(35Å)/AlAs(80Å) superlattices from the electric-field-induced Γ–X anticrossing. On the other hand the interaction potential for the quantum dots (9 ± 1 meV) is significantly larger than that of the monolayer. This is direct experimental evidence that the mixing between Γ and X_z states is enhanced by the potential step at the edges of the dot, which also cause the breaking of translational invariance in the plane perpendicular to the z direction.

VIII. Photomodulated Reflectance

8. PHOTOMODULATED OPTICAL RESPONSE

Photoluminescence spectroscopy is sensitive to spectral features associated with states very close to the valence- and conduction-band edges. This is a consequence of the rapid thermalization of excited carriers. Photoluminescence excitation (PLE) spectroscopy under pressure is often limited by the availability of tunable lasers suitable for tracing the large pressure shifts of direct interband transitions. This also applies to resonance Raman spectroscopy (see, e.g., Refs. [281, 404, 405] for high-pressure studies), which is not a standard method for investigating the pressure dependence of electronic resonance energies. Photomodulated reflectance (PR) spectrscopy, on the other hand, can provide rich spectra due to allowed direct optical transitions over a wide energy range above the fundamental gap in bulk materials and in particular in semiconductor quantum wells [64, 406–408]. The experiments can be carried out easily in combination with a diamond anvil cell [200, 409], and good resolution of transition energies may be obtained even at room temperature because of the derivative nature of the method.

The transmittance can be photomodulated as well. Phototransmittance spectroscopy is a method to detect weak optical transitions in a spectral region where a sample is largely transparent. This method has been applied in high pressure studies of InGaAs/GaAs multiple quantum well (MQW) systems [410, 371], where the subband transitions in the InGaAs wells are much lower in energy compared to the fundamental gap energy of the GaAs substrate and barrier material. It would be interesting to find out if fine structure associated with an indirect absorption edge of a bulk material can be resolved at high pressure using photomodulated transmittance.

The mechanism leading to a photomodulated optical response is closely related [411, 412] to the electro-optical effect which is utilized in electroreflectance spectroscopy [11, 12, 413]. In the latter method the optical properties of a crystal are modified by a modulated external electric field. In PR the sample is illuminated by a modulated laser with photon energy above the lowest direct gap. The photoexcited electron–hole pairs separate in the build-in potential gradient near the surface. This leads to a modulation of the electric field in the surface region. The advantage of photomodulation over electromodulation in high-pressure experiments is clear: there are no electrical contacts required. Electroreflectance spectroscopy, however, can sometimes give more detailed information, partly because of a better signal-to-noise ratio. The electroreflectance technique has been used to pressures less than 1 GPa in an investigation of interband transitions of GaAs [414].

The change of the optical properties near interband critical points under the influence of an electric field is due to the Franz–Keldysh effect. For illustration we consider band edge states in real space, as shown in the left part of Fig. 42. The electric field causes a "tilting" of valence- and conduction-band edges and the electronic wave functions penetrate into the gap region. In this way a finite optical transition probability occurs at an energy below that of the unperturbed gap. Because the phases of valence- and conduction-band wave functions become fixed with respect to each other,

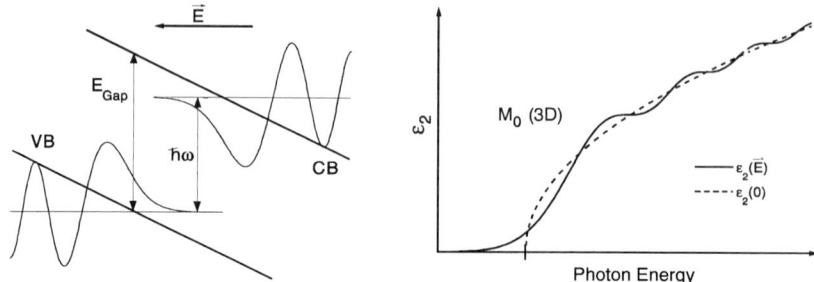

FIG. 42. On the Franz–Keldysh effect: In a potential gradient the electronic wave functions of band edge states extend into the forbidden zone. This results in a change of the interband absorption in the vicinity of an interband critical point, as indicated at right.

there also appear oscillations of $\epsilon_2(\omega)$ at energies above the gap, as is schematically shown for an M_0-type critical point in the right part of Fig. 42. In the low-field limit the oscillations are damped and the main field-induced change of the dielectric response occurs just at the energy of the interband critical point, leading to a sharp spectral feature in a modulation experiment.

As shown by Aspnes [415] the change in the dielectric function with applied electric field $\vec{E}$ is proportional to the third derivative of the unperturbed dielectric function and to the square of the electric field strength:

$$\Delta \epsilon(\omega, \vec{E}) \propto \frac{E^2}{\omega^2} \frac{\partial^3 (\omega^2 \epsilon(\omega))}{\partial \omega^3}. \tag{8.1}$$

By inserting $\epsilon(\omega)$ from Eq. (A.26) one arrives at

$$\Delta \epsilon(\hbar \omega) = A \, \Gamma^n \, e^{i\theta} \, (\hbar \omega - E_{\text{Gap}} + i\Gamma)^{-n} \tag{8.2}$$

which leads to the Aspnes "third-derivative functional form" for the change in reflectance:

$$\frac{\Delta R}{R} = \text{Re} \left[C \, \Gamma^n e^{i\theta} (\hbar \omega - E_{\text{Gap}} + i\Gamma)^{-n} \right]. \tag{8.3}$$

Here, $n = 7/2$, 3, or $5/2$ for a 1-D, 2-D, or 3-D critical point, respectively.

In two-dimensional MQWs the states are confined in the wells and the Franz–Keldysh effect cannot explain a modulation of the dielectric response for a field oriented perpendicular to the wells (this also applies to strongly bound excitons in three dimensions). In an electrical field, however, the subbands are shifted slightly in energy because of the potential change across the wells. Furthermore, electron and hole states of a well are displaced with respect to each other, which results in a change of the dipole transition matrix elements. Thirdly, any change in tunneling probability will result in a linewidth increase of optical transitions. Thus, for MQWs the change $\Delta\epsilon(\omega)$ is given by the first derivative of the unperturbed $\epsilon(\omega)$ with respect to the perturbation [407, 408]:

$$\Delta\epsilon = \left[\frac{\partial\epsilon}{\partial E_n} \frac{\partial E_n}{\partial F} + \frac{\partial\epsilon}{\partial A_n} \frac{\partial A_n}{\partial F} + \frac{\partial\epsilon}{\partial \Gamma_n} \frac{\partial \Gamma_n}{\partial F} \right] \Delta F. \quad (8.4)$$

Here F corresponds to the electric field; A_n is the strength of the transition with energy E_n and width Γ_n. The dielectric function of a homogeneously broadened excitonic transition is [408]

$$\epsilon = C + \frac{A_n}{\hbar\omega - E_n + i\Gamma_n}. \quad (8.5)$$

Taking the first derivative shows that modulation spectra of MQWs (as well as excitonic transitions in three dimensions) can again be described by the Aspnes formula Eq. (8.3), with an exponent $n = 2$ for MQWs. Thus, for PR spectra of bulk and MQW samples the same procedure can be used to extract the optical transition energies, which are the parameters of primary interest in a high-pressure PR experiment.

2. RESULTS FOR BULK MATERIALS

There are only a few PR studies reported of bulk materials under pressure. Hanfland *et al.* [200] measured the E_1 and $E_1 + \Delta_1$ transitions in GaAs up to ~10 GPa. Based on their equation of state data they showed that the volume dependence is quite linear for the E_1-related transitions. The theoretical volume dependence of these transitions obtained by an LMTO-ASA calculation was shown to be in good agreement with experiment.

Another bulk material studied by PR is the layered semiconductor InSe [253], which because of the anisotropic compressibility shows a strongly nonlinear dependence of the direct fundamental gap. The emphasis in the PR study of InSe was on the determination of interband transition energies above the fundamental gap and their pressure coefficients up to 8 GPa. The results were interpreted in combination with a tight-binding LMTO calculation.

We note that bandgap pressure coefficients of bulk materials are obtained as a by-product of PR experiments which are primarily devoted to the study of layered heterostructures.

3. QUANTUM WELLS AND EPILAYERS

Photoreflectance spectroscopy has found wide application in the optical characterization of the subband energy spectrum in 2-D quantum wells [64, 406–408]. High-pressure PR studies of MQWs were first reported by Kangarlu et al. [409]. They studied the effect of hydrostatic pressure on the subband transitions of a GaAs/AlGaAs MQW system at cryogenic temperatures. The same group has subsequently performed more detailed PR studies of GaAs/AlGaAs and GaSb/AlSb quantum wells [65, 416–418].

The energies of the subbands in a MQW system are determined by parameters such as the carrier effective mass m^*, the width and depth of the potential well, and the quantum number n, as can be seen from Eq. (3.34). For a strained MQW system composed of materials with different lattice constants, the built-in strain causes a further splitting of subbands, in addition to the light–heavy hole splitting due to the confining potential. The effect of hydrostatic pressure on these parameters is expected to result in a deviation of the pressure dependence of the MQW subband transitions from those of the bulk materials forming the well and the barrier. In a wide and deep well and for subbands close to the bottom of the well it is reasonable to expect the pressure coefficient dE_n/dP to be close to that of the well material. For narrow wells and large n, however, the wave function penetrates into the barrier material, and this is expected to result in a dE_n/dP value closer to that of the barrier material.

PR studies of lattice-matched GaAs/AlGaAs QW systems [416, 417, 419] in combination with theoretical calculations [420, 421] show that there are mainly two effects which affect the dE_n/dP values in a type-I QW system: (1) As the Γ conduction-band minimum moves up in energy, the effective masses increase, which leads to a reduction of the subband spacing. (2) Because of the difference in pressure coefficients of the well and barrier materials (about 10% in the GaAs/AlGaAs system), the well depth changes with pressure, which changes the penetration of the well states into the barrier. The exciton binding energies also depend on pressure (see Section 6), but this effect as well as the change in the QW width with pressure (typically less than 3% at pressures up to 10 GPa) is negligible.

In order to demonstrate the typical effects of pressure on subband transitions in lattice-matched QWs we discuss results [419] obtained for a modulation-doped GaAs/GaAlAs *single* quantum well sample (well width 24.5 nm, carrier density in the well 5.6×10^{11} cm^{-2}). The properties of the high-mobility 2-D electron gas of this sample are discussed in Section IX. Figure 43 shows room-temperature PR spectra measured at ambient pressure and at 3.1 GPa. According to the selection rules for QW structures, only transitions between subbands with the same index are allowed. The spectra show features due to heavy hole transitions (denoted H_{nn} in Fig. 43) up to subband index $n = 6$. The light hole states give rise to a second set of transitions, but these are expected to be a factor of 3 weaker [422]. Light hole transitions can be identified in the spectra at ambient pressure, but transitions with $n > 1$ are not well resolved under pressure. Additional features are observed in the PR spectra which arise from the E_0 gaps of the AlGaAs barrier material and of GaAs layers deeper in the sample.

The energies of the subband transitions are obtained by fitting a superposition of Aspnes terms [see Eq. (8.3)] to the experimental spectra (dashed line in Fig. 43). Figure 44 shows the transition energies as a function of pressure, and the linear pressure coefficients obtained by fitting second-order polynomials to the $E_n(P)$ data are plotted in Fig. 45 as a function of zero-pressure transition energy. Although the error bars are relatively large

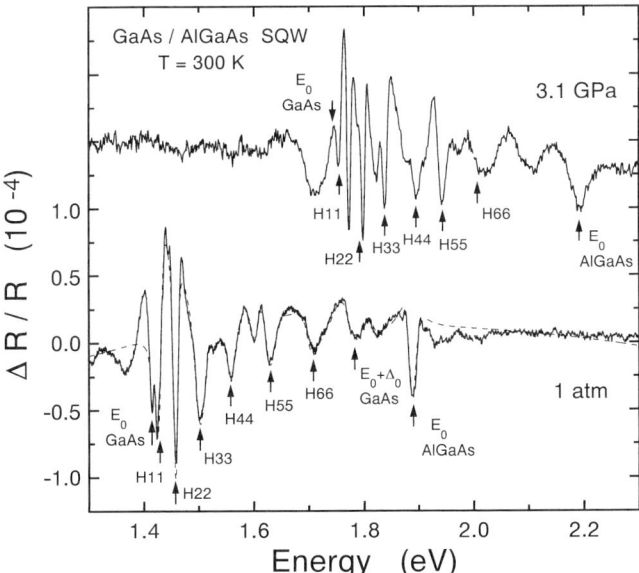

FIG. 43. Photoreflectance spectra of a modulation-doped GaAs single quantum well at two different pressures. Heavy hole quantum well transitions up to subband index $n = 6$ are seen above the fundamental gap with good resolution even at room temperature. After Ref. [419].

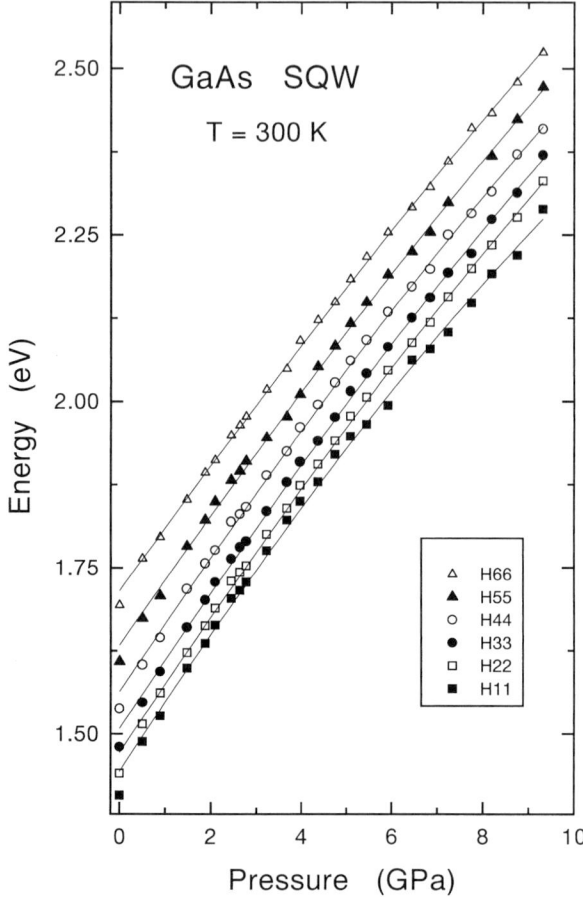

FIG. 44. Heavy hole transition energies between quantum well states of a modulation-doped SQW as a function of pressure. The lines represent results of least-squares fits to the data points above 1.5 GPa. After Refs. [419, 61].

because of the correlation between fitted linear and quadratic coefficients, there is clearly a decrease of the dE_n/dP values with increasing subband index; the higher the index, the closer is the pressure coefficient to that of the barrier material. The decrease of the pressure coefficients with increasing subband index is more directly seen in Fig. 46, which shows, as a function of pressure, the subband transition energies measured relative to that of the E_0 bandgap of GaAs [423]. The initial *increase* of all subband energies seen below 1.5 GPa is attributed to the pressure-driven loss of carriers from

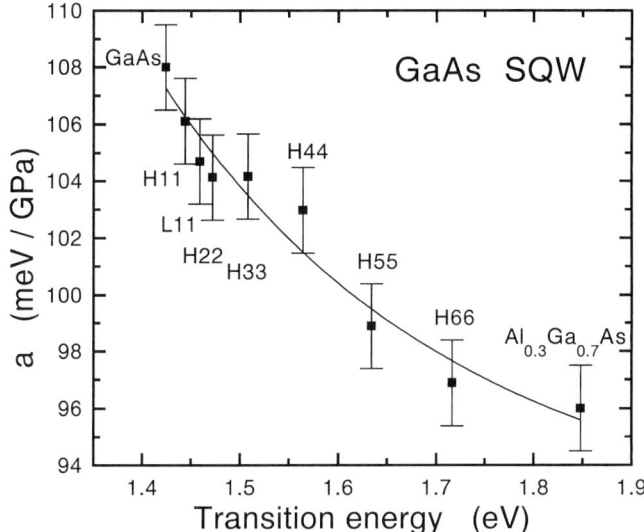

FIG. 45. Linear pressure coefficients of heavy-hole subband transition energies of a GaAs/AlGaAs single quantum well as a function of transition energy at zero pressure. The solid line is a guide to the eye. After Refs. [419, 61].

the QW and the related decrease of the gap renormalization effect (see Section IX).

The differences in the dE_n/dP values of the single quantum well are reproduced by a calculation of the subband energies in the envelope function approximation. The calculation takes into account the known pressure dependencies of the bandgap energies of the bulk barrier and well materials and a change of the effective masses as given by the $\vec{k}\cdot\vec{p}$ method (see Section III). The assumption of a constant relative valence-band discontinuity is supported by experimental evidence obtained for related heterostructures [340, 127]. The result of the calculation is shown by the solid lines in Fig. 46. The agreement with experiment is excellent. The main conclusion is that the pressure dependence of the effective masses derived from the bandgap changes of the bulk material fully accounts for the pressure dependence of subband energy levels in this particular QW system. This is in accordance with the other PR studies of GaAs/AlGaAs QW systems [416, 417].

It has been reported that indirect gap transitions give rise to structure in PR experiments at pressures above the Γ–X crossover in GaAs/AlGaAs QWs [416, 417]. This has not been confirmed for the SQW sample discussed earlier, for which all features seen below the lowest direct transition of GaAs were definitely due to optical interference effects [61, 419].

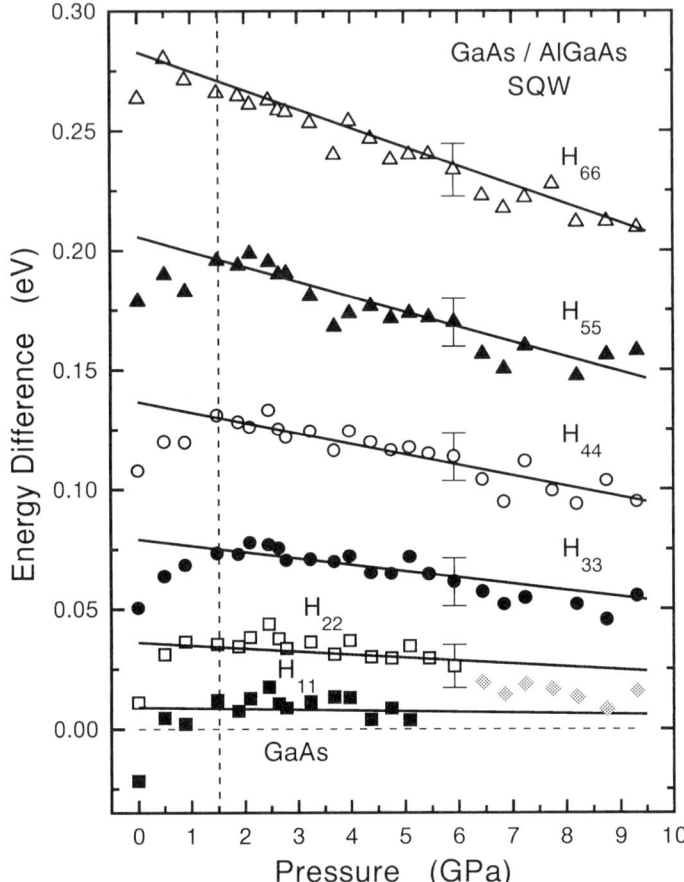

FIG. 46. Energy difference between heavy hole transition energies of a modulation-doped GaAs SQW and the bandgap energy of bulk GaAs as a function of pressure. The bandgap renormalization of the QW transitions due to the presence of free carriers [214] is reduced with increasing pressure, until at about 2 GPa the carrier density in the QW becomes essentially zero. Lines represent theoretical predictions of a $\vec{k} \cdot \vec{p}$ calculation within the envelope function approximation. After Ref. [419].

The effects of quantum confinement on the pressure coefficients were shown to be more pronounced in a PR study of a GaSb/AlSb QW system [418], which has a small lattice mismatch of 0.6% at ambient conditions. GaSb has a much smaller direct gap compared to AlSb (0.81 eV versus about 2.3 eV at 80 K) leading to deep wells compared to other material systems and consequently a small penetration of wave functions into the barrier. Furthermore, the pressure coefficient of the GaSb gap is very large (140 meV/GPa). This leads to a large change in the effective mass under pressure, such that the reduction of the confinement energies in GaSb/AlSb QW systems is larger compared to GaAs/AlGaAs QWs. For large-gap QW systems the change of the effective mass is expected to play a minor role, and strain tuning under hydrostatic pressure is a more important effect.

Chandrasekhar and co-workers applied hydrostatic pressure to tune the electronic structure of a variety of pseudomorphically grown strained epilayer systems, such as ZnSe on GaAs [24], CdTe on InSb [425], ZnTe on InAs [426], and InGaP on GaAs [307, 427]. Depending on the relative signs of the lattice mismatch strain and the pressure-induced strain, the net strain in the QWs increases [425] or decreases [424, 426], with the possibility of inducing exact lattice matching in the latter case. Biaxial strain affects, in particular, the splitting of heavy- and light-hole valence-band states. For instance, in the ZnSe-on-GaAs and ZnTe-on-InAs systems, the pressure-tuned crossing of the $n = 1$ heavy and light hole transitions was demonstrated as well as a strongly nonlinear dependence of the splitting on pressure [424, 426]. The latter effect has been interpreted in terms of a pressure dependence of the shear deformation potential. Part of this work has been summarized in Refs. [428, 429].

IX. Optical Properties of Electron Gases under Pressure

Free carriers in doped semiconductors form electron gases whose densities can be changed by many orders of magnitude, in contrast to normal metals, just by adjusting the doping level. The progress in epitaxial growth techniques enables the fabrication of modulation-doped heterostructures with extremely high electron mobilities. Because of the very low scattering rates achieved for electrons confined in quantum wells, these systems are particularly suitable for studies of fundamental electron–electron interactions. Elementary excitations such as plasmons or spin-density excitations, which are sustained by the electron gas, have great impact on the low-frequency dielectric properties of semiconductors [143, 430]. Application of hydrostatic pressure plays an important role here, because by these means it is possible to vary the electron density as well as the carrier effective mass with profound consequences for the energies of the elementary excitations [94, 431, 273]. Inelastic light scattering has proved to be a powerful tool for probing the energy spectrum of excitations for electron gases of diverse dimensionalities [432]. Therefore, we focus here on recent results from light-scattering experiments on electron gases formed in bulk and quantum well semiconductor samples under high hydrostatic pressure.

1. ELEMENTARY EXCITATIONS OF THE ELECTRON GAS

The elementary excitations of a *noninteracting* electron gas are electron–hole pairs or single-particle excitations (SPEs) of the type depicted in the inset to Fig. 47. They are characterized by an energy δE and a wave vector $\vec{q}$. The range of allowed electron–hole pair energies as a function of wave vector forms the so-called SPE continuum, as illustrated by the shaded area in Fig. 47. For a parabolic band filled with electrons up to the Fermi energy E_F, the SPE continuum is delimited by the energies

$$\delta E^{\pm} = \frac{\hbar^2 q^2}{2m_e^*} \pm \frac{\hbar^2 k_F q}{m_e^*}, \qquad (9.1)$$

where k_F is the Fermi wave vector and m_e^* the electron effective mass.

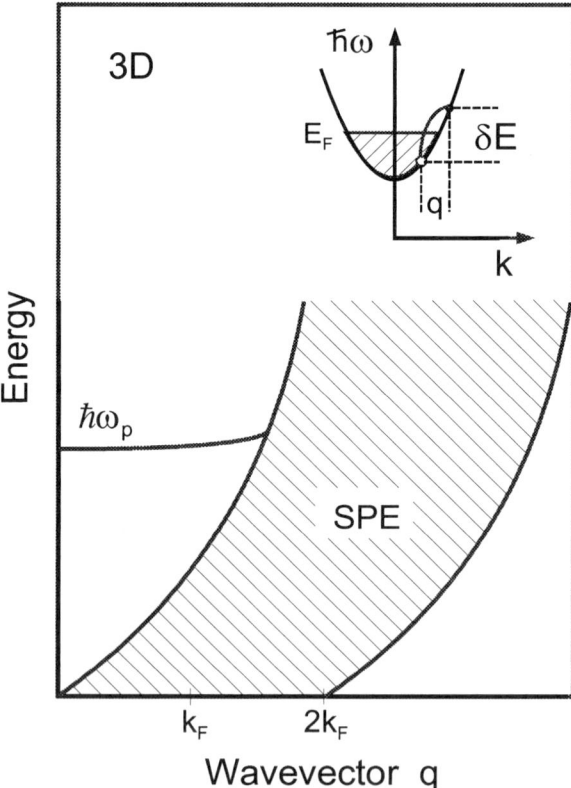

FIG. 47. Wave-vector dispersion of the collective 3-D plasma excitation and the single-particle continuum. The inset shows a characteristic intrasubband electron–hole pair excitation.

4 Optical Properties of Semiconductors under Pressure

The frequency-dependent dielectric function for free electrons in bulk materials is simply given by the Drude model

$$\epsilon(\omega) = \epsilon(0) \cdot \left(1 - \frac{\omega_p^2}{\omega^2}\right) \tag{9.2}$$

with a screened plasma frequency

$$\omega_p^2 = \frac{4\pi n_{3D} e^2}{\epsilon(0) m_e^*}, \tag{9.3}$$

which is proportional to the square root of the 3-D electron density n_{3D} and to the inverse of the electron effective mass. Since the excitation energies of the electron gas are typically below that of the optical phonons, the screening of the electrostatic potential in the material is accounted for by using the static dielectric constant $\epsilon(0)$.

The most important manifestation of collective behavior of the electron gas due to the Coulomb interaction between the electrons are the screening effects, that is, the way in which the electronic charge readjusts itself in response to an external perturbation. Any fluctuation of the charge (spin) distribution induces an electrostatic potential which adds to the external one, leading to self-sustained collective excitations. Here we distinguish between collective charge-density excitations (CDEs), which correspond to in-phase fluctuations of the electron density, and spin-density excitations (SDEs), in which the number of electrons with spin-up or spin-down components is spatially modulated. The best-known example of collective charge-density oscillations are the plasmons. They represent longitudinal compression waves of the charge density with characteristic frequency ω_p corresponding to the zeros of the real part of the dielectric function $\epsilon_1(\omega)$ [168].

The Coulomb interaction contains two terms: the direct or Hartree term and the exchange interaction as a consequence of the Pauli exclusion principle. The Hartree interaction can be represented by a *pair-bubble* Feynman diagram like the one shown in Fig. 48a, in which an electron as it propagates (represented by the solid line) interacts with the rest of the Fermi sea by creating virtual electron–hole pairs. The diagram (b) represents the exchange interaction between two electrons with the same spin, which in consequence interchange sites [433]. In order to take the screening into account we have to sum up the bubble diagrams (a), which leads to the so-called random phase approximation (RPA). Since RPA is a mean-field approximation, local-field effects are neglected, which is a sensible approximation for valence electrons in semiconductors.

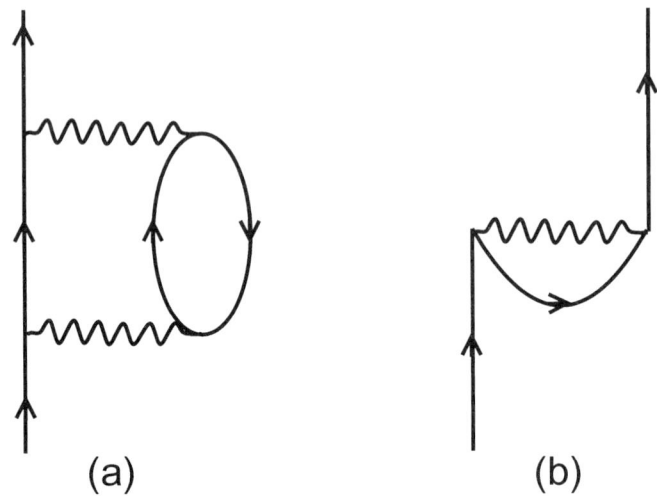

FIG. 48. (a) Particle-hole *pair bubble* Feynman diagram representing the self-energy part of the Coulomb interaction. (b) Exchange diagram.

In the framework of the RPA, the dielectric function of the 3-D electron gas is described by [430]

$$\epsilon(q, \omega) = \epsilon_\infty + 4\pi\chi(q, \omega) + 4\pi\chi_L(q, \omega). \tag{9.4}$$

Here, $\chi(q, \omega)$ describes the electronic contribution and $\chi_L(q, \omega)$ the contribution of the lattice, which is important for polar semiconductors:

$$4\pi\chi_L(q, \omega) = \epsilon_\infty \frac{\omega_{LO}^2 - \omega_{TO}^2}{\omega_{TO}^2 - \omega^2}. \tag{9.5}$$

The calculation of the electronic part starts from the temperature-dependent Lindhard expression [434] for the dielectric susceptibility,

$$4\pi\chi^0(q, \omega + i\Gamma) = \frac{e^2}{\pi^2 q^2} \int_0^{k_F} f(k, T) \times \left[\frac{1}{\frac{\hbar^2 q^2}{2m^*} + \frac{\hbar^2 \vec{q}\cdot\vec{k}}{m^*} - \hbar(\omega + i\Gamma)} \right.$$

$$\left. + \frac{1}{\frac{\hbar^2 q^2}{2m^*} - \frac{\hbar^2 \vec{q}\cdot\vec{k}}{m^*} + \hbar(\omega + i\Gamma)} \right] d^3k, \tag{9.6}$$

where $f(k, T)$ is the Fermi distribution function at finite temperature T and Γ accounts for collision damping effects. After the formal integration of Eq. (9.6) for $T = 0$ K, we achieve analytical expressions for the real and imaginary parts of the dielectric function [271, 270]. To include collision broadening into the Lindhard expression it is not correct to replace ω by $\omega + i\Gamma$. Such a procedure fails to conserve the local electron number. Mermin used a relaxation time approximation in which the collisions relax the electron density matrix to a local equilibrium density matrix [435]. This results in the Lindhard–Mermin function

$$\chi(q, \omega + i\Gamma) = \frac{\chi^0(q, 0)(\omega + i\Gamma)[\chi^0(q, \omega + i\Gamma)]}{\omega\chi^0(q, 0) + i\Gamma[\chi^0(q, \omega + i\Gamma)]}. \tag{9.7}$$

Electron gases formed in semiconductor quantum wells or heterointerfaces possess two-dimensional character. This reduction of the dimensionality of the electronic system with the consequent formation of discrete energy levels affects the spectrum of elementary excitations in a fundamental way: On the one hand, the intrasubband plasmon has zero frequency at $q = 0$ and it exhibits a square-root dispersion $\omega_p \propto \sqrt{q}$; on the other hand, there exist intersubband excitations where electron and hole of the electron–hole pair belong to different subbands [436]. The wave-vector dispersions of 2-D intersubband excitations associated with electronic transitions between an occupied subband i with 2-D density n_i and an empty one j are shown schematically in Fig. 49. The intersubband SPE continuum is centered at the intersubband spacing energy E_{ij} for $q = 0$. For small wave vectors compared to k_F the width of the SPE continuum increases linearly with q. The corresponding collective excitations of charge and spin density are shifted in energy with respect to the SPEs by many-body effects.

When electron interactions are included within RPA, the intersubband susceptibility χ_{ij} is obtained in the form of a Dyson equation [437, 438]

$$\chi_{ij}(q, \omega) = \frac{\chi_0(q, \omega)}{1 - \gamma_{ij}(q) \cdot \chi_0(q, \omega)}, \tag{9.8}$$

where χ_0 is the noninteracting susceptibility. Collective excitations occur at the poles of $\chi_{ij}(q, \omega)$. The factor γ_{ij} represents the Coulomb interaction, as diagrammatically shown in Fig. 50a, which can be written in terms of the direct or Hartree term $\alpha_{ij}(q)$ and the vertex correction $\beta_{ij}(q)$ due to exchange and correlation. For spin-density excitations we set

$$\gamma_{\text{SDE}} = -\beta_{ij}, \tag{9.9}$$

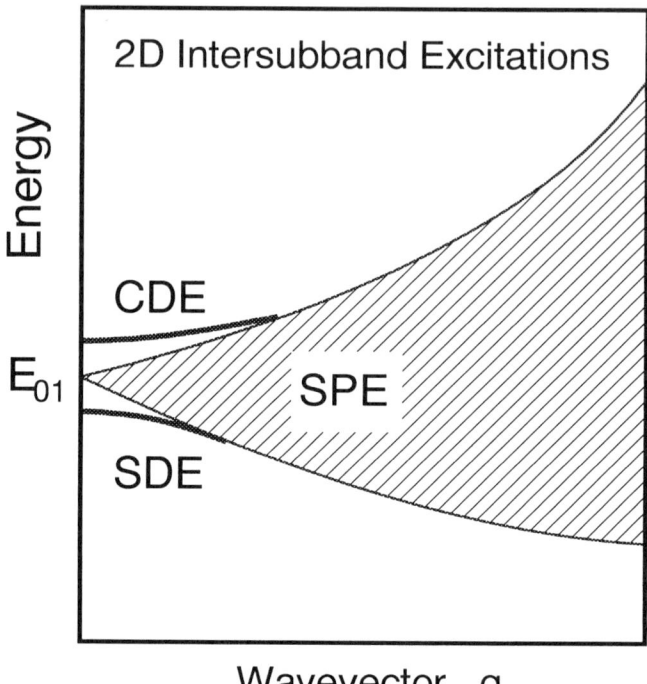

FIG. 49. Wave-vector dispersion of 2-D intersubband excitations. Solid lines correspond to collective charge-density and spin-density excitations. The shaded area represents the single-particle continuum.

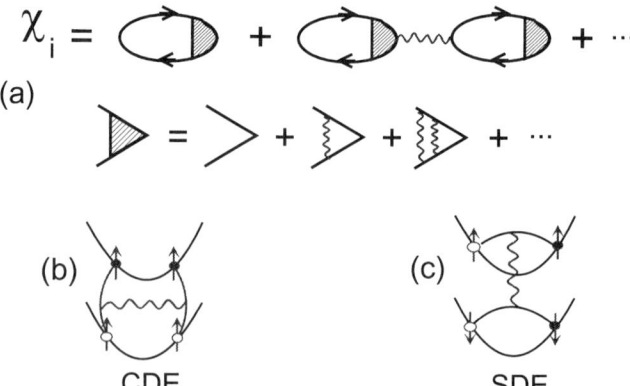

FIG. 50. (a) Feynman diagrams for the calculation of the electronic susceptibility within RPA including exchange-vertex corrections. Graphs for representing the coupling between electron–hole pairs due to (b) the Hartree term and (c) the exchange term of the Coulomb interaction in collective 2-D intersubband excitations.

because only the vertex corrections are of importance for spin-flip excitations without charge fluctuations. The energy of the SDE at $q = 0$ is thus given by [439, 440]

$$\omega_{SDE}^2 = E_{ij}^2 - 2n_i E_{ij} \beta_{ij}. \tag{9.10}$$

As illustrated by the graph (c) of Fig. 50 the so-called *excitonic* shift β_{ij} represents an exciton-like binding between the excited electron in the upper subband and the hole left behind in the lower subband. Because of the orthogonality of the wave functions of different subbands, the exchange-correlation interaction with all other electrons is missing for the excited one, thus leading to a lowering of the energy of collective SDE with respect to E_{ij}.

For charge-density excitations there is an additional contribution from the Hartree term, and the Coulomb interaction is written as

$$\gamma_{CDE} = \frac{\alpha_{ij}}{\tilde{\epsilon}} - \beta_{ij} \tag{9.11}$$

with

$$\tilde{\epsilon}(\omega) = \frac{\omega^2 - \omega_{LO}^2}{\omega^2 - \omega_{TO}^2} \tag{9.12}$$

being the phonon contribution to the dielectric function of the polar lattice. The energy of the CDE is, for long wavelengths,

$$\omega_{CDE}^2 = E_{ij}^2 + 2n_i E_{ij} \left(\frac{\alpha_{ij}}{\tilde{\epsilon}} - \beta_{ij} \right). \tag{9.13}$$

The CDE is shifted upwards from the SDE because of the Hartree interaction or *depolarization* shift α_{ij}. This shift is associated with a macroscopic electric field induced by the collective charge-density fluctuation itself as indicated pictorially by the graph (b) of Fig. 50. It is determined by the electric dipole moment associated with intersubband electron–hole pair transitions, and it is given by [436]

$$\alpha_{ij}(q) = \frac{2\pi e^2}{\epsilon_\infty q} \int dz \int dz' F_i(z) F_j(z) e^{-q|z-z'|} F_i(z') F_j(z'), \tag{9.14}$$

where $F_{i,j}(z)$ are the envelope functions of the different subband states, which for modulation-doped QWs have no definite parity. These dipole moments add up coherently, producing a macroscopic depolarization field which is opposite to the external one, thus increasing the frequency of the collective CDE.

A further reduction of dimensionality as for quantum wires and dots would introduce similar changes in the properties of elementary excitations of the electron gas as discussed above in going from 3-D to 2-D. Probably the most interesting effect, however, concerns the enhancement of exchange and correlation against the Hartree interaction. This is just a consequence of the increasing confinement of the particles in reduced dimensions. One example is the clear observation of SDEs in modulation-doped quantum wells [441, 442]. Here we will show other interesting phenomena arising from electronic correlations when the electron density is reduced by orders of magnitude by applying pressure.

2. Inelastic Light Scattering by Elementary Excitations of the Electron Gas

The coupling between the electronic system and the radiation field is described by the electron–photon interaction Hamiltonian [168]

$$\mathcal{H}'_R = \frac{e}{2m_0 c} \sum_i (\vec{p}_i \cdot \vec{A}(\vec{r}_i) + \vec{A}(\vec{r}_i) \cdot \vec{p}_i) + \frac{e^2}{2m_0 c^2} \sum_i A^2(\vec{r}_i), \quad (9.15)$$

where $\vec{p}_i$ is the one-electron momentum operator and $\vec{A}$ the vector potential. The second term in the Hamiltonian (9.15) contains the field operator squared, thus leading to light-scattering processes involving two photons in first-order perturbation theory. The first term of (9.15) is linear in $\vec{A}$ and has to be taken to second-order. It produces virtual interband transitions with a transition probability which contains an energy denominator $(E_g - \hbar\omega_L)$, where E_g is the bandgap and ω_L is the frequency of the light. Therefore, we have to distinguish between the nonresonant case, that is, $\hbar\omega_L \ll E_g$, and that of a resonance $\hbar\omega_L \approx E_g$.

In the nonresonant case both terms of the electron–photon interaction can be collected into an effective A^2 Hamiltonian [443]

$$\mathcal{H}'_{R,\text{eff}} = \frac{e^2}{c^2} \hat{e}_L \cdot \frac{1}{\tilde{m}^*} \cdot \hat{e}_S \gamma N(\vec{q}) A(\omega_L) A(\omega_S), \quad (9.16)$$

where $\tilde{m}^*$ is the effective mass tensor, $\hat{e}_{L,S}$ are the unit polarization vectors of the incident and scattered photons with wave vectors $\vec{k}_{L,S}$ and $\vec{q} = \vec{k}_L -$

4 OPTICAL PROPERTIES OF SEMICONDUCTORS UNDER PRESSURE   359

$\vec{k}_S$ is the scattering wave vector. In addition, the vector potential amplitude is given by

$$A(\omega) = c\sqrt{\frac{2\pi\hbar}{\mathcal{V}\omega\epsilon(\omega)}} \qquad (9.17)$$

and the Fourier transform of the electron density is

$$N(\vec{q}) = \frac{1}{\mathcal{V}}\sum_j e^{i\vec{q}\cdot\vec{r}_j} \qquad (9.18)$$

with $\mathcal{V}$ the volume of the crystal. This interaction Hamiltonian leads to a differential scattering cross-section per unit frequency and unit solid angle [444]

$$\frac{\partial^2\sigma}{\partial\Omega\partial\omega} = r_0^2 m_0^2 \mathcal{V}^2 \left|\hat{e}_L \cdot \frac{1}{\tilde{m}^*} \cdot e_S\right|^2 S(\vec{q},\omega), \qquad (9.19)$$

where $r_0 = e^2/m_0 c^2$ is the classical radius of the electron, $\omega = \omega_L - \omega_S$ is the scattering frequency, and $S(\vec{q},\omega)$ is the dynamical structure factor defined as

$$S(\vec{q},\omega) = \frac{1}{2\pi}\int_{-\infty}^{\infty} e^{i\omega t}\langle N(\vec{q},t)N^\dagger(\vec{q},0)\rangle\, dt. \qquad (9.20)$$

Here the angular brackets denote thermal average and $N(\vec{q},t)$ is the time-dependent density operator in the Heisenberg representation. The important result expressed by Eqs. (9.19) and (9.20) is that inelastic scattering of light by the electron gas occurs due to thermal fluctuations of the electronic density.

In order to relate the scattering cross-section with the optical response of the system given by the dielectric function we write the density fluctuations in terms of the macroscopic polarization $\vec{P}(\vec{q},t)$ of the free carriers

$$N(\vec{q},t) = -\frac{i}{e}\vec{q}\cdot\vec{P}(\vec{q},t), \qquad (9.21)$$

which allows us to write the dynamical structure factor in terms of the fluctuations of the electron gas polarizability

$$S(\vec{q},\omega) = \frac{q^2}{2\pi e^2}\int_{-\infty}^{\infty} e^{i\omega t}\langle \vec{P}(\vec{q},t)\cdot\vec{P}^\dagger(\vec{q},0)\rangle\, dt. \qquad (9.22)$$

We now make use of the fluctuation-dissipation theorem [445] to calculate $S(\vec{q}, \omega)$ for light scattering by the electron gas. For that purpose we define a *generalized force* $F(\vec{q}, t)$ which acts on the system producing a *generalized displacement* $X(\vec{q}, t)$. In linear response theory both variables are proportional to each other and the proportionality factor is the so-called linear response function $T(\vec{q}, t)$. In our particular case we set the electric displacement vector $\vec{D}$ as the driving force F, the polarization $\vec{P}$ as X, and χ/ϵ as the response function T. Since the Fourier-transformed correlation function $\langle XX^\dagger \rangle_\omega$ is linked through the fluctuation-dissipation theorem to the imaginary part of $T(\vec{q}, \omega)$, we find for the dynamical structure factor

$$S(\vec{q}, \omega) = \frac{\hbar q^2}{\pi e^2} \frac{1}{1 - e^{-\hbar\omega/k_B T}} \operatorname{Im}\left\{\frac{\chi(\vec{q}, \omega)}{\epsilon(\vec{q}, \omega)}\right\}. \quad (9.23)$$

In this way we arrive at a description of light scattering in terms of the plasma dielectric function.

We now like to consider two important physical situations, the scattering by single-particle and collective excitations. In the presence of light scattering by single-particle excitations, one can assume that $\chi \ll \epsilon_\infty$ because screening effects are negligible and the electron gas behaves as noninteracting. Thus, it holds that

$$\operatorname{Im}\left\{\frac{\chi(\vec{q}, \omega)}{\epsilon(\vec{q}, \omega)}\right\} = \frac{1}{\epsilon_\infty} \operatorname{Im}\{\chi_0(\vec{q}, \omega)\}. \quad (9.24)$$

This shows that the inelastic light-scattering cross-section for SPE is proportional to the density of states of electron-hole pairs given by the imaginary part of the free-electron susceptibility χ_0. We will come back to the scattering by SPEs when discussing the resonant case.

For charge-density excitations screening effects are essential, having a significant contribution to the electronic susceptibility ($\chi = (\epsilon - \epsilon_\infty)/4\pi$). The response function can be written as

$$\operatorname{Im}\left\{\frac{\chi(\vec{q}, \omega)}{\epsilon(\vec{q}, \omega)}\right\} = \frac{\epsilon_\infty}{4\pi} \operatorname{Im}\left\{\frac{-1}{\epsilon(\vec{q}, \omega)}\right\}. \quad (9.25)$$

This means that features in light-scattering spectra corresponding to collective excitations will occur at the energies of the zeros of the complex dielectric function.

We now turn to the discussion of the resonant case, which is the most relevant one from the experimental point of view. When the laser frequency is comparable to a bandgap energy, then the intermediate states in virtual transitions become real and there is a huge enhancement of the scattering efficiency caused by the divergent energy denominator. Another important result concerns the activation of an additional light-scattering mechanism associated with spin-density fluctuations. Figure 51 shows two diagrams illustrating schematically the virtual valence-to-conduction band transitions which contribute to resonant light scattering by a 2-D electron gas. Spin-flip scattering processes, in which the intermediate states for the optical interband transitions denoted as (1) and (2) have different spin components, are allowed only if the valence wave function is not an eigenfunction of spin. This is the case for the Γ-point light and split-off hole bands of tetrahedral semiconductors (see Table I), which possess mixed spin character due to spin–orbit interaction.

Using the $\vec{k}\cdot\vec{p}$ wave functions for the valence- and conduction-band extrema at Γ, Hamilton and McWhorter have derived an effective electron–photon Hamiltonian valid near resonance conditions [446, 432],

$$\mathcal{H}'_{R,\text{res}} = \frac{e^2}{2m_0 c^2} \sum_q A(\omega_L) A(\omega_S) \tilde{N}(q), \qquad (9.26)$$

with

$$\tilde{N}(q) = \sum_{\alpha,\beta} c_\beta^\dagger c_\alpha \langle \alpha | e^{i\vec{q}\cdot\vec{r}} | \beta \rangle [\hat{e}_L \cdot \tilde{A} \cdot \hat{e}_S + i(\hat{e}_L \times \hat{e}_S) \cdot \tilde{B} \cdot \langle S_\alpha | \hat{\sigma} | S_\beta \rangle], \qquad (9.27)$$

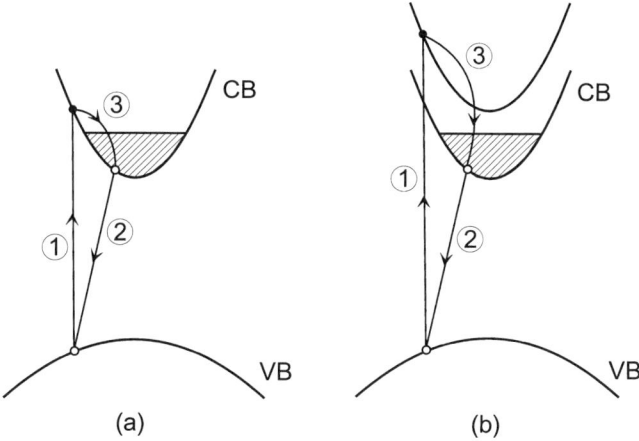

FIG. 51. Schematic diagrams showing the virtual valence-to-conduction band transitions that contribute to resonant inelastic light scattering by the 2-D electron gas for (a) intrasubband and (b) intersubband excitations.

where $c_\beta^\dagger(c_\alpha)$ is the creation (destruction) operator for an electron with envelope function $|\beta\rangle(|\alpha\rangle)$ and spin projection $|S_\beta\rangle(|S_\alpha\rangle)$ and $\tilde{\sigma}$ represents the Pauli matrices. The first term of (9.27), which is symmetric in the polarization of incident and scattered light, is related to the charge-density fluctuations. Thus, light scattering by CDEs is apparent in polarized spectra. The tensor $\tilde{A}$ has a remarkable simple form,

$$\tilde{A} = \tilde{I}\left[1 + \frac{2p^2}{3m_0}\left(\frac{2E_0}{E_0^2 - (\hbar\omega_L)^2} + \frac{E_0 + \Delta_0}{(E_0 + \Delta_0)^2 - (\hbar\omega_L)^2}\right)\right], \quad (9.28)$$

where $\tilde{I}$ is the 2×2 unitary matrix. We note that out of resonance, that is, for $\hbar\omega_L \ll E_0$, Eq. (9.28) becomes identical to the inverse effective mass tensor of the Kane model [see Eq. (3.19)].

The second term of Eq. (9.27) now describes the coupling of light with excitations of spin. By considering only the terms related to the $\tilde{\sigma}_z$ spin component, this part of the Hamiltonian can be rewritten as [432]

$$\mathcal{H}'(\tilde{\sigma}_z) = i\frac{e^2\gamma}{2m_0c^2}(\hat{e}_L \times \hat{e}_S) \cdot \tilde{B} \cdot \hat{e}_z \left(\frac{N_\uparrow(q) - N_\downarrow(q)}{2}\right) A(\omega_L)A(\omega_S), \quad (9.29)$$

where the connection with the spin-density fluctuations is explicitly shown. This term contains the cross product of the light polarization vectors, indicating that SDEs are active in depolarized spectra. These selection rules enable the clear distinction between both types of collective excitations in light-scattering experiments. The operator $\tilde{B}$ takes the form

$$\tilde{B} = \tilde{I}\frac{2p^2}{3m_0}\hbar\omega_L\left(\frac{E_0}{E_0^2 - (\hbar\omega_L)^2} - \frac{E_0 + \Delta_0}{(E_0 + \Delta_0)^2 - (\hbar\omega_L)^2}\right). \quad (9.30)$$

It is easy to see that $\tilde{B}$ vanishes in the nonresonant limit or if there is no spin–orbit interaction ($\Delta_0 = 0$).

We close this section with some remarks about a still-controversial problem, which involves the origin of the unscreened inelastic light scattering by single-component plasmas. Wolff [447] has considered the case of scattering by energy-density fluctuations without an associated net charge, which takes place provided nonparabolicity is present. Nevertheless, the calculated cross-sections are too small to account for the observed intensities [448, 449]. A large body of experiments [441, 442, 450–452] seems to indicate that single-particle scattering might be activated in extreme resonance conditions by residual disorder, but up to now this has not been treated theoretically. A microscopic theory has been developed in which SPE scattering due to second-order intersubband $\vec{p}\cdot\vec{A}$ terms comes out in a natural way when the broadening of intermediate and final states is taken into account [453]. In other words, if the Landau damping of collective excitations is substantial, as is the case within the SPE continuum, then the light is scattered by unscreened electron–hole pair excitations. We have to point out, however, that no prediction about absolute cross-sections is made by this theory.

3. LO-PHONON–PLASMON COUPLED MODES IN BULK SEMICONDUCTORS

The plasmon frequency ω_p in heavily n-doped bulk semiconductors increases as the square root of the electron density [see Eq. (9.3)]. In direct-gap semiconductors and for values of $n_{3D} \geq 5 \times 10^{17}$ cm^{-3}, ω_p typically is comparable to the optical phonon frequencies, ω_{TO} and ω_{LO}. Because of the polar character of the zincblende lattice, the longitudinal optical (LO) phonon interacts with the macroscopic electric field of the plasmon, giving rise to the formation of coupled L_+ and L_- modes. The frequencies of the coupled modes are to be obtained from the zeros of the dielectric function (9.4) containing the contributions from the lattice and the free electrons. If we use for the latter the Drude expression (47), the equation $\epsilon(\omega) = 0$ can be readily solved for the coupled-mode frequencies, giving

$$\omega_{\pm}^2 = \frac{1}{2}\left\{\left(\omega_p^2 + \omega_{LO}^2\right) \pm \left[(\omega_p^2 + \omega_{LO}^2)^2 - 4\omega_p^2\omega_{TO}^2\right]^{1/2}\right\}. \quad (9.31)$$

The coupled-mode frequencies depend strongly on the free carrier density and display a characteristic anticrossing behavior of the two branches (see Fig. 52). For high carrier densities the upper mode (L_+) is plasmon-like and the lower mode (L_-) shows a phonon-like behavior, and vice versa at low densities.

The first systematic study by inelastic light scattering of the plasmon–phonon coupled modes was carried out by Mooradian and Wright [448, 449] in GaAs, InP, and CdTe using samples with different carrier concentrations. The use of hydrostatic pressure constitutes an interesting alternative for changing the electron density, provided the semiconductor undergoes a Γ–X conduction-band crossover in the stability range of the zincblende phase [454]. As illustrated by the inset to Fig. 55, when pressure is tuned across the band crossing, a transfer of electrons between conduction band valleys is established. Because of the large differences in electron effective masses between the Γ and X minima, clear changes are observed in light-scattering spectra.

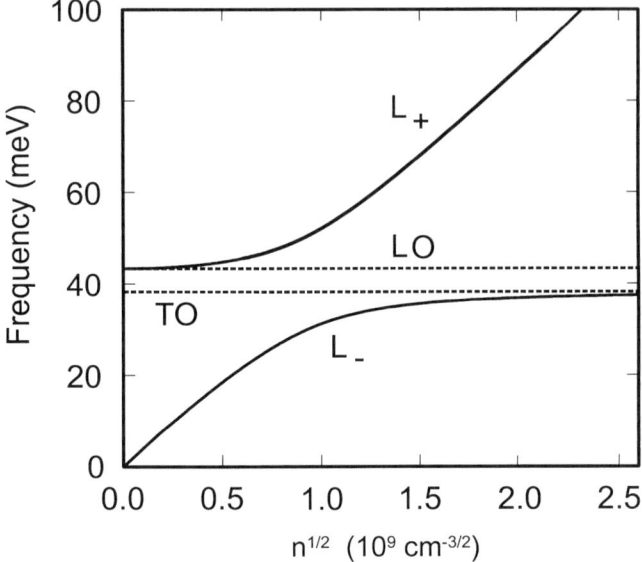

FIG. 52. Energies of the coupled L_ and L_+ modes as a function of charge density. The energies of the modes are calculated from the coupled-mode equation. Parameters corresponding to InP were used in the calculation.

Figure 53 shows a representative Raman spectrum of a Si-doped GaAs sample with a concentration of about 7×10^{17} cm^{-3} measured at low temperatures and at 3.1 GPa [455]. In addition to the L_+ and L_- coupled LO-phonon-plasmon modes of the bulk, the spectrum shows the unscreened LO phonon from the surface depletion region. The solid curve represents the theoretical lineshape of the Raman spectrum for the coupled modes, which is given by an expression similar to (9.25), but where $\epsilon(q, \omega)$ is the total dielectric function of the coupled plasmon–optical-phonon system [432]:

$$L(q, \omega) \sim \left(\frac{\omega_{LO}^2 - \omega^2}{\omega_{TO}^2 - \omega^2}\right)^2 \operatorname{Im}\left[-\frac{1}{\epsilon(q, \omega)}\right]. \qquad (9.32)$$

The electronic contribution to the longitudinal response function $\operatorname{Im}[-1/\epsilon(q, \omega)]$ is calculated using the Lindhard expression [434]. At carrier densities above 1×10^{17} cm^{-3} and for scattering wave vectors $q \geq 5 \times 10^5$ cm^{-1}, Landau damping effects of the plasmons become important. Thus, collision damping is included in the Lindhard function in the relaxation-time approximation of Mermin [435, 456]. Damping of phonons is neglected.

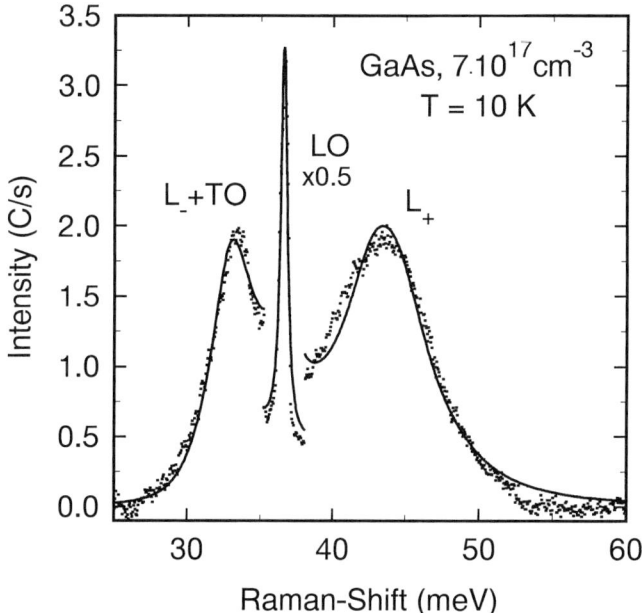

FIG. 53. Raman spectrum of n-doped GaAs at 3.1 GPa showing the L_+ and L_- modes as well as the LO phonon. The solid line represents the result of a fit using Eq. (9.32) and an additional Lorentzian for the LO phonon. After Ref. [455].

Details about the corresponding analytical expression used for fitting experimental spectra are given in Ref. [270].

The spectral function $L(q, \omega)$ depends on two parameters, the free carrier density n_{3D} and the effective electron mass m^*. These parameters cannot be determined uniquely using only Raman spectra. Furthermore, they are not independent of each other. Because of the heavy doping one has to take into account the effect of band nonparabolicity arising from the filling of the conduction band up to the Fermi energy E_F. The effective mass m^* is then given by [457]

$$\frac{m^*}{m_0} = \frac{m_0^*(P)}{m_0}\left(1 - 2\alpha(3\pi^2)^{2/3}\left[\frac{\hbar^2}{2m_0^*(P)}\right]\frac{n_{3D}^{2/3}}{E_g(n_{3D}, P)}\right)^{-1}. \quad (9.33)$$

Here, $E_g(P)$ is the renormalized gap energy at given pressure, m_0^* is the effective mass in the *undoped* material, and α is the nonparabolicity factor which includes terms up to order k^6 [458]. Following $\vec{k}\cdot\vec{p}$ theory [see Eq. (3.19)] [109, 110], a linear dependence of the bottom-of-the-band effective mass parameter is assumed in the undoped case:

$$m_0^*(P) = m_0^* \cdot \frac{E_g(P)}{E_g(0)}. \quad (9.34)$$

One still needs a second equation relating the electron density and the effective mass in order to be able to obtain one of these two parameters from lineshape fits under pressure. For that one can use the values of $E_F(P) = \hbar^2(2\pi^2 n_{3D}(P))^{2/3}/2m^*(P)$ and $E_g(P)$, which can be determined independently from the luminescence data [270]. The renormalized gap corresponds to the onset of PL emission, whereas the Fermi energy is proportional to the PL linewidth. Now $L(q, \omega)$ can be fitted to the Raman spectra with basically two adjustable parameters, the free carrier density n_{3D} and the collision broadening Γ, which is essentially independent of pressure. Furthermore, the pressure-dependent phonon frequencies were used as fixed parameters, calculated using the Murnaghan equation [161] and the Grüneisen relation with the mode Grüneisen parameters and zero pressure frequencies from the literature [459, 77, 78]. We would like to emphasize that, when the Landau damping effects are important, the coupled-mode equation (9.31) is no longer valid for determining the mode frequencies, since it has been derived by using the Drude dielectric function. Secondly, for semiconductors with small electron effective masses such as GaAs or InP, the Fermi energy is of the order of 100 meV and the nonparabolicity effects ought to be taken into account for an accurate determination of the 3-D density.

The first material to be studied for the transfer of electrons between conduction-band valleys by applying pressure was GaAs [454, 270], for which the Γ–X crossover occurs at 4.2 GPa [100, 102]. Figure 54 shows Raman spectra of n-GaAs for different pressures below 4 GPa. The frequencies of the coupled LO-phonon–plasmon modes, L_+ and L_-, remain constant up to an onset pressure of 3.4 GPa. The electron transfer starts before the actual band crossing occurs, because of the band filling up to the Fermi energy. For higher pressures the coupled modes decrease in frequency. The L_+ mode approaches the LO-phonon frequency, whereas the L_- mode vanishes. This behavior is attributed to the reduction of the free carrier density while electrons are transferred from the Γ minimum to X-related donor states. These donors are characterized by a large effective mass and a binding energy of several tens of milli-electron volts, which leads to a freeze-out of carriers into the donor levels. For that reason the coupled modes are no longer observable above the Γ–X crossover.

In Fig. 55 we summarize the results obtained from the Raman measurements for the coupled-mode frequencies (data points) as a function of the carrier density. For comparison we also show (solid lines) the calculated

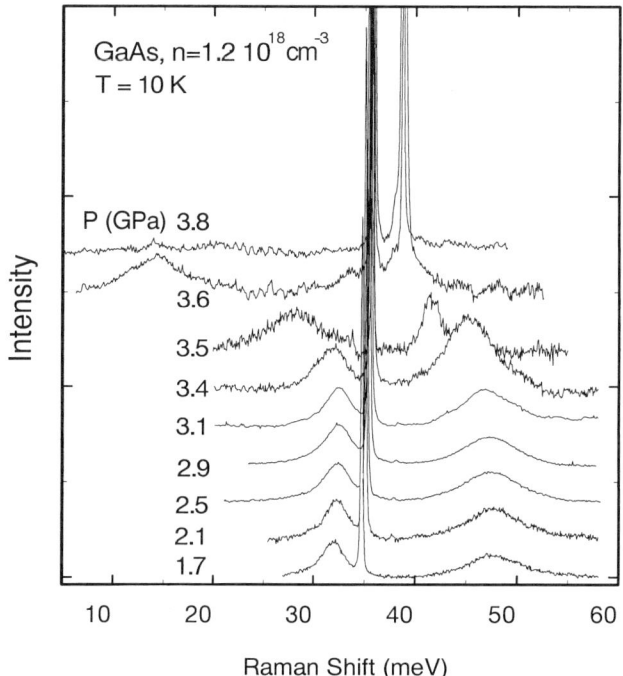

FIG. 54. Raman spectra of the coupled plasmon–phonon modes of n-GaAs for different pressures below 4 GPa. After Ref. [271].

frequencies using the coupled-mode equation (9.31). We notice that the discrepancies between both results are, as expected, more pronounced at higher densities.

The electron effective mass in n-GaAs can now be determined as a function of pressure from the electron density data. Figure 56 displays the dependence of $m^*(P)$ on bandgap energy of an n-GaAs sample with a nominal carrier concentration of $(1-2) \times 10^{18}$ cm^{-3} at ambient pressure. The solid line represents the expected variation of the effective mass within $\vec{k} \cdot \vec{p}$ theory as given by Eq. (9.34). In the doped material there is an additional enhancement of the effective mass due to band nonparabolicity. However, m^* exhibits the same pressure dependence as in the undoped case up to pressures close to the band crossover. The sudden drop of the effective mass at a gap energy of 1.78 eV, which is associated with the onset of electron transfer, results from the vanishing of nonparabolicity effects when the Γ minimum is depopulated.

An application of this effect has made use of the demonstrated sensitivity of the coupled plasmon–phonon modes on changes in the free carrier concentration to determine the $\Gamma-X$ crossover pressure P_c in InP. A wide

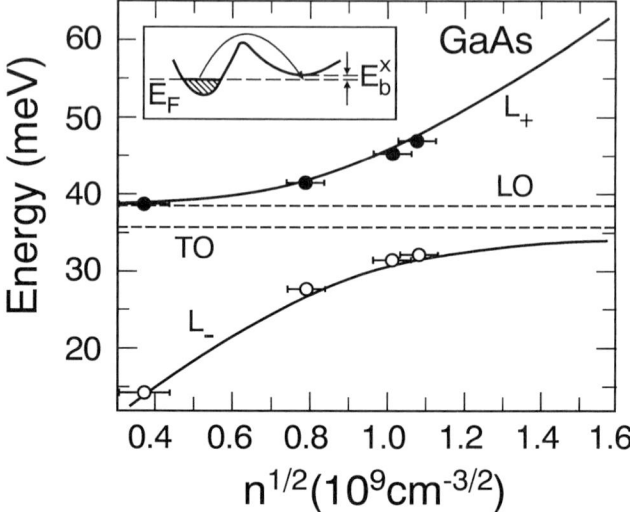

FIG. 55. Energies of the coupled LO-phonon–plasmon modes in n-GaAs as a function of the free carrier density. Solid lines correspond to the calculated L_- and L_+ mode frequencies using the coupled mode equation (9.31). Data points represent the results of the analysis of the Raman and photoluminescence lineshapes. The inset illustrates the transfer of electrons from the Γ minimum to X-related states near the pressure-induced conduction-band crossover. After Ref. [271].

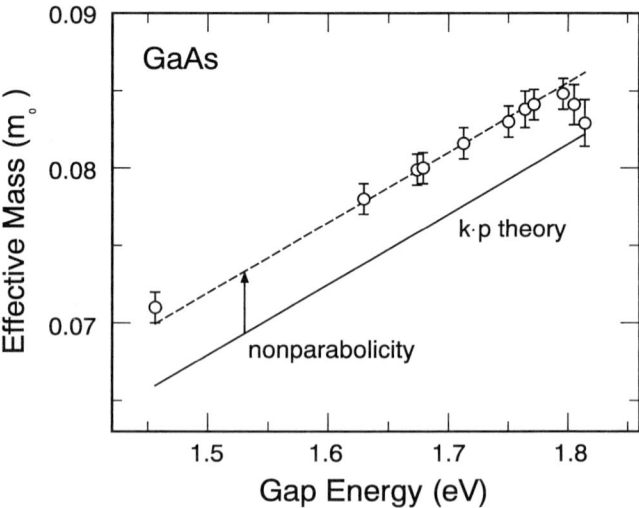

FIG. 56. Effective electron mass of n-doped GaAs versus bandgap energy. The solid line represents Eq. (9.34). After Ref. [271].

range of values for P_c are given in the literature [286, 177, 288, 291, 179, 174, 460–462]. There is a particular complication for its determination: InP undergoes a structural phase transition into the rock-salt structure at a relatively low pressure of 9.8–10.9 GPa [461, 174, 462]. Therefore, Raman scattering and luminescence have been measured in heavily doped n-InP for pressures up to 11 GPa. Figure 57 shows the measured frequency of the L_+ mode together with the carrier density as obtained from the combined lineshape analysis of the Raman and PL data [270]. The onset for the decrease in frequency and density is observed at $P_o = 10.3$ GPa, which is

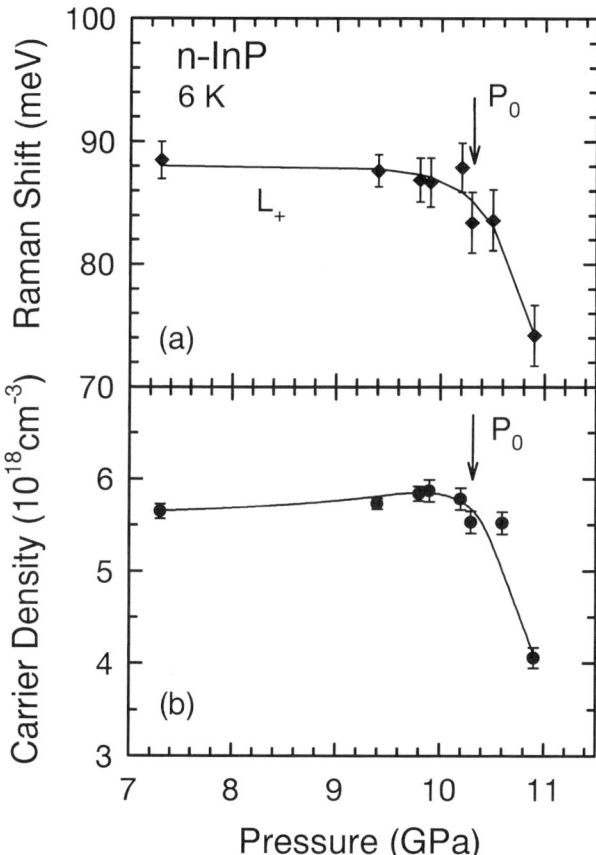

FIG. 57. Pressure dependence (a) of the L_+ mode frequency and (b) of the free carrier density in heavily doped n-InP. P_0 is the onset pressure for the electron transfer between Γ and X conduction-band minima. The lines are guides to the eye. After Ref. [270].

taken as evidence for the onset of electron transfer between conduction-band minima.

The energy of the X minimum at P_o is estimated from the values of $E_g(P_o) = 2.00(1)$ eV and $E_F = 100(5)$ meV from PL measurements and assuming a binding energy of 60(30) meV for the shallow sulfur donors at the X-point. The energy separation between the bottom of the conduction band in Γ and the lowest states in the X-point at the onset pressure P_o is the sum of the Fermi energy and of the donor binding energy at the X-point. Using the literature value for the Γ–X gap energy at ambient pressure (the average value is 2.33 eV [291, 179, 463]), one obtains the linear pressure coefficient of -17 ± 3 meV/GPa for the X-gap of InP. Now one has all the necessary information for the determination of the crossover pressure. For *undoped* InP one finds $P_c = 11.2 \pm 0.4$ GPa, that is, the transition from direct to indirect behavior occurs directly above the structural phase transition, and for that reason it could not be observed before by spectroscopic means.

Perlin *et al.* [464, 337] have used high-pressure Raman spectroscopy of coupled LO-phonon-plasmon modes to investigate the properties of donor states in n-GaN, a material of technological importance for short-wavelength (blue) optoelectronic devices. As-grown but unintentionally doped bulk GaN crystals are always characterized by a high background electron concentration on the order of 10^{19} to 10^{20} cm^{-3}, which is due to native defects. At high pressure n-GaN undergoes a metal–insulator transition which is attributed to the emergence of a localized donor state in the bandgap, which is resonant with the conduction band at ambient conditions [464, 337]. The evidence for that is found in the abrupt change of the L_--mode frequency (it decreases by about 15 cm^{-1} in 1 GPa) at a pressure of 23 GPa. The estimated change in carrier concentration due to the freeze-out of electrons in the donor level is from 5×10^{19} cm^{-3} down to around 3×10^{18} cm^{-3}. This indicates that a deep donor resonant with the conduction band at ambient pressure is responsible for the n-type character of bulk GaN samples.

4. Elementary Excitations of the 2-D Electron Gas under Pressure

The many-body behavior of the 2-D electron gas is characterized by new physical phenomena such as the quantum Hall effect, arising from enhanced electronic correlations in reduced dimensions [436]. As discussed previously, Coulomb interactions manifest themselves in the spectrum of elementary excitations of the 2-D electron gas. The energies of the excitations strongly depend on the electron density; the very dilute regime ($n_{2D} \leq 5 \times 10^{10}$ cm^{-2}) is particularly interesting, because it is in this limit that electron correlations play the dominant role [465–468]. One of the most exciting phenomena, which is expected to occur at low densities but zero magnetic field, is the transition of the electronic system from the gas to a solid phase, the so-called Wigner crystal [465, 466, 469]. In such state the electrons are held together in a triangular type of lattice by Coulomb interactions.

Two-dimensional electron gases of extremely high mobilities in excess of 1×10^6 cm^2/V sec are obtained with current epitaxial techniques, for example, in modulation-doped GaAs/AlGaAs single quantum well (SQW) heterostructures. In these samples the ionized donors are set back from the free carriers concentrated in the GaAs QW by remote doping of the AlGaAs barriers. A sketch of the conduction band profile of such a structure is displayed in Fig. 58. In the sample used in this work, the QW was 245 Å wide, having a starting density n_{2D} of about 5.4×10^{11} cm^{-2} at 2 K. At ambient pressure only the lowest subband is occupied with Fermi energy $E_F = 19.2(8)$ meV and the separation to the second subband being $E_{01} = 28.0(3)$ meV. Pressure enters as a fundamental parameter here because by this means it is possible to reduce the 2-D density in the GaAs QW by more than two orders of magnitude [279, 470–472].

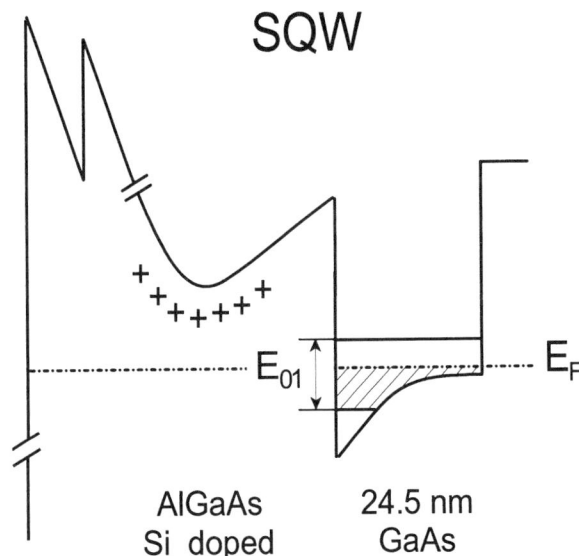

FIG. 58. Sketch of the conduction band profile of a modulation-doped SQW. The sample has only one occupied subband up to the Fermi energy E_F. The subband spacing is E_{01}. After Ref. [273].

This dependence of n_{2D} on pressure has been determined from a quantitative analysis of PL lineshapes. Figure 59 shows some PL spectra of the SQW sample recorded at 6 K and at different pressures. The emission spectra are dominated by intrinsic recombination processes involving vertical ($\vec{k}$-conserving) transitions between conduction- and valence-band states. The spectral lineshapes are thus interpreted in terms of 2-D densities of states (a steplike function) with thermal occupation factors for electrons and holes. In addition, a Gaussian function at the bandgap accounts for excitonic effects in the Fermi sea. The PL emission maximum occurs at the bandgap, whereas the high-energy cutoff corresponds to the energy $E'_F = E_F \cdot (1 + m^*_e/m^*_h)$, where E_F is the Fermi energy [274]. The factor which contains the ratio between the electron and hole effective masses accounts for the curvature of the valence band. Thus, E_F (i.e., the 2-D density) is

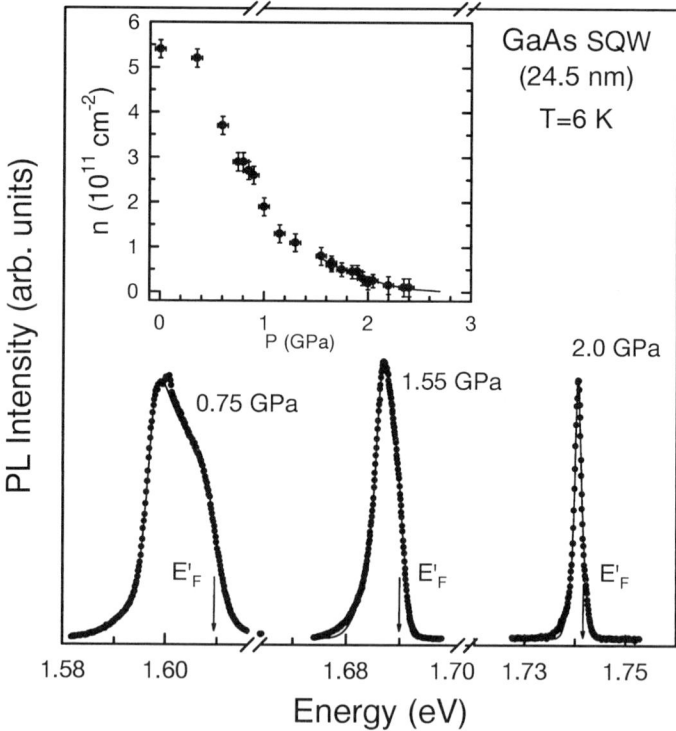

FIG. 59. Photoluminescence spectra of the modulation-doped SQW sample at different pressures. Solid curves represent the lineshape fits. Arrows indicate the Fermi energy E'_F (see text). The inset shows the variation of the 2-D electron density as a function of pressure. After Ref. [273].

proportional to the PL bandwidth. The solid curves in Fig. 59 represent fitting examples at different pressures obtained with a finite broadening parameter of the Gaussian and a carrier temperature of 15 K.

As the pressure increases the PL emission becomes narrower, indicating the reduction in electron density. The values for n_{2D} obtained from fits to the PL data are shown in the inset to Fig. 59 as a function of pressure. Because of the very narrow intrinsic emission peaks (1.3 meV FWHM) of the undoped SQW control sample, it is possible to determine by this method densities down to 1×10^{10} cm^{-2}. This pressure tuning of the 2-D density is fully reversible for decreasing pressure.

The pressure-induced depletion of the QW in these modulation-doped structures results from the pinning of the Fermi level near mid-gap due to the formation of acceptor-like surface states [473] and the tendency of the Γ conduction-band minimum to shift up in energy with pressure. Because acceptor-like levels are tied to the top of the valence band, they are fairly insensitive to the applied pressure. As a consequence, the bottom of the conduction band moves up in energy with respect to the Fermi level, causing a reduction of the carrier density in the well. In addition to this effect, pressure also increases the binding energy of the Si donors. The result is a pressure-induced transfer of electrons, in part to the surface states as required by the larger potential difference between these states and the conduction-band edge at the surface, but also to the deeper donor levels.

In this way it is possible to reduce the density of the 2-D electron gas in a continuous manner for light scattering studies of Coulomb interactions in the very dilute regime. An example of inelastic light scattering spectra of 2-D intersubband excitations measured in the SQW at two different pressures is shown in Fig. 60 (for more details, see Ref. [273]). The sharp peak (FWHM $\approx$ 0.35 meV) which appears for parallel polarization of the incident and scattered beams is assigned to the collective CDE, whereas in depolarized spectra the SDE is active [441]. In addition to the collective modes, single-particle excitations (SPE) at the subband spacing energy E_{01} are observed as a broader feature in spectra for both polarizations [441]. A striking result is that at high pressures the SPE is observed above the energy of the collective depolarization-shifted CDE.

In Fig. 61 we have plotted the energy positions of intersubband excitations as a function of density. With increasing pressure the energy of all excitations and the energy separation between the collective modes and the SPE decrease because of the reduction in electron density. The dashed curves in Fig. 61 represent the theoretical results of Ref. [474] for the energies of the collective modes calculated by numerically solving the Bethe–Salpeter equation for the polarization function. At intermediate carrier densities, $5 \times 10^{10} < n_{2D} < 5 \times 10^{11}$ cm^{-2}, there is remarkably

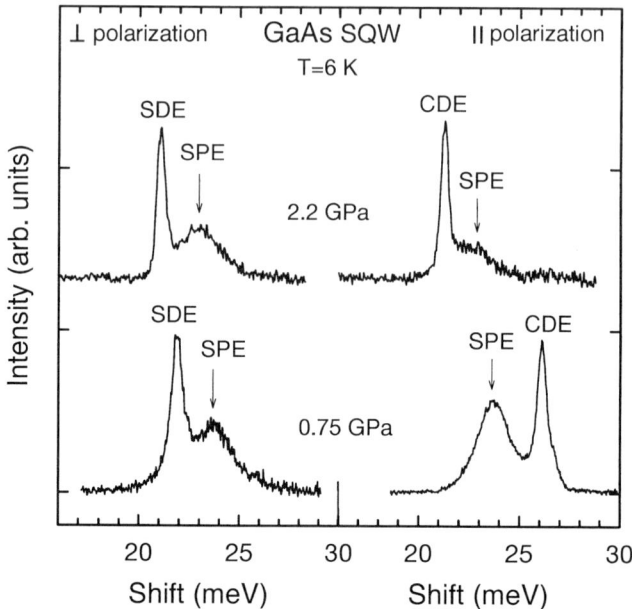

FIG. 60. Depolarized ($\perp$) and polarized ($\|$) light scattering spectra of intersubband excitations of the SQW for two different pressures. The assignment of peaks to charge-density (CDE), spin-density (SDE), and single-particle excitations (SPE) is indicated.

good agreement between the experimental and theoretical results for the CDE and SDE.

This crossing between the energies of intersubband SPE and CDE, which for the SQW structure occurs at a density of 4×10^{10} cm^{-2}, has been predicted within the time-dependent local-density approximation to take place at low densities because of a novel many-body coupling between those excitations [467, 468]. From Eq. (9.13) we recall that the energy shift of the collective CDE with respect to the intersubband spacing E_{01} depends on the relative magnitude of depolarization and excitonic shifts. We therefore conclude that at this crossing the excitonic vertex corrections in the 2-D electron gas are overcoming the Hartree terms of the Coulomb interaction.

From the measured energies of the excitations one obtains, using Eqs. (9.13) and (9.10), the dependence on electron density of the depolarization and excitonic shifts for intersubband transitions. The parameters α_{01} and β_{01} of the SQW, which account for the Hartree and exchange Coulomb interaction, are plotted in Fig. 62 as a function of density. At intermediate densities they display an overall linear dependence in agreement with previ-

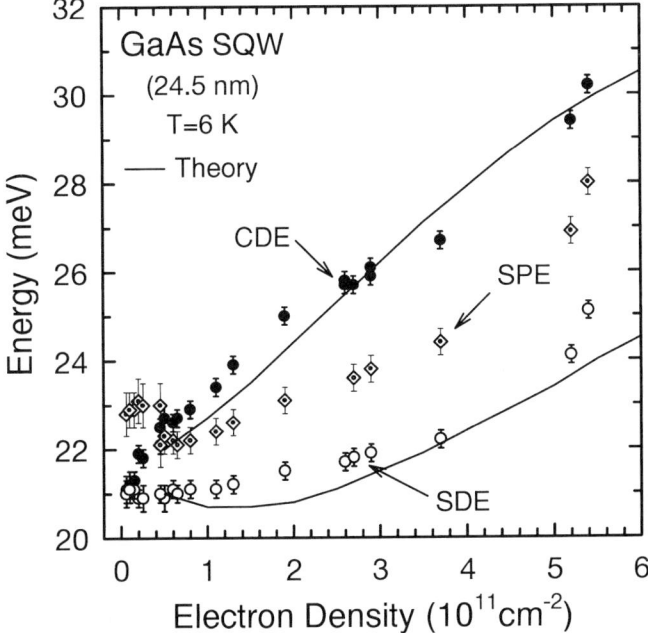

FIG. 61. Measured energies of charge-density (full circles), spin-density (open circles), and single-particle excitations (open diamonds) as a function of 2-D density. The solid curves are the theoretical results of Ref. [474]. After Ref. [273].

ous experimental work on different samples [441, 442]. An important result of this work concerns the collapse of the Hartree term of the Coulomb interaction α_{01} observed at a density of 2×10^{10} cm^{-2} (see Fig. 62). At such low densities the energy difference between collective CDE and SDE, that is, the depolarization shift, almost vanishes.

We interpret the observed collapse of the depolarization shift as evidence that the 2-D electron gas undergoes a metal–nonmetal transition at low carrier densities. The depolarization shift arises from a macroscopic field associated with the charge fluctuations produced by the excitation itself. Screening of the long-range order of the Hartree interaction sets in when the coherence length λ_c, which describes the decay of spatial correlations, becomes smaller than the mean interparticle distance or Wigner–Seitz radius r_s. There is a critical density n_c such that $\lambda_c(n_c) < r_s$. Below this density the depolarization field vanishes. In the case of the SQW we have determined this critical 2-D density n_c as 2×10^{10} cm^{-2}, which corresponds to $r_s \approx 5$. This transition, which is driven by remote impurity scattering

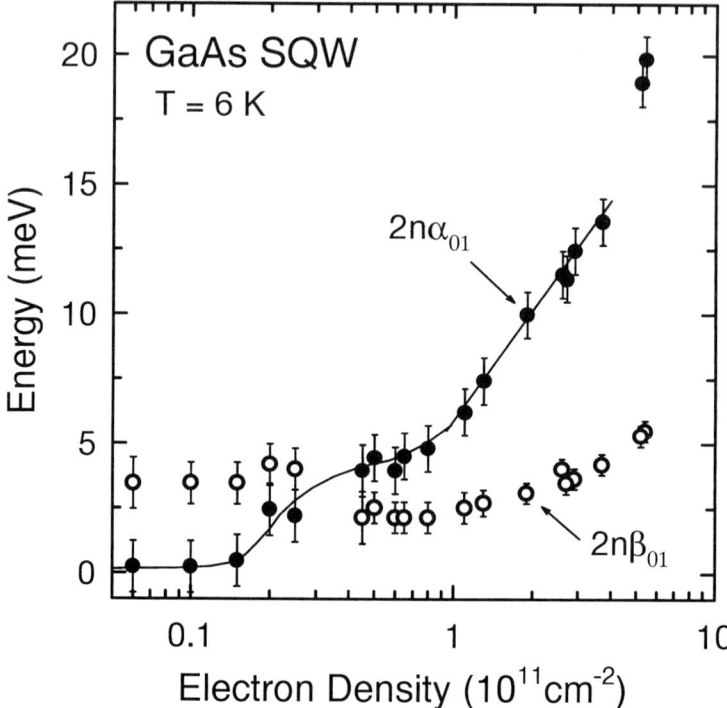

FIG. 62. Dependence on electron density of the parameters $2n\alpha_{01}$ and $2n\beta_{01}$ which represent the Hartree and exchange-correlation terms of the Coulomb interaction, respectively. The solid line is a guide to the eye.

[475, 476], represents an important manifestation of the role played by disorder in the physical properties of the electron gas.

It has been predicted from Monte-Carlo simulations that the disorder potential might favor Wigner-crystal formation in the dilute regime at zero magnetic field [469]. The physical picture for such effect is as follows. The density of impurities is fixed. At small r_s, that is, high electron densities, the number of impurities per electron is small and the disorder potential is effectively screened out. As r_s is increased the electron gas is driven through the metal–insulator transition and the electrons become localized by the random potential. If, however, the impurity density is much smaller than the critical density n_c, then single minima of the disorder potential can act as seeds for Wigner crystallization. Although in this case the solid phase does not have long-range order, it might be characterized by having a shear modulus, which leads to the appearance of additional transverse acoustical excitations. There are at present investigations of the possibility of detecting such modes by inelastic light scattering in samples where the coupling between the 2-D electron gas and the light is produced by a metallic grating fabricated lithographically on the top surface [477].

Exchange and correlation effects, in contrast, remain significant in the very dilute regime. The remarkably different behavior of the exchange interaction in dilute electron gases is a direct consequence of its short-range character and the nature of the exchange-correlation hole whose spatial extension is approximately r_s. This suggests that at very low electron densities the behavior of the electron gas is mainly determined by electronic correlations. These are expected to come particularly into play in double-layer systems, as we will show next.

5. Electron–Electron Interactions in Double-Layer 2-D Electron Gases

Double-layer electron gases produced in modulation-doped double quantum wells (DQW) exhibit a variety of new phenomena associated with interlayer electronic correlations between particles in different quantum wells. For example, at low electron densities and zero magnetic field, an intersubband spin-density instability has been predicted to occur in DQWs induced by enhanced exchange vertex corrections [478–480]. It has also been suggested that Wigner crystallization would be readily achieved in double layers because of interlayer correlations [481]. For this study an asymmetric and a symmetric DQW structure have been chosen for which the relative importance of intra- and interlayer correlations is very different [276, 277].

a. The Asymmetric DQW Sample

The conduction-band profile of the asymmetric DQW sample is depicted schematically in Fig. 63. The DQW consists of 300 and 250 Å wide GaAs wells separated by 50 Å of $Al_{0.3}Ga_{0.7}As$ as barrier layer. The Si doping is incorporated from both sides in the left and right AlGaAs layers with total electron density of $7.9 \times 10^{11} cm^{-2}$. At ambient pressure there are two subbands occupied with electrons, the lowest state being confined in the thinner well, which is farther away from the sample surface, as indicated in Fig. 63. The subband indices are chosen to increase with energy.

For all the QW structures studied thus far, one finds that the effect of increasing pressure is a continuous reduction of the 2-D electron density n_{2D}. For the asymmetric DQW sample, however, it is not possible to determine reliable values for n_{2D} from a quantitative analysis of the PL line-

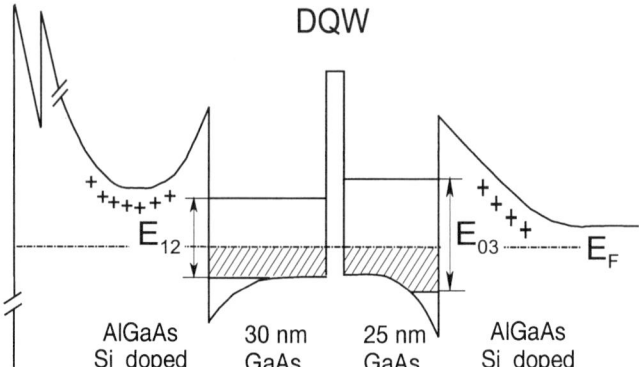

FIG. 63. Sketch of the conduction-band profile of a modulation-doped DQW structure. E_F is the Fermi energy and E_{ij}, $i, j = 0, 1, 2, \ldots$, is the energy spacing between the ith and jth subbands. After Ref. [275].

shapes. This is because of the spectral superposition of the PL emission arising from optical transitions between different electron and hole states of the DQW. Nevertheless, the observed decrease of the total PL bandwidth with pressure clearly indicates the reduction of the carrier density. From the combined PL and light-scattering data one infers that with increasing pressure the overall electron density is reduced in such a way that the 300-Å-wide well is depleted first [275, 276, 277].

The intersubband excitations studied here by inelastic light scattering are associated with electronic transitions labeled as (03) and (12) between confined states of the thinner and the thicker well, respectively. At low pressures the light scattering spectra exhibit two sets of excitations, consisting of a broad SPE peak and sharp collective CDE and SDE features at higher and lower energy, respectively [275]. The disappearance of the features related to the 300-Å-wide well at about 0.7 GPa clearly indicates its early depopulation with pressure. In Fig. 64 we summarize the pressure dependence of the energies of the excitations for the asymmetric DQW. The overall decrease of the excitation energies with increasing pressure is a direct consequence of the pressure-induced reduction of the 2-D density. An interesting result concerns the predicted crossings between the SPEs and the collective CDEs [467, 468], which occur at about 0.5 GPa and 2.0 GPa for the 300-Å and 250-Å well, respectively. This demonstrates that such many-body behavior is a general characteristic of 2-D intersubband excitations in very dilute electron gases, which results from the fact that at low densities the excitonic vertex corrections become more significant compared to the Hartree interaction.

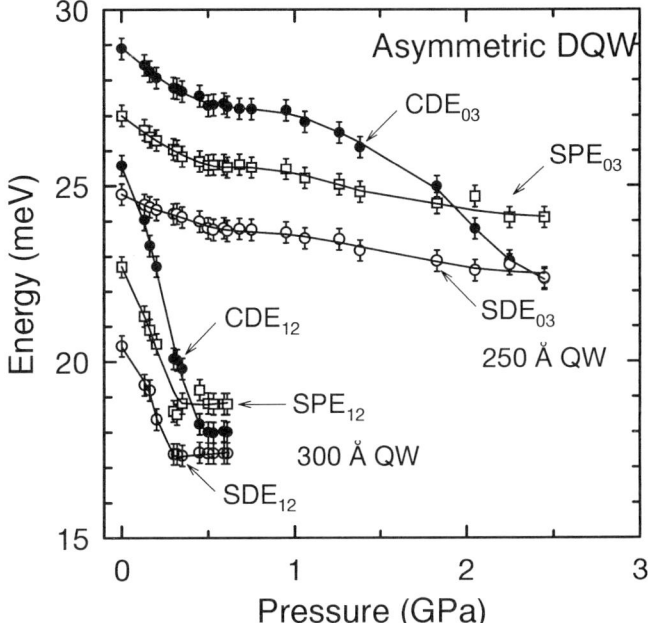

FIG. 64. Dependence on pressure of the charge density, spin density, and single-particle excitation energies in the asymmetric DQW. Lines are a guide to the eye. After Ref. [276].

Another striking result is that the metal–nonmetal transition of the 2-D electron gas revealed by the collapse of the Hartree term (i.e., the energy difference $\omega_{CDE} - \omega_{SDE}$ between the collective excitations of charge and spin density) takes place only for the 250-Å-wide well. For very low densities in the 300-Å well, the energy separation between CDE and SDE remains constant but finite (see Fig. 64). We interpret this behavior as direct evidence for enhanced screening effects in the bilayer system. By comparing with the 250-Å SQW sample, one infers that although the density n_1 of the wider well falls below the critical value determined for the single layer, the disorder potential produced by remote ionized donors is still effectively screened by the *second* layer of electrons due to interlayer correlations. This prevents the collapse of the macroscopic depolarization shift α_{12}.

b. The Symmetric DQW Sample

The interlayer correlations discussed in the previous section are expected to be the largest in the case of a symmetric DQW. Degenerate states confined to different layers mix with each other forming symmetric (S) or antisymmetric (AS) linear combinations. Coupling between these levels opens up the symmetric–antisymmetric gap Δ_{SAS}, which is larger for higher excited states. Figure 65 schematically shows the electronic density along the growth direction z for the first four subbands of the symmetric DQW given by the square modulus of the corresponding envelope wave functions $\zeta_i(z)$. The subband indices are again chosen to increase with energy. For the S states the electronic charge density is more concentrated in the center region of the DQW as compared with the AS states, the difference becoming more pronounced with increasing subband index.

The symmetric DQW structure is similar to the asymmetric one but consists of two 250-Å-wide GaAs wells separated by only 35 Å of $Al_{0.33}Ga_{0.67}As$. At ambient pressure the sample was first characterized by inelastic light scattering. The measurements were performed in backscattering geometry at different angles of incidence by rotating the sample around an axis perpendicular to the plane of incidence. This enables measurement of excitations having a finite in-plane vector $\vec{q}$. Representative spectra measured in the energy region of *intrasubband* excitations are shown in Fig.

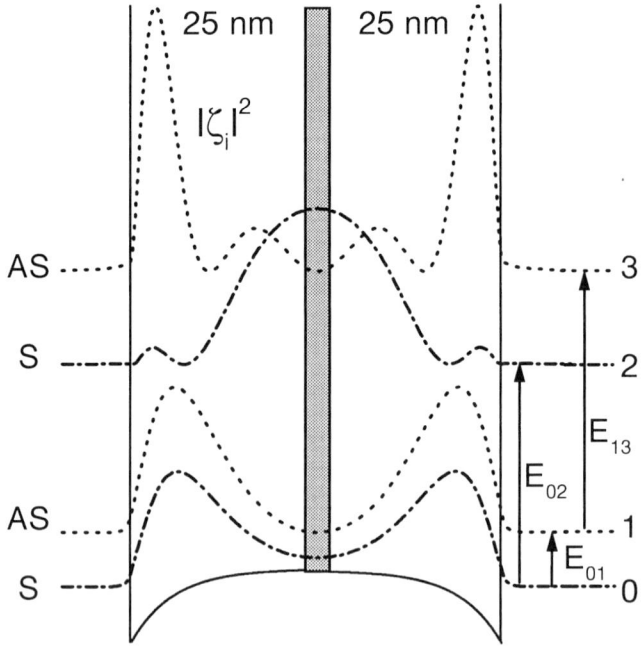

FIG. 65. Confining potential of the symmetric DQW structure, energy-level scheme, and electronic charge densities given by the square modulus of the self-consistent envelope functions $\zeta_i(z)$.

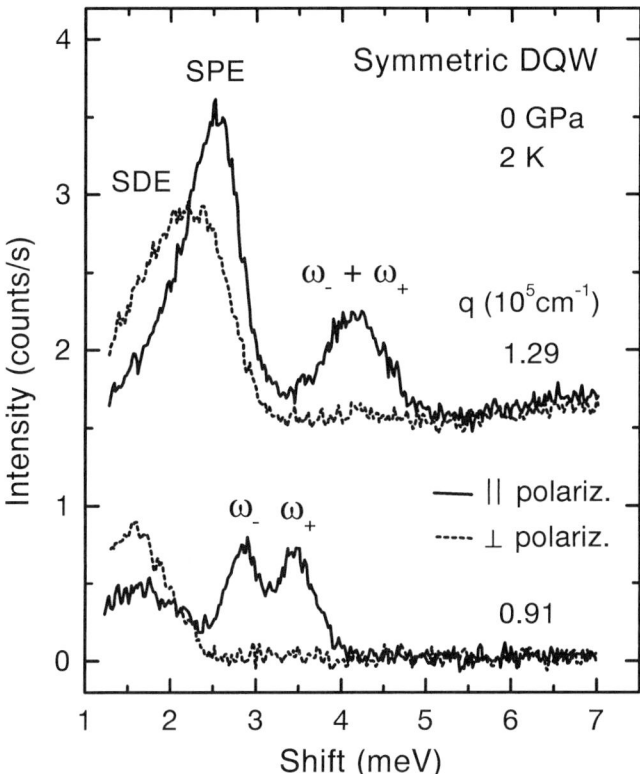

FIG. 66. Inelastic light-scattering spectra of the symmetric DQW measured at 2 K and ambient pressure for two different in-plane wave vectors q. The frequencies of the coupled intrasubband plasmons are denoted as ω_+, ω_-.

66 for two different scattering wave vectors. In spectra with parallel linear polarization of incident and scattered light, we observed two peaks which exhibit a strong wave-vector dispersion. These feature, which are assigned to coupled intrasubband plasmon excitations in the DQW, clearly indicate that the two lowest subbands are occupied with electrons. The features at lower energy with a somewhat weaker dispersive behavior observed in polarized and depolarized spectra are identified as due to intrasubband single-particle and collective spin-density excitations, respectively. The SPE feature exhibits the typical lineshape corresponding to the density of states of electron–hole pair excitations, which has a maximum at the cutoff energy $\hbar^2 k_F q/m^*$ [see Eq. (9.1)] [482]. Furthermore, the intrasubband SDE appears in crossed polarization clearly shifted down in energy with respect to the SPE because of exchange-correlation vertex corrections [438, 484].

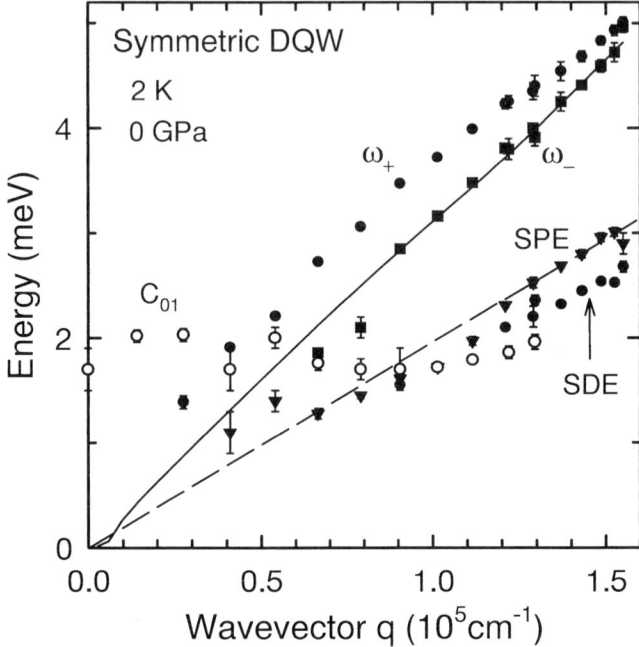

FIG. 67. Wave-vector dispersions of the low-energy excitations of the symmetric DQW. The solid line represents the dispersion of the ω_- plasmon calculated within RPA. The dashed line corresponds to the cutoff energy of the intrasubband SPE continuum.

In Fig. 67 we summarize the wave-vector dependence of the light-scattering peaks observed at low energies. We also show data for a dispersionless excitation observed in polarized spectra, which is tentatively assigned to the depolarization-shifted CDE_{01} for electronic transitions at the Δ_{SAS} gap. The solid line represents the dispersion of the ω_- intrasubband plasmon calculated within RPA, but neglecting the coupling with the intersubband CDE_{01}. This seems to be a good approximation at large q values. From this fit a value of 7.7×10^{11} cm^{-2} has been obtained for the total electron density $n_{2D} = n_0 + n_1$, in very good agreement with the PL results at ambient pressure. The dashed line corresponds to the energy cutoff of the intrasubband SPE continuum as given by Eq. (9.1). The shift of the SDE due to many-body corrections is about 0.35 meV, which is comparable to that measured in SQWs and quantum wires [483, 484].

For the symmetric DQW it was possible to obtain the 2-D densities n_0 and n_1 of the two occupied subbands as a function of pressure from lineshape fits to the PL data. In this case a quantitative analysis of the PL data is possible because the energy difference $E_{g1} - E_{g0} \approx \Delta_{SAS}$ has been experimentally determined by light scattering, as discussed later. The results for the 2-D densities as a function of pressure are displayed in Fig. 68. A very similar pressure dependence of the carrier density has been obtained for a 250-Å SQW sample. In contrast to the case of the asymmetric DQW, here the two densities are reduced with pressure by the same amount, in such a way that the inversion symmetry of the structure is essentially retained in the whole pressure range of the experiment.

Figure 69 shows inelastic light-scattering spectra of the symmetric DQW for both polarizations at 0.45 GPa. A rich structure is apparent at around 22 meV corresponding to intersubband excitations associated with electronic transitions between symmetric or antisymmetric states (see Fig. 65), as allowed by symmetry. At very low energies (1 meV) a well-defined peak is observed for both polarizations. This feature is interpreted as the SPE between the two lowest subbands of the DQW with energy E_{01}, that is, it

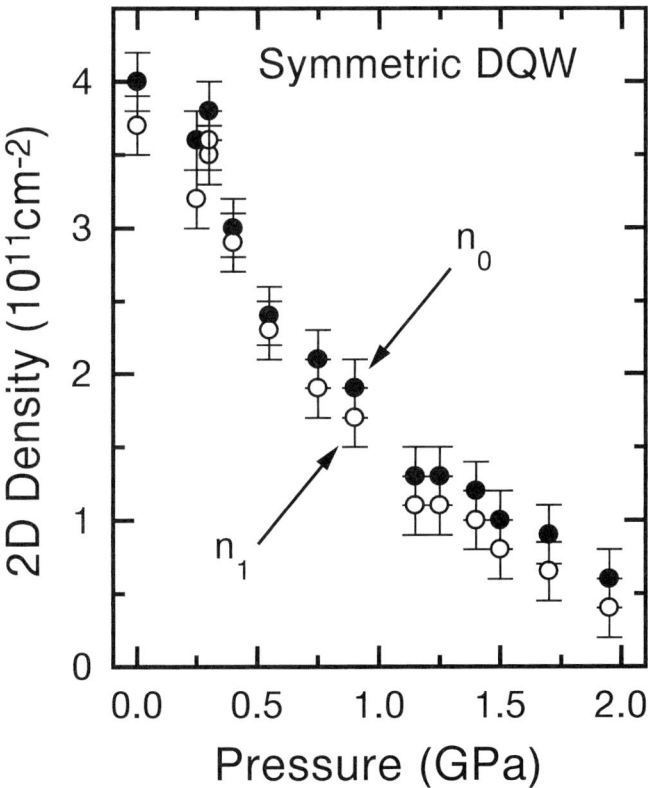

FIG. 68. Dependence on pressure of the 2-D electron densities of the two occupied subbands in the symmetric DQW. After Ref. [277].

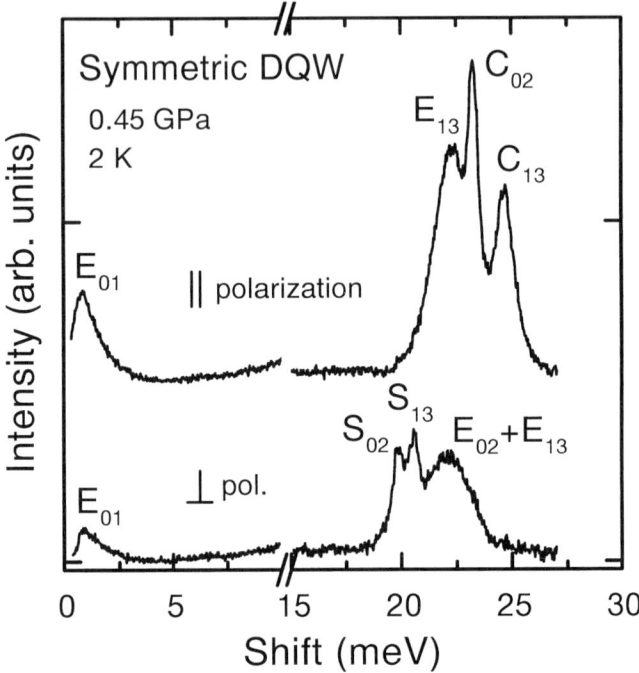

FIG. 69. Polarized and depolarized light-scattering spectra of the symmetric DQW at 0.45 GPa. The assignment of the peaks to charge-density (C_{ij}), spin-density (S_{ij}), and single-particle excitations (E_{ij}) is indicated. After Ref. [276].

appears at the gap energy Δ_{SAS}. Such an excitation together with the collective ones has been previously observed for a similar modulation-doped DQW by Decca et al. [452]. Although this transition is forbidden by symmetry, it might be activated by interface roughness. With increasing pressure the excitations at higher energy exhibit a redshift according to the reduction of the electron density, but they are observable in the entire pressure range up to 2 GPa. The position of the low-energy peak, in contrast, is fairly independent of pressure. This behavior is consistent with the fact that the Fermi energy is always larger than the Δ_{SAS} splitting and that the two lowest subbands remain occupied with electrons [452].

From the differences in the excitation energies one obtains the depolarization and excitonic shifts which are plotted in Fig. 70 as a function of electron density. Absolute values and the almost linear dependence at intermediate densities compare well with the results obtained for the 250-Å SQW. The depolarization shift α_{02} is an exception, showing a pronounced decrease in its magnitude at a relatively high electron density of 2.5×10^{11}

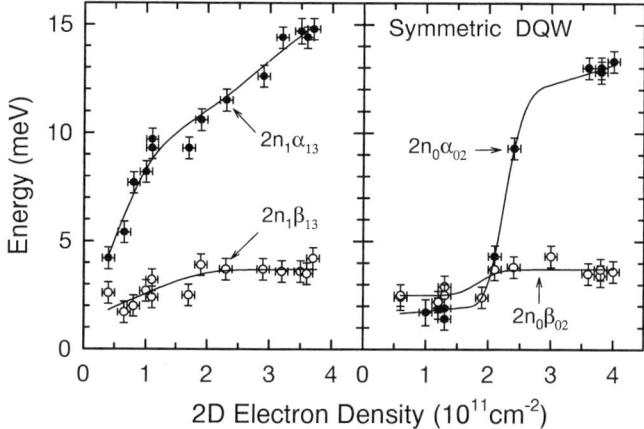

FIG. 70. Dependence on electron density of the depolarization (α_{ij}) and excitonic (β_{ij}) shifts for the symmetric DQW. Lines are a guide to the eye. After Ref. [277].

cm^{-2}. Such behavior is a unique characteristic of symmetric DQWs, and it cannot be ascribed to the same many-body coupling which causes the crossing between SPE and CDE at much lower densities in other QW structures. The quantity α_{02} [see Eq. (9.14)] is a measure of the effects of the depolarization field created by coherent addition of the dipole moments associated with transitions between both *symmetric* states. From the electronic charge densities of Fig. 65 it can be seen that the promotion of an electron to the excited S subband, leaving a hole in the lower one, induces two electric dipoles pointing in opposite directions roughly 280 Å apart (approximately the distance between the quantum well centers). The mean particle separation within each well, which is given by twice the Wigner–Seitz radius r_s, increases with decreasing electron density. For $n_0 = 2.2 \times 10^{11}$ cm^{-2} the interparticle distance $2r_{s0} = 250$ Å is of the same order of the dipole separation, thus leading to a mutual cancellation of the dipole electric fields for other electrons in the wells. This explains the abrupt decrease of α_{02} at high densities. A similar effect occurs for α_{13} but at much lower densities, because the induced dipoles are further apart for intersubband transitions between the antisymmetric states.

X. High-Pressure Phases

1. GENERAL REMARKS

The pressure-induced structural phase transitions of tetrahedrally coordinated semiconductors are in general reconstructive and characterized by a discontinuous decrease in specific volume and an increase in atomic coordination number [485, 486]. The higher coordination implies a different bonding situation in the high-pressure phases, often resulting in metallic behavior. In this context the term *metallic* means that free carriers originate from an intrinsic property of the electronic band structure, that is, from partially filled energy bands.

In early pressure studies it was shown that structural phase transitions in the average valence 4 materials result in changes of the electrical resistivity by several orders of magnitude (see e.g., Refs. [487, 488, 489]). Reconstructive phase transitions cause the formation of a high density of defect states, and a major drop in resistivity is not always sufficient evidence for metallization in the foregoing sense. Examples would be wurtzite CdS and zincblende ZnS, which transform to the sixfold coordinated NaCl-type structure near 3 and 15 GPa, respectively. Optical absorption measurements of CdS [490, 491] and ZnS [492] show that the high-pressure NaCl-type phases have large optical gaps of about 1.5 eV (CdS) and 1.8 eV (ZnS). These gap values, however, are significantly smaller than those of the tetrahedrally coordinated phases at ambient pressure (2.4 eV for CdS, 3.66 eV for zincblende ZnS). Furthermore, the optical absorption spectra indicate an indirect gap for the NaCl-type phases as is also suggested by the calculated energy band structures [492, 493].

From a chemical point of view, the presence of a finite bandgap in the sixfold coordinated phases of CdS and ZnS can be related to the large electronegativity difference between the cations and sulfur, and, due to a small spatial overlap between sulfur orbitals, to a relatively narrow valence band. With decreasing ionicity or increasing overlap between anion p orbitals, the fundamental gap of a six-fold coordinated structure is expected to shrink, resulting in a gap closure and valence–conduction-band overlap. For a purely homonuclear situation (Si, Ge) one arrives at metallic bonding typical for nearly free electron metals [494]. A few selected examples for these two limiting cases, namely ionic materials and elemental semiconductors, are discussed next.

2. RESULTS FOR POLAR COMPOUNDS

With respect to a pressure-driven metallization, ZnTe is just a borderline case, as can be demonstrated by optical reflectivity measurements. ZnTe transforms to the cinnabar-like structure near 9 GPa, followed by a transition to the so-called *Cmcm* structure at 12 GPa [495, 486]. Although the cinnabar phase is closer to fourfold coordination, the *Cmcm* structure can be considered as a distorted NaCl-type structure.

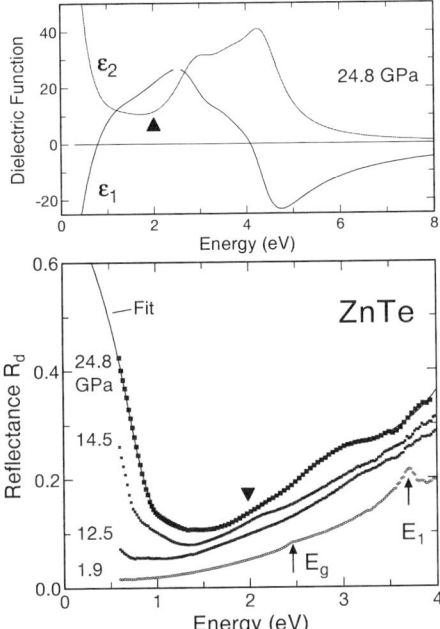

FIG. 71. Optical reflectance spectra of ZnTe in the zincblende phase ($P = 1.9$ GPa), near the transition pressure from the cinnabar to the *Cmcm* (distorted NaCl-type) phase ($P = 12.5$ GPa), and in the *Cmcm* phase (14.5, 24.8 GPa). The top frame shows the real and imaginary parts of the dielectric function obtained from a Lorentz–Drude fit to the reflectance spectrum at 24.8 GPa. The large triangles indicate the onset energy for strong interband transitions at this pressure. Small arrows point to the E_g and E_1 gap energies in the zincblende phase. After Refs. [496, 50].

Figure 71 shows optical reflectivity spectra of ZnTe measured at different pressures [496]. The reflectance values are for the sample in direct optical contact with the diamond window. Because of the large refractive index of diamond, the refractive-index step at the sample–window interface is relatively small. This explains the low reflectance values of the semiconducting phase [see also the Fresnel relation, Eq. (A.29)]. Note that the E_0 gap of the tetrahedral phase shows as a pronounced excitonic feature in reflectivity. The near-infrared reflectivity remains low up to 12 GPa, indicating a semiconducting gap in the cinnabar phase in accordance with theoretical predictions [497, 498]. Starting near 12 GPa, at the transition from the cinnabar to the *Cmcm* structure, we see a Drude-like reflectivity edge developing in the near-IR spectral range. The reflectance continues to increase up to the highest pressure investigated.

The qualitative interpretation of this optical behavior is as follows: Valence- and conduction-band edges are almost degenerate when the *Cmcm* phase is formed. Pressure-driven band broadening results in a continuously increasing band overlap and consequently in the growth of electron and hole pockets. A simplified sketch of the basic features of the band structure near the Fermi level is shown in Fig. 72. The free carriers as well as low-energy interband transitions between split valence bands give rise to the broad Drude-like reflectivity edge seen in the experimental spectra. If the band overlap occurs at the same point in the Brillouin zone, as is suggested by a band-structure calculation of ZnTe in the *Cmcm* phase [498], interband absorption can even start at zero energy. In such a case one cannot separate the interband and intraband contributions to the dielectric response at low energy.

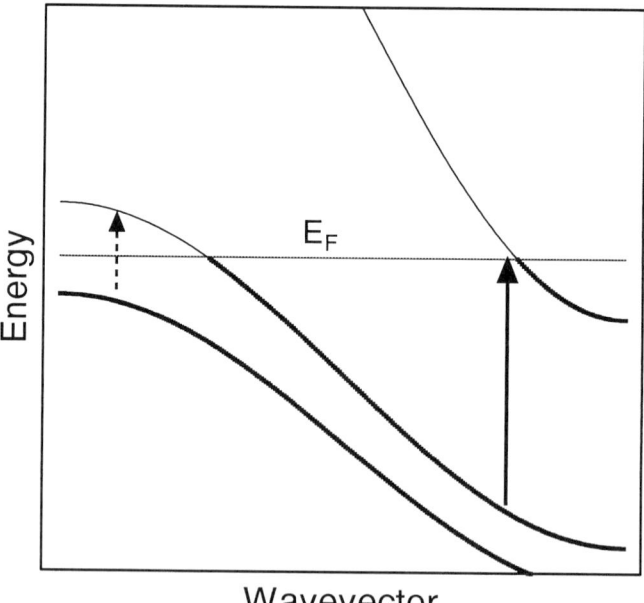

FIG. 72. Schematic representation of the energy-band dispersion for a metallic high-pressure phase of a polar compound with an average valence of 4. The large vertical arrow indicates the onset of direct interband transitions between valence- and conduction-band states. Low-energy interband transitions (dashed arrow) are expected also because of the splitting of valence bands.

An analysis of the reflectivity spectra provides information on the spectral dependence of the dielectric function $\epsilon(\omega)$ in the metallic phase. Based on the procedure described in the Appendix, the reflectance of ZnTe at 24.8 GPa was simulated by an empirical Lorentz–Drude model for $\epsilon(\omega)$. An effective number of four valence electrons per atom was assumed, which, at the extrapolated specific volume, corresponds to a squared valence electron plasma frequency of $(\hbar\Omega)^2 = 290$ eV^2. The upper frame of Fig. 71 shows the obtained real and imaginary parts of $\epsilon(\omega)$. The screened plasma frequency given by the low-energy zero crossing of $\epsilon_1(\omega)$ is 0.8 eV. At around 2 eV the imaginary part $\epsilon_2(\omega)$ shows a steplike increase attributed to onset of direct interband transitions between the topmost valence band and the lowest conduction band (see Fig. 72). About 7% of the total oscillator strength goes to the region below 1 eV. If this fraction is fully attributed to a free carrier response, the carrier concentration would be of the order of 1.5×10^{22} cm^{-1}.

We note that an analysis as just demonstrated is semiquantitative, because it is based on a plausible assumption of the effective number of valence electrons contributing to the optical response. Furthermore, a scalar dielectric function is assumed which does not take into account optical anisotropy.

Figure 73 shows high-pressure reflectivity spectra of ZnSe [499]. This material transforms to the NaCl-type structure near 13.5 GPa [500, 501]. Spectra for the zincblende phase (below 13.5 GPa) show weak features originating from the E_0 and $E_0 + \Delta_0$ transitions. Comparing the first derivatives of these spectra with respect to photon energy at different pressures clearly illustrates their blueshift with increasing pressure (see inset of Fig. 73). The phase transition results in an overall increase of the reflectance which is attributed to a shift of direct interband transitions to lower energy compared to the zincblende phase. Up to 21 GPa there is no evidence of a low-energy oscillator strength due to free carriers. On the other hand, the positive temperature slope of the electrical resistance [502] was interpreted in terms of an intrinsic metallic phase. This interpretation would only be consistent with the optical data if the carrier concentration is less than 1×10^{21} cm^{-1}, falling into the semimetallic or high-doping regime.

Reflectivity studies of phase transitions have been reported for a few other average-valence-4 materials. In the case of HgO [503] the results indicate a continuous transition to metallic behavior of the high-pressure tetragonal phase (a distorted NaCl-type structure) near 28 GPa. This metallization pressure is significantly larger than the previously reported value of about 10 GPa [504], which was based on transport measurements. Another example is GaAs, whose reflectivity has been measured up to 107 GPa and after pressure release [505]. The reflectance spectra of the metastable amorphous phase produced upon unloading were interpreted in terms of

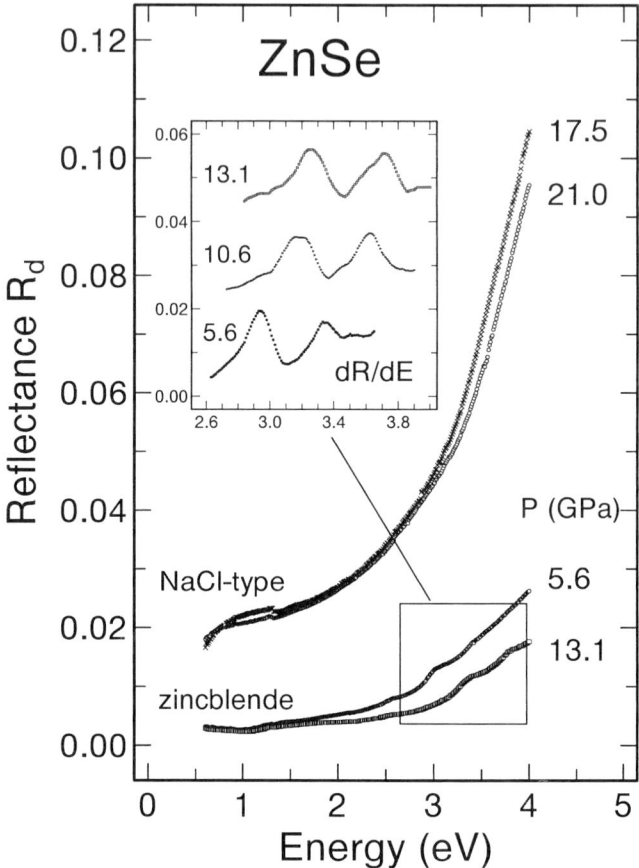

FIG. 73. Optical reflectance spectra of ZnSe in the semiconducting phase ($P < 13.5$ GPa) and in the high-pressure NaCl-type phase ($P > 13.5$ GPa). The inset shows the first derivative of the reflectance spectra of the semiconducting phase. Spectral features are due to the E_0 and $E_0 + \Delta_0$ transitions. After Ref. [499].

a semiconducting behavior. In the case of AlP, reflectance spectra suggest that the NiAs-type high-pressure phase stable above 9.5 GPa is metallic [506].

Semiconductor-to-metal transitions in several group III–VI semiconductors (InSe, GaTe, InTe) were investigated by reflectance spectroscopy [507, 508]. In their low-symmetry ambient-pressure phases these materials are characterized by a localization of an electron pair in a covalent metal–metal bond. Pressure-induced phase transitions to NaCl-type phases are observed between 5 and 11 GPa. The structural change results in a delocalization of the electron pairs and leads to a fairly abrupt change of the optical response. For InSe this is demonstrated in Fig. 74, where the reflectance is scanned at constant photon energy of 0.6 eV. The optical response of the NaCl-type phases of these materials is characterized by screened plasma frequencies near 2 eV, which is significantly higher compared to ZnTe discussed earlier and which is a consequence of the extra delocalized electron per ion pair.

3. Results for the Elemental Semiconductors

The transformation with temperature from gray tin, which has the diamond structure, to white tin, which has the tetragonal β-tin structure, is the classical example for a semiconductor-to-metal transition in a group-IV material. Pressure-driven transitions of Si and Ge to metallic phases were first reported by Minomura and Drickamer in 1962 [487] and superconducting behavior of the high-pressure β-tin phases [509] was discovered a few years later by Wittig [510]. Optical reflectivity studies of metallic Si measured up to 25 GPa [511] were difficult to interpret on the basis of a β-tin structure. Later structural studies have shown that the stability range of Si in β-tin is quite narrow (from 12 to 14 GPa). At higher pressures it

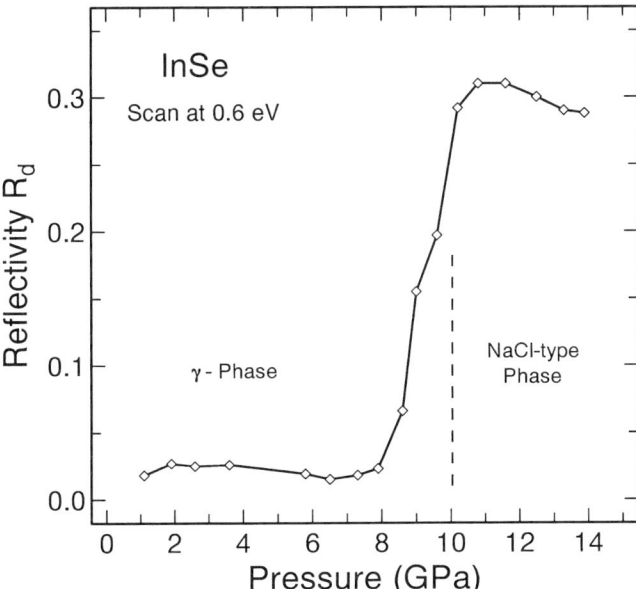

FIG. 74. Variation with pressure of the reflectivity of InSe at a constant photon energy of 0.6 eV. After Ref. [507].

transforms via an intermediate orthorhombic phase [512] to a primitive hexagonal phase (stability range from 16 to 38 GPa) followed by a transition to a hexagonal close-packed structure [513–515]. The superconducting properties of these higher pressure phases have been investigated up to 40 GPa using DAC techniques [516–518] and a theoretical description has been given of the electron–phonon coupling in metallic Si [518].

Figure 75a compares reflectance spectra of Si in the diamond, β-tin, primitive hexagonal (PH), and hexagonal close-packed structures [494, 519]. All three high-pressure phases of Si are characterized by high reflectance values throughout the visible spectral range. The β-tin and PH phases show pronounced absorption bands in the visible range. In Ref. [494] the reflectance spectra of the PH phase were interpreted on the basis of a very simple empty-core pseudopotential description of the band structure and a model for interband absorption between parallel bands in nearly-free electron metals [520, 521]. The model was fitted to the experimental spectra. A result of such a fit is shown in Fig. 75b, which also illustrates how the reflectance spectrum would look for the hypothetical case of the sample being in air. The analysis of the reflectance spectra provides information on the frequency dependence of the dielectric function. The interband part of $\epsilon_2(\omega)$ of the PH phase (Drude-type contribution not included) and its change with pressure is shown in Fig. 75c. Characteristic energies related to maxima in $\epsilon_2(\omega)$ have been attributed to particular band splittings. The optical response of PH-Si was also calculated based on a first-principle linear muffin-tin orbital band structure in the atomic sphere approximation [494]. Figure 75d shows a corresponding result, which is in excellent agreement with the experimental $\epsilon_2(\omega)$ spectrum. One of the main conclusions from the optical study of the metallic PH phase of Si is, that it can be viewed as a polyvalent nearly free electron metal with optical properties explained by parallel band absorption, which is similar to the case of its neighbor element Al.

A. Appendix: Linear Optical Response of Solids

The Appendix is a collection of relations used in the description of the linear optical response of solids.[1] For detailed treatments we refer to stan-

---

[1] For translation from Gaussian units used here to SI units (symbols written, for distinction, with a star): $1/(4\pi) \to \epsilon_0$, $4\pi/c^2 \to \mu_0$, where $\epsilon_0 = 8.854 \times 10^{-12}$ As/Vm and $\mu_0 = 12.57 \times 10^{-7}$ Vs/Am are the vacuum permittivity and permeability, respectively, $\vec{D}^* \to \vec{D}/(4\pi)$, $\chi^* \to 4\pi\chi$, $\vec{B}^* \to \vec{B}/c$, $\vec{H}^* \to c\vec{H}/(4\pi)$, $\vec{M}^* \to 4\pi\vec{M}/c$. Other quantities (e.g., $\vec{E}$ and $\vec{P}$) remain unchanged.

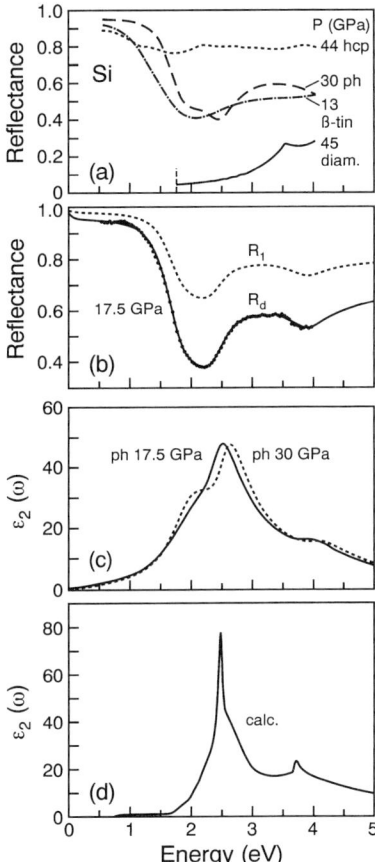

FIG. 75. Optical properties of Si under pressure: (a) Reflectance spectra of different crystalline phases, (b) reflectance spectrum of the primitive hexagonal (PH) phase at 17.5 GPa, result of a least-squares fit (solid line), and reflectance converted to a sample/vacuum interface (dashed line), (c) the interband part of the imaginary dielectric function (ϵ_2) at two different pressures in the PH phase, and (d) the interband part of ϵ_2 calculated within a *first–principles* LMTO-ASA band-structure model. After [494].

dard textbooks [76, 168, 445, 522] and review-type articles [159, 143, 523]. Where appropriate, we comment on the specific case of optical spectroscopy in a high-pressure environment.

A1. Electromagnetic Waves in a Medium

We consider the polarization $\vec{P}(\vec{r}, t)$ (dipole moment per unit volume) of an *isotropic* and *homogeneous* insulating system under an applied electric field $\vec{E}(\vec{r}, t)$, assuming translational invariance in space *and* time. We are interested in the *linear response* regime, corresponding to the weak-field limit. Furthermore, in the *local approximation* the polarization at a location $\vec{r}$ depends on the field at $\vec{r}$ only, that is, spatial dispersion will be neglected. With these assumptions the relation between electric field and polarization of the bound charges is [445]

$$\vec{P}(\vec{r}, t) = \int_{-\infty}^{\infty} dt' \chi^b(t - t') \vec{E}(\vec{r}, t'). \quad \text{(A.1)}$$

The dimensionless susceptibility $\chi^b(t - t')$ (real and scalar) maps the field at $\vec{r}$ and time t' into a polarization (the response) at the same location at a later time t. Causality implies that there can be no response before a driving field is applied; therefore,

$$\chi^b(t - t') = 0 \text{ for } t - t' < 0. \quad \text{(A.2)}$$

An arbitrary field or response in the space/time domain is described by its Fourier components in the wave-vector/frequency domain, which are monochromatic plane waves. The electric field of a plane wave is given by

$$\vec{E}(\omega, \vec{k}) = \vec{E}_0(\omega, \vec{k}) \exp[i(\vec{k} \cdot \vec{r} - \omega t)]. \quad \text{(A.3)}$$

A corresponding relation describes the magnetic-field component. The vector $\vec{E}_0$ is a complex quantity, which allows for an independent phase shift between electric and magnetic field of the electromagnetic plane wave. As a reminder: Only the real part $(\vec{E} + \vec{E}^*)/2$ is a measurable quantity. The advantage of a Fourier transformation is that the convolution in Eq. (A.1) converts into simple products:

$$\vec{P}(\vec{k}, \omega) = \chi^b(\omega) \vec{E}(\vec{k}, \omega). \quad \text{(A.4)}$$

4 OPTICAL PROPERTIES OF SEMICONDUCTORS UNDER PRESSURE

In general $\chi^b(\omega)$ is a complex function. The polarization is therefore composed of an in-phase and an out-of-phase component. The phase shift between driving field and polarization implies absorption. For a nonlocal response the susceptibility would depend on wave vector $\vec{k}$ as well. From this point on it is understood that all optical response functions mentioned depend on frequency without being explicitly written so.

For comparison with experiments it is often more convenient to use the dimensionless complex dielectric function $\epsilon^b = \epsilon_1^b + i\epsilon_2^b = 1 + 4\pi\chi^b$, which relates the dielectric displacement $\vec{D}$ and the electric field through

$$\vec{D} = \vec{E} + 4\pi\vec{P} = (1 + 4\pi\chi^b)\vec{E} = \epsilon^b\vec{E}. \quad (A.5)$$

The magnetic-field component $\vec{H}$ of an electromagnetic wave induces a magnetization $\vec{M}$ such that the $\vec{B}$ field is

$$\vec{B} = \vec{H} + 4\pi\vec{M} = \mu^b\vec{H}. \quad (A.6)$$

Here μ^b is the dimensionless magnetic permeability of the bound charges. Because for nonmagnetic materials $4\pi\vec{M}$ is very small compared to $\vec{H}$, it is justified to set $\mu^b = 1$. For metallic phases or extrinsic semiconductors we have to take into account the contribution of the free carriers to the free current density $\vec{J}^f$ which is given by Ohm's law,

$$\vec{J}^f = \sigma^f \vec{E}, \quad (A.7)$$

where $\sigma^f = \sigma_1^f + i\sigma_2^f$ is the complex electrical conductivity. Any effect of the magnetic field on the free current density will be neglected also.

Thus, two material-specific response functions — $\epsilon^b(\omega)$ and $\sigma^f(\omega)$ — enter into the Maxwell equations.[2] Solving them for $\vec{E}$ ($\vec{B}$), we obtain the wave equation for *transverse* plane waves

$$\vec{k} \cdot \vec{k} = \frac{1}{c^2}\epsilon^b\omega^2 + i\frac{4\pi}{c^2}\sigma^f\omega, \quad (A.8)$$

where c is the speed of light in vacuum. At this point it is convenient to define a total dielectric constant ϵ and an optical conductivity σ which each

---

[2] With the assumptions made here, the Maxwell equations are

$\epsilon^b\vec{\nabla} \cdot \vec{E} = 0, \quad \vec{\nabla} \cdot \vec{B} = 0, \quad \vec{\nabla} \times \vec{E} = -(1/c)\partial\vec{B}/\partial t, \quad \vec{\nabla} \times \vec{B} = (4\pi/c)\sigma^f\vec{E} + (1/c)\epsilon^b(\partial\vec{E}/\partial t).$

include the response of both free carriers *and* bound charges and which can be used alternatively:

$$\epsilon = 1 + \frac{4\pi i}{\omega}\sigma = \epsilon^b + \frac{4\pi i}{\omega}\sigma^f. \quad (A.9)$$

The low-energy optical response of conducting materials is preferentially discussed in terms of $\sigma(\omega)$, which for $\omega \to 0$ extrapolates to the finite longitudinal conductivity σ_{dc}, whereas ϵ_2 diverges. Otherwise, it is more convenient to work with $\epsilon(\omega)$, which is dimensionless.

Inserting the total dielectric function into Eq. (A.8), the frequency versus wave-vector relation becomes

$$\omega^2 = c^2 k^2 / \epsilon \quad (A.10)$$

In spectral regions where ϵ is a complex function, the wave vector $\vec{k}$ is complex as well ($\vec{k} = \vec{k}_1 + i\vec{k}_2$). For homogeneous plane waves the planes of constant amplitude are parallel to planes of constant phase, that is, $\vec{k}_1 \parallel \vec{k}_2$. The wavelength is $\lambda = 2\pi/k_1$. By definition the complex phase velocity is $v_{ph} = \omega/k = c/\sqrt{\epsilon}$.

Wave propagation is commonly described in terms of a complex refractive index n rather than a wave vector. The refractive index is given by

$$n = n_1 + i n_2 = \frac{c}{\omega}\sqrt{k^2} = \sqrt{\epsilon}. \quad (A.11)$$

For a plane wave with complex k propagating along the z direction, we write

$$\exp(ikz) = \exp(i\omega n_1 z/c) \cdot \exp(-\omega n_2 z/c).$$

Thus, n_1 and n_2 determine the dispersion and attenuation of a wave, respectively. For this reason n_2 is called the "extinction coefficient." Conversion relations among the real and imaginary parts of n, ϵ, and σ are

$$n_1 = \sqrt{(|\epsilon| + \epsilon_1)/2}, \quad n_2 = \sqrt{(|\epsilon| - \epsilon_1)/2}$$
$$\epsilon_1 = n_1^2 - n_2^2, \quad \epsilon_2 = 2 n_1 n_2$$
$$\sigma_1 = \frac{\omega}{4\pi}\epsilon_2, \quad \sigma_2 = -\frac{\omega}{4\pi}(\epsilon_1 - 1).$$

At a frequency corresponding to $\hbar\omega = 1$ eV the prefactor $\omega/(4\pi)$ is 1.2090×10^{14} s^{-1} (CGS) or 134.54 siemens (SI units, 1 siemens = 1 A/Vcm).

4 OPTICAL PROPERTIES OF SEMICONDUCTORS UNDER PRESSURE

The absorption coefficient α is defined as the fractional dissipation of time-averaged energy density $\overline{W} = n_1^2 |\tilde{E}_0|^2/(8\pi)$ with distance z along the direction of propagation

$$\alpha = -\frac{1}{\overline{W}} \frac{d\overline{W}}{dz} = -\frac{1}{\overline{W}} \frac{-d\overline{W}}{dt} \frac{dt}{dz}. \qquad (A.12)$$

Integration of the left-hand part of Eq. (A.12) gives

$$\overline{W}(z) = \overline{W}_{z=0} \exp(-\alpha z). \qquad (A.13)$$

Equation (A.12), where we have introduced the energy velocity dz/dt, leads to an interpretation of the absorption coefficient as

$$\alpha = \frac{\text{Mean power dissipation per unit volume}}{\text{Mean energy density} \times \text{Energy velocity}}.$$

With the mean power dissipation

$$\overline{P_{\text{abs}}} = \overline{\text{Re}(\vec{J} \cdot \vec{E})} = \sigma_1 \overline{\vec{E}^2} = \sigma_1 |\tilde{E}_0|^2/2 = 4\pi\sigma_1 \overline{W}/n_1^2, \qquad (A.14)$$

α is given by

$$\alpha = \frac{\sigma_1 \overline{\vec{E}^2}}{(c/n_1)\overline{W}} = \frac{4\pi\sigma_1}{n_1 c} = \frac{\omega \epsilon_2}{n_1 c} = \frac{2\omega n_2}{c}, \qquad (A.15)$$

where we have also given conversion relations.[3] In "practical" units,

$$\alpha\,[\text{cm}^{-1}] = 1.0128912 \cdot 10^5 \cdot n_2 \cdot \hbar\omega\,[\text{eV}]. \qquad (A.16)$$

A2. INTERBAND ABSORPTION

Equation (A.14) provides the link to a microscopic description of the optical properties in the interband regime. The mean absorbed power

---

[3] The preceding derivation for α is strictly valid for nondispersive media, because the relation between energy density and electric field used in Eq. (A.14) holds only for $n_2 = 0$ and because the phase velocity has been used for the energy propagation in Eq. (A.15). The same relations are obtained, however, if one uses the group velocity and the appropriate expression for energy density in a dispersive medium [522].

density $\overline{P_{abs}}$ is given by the absorption transition rate of photons per unit volume $R_{abs}(\omega)$ multiplied by the energy of the photons:

$$\overline{P_{abs}(\omega)} = \sigma_1(\omega)\overline{\vec{E}^2} = \frac{1}{8\pi}\omega\epsilon_2|E_0|^2 = \hbar\omega R_{abs}(\omega). \tag{A.17}$$

In the weak-field limit $R_{abs}(\omega)$ is calculated by applying time-dependent perturbation theory. The transition rate r_{ij} between an occupied initial state $|i>$ and an unoccupied excited state $|j>$ is given by the Fermi Golden Rule. In a crystal, the Bloch states are characterized by a wave vector $\vec{k}$ and a band index. From wave-vector conservation, optical transitions involve states with the same $\vec{k}$ but different band index. Thus, we write

$$r_{ij\vec{k}}(\omega) = \frac{2\pi}{\hbar}|\mathcal{H}'_{ij}(\vec{k})|^2 \delta(\hbar\omega + E_i(\vec{k}) - E_j(\vec{k})). \tag{A.18}$$

The δ function takes care of energy conservation assuming zero lifetime broadening. The matrix element $\mathcal{H}'_{ij}(\vec{k})$ of the electron–radiation interaction Hamiltonian is given by $(e/cm_0)\vec{A}\cdot\vec{p}$. The momentum operator is $\vec{p} = -i\hbar\vec{\nabla}$. For the vector potential $\vec{A}$ of the incident light [in Coulomb gauge $\vec{E} = -(1/c)\partial\vec{A}/\partial t$], we write

$$\vec{A}(\vec{r},t) = \vec{e}A_0 \exp[i(\vec{k}_L \cdot \vec{r} - \omega t)] + \text{c.c.}, \tag{A.19}$$

where $\vec{A}$ is real and polarized along unit vector $\vec{e}$. In the dipole approximation, that is, $\vec{k}_{light} \to 0$, the matrix element $\mathcal{H}'_{ij}(\vec{k})$ is thus given by

$$\begin{aligned}|\mathcal{H}'_{ij}(\vec{k})|^2 &= \left(\frac{e}{m_0 c}\right)^2 |A_0|^2 |<j,\vec{k}|\vec{e}\cdot\vec{p}|i,\vec{k}>|^2 \\ &= \left(\frac{e}{m_0}\right)^2 \frac{|E_0|^2}{\omega^2}|M_{ij}(\vec{k})|^2.\end{aligned} \tag{A.20}$$

Here we have introduced the dipole matrix element $M_{ij}(\vec{k})$. The total transition rate per unit volume (V) is obtained by summing over all occupied (unoccupied) bands i (j) and $\vec{k}$ vectors throughout the Brillouin zone:

$$R_{abs}(\omega) = \frac{2\pi}{\hbar}\left(\frac{e}{m_0}\right)^2 \frac{|E_0|^2}{\omega^2}\frac{1}{V}\sum_{i,j}\sum_{\vec{k}}|M_{ij}(\vec{k})|^2 \delta[\hbar\omega + E_i(\vec{k}) - E_j(\vec{k})]. \tag{A.21}$$

4 OPTICAL PROPERTIES OF SEMICONDUCTORS UNDER PRESSURE

For $\epsilon_2(\omega)$ we then have

$$\epsilon_2(\omega) = \frac{16\pi^2}{\omega^2} \left(\frac{e}{m_0}\right)^2 \frac{1}{V} \sum_{i,j} \sum_{\vec{k}} |M_{ij}(\vec{k})|^2 \, \delta[\hbar\omega + E_i(\vec{k}) - E_j(\vec{k})] \quad (A.22)$$

and the real part $\epsilon_1(\omega)$ is obtained by KK transformation. For semiconductors one is often interested in transition that only involve one of the upper valence bands and the lowest conduction band. For the two-band case and with the assumption that the matrix element is independent of $\vec{k}$, one can simplify Eq. (A.22) to

$$\epsilon_2(\omega) = \left(\frac{4\pi e}{m_0 \omega}\right)^2 |M_{vc}|^2 J_{vc}(\hbar\omega). \quad (A.23)$$

The quantity

$$J_{vc}(\hbar\omega) = \frac{1}{V} \sum_{\vec{k}} \delta[\hbar\omega + E_v(\vec{k}) - E_c(\vec{k})] \quad (A.24)$$

is the so-called *joint density-of-states*, which gives the frequency-dependent density (per unit energy and volume) of pairs of states with a "vertical" (same $\vec{k}$) energy separation $\hbar\omega$. The summation over $\vec{k}$ can be converted into an integral over the first Brillouin zone and further — because of the δ function — to an integral over a surface of constant energy difference:

$$J_{vc}(\hbar\omega) = \frac{1}{V} \sum_{\vec{k}} \delta[\ldots] = \frac{2}{8\pi^3} \int d\vec{k}\, \delta[\ldots]$$

$$= \frac{1}{4\pi^3} \int_{\hbar\omega = E_c(\vec{k}) - E_v(\vec{k})} \frac{dS}{|\vec{\nabla}_{\vec{k}}[E_c(\vec{k}) - E_v(\vec{k})]|}. \quad (A.25)$$

Spin degeneracy is taken into account. Regions of parallel $E(\vec{k})$ dispersion cause singularities in J_{vc} known as Van Hove–type critical points (CPS). It is through the corresponding features in $\epsilon(\omega)$ that optical spectroscopy provides information on the electronic structure at energies above the fundamental bandgaps.

The spectral dependence of the joint density of states near a critical point, around which the energy difference between conduction and valence band exhibits a parabolic disperson $E_c(k) - E_v(k) = \hbar\omega_{vc} + \sum_{l=1}^{3} \frac{\hbar^2 k_l^2}{2m_l^*}$, is determined by the dimensionality of the system and the combined effective mass parameters m_l^* for the principal-axes [64, 69]:

TABLE III

FUNCTIONAL FORM OF THE JOINT DENSITY OF STATES $J^{vc}(\omega)$ FOR DIFFERENT CRITICAL POINT (CP) TYPES IN THREE, TWO, AND ONE DIMENSIONS. CP'S ARE CLASSIFIED ACCORDING TO THE NUMBER OF COMBINED EFFECTIVE MASS PARAMETERS WHICH ARE NEGATIVE (SEE TEXT FOR DETAILS).

		J_{vc}	
Dimensions	CP type	$\omega < \omega_{vc}$	$\omega > \omega_{vc}$
3D	M_0	0	$(\omega - \omega_{vc})^{1/2}$
	M_1	$C - (\omega_{vc} - \omega)^{1/2}$	C
	M_2	C	$C - (\omega - \omega_{vc})^{1/2}$
	M_3	$(\omega_{vc} - \omega)^{1/2}$	0
2D	P_0	0	C
	P_1	$-\ln(\omega_{vc} - \omega)$	$-\ln(\omega - \omega_{vc})$
	P_2	C	0
1D	Q_0	0	$(\omega - \omega_{vc})^{-1/2}$
	Q_1	$(\omega_{vc} - \omega)^{-1/2}$	0

The dielectric function is a superposition of contributions from different critical points in the total joint density of states. Aspnes [413] has given an expression which covers the various types of critical points and the different dimensionalities. With lifetime broadening Γ taken into account, the expression is

$$\epsilon(\hbar\omega) = C - A\,\Gamma^{-n}\,e^{i\theta}\,(\hbar\omega - E_{vc} + i\Gamma)^n. \tag{A.26}$$

The amplitude A is determined by the transition matrix element and the effective mass. The parameter n varies with the dimensionality of the critical point ($n = -\frac{1}{2}$, 0 (logarithmic), and $+\frac{1}{2}$ for a 1-D, 2-D, and 3-D critical point, respectively). The phase θ is an integer multiple of $\pi/2$, with the integer depending on the type of critical point. In a real sample the phase adopts intermediate values because of carrier scattering, sample inhomogeneities, and surface effects.

A3. REFLECTANCE AND TRANSMITTANCE

Reflectance and transmission of electromagnetic plane waves at the interface between two media are determined by the continuity requirements for electric and magnetic fields. The boundary conditions originate from the Maxwell equations and lead to Fresnel's relations. We summarize some relations which apply to high-pressure optical absorption and reflectance spectroscopy at normal incidence. The experimental configuration is that

4 OPTICAL PROPERTIES OF SEMICONDUCTORS UNDER PRESSURE

of an absorbing medium (the sample) either immersed in a transparent dielectric (the pressure medium) or in direct optical contact with the window of the pressure cell. We denote the refractive index of the outer dielectric/window by n_{med}, which in general is a pressure-dependent spectral function.

The complex amplitude reflection coefficient at a single interface is

$$r = -\frac{n_1 + in_2 - n_{\text{med}}}{n_1 + in_2 + n_{\text{med}}} = |r| \exp(i\theta), \tag{A.27}$$

where the phase change θ upon reflection is

$$\tan\theta = \frac{2n_2 n_{\text{med}}}{n_1^2 + n_2^2 - n_{\text{med}}^2}. \tag{A.28}$$

The normal-incidence reflectance (often called reflectivity) is

$$R = r^*r = \frac{(n_1 - n_{\text{med}})^2 + n_2^2}{(n_1 + n_{\text{med}})^2 + n_2^2}. \tag{A.29}$$

The amplitude transmission coefficient and the transmittance are

$$t = \frac{2n_{\text{med}}}{n_1 + in_2 + n_{\text{med}}} \tag{A.30}$$

$$T = tt^* \frac{n_1}{n_{\text{med}}} = \frac{4n_1 n_{\text{med}}}{(n_1 + n_{\text{med}})^2 + n_2^2}. \tag{A.31}$$

The factor n_1/n_{med} in the last equation accounts for the difference in mean energy density flow in the dielectric and the sample. Conservation of energy requires $T + R = 1$.

A case of experimental interest is that of light transmitted through a plane parallel plate of thickness d. Taking into account multiple reflected light the external transmittance T^{ext} (transmitted intensity I_t normalized to incident intensity I_o) is given by [524]

$$T^{\text{ext}} = \frac{I_t}{I_o} = \frac{\exp(-\alpha d) \cdot [(1 - R)^2 + 4R \sin^2\theta]}{[1 - R\exp(-\alpha d)]^2 + 4R\exp(-\alpha d)\sin^2(\theta + \phi)}. \tag{A.32}$$

The second term in the denominator stands for interference effects, which can be observed if $R \neq 0$ ($n_1 \neq n_{\text{med}}$). The quantity

$$\phi = 2\pi n_1 d/\lambda \tag{A.33}$$

is the phase change on one pass within the sample of thickness d. Thus, for a transparent or weakly absorbing sample of known thickness d, the real part of the refractive index can be determined from the spectral separation of interference features in transmission spectra. In practice, the thickness is often not known exactly. In this case the order of the interference fringes must also be determined. Averaging over the interference oscillations in Eq. (A.32) (corresponding to rough or inclined surfaces) gives

$$T^{\text{ext}} = \frac{(1-R)^2 \cdot \exp(-\alpha d)}{1 - R^2 \exp - 2\alpha d} \cdot (1 - n_2^2/n_1^2). \tag{A.34}$$

For most transmission experiments $n_2^2 \ll n_1^2$. If d is further chosen sufficiently large to ensure that $R^2 \exp(-2\alpha d) \ll 1$ (no multiple reflection), the external transmittance simply is

$$T^{\text{ext}} = (1 - R)^2 \cdot \exp(-\alpha d). \tag{A.35}$$

If the experimentally determined external transmittance cannot be converted to an absorption coefficient because the sample thickness is unknown, the optical density, defined as

$$\text{OD} = \log_{10}(1/T^{\text{ext}}), \tag{A.36}$$

is often used to represent the frequency-dependent absorption behavior of a material on a logarithmic scale.

A4. Lorentz–Drude Oscillators

A general result of linear response theory is that any linear response function, such as $\epsilon(\omega)$, can be described by a superposition of so-called Lorentz oscillator expressions [445]. Classically, we arrive at the Lorentz oscillator expression by calculating the response of a harmonic oscillator of mass m, charge q, resonance frequency ω_0, and radiation damping constant $\Gamma > 0$ under the influence of an electric field of variable frequency ω. The linear polarizability is given by

$$\alpha_p(\omega) = \frac{q^2/m}{\omega_0^2 - \omega^2 - i\Gamma\omega}. \tag{A.37}$$

The same resonance profile is obtained in time-dependent perturbation theory of the electron-photon interaction in a system of quantized states with energy difference $\hbar\omega_0$. When considering electronic excitations in a crystal, the difference between macroscopic and local electric field is often neglected. For N electronic "oscillators" per unit volume, the dielectric function simply is

$$\epsilon(\omega) - 1 = 4\pi N \alpha_\mathrm{p}(\omega). \tag{A.38}$$

If excitation energies are distributed over different frequencies ω_i, we write

$$\epsilon(\omega) = 1 + \Omega_\mathrm{p}^2 \sum_i \frac{f_i}{\omega_i^2 - \omega^2 - i\Gamma_i \omega} \tag{A.39}$$

$$\Omega_\mathrm{p}^2 = \frac{4\pi N e^2}{m_0}, \tag{A.40}$$

where m_0 is the free electron mass and Ω_p is the all-electron plasma frequency. By definition, all *relative* oscillator strengths f_i add up to 1 ($\sum_i f_i = 1$). For electronic excitations of an isolated atom, N is replaced by the number of electrons per atom, such that again all oscillator strengths add up to 1.

In the presence of free carriers obeying Fermi statistics, the frequency-dependent transport properties are treated by solving the Boltzmann transport equation for the Fermi distribution function within the mean relaxation time approximation. Formally, the result obtained is the same as if we simply set the resonance frequency ω_0 in the Lorentz oscillator expression equal to zero. With the unscreened plasma frequency

$$\omega_\mathrm{p}^2 = \frac{4\pi N e^2}{m^*}, \tag{A.41}$$

the Drude expression for the transverse $\epsilon(\omega)$ of an electron gas with density N is

$$\epsilon(\omega) = \epsilon_\infty - \frac{\omega_\mathrm{p}^2}{\omega^2 + \Gamma^2} + i\frac{\omega_\mathrm{p}^2 \Gamma}{\omega^3 + \Gamma^2 \omega}. \tag{A.42}$$

Here m^* is the electron effective mass and ϵ_∞ is the background dielectric constant arising from transitions at frequencies larger than ω_p. In the limit $\Gamma = 0$, the dielectric constant becomes zero at the screened plasma frequency $\omega_\mathrm{L} = \omega_\mathrm{p}/\sqrt{\epsilon_\infty}$, that is, the frequency of the longitudinal plasma oscillation at $\vec{k} = 0$. Extrapolating the real part of the optical conductivity $\sigma_1(\omega) = \omega \epsilon_2(\omega)/4\pi$ to zero frequency yields $\sigma_1(0) = N e^2/(m^* \Gamma)$. This expression is identical to the Drude relation for the longitudinal dc conductivity, provided that the relaxation time in transport is identical to $1/\Gamma$ [525]. In the high-frequency limit $\omega \gg \omega_p$ and in the absence of other electronic excitations ($\epsilon_\infty = 1$), we have

$$\epsilon(\omega) = 1 - \frac{4\pi N e^2}{m_0 \omega^2}. \qquad (A.43)$$

Note that we have substituted in Eq. (A.43) the effective mass m^* by the fundamental electron mass m_0 because at sufficiently high frequencies restitutive forces are negligible and electrons behave like free particles. The same relation obviously applies to the high-frequency limit of Eq. (A.42).

For lattice modes of a polar (ionic) crystal, there is an additional contribution to the dielectric function arising from the polarization of the lattice induced by the electric field of the light. The resonance frequency corresponds to that of the transverse optical modes, which within the Lorentz-oscillator model is given by

$$\omega_{TO}^2 = \frac{\Omega^2}{\epsilon(0) - \epsilon_\infty}. \qquad (A.44)$$

Here we have introduced the static dielectric constant $\epsilon(0) = \epsilon(\omega = 0)$. The longitudinal mode frequency is then shifted up in energy because of the effect of the depolarization field associated with the vibrations of the polar lattice themselves (Lyddane–Sachs–Teller relation):

$$\omega_{LO}^2 = \frac{\epsilon(0)}{\epsilon_\infty} \omega_{TO}^2. \qquad (A.45)$$

The dielectric constant is then given by

$$\epsilon(\omega) = \epsilon_\infty + \frac{\omega_{TO}^2 \cdot [\epsilon(0) - \epsilon_\infty]}{\omega_{TO}^2 - \omega^2 - i\omega\Gamma} = \epsilon_\infty \left(1 + \frac{\omega_{LO}^2 - \omega_{TO}^2}{\omega_{TO}^2 - \omega^2 - i\Gamma\omega}\right). \qquad (A.46)$$

This leads to the definition of Born's transverse dynamic effective charge e_T^* through

$$\epsilon_\infty(\omega_{LO}^2 - \omega_{TO}^2) = 4\pi N_{ion} e_T^{*2} / M_{red}. \qquad (A.47)$$

Here, N_{ion} is the density of cation–anion pairs and M_{red} is the reduced ionic mass. Inserting Eq. (A.46) into the wave equation (A.10) leads to a dispersion relation known as phonon polariton dispersion, which has the characteristic feature that for $\Gamma = 0$ there is no wave propagation in the frequency regime between ω_{TO} and ω_{LO}, leading to the Reststrahlen band in reflectance.

A frequency dependence of $\epsilon(\omega)$ similar to Eq. (A.46) also applies to other elementary excitations in solids. For instance, the interaction of light with an excitonic transition can lead to a mixed state known as exciton polariton. The polariton is again characterized by a transverse and longitudinal resonance frequency at $k = 0$. In the case of strong photon–exciton interaction, spatial dispersion cannot be neglected [526, 527] and the dielectric constant depends not only on frequency, but also on the wave vector of the mixed state. An additional k-dependent term reflecting the exciton dispersion appears in the denominator of Eq. (A.46), which has important consequences for the wave propagation in the vicinity of an exciton resonance.

A5. Dispersion Relation and Sum Rules

An important property of linear response functions in physics is that their frequency-dependent real and imaginary parts are related via dispersion relations [528, 159] which are often referred to as Kramers–Kronig relations. The dependence arises from the requirement of causality [see Eq. (A.2)]. A plausibility argument for the necessity of dispersion relations in a causal system may be found in Refs. [528, 529]. For the complex $\epsilon(\omega)$ the Kramers–Kronig relations are

$$\epsilon_1(\omega) - 1 = \frac{2}{\pi} \mathcal{P} \int_0^\infty \frac{\omega' \epsilon_2(\omega')}{\omega'^2 - \omega^2} d\omega' \qquad (A.48)$$

$$\epsilon_2(\omega) = -\frac{2\omega}{\pi} \mathcal{P} \int_0^\infty \frac{\epsilon_1(\omega')}{\omega'^2 - \omega^2} d\omega' + \frac{4\pi \sigma_{dc}}{\omega}, \qquad (A.49)$$

where $\mathcal{P}$ means the Cauchy principal value of the integral. Mathematically, these relations represent Hilbert transforms. The second term on the right-hand side of Eq. (A.49) is nonzero only for materials with finite dc conductivity [528].

Various sum rules apply to optical response functions of solids [159]. For instance, the zero frequency limit $\omega \to 0$ of $\epsilon(\omega)$ is given by

$$\epsilon(0) = \epsilon_1(0) = 1 + \frac{2}{\pi} \int_0^\infty \frac{\epsilon_2(\omega)}{\omega} d\omega. \qquad (A.50)$$

This relation illustrates that the static dielectric constant $\epsilon_1(0)$ is determined by $\epsilon_2(\omega)$ over the entire spectral region. Particularly high values of the electronic contribution to $\epsilon_1(0)$ arise in semiconductors with strong interband transitions at low energy. This sum rule plays an important role in the dielectric theory of the chemical bond of semiconductors.

Another important sum rule is known as the *f*-sum rule. It is derived as follows: Consider a limiting frequency ω_c such that there is no absorption above this frequency. In this case the upper limit of the integral in Eq. (A.48) can be set at ω_c, which implies $\omega' \leq \omega_c$. For high frequency $\omega \gg \omega_c$, the Drude formula Eq. (A.43) holds for $\epsilon_1(\omega)$, which leads to

$$\int_0^{\omega_c} \omega \epsilon_2(\omega) d\omega = \frac{2\pi^2 N e^2}{m_0} = \frac{\pi}{2} \Omega_p^2, \tag{A.51}$$

where N is the electron density. We note that this sum rule may be rewritten in terms of an optical oscillator strength density:

$$f(\omega)|_\epsilon = \frac{m_0}{2\pi^2 e^2} \omega \epsilon_2(\omega). \tag{A.52}$$

If Eq. (A.51) is evaluated out to a frequency $\omega_v < \omega_c$, the electron density obtained is interpreted as the *effective* density N_{eff} of electrons contributing to the optical response up to ω_v. By convention, the effective density is often given in terms of the number of electrons per atom,

$$N_{eff}(\omega_v)|_\epsilon = N_{at} n_{eff}(\omega_v) = \frac{m_0}{2\pi^2 e^2} \int_0^{\omega_v} \omega \epsilon_2(\omega) d\omega, \tag{A.53}$$

where N_{at} is the atom density. A common situation is that valence and core excitations are well separated in energy. If ω_v is chosen to fall into this excitation gap (typically starting near the valence electron plasma frequency), one can regard $n_{eff}(\omega_v)$ as the number of valence electrons per atom contributing to the optical constants in the frequency interval from zero to ω_v [105]. In general, the *f*-sum rule provides an overall consistency check for experimentally determined optical constants.

Another form of dispersion relation applies to the reflectance and the phase angle Θ. The complex amplitude reflection coefficient $r = |r| \exp i\Theta$ expresses a linear relationship between the amplitudes of reflected and incident light. For a sample–vacuum interface the dispersion relation connecting the real and imaginary parts of

$$\ln r = \ln|r| + i\Theta = \frac{1}{2} \ln R + i\Theta$$

is given by [528]

$$\Theta(\omega) = \frac{1}{2\pi} \int_0^\infty \ln \left| \frac{\omega' - \omega}{\omega' + \omega} \right| \frac{d \ln R(\omega')}{d\omega'} d\omega'. \quad (A.54)$$

The phase angle $\Theta(\omega)$ can thus be calculated if $R(\omega)$ is known over the entire spectral region, and other optical constants can then be determined from R and Θ.

Given a measured reflectivity change $\Delta R/R$ under an external perturbation, such as stress or electric field, the change in the phase $\Delta\Theta$ of the complex amplitude reflection coefficient can be determined by a differential Kramers–Kronig transformation [12]:

$$\Delta\Theta(\omega_0) = \frac{\omega_0}{\pi} \mathcal{P} \int_0^\infty \left\{ \frac{\Delta R}{R}(\omega) - \frac{\Delta R}{R}(\omega_0) \right\} \frac{d\omega}{\omega^2 - \omega_0^2}. \quad (A.55)$$

With the known unperturbed $\epsilon(\omega)$, the changes $\Delta\epsilon_1(\omega)$ and $\Delta\epsilon_2(\omega)$ can be calculated directly [523, 12, 530]. For a weak perturbation, the change of the sample reflectance $\Delta R/R$ can be linearized as follows [531, 530]:

$$\frac{\Delta R}{R} = \text{Re}[(\alpha - i\beta) \cdot \Delta\epsilon] = \alpha \, \Delta\epsilon_1 + \beta \, \Delta\epsilon_2. \quad (A.56)$$

The frequency-dependent quantities $\alpha = \partial \ln R/\partial \epsilon_1$ and $\beta = \partial \ln R/\partial \epsilon_2$ are known as the two Seraphin coefficients. These in turn depend on the unperturbed ϵ_1 and ϵ_2. Under certain conditions $\Delta R/R$ is dominated by either $\Delta\epsilon_1$ or $\Delta\epsilon_2$.

In high-pressure experiments the reflectance is measured through a covering transparent medium. Therefore, Eq. (A.54) needs to be modified by introducing a phase shift correction [529, 532]. A major difficulty in the phase determination from high-pressure reflectance spectra is that the experimentally accessible spectral range is quite limited (about 5 eV for diamond windows) compared to typical valence-electron plasma frequencies, and the extrapolation of experimental reflectance spectra becomes a delicate procedure in most cases. Therefore, applying the phase-shift dispersion relation is not straightforward, unless one is looking at isolated reflectivity bands within the experimental spectral range.

A6. Analysis of High-Pressure Reflectance Spectra

An alternative approach for the analysis of reflectance spectra, in particular for high-pressure phases of semiconductors, is to set up an empirical model for the frequency dependence of the dielectric constant and to optimize the model parameters until the reflectance calculated via the Fresnel relation (A.29) gives a satisfactory fit to the experimental spectrum. Lorentz–Drude-type expressions for $\epsilon(\omega)$ are easily handled in fitting procedures. At least one absorption band is usually introduced to account for oscillator strength at frequencies above the experimental range. A feature which makes Lorentz–Drude-type expressions attractive for empirical modeling is that Kramers–Kronig relations are automatically satisfied for $\epsilon_1(\omega)$ and $\epsilon_2(\omega)$. We note that spectral functions for interband absorption between parallel bands in nearly free electron metals [520, 521] also may be found suitable for an empirical modeling of $\epsilon(\omega)$ in the interband absorption regime. For the Kramers–Kronig counterpart of a Gaussian line, see Refs. [225, 533].

The preceding approach will in general not converge to a unique solution for the $\epsilon(\omega)$ spectrum. The situation is improved considerably if a plausible total oscillator strength (f-sum rule) is imposed as a constraint to the fit. For this purpose one estimates the effective number of valence electrons per atom, converts to an effective valence electron density N, and uses the corresponding squared plasma frequency as the prefactor in Eq. (A.39) for $\epsilon(\omega)$ or corresponding expression for other spectral dependences. With N given in units of nm^{-3} and energy in electron volts, the numerical value is

$$(\hbar \Omega_p)^2 = \hbar^2 \frac{4\pi N e^2}{m_0} = 1.379 \, N. \qquad (A.57)$$

In this type of analysis one is usually not interested so much in the numerical values of fitted line positions and widths, but rather in the overall spectral dependence of $\epsilon(\omega)$ generated from the fitted parameters.

We illustrate the modeling of an optical response which resembles the case of an *indirect*-gap material undergoing a transition to a metallic state as a consequence of energetic overlap between valence-band maximum and conduction-band minimum in a single-phase regime. Such a model would, for instance, apply to partly ionic NaCl-type phases. Figure 76 shows simulated spectra of $\hbar\omega \, \epsilon_2$ (corresponding to the oscillator strength spectrum) and the corresponding reflectance spectra. The spectra were calculated for the same total oscillator strength corresponding to a squared plasma frequency of $(\hbar\Omega)^2 = 300$ eV^2. This value is typical for the high-pressure phases of the III–V and II–VI semiconductors. The main optical absorption in the visible and UV spectral range is modeled by two parallel-band terms [520, 521]. The spectra differ in the fraction of total oscillator strength attributed to a single Drude-type term ($\Gamma = 0.1$ eV) which covers both free electrons and holes. The reflectance R_{med} refers to an interface

4 OPTICAL PROPERTIES OF SEMICONDUCTORS UNDER PRESSURE

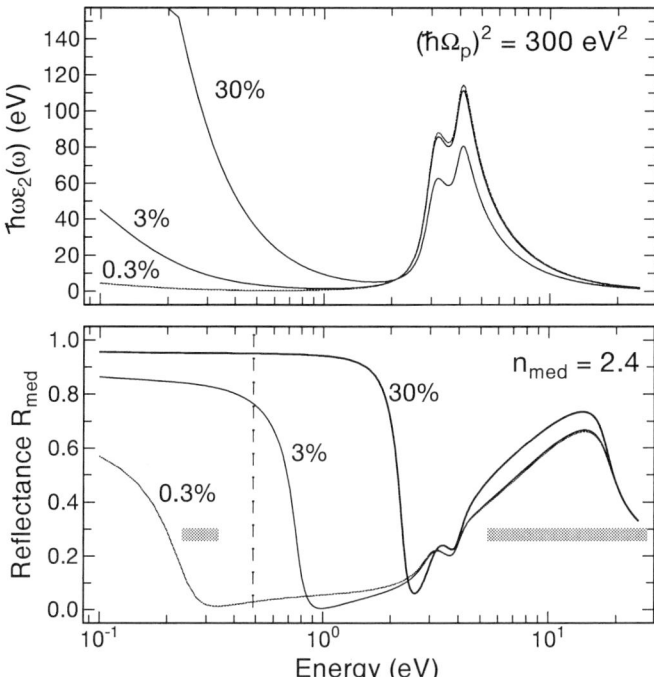

FIG. 76. Simulated spectra of $\hbar\omega\epsilon_2(\omega)$ (top) and corresponding reflectance spectra (bottom). The reflectance R_{med} refers to an outer dielectric medium with constant refractive index $n_{med} = 2.4$, which is close to that of diamond. All spectra are calculated for the same total oscillator strength corresponding to an all-electron squared plasma frequency $(\hbar\Omega)^2 = 300$ eV^2. The spectra differ in the percentage of oscillator strength attributed to a Drude-type contribution. The dashed vertical line near 0.5 eV indicates the borderline between IR and NIR optics, and the gray bars mark the absorbing spectral regions of diamond.

between "sample" and a transparent dielectric with constant refractive index $n_{med} = 2.4$ which is close to that of diamond. We note that a low-energy Drude-like reflectance tail becomes observable above 0.5 eV, if about 1% of the total oscillator strength goes to the Drude part. This corresponds to roughly 10^{21} free carriers per cm^3, assuming an effective mass $m^* = m_0$.

Acknowledgments

With pleasure we acknowledge many stimulating discussions with Manuel Cardona. We thank Klaus Reimann for helpful comments and for making available a database. Niels Christensen has kindly provided us with the calculated band structures of GaAs (Fig. 8). We express our thanks to Regine Noack, Werner Dieterich, and Ulrich Oelke for assistance in art work and to Brigitte Prauser for support in updating our references.

REFERENCES

[1] P. W. Bridgman, *Proc. Amer. Acad. Arts Sci. USA* **81,** 167 (1952).
[2] W. Paul and D. M. Warschauer, *J. Phys. Chem. Solids* **5,** 89 (1958); **5,** 102 (1958); **6,** 6 (1958).
[3] W. Paul, *J. Phys. Chem. Solids* **8,** 196 (1959).
[4] M. Cardona, W. Paul, and H. Brooks, *J. Phys. Chem. Solids* **8,** 204 (1959).
[5] T. E. Slykhouse and H. G. Drickamer, *J. Phys. Chem. Solids* **7,** 207 (1958).
[6] A. L. Edwards, T. E. Slykhouse, and H. G. Drickamer, *J. Phys. Chem. Solids* **11,** 140 (1959).
[7] A. L. Edwards and H. G. Drickamer, *Phys. Rev.* **122,** 1149 (1961).
[8] R. Zallen and W. Paul, *Phys. Rev.* **134,** A1628 (1964).
[9] R. Zallen and W. Paul, *Phys. Rev.* **155,** 703 (1967).
[10] F. H. Pollak, M. Cardona, and K. L. Shaklee, *Phys. Rev. Lett.* **16,** 942 (1966).
[11] F. H. Pollak and M. Cardona, *Phys. Rev.* **172,** 816 (1968).
[12] M. Cardona, *Modulation Spectroscopy, Solid State Physics, Suppl. 11* (Academic Press, New York, 1969).
[13] J. C. Phillips, *Rev. Mod. Phys.* **42,** 317 (1970).
[14] J. A. Van Vechten, *Phys. Rev.* **182,** 891 (1969).
[15] J. A. Van Vechten, *Phys. Rev.* **187,** 1007 (1969).
[16] D. L. Camphausen, G. A. N. Connell, and W. Paul, *Phys. Rev. Lett.* **26,** 184 (1971).
[17] S. S. Mitra, C. Postmus, and J. R. Ferraro, *Phys. Rev. Lett.* **18,** 455 (1967).
[18] S. S. Mitra, O. Brafman, W. B. Daniels, and R. K. Crawford, *Phys. Rev.* **185,** 942 (1969).
[19] A. Van Valkenburg, *Diamond Research* **1964,** 17.
[20] W. A. Bassett, T. Takahashi, and P. W. Stock, *Rev. Sci. Instrum.* **38,** 37 (1967).
[21] R. A. Forman, G. J. Piermarini, J. D. Barnett, and S. Block, *Science* **176,** 284 (1972).
[22] G. J. Piermarini, S. Block, and J. D. Barnett, *J. Appl. Phys.* **44,** 5377 (1973).
[23] G. J. Piermarini, S. Block, J. D. Barnett, and R. A. Forman, *J. Appl. Phys.* **46,** 2774 (1975).
[24] G. J. Piermarini and S. Block, *Rev. Sci. Instrum.* **46,** 973 (1975).
[25] M. Cardona, *Phys. Status Solidi B* **198,** 5 (1996).
[26] F. H. Pollak, in *Semiconductors and Semimetals,* Vol. 32, edited by H. Ehrenreich and D. Turnbull (Academic Press, New York, 1990), p. 17.
[27] K. Syassen, R. A. Stradling, and A. R. Goñi, *Proc. 7th Int. Conf. on High Pressure Semiconductor Physics,* Schwäbisch-Gmünd, Germany, 1996, *Phys. Stat. Sol. B,* **198,** 1 (1996).
[28] B. A. Weinstein and W. Paul, *Proc. 6th Int. Conf. on High Pressure Semiconductor Physics,* Vancouver, Canada 1994, *J. Phys. Chem. Sol.* **56,** 311 (1995).
[29] T. Arai and S. Onari, *Proc. 5th Int. Conf. on High Pressure Semiconductor Physics,* Kyoto, Japan 1992, *Jpn. J. Appl. Phys.* **32,** Suppl. 32-1, 1 (1993).
[30] D. D. Kyriakos and O. E. Valassiades, *Proc. 4th Int. Conf. on High Pressure Semiconductor Physics,* Porto Carras, Greece 1990 (Aristotle University, Thessaloniki, 1990).
[31] T. Suski, *3rd Int. Conf. on High Pressure Semiconductor Physics,* Tomaszow, Poland, *Semicond. Sci. Technol.* **4,** pp. 211 (1989).

[32] J. M. Besson and J. P. Pincaeaux, *Science* **206,** 1073 (1979).
[33] H. K. Mao and P. M. Bell, *Carnegie Institute Washington Year Book 1978,* p. 659.
[34] A. Jayaraman, *Rev. Mod. Phys.* **55,** 65 (1983).
[35] A. Jayaraman, *Rev. Sci. Instrum.* **57,** 1013 (1986).
[36] M. I. Eremets, *High Pressure Experimental Methods* (Oxford University Press, Oxford, 1996).
[37] *High Pressure Techniques in Chemistry and Physics,* edited by W. B. Holzapfel and N. S. Isaacs (Oxford University Press, Oxford, 1997).
[38] K. Syassen, *Phys. Rev. B* **25,** 6548 (1982).
[39] P. E. Van Camp, V. E. Van Doren, and J. T. Devreese, *Solid State Commun.* **84,** 731 (1992).
[40] M. P. Surh, S. G. Louie, and M. L. Cohen, *Phys. Rev. B* **45,** 8239 (1992).
[41] J. Walker, *Rep. Prog. Phys.* **42,** 1605 (1979).
[42] A. T. Collins, *Physica B* **185,** 284 (1993).
[43] G. Huber, K. Syassen, and W. B. Holzapfel, *Phys. Rev. B* **15,** 5123 (1977).
[44] J. H. Burnett, H. M. Cheong, and W. Paul, *Rev. Sci. Instrum.* **61,** 3904 (1990).
[45] K. Reimann, *High Pressure Res.* **15,** 73 (1996).
[46] K. Reimann and K. Syassen, *Semicond. Sci. Technol.* **4,** 263 (1989).
[47] K. Reimann and K. Syassen, *Phys. Rev. B* **39,** 11113 (1989).
[48] P. Loubeyre, J. M. Besson, J. P. Pincaeaux, and J. P. Hansen, *Phys. Rev. Lett.* **49,** 1172 (1982).
[49] W. L. Vos, M. G. E. van Hinsberg, and J. A. Schouten, *Phys. Rev. B* **42,** 6106 (1990).
[50] C. Abraham, Ph.D. Thesis, Universität Stuttgart, 1992.
[51] M. Lindner, G. F. Schötz, P. Link, H. P. Wagner, W. Kuhn, and W. Gebhardt, *J. Phys.: Condens. Matter* **4,** 6401 (1992).
[52] H. K. Mao, P. M. Bell, J. W. Shaner, and D. J. Steinberg, *J. Appl. Phys.* **49,** 3276 (1978).
[53] H. K. Mao, J. Xu, and P. M. Bell, *J. Geophys. Res.* **91,** 4673 (1986).
[54] R. J. Hemley, C. S. Zha, A. P. Jephcoat, H. K. Mao, L. W. Finger, and D. E. Cox, *Phys. Rev. B* **39,** 11820 (1989).
[55] R. A. Noack and W. B. Holzapfel, in *High Pressure Science and Technology,* edited by K. D. Timmerhaus and M. S. Barber (Plenum, New York, 1979), p. 748.
[56] S. Yamaoka, O. Shimomura, and O. Fukunaga, *Proc. Jpn. Acad. B* **56,** 103 (1980).
[57] S. L. Wunder and P. E. Schoen, *J. Appl. Phys.* **52,** 3775 (1981).
[58] P. D. Horn and Y. M. Gupta, *Phys. Rev. B* **39,** 973 (1989).
[59] D. E. McCumber and M. D. Sturge, *J. Appl. Phys.* **34,** 1682 (1963).
[60] D. D. Ragan, R. Gustavsen, and D. Schiferl, *J. Appl. Phys.* **72,** 5539 (1992).
[61] C. Ulrich, Ph.D. Thesis, Universität Stuttgart, 1997.
[62] L. Viña, S. Logothetidis, and M. Cardona, *Phys. Rev. B* **30,** 1979 (1984).
[63] W. B. Holzapfel, in *High Pressure Techniques in Chemistry and Physics,* edited by W. B. Holzapfel and N. S. Isaacs (Oxford University Press, Oxford, 1997), p. 47.
[64] F. H. Pollak and H. Shen, *J. Crystal Growth* **98,** 53 (1989).
[65] H. R. Chandrasekhar and M. Chandrasekhar, *Proc. SPIE* **1286,** 207 (1990).
[66] H. Shen, P. Parayanthal, Y. F. Liu, and F. H. Pollak, *Rev. Sci. Instrum.* **58,** 1429 (1987).
[67] H. Shen, X. C. Shen, F. H. Pollak, and R. N. Sacks, *Phys. Rev. B.* **36,** 3487 (1987).
[68] H. R. Philipp and E. A. Taft, *Phys. Rev.* **127,** 159 (1962).
[69] M. I. Eremets, V. V. Struzhkin, J. A. Timofeev, I. A. Trojan, A. N. Utjuzh, and A. M. Shirokov, *High Pressure Res.* **9,** 347 (1992).
[70] A. Ruoff and K. Ghandehari, in *High-Pressure Science and Technology — 1993,* edited by S. C. Schmidt, J. W. Shaner, G. A. Samara, and M. Ross (AIP Conference Proceedings 309, American Institute of Physics, New York, 1994), p. 1523.

[71] J. Nunnenkamp, K. Reimann, J. Kuhl, and K. Ploog, *Phys. Rev. B* **44**, 8129 (1991).
[72] J. Nunnenkamp, K. Reimann, J. Kuhl, and K. Ploog, *Surf. Sci.* **263**, 553 (1992).
[73] K. Reimann, M. Holtz, K. Syassen, J. Nunnenkamp, J. Kuhl, R. Nötzel, A. J. Shields, and K. Ploog, in *Recent Trends in High Pressure Research — Proceedings of the XIIIth AIRAPT Conference, Bangalore 1991*, edited by A. K. Singh (Oxford & IBH Publishing, New Delhi, 1992), p. 41.
[74] K. Reimann, M. Holtz, K. Syassen, J. Nunnenkamp, J. Kuhl, R. Nötzel, A. J. Shields, and P. Ploog, *High Pressure Res.* **9**, 83 (1992).
[75] W. A. Harrison, *Electronic Structure and the Properties of Solids* (W. H. Freeman and Company, San Francisco, 1980).
[76] P. Y. Yu and M. Cardona, *Fundamentals of Semiconductors* (Springer, Berlin, 1996).
[77] *Landolt–Börnstein — Zahlenwerte und Funktionen aus Naturwissenschaften und Technik, New Series, Physics of Group IV Elements and III–V Compounds*, edited by O. Madelung, M. Schulz, and H. Weis (Springer, Heidelberg, 1982), Vol. III/17a.
[78] *Landolt–Börnstein — Zahlenwerte und Funktionen aus Naturwissenschaften und Technik, Intrinsic Properties of Group IV Elements and III–V, II–VI, and I–VII Compounds*, edited by O. Madelung, M. Schulz, and H. Weis (Springer, Heidelberg, 1987), Vol. III/22a.
[79] D. Chadi and M. Cohen, *Phys. Status Solidi B* **68**, 405 (1975).
[80] G. Sai-Halasz, L. Esaki, and W. Harrison, *Phys. Rev. B* **18**, 2812 (1978).
[81] P. Vogl, H. P. Hjalmarson, and J. D. Dow, *J. Phys. Chem. Solids* **44**, 365 (1983).
[82] K. C. Hass, H. Ehrenreich, and B. Velicky, *Phys. Rev. B* **27**, 1088 (1983).
[83] J. C. Slater and G. F. Koster, *Phys. Rev.* **94**, 1498 (1954).
[84] J. Robertson, *J. Phys. C* **12**, 4777 (1979).
[85] P. V. Smith and D. McMahon, *J. Phys. C* **16**, 6947 (1983).
[86] N. E. Christensen, *Physica Scripta* **T19A**, 298 (1987).
[87] G. Grosso and C. Piermarocchi, *Phys. Rev. B* **51**, 16772 (1995).
[88] M. C. Muñoz and G. Armelles, *Phys. Rev. B* **48**, 2839 (1993).
[89] D. Bertho, J.-M. Jancu, and C. Jouanin, *Phys. Rev. B* **50**, 16956 (1994).
[90] G. H. Li, A. R. Goñi, C. Abraham, K. Syassen, P. V. Santos, A. Cantarero, O. Brandt, and K. Ploog, *Phys. Rev. B* **50**, 1575 (1994).
[91] M. Kumagai, T. Takagahara, and E. Hanamura, *Phys. Rev. B* **37**, 898 (1988).
[92] Y. Gu and Z. Xu, *J. Phys. C* **3**, 6553 (1991).
[93] T. Yamauchi, Y. Arakawa, and J. N. Schulman, *Surf. Sci.* **257**, 291 (1992).
[94] G. Martinez, in *Optical Properties of Solids (Handbook on Semiconductors, Vol. 2)*, edited by M. Balkanski (North Holland, Amsterdam, 1980), p. 181.
[95] K. J. Chang, S. Froyen, and M. L. Cohen, *Solid State Commun.* **50**, 105 (1984).
[96] M. Alouani, L. Brey, and N. Christensen, *Phys. Rev. B* **37**, 1167 (1988).
[97] P. Boguslawski and I. Gorczyca, *Semicond. Sci. Technol.* **9**, 2169 (1994).
[98] N. E. Christensen, private communication.
[99] B. Welber, M. Cardona, C. K. Kim, and S. Rodríguez, *Phys. Rev. B* **12**, 5729 (1975).
[100] D. J. Wolford and J. A. Bradley, *Solid State Commun.* **53**, 1069 (1985).
[101] M. Leroux, G. Pelous, F. Raymond, and C. Verie, *Appl. Phys. Lett.* **46**, 288 (1985).
[102] A. R. Goñi, K. Strössner, K. Syassen, and M. Cardona, *Phys. Rev. B* **36**, 1581 (1987).
[103] D. R. Penn, *Phys. Rev.* **128**, 2093 (1962).
[104] V. Heine and R. O. Jones, *J. Phys. C* **2**, 719 (1969).
[105] H. R. Philipp and H. Ehrenreich, *Phys. Rev.* **129**, 1550 (1963).
[106] A. R. Goñi, K. Syassen, and M. Cardona, *Phys. Rev. B* **39**, 12921 (1989).
[107] A. Mang, K. Reimann, and S. Rübenacke, in *Proc. 22nd Int. Conf. on the Physics of*

Semiconductors, Vancouver 1994, edited by D. J. Lockwood (World Scientific, Singapore, 1995), Vol. 1, p. 317.
[108] G. Dresselhaus, A. F. Kip, and C. Kittel, *Phys. Rev.* **98,** 368 (1955).
[109] E. O. Kane, *J. Phys. Chem. Solids* **1,** 249 (1957).
[110] M. Cardona, *J. Phys. Chem. Solids* **24,** 1543 (1963).
[111] U. Rössler, *Solid State Commun.* **49,** 943 (1984).
[112] W. Zawadzki, P. Pfeffer, and H. Sigg, *Solid State Commun.* **53,** 777 (1985).
[113] M. Cardona, in *Atomic Structure and Properties of Solids,* edited by E. Burstein (Academic Press, New York, 1972), Vol. LII.
[114] A. R. Goñi, K. Syassen, Y. Zhang, K. Ploog, A. Cantarero, and A. Cros, *Phys. Rev. B* **45,** 6809 (1992).
[115] H. Siegle, A. R. Goñi, C. Thomsen, C. Ulrich, K. Syassen, B. Schöttker, D. J. As, and D. Schikora, in GaN and Related Materials II, Proc. MRS 1997, Spring Meeting, San Francisco, Vol. 486, p. 225 (1997).
[116] J. F. Nye, *Physical Properties of Crystals* (Oxford University, Oxford, 1969).
[117] G. L. Bir and G. E. Pikus, *Symmetry and Strain-Induced Effects in Semiconductors* (Wiley, New York, 1974).
[118] M. Cuevas and H. Fritsche, *Phys. Rev.* **137,** A1847 (1965).
[119] I. Balslev, *Phys. Rev.* **143,** 636 (1966).
[120] I. Balslev, *J. Phys. Soc. Jpn.* **21,** 101 (1966).
[121] E. O. Kane, *Phys. Rev.* **178,** 1368 (1969).
[122] H. Vogelman and T. Fjeldly, *Rev. Sci. Instrum.* **45,** 3096 (1974).
[123] P. Etchegoin, J. Kircher, M. Cardona, and C. Grein, *Phys. Rev. B* **45,** 11721 (1992).
[124] A. Blacha, H. Presting, and M. Cardona, *Phys. Status Solidi B* **126,** 11 (1984).
[125] G. Bastard, J. A. Brum, and R. Ferreira, in *Solid State Physics,* edited by H. Ehrenreich and D. Turnball (Academic Press, San Diego, 1991), Vol. 44, p. 229.
[126] M. Altarelli, in *Heterojunctions and Semiconductors Superlattices,* edited by G. Allan, G. Bastard, N. Boccara, M. Lanoo, and M. Voos (Springer, Heidelberg, 1985), p. 12.
[127] D. J. Wolford, T. F. Kuech, J. A. Bradley, M. A. Gell, D. Ninno, and M. Jaros, *J. Vac. Sci. Technol. B* **4,** 1043 (1986).
[128] J. Chen, J. R. Sites, I. L. Spain, M. J. Hafich, and G. Y. Robinson, *Appl. Phys. Lett.* **58,** 744 (1991).
[129] A. D. Prins, J. L. Sly, A. T. Meney, D. J. Dunstan, E. P. O'Reilly, A. R. Adams, and A. Valster, *J. Phys. Chem. Solids* **56,** 423 (1995).
[130] J. C. Maan, *Surf. Sci.* **196,** 518 (1988).
[131] L. M. Claessen, J. C. Maan, M. Altarelli, P. Wyder, L. L. Chang, and L. Esaki, *Phys. Rev. Lett.* **57,** 2556 (1986).
[132] H. M. Cheong, W. Paul, M. E. Flatte, and R. H. Miles, *Phys. Rev. B* **55,** 4477 (1997).
[133] G. A. Samara, *Phys. Rev. B* **27,** 3494 (1983).
[134] M. Kastner, *Phys. Rev. B* **6,** 2273 (1972).
[135] Y. F. Tsay, S. S. Mitra, and B. Bendow, *Phys. Rev. B* **10,** 1476 (1974).
[136] B. Bendow, P. D. Gianino, Y.-F. Tsay, and S. S. Mitra, *Appl. Optics* **13,** 2382 (1974).
[137] Z. H. Levine and D. Allan, *Phys. Rev. B* **43,** 4187 (1991).
[138] Z. H. Levine and D. Allan, *Phys. Rev. Lett.* **66,** 41 (1991).
[139] A. Dal Corso, S. Baroni, and R. Resta, *Phys. Rev. B* **49,** 5323 (1994).
[140] S. Baroni and R. Resta, *Phys. Rev. B* **33,** 7017 (1986).
[141] M. Alouani and J. M. Wills, *Phys. Rev. B* **54,** 2480 (1996).
[142] N. E. Christensen, I. Wenneker, A. Svane, and M. Fanciulli, *Phys. Status Solidi B* **198,** 23 (1996).

[143] F. Bassani and M. Altarelli, in *Handbook of Synchrotron Radiation,* Vol. 1, edited by E. E. Koch (North Holland, Amsterdam, 1983).
[144] N. M. Balzaretti and J. A. H. da Jornada, *Solid State Commun.* **99,** 943 (1996).
[145] R. Vetter, *Phys. Status Solidi A* **9,** 443 (1971).
[146] N. J. Trappeniers, R. Vetter, and H. A. R. de Bruin, *Physica* **45,** 619 (1970).
[147] A. R. Goñi, K. Syassen, and M. Cardona, *Phys. Rev. B* **41,** 10104 (1990).
[148] P. Perlin, I. Gorczyca, N. E. Christensen, I. Grzegory, H. Teisseyre, and T. Suski, *Phys. Rev. B* **45,** 13307 (1992).
[149] K. Strössner, S. Ves, and M. Cardona, *Phys. Rev. B* **32,** 6614 (1985).
[150] W. M. DeMeis and G. McKay, Lab. Technical Report No. HP-15, Harvard University, unpublished.
[151] M. Hanfland, Master's Thesis, Universität Düsseldorf, 1985.
[152] P. Wiesniewski, C. Skierbiszewski, and T. Suski, *Acta Phys. Pol. A* **69,** 889 (1986).
[153] G. F. Schötz, E. Griebl, H. Stanzl, T. Reisinger, and W. Gebhardt, *Mater. Sci. Forum* **182–184,** 271 (1995).
[154] M. Gauthier, A. Polian, J. M. Besson, and A. Chevy, *Phys. Rev. B* **40,** 3837 (1989).
[155] N. Kuroda, O. Ueno, and Y. Nishina, *Phys. Rev. B* **35,** 3860 (1987).
[156] P. G. Johannsen, G. Reiss, U. Bohle, and W. B. Holzapfel, *Phys. Status Solidi B* **198,** 93 (1996).
[157] D. E. Aspnes and A. A. Studna, *Phys. Rev. B* **27,** 985 (1983).
[158] S. Adachi, *Phys. Rev. B* **35,** 7454 (1987).
[159] D. Y. Smith, in *Handbook of Optical Constants of Solids,* edited by E. D. Palik (Academic Press, New York, 1985), p. 35.
[160] *Handbook of Optical Constants of Solids,* edited by E. D. Palik (Academic Press, New York, 1985).
[161] F. D. Murnaghan, *Proc. Natl. Acad. Sci. USA* **44,** 244 (1944).
[162] P. G. Johannsen, *Meas. Science Technol.* **4,** 237 (1993).
[163] P. G. Johannsen, G. Reiss, U. Bohle, J. Magiera, R. Müller, H. Spiekermann, and W. Holzapfel, *Phys. Rev. B* **55,** 6865 (1997).
[164] R. Le Toullec, P. Loubeyre, and J.-P. Pinceaux, *Phys. Rev. B* **40,** 2368 (1989).
[165] J. P. Itié and R. Le Toullec, *J. Physique Colloq.* **45,** C8 (1984).
[166] R. W. Godby, M. Schlüter, and L. J. Sham, *Phys. Rev. B* **37,** 10159 (1988).
[167] F. Gygi and A. Baldareschi, *Phys. Rev. Lett.* **62,** 2160 (1989).
[168] O. Madelung, *Introduction to Solid State Theory* (Springer, Berlin, 1982).
[169] R. L. Hartman, *Phys. Rev.* **127,** 765 (1962).
[170] W. P. Dumke, M. R. Lorenz, and G. Pettit, *Phys. Rev. B* **5,** 2978 (1972).
[171] B. Welber, M. Cardona, Y.-F. Tsay, and B. Bendow, *Phys. Rev. B* **15,** 875 (1977).
[172] M. Kobayashi, M. Yamanaka, and M. Shinohara, *J. Phys. Soc. Jpn.* **58,** 2673 (1989).
[173] S. Ernst, M. Rosenbauer, U. Schwarz, P. Deák, K. Syassen, M. Stutzmann, and M. Cardona, *Phys. Rev. B* **49,** 5362 (1994).
[174] H. Müller, R. Trommer, M. Cardona, and P. Vogl, *Phys. Rev. B* **21,** 4879 (1980).
[175] S. Ves, K. Strössner, C. Kim, and M. Cardona, *Solid State Commun.* **55,** 327 (1985).
[176] K. Strössner, S. Ves, C. Kim, and M. Cardona, *Phys. Rev. B* **33,** 4044 (1986).
[177] T. Kobayashi, K. Aoki, and K. Yamamoto, *Physica B* **139–140,** 537 (1986).
[178] T. Kobayashi, *Semicond. Sci. Technol.* **4,** 248 (1989).
[179] A. R. Goñi, K. Syassen, K. Strössner, and M. Cardona, *Phys. Rev. B* **39,** 3178 (1989).
[180] G. Abrazevicius, G. Babonas, S. Marcinkevicius, V. D. Prochukhan, and Y. V. Rud, *Solid State Commun,* **49,** 651 (1984).
[181] S. Ves, K. Strössner, N. E. Christensen, C. K. Kim, and M. Cardona, *Solid State Commun.* **56,** 479 (1985).

[182] S. Ves, K. Strössner, W. Gebhardt, and M. Cardona, *Phys. Rev. B* **33,** 4077 (1986).
[183] K. Strössner, S. Ves, C. K. Kim, and M. Cardona, *Solid State Commun.* **61,** 275 (1987).
[184] S. Jiang, S. C. Shen, Q. G. Li, H. R. Zhu, G. L. Lu, and W. Giriat, *Phys. Rev. B* **40,** 8017 (1989).
[185] T. Zengju and S. Xuechu, *J. Phys.: Condens. Matter* **2,** 6293 (1990).
[186] G. Schötz, Ph.D. Thesis, Universität Regensburg, 1995.
[187] J. González, F. V. Pérez, E. Moya, and J. C. Chervin, *Phys. Chem. Solids* **56,** 1995 (1995).
[188] P. Cervantes, Q. Williams, M. Cote, O. Zakharov, and M. L. Cohen, *Phys. Rev. B* **54,** 17585 (1996).
[189] S. Ves, D. Glötzel, M. Cardona, and H. Overhof, *Phys. Rev. B* **24,** 3073 (1981).
[190] Y. Sasaki, K. Hoshi, S. Saito, K. Yamaguchi, and Y. Nishina, *J. Phys. Soc. Jpn.* **52,** 3706 (1983).
[191] G. Valiukonis, D. A. Guseinova, G. Krivaite, and A. Sileika, *Phys. Status Solidi B* **135,** 299 (1986).
[192] N. Kuroda, O. Ueno, and Y. Nishina, *J. Phys. Soc. Jpn.* **55,** 581 (1986).
[193] J. González and C. Rincón, *J. Appl. Phys.* **65,** 2031 (1989).
[194] L.-H. Choi and P. Y. Yu, *Phys. Status Solidi B* **198,** 251 (1996).
[195] L. Roa, J. C. Chervin, A. Chevy, M. Dávila, P. Grima, and J. González, *Phys. Status Solidi B* **198,** 99 (1996).
[196] N. E. Christensen, *Phys. Rev. B* **30,** 5753 (1984).
[197] C. O. Rodríguez, E. L. Peltzer y Blanca, and O. M. Capannini, *Phys. Rev. B* **33,** 8436 (1986).
[198] A. D. Prins, J. Sly, and D. J. Dunstan, *Phys. Status Solidi B* **198,** 57 (1996).
[199] W. Hanke and L. J. Sham, *Phys. Rev. B* **38,** 13361 (1988).
[200] M. Hanfland, K. Syassen, and N. Christensen, *J. Physique Colloq.* **45,** C8 (1984).
[201] H. J. McSkimin and P. Andreatch, Jr., *J. Appl. Phys.* **34,** 651 (1963).
[202] B. H. Lee, *J. Appl. Phys.* **41,** 2984 (1970).
[203] G. H. Li, A. R. Goñi, K. Syassen, and M. Cardona, *Phys. Rev. B* **49,** 8017 (1994).
[204] P. J. Melz and I. B. Ortenburger, *Phys. Rev. B* **3,** 3257 (1971).
[205] A. R. Goñi, K. Syassen, K. Strössner, and M. Cardona, *Semicond. Sci. Technol.* **4,** 246 (1989).
[206] G. G. Macfarlane, T. P. McLean, J. E. Quarrington, and V. Roberts, *Phys. Rev.* **111,** 1245 (1958).
[207] G. G. Macfarlane, T. P. McLean, J. E. Quarrington, and V. Roberts, *Phys. Rev.* **108,** 1377 (1957).
[208] A. R. Goñi, Ph.D. Thesis, Universität Stuttgart, 1989.
[209] R. J. Elliott, *Phys. Rev.* **108,** 1384 (1957).
[210] R. J. Elliott, in *Theory of Excitons I,* edited by C. G. Kuper and G. Whitfield (Oliver and Boyd, Edinburgh, 1963).
[211] Y. Toyozawa, *Prog. Theor. Phys.* **20,** 53 (1958).
[212] M. D. Sturge, *Phys. Rev.* **127,** 768 (1962).
[213] P. J. Dean and D. C. Herbert, in *Bound Excitons in Semiconductors, Topics in Current Physics Series,* edited by K. Cho (Springer, Berlin, 1979), p. 55.
[214] A. R. Goñi, A. Cantarero, K. Syassen, and M. Cardona, *Phys. Rev. B* **41,** 10111 (1990).
[215] B. Gerlach and J. Pollmann, *Phys. Status Solidi B* **67,** 93 (1975).
[216] M. Shinada and S. Sugano, *J. Phys. Soc. Jpn.* **21,** 1936 (1966).
[217] G. E. W. Bauer and T. Ando, *Phys. Rev. B* **38,** 6015 (1988).
[218] N. H. Lu, P. M. Hui, and T. M. Hsu, *Solid State Commun.* **78,** 145 (1991).
[219] P. W. Baumeister, *Phys. Rev.* **121,** 359 (1961).
[220] D. D. Sell, *Phys. Rev. B* **6,** 3750 (1972).

[221] J. Camassel, P. Merle, H. Mathieu, and A. Chevy, *Phys. Rev. B* **17**, 418 (1978).
[222] E. F. Schubert, E. O. Göbel, Y. Korikoshi, K. Ploog, and H. Queisser, *Phys. Rev. B* **30**, 813 (1984).
[223] D. D. Sell and P. Lawaetz, *Phys. Rev. Lett.* **26**, 311 (1971).
[224] H. Iwamura, H. Kobayashi, and H. Okamoto, *Jpn. J. Appl. Phys.* **23**, L795 (1984).
[225] W. Kuhn and E. Braun, *Z. Phys. Chem.* **8**, 281 (1930).
[226] A. R. Goñi, A. Cantarero, U. Schwarz, K. Syassen, and A. Chevy, *Phys. Rev. B* **45**, 4221 (1992).
[227] M. Lindner, S. H. L. Zott, G. F. Schötz, W. Gebhardt, P. Perlin, and P. Wisniewski, *High Pressure Res.* **10**, 408 (1992).
[228] A. Mang, K. Reimann, and S. Rübenacke, *Solid State Commun.* **94**, 251 (1995).
[229] A. Mang, K. Reimann, S. Rübenacke, and M. Steube, *Phys. Rev. B* **53**, 16283 (1996).
[230] K. Reimann and S. Rübenacke, *Phys. Rev. B* **49**, 11021 (1994).
[231] E. Griebl, B. Haserer, T. Frey, T. Reisinger, and W. Gebhardt, *Phys. Status Solidi B* **198**, 355 (1996).
[232] G. H. Li, A. R. Goñi, K. Syassen, H. Q. Hou, W. Feng, and J. Zhou, *Phys. Rev. B* **54**, 13820 (1996).
[233] K. Reimann, M. Haselhoff, S. Rübenacke, and M. Steube, *Phys. Status Solidi B* **198**, 71 (1996).
[234] A. Baldereschi and N. O. Lipari, *Phys. Rev. B* **3**, 439 (1971).
[235] A. Bourdon, A. Chevy, and J. M. Besson, in *Proc. 14th Int. Conf. Phys. Semicond.*, edited by B. L. H. Wilson (The Institute of Physics, London, 1979), p. 1371.
[236] N. Piccioli, R. Le Toullec, F. Bertrand, and J. C. Chervin, *J. Phys. (Paris)* **42**, 1129 (1981).
[237] K. F. Brennan, D. H. Park, K. Hess, and M. A. Littlejohn, *J. Appl. Phys.* **63**, 5004 (1988).
[238] M. V. Fischetti and S. E. Laux, *Phys. Rev. B* **38**, 9721 (1988).
[239] D. Kim and P. Y. Yu, *Phys. Rev. Lett.* **64**, 946 (1990).
[240] D. Kim and P. Y. Yu, *Phys. Rev. B* **43**, 4158 (1991).
[241] H. Kalt, W. W. Rühle, K. Reimann, M. Rinker, and E. Bauser, *Phys. Rev. B* **43**, 12364 (1991).
[242] J. A. Kash, *Phys. Rev. B* **47**, 1221 (1993).
[243] J. A. Kash and J. C. Tsang, in *Light Scattering in Solids VI*, edited by M. Cardona and G. Güntherodt (Springer, Berlin, 1991), p. 423.
[244] E. M. Conwell, *High Field Transport in Semiconductors* (Academic Press, New York, 1967).
[245] S. Zollner, S. Gopalan, and M. Cardona, *J. Appl. Phys.* **68**, 1682 (1990).
[246] S. Zollner, S. Gopalan, and M. Cardona, *Solid State Commun.* **76**, 877 (1990).
[247] S. Satpathy, M. Chandrasekhar, H. R. Chandrasekhar, and U. D. Venkateswaran, *Phys. Rev. B* **44**, 11339 (1991).
[248] X. Q. Zhou, H. M. V. Driel, and G. Mak, *Phys. Rev. B* **50**, 5226 (1994).
[249] S. Krishnamurthy and M. Cardona, *J. Appl. Phys.* **74**, 2117 (1993).
[250] K. Tanaka, H. Ohtake, and T. Suemoto, *Phys. Rev. Lett.* **71**, 1935 (1993).
[251] N. E. Christensen, private communication.
[252] A. Segura, J. M. Besson, A. Chevy, and M. Martin, *Nuovo Cimento B* **38B**, 345 (1977).
[253] C. Ulrich, A. R. Goñi, K. Syassen, O. Jepsen, A. Cantarero, and V. Muñoz, in *High Pressure Science and Technology — 1995 (Proc. AIRAPT Conf., Warsaw 1995)*, edited by W. A. Trzciakowski (World Scientific, Singapore, 1996), p. 612.
[254] Y. Onodera and Y. Toyozawa, *J. Phys. Soc. Jpn.* **22**, 833 (1967).
[255] M. Steube, S. Rübenacke, and K. Reimann, in *Proc. 23nd Int. Conf. on the Physics of Semiconductors, Berlin 1996*, edited by M. Scheffler and R. Zimmerman (World Scientific, Singapore, 1996), Vol. 1, p. 337.

[256] K. Reimann, S. Rübenacke, and M. Steube, *Solid State Commun.* **96,** 279 (1995).
[257] N. O. Lipari and A. Baldereschi, *Phys. Rev. B* **6,** 3764 (1972).
[258] M. Sondergeld and R. G. Stafford, *Phys. Rev. Lett.* **35,** 1529 (1975).
[259] A. R. Adams, M. Silver, E. P. O'Reilly, B. Gonul, A. F. Phillips, S. J. Sweeney, and P. J. A. Thijy, *Phys. Status Solidi B* **198,** 381 (1996).
[260] D. Patel, C. S. Menoni, A. A. Bernussi, and H. Temkin, *Phys. Status Solidi B* **198,** 375 (1996).
[261] P. Stepinski, Y. Tyagur, T. Sosin, and W. A. Trzeciakowski, in *High Pressure Science and Technology — 1995 (Proc. AIRAPT Conf., Warsaw 1995),* edited by W. A. Trzciakowski (World Scientific, Singapore, 1996), p. 651.
[262] D. Patel, C. S. Menoni, H. Temkin, C. Tome, R. A. Logan, and D. Coblentz, *J. Appl. Phys.* **74,** 737 (1993).
[263] A. R. Adams, *High Pressure Res.* **3,** 43 (1990).
[264] J. E. Epler, R. W. Kaliski, N. Holonyak, Jr., M. J. Peanasky, G. A. Herrmannsfeldt, H. G. Drickamer, R. D. Burnham, and R. L. Thornton, *J. Appl. Phys.* **57,** 1495 (1985).
[265] H. B. Bebb and E. W. Williams, in *Semiconductors and Semimetals,* Vol. 8, *Transport and Optical Phenomena,* edited by R. K. Willardson and A. C. Beer (Academic Press, New York, 1972), p. 181.
[266] J. C. Inkson, *J. Phys. C* **9,** 1177 (1976).
[267] A. R. Goñi, unpublished.
[268] V. A. Vilkotskii, D. S. Domansvskii, R. D. Kakanov, V. V. Krasovaskii, and V. D. Tkarchev, *Phys. Status Solidi B* **91,** 71 (1979).
[269] S. Bendapudi and D. Bose, *Appl. Phys. Lett.* **42,** 287 (1982).
[270] S. Ernst, A. R. Goñi, K. Syassen, and M. Cardona, *Phys. Rev. B* **53,** 1287 (1996).
[271] S. Ernst, A. R. Goñi, K. Syassen, and M. Cardona, *J. Phys. Chem. Solids* **56,** 567 (1995).
[272] D. Olego, M. Cardona, and H. Müller, *Phys. Rev. B* **22,** 894 (1980).
[273] S. Ernst, A. R. Goñi, K. Syassen, and K. Eberl, *Phys. Rev. Lett.* **72,** 4029 (1994).
[274] R. Sooryakumar, A. Pinczuk, A. C. Gossard, D. S. Chemla, and L. J. Sham, *Phys. Rev. Lett.* **58,** 1150 (1987).
[275] A. R. Goñi, S. Ernst, K. Syassen, and K. Eberl, *J. Phys. Chem. Solids* **56,** 367 (1995).
[276] A. R. Goñi, S. Ernst, K. Syassen, and K. Eberl, in *High Pressure Science and Technology — 1995 (Proc. AIRAPT Conf., Warsaw 1995),* edited by W. A. Trzciakowski (World Scientific, Singapore, 1996), p. 634.
[277] A. R. Goñi, K. Syassen, and K. Eberl, in *Proc. 23nd Int. Conf. on the Physics of Semiconductors, Berlin 1996,* edited by M. Scheffler and R. Zimmermann (World Scientific, Singapore, 1996), p. 2303.
[278] B. A. Weinstein, S. K. Hark, C. Mailhiot, and C. H. Perry, *Superlatt, Microstruct.* **3,** 273 (1987).
[279] W. Zhou, C. H. Perry, and J. M. Worlock, *Phys. Rev. B* **42,** 9657 (1990).
[280] F. Engelbrecht, J. Zeman, G. Wellenhofer, C. Peppermüller, R. Helbig, G. Martinez, and U. Rössler, *Phys. Status Solidi B* **198,** 81 (1996).
[281] P. Y. Yu and B. Welber, *Solid State Commun.* **25,** 209 (1978).
[282] D. Patel, J. Chen, S. R. Kurtz, J. M. Olson, J. H. Quigley, M. J. Hafich, and G. Y. Robinson, *Phys. Rev. B* **39,** 10978 (1989).
[283] S. J. Hwang, W. Shan, R. J. Hauenstein, J. J. Song, M.-E. Lin, S. Strite, B. N. Sverdlov, and H. Morkoç, *Appl. Phys. Lett.* **64,** 2928 (1994).
[284] S. Kim, I. P. Herman, J. A. Tuchman, K. Doverspike, L. B. Rowland, and D. K. Gaskill, *Appl. Phys. Lett.* **67,** 380 (1995).
[285] W. Shan, T. J. Schmidt, R. J. Hauenstein, J. J. Song, and B. Goldenberg, *Appl. Phys. Lett.* **66,** 3492 (1995).

[286] T. Kobayashi, T. Tei, K. Aoki, K. Yamamoto, and K. Abe, *J. Lumin.* **24–25,** 347 (1981).
[287] C. S. Menoni, H. D. Hochheimer, and I. L. Spain, *Phys. Rev. B* **33,** 5896 (1986).
[288] M. Leroux, *Semicond. Sci. Technol.* **4,** 231 (1989).
[289] R. People, A. Jayaraman, K. W. Wecht, D. L. Sisco, and A. Y. Cho, *Appl. Phys. Lett.* **52,** 2124 (1988).
[290] J. D. Lambkin and D. J. Dunstan, *Solid State Commun.* **67,** 827 (1988).
[291] S. W. Tozer, D. J. Wolford, J. A. Bradley, D. Bour, and G. B. Stringfellow, in *Proc. 19th Int. Conf. on the Physics of Semiconductors, Warsaw 1988,* edited by W. Zawadzki (Institute of Physics, Warsaw, 1988), p. 881.
[292] A. D. Prins and D. J. Dunstan, *Semicond. Sci. Technol.* **4,** 239 (1989).
[293] I. T. Ferguson, T. P. Beales, T. S. Cheng, C. M. Sotomayor-Torres, and E. G. Scott, *Semicond. Sci. Technol.* **4,** 243 (1989).
[294] H. Kalt, K. Reimann, M. Rinker, and E. Bauser, in *Proc. 20th Int. Conf. on the Physics of Semiconductors, Thessaloniki 1990,* edited by E. M. Anastassakis and J. D. Joannopoulos (World Scientific, Singapore, 1990), p. 2498.
[295] K. Reimann, M. Holtz, K. Syassen, Y. C. Lu, and E. Bauser, *Phys. Rev. B* **44,** 2985 (1991).
[296] J. A. Tuchman, S. Kim, Z. Sui, and I. P. Herman, *Phys. Rev. B* **46,** 13371 (1992).
[297] A. Arora and T. Sakuntala, *Phys. Rev. B* **52,** 11052 (1995).
[298] A. Anastassiadou, E. Liarokapis, S. Stoyanov, and E. Anastassakis, *Solid State Commun.* **67,** 633 (1988).
[299] U. D. Venkateswaran and M. Chandrasekhar, *Phys. Rev. B* **31,** 1219 (1985).
[300] D. J. Dunstan, B. Gil, C. Priester, and K. P. Homewood, *Semicond. Sci. Technol.* **4,** 241 (1989).
[301] H. M. Cheong, J. H. Burnett, and W. Paul, *Solid State Commun.* **77,** 565 (1991).
[302] V. Lemos, J. R. Moro, Q. A. G. de Souza, and P. Motisuke, *Solid State Commun.* **60,** 853 (1986).
[303] J. Gregus, J. Nakahara, K. Takamura, J. Watanabe, K. Ando, and H. Akinaga, *J. Phys. Chem. Solids* **56,** 407 (1995).
[304] N. Kuroda and Y. Matsuda, *Phys. Status Solidi B* **198,** 61 (1996).
[305] T. Kobayashi, M. Ohtsuji, and R. S. Deol, *J. Appl. Phys.* **74,** 2752 (1993).
[306] K. Uchida, P. Y. Yu, N. Noto, and E. R. Weber, *Appl. Phys. Lett.* **64,** 2858 (1994).
[307] H. R. Chandrasekhar, M. Chandrasekhar, R. J. Thomas, E. D. Jones, and R. P. Schneider, *Superlatt. Microstruct.* **18,** 131 (1995).
[308] H. Kojima, H. Kayama, T. Kobayashi, K. Uchida, and J. Nakahara, *J. Phys. Chem. Solids* **56,** 345 (1995).
[309] J. Dong, Z. Wang, D. Lu, X. Liu, D. Sun, Z. Wang, M. Kong, and G. Li, *Appl. Phys. Lett.* **68,** 1711 (1996).
[310] J. Dong, G. Li, Z. Wang, D. Lu, X. Liu, X. Li, D. Sun, M. Kong, and Z. Wang, *J. Appl. Phys.* **79,** 7177 (1996).
[311] E. M. Baugher, M. Chandrasekhar, H. R. Chandrasekhar, H. Luo, J. K. Furdyna, and L. R. Ram-Mohan, *J. Phys. Chem. Solids* **56,** 323 (1995).
[312] S. H. Wei and A. Zunger, *Appl. Phys. Lett.* **64,** 1676 (1994).
[313] A. Niilisk, A. Laisaar, I. S. Gorban, and A. V. Slobodyanyuk, *J. Phys. Chem. Solids* **56,** 603 (1995).
[314] M. Leroux, G. Neu, and C. Verie, *Solid State Commun.* **58,** 289 (1986).
[315] H. Tanino and M. Tajima, *Phys. Rev. B* **33,** 5965 (1986).
[316] D. J. Wolford, in *Proc. 18th Int. Conf. on the Physics of Semiconductors, Stockholm,* edited by O. Engström (World Scientific, Singapore, 1986), p. 1115.
[317] M. Gerling, X. Liu, S. Nilsson, M. E. Pistol, and L. Samuelson, *Semicond. Sci. Technol.* **4,** 257 (1989).

[318] J. Leymarie, M. Leroux, and G. Neu, *Semicond. Sci. Technol.* **4**, 235 (1989).
[319] G. A. R. Lima, M. R. Sardela, Jr., A. Fazzio, and R. Mota, *Solid State Commun.* **69**, 461 (1989).
[320] X. Liu, L. Samuelson, M. E. Pistol, M. Gerling, and S. Nilsson, *Phys. Rev. B* **42**, 11791 (1990).
[321] M. Leroux, J. M. Sallese, J. Leymarie, G. Neu, and P. Gibart, *Semicond. Sci. Technol.* **6**, 514 (1991).
[322] A. Kangarlu, H. Guarriello, R. Berney, and P. Y. Yu, *Appl. Phys. Lett.* **59**, 2290 (1991).
[323] M. Holtz, T. Sauncy, T. Dallas, M. Seon, C. P. Palsule, and S. Gangopadhyay, *Phys. Status Solidi B* **198**, 199 (1996).
[324] W. P. Roach, M. Chandrasekhar, H. R. Chandrasekhar, and F. A. Chambers, *Phys. Rev. B* **44**, 13404 (1991).
[325] K. Uchida, P. Seguy, H. Wong, P. L. Souza, P. Y. Yu, E. R. Weber, and K. Matsumoto, *Jpn. J. Appl. Phys.* **32**, 246 (1993).
[326] M. Gerling and L. P. Tilly, *Appl. Phys. Lett.* **62**, 2839 (1993).
[327] M. I. Eremets, O. A. Krasnovskij, V. V. Struzhkin, and A. M. Shirokov, *Semicond. Sci. Technol.* **4**, 267 (1989).
[328] H. Tisseyre, G. Nowak, M. Leszczynski, I. Grzegory, M. Bockowski, S. Krukowski, S. Porowski, M. Mayer, A. Pelzmann, M. Kamp, K. J. Ebeling, and G. Karczewski, *MRS Internet J. Nitride Semicond. Res.* **1**, 13 (1996).
[329] H. Tesseyre, B. Kozankiewicz, M. Leszczynski, I. Grzegory, T. Suski, M. Bochowski, S. Porowski, K. Pakula, P. Mensz, and I. Bhat, *Phys. Status Solidi B* **198**, 235 (1996).
[330] Z. X. Liu, A. R. Goñi, K. Syassen, H. Siegle, C. Thomsen, B. Schöttker, D. J.As, and D. Schikora, Pressure and temperature effects on the optical transitions in cubic GaN, unpublished.
[331] A. Stapor, A. Kozanecki, K. Reimann, K. Syassen, J. Weber, M. Moser, and F. Scholz, *Acta Phys. Pol. A* **79**, 315 (1991).
[332] K. Takarabe, *Phys. Status Solidi B* **198**, 211 (1996).
[333] B. A. Weinstein, T. M. Ritter, D. Strachan, M. Li, H. Luo, M. Tamargo, and R. Park, *Phys. Status Solidi B* **198**, 167 (1996).
[334] J. M. Lang, Z. A. Dreger, and H. G. Drickamer, *J. Solid State Chem.* **106**, 144 (1993).
[335] I.-H. Choi and P. Y. Yu, Pressure dependence of defects and $p-d$ hybridization in chalcopyrite semiconductors, submitted.
[336] F. Bassani, *Electronic States and Optical Transitions in Solids* (Pergamon, London, 1975).
[337] P. Perlin, W. Knap, J. Camassel, A. Polian, J. C. Chervin, T. Suski, I. Grzegory, and S. Porowski, *Phys. Status Solidi B* **198**, 223 (1996).
[338] S. Y. Ren, J. D. Dow, and D. J. Wolford, *Phys. Rev. B* **25**, 664 (1982).
[339] D. J. Wolford, J. A. Bradley, K. Fry, J. Thompson, and H. E. King, in *Proc. Int. Symp. GaAs and Related Compounds, Albuquerque, USA 1982*, (The Institute of Physics, New York, 1983), p. 477.
[340] U. D. Venkateswaran, M. Chandrasekhar, H. R. Chandrasekhar, B. A. Vojak, F. A. Chambers, and J. M. Meese, *Phys. Rev. B* **33**, 8416 (1986).
[341] S. Minomura, *AIP Conf. Proc.* **309**, Pt. 1, p. 569.
[342] Z. X. Liu, A. R. Goñi, K. Brunner, K. Eberl, and K. Syassen, *Phys. Status Solidi B* **198**, 315 (1996).
[343] U. D. Venkateswaran, M. Chandrasekhar, H. R. Chandrasekhar, T. Wolfram, R. Fischer, W. T. Masselink, and H. Morkoç, *Phys. Rev. B* **31**, 4106 (1985).
[344] B. A. Weinstein, S. K. Hark, and R. D. Burnham, *J. Appl. Phys.* **58**, 4662 (1985).
[345] U. D. Venkateswaran, M. Chandrasekhar, H. R. Chandrasekhar, B. A. Vojak, F. A. Chambers, and J. M. Meese, *Superlatt. Microstruct.* **3**, 217 (1987).

[346] M. A. Gell, D. Ninno, M. Jaros, D. J. Wolford, T. F. Keuch, and J. A. Bradley, *Phys. Rev. B* **35,** 1196 (1987).
[347] D. J. Wolford, T. F. Kuech, T. W. Steiner, J. A. Bradley, M. A. Gell, D. Ninno, and M. Jaros, *Superlatt. Microstruct.* **4,** 525 (1988).
[348] M. G. W. Alexander, M. Nido, K. Reimann, and W. W. Rühle, *Appl. Phys. Lett.* **55,** 2517 (1989).
[349] V. A. Wilkinson, J. D. Lambkin, A. D. Prins, and D. J. Dunstan, *High Pressure Res.* **3,** 57 (1990).
[350] M. Nido, M. G. W. Alexander, K. Reimann, K. Ploog, W. W. Rühle, and K. Köhler, *Surf. Sci.* **229,** 195 (1990).
[351] P. Harrison, I. Morrison, J. P. Hagon, and M. Jaros, *Superlatt. Microstruct.* **10,** 421 (1991).
[352] Z. X. Liu, G. H. Li, H. X. Han, Z. P. Wang, and D. S. Jiang, *Jpn. J. Appl. Phys.* **32,** 125 (1993).
[353] H. M. Cheong, J. H. Burnett, W. Paul, P. F. Hopkins, and A. C. Gossard, *Phys. Rev. B* **49,** 10444 (1994).
[354] P. Perlin, T. P. Sosin, W. Trzeciakowski, and E. Litwin-Staszewska, *J. Phys. Chem. Solids* **56,** 411 (1995).
[355] G. W. Smith, M. S. Skolnick, A. D. Pitt, I. L. Spain, C. R. Whitehouse, and D. C. Herbert, *J. Vac. Sci. Technol. B* **7,** 306 (1989).
[356] J. D. Lambkin, A. R. Adams, D. J. Dunstan, P. Dawson, and C. T. Foxon, *Phys. Rev. B* **39,** 5546 (1989).
[357] G. Li, D. S. Jiang, H. Han, Z. Wang, and K. Ploog, *Phys. Rev. B* **40,** 10430 (1989).
[358] M. S. Skolnick, G. W. Smith, I. L. Spain, C. R. Whitehouse, D. C. Herbert, D. M. Whittaker, and L. J. Reed, *Phys. Rev. B* **39,** 11191 (1989).
[359] Y. Masumoto, Y. Kinoshita, O. Shimomura, and K. Takemura, *Phys. Rev. B* **40,** 11772 (1989).
[360] M. Holtz, K. Syassen, R. Muralidharan, and K. Ploog, *Phys. Rev. B* **41,** 7647 (1990).
[361] R. Cingolani, M. Holtz, R. Muralidharan, K. Ploog, K. Reimann, and K. Syassen, *Surf. Sci.* **228,** 217 (1990).
[362] G. H. Li, D. S. Jiang, H. Han, Z. Wang, and K. Ploog, *J. Lumin.* **46,** 261 (1990).
[363] M. Holtz, R. Cingolani, K. Reimann, R. Muralidharan, K. Syassen, and K. Ploog, *Phys. Rev. B* **41,** 3641 (1990).
[364] K. Takarabe, S. Hitomi, T. Yoshimura, S. Minomura, H. Kato, Y. Watanabe, and M. Nakayama, *Semicond. Sci. Technol.* **6,** 465 (1991).
[365] M. Jaros, L. D. Brown, and I. Morrison, *Semicond. Sci. Technol.* **6,** 417 (1991).
[366] M. Leroux, N. Grandjean, B. Chastaingt, C. Deparis, G. Neu, and J. Massies, *Phys. Rev. B* **45,** 11846 (1992).
[367] M. Nakayama, K. Imazawa, K. Suyama, I. Tanaka, and H. Nishimura, *Phys. Rev. B* **49,** 13564 (1994).
[368] W. R. Tribe, P. C. Klipstein, R. Grey, J. S. Robert, and G. W. Smith, *J. Phys. Chem. Solids* **56,** 429 (1995).
[369] V. A. Wilkinson, A. D. Prins, J. D. Lambkin, E. P. O'Reilly, D. J. Dunstan, L. K. Howard, and M. T. Emeny, *Phys. Rev. B* **42,** 3113 (1990).
[370] H. Q. Hou, L. J. Wang, R. M. Tang, and J. M. Zhou, *Phys. Rev. B* **42,** 2926 (1990).
[371] W. Shan, X. M. Fang, D. Li, S. Jiang, S. C. Shen, H. Q. Hou, W. Feng, and J. M. Zhou, *Phys. Rev. B* **43,** 14615 (1991).
[372] S. G. Lyapin, M. I. Eremets, O. A. Krasnovskii, A. M. Shirokov, V. D. Kulakovskii, T. G. Andersson, and Z. G. Chen, *Superlatt. Microstruct.* **10,** 301 (1991).
[373] G. H. Li, A. R. Goñi, K. Syassen, O. Brandt, and K. Ploog, *Phys. Rev. B* **50,** 18420 (1994).

[374] R. People, A. Jayaraman, S. K. Sputz, J. M. Vandenberg, D. L. Sivco, and A. Y. Cho, *Phys. Rev. B* **45,** 6031 (1992).
[375] Z. X. Liu, G. H. Li, H. X. Han, and Z. P. Wang, *Solid State Electron.* **37,** 885 (1994).
[376] M. F. Whitaker, D. J. Dunstan, M. Missous, and L. Gonzales, *Phys. Status Solidi B* **198,** 349 (1996).
[377] J. L. Sly and D. J. Dunstan, *Phys. Rev. B* **53,** 10116 (1996).
[378] M. Leroux, M. L. Fille, B. Gil, J. P. Landesman, and J. C. Garcia, *Phys. Rev. B* **47,** 6465 (1993).
[379] W. Shan, S. J. Hwang, J. J. Song, H. Q. Hou, and C. W. Tu, *Phys. Rev. B* **47,** 3765 (1993).
[380] J. S. Lambkin, D. J. Dunstan, E. P. O'Reilly, and B. R. Butler, *J. Crystal Growth* **93,** 323 (1988).
[381] C. N. Yeh, L. E. McNeil, R. E. Nahory, and R. Bhat, *Phys. Rev. B* **52,** 14682 (1995).
[382] A. D. Prins, J. L. Sly, A. T. Meney, D. J. Dunstan, E. P. O'Reilly, A. R. Adams, and A Valster, in *Proc. 22nd Int. Conf. Physics of Semiconductors, Vancouver 1994,* edited by D. J. Lockwood (World Scientific, Singapore, 1995), Vol. 1, p. 719.
[383] J. W. Cockburn, O. P. Kowalski, D. J. Mowbray, M. S. Skolnick, R. Teissier, and M. Hopkinson, in *Proc. 22nd Int. Conf. Physics of Semiconductors, Vancouver 1994,* edited by D. J. Lockwood (World Scientific, Singapore, 1995), Vol. 1, p. 747.
[384] D. Patel, K. Interholzinger, P. Thiagarajan, G. Y. Robinson, and C. S. Menoni, *Phys. Status Solidi B* **198,** 337 (1996).
[385] C. Ulrich, S. Ves, A. R. Goñi, A. Kurtenbach, K. Syassen, and K. Eberl, *Phys. Rev. B* **52,** 12212 (1995).
[386] K. Uchida, N. Miura, and H. Kukimoto. *Physica B* **227,** 352 (1996).
[387] R. J. Warburton, T. P. Beales, N. J. Mason, R. J. Nicholas, and P. J. Walker, *Semicond. Sci. Technol.* **6,** 527 (1991).
[388] R. J. Warburton, R. J. Nicholas, N. J. Mason, P. J. Walker, A. D. Prins, and D. J. Dunstan, *Phys. Rev. B* **43,** 4994 (1991).
[389] Y. Yamada, Y. Masumoto, T. Taguchi, and K. Takemura, *Phys. Rev. B* **44,** 1801 (1991).
[390] M. Lomascolo, G. H. Li, K. Syassen, R. Cingolani, and I. Suemune, *Phys. Rev. B* **50,** 14635 (1994).
[391] B. Gil, D. J. Dunstan, J. Calatayud, H. Mathieu, and J. P. Faurie, *Phys. Rev. B* **40,** 5522 (1989).
[392] A. D. Prins, B. Gil, D. J. Dunstan, and J. P. Faurie, *High Pressure Res.* **3,** 63 (1990).
[393] M. Zigone, H. Roux-Buisson, H. Tuffigo, N. Magnea, and H. Mariette, *Semicond. Sci. Technol.* **6,** 454 (1991).
[394] V. A. Wilkinson, D. E. Ashenford, B. Lunn, and D. J. Dunstan, *High Pressure Res.* **3,** 72 (1990).
[395] H. P. Zhou, C. M. Sotomayor-Torres, B. Lunn, and D. E. Ashenford, *Proc. SPIE Int. Soc. Opt. Eng.* **1675,** 203 (1992).
[396] R. Meyer, M. Dahl, G. Schaak, A. Waag, and R. Boehler, *Solid State Commun.* **96,** 271 (1995).
[397] H. X. Han, Z. X. Liu, Z. P. Wang, J. Q. Zhang, Z. L. Peng, and S. X. Yuang, *J. Phys. Chem. Solids* **56,** 389 (1995).
[398] H. M. Cheong, J. H. Burnett, W. Paul, P. M. Young, Y. Lansari, and J. F. Schetzina, *Phys. Rev. B* **48,** 4460 (1993).
[399] T. Arai, T. Inokuma, T. Makino, and S. Onari, Jpn. *J. Appl. Phys.* **32,** Suppl. 32 (1993).
[400] J. Schröder and P. Pearsans, *J. Lumin.* **70,** 69 (1996).
[401] G. S. Spencer, J. Menéndez, L. N. Pfeiffer, and K. W. West, *Phys. Rev. B* **52,** 8205 (1995).
[402] W. Ge, M. D. Sturge, W. D. Schmidt, L. N. Pfeiffer, and K. W. West, *Appl. Phys. Lett.* **57,** 55 (1990).

[403] M.-H. Meynadier, R. E. Nahory, J. M. Worlock, M. C. Tamargo, J. L. de Miguel, and M. D. Sturge, *Phys. Rev. Lett.* **60,** 1338 (1988).
[404] K. Aoki, E. Anastassakis, and M. Cardona, *Phys. Rev. B* **30,** 681 (1984).
[405] M. Holtz, U. D. Venkateswaran, K. Syassen, and K. Ploog, *Phys. Rev. B* **39,** 8458 (1989).
[406] F. H. Pollak and H. Shen, *Superlatt. Microstruct.* **6,** 203 (1989).
[407] O. J. Glembocki, *Proc. SPIE* **1286,** 2 (1990).
[408] O. J. Glembocki and B. V. Shanabrook, in *Semiconductors and Semimetals,* Vol. 36, edited by D. G. Sailer (Academic Press, New York, 1992), p. 221.
[409] A. Kangarlu, H. R. Chandrasekhar, M. Chandrasekhar, F. A. Chambers, B. A. Vojak, and J. M. Meese, *Superlatt. Microstruct.* **2,** 569 (1986).
[410] W. Shan, X. M. Fang, D. Li, S. Jiang, S. C. Shen, H. Q. Hou, W. Feng, and J. M. Zhou, *Appl. Phys. Lett.* **57,** 475 (1990).
[411] F. Cerdeira and M. Cardona, *Solid State Commun.* **7,** 879 (1969).
[412] D. E. Aspnes, *Solid State Commun.* **8,** 267 (1970).
[413] D. E. Aspnes, in *Optical Properties of Solids (Handbook on Semiconductors, Vol. 2),* edited by M. Balkanski (North Holland, Amsterdam, 1980).
[414] R. Bendorius and A. Shileika, *Solid State Commun.* **8,** 1111 (1970).
[415] D. E. Aspnes, *Surf. Sci.* **37,** 418 (1973).
[416] M. Chandrasekhar, H. R. Chandrasekhar, A. Kangarlu, and U. D. Venkateswaran, *Superlatt. Microstruct.* **4,** 107 (1988).
[417] A. Kangarlu, H. R. Chandrasekhar, M. Chandrasekhar, Y. M. Kapoor, F. A. Chambers, B. A. Vojak, and J. M. Meese, *Phys. Rev. B* **38,** 9790 (1988).
[418] B. Rockwell, H. R. Chandrasekhar, M. Chandrasekhar, F. H. Pollak, H. Shen, L. L. Chang, W. I. Wang, and L. Esaki, *Surf. Sci.* **228,** 322 (1990).
[419] C. Ulrich, A. R. Goñi, K. Eberl, and K. Syassen, in *High Pressure Science and Technology — 1995 (Proc. AIRAPT Conf., Warsaw 1995),* edited by W. A. Trzciakowski (World Scientific, Singapore, 1996), p. 647.
[420] D. Z.-Y. Ting and Y.-C. Chang, *Phys. Rev. B* **36,** 4359 (1987).
[421] P. Lefebvre, B. Gil, and H. Mathieu, *Phys. Rev. B* **35,** 5630 (1987).
[422] Y. R. Lee, A. K. Ramdas, L. A. Kolodziejksi, and R. L. Gunshor, *Phys. Rev. B* **38,** 13143 (1988).
[423] A. R. Goñi, A. Cantarero, K. Syassen, and M. Cardona, *Phys. Rev. B* **41,** 10111 (1990).
[424] B. Rockwell, H. R. Chandrasekhar, M. Chandrasekhar, A. K. Ramdas, M. Kobayashi, and R. L. Gunshor, *Phys. Rev. B* **44,** 11307 (1991).
[425] M. S. Boley, R. J. Thomas, M. Chandrasekhar, H. R. Chandrasekhar, A. K. Ramdas, M. Kobayashi, and R. L. Gunshor, *J. Appl. Phys.* **74,** 4136 (1993).
[426] R. J. Thomas, M. S. Boley, H. R. Chandrasekhar, M. Chandrasekhar, C. Parks, A. K. Ramdas, J. Han, M. Kobayashi, and R. L. Gunshor, *Phys. Rev. B* **49,** 2181 (1994).
[427] R. J. Thomas, H. R. Chandrasekhar, M. Chandrasekhar, E. Jones, and R. P. Schneider, Jr., *J. Phys. Chem. Solids* **56,** 357 (1995).
[428] M. Chandrasekhar and H. R. Chandrasekhar, *High Pressure Res.* **9,** 57 (1992).
[429] M. Chandrasekhar and H. R. Chandrasekhar, *Philos. Mag. B* **70,** 369 (1994).
[430] G. Martinez, in *Optical Properties of Semiconductors (NATO ASI Series, Series E: Appl. Sci, V. 228),* edited by G. Martinez (Kluver, Dordrecht, 1993), p. 1.
[431] A. Raymond, C. Chaubet, D. Dur, W. Knap, W. Zawadzki, and J. P. Andre, *Jpn. J. Appl. Phys.* **32,** 78 (1993).
[432] G. Abstreiter, M. Cardona, and A. Pinczuk, in *Light Scattering in Solids IV,* edited by M. Cardona and G. Güntherodt (Springer, Berlin, 1984), p. 5.
[433] R. D. Mattuck, *A Guide to Feynman Diagrams in the Many-Body Problem* (Dover, New York, 1992).

[434] T. Lindhard, *Dan. Mat. Fys. Medd.* **28,** 881 (1954).
[435] N. D. Mermin, *Phys. Rev. B* **1,** 2362 (1970).
[436] T. Ando, A. B. Fowler, and F. Stern, *Rev. Mod. Phys.* **54,** 457 (1982).
[437] T. Ando, Z. Physik B **26,** 263 (1977).
[438] S. Katayama and T. Ando, *J. Phys. Soc. Jpn.* **54,** 1615 (1985).
[439] D. A. Dahl and L. J. Sham, *Phys. Rev. B* **16,** 651 (1977).
[440] A. C. Tselis and J. J. Quinn, *Phys. Rev. B* **29,** 3318 (1984).
[441] A. Pinczuk, S. Schmitt-Rink, G. Danan, J. P. Valladares, L. N. Pfeiffer, and K. W. West, *Phys. Rev. Lett.* **63,** 1633 (1989).
[442] D. Gammon, B. V. Shanabrook, J. C. Ryan, and D. S. Katzer, *Phys. Rev. B* **41,** 12311 (1990).
[443] M. V. Klein, in *Light Scattering in Solids I, Introductory Concepts,* edited by M. Cardona (Springer, Berlin, 1983), p. 147.
[444] W. Hayes and R. Loudon, *Scattering of Light by Crystals* (Wiley, New York, 1978).
[445] F. Wooten, *Optical Properties of Solids* (Academic Press, New York, 1972).
[446] D. C. Hamilton and A. L. McWhorter, in *Light Scattering Spectra of Solids,* edited by G. B. Wright (Springer, Berlin, 1969), p. 309.
[447] P. Wolff, *Phys. Rev.* **171,** 436 (1968).
[448] A. Mooradian and G. B. Wright, *Phys. Rev. Lett.* **16,** 999 (1966).
[449] A. Mooradian, in *Festkörperprobleme IX* (Pergamon-Vieweg, Braunschweig, 1969), p. 74.
[450] A. Pinczuk and G. Abstreiter, in *Light Scattering in Solids V,* edited by M. Cardona and G. Güntherodt (Springer-Verlag, Berlin, 1989), p. 153.
[451] A. R. Goñi, A. Pinczuk, J. S. Weiner, J. M. Calleja, B. S. Dennis, L. N. Pfeiffer, and K. W. West, *Phys. Rev. Lett.* **67,** 3298 (1991).
[452] R. Decca, A. Pinczuk, S. Das Sarma, B. Dennis, L. N. Pfeiffer, and K. W. West, *Phys. Rev. Lett.* **72,** 1506 (1994).
[453] V. I. Belitsky, A. Cantarero, and S. T. Pavlov, *Solid State Commun.* **94,** 589 (1995).
[454] A. Pinczuk, S. G. Louie, B. Welber, J. C. Tsang, and J. A. Bradley, in *Proc. 14th Int. Conf. Phys. Semicond.,* edited by B. L. H. Wilson (The Institute of Physics, London, 1979), p. 1191.
[455] S. Ernst, Ph.D. Thesis, Universität Stuttgart, 1994.
[456] G. Abstreiter, R. Trommer, M. Cardona, and A. Pinczuk, *Solid State Commun.* **30,** 703 (1979).
[457] M. Bugajaski and W. Lewandowski, *J. Appl. Phys.* **57,** 521 (1985).
[458] A. Raymond, J. L. Robert, and C. Bernard, *J. Phys. C* **12,** 2289 (1979).
[459] B. A. Weinstein and R. Zallen, in *Light Scattering in Solids IV,* edited by M. Cardona and G. Güntherodt (Springer, Berlin, 1984), p. 463.
[460] G. D. Pitt, *J. Phys. C* **6,** 1586 (1973).
[461] I. Gorczyca, N. E. Christensen, and M. Alouani, *Phys. Rev. B* **39,** 7705 (1989).
[462] C. S. Menoni and I. L. Spain, *Phys. Rev. B* **35,** 7520 (1987).
[463] A. Onton, R. J. Chiotka, and Y. Yacoby, in *Proceedings 11th Int. Conf. Phys. Semiconductors* (Polish Scientific, Warsaw, 1972), p. 1023.
[464] P. Perlin, T. Suski, H. Teisseyre, M. Leszczynski, I. Grzegory, J. Jun, S. Porowski, P. Boguslawski, J. Bernholc, J. C. Chervin, A. Polian, and T. D. Moustakas, *Phys. Rev. Lett.* **75,** 296 (1995).
[465] R. Price, P. M. Platzman, and S. He, *Phys. Rev. Lett.* **70,** 339 (1993).
[466] B. Tanatar and D. Ceperly, *Phys. Rev. B* **39,** 5005 (1989).
[467] S. Das Sarma and I. K. Marmorkos, *Phys. Rev. B* **47,** 16343 (1993).
[468] I. K. Marmorkos and S. Das Sarma, *Phys. Rev. B* **48,** 1544 (1993).

[469] S. T. Chui and B. Tanatar, *Phys. Rev. Lett.* **74,** 458 (1995).
[470] L. Robert, *Physica Scripta* **T39,** 235 (1991).
[471] G. Berthold, T. Suski, J. Smoliner, R. Maschek, E. Gornik, G. Bohm, and G. Weimann, *Semicond. Sci. Technol.* **8,** 1512 (1993).
[472] L. H. Dmowski, A. Zduniak, E. Litwin-Staszewska, S. Contreras, W. Knap, and J. L. Robert, *Phys. Status Solidi B* **198,** 283 (1996).
[473] M. D. Pashley, K. W. Heberern, and R. M. Feenstra, *J. Vac. Sci. Technol. B* **10,** 1874 (1992).
[474] S. L. Chuang, M. S. C. Luo, S. Schmitt-Rink, and A. Pinczuk, *Phys. Rev. B* **46,** 1897 (1992).
[475] A. Gold, *Appl. Phys. Lett.* **54,** 2100 (1989).
[476] A. L. Efros, *Solid State Commun.* **70,** 253 (1989).
[477] A. R. Goñi, K. Syassen, P. Grambow, and K. Eberl, unpublished.
[478] S. Das Sarma and P. I. Tamborenea, *Phys. Rev. Lett.* **73,** 1971 (1994).
[479] P. I. Tamborenea and S. Das Sarma, *Phys. Rev. B* **49,** 16821 (1994).
[480] F. A. Reboredo and C. R. Proetto, *Phys. Rev. Lett.* **79,** 463 (1997).
[481] L. Swierkowski and D. Neilson, *Phys. Rev. Lett.* **67,** 240 (1991).
[482] J. K. Jain and S. Das Sarma, *Phys. Rev. B* **36,** 5949 (1987).
[483] M. Berz, J. M. Walker, P. von Allmen, E. F. Steigmeier, and F. K. Reinhart, *Phys. Rev. B* **42,** 11957 (1990).
[484] A. Schmeller, A. R. Goñi, A. Pinczuk, J. S. Weiner, J. M. Calleja, B. S. Dennis, L. N. Pfeiffer, and K. W. West, *Phys. Rev. B* **49,** 14778 (1994).
[485] E. Y. Tonkov, *High Pressure Phase Transformations — A Handbook* (Gordon and Breach, Philadelphia, 1992).
[486] M. I. McMahon, R. J. Nelmes, H. Liu, and S. A. Belmonte, *Phys. Rev. Lett.* **77,** 1781 (1996).
[487] S. Minomura and H. G. Drickamer, *J. Phys. Chem. Solids* **23,** 451 (1962).
[488] G. A. Samara and H. G. Drickamer, *J. Phys. Chem. Solids* **23,** 457 (1962).
[489] A. Ohtani, M. Motobayashi, and A. Onodera, *Phys. Lett. A* **75,** 435 (1980).
[490] B. Battlog, A. Jayaraman, J. E. Van Cleve, and R. G. Maines, *Phys. Rev. B* **27,** 3920 (1983).
[491] X. S. Zhao, J. Schroeder, T. G. Bilodeau, and L. G. Hwa, *Phys. Rev. B* **40,** 1257 (1989).
[492] S. Ves, U. Schwarz, N. E. Christensen, K. Syassen, and M. Cardona, *Phys. Rev. B* **42,** 9113 (1990).
[493] W. Andreoni and K. Maschke, *Phys. Rev. B* **22,** 4816 (1980).
[494] M. Hanfland, M. Alouani, K. Syassen, and N. E. Christensen, *Phys. Rev. B* **38,** 12864 (1988).
[495] R. J. Nelmes, M. I. McMahon, N. G. Wright, and D. R. Allan, *Phys. Rev. Lett.* **73,** 1805 (1994).
[496] Z. Wang and K. Syassen, unpublished.
[497] G. D. Lee and J. Ihm, *Phys. Rev. B* **53,** 7622 (1996).
[498] G. D. Lee, Chiduck Hwang, M. H. Lee, and Jisoon Ihm, *J. Phys.: Condens. Matter* **9,** 6619 (1997).
[499] U. Oelke, Z. Wang, and K. Syassen, unpublished.
[500] P. L. Smith and J. E. Martin, *Phys. Rev. Lett.* **19,** 541 (1965).
[501] R. G. Greene, H. Luo, and A. L. Ruoff, *J. Phys. Chem. Solids* **56,** 521 (1995).
[502] G. Itkin, G. R. Hearne, E. Sterer, and M. P. Pasternak, *Phys. Rev. B* **51,** 3195 (1995).
[503] T. Zhou, U. Schwarz, M. Hanfland, Z. X. Liu, K. Syassen, and M. Cardona, *Phys. Rev. B,* in press.
[504] V. V. Shchennikov, *Phys. Solid State* **35,** 401 (1993).
[505] Y. K. Vohra, Hui Xia, and A. L. Ruoff, *Appl. Phys. Lett.* **57,** 2666 (1990).

[506] R. G. Greene, H. Luo, and A. L. Ruoff, *J. Appl. Phys.* **76,** 7296 (1994).
[507] U. Schwarz, A. R. Goñi, K. Syassen, and A. Chevy, *High Pressure Res.* **8,** 396 (1991).
[508] U. Schwarz, K. Syassen, and R. Kniep, *J. Alloys and Compounds* **224,** 212 (1995).
[509] J. Jamieson, *Science* **139,** 762 (1963).
[510] J. Wittig, *Z. Phys.* **195,** 215 (1966).
[511] H. G. Zimmer, Master's Thesis, Universität Düsseldorf, 1983.
[512] M. I. McMahon and R. J. Nelmes, *Phys. Rev. B* **47,** 8337 (1993).
[513] H. Olijnyk, S. K. Sikka, and W. B. Holzapfel, *Phys. Lett.* **103A,** 137 (1984).
[514] J. Z. Hu, L. D. Merkle, C. S. Menoni, and I. L. Spain, *Phys. Rev. B* **34,** 4679 (1986).
[515] S. J. Duclos, Y. K. Vohra, and A. L. Ruoff, *Phys. Rev. Lett.* **58,** 775 (1987).
[516] K. J. Chang, M. M. Dacorogna, M. L. Cohen, J. M. Mignot, G. Chouteau, and G. Martinez, *Phys. Rev. Lett.* **54,** 2375 (1985).
[517] T. H. Lin, W. Y. Dong, K. J. Dunn, and C. N. J. Wagner, *Phys. Rev. B* **33,** 7820 (1985).
[518] D. Erskine, P. Y. Yu, K. J. Chang, and M. L. Cohen, *Phys. Rev. Lett.* **57,** 2741 (1986).
[519] M. Hanfland, K. Syassen, M. Alouani, and N. E. Christensen, *Semicond. Sci. Technol.* **4,** 250 (1989).
[520] N. W. Ashcroft and K. Sturm, *Phys. Rev. B* **3,** 1898 (1971).
[521] N. W. Ashcroft and K. Sturm, *Phys. Rev. B* **24,** 2315 (1981).
[522] L. D. Landau and E. M. Lifshitz, *Electrodynamics of Continuous Media* (Pergamon Press, New York, 1963).
[523] F. Stern, *Optical Properties of Solids, Solid State Physics* Vol. 15 (Academic Press, New York, 1963), p. 299.
[524] J. A. Stratton, *Electromagnetic Theory* (McGraw-Hill, New York, 1941).
[525] N. W. Ashcroft and N. D. Mermin, *Solid State Physics* (Holt, Rinehart and Winston, New York, 1976).
[526] J. J. Hopfield and D. G. Thomas, Phys. Rev. **132,** 563 (1963).
[527] V. M. Agranovich and V. L. Ginzburg, *Crystal Optics with Spatial Dispersion and Excitons* (Springer, Berlin, 1984).
[528] J. S. Toll, *Phys. Rev.* **104,** 1760 (1956).
[529] M. Cardona, in *Optical Properties of Solids,* edited by S. Nudelmann and S. Mitra (Plenum Press, New York, 1969), p. 137.
[530] B. O. Seraphin, in *Semiconductors and Semimetals,* Vol. 9, *Modulation Techniques,* edited by R. K. Willardson and A. Beer (Academic Press, New York, 1972).
[531] B. O. Seraphin and N. Bottka, *Phys. Rev.* **145,** 628 (1966).
[532] J. S. Plaskett and P. Schatz, *J. Chem. Phys.* **38,** 612 (1963).
[533] J. Humlicek, E. Schmidt, L. Bocank, R. Svehla, and K. Ploog, *Phys. Rev. B* **48,** 5241 (1993).

CHAPTER 5.1

Hydrostatic Pressure and Uniaxial Stress in Investigations of the EL2 Defect in GaAs

Pawel Trautman, Michal Baj, and Jacek M. Baranowski

INSTITUTE OF EXPERIMENTAL PHYSICS
WARSAW UNIVERSITY
WARSAW, POLAND

I. INTRODUCTION 427
II. BASIC PROPERTIES OF THE EL2 DEFECT IN GaAs 428
   1. Occurrence of the EL2 Defect in GaAs 428
   2. Optical Absorption due to EL2 430
   3. Metastable Properties of the EL2 Defect in GaAs 433
   4. Discovery of As_{Ga} Antisite by EPR Spectroscopy 435
III. PIEZOSPECTROSCOPIC INVESTIGATIONS OF NO-PHONON LINES OF EL2 ... 436
IV. STUDIES OF THE EL2 DEFECT UNDER HYDROSTATIC PRESSURE 440
V. MICROSCOPIC NATURE OF THE METASTABLE STATE OF EL2 445
   1. Theoretical Model of the Metastability of EL2 445
   2. Determination of Symmetry of EL2 in the Metastable State 447
VI. CONCLUSIONS 453
   References 453

I. Introduction

The EL2 defect is a dominant native deep donor in GaAs crystals grown by both Czochralski and Bridgman methods. It is one of the few observed intrinsic defects in III–V semiconducting compounds. The most characteristic feature of this defect is that it can be transformed to an excited metastable state through illumination of the crystal at low temperature.

EL2 is the most important native point defect in bulk GaAs crystals. The reasons for it are both technical and scientific.

The presence of EL2 in GaAs crystals makes possible fabrication of semi-insulating (SI) GaAs material, that is, crystals of high resistivity ($\approx 10^8$ Ω cm) with a free carrier concentration of about 10^7 cm^{-3} at room temperature. Such a low carrier concentration cannot be achieved solely by purification. The semi-insulating property of undoped GaAs results from pinning of the Fermi level by the midgap donor level of EL2. Semi-insulating GaAs serves as substrate material for high-speed digital and microwave monolithic integrated circuits (MMICs). GaAs semiconductor devices have essentially replaced other technologies and materials for low-power devices operating in the 2–40 GHz frequency range. A review of applications of GaAs and in particular of EL2 compensated semi-insulating GaAs can be found in Chapters 20, 21, and 22 of *Properties of Gallium Arsenide,* 3rd edition (Brozel and Stillman, 1996).

The scientific interest in EL2 comes mainly from its rare property of having an excited metastable state, into which EL2 can be transformed through illumination of the crystal at low temperature. The main purpose of scientific research on EL2 was to determine its microscopic structure and to reveal the mechanism of metastability. The puzzling property of metastability together with technological importance have made this defect one of the most intensively studied defects in any semiconductor.

In this chapter, we review the information on EL2 obtained from studies of its properties with application of hydrostatic pressure and uniaxial stress. These investigations contributed greatly to the determination of the microscopic structure of EL2 and to the explanation of the mechanism of its metastability. The remainder of this chapter is organized as follows. Part II provides a brief introduction to the basic properties of EL2. In Part III, we describe how symmetry of EL2 in the normal state was determined by standard piezospectroscopic measurements performed on a weak no-phonon line of the EL2 intracenter absorption and later confirmed by piezospectroscopy of a sharp luminescence line. Part IV outlines the discovery of the acceptor level of EL2 in the metastable state by investigations of EL2 under hydrostatic pressure and briefly describes properties of EL2 connected with the presence of this acceptor level. Part V presents the determination of symmetry of EL2 in the metastable state by a novel method consisting of investigation of the thermal recovery of EL2 from its metastable state under uniaxial stress. Part VI concludes this chapter.

II. Basic Properties of the EL2 Defect in GaAs

1. OCCURRENCE OF THE EL2 DEFECT IN GaAs

The name EL2, short for electron level number 2, was given in 1977 to a deep trap having characteristic electron emission parameters E_a (activation energy) and σ_a (emission cross-section) determined by deep-level transient spectroscopy (DLTS) experiments (Martin *et al.*, 1977). These parameters for EL2 are $E_a = 0.82$ eV and $\sigma_a \approx 10^{-13}$ cm^2. The capture cross-section for electrons σ_n was directly measured (Mitonneau *et al.*, 1979) in a large temperature range (50–270 K) and is given by

$$\sigma_n(\text{cm}^2) = 5 \times 10^{-19} + 6 \times 10^{-15} \times \exp(-0.066 \text{ eV}/kT).$$

Thus, it is strongly temperature dependent and, in order to determine the true value of the EL2 thermal ionization energy E_i, one has to correct the DLTS activation energy by subtracting the activation energy of σ_n (Lang and Logan, 1975). This gives $E_i = 0.82 - 0.07 \text{ eV} = 0.75 \text{ eV}$, that is, a value equal to the ionization energy obtained from measurements (Gooch *et al.*, 1961) of the Hall effect on semi-insulating GaAs. EL2 is commonly present in a concentration of about 10^{16} cm^{-3} in GaAs crystals grown by both liquid-encapsulated Czochralski and Bridgman methods. The midgap level due to EL2 was, for the first time, directly detected by Williams in 1966 at concentrations around 10^{16} cm^{-3} in measurements of capacitance of Schottky barriers produced on *n*-type GaAs.

Semi-insulating GaAs is now commonly obtained by the LEC technique without intentional doping (Holmes *et al.*, 1982). The concentration of EL2 determined by measurements of optical absorption in crystals grown from pyrolytic boron nitride crucibles, starting with a charge of high-purity Ga and As, and using dry B_2O_3 encapsulant, was found to increase from 5×10^{15} to 1.7×10^{16} cm^{-3} as the As atom fraction in the melt increases from 0.48 to 0.52 (Holmes *et al.*, 1982). A similar dependence of EL2 concentration on growth conditions was observed in crystals grown by the horizontal Bridgman technique; the concentration of EL2, determined by the DLTS measurements, was found to increase from 10^{16} to 3×10^{16} cm^{-3} as the As source temperature increases from 613 to 619°C (Lagowski *et al.*, 1982). The EL2 defect is also present in GaAs layers grown by vapor-phase epitaxy (VPE) (Miller *et al.*, 1977; Bhattacharya *et al.*, 1980; Samuelson *et al.*, 1981). Its concentration in these layers was observed to be roughly proportional to the As/Ga ratio in the vapor phase during growth, ranging from 2×10^{13} cm^{-3} for an As/Ga ratio of 1/3 to 4×10^{14} cm^{-3} for an As/Ga ratio of 12.

It is now well established that the EL2 defect is not related to any chemical impurity and that it is more abundant in crystals grown under arsenic-rich conditions.

The fact that EL2 is a native defect is confirmed by an observation of EL2 in GaAs layers grown by molecular beam epitaxy (MBE) at low substrate temperatures (LT GaAs). Measurements of near-infrared absorption demonstrated the presence of EL2 in concentrations of about 10^{20} cm^{-3} in layers grown at 200°C (Kaminska and Weber, 1990). This absorption spectrum has a shape characteristic of EL2 and the property of optical quenching at helium temperatures and recovery during heating of the crystal at the characteristic for EL2 temperature of approximately 120 K. The concentration of EL2 in these layers is many orders of magnitude higher than the concentration of any chemical impurity.

2. Optical Absorption Due to EL2

The spectrum of near infrared absorption due to the EL2 defect was first reported by Martin in 1981. Two years later, Kaminska *et al.* (1983) found that the central band of this absorption begins with a fine structure consisting of a no-phonon line at 1.039 eV (8378 cm^{-1}) followed by phonon replicas separated by about 11 meV. These spectra measured at 10 K are shown in Figs. 1 and 2.

The absorption due to EL2 begins at 0.8 eV with photoionization transitions from the midgap donor level of EL2 to the Γ valley of the conduction band (see curve (a) in Fig. 1). At 1.1 eV, there is onset of transitions to the L valleys of the conduction band resulting in a rapid increase of the slope of absorption coefficient versus photon energy. A band of intracenter absorption beginning at 1.039 eV with the no-phonon line, and extending to about 1.3 eV, is superimposed on this photoionization background. The most characteristic feature of the optical absorption due to EL2 is that it

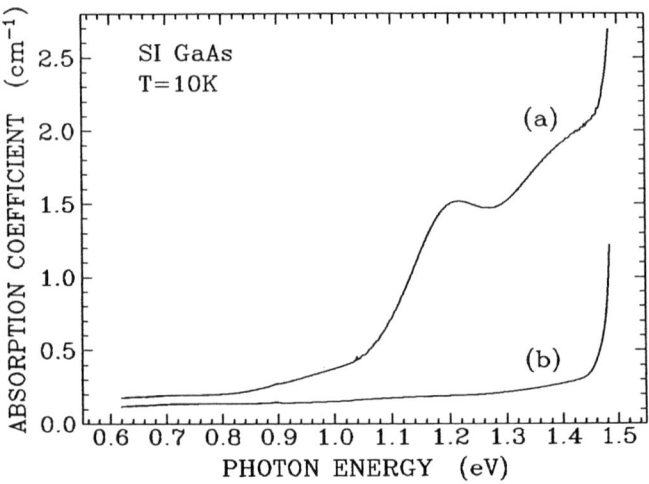

Fig. 1. Near-infrared absorption spectra measured at 10 K in semi-insulating GaAs containing the EL2 defect in concentration of about 2×10^{16} cm^{-3}. Curve (a) was recorded after cooling the sample in the dark. Curve (b) was obtained after illumination of the sample with white light for 10 min.

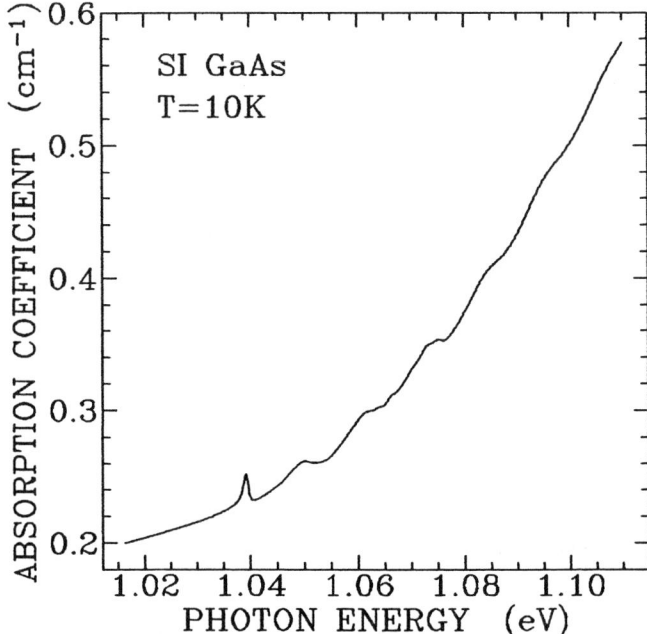

FIG. 2. Fine structure of the intracenter absorption due to the EL2 defect in GaAs. The fine structure consists of a no-phonon line at 1.039 eV and its phonon replicas. This spectrum was measured in the same sample as that shown in Fig. 1.

can be quenched by illumination of the crystal at low temperature (Martin, 1981). The absorption disappears completely after illumination of the crystal with white light for about 10 min (see curve (b) in Fig. 1).

The spectra shown in Figs. 1 and 2 are commonly observed in n-type and undoped semi-insulating GaAs. These spectra are attributed to the neutral charge state of EL2. When EL2 is partially quenched, the whole absorption band is quenched homogeneously, that is, by the same factor for each photon energy. The magnitude of the no-phonon line was observed to be proportional to that of the entire band. The magnitude of the absorption shown by curve (a) in Fig. 1 and the intensity of the no-phonon line were shown to be proportional to the concentration of EL2 determined by DLTS in n-type GaAs (Martin, 1981; Skowronski et al., 1986). Therefore, the entire absorption band including the fine structure is due to the EL2 defect.

The EL2 defect, being a double donor, may exist also in two other charge states: singly positively charged (+) and doubly positively charged (++).

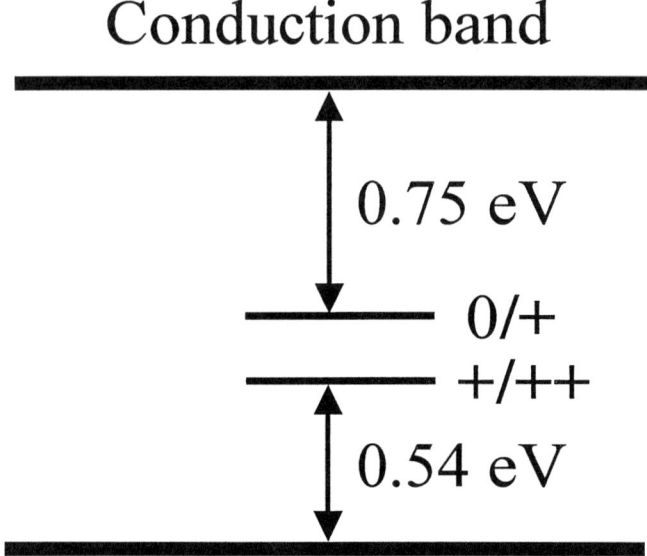

FIG. 3. Energy levels of the EL2 defect determined by photo-EPR, DLTS, and photocapacitance spectroscopies. The EL2 defect being a double donor has three (0, +, ++) charge states in the energy gap.

The positions of the two donor levels of EL2, shown in Fig. 3, were determined by photo-EPR experiments (Weber et al., 1982) and by means of DLTS and photocapacitance spectroscopies (Lagowski et al., 1985; Omling et al., 1988; Silverberg et al., 1988). The optical properties of EL2$^+$ and EL2^{++} are less well known than those of EL2°. For EL2^{++}, only transitions from the valence band to the empty +/++ level of EL2 are expected to occur, and these were observed by Skowronski (1990). The situation concerning the nature of transitions contributing to the optical absorption of EL2 is most complicated in the case of the + charge state. For this charge state, three kinds of optical transitions can occur: (i) transitions from the valence band to the empty 0/+ level of EL2, (ii) transitions from the filled +/++ level of EL2 to the conduction band, and (iii) intracenter $A_1 \rightarrow T_2$ transitions. The magnitudes of the relative contributions of these transitions to the experimentally observed spectrum of magnetic circular dichroism (MCD) due to EL2$^+$ are a matter of controversy (Meyer et al., 1984, 1985; Kaufmann, 1985; Kaufmann and Windscheif, 1988; Krambrock et al., 1992). The rare occurrence of GaAs crystals with most EL2 in the

+ charge state is one of the reasons why the optical properties of EL2$^+$ are poorly understood.

3. METASTABLE PROPERTIES OF THE EL2 DEFECT IN GaAs

The unusual metastable property of high-resistivity GaAs crystals, which is now known to be due to EL2, was first reported in a paper by Lin *et al.* (1976) on persistent quenching of photoconductivity. At the low temperature of 82 K, the crystal was converted from a high-photosensitivity state with *n*-type photoconductivity to a low-photosensitivity state with *p*-type photoconductivity by exposure to photons in the 1.0–1.3 eV energy range. Recovery from the low-sensitivity quenched state to the high-sensitivity normal state occurred during heating of the crystal at about 120 K. The observed range of photon energies inducing quenching and the temperature of recovery are characteristic of the EL2 defect. In semi-insulating GaAs the transformation of EL2 into the metastable state shifts down the position of the Fermi level in the energy gap.

Systematic studies of the metastability of EL2 began in 1977 after quenching of photocapacitance of Schottky barriers produced on *n*-type GaAs had been discovered (Bois and Vincent, 1977; Vincent and Bois, 1978). At low temperature ($\approx$77 K) and for photon energy around 1.15 eV, a peculiar behavior of transient photocapacitance associated with EL2 was observed. The measurement of photocapacitance transient begins with filling the EL2 traps with electrons by forward biasing the Schottky diode; then reverse bias is applied in the dark. After the light is switched on, a fast increase of capacitance is observed, followed by a decrease to a final value that is nearly equal to the initial value. The fall time is about 100 times longer than the rise time. In the case of a usual photocapacitance transient, the capacitance increases monotonically and then saturates as the result of emptying of the EL2 level. It is observed for EL2 for photon energy outside of the photoquenching band (e.g., for $h\nu = 1.4$ eV) and for any photon energy inducing photoionization of EL2 when the temperature is above 120 K.

The peculiar photocapacitance transient is explained by the transformation of EL2 from the normal photoactive state into a passive metastable state EL2* in which the defect has the same charge state as in the normal state. EL2* must have the same charge state as EL2^0, since the value of capacitance after completion of photoquenching, when EL2 is completely converted to the metastable state, is the same as the initial value.

The normal photoactive state can be recovered by heating the sample. There are two distinct recovery processes. The first one is observed in the absence of free electrons, that is when the diode is reverse polarized. In this case, the recovery rate is dependent only on the temperature and is given by the formula (Mitonneau and Mircea, 1979)

$$r_{th}(\text{sec}^{-1}) = 8.6 \times 10^{11} \times \exp(-0.34 \text{ eV}/kT).$$

The recovery rate is greatly accelerated when free electrons are present, that is, when forward bias is applied. In this case, it is proportional to the concentration n of free electrons and is given by the formula (Mitonneau and Mircea, 1979)

$$r_n(\text{sec}^{-1}) = 1.9 \times 10^{-14} \times n v_{th} \times \exp(-0.107 \text{ eV}/kT),$$

where v_{th} is the thermal velocity of electrons in the conduction band equal to $\approx 2 \times 10^7$ cm/sec at 77 K. This recovery is also thermally activated, but with a smaller activation energy of 0.107 eV. The total recovery rate may be written as the sum of r_{th} and r_n,

$$r(\text{sec}^{-1}) = 8.6 \times 10^{11} \times \exp(-0.34 \text{ eV}/kT) \\ + 1.9 \times 10^{-14} \times n v_{th} \times \exp(-0.107 \text{ eV}/kT).$$

In optical measurements the metastability of EL2 manifests itself as a quenching of the near-infrared absorption (see Fig. 1). The quenching is induced by illumination of the crystal at low temperature. Similarly to the photocapacitance signal, the absorption can be recovered by heating of the crystal (see Fig. 4) (Trautman et al., 1988). The recovery of EL2 from the

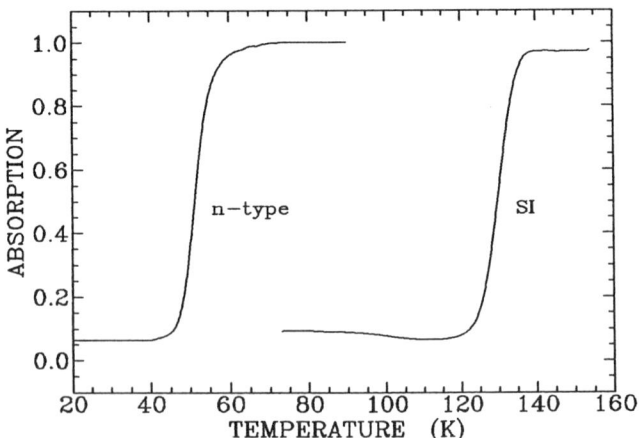

FIG. 4. Thermal recovery of absorption due to the EL2 defect measured at photon energy of 1.2 eV for n-type ($n \approx 10^{16}$ cm^{-3}) and semi-insulating (SI) GaAs during slow (≈ 2 K/min) heating of the crystal after previous quenching of the absorption with white light at 10 K.

metastable state occurs at ≈50 K in *n*-type GaAs and at ≈130 K in semi-insulating GaAs. Analysis of the rate of EL2 recovery as a function of temperature by means of Arrhenius plots for several *n*-type and semi-insulating GaAs crystals allowed the recovery rate to be expressed by

$$r(\text{sec}^{-1}) = 1.7 \times 10^{12} \exp(-0.36/kT) + 1.6 \times 10^{-9} \times n \times \exp(-0.085/kT).$$

Similar results were obtained by Fuchs and Dischler (1987). The formula for the rate *r* of EL2 recovery obtained from measurements of recovery of absorption is consistent with that obtained from measurements of the recovery of photocapacitance. Differences between experimentally determined parameters in the two formulas are in part due to experimental errors, such as errors in calibration of temperature, and in part due to different experimental conditions. In photocapacitance experiments, measurements are performed in a thin layer of semiconductor in which a gradient of free electron concentration is present and faster recovery is measured, that is, the recovery is investigated at higher temperature than in optical experiments.

A very important property of the photoquenching process is the spectral dependence of the optical cross-section σ^* for the bleaching transitions. Vincent, Bois, and Chantre (1982) determined from photocapacitance measurements that σ^* is in the form of a Gaussian band extending between about 1.0 and 1.3 eV. This covers the same energy range as the intracenter absorption band observed in optical absorption measurements.

Metastability of EL2 was also observed as a quenching of the luminescence bands peaked at ≈0.62 and ≈0.68 eV (Leyral *et al.*, 1982; Yu, 1984; Nissen *et al.*, 1990). The excitation spectrum of quenching of luminescence extends over the same 1.0–1.3 eV energy range as the band of intracenter transitions in the absorption spectrum. The luminescence signal returns to its initial value after heating of the crystal above ≈120 K.

4. DISCOVERY OF As_{Ga} ANTISITE BY EPR SPECTROSCOPY

In 1980, the arsenic antisite was discovered in as-grown semi-insulating GaAs by submillimeter electron paramagnetic resonance (EPR) spectroscopy (Wagner *et al.*, 1980). The corresponding spectrum consists of four nearly equally spaced lines and is isotropic. The four lines were interpreted as originating from the hyperfine interaction of an unpaired electron with a central nucleus of spin $I = 3/2$. The spectrum can be described by the spin Hamiltonian

$$H = g\mu_B \mathbf{B} \cdot \mathbf{S} + A\mathbf{S} \cdot \mathbf{I}.$$

The experimentally determined (Wagner et al., 1980) parameters are $g = 2.04 \pm 0.01$ and $|A| = 0.090 \pm 0.001$ cm^{-1}. As the arsenic isotope ^{75}As is 100% abundant and has nuclear spin $I = 3/2$, the observed spectrum may be due to the positively charged arsenic antisite As$_{Ga}^+$. This identification was confirmed by comparison, with appropriate scaling, of the hyperfine interaction parameter A with that for the phosphorus antisite P$_{Ga}$ in GaP, where super-hyperfine interaction with ligands is resolved, allowing unambiguous identification of the defect.

The identification of the As$_{Ga}$ antisite with EL2 was postulated on the basis of correlation between the intensity of the EPR spectrum and the concentration of carbon acceptor (Elliott et al., 1984). It was based on the following argument: If EL2 and As$_{Ga}$ were not the same defect, then the EPR signal would be independent of the changes in residual acceptor content, since these would change the occupation of the EL2 defect, the compensating center, but not the occupation of As$_{Ga}$. The identity of EL2 with As$_{Ga}$ was further supported by the similarity between the enhancement and quenching spectra of the EPR signal due to As$_{Ga}^+$ and the photoionization spectra of EL2 (Baeumler et al., 1985). The presence of the As$_{Ga}^+$ EPR signal in undoped semi-insulating GaAs and its ability to be enhanced and quenched by illumination of the crystal indicates that the $0/+$ level of As$_{Ga}$ coincides, within a few milli-electron volts, with the $0/+$ midgap level of EL2. Indeed, if this were not the case, then all the As$_{Ga}$ antisites would be in the same charge state and either the entire EPR signal or no signal at all would be observed before illumination. The coincidence of the levels of EL2 and As$_{Ga}$ is a strong argument for the identity of the two defects.

III. Piezospectroscopic Investigations of No-Phonon Lines of EL2

The large width of the EPR lines due to As$_{Ga}^+$ did not allow the unambiguous determination that EL2 is due to the isolated arsenic antisite. The symmetry of EL2 in the normal state was first determined by Kaminska, Skowronski, and Kuszko (1985) on the basis of measurements of splittings of the no-phonon line (also termed zero-phonon line, ZPL) of EL2 under uniaxial stress applied in several high-symmetry crystallographic directions. The results of their experiment indicated that the no-phonon line is due to the $A_1 \rightarrow T_2$ transition in a center of tetrahedral T_d symmetry. Therefore, EL2 has to be a simple isolated point defect, but not a complex of lower than tetrahedral symmetry. Combining this determination of symmetry with previous experimental and theoretical data, Kaminska et al. (1985) identified the EL2 defect with the isolated arsenic antisite.

5.1 HYDROSTATIC PRESSURE AND UNIAXIAL STRESS INVESTIGATIONS

On the other hand, on the basis of EPR and DLTS investigations performed on GaAs samples subjected to electron irradiation and heat treatments, von Bardeleben et al. (1985, 1986) suggested that EL2 is a complex of an As$_{Ga}$ antisite and an arsenic interstitial As$_i$. This suggestion found strong experimental support in measurements of the optically detected electron nuclear double resonance (ODENDOR) (Meyer et al., 1987a, b). The conclusions of Kaminska et al. (1985) concerning the symmetry of EL2 were questioned by Figielski and Wosinski (1987). An inconsistency of the experimentally observed selection rules with those predicted theoretically for a transition within a defect of T_d symmetry was pointed out, and an alternative interpretation of the observed splittings was proposed, leading to the conclusion that EL2 may have the orthorhombic I C_{2v} symmetry. In the light of these facts, an independent uniaxial stress experiment on the no-phonon line of EL2 was performed by Trautman, Walczak, and Baranowski (1990).

Optical transmission of uniaxially stressed GaAs crystals containing the EL2 defect in concentration of $\approx 10^{16}$ cm^{-3} was measured at 10 K using monochromatic polarized light with electric field parallel (π spectrum) or perpendicular (σ spectrum) to the direction of stress. The 8378 cm^{-1} no-phonon line of EL2 was observed to split into two, two, and three components under uniaxial stress applied along the [100], [111], and [110] crystallographic directions, respectively (see Figs. 5 and 6). Exactly one stress-split component of the no-phonon line was observed in each polarization, including three nonequivalent polarizations in the case of [110] stress. The splitting of an optical transition results from the removal of orbital (electronic) or orientational degeneracy of the defect due to lowering of the cubic symmetry of the crystal by uniaxial stress. In centers having trigonal or tetragonal symmetry, combination of these two degeneracies is possible. The splitting patterns depend on the spatial symmetry of the defect and on the symmetry of the electronic states involved in the optical transition. Therefore, piezospectroscopy gives information on these symmetry properties, which, in conjunction with theoretical predictions, helps to establish the microscopic nature of the investigated defect. The observed pattern of splitting of the no-phonon line of EL2 indicates unequivocally that it is due to $A_1 \rightarrow T_2$ (or $A_2 \rightarrow T_1$) electric dipole transitions in a defect of the tetrahedral T_d symmetry (Trautman et al., 1990; Trautman and Baranowski, 1995). Uniaxial stress experiments cannot distinguish between $A_1 \rightarrow T_2$ and $A_2 \rightarrow T_1$ transitions. We will refer to the $A_1 \rightarrow T_2$ transition, as it is in agreement with theoretical predictions for the As$_{Ga}$ defect. The results of Trautman et al. (1990) are essentially in agreement with those of Kaminska et al. (1985); the main difference is that the viewing directions for [110] stress in Fig. 1 of paper by Kaminska et al. (1985) are

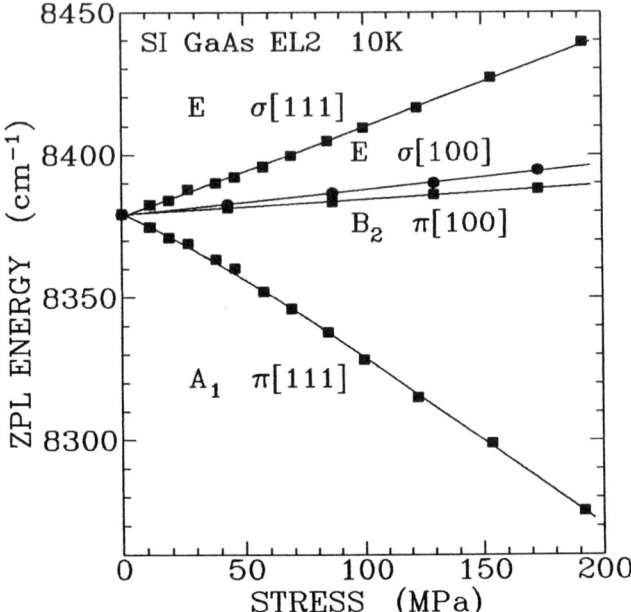

FIG. 5. The splitting of the 8378 cm^{-1} zero-phonon line (ZPL) of the EL2 intracenter absorption under [100] and [111] uniaxial stresses. The points were experimentally determined. The solid lines are theoretical fits to the experimental points. Symbols π and σ indicate components present in π and σ polarizations, respectively. Indices [100] and [111] denote components observed under [100] and [111] stress, respectively.

interchanged, as was pointed out by Figielski and Wosinski (1987) and Bergman et al. (1988).

The excited T_2 state, transitions to which give rise to the no-phonon line, is triply orbitally degenerate. This is clearly seen from the splitting pattern of the no-phonon line under [110] stress, in which case exactly one component is observed for light polarized along each of the three principal axes ([110], [001], and [1$\bar{1}$0]) of the crystal stressed in the [110] direction (see Fig. 6).

Threefold orbital degeneracy does not exist in defects of lower than T_d symmetry. Namely, in centers of trigonal or tetragonal (axial) symmetry only doubly orbitally degenerate levels are possible. In centers of still lower symmetry, only orbital singlets exist. Therefore, the EL2 defect has to have unperturbed tetrahedral T_d symmetry.

The stress-split components are due to transitions from the A_1 ground state, which does not split, to sublevels of the split T_2 state. These sublevels

5.1 Hydrostatic Pressure and Uniaxial Stress Investigations

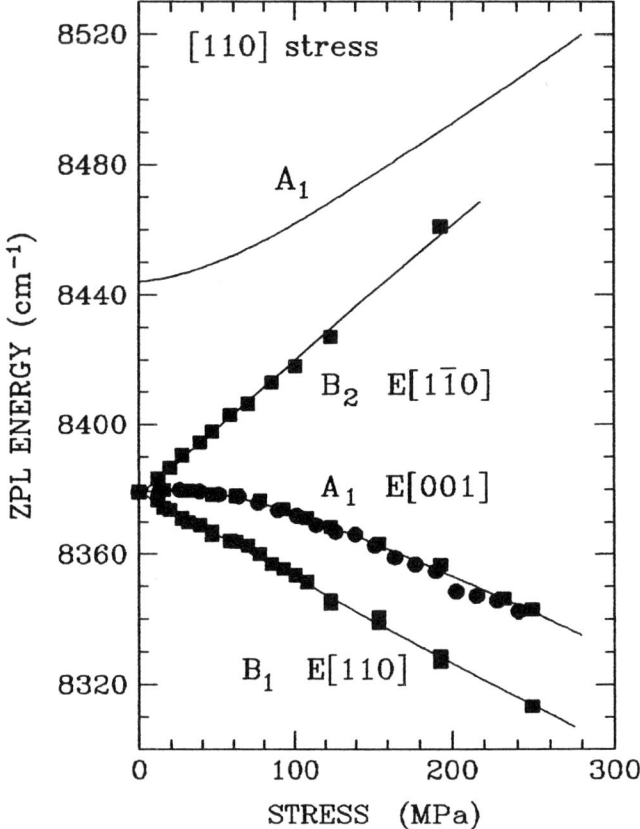

FIG. 6. The splitting of the 8378 cm^{-1} zero-phonon line (ZPL) of the EL2 intracenter absorption under [110] stress. The symbol E[hkl] indicates that incident light was polarized in the [hkl] crystallographic direction. The points represent experimental results. The solid lines are theoretical fits to the experimental points. The presence of the upper A_1 state was inferred from nonlinear variation of energy of the experimentally observed lower A_1 component.

transform as A_1, E, B_2, and B_1 irreducible representations of the point symmetry groups of stressed crystal, which are D_{2d}, C_{3v}, and C_{2v} for stress applied in the [100], [111], and [110] directions, respectively. With the help of polarization selection rules one can attribute these representations to the corresponding stress-split components as is shown in Figs. 5 and 6. The components that were observed to move nonlinearly with the stress are due to transitions from the ground A_1 state to A_1 symmetry sublevels of the split T_2 state. This indicates the presence of an A_1 state lying above the T_2 state. Stress-induced interaction with this state produces bending of the A_1 components. However, attempts to detect optical transitions to the upper A_1 state in experiments have been unsuccessful.

The A_1 and T_2 symmetry of electronic states participating in the intracenter transition of EL2, determined by the piezospectroscopic experiments, is in agreement with the theoretically predicted symmetry of electronic states of the arsenic antisite (Scherz and Scheffler, 1993), which supports the identification of EL2 with the As_{Ga} antisite.

Clear confirmation of the T_d symmetry of the EL2 defect was provided by Nissen et al. (1991) from measurements of a sharp luminescence line at 0.7028 eV under uniaxial stress. This line remained unresolved under [100], [110], and [111] stresses up to 500 MPa. The only effect of uniaxial stress on this line is a hydrostatic shift of 30 meV/GPa. The luminescence is due to transitions of an electron from an A_1 symmetry hydrogenic state bound under the Γ point minimum of the conduction band to the midgap A_1 state of EL2. The lack of any splitting of the 0.7028 eV line under uniaxial stress strongly supports the isolated As_{Ga} model of EL2, since any complex of lower than T_d symmetry should exhibit a splitting due to removal of the orientational degeneracy.

IV. Studies of the EL2 Defect under Hydrostatic Pressure

Detailed studies of the EL2 defect and its metastability under hydrostatic pressure were made by Baj, Dreszer, and Babinski (1989a, 1989b, 1991). These investigations resulted in the discovery of a great wealth of properties of EL2; in particular, the mechanism of acceleration of EL2 recovery in the presence of free electrons was explained (Dreszer and Baj, 1991). The most important of their findings is that EL2 in the metastable state has an acceptor level EL2*$^{-/0}$ lying about 15 meV above the bottom of the conduction band and that this level moves down under hydrostatic pressure entering the band-gap at about 200 MPa (Baj et al., 1991).

At atmospheric pressure the transformation of EL2 from the normal to the metastable state does not affect the electrical conduction or the concentration of free electrons in n-type GaAs. The electrical conduction of n-type GaAs was observed to decrease significantly when EL2 was transformed into its metastable state under a hydrostatic pressure of the order of a few hundred megapascals (Baj et al., 1991), which is shown in Figs. 7 and 8. The magnitude of this decrease in electrical conduction was observed to increase with increasing pressure and with increasing number of EL2 centers transformed into the metastable state. In sample no. 2,

5.1 HYDROSTATIC PRESSURE AND UNIAXIAL STRESS INVESTIGATIONS

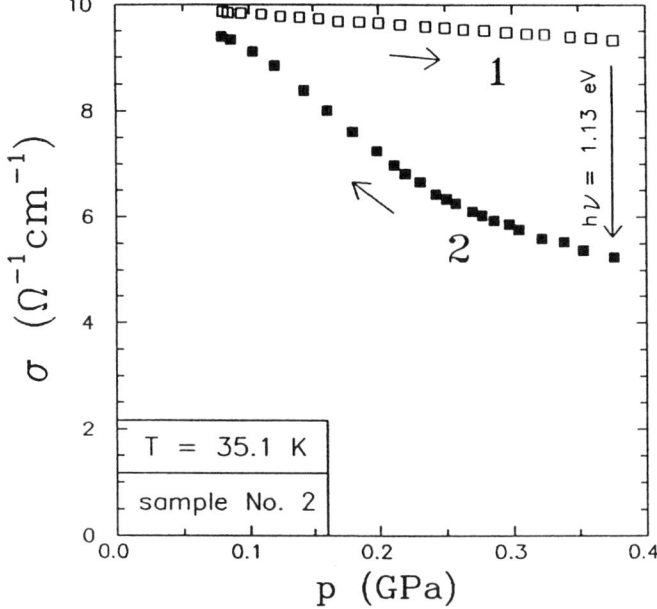

FIG. 7. Electrical conductivity of sample no. 2 versus pressure measured at $T = 35.1$ K before (curve 1) and after (curve 2) photoquenching of EL2 with $h\nu = 1.13$ eV for 10 min. Arrows indicate directions of changes of pressure.

having approximately twice as high concentration of free electrons as that of EL2, the maximum decrease of the electrical conduction was equal to about half of its value at atmospheric pressure (see Fig. 7). On the other hand, in sample no. 1, having twice as many EL2 defects as free electrons, the conduction decreased by over two orders of magnitude under 400 MPa and at 35 K (see Fig. 8). These results indicate that EL2 in the metastable state has an acceptor level EL2*$^{-/0}$ resonant with the conduction band at atmospheric pressure and that this level enters the bandgap under hydrostatic pressure and traps electrons from the conduction band.

Another property of EL2 in the metastable state, which is observed under hydrostatic pressure only in n-type GaAs, is a broad absorption band due to EL2* (Baj and Dreszer, 1989a). This absorption band most likely arises from a photoionization of EL2*$^-$ to the conduction band, because this absorption is not observed in semi-insulating GaAs, in which material EL2* remains in its neutral charge state under pressure because of the lack of free electrons that could be trapped by the EL2*$^{-/0}$ level. The strength of this absorption is similar to that of EL2^0, but its spectral dependence is distinctly different.

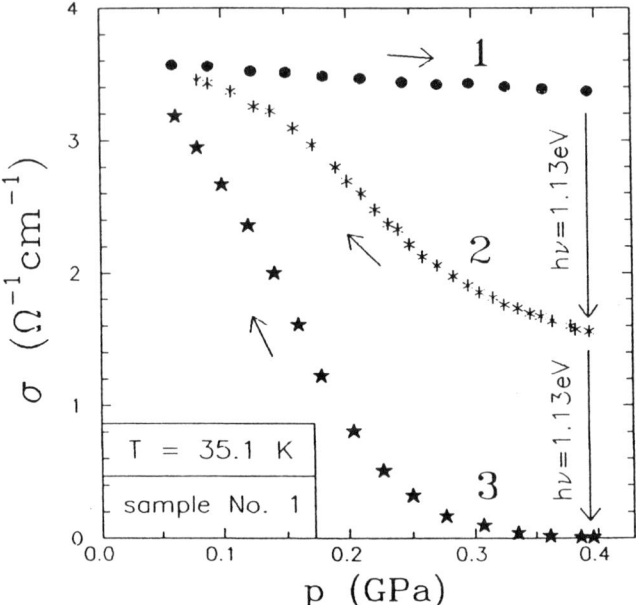

FIG. 8. Electrical conductivity of sample no. 1 versus pressure measured at $T = 35.1$ K before (curve 1), after partial (curve 2), and after complete (curve 3) photoquenching of EL2 with $h\nu = 1.13$ eV. Arrows indicate directions of changes of pressure.

Under hydrostatic pressure $p > 300$ MPa, EL2* can be effectively transformed back into its normal state without heating of the sample in both n-type and semi-insulating GaAs by illumination of the crystal with photons of energy close to the onset of band to band transitions (Baj and Dreszer, 1989a; Dreszer *et al.*, 1992). This effect for samples nos. 2 and 1 is shown in Figs. 9 and 10, respectively. This photorecovery of EL2 is most likely resulting from a capture of a photogenerated hole by a negatively charged EL2 in the metastable state EL2*$^-$ on a hydrogenic orbit. The system consisting of EL2*$^-$ and a hole bound to it by the Coulomb attraction can be considered to be equivalent to an electronically excited (by about 1.5 eV) state of EL2*^{0} because the total charge of this system is zero and the energies of binding of a free electron and then a hole by EL2*^{0} are of the order of 10 meV. This electronically excited state of EL2*^{0} may use this energy to overcome the energy barrier separating it from the normal state: EL2*$^-$ + h^+ = *excited state of* EL2*^{0} → EL2^0, in analogy with the photoinduced transformation of EL2^0 into EL2*: EL2^0 + $h\nu$ → *excited state of* EL2^0 → EL2*. Without hydrostatic pressure this mechanism of photorecov-

5.1 HYDROSTATIC PRESSURE AND UNIAXIAL STRESS INVESTIGATIONS

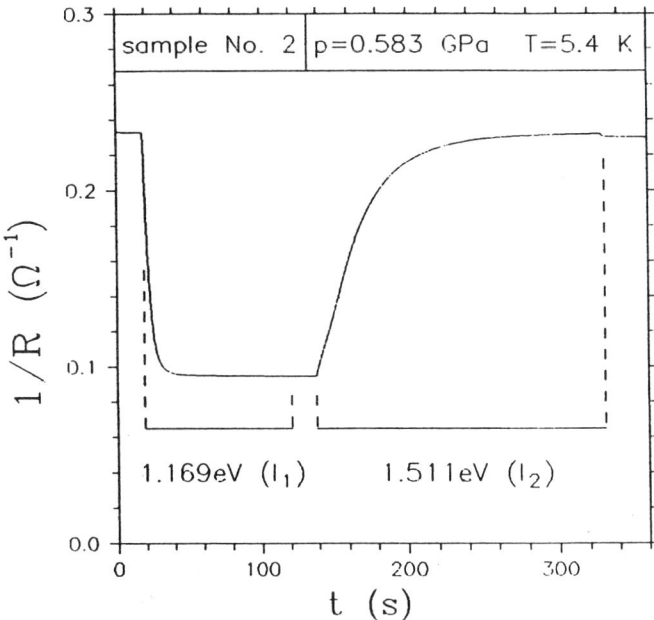

FIG. 9. Changes of electrical conductance of sample no. 2 recorded at $p = 0.583$ GPa and $T = 5.4$ K during successive optical quenching ($hv_1 = 1.169$ eV) and optical recovery ($hv_2 = 1.511$ eV) of the EL2 defect. The ratio of light intensities was $I_1 : I_2 = 5 : 1$.

ery is ineffective because the EL2$^{*-/0}$ level lying approximately 15 meV above the bottom of the conduction band is not populated with an electron.

In n-type GaAs, the EL2 defect recovers at much lower temperature (45 K) than in semi-insulating GaAs (125 K). This difference is undoubtedly connected with the presence of free electrons, since EL2 recovers at the higher temperature in Schottky barriers produced on n-type GaAs depleted from electrons by a reverse polarization, and it recovers at the lower temperature when electrons are brought back by direct polarization. The reduction of the temperature of EL2 recovery by the presence of free electrons is readily explained when the existence of the EL2$^{*-/0}$ level is taken into account. In the presence of free electrons, a fraction of the EL2$^{*-/0}$ level is filled with electrons because this level lies only 15 meV above the bottom of the conduction band. When EL2* is in the negative charge state, the energy barrier separating it from the normal configuration is smaller than when it is in the neutral charge state and EL2 recovers at the lower temperature. Therefore, the location of the EL2$^{*-/0}$ level only slightly above the bottom of the conduction band accounts for the apparently inexplicable phenomenon of thermal recovery of the same defect in the same charge

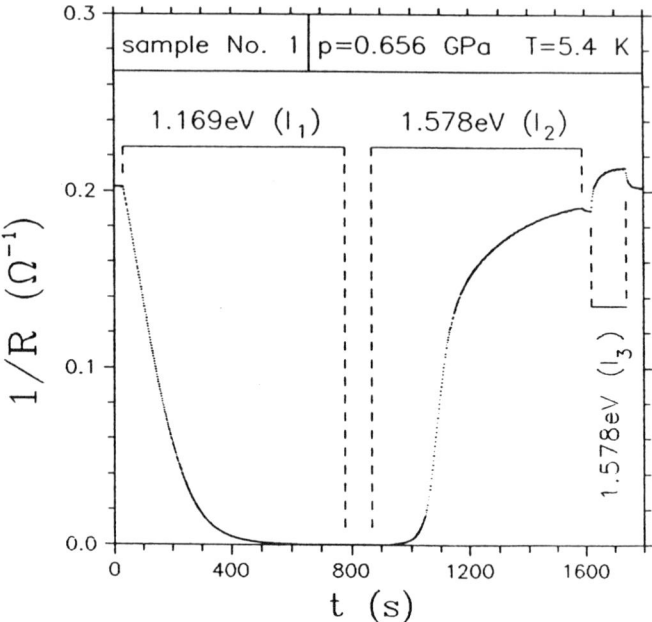

FIG. 10. Changes of electrical conductance of sample no. 1 recorded at $p = 0.656$ GPa and $T = 5.4$ K during successive optical quenching ($hv_1 = 1.169$ eV) and optical recovery ($hv_2 = 1.578$ eV) of the EL2 defect. The ratio of light intensities was $I_1:I_2:I_3 = 5:1:11$.

state at different temperatures in n-type and in semi-insulating GaAs. This difference results from a temporary population of the EL2*$^{-/0}$ level by electrons from the conduction band in n-type GaAs and complete lack of population of this level in semi-insulating GaAs.

The properties of the EL2 defect are significantly altered when a hydrostatic pressure of moderate value, on the order of 300 MPa, is applied to the sample. Under hydrostatic pressure, EL2 can be effectively photorecovered, and in n-type GaAs there appears a broad absorption band due to the negative charge state of the metastable state of EL2. These properties are contrary to those of EL2 under atmospheric pressure. These alterations of properties of EL2 by the presence of hydrostatic pressure may be the basis for explanation of the apparently different properties of EL2 in GaAs irradiated with high-energy particles and in GaAs grown at low temperature, since the local strain fields present in these materials may induce changes in the properties of EL2 similar to those due to the application of hydrostatic pressure.

The discovery of the EL2*$^{-/0}$ acceptor level paved the way for a series of novel experiments on EL2 (Przybytek et al., 1995, 1996a, 1996b). The basic idea underlying these experiments was that at sufficiently high hydrostatic pressure the gain in energy connected with the capture of an electron from the conduction band can be larger than the difference in total energy between the metastable EL2* and the normal EL2 configurations. Thus, at high pressure, the metastable EL2* configuration of C_{3v} symmetry may become the fundamental one in n-type GaAs, and the usually ground configuration of EL2 of T_d symmetry may be the excited metastable one. In order to reduce the necessary pressure to the values obtainable experimentally, Przybytek *et al.* (1995, 1996a, 1996b) have used a series of $GaAs_{1-x}P_x$ crystals with x ranging from 0 to 0.35. Alloying GaAs with P widens the energy gap of the mixed crystal, which makes the value of the pressure required in these experiments substantially smaller. One of the most spectacular of their results is that by cooling down to 28 K in the dark a $GaAs_{0.8}P_{0.2}$ sample under pressure of 0.8 GPa, they obtained EL2 in the metastable (distorted) configuration, which was manifested by the appearance of the EL2 recovery step at about 50 K during heating of the sample after the pressure was released (Przybytek *et al.*, 1995). Therefore, they accomplished the transformation of EL2 into the metastable state without the use of light, or more precisely realized conditions in which the metastable distorted configuration of EL2 is the fundamental one. Later, by studying the transformation of EL2 between its two atomic configurations as a function of pressure in a series of samples with different phosphorus content, Przybytek *et al.* (1996b) determined the difference in total energy between the metastable (C_{3v}) and normal (T_d) configurations of EL2 in GaAs to be 105 meV. This value was the last important energy parameter of the configuration coordinate diagram of EL2 to be determined experimentally. The small value of this energy at least partly accounts for the high efficiency of the optical transformation of EL2 into the metastable state.

V. Microscopic Nature of the Metastable State of EL2

1. Theoretical Model of the Metastability of EL2

When the identification of EL2 with the isolated As_{Ga}, a simple substitutional defect, was proposed, this defect was considered to be too simple to account for the metastable properties of EL2. This view was supported by the results of early theoretical calculations of lattice relaxation around As_{Ga} (Bachelet and Scheffler, 1984; Scheffler *et al.*, 1985). The turning point in the search for a model of EL2 metastability was the suggestion of Baranowski *et al.* (1986) that the transformation of EL2 into the metastable state is connected with a displacement of the central As atom of arsenic antisite to an interstitial position. Within this model, the transformation of EL2 into the metastable state more closely resembles a photodissociation of As_{Ga} into a gallium vacancy and an arsenic interstitial; $As_{Ga} \rightarrow V_{Ga}–As_i$, than the lattice relaxation usually accompanying the change of electronic state of a defect.

This new idea found strong support in the state-of-the-art theoretical calculations performed by three independent research groups: Dabrowski and Scheffler (1988, 1989), Chadi and Chang (1988), and Kaxiras and Pandey (1989). These calculations showed that the arsenic antisite defect indeed has an excited metastable configuration in which the total energy of the defect is only about 0.2 eV higher than in the ground state. During the transformation into the metastable state the central As atom breaks one of its four As–As bonds and is displaced by approximately 1.3 Å along the antibonding [111] direction to an interstitial position (see Fig. 11). Therefore, the metastable state of As_{Ga} is the tightly bound gallium-vacancy–arsenic-interstitial $V_{Ga}As_i$ defect pair. As a result of the transformation into the metastable state, two As atoms change their coordination from fourfold to threefold.

The relative stability of the metastable configuration is a consequence of the ability of group V atoms to be threefold bonded as in the case of arsine AsH_3 or various forms of solid arsenic. In consequence of the transformation into the metastable state, the symmetry of As_{Ga} is reduced from tetrahedral T_d to trigonal C_{3v}. The metastable state $V_{Ga}As_i$ is fourfold orientationally degenerate because the transformation into the metastable state may take place in four ways by breaking any one of the four bonds between the central As atom and its ligands. The transformation of As_{Ga}

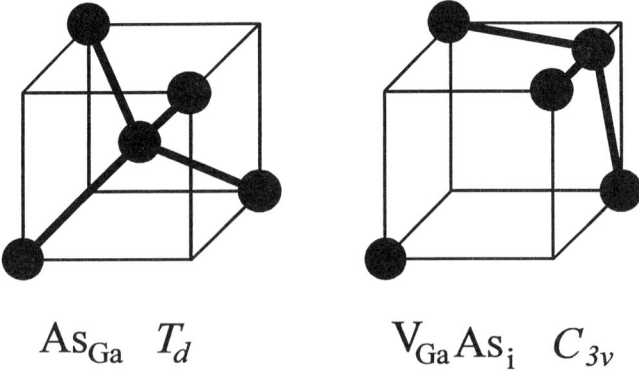

FIG. 11. Local atomic structure of the undistorted arsenic antisite As_{Ga} and the gallium-vacancy–arsenic-interstitial $V_{Ga}As_i$ defect pair that is considered to be the excited metastable state of As_{Ga} responsible for the metastability of EL2.

into the metastable state is initiated by the Jahn–Teller distortion following the excitation of As_{Ga} to the electronically degenerate T_2 state. From this distorted excited state the defect may reach the metastable state by tunneling. The theoretically calculated values of the energy barrier between the metastable and the normal configuration of As_{Ga} strongly depend on the particular method of calculation and lie in a rather broad range from 0.16 to 0.92 eV, that is, these theoretical values range from less than half to more than twice the experimental value of 0.36 eV.

6. DETERMINATION OF SYMMETRY OF EL2 IN THE METASTABLE STATE

When the transformation of As_{Ga} into $V_{Ga}As_i$ was proposed as the mechanism of EL2 metastability, the only experimental support for this model was that it successfully explained the metastability of the isolated arsenic antisite. There was no experimental evidence for this particular structure or symmetry of the metastable state.

The metastable state of EL2 could not be studied by conventional methods of investigation of defects in semiconductors because of the lack of absorption, EPR, and luminescence signals due to this state. The process of thermally activated recovery of EL2 from its metastable state is one of the few experimentally observed phenomena in which the metastable state is involved. Study of the thermal recovery of optical absorption due to the EL2 defect under uniaxial stress applied along the [111] and [100] crystallographic directions was performed by Trautman and Baranowski (1992, 1995). The idea of the experiment was that if EL2 in the metastable state (i.e., EL2*) has a symmetry lower than tetrahedral, then EL2* centers having different orientations with respect to the stress may recover at different temperatures, that is, the recovery step (see Fig. 4) may split under the uniaxial stress. From the dependence of splitting on the direction of the stress in the crystal, the symmetry of the metastable state can be determined.

The experimental procedure was as follows. The sample was first cooled to 10 K in the dark, after which the EL2 defect was excited to the metastable state by illumination with white unpolarized light. Uniaxial compressive stress was then applied. Finally, the recovery of absorption of the normal state of EL2 was measured during heating of the sample at a rate of approximately 5 and 2.6 K/min for n-type and semi-insulating GaAs samples, respectively. After the EL2 defect had completely recovered, the stress was removed.

Figure 12 shows the thermal recovery of the absorption due to the EL2 defect in semi-insulating GaAs under [111] and [100] stresses of several values from 0 to 600 MPa. The recovery step starting at about 120 K evidently splits into two components for large values of [111] stress, but no splitting is observed under [100] stress. Qualitatively the same behavior is observed for the recovery step starting at about 45 K in n-type GaAs (see Fig. 13). The presence of splitting under [111] stress and lack of splitting under [100] stress is more evident after computation of numerical derivatives of the recovery curves with respect to temperature, as is shown in Figs. 12 and 13.

The observed splitting of the recovery step under [111] stress is clear evidence for an orientational degeneracy of the EL2 defect in the metastable state (EL2*). This means that EL2 in the metastable state has lower than tetrahedral (noncubic) symmetry. This follows from the fact that if the symmetry of the EL2* was tetrahedral, then all of them would be equivalent with respect to the stress and no splitting of the recovery step would be possible. The EL2* defects are equivalent with respect to the [100] stress because there is no splitting or even broadening of the recovery step up to 600 MPa of [100] stress. Among defects having orientational degeneracy, only those having trigonal C_{3v} symmetry do not split under the [100] stress. The recovery step splits into two components under [111] stress as expected for a defect of C_{3v} symmetry. Therefore, EL2 in the metastable state has trigonal C_{3v} symmetry. The splitting of the recovery step cannot result from a removal of an electronic (orbital) degeneracy by the uniaxial stress, since the removal of the electronic degeneracy alone does not divide the EL2* defects into groups of defects which may recover at different temperatures.

A given EL2 center in the metastable state is aligned along one of the four equivalent (in the unstressed crystal) $\langle 111 \rangle$ directions. The [111] stress differentiates one ([111]) of these directions from the other three ([$\bar{1}\bar{1}1$], [$\bar{1}1\bar{1}$], and [$1\bar{1}\bar{1}$]), which results in the splitting of the recovery step. On the other hand, all four $\langle 111 \rangle$ directions are equivalent under [100] stress, which accounts for the lack of splitting under stress applied in this direction. There is no systematic shift of the recovery step under [100] stress in semi-insulating GaAs and only a small shift of about 2 K to lower temperatures in n-type GaAs, which is in agreement with the results of studies of EL2 recovery under hydrostatic pressure (Dreszer and Baj, 1991).

The numbers of EL2 defects recovering at lower and higher temperature under [111] stress amount to about one-fourth and three-fourths of the total number of the EL2 defects, as judged from the heights of the recovery steps occurring at lower and higher temperature, respectively. This implies that the defects recovering at lower temperature are oriented along the direction of the stress and those recovering at higher temperature are

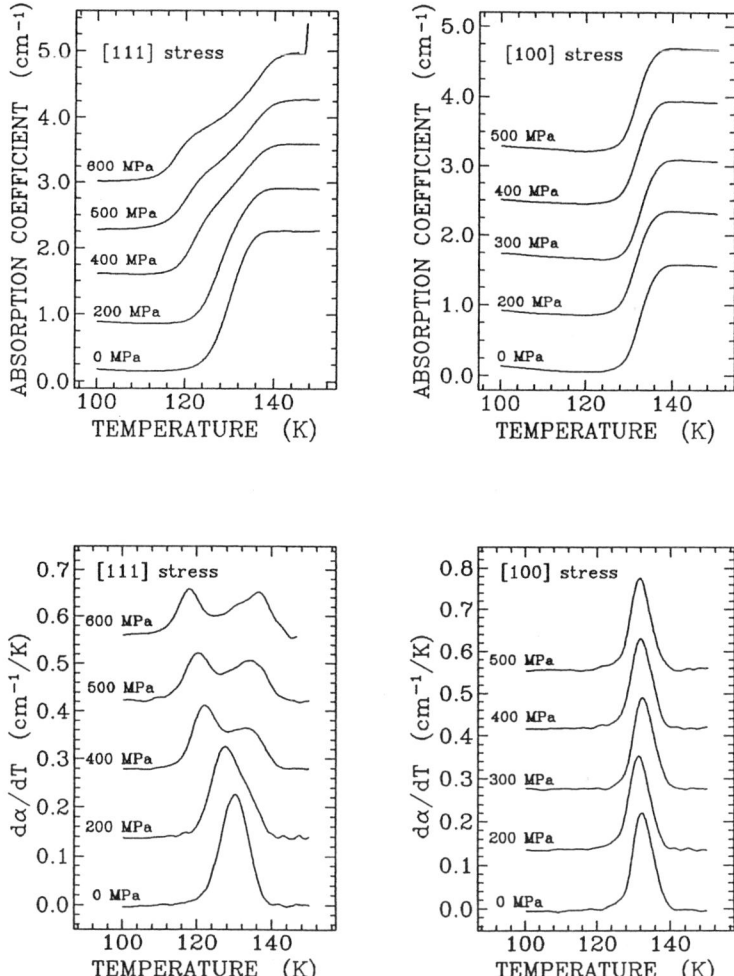

FIG. 12. Thermal recovery of absorption due to the EL2 defect in semi-insulating GaAs measured during slow heating of the crystal under uniaxial stress applied in the [111] and [100] directions (top) and computed first derivatives of these curves with respect to temperature (bottom). Curves for different values of the stress are shifted vertically for clarity. The abrupt increase of absorption at the end of the curve measured under [111] stress of 600 MPa was caused by a fracture of the sample. The recovery step splits into two components under [111] stress, but no splitting is observed under [100] stress; this indicates that the metastable state of EL2 has trigonal C_{3v} symmetry.

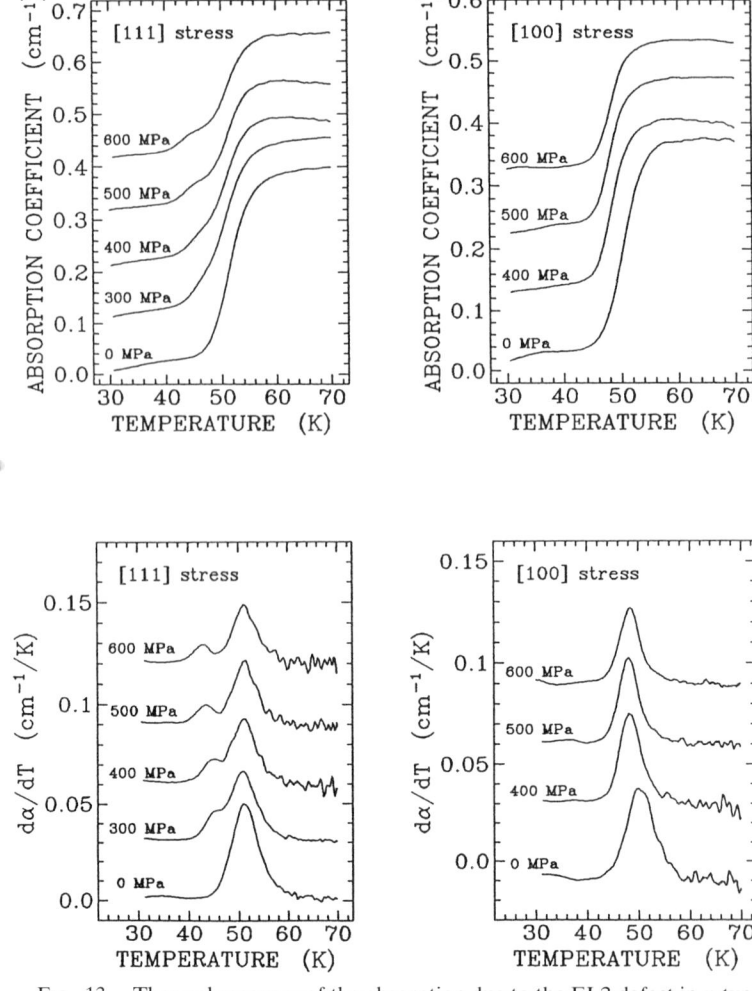

FIG. 13. Thermal recovery of the absorption due to the EL2 defect in *n*-type GaAs measured during slow heating of the crystal under uniaxial stress applied in the [111] and [100] directions (top) and computed first derivatives of these curves with respect to temperature (bottom). Curves for different values of the stress are shifted vertically for clarity.

oriented at a slant to the direction of the stress, assuming that the EL2* defects approximately equally populate different orientations. The EL2 defects in the metastable state do not reorient during the heating of the crystal under stress, since we observe the splitting of the recovery step. This means that the barrier for reorientation of the metastable state is larger than the barrier between the metastable and the normal configurations. The total increase of absorption in *n*-type GaAs during the recovery under stress is significantly smaller than that with no stress applied (see Fig. 13). This is most likely due to the fact that EL2 in the metastable state in *n*-type GaAs absorbs light when the sample is under pressure (Baj and Dreszer, 1989a). This effect is absent in semi-insulating GaAs (see Fig. 12).

Trautman and Baranowski (1992, 1995) have also studied the effect of different methods of excitation of EL2 to the metastable state on the splitting of the recovery step in semi-insulating GaAs (see Fig. 14). The curves (a–f) were all measured under [111] stress of 500 MPa; thus, the magnitude of the splitting of the recovery step should be the same in each of these curves. It is clearly seen that the ratio of the number of EL2 defects recovering at lower temperature to that recovering at higher temperature strongly depends on the method of excitation of EL2 to the metastable state. Curve (a) shows, for reference, the recovery after excitation of EL2

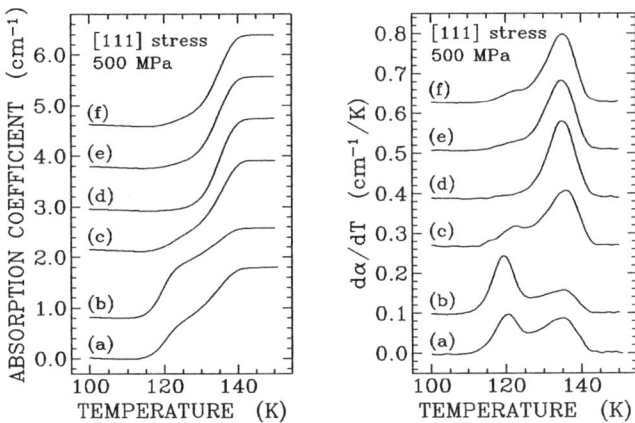

FIG. 14. Thermal recovery of the absorption due to the EL2 defect in semi-insulating GaAs measured under [111] uniaxial stress of 500 MPa (on the left) and the computed first derivatives of these curves with respect to temperature (on the right). Curves (a) show the recovery after excitation of EL2 to the metastable state with unpolarized light; curves (b) and (c) were obtained after excitation with light polarized parallel (b) and perpendicularly (c) to the direction of the stress, which was applied after the excitation; curve (d) was recorded after excitation performed under [111] stress of 500 MPa with unpolarized light. During excitation in cases (a), (b), and (c) no stress was applied. Curve (e) was obtained after excitation under [111] stress of 500 MPa followed by excitation with no stress applied. Curve (f) was measured after excitation without stress followed by excitation under [111] stress of 500 MPa. In the cases (e) and (f), unpolarized light was used for excitation. The curves are shifted vertically for clarity.

to the metastable state with unpolarized light and with no stress applied. Excitation with light polarized parallel to the [111] direction results in an increased number of EL2* centers oriented along this direction, whereas excitation with light polarized perpendicularly results in a decreased number of EL2* centers oriented along the [111] direction (see curves (b) and (c) in Fig. 14). All the EL2* centers became oriented at a slant to the [111] direction of the stress after excitation performed under [111] stress of 500 MPa (see curve (d) in Fig. 14).

When the EL2 centers were first transferred to the metastable state by illumination with unpolarized light, and therefore became oriented both along and at a slant to the [111] direction, and the crystal was then illuminated under stress of 500 MPa, the centers finally became all oriented at a slant to the direction of the stress (see curve (f) in Fig. 14). When the order of these two illuminations was reversed, the centers were also oriented aslant the [111] direction (see curve (e) in Fig. 14), that is, the second illumination performed without stress had no effect on the orientation of EL2* centers. This is an indication that photorecovery of EL2 is present under [111] stress of 500 MPa, and that it is absent when the stress is not applied.

The fact that after excitation of EL2 to the metastable state under [111] stress of 500 MPa no EL2* centers are oriented along the direction of the stress finds an explanation in the splitting of the acceptor level of EL2*. The process of photorecovery of EL2 becomes effective under pressure when the acceptor level of EL2* enters the energy gap (Baj and Dreszer, 1989a; Baj et al., 1991). The EL2* centers oriented along the [111] direction of the stress have the acceptor level lying at a lower energy than the centers oriented at a slant, which results in a more efficient optical recovery of these centers than of those oriented at a slant. This difference in the rates of optical recovery results in a preferential orientation of the EL2* centers at a slant to the direction of the stress after excitation performed under stress. The phenomenon of optical recovery also explains the transformation of EL2* centers oriented along the [111] direction into centers oriented at a slant as the result of illumination under [111] stress of 500 MPa. This effect is represented by curve (f) in Fig. 14. Photorecovery is negligible without stress, which results in the lack of an effect of illumination without stress on the orientation of EL2* centers, as evidenced by the similarity of curves (e) and (d) in Fig. 14.

In conclusion, Trautman and Baranowski (1992, 1995) showed that the EL2 defect in the metastable state has trigonal C_{3v} symmetry. It was observed that one and the same EL2 defect in the metastable state can be oriented in different directions depending on the method of transformation of EL2 into the metastable state. Namely, the EL2* defects can be oriented mostly along the [111] direction by excitation of EL2 to the metastable state with light polarized parallel to this direction, and they can be oriented completely at a slant to the [111] direction by excitation under [111] stress of 500 MPa. It is much easier to understand the ability of the EL2 defect to assume different orientations in the metastable state assuming that EL2 in the normal state has perfect tetrahedral symmetry than when EL2 is attributed to a complex of lower symmetry because the orientation of the complex in the metastable state is expected to be predetermined by its orientation in the normal state rather than by the polarization of light used to excite EL2 to the metastable state. Therefore, the investigation of recovery of absorption due to EL2 under uniaxial stress not only allowed the determination of the C_{3v} symmetry of EL2 in the metastable state, but also supports the identification of the EL2 defect with a defect of unperturbed tetrahedral symmetry.

VI. Conclusions

Much progress has been made in the understanding of the electrical, optical, and structural properties of the EL2 defect in GaAs since the discovery of its unusual metastable properties in 1977 began intensive studies of this defect. In the early 1980s, EL2 was probably the most puzzling defect in a semiconductor because its universally occurring metastability could not be reconciled with the simple microscopic structure of this defect. Now, it is well established that EL2 is due to the arsenic antisite As_{Ga} and the nature of its metastability is well understood. Much information on EL2, in particular that related to its microscopic structure and the mechanism of metastability, was obtained from the study of its properties with application of hydrostatic pressure and uniaxial stress. These investigations helped to establish that the metastability of EL2 is due to the optically inducible transformation $As_{Ga} \rightarrow V_{Ga}As_i$ of an arsenic antisite into a gallium-vacancy–arsenic-interstitial defect of trigonal symmetry.

References

Bachelet, G. B., and Scheffler, M. (1984). *Proc. 17th Int. Conf. Phys. of Semiconductors*, p. 755 (eds. Chadi, J. D., and Harrison, W. A.). Springer-Verlag, New York.
Baeumler, M., Kaufmann, U., and Windscheif, J. (1985). *Appl. Phys. Lett.* **46,** 781.
Baj, M., and Dreszer, P. (1989a). *Phys. Rev.* **B39,** 10470.
Baj, M., and Dreszer, P. (1989b). *Proc. 15th Int. Conf. Defects in Semiconductors* (ed. Ferenczi, G.), *Mater. Sci. Forum* **38–41,** 101.
Baj, M., Dreszer, P., and Babinski, A. (1991). *Phys. Rev.* **B43,** 2070.

Baranowski, J. M., Kaminska, M., Kuszko, W., Walczak, J. P., Trautman, P., and Jezewski, M. (1986). *Proc. 14th Int. Conf. Defects in Semiconductors* (ed. von Bardeleben, H. J.), *Mater. Sci. Forum* **10–12,** 317.
Bergman, K., Omling, P., Samuelson, L., and Grimmeiss, H. G. (1988). *Proc. 5th Conf. Semi-insulating III–V Materials,* p. 397 (eds. Grossmann, G., and Ledebo, L.). Adam Hilger, Bristol and Philadelphia.
Bhattacharya, P. K., Ku, J. W., Owen, S. J. T. Aebi, V., Cooper, III, C. B., and Moon, R. L. (1980). *Appl. Phys. Lett.* **36,** 304.
Bois, D., and Vincent, G. (1977). *J. Phys. Lett. (France)* **38,** L351.
Brozel, M. R., and Stillman, G. E. (eds.) (1996). *Properties of Gallium Arsenide,* 3rd ed., EMIS Datareviews Series No. 16. INSPEC, IEE, London.
Chadi, D. J., and Chang, K. J. (1988). *Phys. Rev. Lett.* **60,** 2187.
Dabrowski, J., and Scheffler, M. (1988). *Phys. Rev. Lett.* **60,** 2183.
Dabrowski, J., and Scheffler, M. (1989). *Phys. Rev. B* **40,** 10391.
Dreszer, P., and Baj, M. (1991). *J. Appl. Phys.* **70,** 2679.
Dreszer, P., Baj, M., and Korzeniewski, K. (1992). *Proc. 16th Int. Conf. Defects in Semiconductors* (eds. Davies, G., Deleo, G., and Stavola, M.), *Mater. Sci. Forum* **83–87,** 875.
Elliott, K., Chen, R. T., Greenbaum, S. G., and Wagner, R. J. (1984). *Appl. Phys. Lett.* **44,** 907.
Figielski, T., and Wosinski, T. (1987). *Phys. Rev. B* **36,** 1269.
Fuchs, F., and Dischler, B. (1987). *Appl. Phys. Lett.* **51,** 679.
Gooch, C. H., Hilsum, C., and Holeman, B. R. (1961). *J. Appl. Phys. Suppl.* **32,** 2069.
Holmes, D. E., Chen, R. T., Elliot, K. R., and Kirkpatrick, C. G. (1982). *Appl. Phys. Lett.* **40,** 46.
Kaminska, M., and Weber, E. R. (1990). *Proc. 20th Int. Conf. Phys. of Semiconductors,* p. 473 (eds. Anastassakis, E. M., and Joannopoulos, J. D.). World Scientific, Singapore.
Kaminska, M., Skowronski, M., Lagowski, J., Parsey, J. M., and Gatos, H. C. (1983). *Appl. Phys. Lett.* **43,** 302.
Kaminska, M., Skowronski, M., and Kuszko, W. (1985). *Phys. Rev. Lett.* **55,** 2204.
Kaufmann, U. (1985). *Phys. Rev. Lett.* **54,** 1332.
Kaufmann, U., and Windscheif, J. (1988). *Phys. Rev. B* **38,** 10060.
Kaxiras, E., and Pandey, K. C. (1989). *Phys. Rev. B* **40,** 8020.
Krambrock, K., Spaeth, J.-M., Delerue, C., Allan, G., and Lannoo, M. (1992). *Phys. Rev. B* **45,** 1481.
Lagowski, J., Gatos, H. C., Parsey, J. M., Wada, K., Kaminska, M., and Walukiewicz, W. (1982). *Appl. Phys. Lett.* **40,** 342.
Lagowski, J., Lin, D. G., Chen, T.-P., Skowronski, M., and Gatos, H. C. (1985). *Appl. Phys. Lett.* **47,** 929.
Lang, D. V., and Logan, R. A. (1975). *J. Electron. Mater.* **4,** 1053.
Leyral, P., Vincent, G., Nouailhat, A., and Guillot, G. (1982). *Solid State Commun.* **42,** 67.
Lin, A. L., Omelianovski, E., and Bube, R. H. (1976). *J. Appl. Phys.* **47,** 1852.
Martin, G. M. (1981). *Appl. Phys. Lett.* **39,** 747.
Martin, G. M., Mitonneau, A., and Mircea, A. (1977). *Elektron. Lett.* **13,** 191.
Meyer, B. K., Spaeth, J.-M., and Scheffler, M. (1984). *Phys. Rev. Lett.* **52,** 851.
Meyer, B. K., Spaeth, J.-M., and Scheffler, M. (1985). *Phys. Rev. Lett.* **54,** 1333.
Meyer, B. K., Hofmann, D. M., Niklas, J. R., and Spaeth, J.-M. (1987a). *Phys. Rev. B* **36,** 1332.
Meyer, B. K., Hofmann, D. M., and Spaeth, J.-M. (1987b). *J. Phys. C* **20,** 2445.
Miller, M. D., Olsen, G. H., and Ettenberg, M. (1977). *Appl. Phys. Lett.* **31,** 538.
Mitonneau, A., and Mircea, A. (1979). *Solid State Commun.* **30,** 157.
Mitonneau, A., Mircea, A., Martin, G. M., and Pons, D. (1979). *Rev. Phys. Appl. (France)* **14,** 853.

5.1 HYDROSTATIC PRESSURE AND UNIAXIAL STRESS INVESTIGATIONS

Nissen, M. K., Steiner, T., Beckett, D. J. S., and Thewalt, M. L. W. (1990). *Phys. Rev. Lett.* **65,** 2282.
Nissen, M. K., Villemaire, A., and Thewalt, M. L. W. (1991). *Phys. Rev. Lett.* **67,** 112.
Omling, P., Silverberg, P., and Samuelson, L. (1988). *Phys. Rev. B* **38,** 3606.
Przybytek, J., Baj, M., and Slupinski, T. (1995). *Acta Phys. Pol.* **A88,** 881.
Przybytek, J., Baj, M., Slupinski, T., and Li, Ming-Fu (1996a). *Phys. Stat. Sol. (b)* **198,** 193.
Przybytek, J., Baj, M., Slupinski, T., and Mikucki, J. (1996b). *Proc. 23rd Int. Conf. Phys. Semiconductors,* p. 2749 (eds. Scheffler, M., and Zimmermann, R.). World Scientific, Singapore.
Samuelson, L., Omling, P., Titze, H., and Grimmeiss, H. G. (1981). *J. Cryst. Growth.* **55,** 164.
Scheffler, M., Beeler, F., Jepsen, O., Gunnarsson, O., Andersen, O. K., and Bachelet, G. B. (1985). *Proc. 13th Int. Conf. Defects in Semiconductors,* p. 45. (eds. Kimerling, L. C., and Parsey, J. M., Jr.). AIME, Warrendale, PA.
Scherz, U., and Scheffler, M. (1993). *Semiconductors and Semimetals* **38,** 1 (eds. Willardson, R. K., Beer, A. C., and Weber, E. R.). Academic Press, New York.
Silverberg, P., Omling, P., and Samuelson, L. (1988). *Appl. Phys. Lett.* **52,** 1689.
Skowronski, M. (1990). *J. Appl. Phys.* **68,** 3741.
Skowronski, M., Lagowski, J., and Gatos, H. C. (1986). *J. Appl. Phys.* **59,** 2451.
Trautman, P., and Baranowski, J. M. (1992). *Phys. Rev. Lett.* **69,** 664.
Trautman, P., and Baranowski, J. M. (1995). *Int. J. Mod. Phys. B* **9,** 1263.
Trautman, P., Kaminska, M., and Baranowski, J. M. (1988). *Cryst. Res. Technol.* **23,** 413.
Trautman, P., Walczak, J. P., and Baranowski, J. M. (1990). *Phys. Rev. B* **41,** 3074.
Vincent, G., and Bois, D. (1978). *Solid State Commun.* **27,** 431.
Vincent, G., Bois, D., and Chantre, A. (1982). *J. Appl. Phys.* **53,** 3643.
von Bardeleben, H. J., Stivenard, D., Bourgoin, J. C., and Huber, A. (1985). *Appl. Phys. Lett.* **47,** 970.
von Bardeleben, H. J., Stivenard, D., Deresmes, D., Huber, A., and Bourgoin, J. C. (1986). *Phys. Rev. B* **34,** 7192.
Wagner, R. J., Krebs, J. J., Strauss, G. H., and White, A. M. (1980). *Solid State Commun.* **36,** 15.
Weber, E. R., Ennen, H., Kaufmann, U., Windscheif, J., Schneider, J., and Wosinski, T. (1982). *J. Appl. Phys.* **53,** 6140.
Williams, R. (1966). *J. Appl. Phys.* **37,** 3411.
Yu, P. W. (1984). *Appl. Phys. Lett.* **44,** 330.

CHAPTER 5.2

High-Pressure Study of DX Centers Using Capacitance Techniques

Ming-fu Li

DEPARTMENT OF ELECTRICAL ENGINEERING
NATIONAL UNIVERSITY OF SINGAPORE
SINGAPORE

Peter Y. Yu

DEPARTMENT OF PHYSICS
UNIVERSITY OF CALIFORNIA
BERKELEY, CA
AND
MATERIALS SCIENCES DIVISION
LAWRENCE BERKELEY NATIONAL LABORATORY
BERKELEY, CA

I. INTRODUCTION 457
II. TECHNIQUES FOR ELECTRICAL MEASUREMENTS ON SAMPLES INSIDE THE DAC   459
   1. *Introducing Wires into the DAC* 459
   2. *Performing Capacitance Measurements Inside the DAC* 461
III. INTRODUCTION TO CAPACITANCE TRANSIENT TECHNIQUES 462
   1. *Capacitance Transients at Constant Temperature* 463
   2. *Capacitance Transient when Scanning Temperature — Deep-Level Transient Spectroscopy* 465
   3. *Photocapacitance Transient Measurements* 466
IV. EXPERIMENTAL STUDIES OF DX CENTERS 467
   1. *Introduction* 467
   2. *Establishment of the DX Center as Due to Substitutional Donors* 469
   3. *Models of the DX Center* 474
V. CONCLUDING REMARKS 481
   Acknowledgments 482
   References 482

I. Introduction

Pressure can change the band structure of a semiconductor without changing its symmetry or composition. Thus, pressure is a powerful technique for studying the influence of electronic band structures on the properties of defects in semiconductors. This is particularly true for zincblende-type semiconductors because the pressure dependence of the various high symmetry critical points in their lowest conduction band are known to obey the so-called Paul's Empirical Rule [1]. For example, the pressure coefficients of the energy of the conduction band at the Brillouin zone center (or Γ point) in this family of semiconductors are all *positive* and of the order of ~100 meV/GPa in magnitude. On the other hand, the pressure coefficients of the conduction band at the zone edge in the [100] direction (or X point) are all *negative* and ~10 meV/GPa in magnitude. As a result, if the X valleys are higher in energy than the Γ valley, then the conduction bandwidth decreases with pressure. In many semiconductors the location of the conduction-band minimum in the Brillouin zone can be made to change from the Γ point to the X point under sufficiently high pressure. This phenomenon is known as the Γ–X *crossover*. For example, this crossover occurs around 4 GPa in GaAs [2].

Defect centers in semiconductors are usually classified as shallow (or hydrogenic) and deep. A defect energy level whose wave function can be constructed out of the nearest band extremum is considered shallow. On the other hand, the wave function of a highly localized center can be expressed only as a linear combination of wave functions from a large region of the Brillouin zone. In some cases many bands may be involved. Such centers are said to be deep. The properties of shallow and deep centers are quite different, their pressure dependence being one of them. So far, this property has proven to be one of the most reliable ways to distinguish between shallow and deep defects. By definition, the wave function of a shallow center is constructed from the wave functions of its nearest band extremum. Hence, the pressure dependence of a shallow center should be identical to that of its nearest band extremum. On the other hand, the pressure dependence of a deep center can be quite different from that of its nearest band extremum. In addition to being a method for distinguishing shallow and deep defects, pressure can change the properties of a defect by changing the host band structure. In some cases pressure can convert a shallow defect into a deep one or vice versa. Thus, high pressure can play an important role in the study of defects, especially in deep centers whose nature is often poorly understood.

With the development of the diamond anvil cell (DAC), it is possible both to achieve high pressures and to obtain optical access to the sample [3]. Many shallow impurities and a few deep centers have been studied with the DAC using optical techniques. Many defects, especially deep centers, are important in semiconductor technology because of their electrical characteristics. Thus, it is desirable to develop methods for carrying out electrical measurements on deep levels inside the DAC. Furthermore, such methods will enable electro-optical measurements to be performed inside the DAC. It should be noted that high-pressure electrical measurements have been performed using either the Bridgman cell or the large piston-cylinder type of high-pressure cell (see references in [3] for further details). The disadvantages of these cells are that the former provides no optical access so pressure calibration cannot be performed easily as with the ruby fluorescence technique, whereas the latter kind of cell is limited to pressures less than 2 GPa. In this article we will concentrate on the DAC because of its versatility and ability to reach the highest pressure. We shall discuss the techniques for performing transient capacitance and photocapacitance measurements inside the DAC developed at the University of California at Berkeley. The usefulness of these techniques will be illustrated by results obtained on DX centers found in GaAs and its alloys.

The organization of this article is as follows. In Section II we concentrate on the technique of introducing wires into the DAC for capacitance and photocapacitance measurements. In Section III we present a short introduction to various capacitance transient techniques. The experimental results on the DX centers obtained with these techniques are presented in Section IV where they are discussed in light of recent models.

II. Techniques for Electrical Measurements on Samples Inside the DAC

1. INTRODUCING WIRES INTO THE DAC

The design of our DAC high-pressure cells has been described extensively in the literature [3] and so will not be repeated here. The major difficulty in performing electrical measurements on samples inside the DAC lies in making electrical contacts to the sample with wires [3]. High pressure is achieved in a DAC by pressing two diamond anvils onto a metal gasket as shown in Fig. 1. The sample and the pressure medium are confined in a hole drilled in the gasket. To make electrical contact to the sample inside the pressure medium, wires have to pass through the contact region between one of the diamond anvils and the gasket. Thus, the wires have to be insulated from the metallic gasket. Because of the small size of the sample to be contacted by the wires, the diameter of the wire is typically 20–50 μm. Such thin wires are necessarily quite fragile. Unless they are protected in some way, they are easily cut by the diamond anvil.

Various methods have been devised to solve these two problems. Insulating gaskets have been used to avoid the problem of shorting the wires through the gasket [4]. However, most insulating materials are hard rather

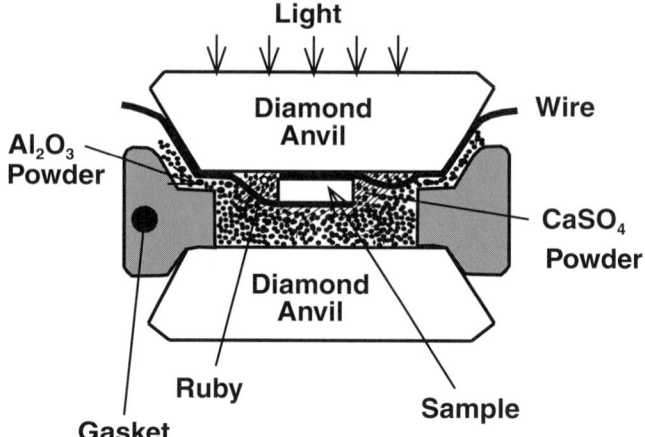

FIG. 1. Schematic diagram of the diamond anvils and sample inside the hole drilled in the metal gasket. In particular note the details of how the wire is insulated from the gasket by Al_2O_3 powders while the sample is surrounded by a softer powder ($CaSO_4$).

than malleable like a metal. Since they do not "flow" under stress, they cannot confine the pressure medium by forming a tight seal with the diamond anvils. Another approach is to insulate the gasket from the wires by coating it with an insulator. Typically, a thin layer of Al_2O_3 is deposited on a metal gasket by plasma spraying or sputtering [5]. The disadvantage of this technique is that most laboratories do not have plasma spraying or sputtering facilities. The gaskets have to be indented and then sent away to be coated with Al_2O_3. As an alternative to spraying and sputtering, we have developed the technique of applying Al_2O_3 in the form of a fine powder (the same powder used in polishing crystals, with grain sizes about 1 μm or less). The powder is compacted into a layer about 30 μm thick on the indented gasket using the diamond anvils themselves as the press. The shortcoming of this method is that the Al_2O_3 powder tends to spill over into the hole in the gasket, and so Al_2O_3 powder becomes the pressure medium. Since Al_2O_3 is a very hard material, it is not a good medium for a homogeneous pressure. Erskine *et al.* [6] alleviated this problem by adding a softer powder, such as $CaSO_4$, as the pressure medium inside the hole. A schematic diagram of the gasket, sample, and wire after installation inside the cell is shown in Fig. 1. Other groups have independently developed this technique [7, 8] or variations of this method, such as using a mixture of epoxy and Al_2O_3 powder to insulate the metal gasket [9]. The latter method has the advantage that the seal between the gasket and the diamond

5.2 STUDY OF DX CENTERS USING CAPACITANCE TECHNIQUES

remains tight at even low pressure. This allows a liquid pressure medium to be used for achieving greater pressure homogeneity.

The pressure homogeneity inside a DAC prepared in this way has been studied by Erskin *et al.* [6] using either the width of the ruby fluorescence line or the sharpness of the superconducting transition in superconductors such as Pb. Typical pressure inhomogeneity in cells prepared this way is less than 10%. Using this technique we have carried out capacitance measurements on semiconductor diodes inside the DAC. Since the resistance of these diodes under reverse bias is usually very high, the pressure medium must be a good insulator. In this respect, powders such as $CaSO_4$ have an advantage over liquid pressure media. However, diodes can also be destroyed by large pressure gradients. So far we have succeeded in making capacitance measurements on diodes at pressures up to about 4 GPa. Assuming that the pressure gradient inside the gasket hole is about 10%, the nonhydrostatic component of the pressure can be as high as 0.4 GPa. This stress is sufficient to damage some materials such as GaAs. In addition to standard capacitance measurements, we have made transient measurements, such as deep-level transient spectroscopy (DLTS), on samples inside the DAC.

2. PERFORMING CAPACITANCE MEASUREMENTS INSIDE THE DAC

Some special consideration are relevant to the application of capacitance techniques to samples inside the DAC. Given the sample configuration in Fig. 1, it is unavoidable that there will be a rather large background stray capacitance. Thus, one advantage of transient capacitance experiments is that one will be measuring a small change in the diode capacitance. Since the background stray capacitance is unchanged during the experiment, it can be easily subtracted out. In DLTS measurements the sample temperature has to be monitored. Since it is difficult to put the thermometer inside the cell next to the sample, we have attached a calibrated Si diode thermometer to the diamond anvil nearest to the sample. To minimize the temperature difference between the sample and the diode sensor, the cell temperature is changed very slowly (usally at the rate of about 2 K/min or less). To achieve a short equilibration time between the sample and thermometer, the thermal mass of the cell should be minimal. We have achieved this by separating the lever system used to apply the pressure from the cylinders holding the anvils. Pressure is applied to the cell via a hydraulic press. Once the desired pressure has been reached, the pressure is maintained by a locking ring on the cell. The linewidth of DLTS spectra is inherently rather broad and therefore the larger pressure inhomogeneity associated with a solid pressure medium is not a major disadvantage.

We mentioned earlier that one advantage of the DAC is that it will be possible to perform electro-optical measurements on samples inside the DAC. One such measurement we have carried out is photocapacitance. The problem with performing photocapacitance experiments inside the DAC results from the metal electrodes on the diodes. The metal overlayer prevents light from reaching the depletion layer of the sample directly. We have solved this problem by scattering the light in the pressure medium so that light can enter the diode from its sides. Unfortunately, it is not possible to determine exactly the amount of light absorbed by the sample.

III. Introduction to Capacitance Transient Techniques

Capacitance transient techniques are among the most important methods for characterizing deep defects in semiconductors [10–15]. Although these techniques have been extensively studied and described in the literature, we shall give a short introduction to them here for the benefit of readers not familiar with them. This also allows us to discuss the advantages and the problems of such measurements on samples inside DAC.

Let us consider a p^+n junction (or an n-Schottky barrier junction) with a reverse bias voltage V_b. In order to simplify the discussion we shall assume that the shallow donors are distributed uniformly throughout the n-type layer. The region $0 < x < W$ is completely depleted of free electrons (i.e., the electron concentration is zero). This assumption is known as the *depletion approximation*. The uniform space charge due to the ionized donors gives rise to a electric field which is linearly dependent on distance x as a result of the Gauss law. Alternatively, the electric potential V varies quadratically with x such that $V = V_b$ at $x = 0$ and $V = 0$ at $x = W$. In the region $x > W$, the potential is assumed to be identically zero. The free electron concentration n_0 is determined by the donor doping concentration N_d. The small signal junction capacitance C can be written as

$$C = \frac{A\epsilon}{W}, \tag{1}$$

where A is the area of the junction and ϵ is the permittivity of the semiconductor. The depletion layer thickness W and hence the capacitance C are both dependent on the bias voltage V_b. Let us assume that there are deep centers in the depletion region with concentration $N_t \ll N_d$. Suppose the bias voltage V_b is fixed but the occupancy of the deep centers is changed. As an illustration, let us assume that electrons are thermally excited from the deep centers into the conduction band. This changes the density of space charges (ρ) in the depletion layer. As a result, the width of the depletion layer will increase by ΔW, which leads to a change ΔC in the junction capacitance. From Eq. (1) the two quantities are related by

$$\Delta C/C = -\Delta W/W. \tag{2}$$

On the other hand, because the bias voltage is kept constant and

$$V_b \sim \rho W^2, \tag{3}$$

we obtain the relation

$$-2\Delta W/W = \Delta \rho/\rho. \tag{4}$$

Under our assumption that $N_d \gg N_t$, ρ is determined mainly by N_d while $\Delta \rho$ is equal to $(-e)\Delta n_t$, where Δn_t is the change in the concentration of electrons trapped on the deep levels. Substituting these results into Eq. (4) and then combining the resultant equation with Eq. (2), we obtain the simple expression

$$\Delta C/C = -\Delta n_t/2N_d. \tag{5}$$

In deriving Eq. (5) we have assumed that the concentration of acceptors in the p^+ region is much higher than N_d so that the change in the depletion width W caused by the change Δn_t occurs entirely in the n-doped region. We shall now consider three ways to apply Eq. (5) to study deep centers.

1. CAPACITANCE TRANSIENTS AT CONSTANT TEMPERATURE

The most obvious way to apply Eq. (5) is to keep the sample temperature constant and superimpose an applied voltage pulse on the bias voltage to induce a transient Δn_t in the deep center population [10]. In this way the rates of emission from and capture of carriers into the deep center can be determined from the transient in the capacitance. Such an experiment can be carried out in two phases. In phase I a *forward* bias (filling) pulse with amplitude V_p and pulse width τ is added to the *reverse* bias voltage V_b so that the total bias is given by

$$V_1 = V_b - V_p. \tag{6}$$

The change in bias from V_b to V_1 results in the depletion width W also changing from W_b to W_1. As V_b is larger than V_1, so is W_b larger than W_1. Hence, the region between W_1 and W_b is no longer depleted of carriers, and electrons will be *captured* from the conduction band into the deep level E_t with a capture rate C_n. By solving a simple rate equation we can easily show that the increase in the deep level population Δn_t is given by

$$\Delta n_t = N_t(1 - e^{-C_n \tau}), \tag{7}$$

assuming that $\Delta n_t = 0$ at the beginning of the applied pulse V_p. In phase II the bias is returned from V_1 to V_b and the depletion width also returns from W_1 to W_b. In the region between W_1 and W_b, electrons are emitted with an emission rate e_n by the deep levels to the conduction band and then swept out of the depletion region. In this phase Δn_t changes with time t as

$$\Delta n_t(t) = N_t(1 - e^{-C_n \tau})e^{-e_n t}, \tag{8}$$

where we have assumed $t = 0$ to be the moment the filling pulse is turned off. Substituting Eq. (8) into Eq. (5), the corresponding change ΔC in the junction capacitance C is

$$\frac{\Delta C}{C} = -\frac{N_t(1 - e^{-C_n \tau})e^{-e_n t}}{2N_d}. \tag{9}$$

Using Eq. (9) we can determine the capture rate C_n by measuring ΔC as a function of the filling pulse width τ. Similarly, the thermal emission rate e_n can be deduced from the time dependence of $\Delta C/C$. The thermal emission and capture rates are further related to each other by the principle of detailed balance [10, 15],

$$e_n = N_c C_n e^{-(E_c - E_t)/kT}, \tag{10}$$

where E_c and N_c are, respectively, the conduction band edge energy and effective density of states, and k is the Boltzmann constant. The energy $E_c - E_t$ is the thermal ionization energy of the deep level. In the literature it is a common practice to define the probability of capture in terms of the capture cross-section σ_n. It is related to the capture rate C_n by

$$C_n = n_0 \langle v \rangle \sigma_n, \tag{11}$$

5.2 STUDY OF DX CENTERS USING CAPACITANCE TECHNIQUES

where $\langle v \rangle$ and n_0 are, respectively, the thermal velocity and concentration of the free carriers. The capture cross-section is usually assumed to be thermally activated and has the temperature dependence

$$\sigma_n(T) = \sigma_n(\infty)e^{-E_B/kT}, \qquad (12)$$

where E_B is defined as the capture barrier height. Substituting Eqs. (12) and (11) into Eq. (10), we obtain

$$e_n(T) = n_o\langle v \rangle \sigma_n(\infty) N_c e^{-(E_c - E_t + E_B)/kT}. \qquad (13)$$

Equations (12) and (13) are commonly used in the literature for determining the capture and emission barrier heights.

2. CAPACITANCE TRANSIENT WHEN SCANNING TEMPERATURE — DEEP-LEVEL TRANSIENT SPECTROSCOPY

According to Eq. (13), one can in principle determine the emission barrier height by measuring the temperature dependence of the emission rate. The latter can be obtained by measuring the capacitance transient after a filling pulse as a function of time while keeping the temperature T constant. An alternative to scanning the time at fixed temperature is to scan the temperature while keeping a "time window" constant [12]. In this approach, a filling pulse is applied first and then the capacitance is measured only at two preset times t_1 and t_2, which are said to define a "time window." The corresponding capacitances are denoted by C_1 and C_2, respectively. Their difference $(C_1 - C_2)$ is then measured as a function of temperature. The idea is that $(C_1 - C_2)$ is nonzero only when there is significant change in the deep occupation during the time window. If t_1 and t_2 are both $\ll 1/e_n(T)$, then both C_1 and C_2 are unchanged so their difference $(C_1 - C_2) \sim 0$. On the other hand, if t_1 and t_2 are both $\gg 1/e_n(T)$, then C_1 and C_2 are both almost equal to the equilibrium value and so again $(C_1 - C_2) \sim 0$. Thus, a plot of $(C_1 - C_2)$ versus temperature will exhibit a peak when $1/e_n(T)$ is approximately equal to the time window $(t_2 - t_1)$. Such a $(C_1 - C_2)$ versus temperature curve is known as a *DLTS spectrum* [12]. The temperature T_m where the DLTS curve shows a maximum allows the emission rate at T_m to be determined:

$$e_n(T_m) \approx 1/(t_2 - t_1). \qquad (14)$$

The DLTS technique takes advantage of the fact that $e_n(T)$ depends on T exponentially so it takes a relatively small change in T to vary $e_n(T)$ by many orders of magnitude. As may be expected, a slow scanning of temperature is desirable in obtaining DLTS spectra. Using different time windows and measuring the corresponding values of T_m, one can obtain the temperature dependence of e_n and hence the *DLTS activation energy* E_{DLTS}, defined as

$$E_{DLTS} = [E_c - E_t] + E_B. \tag{15}$$

From Eq. (13) it is clear that we can also interpret the energy E_{DLTS} as an emission barrier height.

3. Photocapacitance Transient Measurements

From the preceding discussions we see that the emission rate of deep levels can be made negligibly small by lowering the sample temperature so that $kT \ll E_{DLTS}$. Under this condition it may become possible to photoionize the deep level. To carry out such photoionization measurement one fills the deep levels by setting the bias voltage $V_b = 0$ at room temperature so that $n_t = N_t$. The sample is then cooled to low temperature in the dark. A reverse bias voltage V_b is then applied to the junction. As the thermal emission rate e_n is extremely small and negligible at low temperature, a metastable state is created. If the sample is now illuminated with radiation with sufficient energy to optically excite carriers out of the deep level, the photoinduced change of the junction capacitance is a measure of the photoionization of the deep levels. Let $h\nu$ be the incident photon energy and $\Phi(h\nu)$ be its flux density. σ_n^o and σ_p^o are, respectively, the electron and hole photoionization cross-sections for the deep centers. Then dn_t/dt is given by [16]

$$dn_t/dt = \Phi(h\nu)[\sigma_p^o(h\nu)(N_t - n_t) - \sigma_n^o(h\nu)n_t]. \tag{16}$$

Let $t = 0$ be the point when the light is turned on. Since most of the deep centers are occupied before the light is turned on, $n_t(0) = N_t$ and (16) can be approximated at $t \geq 0$ by

$$dn_t/dt \approx -\Phi(h\nu)\sigma_n^o(h\nu)N_t. \tag{17}$$

Substituting Eq. (17) into Eq. (6), we obtain

$$dC/dt \approx (CN_t/2N_d)\Phi(h\nu)\sigma_n^o(h\nu). \tag{18}$$

Thus, by measuring the initial rate of change $dC/dt(t \sim 0)$ and the incident photon flux density, one can measure the electron photoionization cross-section $\sigma_n^o(h\nu)$.

The capacitance transient technique is suitable for measuring samples inside the DAC. First, the DAC allows electric and optical perturbations to be applied simultaneously to the sample. Second, the advantage of such transient experiments is that a small change in the junction capacitance can be more precisely determined on top of a large constant background stray capacitance associated with a sample inside the DAC. In DLTS measurements, the sample temperature is monitored by a calibrated Si diode thermometer attached to the diamond anvil nearest to the sample. To minimize the temperature difference between the sample and the diode sensor, the cell temperature is changed very slowly. To achieve a short equilibration time between the sample and thermometer, the thermal mass of the cell should be minimal.

IV. Experimental Studies of DX Centers

1. INTRODUCTION

The deep trap known as the DX center was discovered in 1979 by Lang and co-workers [17] in *n*-type $Al_xGa_{1-x}As$ with $x > 0.22$. During the past two decades the properties of the DX center have been studied extensively [18–21]. Lang *et al.* named this defect the DX center because they thought that it involved a complex consisting of a donor atom D and an unknown constituent X. Since this center was first observed in alloys of GaAlAs only, it was believed that X is an intrinsic defect found in alloys. Some key characteristics of the DX center which distinguish it from other deep centers are (1) its optical ionization energy E_{op} (~ 1 eV) is an order of magnitude larger than its thermal ionization energy $(E_c - E_t)$; and (2) its capture barrier height is on the order of 0.2 eV, and it has a very small capture cross-section for electrons at 77 K, resulting in a very long lifetime for free electrons. This give rise to the phenomenon known as persistent photoconductivity (PPC). The electron capture cross-section of a number of deep levels in GaAs and GaP has been studied extensively by Henry and Lang [22], as summarized in Fig. 2. These data can be fitted by the multiphonon emission theory (MPT) [22, 23] with a capture barrier height E_B as defined in Eq. (12) when kT is larger than the phonon energy. On the other hand, there is a lower limit of 10^{-21} cm^2 for the capture cross-section at low temperatures due to optical capture. However, the capture cross-section of the DX center can be several orders of magnitude lower than 10^{-21} cm^2 as shown in Fig. 2. This implies that optical capture into the DX center is almost prohibited. Lang explained these properties of the DX centers by the existence of a large lattice relaxation in these centers [17, 18]. Since Lang *et al.*'s original paper, great progress has been made toward developing a microscopic model of the DX center. In this review, we mainly focus on some transient capacitance experiments performed inside the diamond anvil high pressure cells which have helped to elucidate the nature of the DX center.

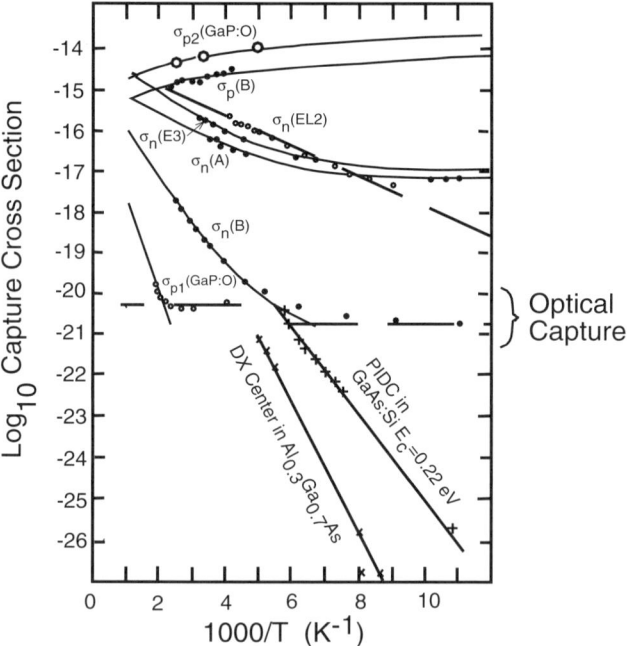

FIG. 2. The capture cross-section of PIDC (+) in GaAs:Si under 2.9 GPa [27] compared to that of the DX center (×) in Al$_{0.3}$Ga$_{0.7}$As:Si [35]. Notice how both of them behave differently from the capture behaviors of most other deep centers in GaAs and GaP. Reproduced from Ref. 22.

2. ESTABLISHMENT OF THE DX CENTER AS DUE TO SUBSTITUTIONAL DONORS

Lifshitz et al. [24] were the first group to notice the existence of a correlation between the effect of pressure and of Al alloying on the band structure properties of AlGaAs. They noticed that 0.1 GPa of pressure has approximately the same effect on the conduction band of GaAs as increasing the Al concentration by 1%. Since Lang et al. [17] have found that the DX center appeared in $Al_xGa_{1-x}As$ only when $x > 0.22$, it was natural to ask whether the DX center will appear in GaAs under a pressure of more than 2 GPa, assuming that the properties of the DX center is determined entirely by the conduction band structure. In 1985, Mizuta et al. [25] applied pressure to n-type GaAs doped with Si using a Bridgman anvil device and discovered that at pressures exceeding 2.4 GPa, a peak appeared in the DLTS spectrum (as shown in Fig. 3), with a DLTS activation energy of 0.31–0.33 eV. Similar results were obtained for n-type GaAs doped with Sn. Mizuta et al. identified this DLTS peak with the DX center peak found by Lang et al. in $Al_xGa_{1-x}As$ for $x > 0.2$. In a subsequent paper, Tachikawa et al. found evidence of PPC in GaAs:Si under pressure at 77 K [26] using an LED loaded into the high-pressure cell as the light source. Their result is shown in the inset of Fig. 3. If this pressure-induced deep center (PIDC) were indeed identical to the DX center, then the pressure experiments would have invalidated Lang's proposal that the DX center in AlGaAs involves an unknown constituent X introduced by alloying. Instead, one has to conclude that the DX center is the result of a shallow-to-deep transformation of substitutional Si donors in GaAs *induced by changes in the conduction-band structure only*. Such changes can be produced either by alloying or by pressure.

A crucial test of whether the PIDC in GaAs:Si and the DX center in AlGaAs are identical is the determination of the photoionization energy E_{op} and the thermal ionization energy $E_c - E_t$ of the PIDC and comparing them with those of the DX centers. Figure 4 shows the capture behavior of the PIDC in GaAs under 2.5–2.9 GPa of pressure. By analyzing the experimental result using Eqs. (10)–(13), the PIDC was found to obey the MPT with a capture barrier height $E_B = 0.22$ eV. Furthermore, the pressure coefficient of the capture barrier height (dE_B/dP) was determined to be -21 meV/GPa. This value of E_B is comparable to the value $E_B = 0.33 \pm 0.05$ eV for Si-doped AlGaAs measured by Lang [18]. Similarly, the values of $E_{DLTS} = 0.30$ eV and the pressure coefficient $dE_{DLTS}/dP = -13$ meV/GPa were obtained by DLTS measurement inside the DAC. These values are in good agreement with the results of Mizuta et al. The thermal ionization energy $E_c - E_t$ obtained from these energies using Eq. (15) was found

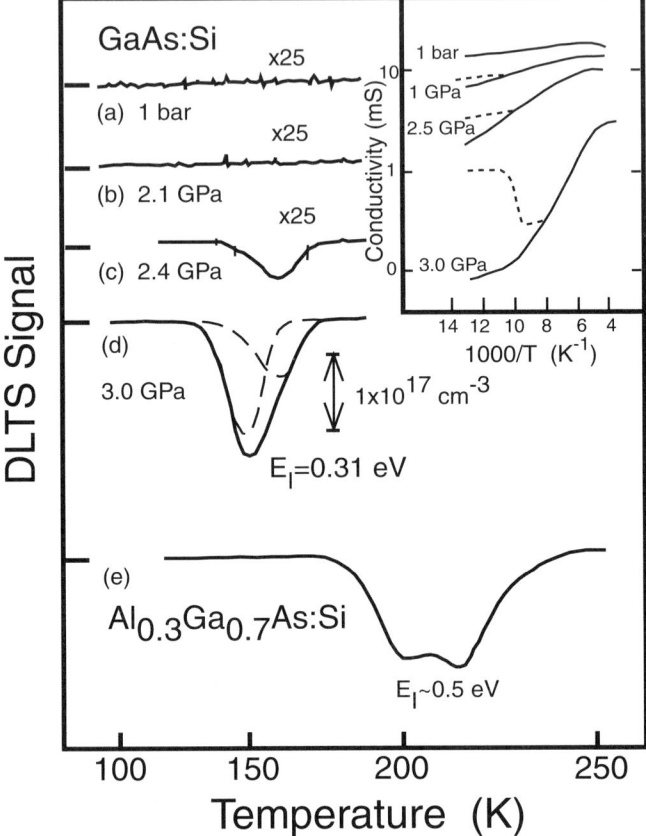

FIG. 3. (a)–(d) The DLTS spectra of GaAs:Si measured at different pressures. The rate window used is 66 sec^{-1} with $t_2/t_1 = 2$. Reproduced from [25]. (e) The corresponding spectrum for the DX center in Al$_{0.3}$Ga$_{0.7}$As:Si. The inset shows persistent photoconductivity effect of GaAs:Si under different pressures. Reproduced from [26]. The solid curves were measured in the dark while the broken curves were measured after light illumination.

to be 0.08 eV, which is comparable with the value of 0.10 ± 0.05 eV measured by Lang for Si-doped AlGaAs [18].

Photoionization experiments are difficult to perform with the Bridgman anvil device used by Mizuta *et al.* [25] since these anvils are made from sintered diamond and are opaque. To overcome this difficulty, we have instead used a DAC to perform photocapacitance transient measurements on this PIDC in Si-doped GaAs [27, 28]. The transparent diamond anvils

5.2 Study of DX Centers Using Capacitance Techniques

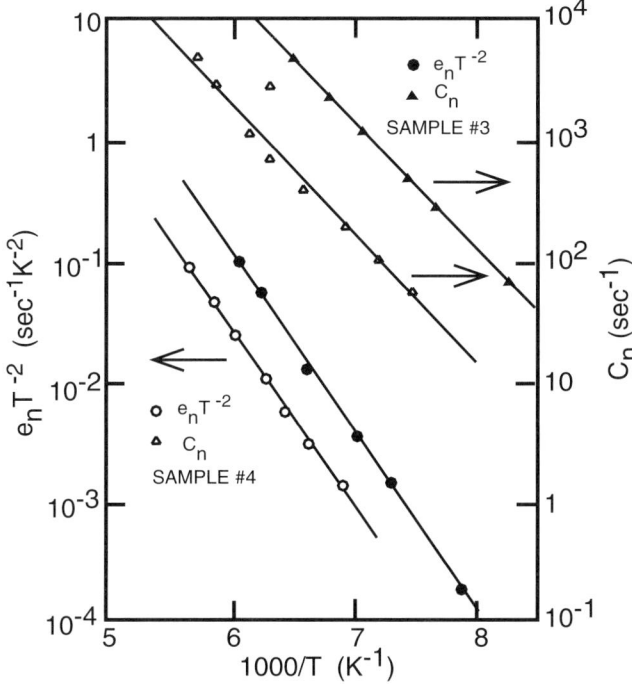

FIG. 4. Plots of the electron emission and capture rates versus 1/temperature for PIDC in two samples of GaAs under 2.9 GPa of pressure. Reproduced from [27].

allowed us to measure the photoionization spectra of the PIDC by applying Eq. (18). The results are shown in Fig. 5. By fitting our experimental data with the theory of Lang, Logan, and Jaros [17], we determined the photoionization threshold energy E_{op} of the PIDC in Si-doped GaAs to be 1.4 eV. This value is larger than its thermal ionization energy 0.08 eV by an order of magnitude. As seen from Fig. 5, there is also good agreement with the DX center's photoionization spectra measured by Legros et al. [29] in Si-doped AlGaAs. These results, together with those of Mizuta et al. [25] in GaAs:Si, thus show conclusively that the PIDC found in GaAs has all the important attributes of the DX centers in AlGaAs. To our knowledge this is the first time pressure has played such a crucial role in revealing the nature of a deep center in semiconductors.

It is interesting to note that on close examination there are actually quantitative differences between the DX centers in GaAs under pressure and those found in AlGaAs at ambient pressure [30]. The origin of this

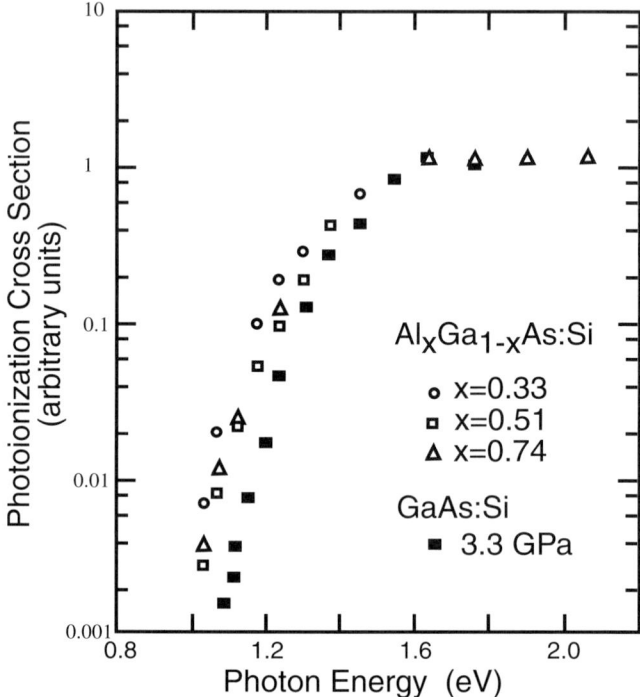

FIG. 5. Photoionization spectrum of the PIDC in GaAs:Si under a pressure of 3.3 GPa [28] at 77 K, compared with that of the DX center in $Al_xGa_{1-x}As$:Si at 84 K for three different values of x [29].

difference has been explained by Mooney *et al.* [31] to be due to the different local environments of the DX center in the alloy. Although a Si DX center has only one possible local environment, the analogous center in AlGaAs can have one, two, or three Al atoms as its neighbors. The convincing experiments in support of this explanation again involve the application of pressure. The experiments were performed by Calleja *et al.* [32] and by Baba *et al.* [33]. The results of Calleja *et al.* [32] are shown in Fig. 6. They measured the DLTS spectra of $Al_xGa_{1-x}As$ samples with $x = 0, 0.04, 0.08$, respectively, under pressure. In Fig. 6, the lower temperature DLTS peak with an activation energy of 0.34 eV is attributed to a DX center without Al atoms as nearest neighbors. This assignment is consistent with high-pressure data on GaAs:Si [27, 34]. The higher DLTS peak with activation energy 0.44 eV corresponds to a DX center with Al atoms as nearest neighbors and is also consistent with previous results in AlGaAs:Si [35]. The 0.1-eV difference in DLTS activation energies between DX cen-

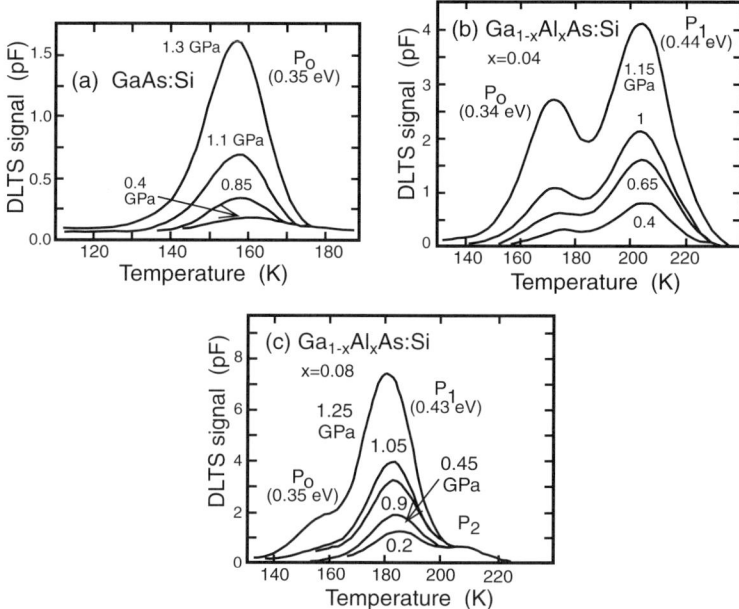

FIG. 6. DLTS spectra of DX centers under various applied pressures: (a) GaAs;Si, (b)Al$_x$Ga$_{1-x}$As ($x = 0.04$), and (c) Al$_x$Ga$_{1-x}$As ($x = 0.08$). Reproduced from [32].

ters with and without Al atoms as nearest neighbors has been confirmed by self-consistent theoretical calculations [36].

In addition to DX centers produced by group IV donors, such as Si, the pressure dependence of DX centers formed by group VI donors, such as Te, has also been studied [37]. The experiments were performed on Al$_x$Ga$_{1-x}$As epilayers, with $x = 0.15, 0.25$, and 0.35 and doped with 5×10^{16} cm^{-3} of Te. The DLTS peak emerges, respectively, at 1.6 GPa, 0.7 GPa, and 1 bar for the $x = 0.15, 0.25$, and 0.35 samples. These experiments showed that Liftshitz et al.'s result that 0.1 GPa of pressure has approximately the same effect on the conduction band of GaAs as increasing the Al concentration by 1% is correct for predicting the energy of the DX level relative to the conduction band edge. Together with the results in GaAs : Si they show quite convincingly that at ambient pressure the DX level associated with donor atoms in GaAs is actually a *resonance state* above the conduction band. As a result of the change in the conduction-band structure caused either by alloying or by pressure, the DX center emerges from the conduction band into the energy gap and becomes the stable ground state of the donor.

3. MODELS OF THE DX CENTER

Since it became clear that the DX center is a simple substitutional donor in GaAs which exhibits a shallow-to-deep transformation as a result of changes in the conduction-band structure induced by either pressure or alloying, many models have been proposed to explain its properties. However, the atomic and electronic configurations of the DX center have remained controversial for some time. The debates center on two areas. On the atomic configuration of the DX center, the question is whether there is large [17, 38, 39] or small lattice relaxation [40–43]. As far as the electronic configuration of the DX center is concerned, the issue is whether the DX center has a negative on-site Coulomb interaction U (abbreviated as $-U$) or a positive U between the two electrons localized on the same impurity. If the former case is correct, then the ground state of the DX center contains two electrons [38, 43], whereas in the latter case it will contain only one electron [41, 42]. Now it is generally accepted that the model proposed by Chadi and Chang in 1988 [38] is correct. The important features of this model, based on their supercell self-consistent pseudopotential calculation, can be summarized as follows:

1. The DX center is a $-U$ center resulting from the reaction

$$2d^0 \rightarrow d^+ + DX^-, \qquad (19)$$

   where d^0 and d^+ represent fourfold-coordinated substitutional donors in the neutral and ionized state, respectively. DX^- is a negative charged donor that has captured two electrons.
2. The DX^- defect formation involves a large bond-rupturing displacement of the host lattice atoms. For donors on cation sites, such as Si_{Ga}, the donor atom is displaced as depicted in Figs. 7a and b. In the case of donors located on anion sites, such as S_{As}, one of its nearest-neighbor Ga (or Al) atoms along a bond axes is displaced as shown in Figs. 7c and d. In other words, the local symmetry of a donor is charge dependent. When the donor electron occupancy is 0 or 1, corresponding to the positively charged d^+ or neutral charge d^0 states, the donor atom symmetry is T_d and there is no lattice relaxation. When the donor electron occupancy is 2, corresponding to a negatively charged DX^- state, the defect symmetry is reduced to C_{3v} as a result of bond-breaking relaxation.

One prediction of the $-U$ model is that the DX center should produce no electron paramagnetic resonance (EPR) signal. The reason is because

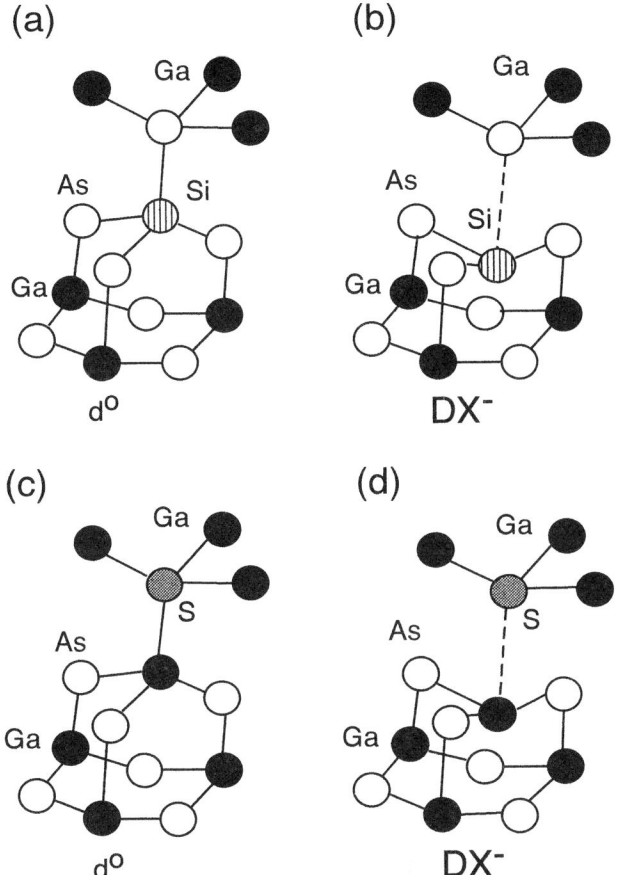

FIG. 7. Schematic diagrams of the normal substitutional d° (a and c) and the broken-bond DX⁻ configurations (b and d) of Si and S donors in GaAs. Reproduced from [38].

these two electrons should have opposite spin in order to satisfy the Pauli Exclusion Principle. Unfortunately, two experimental attempts to test the validity of the $-U$ model turned out to be contradictory [43, 44]. The strongest confirmation of the $-U$ property of the DX center comes from codoping experiments carried out by different groups using a variety of sample sources and measurement techniques.

Fujisawa *et al.* [45] performed the first successful codoping test of the $-U$ model by applying pressure to GaAs codoped with two donors: Ge and Si with different binding energies. At a pressure of 2.2 GPa, Ge in GaAs is converted into DX centers while Si remains as a shallow donor.

If the $-U$ model is correct, then the number of Ge DX centers can be varied by changing the concentration of shallow Si donors while keeping the number of Ge atoms fixed, since the former supplies the second electron to be trapped by the Ge DX state. Fujisawa *et al.* studied several samples in which the Ge concentration is fixed at 1×10^{17} cm^{-3} while the Si doping concentration is varied up to 2.6×10^{17} cm^{-3}. From the DLTS spectra, Fujisawa *et al.* found that the concentration of electrons trapped at the Ge DX centers increases with Si concentration and saturates at a value of 2.3×10^{17} cm^{-3} as shown in Fig. 8. The Ge donor concentration was estimated to be at most 1.5×10^{17} cm^{-3} with a compensating acceptor concentration of 0.5×10^{17} cm^{-3}. Thus, the saturated concentration of 2.3×10^{17} cm^{-3} electrons trapped on Ge cannot be explained by a Ge ground state with only one electron. Instead, one has to assume that each Ge atom can trap two electrons with a concentration of 1.15×10^{17} cm^{-3} Ge atoms.

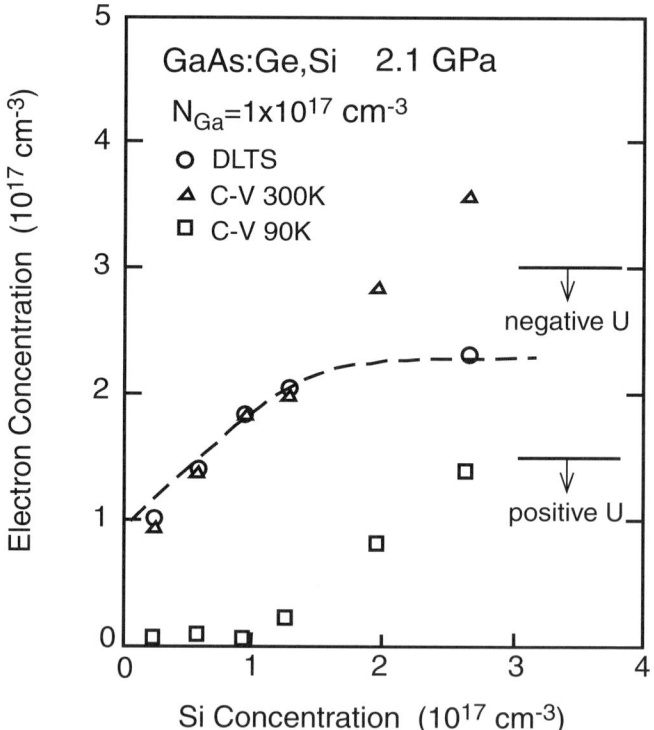

FIG. 8. Plots of the electron concentration at Ge DX centers determined by DLTS (○) and of the free carrier concentration (△) measured by CV at two different temperatures in GaAs codoped with Ge and Si at 2.1 GPa as a function of Si donor concentration. Reproduced from [45].

5.2 STUDY OF DX CENTERS USING CAPACITANCE TECHNIQUES

Unfortunately, Fujisawa et al.'s experiment is not unambiguous. First, if the true Ge concentration in their samples is 2.3×10^{17} cm^{-3} while the concentration of compensating acceptors is 1.3×10^{17} cm^{-3}, then their experimental result is consistent with a $+U$ center model. Second, the concentration of trapped electrons estimated by the DLTS method may not be reliable enough because of the high concentration of deep defects and the very large edge region effect [17]. Third, Fujisawa et al. tried to maintain the Ge doping concentration constant but, due to fluctuation in the growth conditions, the precision in controlling the doping concentration is low.

Baj et al. [46] avoided the difficulties in interpreting the codoping experiment of Fujisawa et al. by using a single GaAs sample codoped with Te and Ge instead. In addition to the large lattice relaxation DX levels, the Ge impurities in GaAs form also a small lattice relaxation A_1 level. These levels lie in the conduction band at ambient pressure but move into the gap at pressures exceeding 1.0 GPa. On the other hand, Te remains a shallow donor level in the gap at pressure less than 1.5 GPa. Thus, the idea behind the experiment of Baj et al. is to use pressure to convert Ge first into the positive U A_1 level impurities (labeled as the D^0 state in some literatures [47]) and then into the deep DX$^-$ center while shallow Te levels provide the electrons to be trapped on the DX levels. Furthermore, instead of measuring the concentration of electrons trapped at the DX centers by DLTS, the free carrier concentration is determined by the Hall effect, which is more precise than DLTS. A combination of control methods, such as irradiating the sample with light and changing its temperature, allow the number of electrons trapped on the DX centers to be varied via PPC. Figure 9 shows the Hall carrier concentration after light illumination measured by Baj et al. as a function of pressure at 77 and 100 K, respectively. Both curves show a step at pressure between 0.5 GPa to 1.0 GPa. The step in the 77 K curve is smaller and has a magnitude 1×10^{17} cm^{-3}. This step is explained by the trapping of electrons from the conduction band into the shallower A_1 level of the Ge centers. Thus, the concentration of Ge impurity is determined accurately to be 1×10^{17} cm^{-3} since each A_1 state captures only one electron. The deeper DX level associated with the Ge impurities does not capture electrons at 77 K because of its large capture barrier height. However, at 100 K the capture rate of the Ge DX center becomes much faster, so if the DX center is a $-U$ center, one expects to see a bigger drop in the carrier concentration due to trapping into the DX state. Indeed, Baj et al. found that the step in the 100 K curve in Fig. 9 is 2×10^{17} cm^{-3}, or exactly twice the concentration of the Ge impurity. This experiment unambiguously demonstrates that each DX level of the Ge impurity in GaAs captures two electrons. The beauty of this experiment is that the

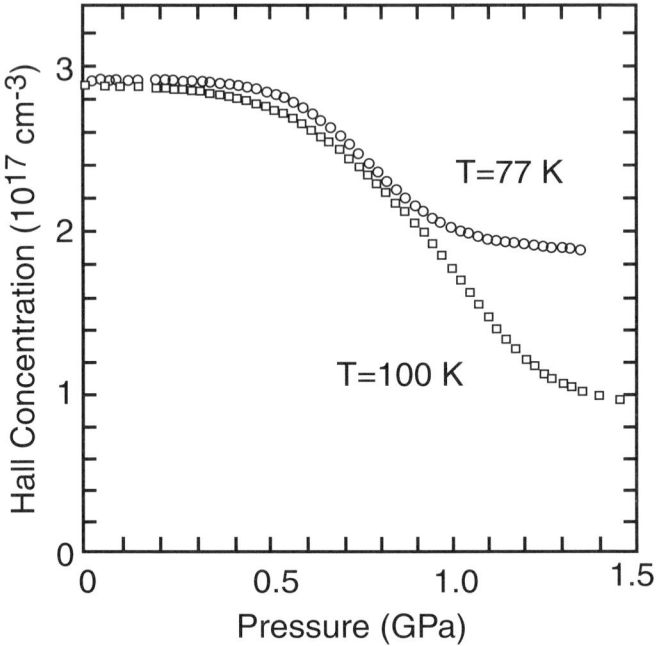

FIG. 9. Plots of the Hall carrier concentration in GaAs codoped with Ge and Te measured as a function of hydrostatic pressure at $T = 77$ and 100 K. Reproduced from [46].

result is independent of any compensating acceptors. Also, no prior information on doping concentrations is needed. The only requirement is that the concentration of Te shallow donor be higher than that of the Ge donors so that Te can provide enough electrons to fill the Ge DX states.

Willke *et al.* [48] also reported a codoping experiment using $Al_xGa_{1-x}As$ samples which contain Si and Sn instead of Ge and Te. Under hydrostatic pressure, both Si and Sn DX levels move into the energy gap. The idea behind their experiment is to use light to selectively photoionize the two deep levels, since the photoionization thresholds of Sn and Si in AlGaAs are different, 0.8 and 1.1 eV, respectively. Figure 10 shows the phototransient carrier concentration (n) of their sample at 4.2 K as measured by Hall and Shubnikov–de Haas effects. In Fig. 10a the sample is first illuminated with 1 eV radiation at $t = 0$ sec. Then the illumination is switched to 1.4-eV radiation starting at $t = 7000$ sec. The transient step at $t = 0$ sec is accounted for by photoionization of the Sn centers and suggests a Sn concentration of 5.6×10^{17} cm^{-3}. The transient step at $t = 7000$ sec is due to Si and indicates that its concentration is 1.21×10^{18} cm^{-3}. The sample is then subjected to a thermal cycling in the dark to 70 K in order to allow

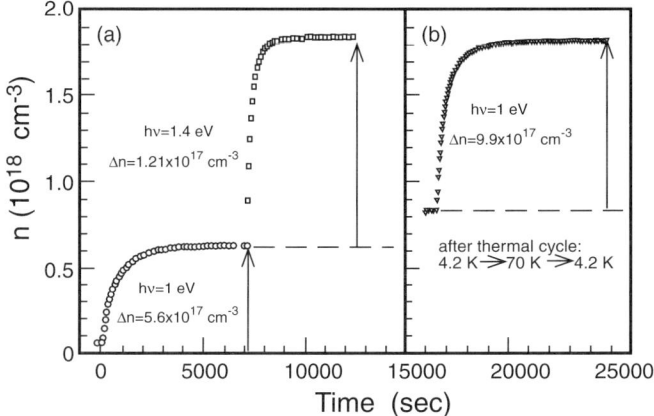

FIG. 10. Phototransients in the carrier concentration (n) measurement at 4.2 K in $Al_xGa_{1-x}As$ sample codoped with Sn and Si. (a) The sample was first slowly cooled down in the dark and then illuminated with 1 eV radiation at $t = 0$ sec followed by irradiation with 1.4-eV light at $t = 7000$ sec. (b) The sample was thermally cycled in the dark to 70 K and then illuminated with 1-eV radiation at 4.2 K. Reproduced from [48].

carriers to be recaptured into the shallower Sn DX centers only. The idea is that PPC of the deeper Si DX center now provides the free carriers to be trapped on the Sn centers in case the latter has a negative U. When the measurement is repeated with 1-eV radiation (result shown in Fig. 10b, the step in the phototransient is now found to be larger, 9.9×10^{17} cm^{-3}. This is almost double the Sn concentration of 5.6×10^{17} cm^{-3} as predicted by the $-U$ model for the DX center. Thus, the evidence again supports that the Sn DX centers have $-U$.

Finally, we briefly describe another codoping experiment based on similar ideas but not using GaAs. In this experiment [49], $GaAs_{0.6}P_{0.4}$ samples with a uniformly doped background of Te with concentration N_{Te} are used. A Gaussian distribution of S with concentration $N_s(x)$, where x is the depth, is introduced by ion implantation. It is well known that Te is a shallow donor while S forms a DX center ground state in GaAsp [49, 50]. The carrier concentration (n) is measured by the CV method. When the sample is illuminated at 77 K by light, the carrier concentration n_{op} is given by $N_{Te} - N_A + N_s$ (shown in Fig. 11), where N_A is the compensating acceptor concentration. If n is measured in the dark instead, the resultant carrier concentration n_{dark} will depend on whether the DX centers trap one or two electrons. If the DX centers trap two electrons, then

$$n_{dark} = N_{Te} - N_A - N_s \tag{20}$$

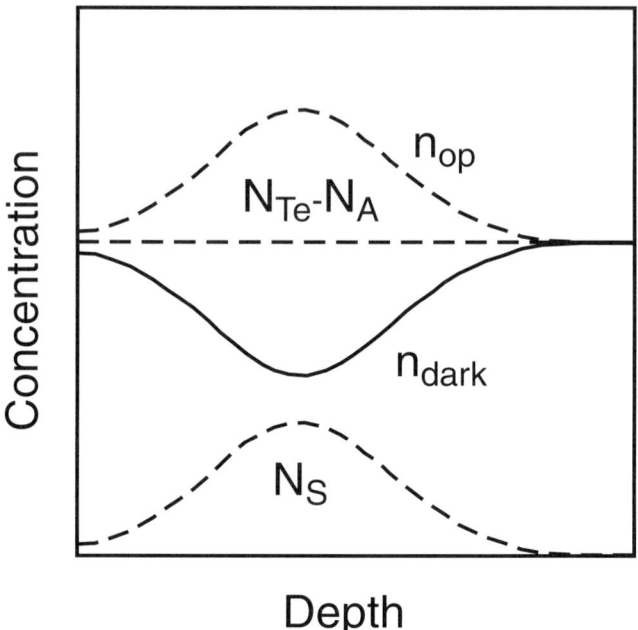

FIG. 11. Schematic diagram of donor distribution as a function of depth in a $GaAs_{0.6}P_{0.4}$ sample codoped with Te and S. N_s represents the implanted S ion spatial profile. $N_{Te} - N_A$ is the free carrier profile in the sample before implantation of S. n_{dark} is the predicted free-carrier profile if the S DX centers capture two electrons. n_{op} is the predicted profile after all the S DX centers have been emptied of electrons via PPC. Reproduced from [49].

will have the shape shown in Fig. 11. This curve should be a mirror image of the curve n_{op} with respect to the horizontal line $N_{Te} - N_A$. On the other hand, if the DX centers trap only one electron, then $n_{dark} = N_{Te} - N_A$ will be a horizonal line as in the unimplanted sample. Other defects produced by ion implantation can complicate this scheme. Fortunately, these defects have no PPC effect and can be distinguished from the DX center signal. Figure 12 shows the carrier spatial profiles measured by the CV technique. In case of the dark profile, n_{dark}, the curve has been corrected for the effects due to ion-implantation-induced defects. The fact that the profiles n_{op} and n_{dark} are roughly mirror images of each other with respect to the unimplanted curve clearly shows that the sulfur DX center traps two electrons. Finally, by taking into consideration the fact that the DLTS spectra in GaAsP exhibit only one peak [51], one can conclude that the two-electron DX state of S in GaAsP is indeed the ground state, and therefore the S DX center in GaAsP is a $-U$ system. If this were not the case, then

5.2 STUDY OF DX CENTERS USING CAPACITANCE TECHNIQUES

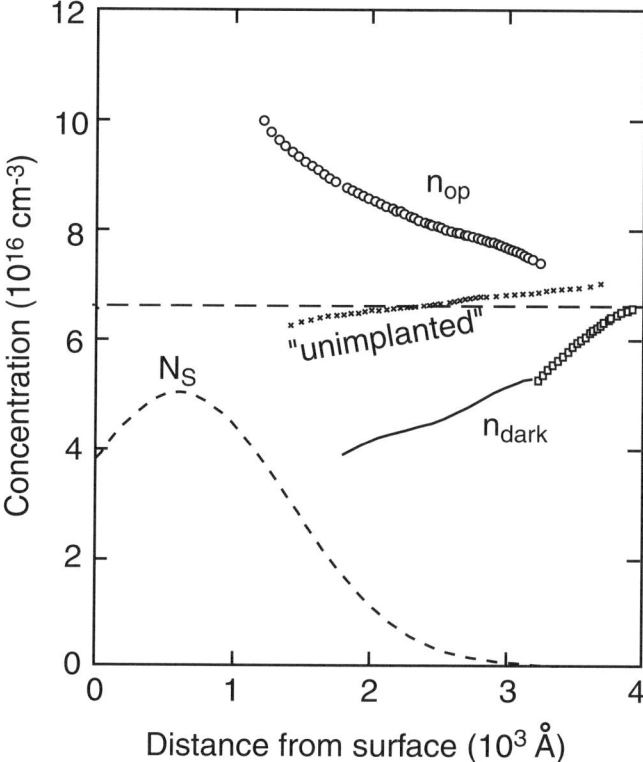

FIG. 12. The spatial profiles of free carrier concentration in the GaAs$_{0.4}$P$_{0.6}$ sample codoped with Te and S measured at 77 K using CV techniques. The broken curve N_s is the calculated S profile based on the ion implantation parameters. The curve labeled "unimplanted" represents the free carrier profile due to the shallow Te donors before the sample was implanted with S. The curve labeled n_{dark} is the free carrier profile measured with the sample in the dark. This curve consists of two parts. The points are the measured data, while the solid curve is obtained after correction for the effects of the ion-implantation-induced defects. The curve labeled n_{op} is the spatial profile obtained under light illumination. Reproduced from [49].

the DLTS specta would show a second peak at a lower temperature, in disagreement with experiment [48, 50].

V. Concluding Remarks

In this chapter we have described how to perform capacitance measurements on samples subjected to high pressure inside the DAC. As an illustration of the usefulness of this technique, we have discussed its application to determine the nature of the DX centers in GaAs and related alloys. Because of space limitations, it is impossible to summarize all the contributions made by high-pressure techniques to our understanding of DX centers. For example, we have omitted mentioning many important high-pressure optical experiments performed on DX centers inside the DAC [52]. We should also point out that the DX center is not the only deep center that capacitance experiments inside the DAC have made significant contributions to understanding [53].

Acknowledgments

The part of this work performed in Singapore was supported by the Singapore NSTB RIC-University Research Funding Project 681305. The work at Berkeley was supported by the Director, Office of Energy Research, Office of Basic Energy Sciences, Materials Sciences Division, of the U.S. Department of Energy under Contract No. DE-AC03-76SF00098.

References

1. W. Paul, *J. Appl. Phys.* **32,** 2082 (1961).
2. P. Y. Yu and B. Welber, *Solid State Commun.* **25,** 209 (1978).
3. See, for example, review article by A. Jayaraman in *Rev. Modern Phys.* **57,** 1013 (1986).
4. H. K. Mao and P. M. Bell, *Rev. Sci. Instrum.* **52,** 615 (1981).
5. R. L. Reichlin, *Rev. Sci. Instrum.* **54,** 1674 (1983).
6. D. Erskine, P. Y. Yu, and G. Martinez. *Rev. Sci. Instrum.* **58,** 406 (1987).
7. D. Patel, T. C. Crumbaker, J. R. Sites, and I. L. Spain, *Rev. Sci. Instrum.* **57,** 2795 (1986).
8. D. Patel and I. L. Spain, *Rev. Sci. Instrum.* **58,** 1317 (1987).
9. H. Huiberts, Ph.D. Thesis, Vrije Universiteit, Amsterdam (1997).
10. C. T. Sah, L. Forbes, L. L. Rosier and A. F. Tasch, *Solid State Electron.* **13,** 759 (1970).
11. C. T. Sah, *Solid State Electron.* **19,** 975 (1975).
12. D. V. Lang, *J. Appl. Phys.* **45,** 3023 (1974).
13. G. L. Miller, D. V. Lang, and L. C. Kimerling, *Ann. Rev. Mater. Sci.* **7,** 377 (1977).
14. H. G. Grimmeis, *Ann. Rev. Mater. Sci.* **7,** 343 (1977).
15. M. F. Li, *Modern Semiconductor Quantum Physics,* World Scientific, Singapore, 1994, Section 360.
16. A. Chantre, G. Vincent, and D. Bois, *Phys. Rev. B* **23,** 5335 (1981).
17. D. V. Lang, R. A. Logan, and M. Jaros, *Phys. Rev. B.* **19,** 1015 (1979).
18. D. V. Lang, in *Deep Centers in Semiconductors,* ed. by S. T. Pantelides. Gordon and Breach, New York, 1985, p. 489.
19. P. M. Mooney, *J. Appl. Phys.* **67,** R1 (1990).
20. J. C. Bourgoin (ed.), *Physics of the DX Centers in AlGaAs Alloys.* Sci. Tech. Publications, Vaduz, 1990.

21. K. J. Malloy and K. Khachaturyan, in *Imperfections in III/V Materials,* Vol. 38, *Semiconductors and Semimetals.* Academic Press, Boston, 1993, p. 235.
22. C. H. Henry and D. V. Lang, *Phys. Rev. B* **15,** 989 (1977).
23. K. Huang and R. Rhys, *Proc. Roy. Soc.* **204,** 406 (1950).
24. N. Lifshitz, A. Jayaraman, and R. A. Logan, *Phys. Rev. B* **21,** 670 (1980).
25. M. Mizuta, M. Tachikawa, H. Kukimoto, and S. Minomura, *Jpn. J. Appl. Phys.* **24,** L143 (1985).
26. M. Tachikawa, T. Fujisawa, H. Kukimoto, A. Shibata, G. Oomi, and S. Minomura, *Jpn. J. Appl. Phys.* **24,** L893 (1985).
27. M. F. Li, P. Y. Yu, E. R. Weber, and W. L. Hansen, *Appl. Phys. Lett.* **51,** 349 (1987).
28. M. F. Li, P. Y. Yu, R. Weber, and W. L. Hansen, *Phys. Rev. B* **35,** 7505 (1987).
29. R. Legros, P. M. Mooney, and S. L. Wright, *Phys. Rev. B* **35,** 7505 (1987).
30. M. F. Li, P. Y. Yu, W. Shan, W. L. Hansen, and E. R. Weber, *Proc. 19th Int. Conf. Physics of Semiconductors,* ed. by W. Zawadzki. Institute of Physics, Polish Academy of Sciences, 1988, p. 1051.
31. P. M. Mooney, T. N. Theis, and S. L. Wright, *Appl. Phys. Lett.* **53,** 2546 (1988).
32. E. Calleja, F. Garcia, A. Gomez, E. Munoz, P. M. Mooney, T. N. Morgan, and S. L. Wright, *Appl. Phys. Lett.* **56,** 934 (1990).
33. T. Baba, M. Mizuta, T. Fujisawa, J. Yoshino, and H. Kukimoto, *Jpn. J. Appl. Phys.* **28,** L891 (1989).
34. T. N. Theis, P. M. Mooney, and S. L. Wright, *Phys. Rev. Lett.* **60,** 3619 (1988).
35. B. L. Zhou, K. Ploog, E. Gmelin, X. O. Zheng, and M. Schultz, *Appl. Phys. A* **28,** 223 (1982).
36. Z. B. Zhang, *Phys. Rev. B* **44,** 3417 (1991).
37. W. Shan, P. Y. Yu, M. F. Li, W. L. Hansen, and E. Bauser, *Phys. Rev. B* **40,** 7831 (1989).
38. D. J. Chadi and K. J. Chang, *Phys. Rev. Lett.* **61,** 873 (1988); *Phys. Rev. B* **39,** 10366 (1989).
39. K. A. Khachaturyan, E. R. Weber, and M. Kaminska, in *Defects in Semiconductors, 15.* Trans. Tech., Switzerland, 1989, p. 1067.
40. J. C. M. Henning and J. P. M. Ansems, *Semicond. Sci. Tech.* **2,** 1 (1987).
41. H. P. Hjarmarson and T. J. Drumond, *Appl. Phys. Lett.* **48,** 656 (1986).
42. J. C. Bourgoin, *Solid State Phenomena* **10,** 253 (1989).
43. E. Yamaguchi, K. Sheraishi, and T. Ohno, *J. Phys. Soc. Japan* **60,** 3093 (1990).
44. K. A. Khachaturyan, D. D. Awschalom, J. R. Rosen, and E. R. Weber, *Phys. Rev. Lett.* **63,** 1311 (1989).
45. T. Fujisawa, J. Yoshino, and H. Kukimoto, *Jpn. J. Appl. Phys.* **29,** L388 (1990); and in *Proc. 20th Int. Conf. Phys. Semiconductors,* ed. by E. M. Anastassakis and J. D. Joannopoulos. World Scientific, Singapore, 1990, Vol. 1, p. 509.
46. M. Baj, L. H. Dmowski, and T. Stupinski, *Phys. Rev. Lett.* **71,** 3529 (1993); M. Baj and L. H. Dmowski, *J. Phys. Chem. Solids* **56,** 589 (1995).
47. For a discussion on the theory of the D^0 state see, for example, J. Dabrowski and M. Scheffler, in *Defects in Semiconductors 16, Materials Science Forum* **83–87** (Trans Tech Publ., Switzerland, 1992), p. 735. For discussions on experimental results see, for example, T. Suski in *Defects in Semiconductors 17, Materials Science Forum* **143–147** (Trans Tech Publ., Switzerland, 1994), p. 975.
48. U. Willke, M. L. Fille, D. K. Maude, J. C. Portal, and P. Gibart, in *Proc. 22nd Int. Conf. Phys. Semiconductors,* ed. by D. J. Lockwood. World Scientific, Singapore, 1994, p. 2295.
49. M. F. Li, Y. Y. Luo, P. Y. Yu, E. R. Weber, H. Fujioka, A. Y. Du, S. J. Chua and Y. T. Lim, *Phys. Rev. B* **50,** 7996 (1994); and in *Proc. 22nd Int. Conf. Phys. Semiconductors,* ed. by D. J. Lockwood. World Scientific, Singapore, 1994, p. 2303.
50. M. G. Craford, G. E. Stillman, N. Holonyak, Jr., and J. A. Rossi, *J. Electronic Materials* **20,** 3 (1991); *Phys. Rev.* **168,** 867 (1968).

51. R. A. Craven and D. Finn, *J. Appl. Phys.* **50,** 6334 (1979).
52. See, for example, J. A. Wolk, W. Walukiewicz, M. L. Thewalt, and E. E. Haller, *Phys. Rev. Lett.* **68,** 3619 (1992) for a description of the infrared vibrational study and J. Zeman, M. Zigone, and G. Martinez, *Phys. Rev. B* **51,** 1755 (1995) for a review of the Raman investigation of the DX center, both under pressure inside the DAC.
53. See, for example, N. M. Johnson, W. Shan, and P. Y. Yu, *Phys. Rev. B* **39,** 3431 (1989) for a study of the pressure dependence of the P_b center at the $\langle 111 \rangle$ Si–SiO$_2$ interface.

CHAPTER 5.3

Spatial Correlations of Impurity Charges in Doped Semiconductors

Tadeusz Suski

UNIPRESS, HIGH PRESSURE RESEARCH CENTER
POLISH ACADEMY OF SCIENCES
WARSAW, POLAND

I. INTRODUCTION	485
II. EXPERIMENTAL TECHNIQUES	487
III. DONOR CHARGES IN HgSe:Fe	487
IV. DONOR STATES IN GaAs AND AlGaAs	490
V. EXPERIMENTAL STUDIES OF SPATIAL CORRELATIONS FOR DX-CENTER CHARGES IN BULK SEMICONDUCTORS	495
VI. SPATIAL CORRELATIONS OF REMOTE IMPURITY CHARGES	502
VII. REMOTE-CHARGE CORRELATIONS AND QUANTUM TRANSPORT OF 2DEG	505
VIII. CONCLUDING REMARKS	508
IX. REFERENCES	510

I. Introduction

The spatial distributions of dopants in semiconductors, and of the dopant-related charges are usually assumed to be random, that is, to follow Poisson statistics (see, for example, [1]). Shklovskii and Efros [2] pointed out that correlated impurity distribution (ionized donors and acceptors) occurs in compensated semiconductors. Gerlach and Rautenberg [3] showed that in a typical semiconductor the scattering of electrons by the ionized impurities cannot be considered as being caused by the random arrangement of charges for a doping level greater than 5×10^{16} cm^{-2}. Schubert [4] reviewed an interesting case of highly Be δ-doped GaAs samples where a significant deviation of the dopant distribution from pure random was found. It has been proposed that repulsive Coulomb interactions of ionized Be acceptors (Be$^-$) act as a driving force to spread dopants out of the narrow, highly doped sheet (δ-doping). At Be concentrations above about 1×10^{14} cm^{-2}, the mean Be separation could be of the order of two cation sites. At the growth temperature, the interacceptor Coulomb interaction is strong enough to induce an increase in the impurity separation. Thus, the width of the doping layer increases from typically 5 nm to over 10 nm along the growth direction. As a result, a kind of ordering is superimposed on the sublattice of sites occupied by impurities. Impurities tend to keep the maximum distance between themselves.

At lower impurity concentrations, in the range of 10^{12} cm^{-2}, the mean dopant-to-dopant separation is large, and Coulomb interactions are insufficient to influence impurity-site arrangement in a semiconductor. Thus, the presence of built-in disorder of the available impurity positions, corresponding to the random distribution of dopants, usually characterizes an impurity distribution. The considered two-dimensional dopant concentration ($\geq 10^{12}$ cm^{-2}) is equivalent to a concentration on the order 10^{18}–10^{19} cm^{-3} in a bulk semiconductor. It represents the doping level useful for most applications. It turns out that in such a semiconductor, Coulomb interactions between charges localized on individual donors can cause their ordering among the available impurity sites, provided that the impurities are only partially occupied. The considered spatial correlations arises when charges of only one sign or both signs are present. The former case can be illustrated by positive charges of ionized donors. Consequently, the correlation caused by Coulomb repulsion resembles that in the Wigner glassy state (see, e.g., [5, 6]). The latter situation can be found in compensated semiconductors or in systems with donors which can localize two electrons. In both cases, the Coulomb attraction between positively charged ionized donors and acceptors/negatively charged donors contributes to the crystal total energy.

In this context, the cases of resonant Fe donors in HgSe and DX centers in GaAs, AlGaAs, and CdTe, remote donors (DX centers as well) in modulation-doped heterostructures of GaAs/AlGaAs, and quantum wells of AlGaAs/InGaAs/GaAs will be discussed.

For establishing the correlated arrangement of charges, the lifetime of the impurity state must be sufficiently long. This requirement is equivalent to the condition of small hybridization between the impurity state and the extended band states. In the two examples discussed in this chapter, the Fe-related resonant donor state in HgSe is very long-living as demonstrated by an extreme narrowness of the EPR line [7], whereas strong coupling to the matrix lattice in case of DX centers leads to a long-living state as well.

Since spatially correlated charged impurities are less effective scatterers of the conduction electrons than in the case of their random distribution, electrical transport phenomena are significantly influenced by the presence of spatial correlations of impurity charges. We will demonstrate how the hydrostatic pressure and classical and quantum transport can be used to visualize and to study the effect of spatial ordering of impurity charges.

II. Experimental Techniques

High-pressure experiments performed for the purpose of studying the phenomena characteristic of interacting impurity charges consist of low-temperature magnetotransport measurements. Sometimes it was necessary to illuminate a sample mounted in a high-pressure cell. In such situations, either a cell equipped with an optical window or a semiconductor light-emitting diode mounted inside the cell were used.

For standard high-pressure experiments at low temperatures, a special liquid clamp cell was used. A schematic view of such a cell is shown in Fig. 1a. The sample space is filled with petroleum spirit. The pressure transmitting medium is compressed at room temperature and then cooled down. During the cool-down, the pressure medium becomes frozen, which causes a pressure drop by a magnitude ΔP. This decrease depends on the P value established at room temperature. The pressure on the sample remains hydrostatic and is measured *in situ* by an n-InSb semiconductor manometer. In this way hydrostatic pressures up to ~2.1 GPa (21 kbar) ($T = 300$ K) and up to ~2.0 GPa ($T = 4.2$ K) can be reached. Since the pressure cell is made of beryllium–copper and/or a special nonmagnetic Ni–Cr alloy, it can be used in magnetic fields as high as 25 T.

A disadvantage of the cell is that the pressure must be changed at room temperature. Therefore, the cell has to be heated up to generate a new pressure value.

In a second type of pressure cell (Fig. 1b), helium gas is used as a pressure transmitting medium. The cell is connected to a gas compressor by a special capillary. Because of the very low temperature of He-gas solidification (e.g., ~77 K at 1.5 GPa), it is possible to vary pressure at low temperatures. The maximum pressure available in this type of cell is 1.5 GPa.

III. Donor Charges in HgSe:Fe

In HgSe the ground state of the Fe^{2+} donor is degenerate with the conduction band. The donor level corresponding to the Fe^{2+} state, E_{Fe}, is located ~210 meV above the conduction band minimum. As a consequence, autoionization processes occur: $Fe^{2+} \rightarrow Fe^{3+} + e^-$. For Fe concentrations less than $N_{Fe} = 4.5 \times 10^{18}$ cm^{-3}, the number of electrons in the band is proportional to the number of impurities. As the concentration of Fe exceeds 4.5×10^{18} cm^{-3}, the Fermi energy of conduction electrons reaches E_{Fe}, and consequently Fe donors become only partly ionized (Fermi-level pinning), exhibiting two charge states. This is the so-called mixed valence

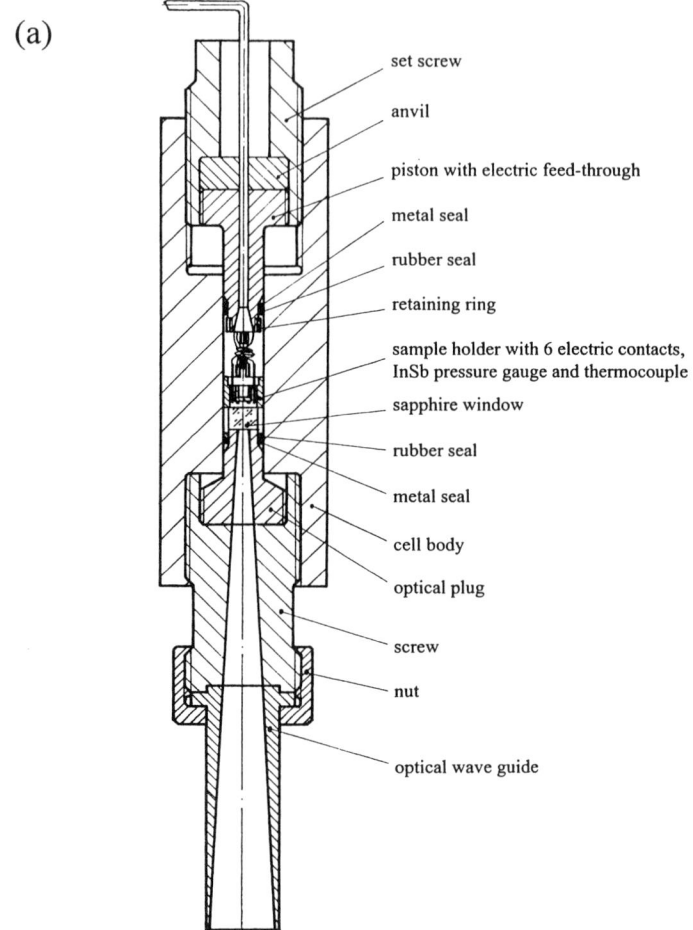

FIG. 1. Schematic view of a UNIPRESS liquid clamp cell (a) with optical access and (b) with a gas cell for magnetotransport measurements.

regime (MVR). In the higher doping range, the concentration of electrons is roughly independent of N_{Fe} (see, e.g., [7]).

In the MVR the mobility of the conduction electrons, μ, at temperatures lower than ~ 100 K was found to exceed greatly that in HgSe doped with Ga to similar doping levels (the Ga donor does not form a resonant localized state). Moreover, theoretical calculations of μ, μ_{th}, underestimated considerably the values observed experimentally, μ_{exp}; $\mu_{exp}/\mu_{th} > 10$ [7]. To explain the reduced scattering rate, the interesting concept of formation of a Wigner-like lattice of impurity charges (Fe^{+3}) has been proposed [8]. The following simple model was used with the purpose of explaining the

5.3 Spatial Correlations of Impurity Charges

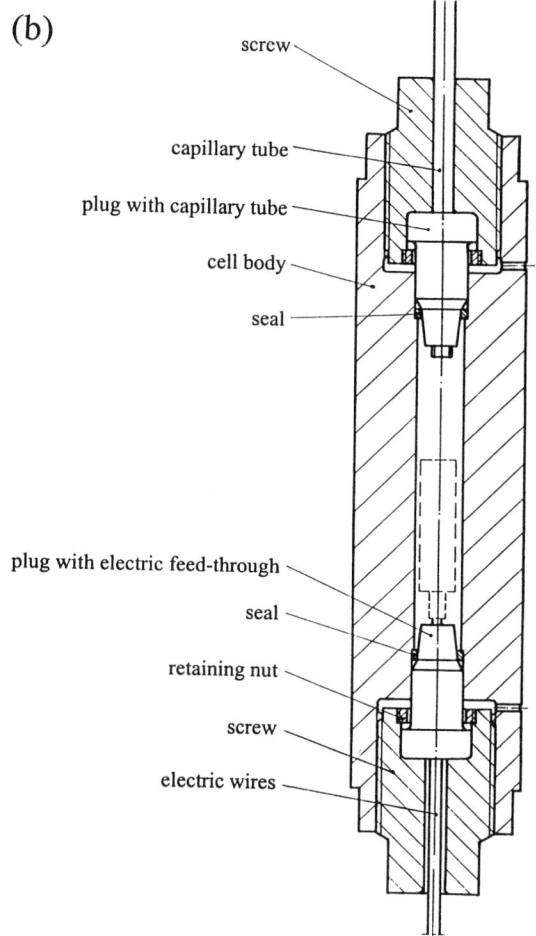

Fig. 1. (*continued*)

mobility enhancement in HgSe:Fe. The Coulomb repulsion between the positive charges localized on Fe^{3+} ionized donors causes the appearance around each Fe^{3+} donor of a region without other Fe^{3+} ions. One can introduce the pair correlation function for the ionized donors in a steplike form: $g_{++}(r) = 0$ for $r < r_c$ and $g_{++}(r) = 1$ for $r > r_c$ [7]. It means that any possible correlations in positions of donor charges more distant than r_c are neglected (short-range correlation model). The correlation radius can be found from the equation

$$nV_c = 1 - \exp(N_{Fe}V_c),$$

where n is the conduction electron concentration and $V_c = 4\pi r_c^3/3$.

The Fourier transform of $g_{++}(r)$ corresponds to the structure factor $S(q)$ of scattering centers, which in turn is decisive in determining the momentum relaxation time, τ_t:

$$\frac{1}{\tau_t} = \frac{m^*}{6\pi\hbar^2} \int_0^{2k_F} q^3 S(q) |V(q)|^2 \, dq.$$

$V(q)$ is the matrix element of the screened Coulomb potential (scattering potential); m^* and k_F are the electron effective mass and wave vector corresponding to the Fermi level. τ_t determines the scattering of the conduction electrons by ionized impurities. This scattering mechanism becomes more important at lower temperatures where the contribution to electron scattering by lattice vibration decreases. τ_t is proportional to the mobility. Calculations of μ which take into account the considered interdonor interactions reproduce well the experimentally observed increase in the mobility. A number of Fe atoms in the cation sublattice of (HgFe)Se become sufficient to induce this kind of scattering mechanism [7, 9].

Application of hydrostatic pressure to the HgSe:Fe with $N_{Fe} \geq 4.5 \times 10^{18}$ cm^{-3} enables verification of the concept of ionized donor superlattice formation [10, 11]. Since HgSe represents the zero-gap semiconductor, a pressure increase causes a significant reduction of the energetic distance between the Γ^6 (light hole) and Γ^8 (conduction) bands and gives rise to a decrease in the density of states in the conduction band. It induces an upward shift of the Fermi level, and at sufficiently high pressure the Fermi level approaches the position of E_{Fe}. Thus, with increasing pressure more Fe ions are occupied by electrons and the transition to the MVR occurs (Fig. 2). Assuming that the concept of the formation of an ionized donor superlattice is correct, the corresponding increase in the electron mobility should exceed the variation in μ resulting from the pressure-induced reduction in the effective mass value. The observed increase in μ is fully consistent with the calculations based on the presented short-range correlation model (Fig. 3). One advantage of using pressure for the verification of the concept based on the formation of the spatial correlation in HgSe highly doped with Fe is the use of only one sample. The equivalent ("ambient pressure") procedure of modifying the ratio of ionized to neutral Fe donors (Fe^{+3}/Fe^{+2}) requires application of several samples with different N_{Fe} values. In this case uncontrolled parameters of the samples come into play, such as different compensation rates. Compensation represents a parameter which influences the mobility magnitude.

IV. Donor States in GaAs and AlGaAs

Donor impurities in GaAs and AlGaAs may form three distinct states: (i) shallow, hydrogen-like, and relatively extended states, (ii) considerably

5.3 SPATIAL CORRELATIONS OF IMPURITY CHARGES

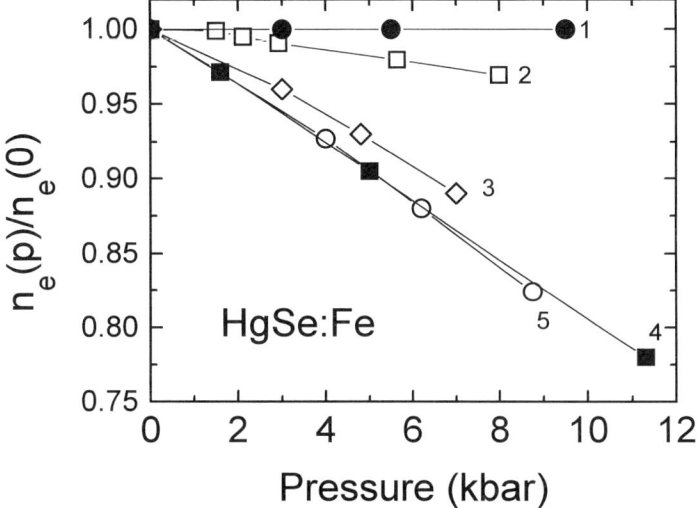

FIG. 2. Relative electron concentration in HgSe:Fe versus hydrostatic pressure at $T = 4.2$ K observed for various Fe concentrations. (1) $N_{Fe} = 1 \times 10^{18}$ cm^{-3}, (2) $N_{Fe} = 2 \times 10^{18}$ cm^{-3}, (3) $N_{Fe} = 4 \times 10^{18}$ cm^{-3}, (4) $N_{Fe} = 8 \times 10^{18}$ cm^{-3}, (5) $N_{Fe} = 20 \times 10^{18}$ cm^{-3} [10].

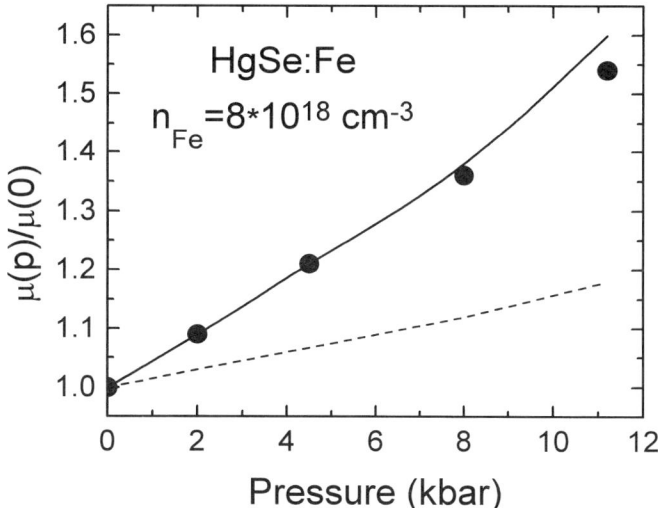

FIG. 3. Relative mobility as a function of hydrostatic pressure in HgSe:Fe sample (with $n = 8 \times 10^{18}$ cm^{-3}) measured at $T = 4.2$ K. The dashed curve shows the result of calculations without and with (solid curve) spatial correlations of Fe-donor charges (experimental data [10], calculations [7]).

more localized DX states accompanied by a strong lattice relaxation in the vicinity of the donor, and (iii) localized states, D^o, with weak coupling to the lattice [12]. The ionized shallow donor represents a positively charged center d^+. It is commonly accepted that state (ii) is doubly occupied (negatively charged DX^-), corresponding to the negative Hubbard correlation energy of electrons on the impurity site, that is, $U < 0$ [13, 14]. The D^o state localizes one electron and represents a neutral donor state.

Now we will make few comments about the availability of these three states in $Al_xGa_{1-x}As$ ($0 < x < 0.4$). For electron concentrations above roughly 10^{18} cm^{-3}, the hydrogenic state d^+ cannot localize electrons (because of the Mott transition). Thus, in highly doped n-type semiconductors, d^+ states exist at all temperatures.

Though the energetic position of DX^- states in GaAs and $Al_xGa_{1-x}As$ depends on the donor species (see, e.g., [12]), it is commonly accepted to consider $x \approx 0.2$ as a critical value below which the electronic level related to DX^- state of a donor is resonant with the conduction band (Fig. 4a). The d^+ state represents a ground state of the donor ($n \leq 10^{18}$ cm^{-3}). For $x > 0.2$, the DX^- becomes a donor ground state, forming a level in the gap of AlGaAs alloy. By applying an external pressure, it is possible to vary the relative position of the DX^- level, E_{DX^-}, into coincidence with the Fermi level in the case of GaAs [15]. In such situation, some of the conduction electrons may occupy DX states, provided that their thermal energy exceeds the energy barrier resulting from the relaxation effect just mentioned (see Chapter 5.2). With decreasing temperature, the characteristic time for capture or emission of electrons by DX centers becomes very long. It takes place below ~100 K for Si, S-, and Te-donors and at about 40 K for Sn-donors [12]. Because of the described phenomena, variation of the applied pressure either leads to changes in the conduction electron

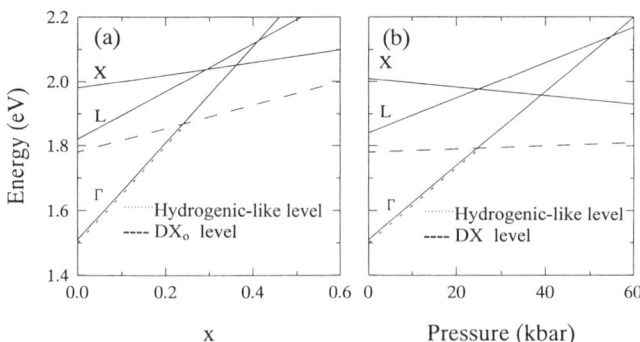

FIG. 4. Energy band minima as a function of aluminum content x in $Al_xGa_{1-x}As$ (a) and hydrostatic pressure in GaAs (b). The energies of the Si DX level and the hydrogenic-like state related to the Γ minimum are also shown.

concentration (high temperatures) or keeps this concentration independent of the pressure (at $T < T_m$, where T_m is the temperature below which barriers for electron capture and emission of electrons become relevant) (Fig. 5).

Optical ionization of DX centers at low temperatures leads to the persistent photoconductivity effect (PPC). After illumination is terminated, the electrons excited from DX states stay in the conduction band. Barriers to electron capture block electron localization on DX centers.

Appearance of the D^0 donor state has been clearly demonstrated in GaAs:Ge (see, e.g., [12] and references therein). Moderate pressures $P \sim 1$ GPa are sufficient to populate both D^0 and DX states of Ge, if temperature is above ~ 100 K. Applying pressures below this temperature enables tuning of the D^0 state population only (Fig. 6). This different behavior of DX and the D^0 states of the Ge-donor in GaAs reflects a strong and weak coupling to the matrix lattice of these two states of the donor. Sufficiently high doping can bring the Fermi level to coincidence with the D^0 and DX states. For other donors (e.g., Si donors), the considered localized states lie much higher in the GaAs conduction band. Therefore, application of $P \sim 3$ GPa is necessary to visualize D^0 states by means of optical experiments [16]. The D^0 states practically do not influence electrical properties of $Al_xGa_{1-x}As$ semiconductor.

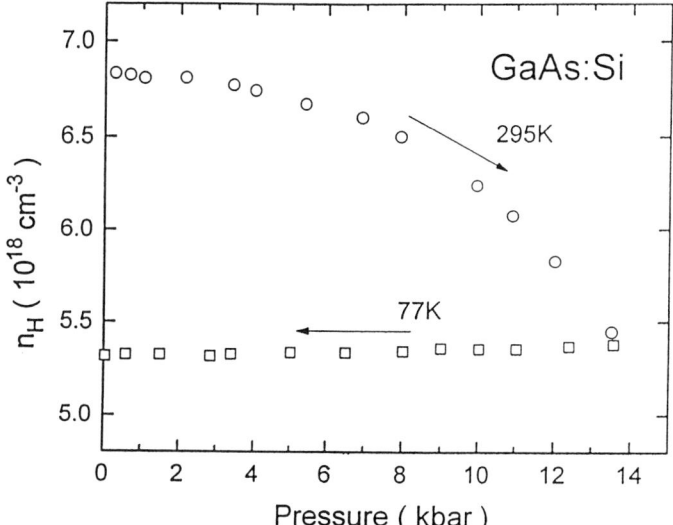

FIG. 5. Electron concentration in MBE GaAs:Si as a function of hydrostatic pressure measured at $T = 295$ K (circles) and at $T = 77$ K during release of pressure from $P = 14$ kbar applied at $T = 295$ K (squares) [21].

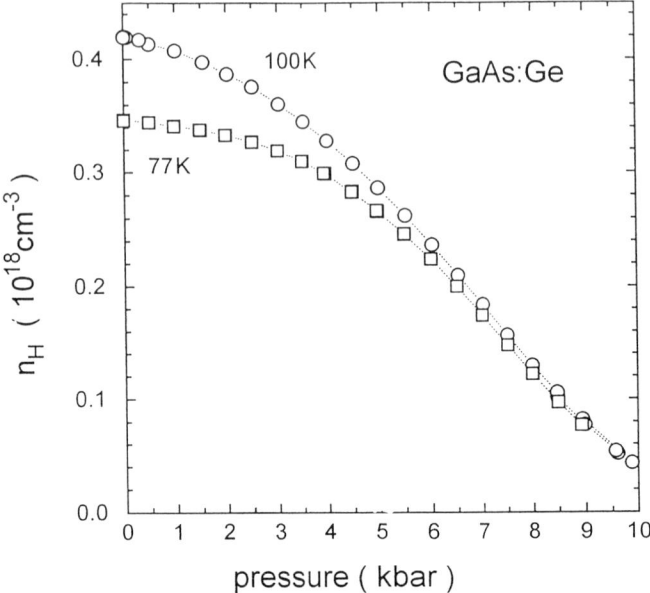

FIG. 6. Electron concentration in MOVPE GaAs:Ge as a function of hydrostatic pressure measured at $T = 100$ K (circles) and at $T = 77$ K during release of pressure from $P = 10$ kbar applied at $T = 100$ K (squares) [12].

In this section we will concentrate on the situation when donors in a $Al_xGa_{1-x}As$ semiconductor are partially occupied with electrons (mixed valence regime). This corresponds to the presence of positively (d^+) and negatively (DX^-) charged objects, both of which contribute to scattering of the conduction electrons. The case of D^o state contribution will not be illustrated here. Because of the electrostatic interactions, in thermal equilibrium conditions, the spatial arrangement of d^+ and DX^- becomes correlated. Attraction between differently charged donor states induces formation of dipole-like objects consisting of closely located impurity pairs. Though of different character than the Fe donors in HgSe, the spatial correlation of donor charges, d^+–DX^-, significantly modifies the carrier mobility [17–19]. This results from the fact that d^+–DX^- dipole-like objects represent much weaker scattering efficiency compared with independent and distant d^+ and DX^- states.

The DX center characterized by a negative correlation energy requires consideration of three pair correlation functions: $g_{++}(r)$, $g_{+-}(r)$, and $g_{--}(r)$, where the subscripts specify which of the possible charge states of two impurities are concerned [7, 20]. The most important of the three correlation functions is $g_{+-}(r)$. Since the charges of opposite sign attract each other, there is a tendency to form spaced pairs of ionized and occupied donors d^+–DX^-. Therefore, it is reasonable to assume a steplike form of $g_{+-}(r)$ with values exceeding unity for $r < r_d$, and equal to 1 (no correlation) for $r > r_d$. The parameter r_d can be interpreted as a maximum dipole length of the d^+–DX^- pair. One can derive an equation for r_d [20],

$$N_- V_d = 1 - \exp(-NV_d)(1 + NV_d),$$

where $V_d = 4\pi r_d^3/3$.

Considering the gain in total crystal energy due to the electrostatic energy related to correlations, E_C, it is worthwhile to point out an important difference between semiconductors with impurities characterized by positive or negative correlation energy, U. In the case of HgSe:Fe (or other spatial correlations realized in the system of positively charged and neutral impurities), the maximum of E_C occurs for high (~0.8) ratios of ionized to all donor centers. Energetic gain, E_c, is very small when this rate tends to zero (all conduction electrons trapped by donors). In the case of systems with impurities of negative-U character, the total freezeout of electrons on the DX states results in formation of charged impurities, half of which are positively charged and the second half negatively charged (assuming the absence of other impurities). Here, a high degree of spatial correlations between d^+ and DX^- can be achieved [20]. This situation corresponds to high values of E_c.

V. Experimental Studies of Spatial Correlations for DX-Center Charges in Bulk Semiconductors

As mentioned in the previous section, at high temperatures ($T > T_m$) the concentration of the conduction electrons, n, decreases with applied pressure because when E_{DX^-} approaches E_F, the transfer of electrons from the band to DX centers takes place. On the other hand, at $T < T_m$, releasing the pressure does not change n because the trapped electrons cannot pass over the energy barrier between DX and band states. Thus, depending on the pressure value at which the temperature is reduced, different n values can be achieved in one sample (and at ambient pressure $T < T_m$). As in the case of HgSe:Fe system, a hydrostatic pressure is a tool which enables study of variation of μ with n. Moreover, for studies of semiconductor systems consisting of DX centers, there is an additional advantage to using pressure. At low temperatures, one can get rid of pressure contributions

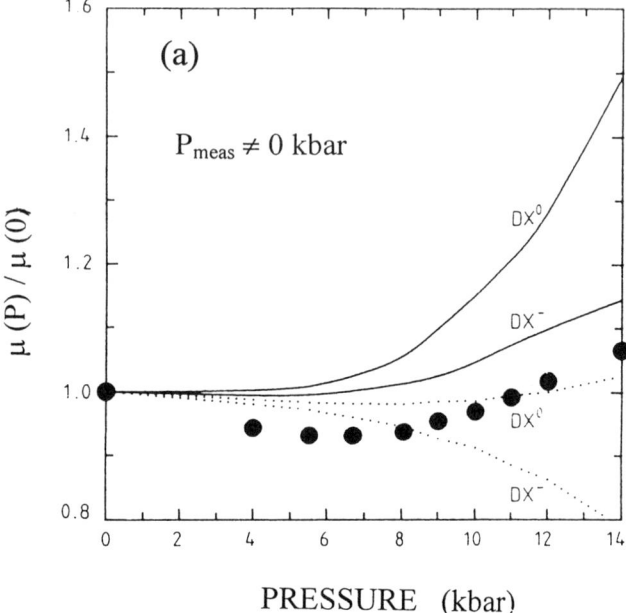

FIG. 7. Relative changes of the electron mobility in MBE GaAs:Si (6.9×10^{18} cm^{-3}) measured at $T = 4.2$ K as a function of the hydrostatic pressure applied at $T = 295$ K (a) and after pressure release at $T = 77$ K (b). Theoretical values are calculated without (dotted curves) and with (solid curves) spatial correlations of donor charges. Results for both D^0 (i.e., $U > 0$) and DX$^-$ (i.e., $U < 0$) models of DX centers are presented [21].

to the band-structure changes. Figure 7 gives a comparison between mobility variation ($T = 4.2$ K) with pressure and the same measured after the pressure release at low temperatures. To reproduce the increase of μ with n, it was necessary to take into account correlations in the spatial distribution of the donor charges, that is, a formation of d$^+$–DX$^-$ dipoles. Calculations performed within the assumption about the random distribution of the donor charges led to the mobility drop accompanying an increase of the applied pressure [21].

A useful idea for experimental testing of the existence of correlations in the arrangement of d$^+$ and DX$^-$ states has been applied in various semiconducting systems. It consists of determining of the spatial-correlation-related contribution to the electron transport by examining μ as a multivalued function of the electron concentration. Accordingly, it has been demonstrated that depending on the method of preparation of the same sample (GaAs:Si, CdTe:In, AlGaAs:Si, modulation-doped hetero-

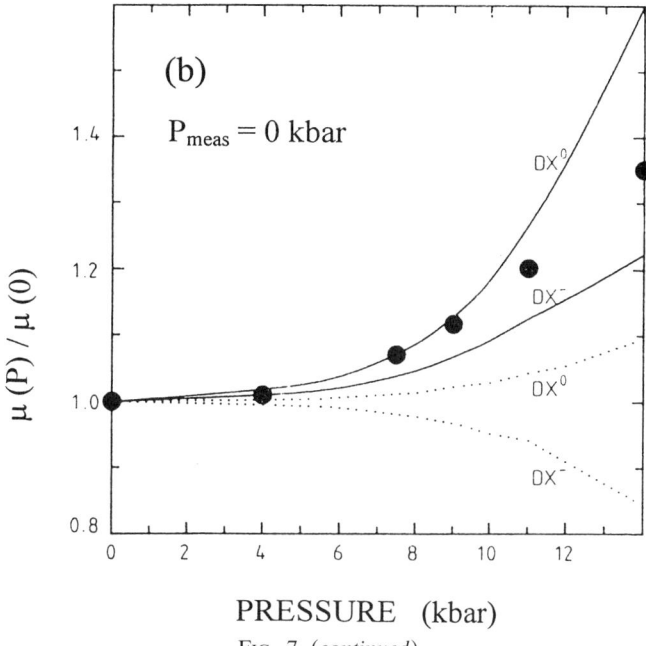

FIG. 7. (*continued*)

structures of AlGaAs/GaAs, and quantum wells of AlGaAs/InGaAs/GaAs), various values of μ can be obtained for the same concentration of conducting electrons. This is interpreted as originating from various distributions of charges among the donor sites. In the performed experiments, mobilities (T = 4.2 K or 77 K) measured for the most favorable spatial correlation of charges for a given amount of occupied donors (obtained during slow cooling of the sample) are compared with mobility values corresponding to the situation with reduced correlations. In general, processes of electron capture on DX centers favor a high degree of spatial correlation. In contrast, processes of electron emission from DX centers induce a reduction in the correlation degree [22].

To alter the electron concentration (i.e., the amount of occupied DX centers), four procedures have been employed:

1. *High-pressure freezeout* (HPFO) of electrons on the metastable DX⁻ states as described earlier. Providing spatial correlations in the distribution of d⁺ and DX⁻ states, the HPFO procedure allows high values of μ for each value of n (determined by the magnitude of the freezeout pressure).

2. *Ionization of DX centers by phonons.* This procedure leads to the destruction of spatial correlations of the donor charges prepared by the HPFO method. The thermal excitation is used to transfer localized electrons to the conduction band (or conducting channel in the case of two-dimensional electron gas, 2DEG) of the examined semiconductor (or semiconductor structure). This consists of subsequent annealing of the sample to temperatures above about 100 K and cooling down to 4.2 K (or 77 K) for measurements of n and μ. Each successive annealing step requires heating of the sample to a higher temperature (above ~110 K). Increasing temperature induces transfer of a portion of electrons from the metastable DX^- state to the conduction band. If the heating temperature is not too high, destruction of the correlations in the impurity-charge distribution is caused by electron emission from randomly chosen DX^- centers. At this temperature range, retrapping of electrons onto DX^- centers and thus rearrangement of their spatial positions is hindered by a capture barrier.
3. *Ionization of DX centers by photons* (persistent photoconductivity, PPC). Having frozen-in the correlated occupation of DX centers (e.g., by the HPFO procedure), one can study its optical destruction. It can be performed by illuminating the sample with a sequence of pulses from an infrared light-emitting diode (excitation energy below the energy gap). Each step of the illumination is followed by n and μ measurement. The incident photons persistently photoionize DX centers (n increases in a steplike manner) independently of the position of the donor with respect to other donors. Hence, the random nature of the photoionization can be used to destroy the spatial correlations.
4. *Temperature-induced capture of electrons on DX centers.* Persistent photoionization of DX^- states (PPC method) produces a reservoir of electrons in the conduction band. Now, the procedure inverse to that described in Z can be employed, with the annealing cycles now leading to the capture of electrons by DX centers. As in the former cases, measurements of n and μ are performed at low temperatures for each annealing step. This annealing-stimulated capture causes an increase in the correlation degree. Thus, high values of μ should be observed.

The first example concerns studies of spatial correlations in GaAs highly doped with Si donors; performed by Dmowski *et al.* [23]. Figure 8 shows the electron mobility versus n obtained after various procedures of the sample preparation. All measurements were performed at $T = 77$ K. The open circles, corresponding to employing HPFO, represent the highest magnitude of μ for each value of n. The respective degree of spatial correlations for this method is the highest. One can see that an increase in n

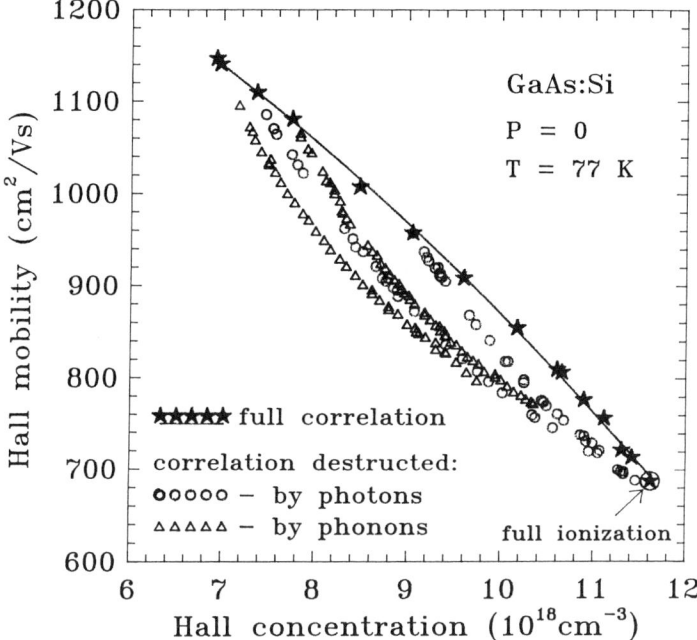

FIG. 8. Mobility versus carrier concentration obtained by different procedures. Open circles represent high-pressure freezeout (HPFO) which results in a high degree of spatial correlations, solid squares and open triangles mark the low correlation path achieved by PPC effect and subsequent annealing steps, respectively [23].

caused by either method 2 or 3, after their employment at various stages of the HPFO procedure, gives much lower values of μ. In agreement with the expectations for processes based on electron emission from DX states, that is 2 and/or 3 [22], spatial correlations of donor charges are subject to destruction.

Figure 9 illustrates the case of CdTe:In [24]. An In or indium donor introduces a resonant DX metastable state whose localized level is situated more than 125 meV above the conduction band minimum. One has to apply pressures above 0.75 GPa to induce electron localization by DX states. Figure 9a shows μ versus n dependence for the HPFO procedure (higher μ values) in comparison with the situation induced by procedure 2. It resembles the behavior observed in GaAs:Si. HPFO favors a high degree of donor-charge correlation. In contrast, using procedure 2 causes a reduction in correlation.

In the case shown in Fig. 9b, the conducting electron concentration was increased stepwise by means of procedure 3 employed at low temperature

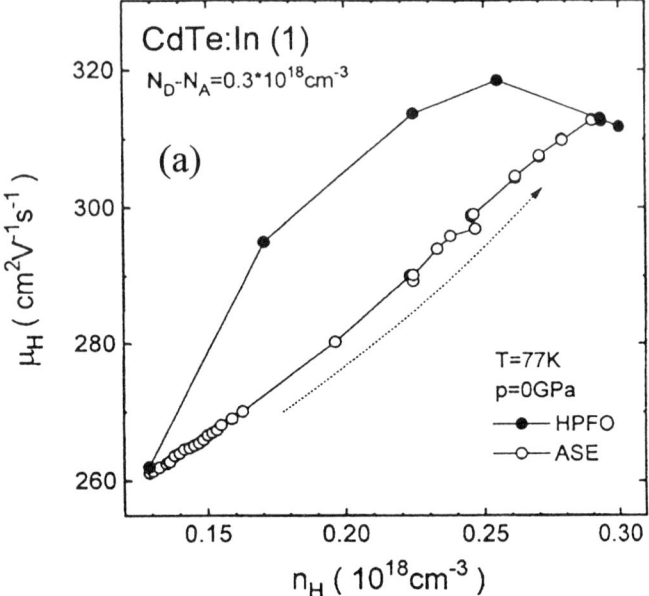

Fig. 9. Mobility of electrons as a function of Hall electron concentration measured at $T = 77$ for CdTe:In. (a) Sample 1. Solid circles represent HPFO, which results in a high degree of spatial correlations; open circles mark the low correlation path achieved by subsequent annealing steps (ASE). (b) Sample 2, measured at $P = 1.35$ GPa. Open squares correspond to the PPC effect and result in a lower degree of correlation; open circles represent subsequent sample annealing (ASC) leading to an increase in the correlation degree (after [24]).

and at pressure of $P = 1.35$ GPa (a partial occupation of DX states and thus a high degree of correlation was achieved by applying pressure at $T = 300$ K and cooling the sample at $P = $ const). The obtained results correspond to the electron emission from DX states and are associated with a reduction in the correlation degree. They are compared with the evolution of μ versus n obtained by method 4. The capture of electrons by DX states leads in the latter case to establishing a high degree of correlation in the donor-charge arrangement. Analysis of the mobility variation with n strongly suggests that the In-DX state in CdTe represents likely a negative U-center [24], and correspondingly the In donor localizes two electrons at high pressures.

In summary, one can say that the electron mobility enhancement caused by the reduction of scattering efficiency related to ionized impurities can occur independently of the semiconductor compound. The donors forming

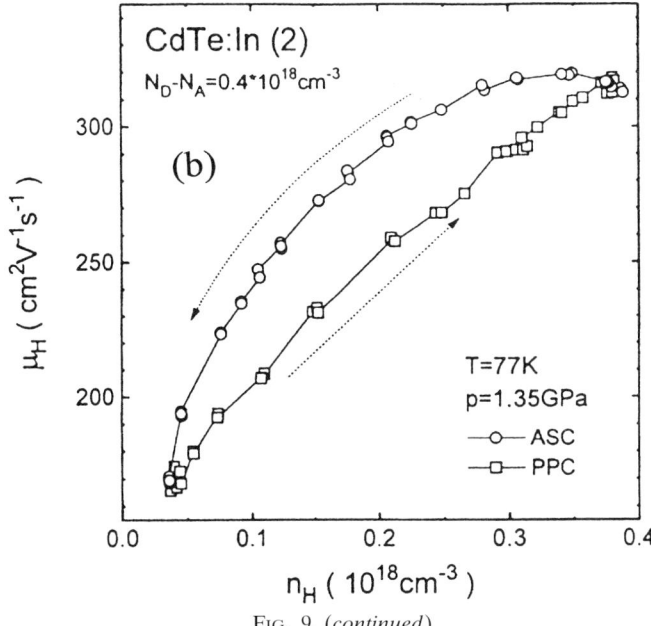

FIG. 9. (continued)

DX centers in both III–V and II–VI compounds exhibit a similar contributions to the electron mobility increase originating from spatial correlations in the arrangement of their charges among the dopant sites.

The multivalued character of $\mu(n)$ dependencies and the hysteresis observed in the capture–emission cycles have also been reported for ternary AlGaAs compounds [25–27]. In the case of AlGaAs, the Si DX centers split into four groups depending on the number of Al neighbors in the close vicinity of the Si dopant [28]. Figure 10a illustrates this phenomenon (see also [26]). First, the HPFO procedure is applied to prepare a reservoir of electrons captured by Si DX centers in $Al_{0.15}Ga_{0.85}As$. Then, the Hall electron concentration is measured while the sample temperature T is slowly increased in a controlled way. Thermal emission of electrons from DX centers yields the observed increase of n. When alloy splitting of the levels occurs, the emission rates of different levels differ. Temperature ranges where emission from a particular level occurs are separated. This results in the appearance of "emission steps" on the $n(T)$ dependence. Three well-defined steps correspond to consecutive emission from the levels DX_0, DX_1, and DX_2 (the subscript denotes the number of Al atoms replacing Ga in the vicinity of the shifted Si atom). By applying pressure, Piotrzkowski et al. [26, 29] demonstrated that the mobility in AlGaAs:Si depends significantly on the distribution of DX centers among various alloy-induced

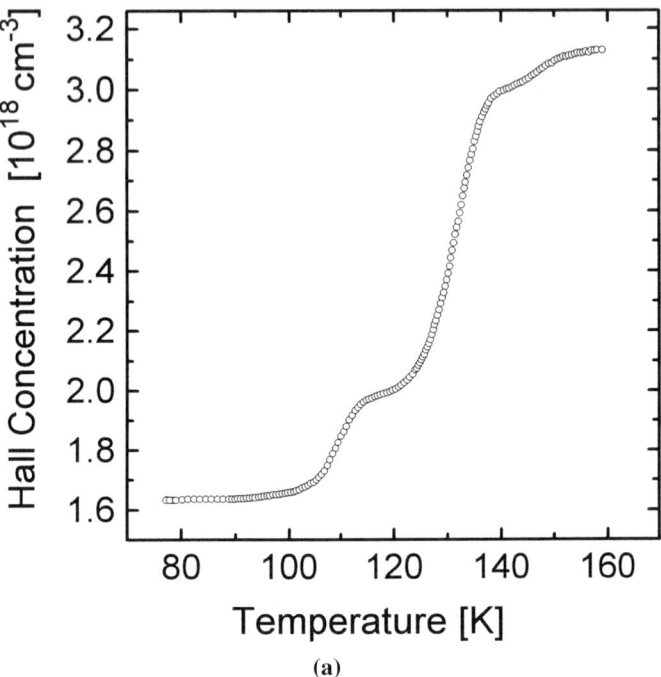

(a)

FIG. 10. Hall electron concentration (a) and mobility (b) in Si: $Al_{0.15}Ga_{0.85}As$ film as a function of increasing temperature (77–170 K) measured after HPFO of electrons onto Si DX centers (prepared according to measurements performed in [26]).

configurations as well as the corresponding spatial correlation of donor charges (Fig. 10b).

VI. Spatial Correlations of Remote Impurity Charges

This section discusses the existence of spatial correlations of impurity charges in 2D semiconductor systems. To illustrate this effect we will use an analysis of two-dimensional electron gas (2DEG) mobility, μ_{2D}, in modulation-doped heterostructures of $GaAs/Al_xGa_{1-x}As$ [30]. Very high mobilities achieved in such 2D systems result mainly from the spatial separation of the remote Si donors and the conducting channel formed at the interface in GaAs material. However, the contribution caused by the spatial correlations of impurity charges to the increase of μ_{2D} is often quite significant. The separation of dopants from the conducting channel is determined by thickness of the spacer, d, that is, the undoped layer of the AlGaAs

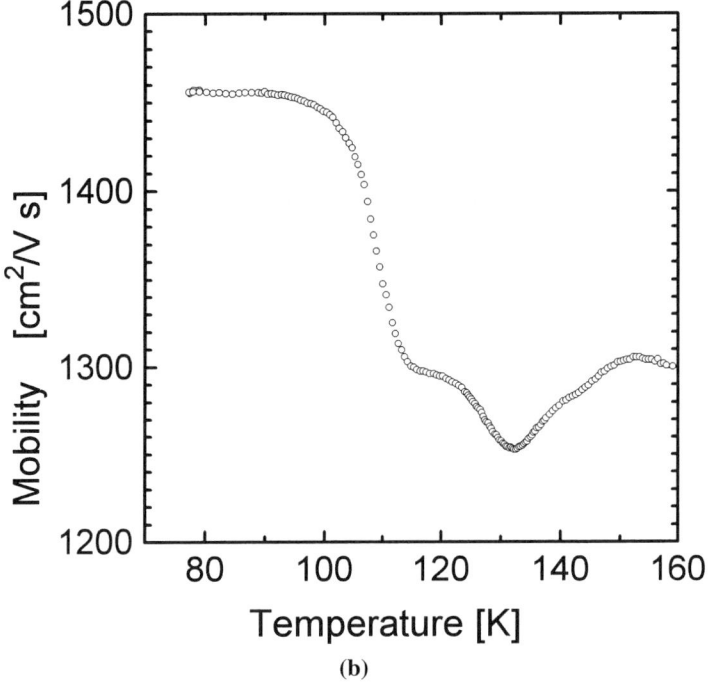

Fig. 10. (*continued*)

barrier. With increasing d, the contribution to the scattering mechanism of 2DEG caused by ionized impurities is significantly reduced (low-temperature μ_{2D} exceeds by a few orders of magnitude the mobility of electrons in bulk GaAs samples). At the same time (when d increases) the concentration of electrons, n_{2D}, transferred from Si donors to the 2D channel decreases. For device-oriented applications the conductivity value (proportional to the product of μ_{2D} and n_{2D}) is decisive for device quality. Thus, the d value has to be chosen as a compromise leading to maximum conductivity. Practically, a d value in the range of 10–20 nm is used in the design of real heterostructures employed in, for example, high-mobility field-effect transistors.

The behavior of remote Si donors is dominated by the DX center properties. One can expect that the effects of correlations within the remote impurity charges contribute considerably to the enhancement of mobility observed in 2D systems [31, 32]. This intuitive expectation has been proved by means of applying pressure [33] and confirmed in gated samples [34, 35].

Figure 11 shows the result of model calculations based on Monte Carlo simulations of correlations [20]. Accordingly, the calculations demonstrate

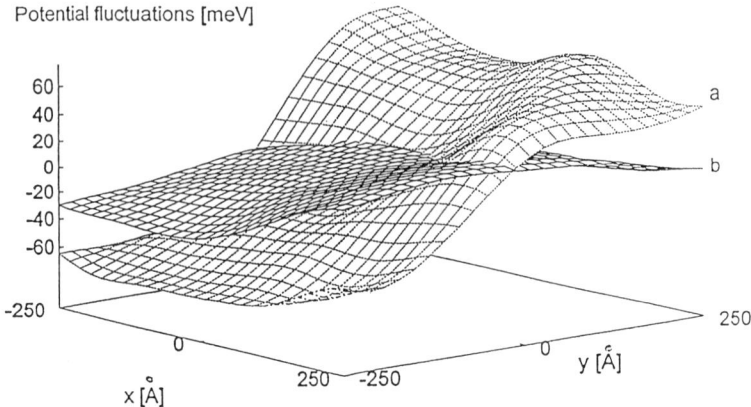

FIG. 11. Potential fluctuations produced by remote impurities (in the x, y plane perpendicular to the growth direction) and seen by 2DEG separated from dopants by 10-nm spacer (a) without spatial correlations of dopant charges, (b) with spatial correlations [36].

that the process of correlations in the population of remote donors by electrons leads to a considerable reduction in the impurity potential fluctuations seem by 2DEG [36].

Enhancement of μ_{2D} due to the correlations has been identified at low temperatures in modulation-doped GaAs/AlGaAs heterostructures and in pseudomorphic AlGaAs/InGaAs/GaAs quantum wells [37]. These findings were based on the same type of observation that was used for bulk semiconductors consisting of DX centers. Depending on the way electrons are distributed among Si donors, different values of μ_{2D} are achieved for the same 2DEG concentration n_{2D} (see Fig. 12, illustrating the situation for three AlGaAs/InGaAs/GaAs quantum wells with different spacer thickness, Si donors introduced in the form of δ-doping to the AlGaAs barrier, and a pseudo-2D conducting channel formed in a triangular well in the InGaAs layer). Changes of n_{2D} have been induced either by the high-pressure freezeout (HPFO) of electrons on the Si DX states of remote donors, or by the consecutive annealing of this reservoir of localized electrons. The former procedure leads to higher correlation degree in the remote charge system, and thus to higher mobility values. Annealing produces the opposite effect. The absolute value of μ_{2D} in the studied samples decreases with decreasing spacer thickness. In the described experiment, all measurements of μ_{2D} and n_{2D} were performed at 77 K. With reduced temperature, the absolute value of μ_{2D} and the contribution to mobility originating in spatial correlations of the remote charges both increase.

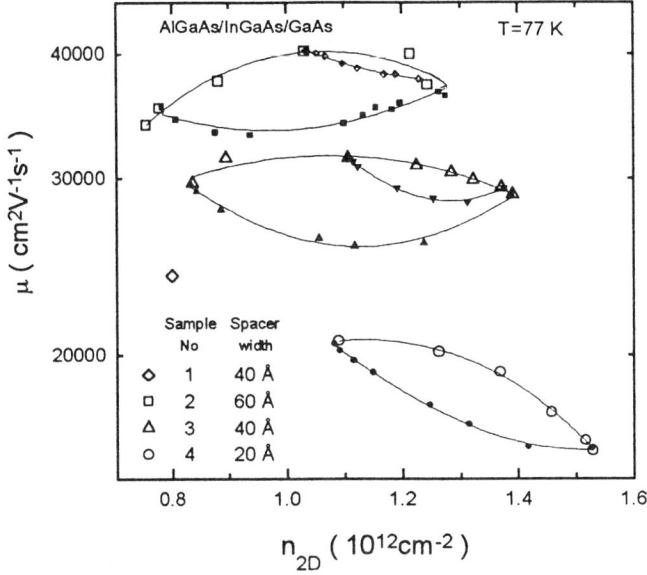

FIG. 12. Hall mobility versus carrier concentration for Si : Al$_{0.32}$As/In$_{0.15}$Ga$_{0.85}$As/GaAs structures with different spacer widths. Measurements were performed at $T = 77$ K and at ambient pressure. Open symbols represent the HPFO procedure performed at different freezout pressures (capture). Solid symbols correspond to the n_{2D} increase induced by an annealing path following HPFO procedure (emission) (after [37]).

VII. Remote-Charge Correlations and Quantum Transport of 2DEG

In the previous sections electron mobility (classical mobility) and related (classical) scattering time τ_t have been examined as parameters reflecting the existence of spatial correlations of impurity charges. There are also other parameters which should be sensitive to the potential fluctuations caused by the remote-donor charges. The natural candidates include the single-particle relaxation time (or quantum lifetime) of 2DEG, τ_s, and the width of the integer quantum Hall effect (IQHE) plateaus. In this paragraph we will demonstrate how changes in the arrangement of remote impurity charges influence these parameters. To keep the relation with the previously discussed 2DEG mobility, a comparison between the contributions of the spatial correlations to τ_s and to the transport scattering time $\tau_t = \mu_{2D} m^*/e$ will be discussed.

When the simplified Boltzmann equation is solved in the relaxation-time approximation, the mean time between collisions is weighted by a factor of $(1 - \cos \theta)$, as in the equation

$$\tau_t^{-1} = \int dk' \, W_{kk'}(1 - \cos \theta),$$

where $W_{kk'}$ is the probability of electron scattering from state k to k' and θ is the scattering angle. Within the relaxation-time approximation the large-angle scattering represents a main factor reducing the mobility value (because of the weighting factor of $(1 - \cos \theta)$). Thus, in a system where small-angle scattering dominates, the transport scattering time represents only a fraction of the actual number of electron collisions. Electron scattering by ionized impurities is a large-angle scattering process in bulk samples. In modulation-doped structures, moving the ionized donors away from the 2D channel means that the remote impurity scattering becomes more of a small-angle process. Consequently, in 2D systems the transport mobility can exceed the single-particle mobility by one or two orders of magnitude. Since the small-angle contribution to the transport mobility, that is, τ_t, is strongly supressed with respect to τ_s, the latter parameter becomes a much more sensitive tool for studying modifications in remote impurity scattering [32, 38, 39]. A comparison of τ_s behavior in a set of samples with different 2DEG and remote donor concentrations led to a suggestion about the possible contribution of the remote-charge correlation to τ_s [32],

$$\tau_s^{-1} = \int dk' \, W_{kk'}.$$

τ_s characterizes the total scattering probability of the electron and can be extracted from the evolution with the magnetic field of the Shubnikov–de Haas oscillations (see, for example, [38]).

Single-particle mobility and the related single-particle life time τ_s can be determined experimentally from so-called Dingle plots. The amplitude ΔR of the Shubnikov–de Haas oscillation is given by

$$\Delta R = 4R_o X(T) \exp(-\pi/\omega_c \tau_s),$$

where R_o is the zero field resistance, ω_c the cyclotron frequency, and $X(T)$ a thermal damping factor given by

$$X(T) = (2\pi^2 kT/\hbar\omega_c)/\sinh(2\pi^2 kT/\hbar\omega_c)$$

If the logarithm of the amplitude, divided by $X(T)$, is plotted against $1/B$, the slope gives $1/\tau_s$ with an intercepts of $4R_o$. The data presented in Fig. 13 (lower hysteresis loop) correspond to the values of τ_s obtained by means of the described procedure. The higher loop illustrates the dependence of τ_t on n_{2D}. Similarly to other semiconductor systems with 2DEG, values of τ_t exceed those of τ_s (for the same n_{2D}). The HPFO procedure used for n_{2D} tuning within the regime of high correlation degree causes higher values of both τ_s and τ_t with respect to annealing path (for the same n_{2D}) [46]. The latter procedure favors reduced degree of the remote charge correlation. A qualitatively similar behavior of τ_s and τ_t has been obtained in AlGaAs/InGaAs/GaAs pseudomorphic heterostructures [40].

Another effect which can be sensitive to the spatial correlations of remote impurity charges (i.e., a reduction in the amplitude of long-range Coulomb potential fluctuations) is the width of the integer quantum Hall (IQH) plateaus. Dependence of plateau width on impurity concentration has been demonstrated (see, e.g., [41]). Much less is known about the influence of impurity charge distribution on the density of states of 2DEG and on the width of IQH plateaus [42]. For a 2DEG system realized in a Si metal oxide semiconductor field-effector transistor, Furneaux and Reinecke [43]

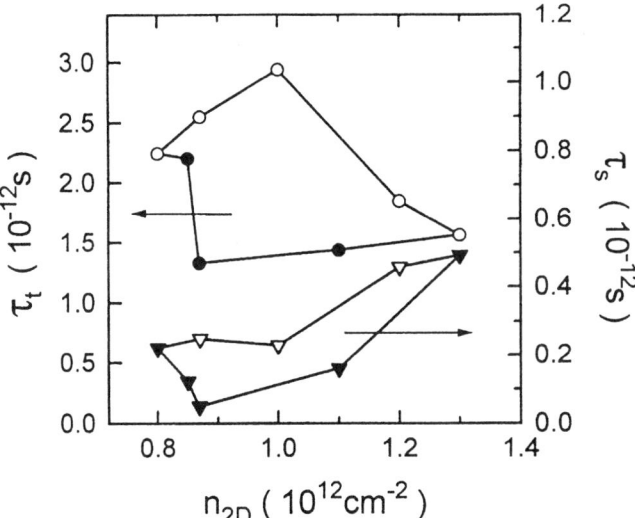

FIG. 13. Transport relaxation time (τ_t) and single-particle relaxation time (τ_s) variations with electron concentration n_{2D} for GaAs/AlGasAs heterostructure. Measurements were performed at $T = 4.2$ K and at ambient pressure. Open symbols correspond to the HPFO procedure used to reduce n_{2D}. Solid symbols represent the n_{2D} increase induced by the annealing path following HPFO of 2DEG [46].

investigated the effects of driftable Na$^+$ ions in the oxide. More ions in the region of the interface between the oxide and the semiconductors result in an increase in the width of the QH plateaus.

Most theories of the IQH effect assume that the localization of electronic states is the reason for the existence of plateaus in the Hall resistance, ρ_{xy}, and broad minima in the magnetoresistance, ρ_{xx}. The quantized Hall resistance is given by $\rho_{xx} = h/ie^2$, where h is the Planck constant and the quantum number i is the number of completely filled Landau levels. The relation between the mobility of carriers and the width of the plateaus indicates that the origin of the Hall plateaus is related to scattering centers which lead to carrier localization in strong magnetic fields. It is generally accepted that the plateaus originate from a mobility gap and not from a density-of-states gap between Landau levels. Because of disorder, the Landau levels are broadened with localized states in the tails of the levels [44, 45]. For long-range potential fluctuations, the motion of electrons can be well described by a percolation picture where the electrons in a strong magnetic field move on equipotential lines. Plateaus are due to Fermi levels situated in localized states due to disorder. The localized fraction of the density of states determines the width of the plateaus.

It has been demonstrated that width of the Hall plateaus increases with decreasing degree of spatial correlation among the remote impurity charges (Fig. 14). The HPFO procedure used for n_{2D} tuning within the regime of high correlation degree causes lower values of plateau width with respect to the annealing path (for the same n_{2D}) [46]. The latter procedure favors a reduced degree of remote charge correlation, that is, a decrease in the amplitude of potential fluctuations caused by charges localized on remote donor impurities.

VIII. Concluding Remarks

High mobility of band carriers and their high density are desirable features of semiconducting materials and their various structures. These requirements, very important from the device-oriented point of view, are in a sense contradictory. Ionized impurities supplying carriers are at the same time among the most efficient scatterers, reducing mobility of carriers. One of the widely used solutions of this problem is the idea of remote doping in semiconductor heterostructures where the two-dimensional electron gas is separated from the region where dopants are located. The other possibility of reducing the scattering efficiency by ionized impurities consists of designing a system with spatially correlated positions of dopants. A strong

5.3 Spatial Correlations of Impurity Charges

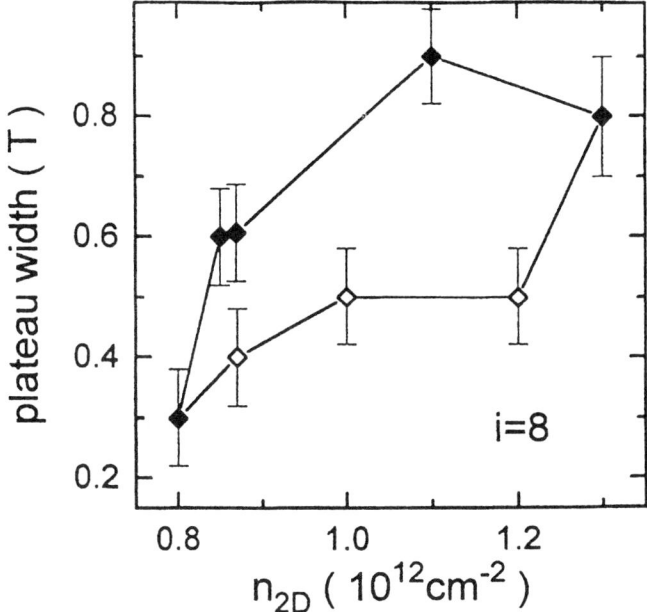

FIG. 14. Integer quantum Hall plateau width (for $i = 8$) versus 2DEG concentration. Solid and open symbols correspond to different procedures used for tuning n_{2D} (see caption of Fig. 11) [46].

Coulomb interaction between impurity charges may lead to establishing a correlated arrangement of their positions in the crystal lattice during growth of bulk semiconductors or low-dimensional structures. Then, a portion of the scattering events has a coherent character and contributes neither to the momentum relaxation rate nor to a shortening of carrier lifetime.

In this paper we concentrated on a different aspect of the ordering within an impurity sublattice, namely, ordering of impurity charges. The positions of impurities are restricted to a randomly distributed and fixed set of sites in the matrix lattice. The importance of the interimpurity interactions appears when under the condition of partial occupancy of impurities by electrons, Coulomb interactions among the impurity charges induce spatial correlations of positions of these charges. This lowers the total energy of the system and enhances carrier mobility. In this paper we demonstrated the contribution of the spatial arrangement of impurity charges to transport and magnetotransport effects (of classical or quantum character) in a variety of semiconductor systems. In this context we discussed bulk materials as well as semiconductor heterostructures. The considered phenomenon has a very general character and can influence many physical effects, some of them important from the application point of view. One may, for example, expect interesting effects caused by spatial correlations of impurity charges in doped superlattices, quantum wires, and quantum dot/antidot structures. In this paper we showed that hydrostatic pressure represents a very useful tool for studying phenomena sensitive to the presence of spatial correlations in the arrangement of charges located on impurities/defects. However, one must be aware of the importance of the considered effects at ambient pressure conditions.

References

1. Shockley, W. (1961). Problems related to $p–n$ junctions in silicon. *Solid State Electron.* **2**, 35–67.
2. Shklovskii, B. I., and Efros, A. L. (1984). *Electronic Properties of Doped Semiconductors,* p. 303. Springer, New York.
3. Gerlach, E., and Rautenberg, M. (1978). Ionized impurity scattering in semiconductors. *Phys. Stat. Sol.(b)* **86**, 479–482.
4. Schubert, E. F. (1994). Delta-doping of semiconductors: Electronic optical and structural properties of materials and devices, in *Semiconductors and Semimetals,* Vol. 40 (A. C. Gossard, ed.), pp. 2–151. Academic Press, New York.
5. Baranovskii, S. D., Efros, A. L., Gelmont, B. L., and Shklovskii, B. I. (1979). Coulomb gap in disordered systems: computer simulations. *J. Phys. C* **12**, 1023–1034.
6. Davies, J. H., Lee, P. A., and Rice, T. M. (1984). Properties of the electron gas. *Phys. Rev. B* **29**, 4260–4271.
7. Kossut, J., Wilamowski, Z., Dietl, T., and Swiatek, K. (1990). Spatial correlation of impurity charges in doped semiconductors. In *Proc. 20th Int. Conf. Physics of Semiconductors* (E. M. Anastassakis and J. D. Joannopoulos, eds.), pp. 613–620. World Scientific, Singapore.
8. Mycielski, J. (1986). *Solid State Commun.* **60**, 165.
9. Tsidlikovski, I. M., and Kuleyev, I. G. (1996). Spatial correlations of impurity charge in gapless semiconductors. *Semicond. Sci. Technol.* **11**, 625–640.
10. Skierbiszewski, C., Suski, T., E. Litwin-Staszewska, E., Dobrowolski, W., Dybko, K., and Mycielski, A. (1989). The influence of hydrostatic pressure on the formation of a donor superlattice in HgSe:Fe. *Semicond. Sci. Technol.* **4**, 293–295.
11. Skierbiszewski, C., Wilamowski, Z., Suski, T., Kossut, J., and Witkowska, B. (1993). Why various types of donors can either enhance or reduce electron mobility in narrow gap semiconductors. *Semicond. Sci. Technol.* **8**, S40–S43.
12. Suski, T. (1994). Hydrostatic pressure investigations of metastable defect states. *Mat. Sci. Forum* **143–147** (G. Heinrich, ed.), pp. 975–982. Trans. Tech. Publications, Switzerland.
13. Chadi, D., and Chang, K. (1988). Theory of the atomic and electronic structure of DX centers in GaAs and AlGaAs alloys. *Phys. Rev. Lett.* **61**, 873–876.
14. Baj, M., Dmowski, L. H., and Slupinski, T. (1993). Direct proof of two-electron occupation of Ge-DX centers in GaAs codoped with Ge and Te. *Phys. Rev. Lett.* **71**, 3529–3532.
15. Mizuta, M., Tachikawa, M., Kukimoto, H., and Minomura, S. (1985). Direct evidence for the DX center being a substitutional donor in AlGaAs alloy system. *Jpn. J. Appl. Phys.* **24**, L143–L146.
16. Dmochowski, J. E., Wang, P. D., Stradling, R. A., and Trzeciakowski, W. (1992). High pressure studies of electronic states with small lattice relaxation of DX-centres in GaAs.

Mat. Sci. Forum **83–87** (M. Stavola, ed.), pp. 751–756. Trans. Tech. Publications, Switzerland.
17. Suski, T., Piotrzkowski, R., Wisniewski, P., Litwin-Staszewska, E., and Dmowski, L. (1989). High pressure and DX centers in heavily doped bulk GaAs. *Phys. Rev. B* **40,** 4012–4021.
18. O'Reilly, E. P. (1989). Pressure dependence of DX center mobility in highly doped GaAs. *Appl. Phys. Lett.* **55,** 1409–1411.
19. Dietl, T., Dmowski, L., Kossut, J., Litwin-Staszewska, E., Piotrzkowski, R., Suski, T., Swiatek, K., and Wilamowski, Z. (1990). Inter-donor interactions — source of electron mobility increase under pressure. *Acta Phys. Polon. A* **77,** 29–32.
20. Sobkowicz, P., Wilamowski, Z., and Kossut, J. (1992). Monte Carlo simulations of spatial correlation effects of charged centres in δ-doping layers. *Semicond. Sci. Technol.* **7,** 1155–1161.
21. Suski, T., Wisniewski, P., Litwin-Staszewska, E., Kossut, J., Wilamowski, Z., Dietl, T., Swiatek, K., Ploog, K., and Knecht, J. (1990). Pressure dependence of electron mobility in GaAs:Si — effects of on-site and inter-site interactions within a system of DX centres. *Semicond. Sci. Technol.* **5,** 261–264.
22. Jantsch, W., Wilamowski, Z., and Ostermayer, G. (1992). Intra- and inter-defect electronic correlation in III–V semiconductors. *Phys. Scr.* **T45,** 140–144.
23. Dmowski, L. H., Goutiers, B., Maude, D. K., Eaves, L., Portal, J. C., and Harris, J. J. (1993). Spatial correlations of occupied DX centres in GaAs and its destruction by thermal and optical excitation. *Jpn. J. Appl. Phys.* **32,** Suppl. 32-1, 221–223.
24. Suski, T., Wisniewski, P., Litwin-Staszewska, E., Wasik, D., Przybytek, J., Baj, M., Karczewski, G., Wojtowicz, T., Zakrzewski, A., and Kossut, J. (1996). Spatial correlations of In-donor charges in CdTe layers. *J. Crystal Growth* **159,** 380–383.
25. Baraldi, A., Ghezzi, C., Parisini, A., Bosachi, A., and Franchi, S. (1991). Low temperature mobility of photoexicted electrons in AlGaAs containing DX centers. *Phys. Rev. B* **44,** 8713–8721.
26. Piotrzkowski, R., Konczewicz, L., Litwin-Staszewska, E., Robert, J. L., and Lorenzini, P. (1991). Dependence of mobility on DX-centre configuration in GaAlAs alloy. *Semicond. Sci. Technol.* **6,** 250–253.
27. Ghezi, C., Parisini, A., and Dallacasa, V. (1994). Electron scattering by spatially correlated DX charges. *Phys. Rev. B* **50,** 2166–2175.
28. Morgan, T. N. (1989). The vacancy-interstitial model of DX centers. *Mat. Sci. Forum* **38–41** (G. Ferenczi, ed.), pp. 1079–1084. Trans. Tech. Publications, Switzerland.
29. Piotrzkowski, R., Litwin-Staszewska, E., and Robert, J. L. (1994). The DX-centers related mobility in AlGaAs: charge correlation and multilevel-structure effects. *Mat. Sci. Forum* **143–147** (G. Heinrich, ed.), pp. 1135–1140. Trans. Tech. Publications, Switzerland.
30. Walukiewicz, W., Ruda, H. E., Lagowski, J., and Gatos, H. C. (1984). Electron mobility in modulation-doped heterostructures. *Phys. Rev. B* **30,** 4571–4582.
31. Efros, A. L., Pikus, F. G., and Samsonidze, G. G. (1990). Maximum low temperature mobility of two-dimensional electrons in heterounctions with a thick spacer layer. *Phys. Rev. B* **41,** 8295–8301.
32. Coleridge, P. T. (1991). Small angle scattering in two-dimensional electron gases. *Phys. Rev. B* **44,** 3793–3801.
33. Suski, T., Wisniewski, P., Gorczyca, I., Dmowski, L. H., Piotrzkowski, R., Sobkowicz, P., Smoliner, J., Gornik, E., Boehm, G., and Weimann, G. (1994). Spatial correlations of remote impurity charges: Mechanism responsible for the high mobility of a two-dimensional electron gas. *Phys. Rev. B* **50,** 2723–2726.
34. Buks, E., Heiblum, M., and Shtrikman, H. (1994). Correlated charged donors and strong mobility enhancement in a two-dimensional electron gas. *Phys. Rev. B* **49,** 14790–14793.

35. Brunthaler, G., Penn, C., Suski, T., Wisniewski, P., Litwin-Staszewska, E., and Koehler, K. (1996). Influence of spatial doping correlation on scattering times studied in gated and ungated aAs/AlGaAs quantum wells under hydrostatic pressure. *Solid State Electron.* **40,** 105–108.
36. Wisniewski, P., Suski, T., Dmowski, L. H., Gorczyca, I., Sobkowicz, P., Smoliner, J., Gornik, E., Boehm, G., and Weimann, G. (1994). Tuning of 2DEG mobility by modification in ordering of remote impurity charges in GaAs/AlaAs heterostructures. *Mat. Sci. Forum* **143–147** (H. Heinrich, ed.), pp. 617–622. Trans. Tech. Publications, Switzerland.
37. Litwin-Staszewska, E., Suski T., Skierbiszewski, C., Kobbi, F., Robert, J. L., and Mosser, V. (1995). Spatial correlations of donor charges in δ-doped AlGaAs/InGaAs/GaAs structures. In *High Pressure Science and Technology* (W. A. Trzeciakowski, ed.), pp. 654–656. World Scientific, Singapore.
38. Harrang, J. P., Higgins, R. J., Goodall, R. K., Jay, P. R., Laviron, M., and Delescluse, P. (1985). Quantum and classical mobility determination of the dominant scattering mechanism in the two-dimensional electron gas of an AlGaAs/GaAs heterojunction. *Phys. Rev. B* **32,** 8126–8135.
39. Gold, A. (1987). Electronic transport properties of a two-dimensional electron gas in a silicon quantum well structure at low temperatures. *Phys. Rev. B* **35,** 723–733.
40. Dmowski, L. H., Zduniak, A., Litwin-Staszewska, E., Contreras, S., Knap, W., Robert, J. L., and Mosser, V. (1996). Study of classical and quantum scattering times in pseudomorphic AlGaAs/InGaAs/GaAs heterostructures by means of pressure. In *High Pressure Science and Technology* (W. A. Trzeciakowski, ed.), pp. 657–659. World Scientific, Singapore.
41. Chang, A. M. (1987). In *The Quantum Hall Effect* (R. E. Prange and S. M. Girvin, eds.), p. 175. Springer, New York.
42. Haug, R. J., Gerhardts, R. R., v. Klitzing, K., and Ploog, K. (1987). Effect of repulsive and attractive scattering centers on the magnetotransport properties of a two-dimensional electron gas. *Phys. Rev. Lett.* **59,** 1349–1352.
43. Furneaux, J. F., and Reinecke, T. L. (1984). Novel features of quantum Hall plateaus for varying interface charge. *Phys. Rev.* **29,** 4792–4795.
44. Aoki, H., and Ando, T. (1981). Effect of localization on the Hall conductivity in the two-dimensional systems in strong magnetic fields. *Solid State Commun.* **38,** 1079–1082.
45. Prange, R. E. (1981). Quantized Hall resistance and the measurement of the fine structure constant. *Phys. Rev.* **23,** 4802–4805.
46. Wisniewski, P., Suski, T., Litwin-Staszewska, E., Brunthaler, and Koehler, K. (1996). Mobility and quantum lifetime in a GaAs/AlGaAs heterostructure: tuning of the remote-charge correlations. *Surface Sci.* **361/362,** 579–582.

CHAPTER 6

Pressure Effects on the Electronic Properties of Diluted Magnetic Semiconductors

Noritaka Kuroda

INSTITUTE FOR MATERIALS RESEARCH
TOHOKU UNIVERSITY
SENDAI, JAPAN

I. INTRODUCTION 513
II. STRUCTURAL PROPERTIES 515
    1. Under Ambient Conditions 515
    2. Elastic Properties 517
    3. Phase Transition 518
III. ELECTRONIC PROPERTIES 523
    1. Energy Gap 523
    2. Exchange Interactions 524
    3. Intraion d–d Optical Transitions 540
IV. SUPERSTRUCTURES 543
    1. $ZnSe/Zn_{1-x}Mn_xSe$ 543
    2. $CdTe/Cd_{1-x}Mn_xTe$ 544
V. MISCELLANEOUS 545
    1. Raman Scattering 545
    2. Galvanomagnetic Effects in $Hg_{1-x}Fe_xSe$ 546
    Acknowledgments 546
    References 547

I. Introduction

Diluted magnetic semiconductors (DMSs) are the materials in which transition metal elements are substituted for a fraction of one-tenth of a percent to several tens of percents of cations of host semiconductors. In the absence of a magnetic field they behave very much as isoelectronic pseudobinary compounds and yield comparable semiconducting properties. If a magnetic field is applied, however, these materials exhibit a variety of unusual galvanomagnetic and magneto-optical properties, which are of potential technological significance. In particular, giant magneto-optical effects in wide-gap DMSs such as $Cd_{1-x}Mn_xTe$ are conceived to be useful for optical devices which operate in the near-infrared or visible region of light. For instance, there is an attempt to apply the giant Verde constant of the Faraday rotation (Bartholomew et al., 1986) in the laser-light isolator for the fiber amplifier of optical telecommunication systems (Onodera et al., 1994).

DMSs are, of course, transition metal compounds. In the majority of group II–VI DMSs the on-site Coulomb energy U and the charge-transfer energy Δ satisfy a condition of $U > \Delta$. According to the classification by Zaanen, Sawatzky, and Allen (1985), those DMSs belong to the regime of charge-transfer semiconductors, as shown in Fig. 1. It is known that many transition metal compounds related to high-T_c superconductors and colossal-magnetoresistance materials also belong to the regime of charge-transfer semiconductors (Ohta et al., 1991). We may envisage, therefore, that DMSs have electronic properties common in various respects to those of highly correlated materials. In fact, most of the unusual galvanomagnetic and magneto-optical properties of DMSs are related to the admixture of the d orbitals of substituted transition metal ions with the s and p orbitals making up the conduction and valance bands of host semiconductors. The intersite

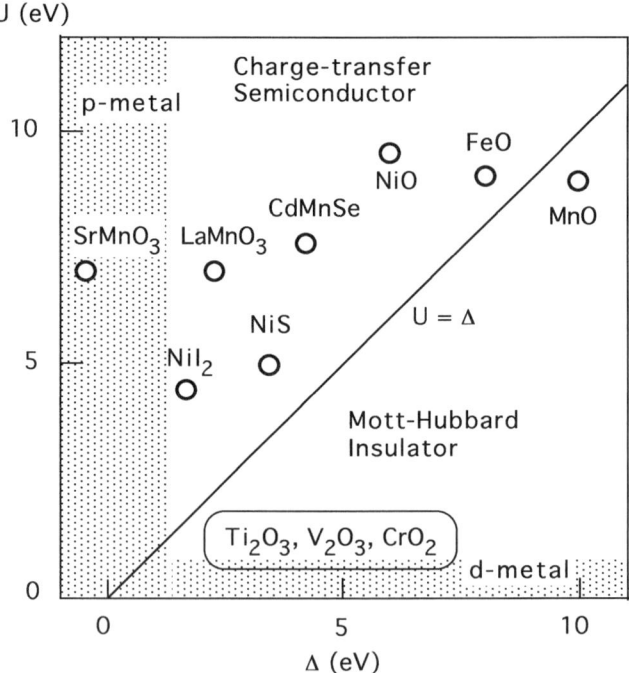

FIG. 1. Classification of transition-metal compounds in terms of the on-site Coulomb energy U and the charge-transfer energy Δ.

hopping, whether real or virtual, of an electron via the hybridized orbitals accompanies charge fluctuations of transition-metal ions and thus is influenced strongly by the electron correlation that originates from many-body Coulomb interactions of the d electrons.

Hydrostatic pressure can compress a solid continuously at a rate as high as 10% per 10 GPa in volume without breaking and alloying. Since this volume change largely affects the electron structure, studies of high-pressure effects provide a wealth of information on the properties of a solid. So far extensive studies have been made on the pressure effects in DMS. From those studies not only the galvanomagnetic and magneto-optical properties, but also the basic semiconductor properties have emerged to be associated with the admixture between the d and sp orbitals. This article reviews those studies with attention focused on how the presence of d electrons influences the electronic properties of DMS.

Section II surveys the structural features, including pressure-induced phase transitions. Previous structural studies are concerned with optical and ultrasonic experiments, as well as X-ray diffraction analyses, mostly in $A^{II}_{1-x}Mn_xB^{VI}$ compounds. In Section III the studies on the fundamental energy gap, exchange interactions, and local d levels of transition metal ions, particularly Mn^{2+}, are reviewed. The properties of the optical energy gap and local d levels have been studied by optical absorption and photoluminescence experiments. As for the mechanisms of exchange interactions, valuable information has been extracted from Shubnikov–de Haas, magnetic, and optical studies. Although many of those works are concerned with II–VI-based, tetrahedrally coordinated compounds, important contributions have also come from magnetic experiments in PbSnMnTe and $(CdZnMn)_3As_2$ systems, the former having a rock-salt structure, and the latter having a rather complicated tetragonal crystal structure. Section IV describes briefly several examples of high-pressure optical studies of superstructures of DMS. In Section V the knowledge of phonons obtained by Raman scattering spectroscopy in several substances and the knowledge of novel galvanomagnetic effects in HgFeSe are presented.

II. Structural Properties

1. UNDER AMBIENT CONDITIONS

Under atmospheric pressure, solid solutions $A^{II}_{1-x}Mn_xB^{VI}$ crystallize in zincblende- or wurtzite-type structures, as shown in Table I. In all the mixed crystals the lattice parameters determined from X-ray diffraction analyses

TABLE I

CRYSTAL STRUCTURE, RANGE OF COMPOSITION x, AND MEAN CATION–CATION DISTANCE IN $A^{II}_{1-x}Mn_xB^{VI}$ COMPOUNDS[a]

Substance	Crystal structure	Range of Mn composition, x	Mean cation–cation distance (Å)
ZnMnS	zincblende	0–0.1	$3.830 + 0.139x$
	wurtzite	0.1–0.45	
ZnMnSe	zincblende	0–0.3	$4.009 + 0.165x$
	wurtzite	0.3–0.57	
ZnMnTe	zincblende	0–0.86	$4.315 + 0.168x$
CdMnS	wurtzite	0–0.45	$4.121 - 0.150x$
CdMnSe	wurtzite	0–0.5	$4.296 - 0.123x$
CdMnTe	zincblende	0–0.77	$4.587 - 0.105x$
HgMnS	zincblende	0–0.37	$4.141 - 0.116x$
HgMnSe	zincblende	0–0.38	$4.301 - 0.123x$
HgMnTe	zincblende	0–0.75	$4.568 - 0.080x$

[a] The table is made from the data of Furdyna (1988) with several data for cation–cation distance replaced by the data of Rodic et al. (1996).

are known to obey Vegard's law well. A lattice parameter d varies linearly with x as $d = xd_{MnB} + (1 - x)d_{AB}$. It has long been known that in an isostructural system with a constant anion, the unit cell volume has a good linear relationship with the ionic volume of cation. Rodic et al. (1996) have noticed that in the $A^{II}_{1-x}Mn_xB^{VI}$ system the unit cell volume is proportional to the volume of cations $r^3 = \{xr_{Mn} + (1 - x)r_A\}^3$ regardless of the difference in the crystal structure, where r_{Mn} and r_A are the radii of Mn^{2+} and A^{2+} ions, respectively.

In the study of electronic properties of DMS we have to bear in mind that the lattice is significantly distorted around substituted transition metal ions because of the difference in radii of the cations. According to EXAFS measurements in CdMnTe (Balzarotti et al., 1984), ZnMnSe (Furdyna, 1988), and ZnCoS and ZnCoSe (Lawniczak-Jablonska and Golacki, 1994), the distribution of the nearest-neighbor anion–cation distances is bimodal, that is, the average lengths of individual A^{II}–B^{VI} and $Mn(Co)$–B^{VI} bonds are rather independent of x. It is worth noting that the bimodal behavior has also been observed in a nonmagnetic compound, CdZnTe (Terauchi et al., 1993).

Detailed reviews on the structural properties have been given by Furdyna (1988) and Jain and Robins (1991).

2. ELASTIC PROPERTIES

In general, solids are hardened by hydrostatic pressure because of volume anharmonicity and thus the bulk modulus increases as pressure increases. Table II shows the values of the bulk modulus $B = (c_{11} + 2C_{12})/3$ and its pressure coefficient B' or the volume compressibility κ and its pressure coefficient κ' in several DMSs. For the compressibility in pure substances one may refer to literature (Montalvo and Langer, 1970; Martinez, 1980). As seen in Table II, B and κ in $A^{II}MnB^{VI}$ compounds do not depend much on substances, consistent with Paul's rule (Paul, 1961) that within a given crystal structure the pressure coefficient of an optical energy gap is rather independent of the chemical composition of the crystal. Since we have $B = C_0 \equiv e^2/b^4$ to a good approximation for II — VI compounds (Phillips, 1973), where e is electron charge and b is the anion–cation bond length, B tends to decrease with increasing bond length.

TABLE II

BULK MODULUS B, $dB/dP = B'$, VOLUME COMPRESSIBILITY κ, AND ITS PRESSURE COEFFICIENT κ'

Substance		B (GPa)	B'	κ (GPa^{-1})	κ' (GPa^{-2})
ZnMnSe	$x = 0$	62^a, 58^b			
	$x = 0.5$	55^a			
ZnMnTe	$x = 0$	48.0^c	4.7^c		
	$0 \leq x \leq 0.3$	$dB/dx \approx -15$	$dB'/dx \approx -4$		
CdMnTe	$x = 0$	42.5^d	4.2^d	0.022 ± 0.001^e	0.0018 ± 0.0005^e
	$x = 0$	42.0 ± 2^f	6.4 ± 0.6^f		
	$0.1 \leq x \leq 0.7$			0.022 ± 0.006^e	0.001 ± 0.002^e
	$x = 0.5$	40^d	4.1^d		
HgSe	$x = 0$	51.7^g	2.6^g		
HgMnTe	$x = 0$	42.3^h	3.8^h		
	$x = 0, 0.09$	$38.5 (= 1/\kappa)$		0.026 ± 0.002^i	0.003 ± 0.001^i
	$x = 0.2$	40.8^j	4.0^j		

[a] Mayanovic et al. (1988).
[b] Becker (1988).
[c] Strössner et al. (1987).
[d] Maheswaranathan et al. (1985).
[e] Qadri et al. (1987).
[f] Strössner et al. (1985).
[g] Ford et al. (1982).
[h] Miller et al. (1981).
[i] Qadri et al. (1989).
[j] Chao and Sladek (1981).

Ultrasonic data of elastic moduli in ZnMnSe (Mayanovic *et al.*, 1988) show that B decreases upon Mn substitution, in accordance with the increase in the bond length. It is also the case in ZnMnTe, in which the bulk modulus of 48.0 GPa in ZnTe is reduced to 43.6 GPa in $Zn_{0.7}Mn_{0.3}Te$ (Strössner *et al.*, 1987). In contrast, in Cd- and Hg-based compounds such as CdMnTe and HgMnTe, in which the mean bond length is shortened by Mn, the effect of Mn substitution on the bulk modulus seems comparatively weak, suggesting that Mn tends to make the zincblende lattice less stable.

3. PHASE TRANSITION

Most of the tetrahedrally coordinated $A^N B^{8-N}$ compounds undergo semiconductor (zincblende or wurtzite) → ionic (NaCl) → metallic phase transitions successively at elevated pressures. In HgS, HgSe and telluride compounds, ZnTe, CdTe, and HgTe, the NaCl phase is preceded by a cinnabar phase. They transform zincblende → cinnabar → NaCl → orthorhombic, except that ZnTe has no NaCl phase at room temperature (Nelmes *et al.*, 1995). In CdTe the cinnabar phase appears only in a very narrow range of pressure before transforming to NaCl phase at ~3.5 GPa, whereas in ZnTe and HgTe the cinnabar phase is stable in the ranges 9.3–11.0 GPa and 1.4–8.6 GPa, respectively.

Using a Born model of the interatomic forces to describe the lattice energy under pressure, Demarest (1972) has proposed as a modification of the Born stability criterion that the ratios C_S/B and C_{44}/B should reach certain critical values for a given structure to transform under pressure, where $C_S \equiv (C_{11} - C_{12})/2$. The cinnabar structure can be regarded as a distorted NaCl structure, the [0001] direction of hexagonal cinnabar corresponding to the [111] direction of NaCl. A topotactic consideration leads one to note that the transition from zincblende structure to cinnabar or NaCl structure can be induced by shear displacement of atoms on {110} planes (Miller *et al.*, 1981). Experimental results of the pressure dependence of the shear elastic constants C_S and C_{44} in HgTe (Miller *et al.*, 1981) and HgSe (Ford *et al.*, 1982) support the criterion of Demarest. For a number of zincblende semiconductors extrapolation of the low-pressure data suggests that at the transition pressure, which is hereafter denoted as P_t, the ratio C_S/B is near 0.2 and C_{44}/B is about 0.5 (Yogurtçu *et al.*, 1981; Ford *et al.*, 1982). Figure 2 shows the results of Miller *et al.* (1981) on the direct experimental assessment of C_S/B up to P_t of 1.6 GPa in HgTe. The value of C_S/B, which is 0.20 at ambient pressure, is reduced to 0.17 at P_t. Correspondingly, the ratio of the bond-bending force constant F_b to the bond-stretching force constant F_s decreases continuously, as shown in Fig. 2.

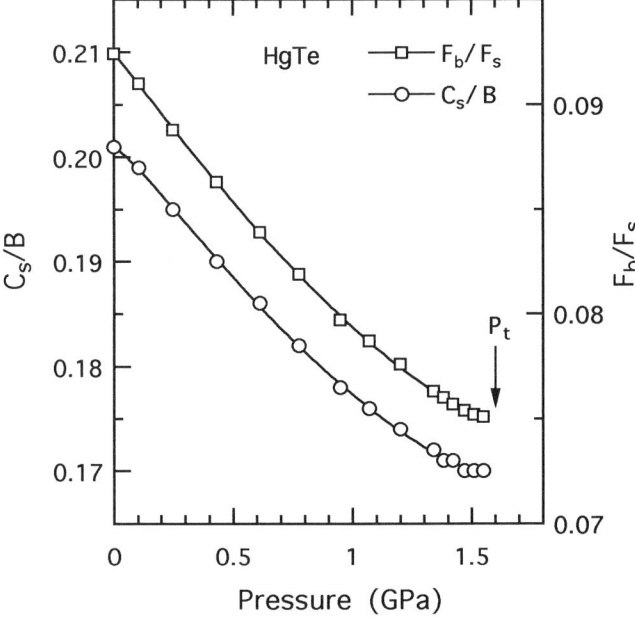

FIG. 2. Pressure dependence of C_s/B and F_b/F_s in HgTe. [Data taken from Miller *et al.* (1981).]

The ultrasonic data in CdMnTe (Maheswaranathan *et al.*, 1985) and ZnMnSe (Mayanovic *et al.*, 1988; Maheswaranathan and Sladek, 1990) show that C_S/C_0 and C_{44}/C_0 decrease significantly as the Mn concentration increases, as seen in Fig. 3. These moduli decrease further with pressure. From their ultrasonic data Maheswaranathan *et al.* have deduced that the critical value of the ratio C_S/B is 0.126 at P_t in CdMnTe. Hence, based on Demarest's stability criterion, the linear extrapolation of the low-pressure data predicts that P_t will be reduced to 2.9 and 2.75 GPa in CdMnTe of $x = 0.45$ and 0.52, respectively. In the same way Maheswaranathan and Sladek have deduced the critical value of C_S/B for ZnMnSe to be 0.21 and consequently deduced the value of P_t, which is 14 GPa in ZnSe, to be 6.4 and 5.2 GPa for their ZnMnSe samples of $x = 0.37$ and 0.53, respectively.

A number of experiments have been conducted to directly measure the dependence of P_t on Mn concentration. Figures 4a and 4b show the representative data for Zn- and Cd-based $A^{II}MnB^{VI}$ compounds, respectively. The values of P_t predicted by Maheswaranathan *et al.* and Maheswaranathan and Sladek agree well with the experimental results. It is evident that the substitution of Mn makes the zincblende or wurtzite lattice less stable not only in CdMnTe and ZnMnSe, but also in all the other Zn- and

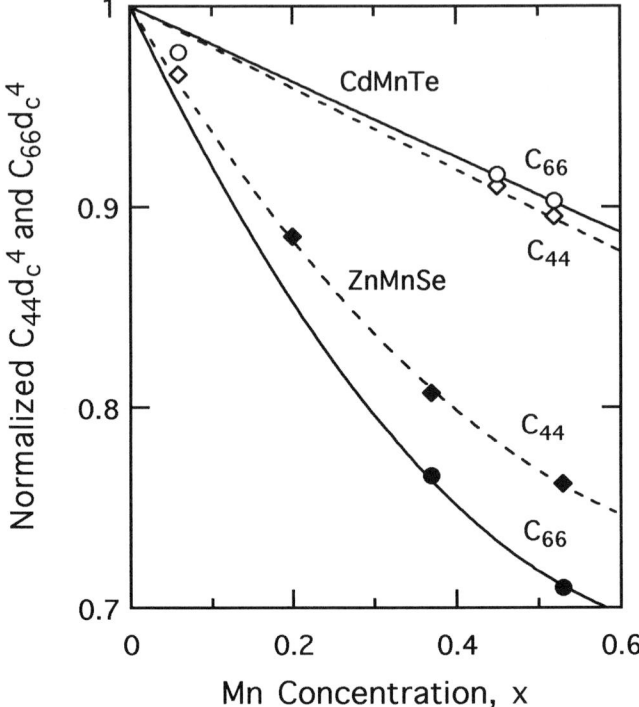

FIG. 3. Normalized $C_{44}d_c^4$ and $C_{66}d_c^4$ versus Mn concentration in CdMnTe and ZnMnTe, where d_c is the mean cation–cation distance. [Data taken from Mayanovic et al. (1988).]

Cd-based $A^{II}MnB^{VI}$ compounds. We note that $dP_t/P_t\,dx$ is of the order of -1. The x-dependence of P_t seems to be enhanced on going from telluride to sulfide in both Zn- and Cd-based compounds.

As a by-product of Raman scattering measurement of the pressure effects on TO and LO phonons, Arora et al. (1988, 1995) have found that in ZnMnSe an intermediate phase, whose structure is unidentified, precedes the NaCl phase in a comparatively low pressure range 2–4 GPa. According to the optical absorption and X-ray diffraction measurements, a similar intermediate phase appears also in ZnMnTe for $x < 0.3$, whereas the zincblende phase transforms directly to the NaCl phase for $x \geq 0.3$ (Strössner et al., 1987).

The tetrahedral structure of zincblende and wurtzite is sustained by the sp^3-hybridized, partially ionic covalent bonds. The pressure-induced decrease seen in C_S/B and F_b/F_s suggests that as pressure increases, the valence electrons depopulate the sp^3 bonding states of the cation to make the tetrahedral bonds more ionic. In view of the fact that Pauling's electro-

6 ELECTRONIC PROPERTIES OF DILUTED MAGNETIC SEMICONDUCTORS 521

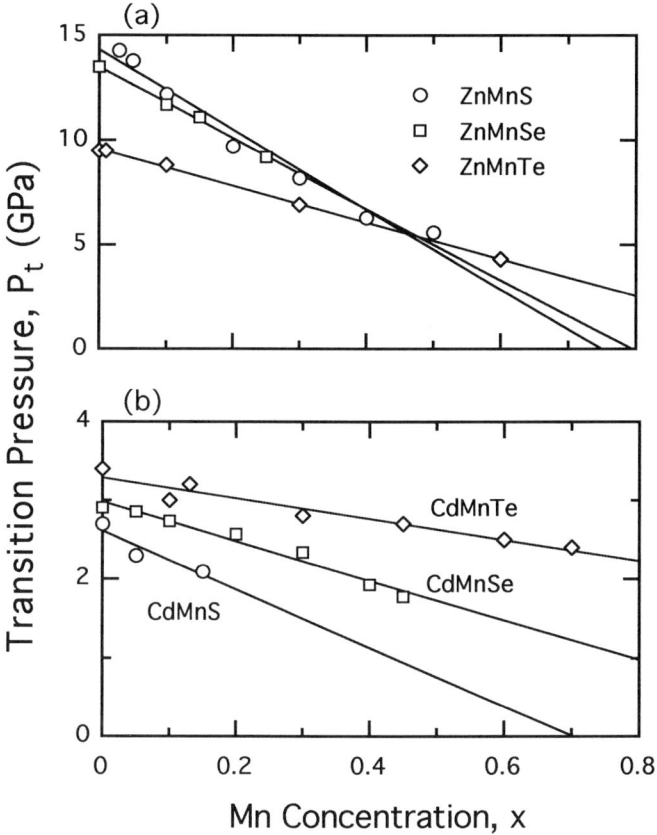

FIG. 4. Dependence of the transition pressure on Mn concentration in Mn-based II–VI DMS. [Data taken from Anastassiadou et al. (1988) for ZnMnS, Ves et al. (1986b) for ZnMnSe, Ves et al. (1986a) for ZnMnTe, Jiang et al. (1990) for CdMnS, Jiang et al. (1989a) for CdMnSe, and Qadri et al. (1987) for CdMnTe.]

negativity of Mn (1.5) is smaller than Zn (1.6), Cd (1.7), and Hg (1.8), it may be plausible to attribute the significant lowering of P_t in DMS (Fig. 4) in part to an increase in ionicity due to Mn substitution.

The spectroscopic ionicity theory of Phillips (1973, 1982) has successfully explained various elastic and electromechanical properties of tetrahedrally coordinated semiconductors. Phillips has related P_t in Zn chalcogenide, Cd chalcogenide, and Cu halide compounds to the spectroscopic ionicity f_i through the expression

$$P_t \Delta V_t / \Delta H_a = 0.7(0.785 - f_i) \quad \text{for } 0.5 < f_i \leq 0.785, \tag{1}$$

where ΔV_t and ΔH_a are the changes in volume and heat of atomization, respectively, per mole upon the zincblende–NaCl transition. The critical ionicity $f_i = 0.785$ signifies the upper limit of the ionicity for zincblende or wurtzite, beyond which a compound can no longer withstand shear and thus crystallizes in a NaCl structure even at ambient pressure. The large difference in P_t between Zn and Cd chalcogenide compounds, which is seen in Fig. 4, is ascribed to the difference in f_i. Phillips has shown that $C_S^* = C_S/C_0$ at ambient pressure scales well with f_i empirically as

$$C_S^* = 0.65 - 0.62 f_i. \tag{2}$$

It follows from Eq. (2) that the critical value of C_S^* is about 0.16. Similarly, there is a trend between f_i and the bending/stretching ratio, F_b/F_s, of force constants as

$$F_b/F_s = 0.3(1 - f_i), \tag{3}$$

which gives the critical value of F_b/F_s to be about 0.06. Recall that the ionicity f_i in Eqs. (1), (2), and (3) is defined at ambient pressure. It is of interest that Eqs. (2) and (3) explain substantially the experimental results (Fig. 2) for HgTe not only at ambient pressure but also under high pressures up to P_t, although it is difficult to theoretically explain the continuous increase of f_i up to the critical value 0.785 or some value near it (Van Vechten, 1969; Martinez, 1980).

In the case of DMS, as pointed out by Sladek (1986), in addition to a rearrangement of charge distribution due to the change in the anion–cation electronegativity difference, the concomitant p–d hybridization between Mn^{2+} $3d$ and anion p orbitals (Taniguchi et al., 1986; Wei and Zunger, 1987) would diminish the sp^3 hybridization. This would weaken still more the inherent directionality of the covalent bonds. In this sense the Mn substitution seems to tend to *metallize* the chemical bonds in Zn- and Cd-based DMS.

Rodic et al. (1996) have found that in HgMnS the zincblende → cinnabar transition takes place at a pressure below 1.8 GPa for $x < 0.1$, but the transition pressure goes beyond 1.8 GPa for $x \geq 0.1$. Moreover, Qadri et al. (1989) have demonstrated from X-ray diffraction experiments that successive zincblende → cinnabar → NaCl → β-Sn transitions occur at 1.6, 8.0, and 11.5 GPa, respectively, in HgTe, whereas transition pressures shift to 2.4, 10.0, and 11.5 GPa in $Hg_{0.91}Mn_{0.09}Te$, in spite of the fact that the bulk modulus of $Hg_{0.91}Mn_{0.09}Te$ is almost identical with that of pure HgTe. Apparently this is contrary to the trend, $dP_t/dx < 0$, of Zn- and Cd-based DMSs. As described in the next section, Hg-based DMSs are zero-gap semiconductors for $x < x_0 \approx 0.07$, while they have a positive energy gap for $x > x_0$. As far as Hg-based DMSs are concerned, Mn ions are likely to *demetallize* the chemical bonds, causing $dP_t/dx > 0$. Anyhow, systematic experimental data of dC_S^*/dx and the x dependence of dC_S^*/dP are desired to provide insight into this problem.

To conclude this section we would mention the theoretical approach of Singh (1991) and Singh and Arora (1993a, b). By incorporating characters of anions into interatomic potentials such as the Madelung energy, they have examined dP_t/dx in ZnMnTe and CdMnTe in terms of the so-called three-body-potential model. It has been shown that if an appropriate set of parameters is used to calculate the Gibbs free energy, dP_t/dx and ΔV_t can be evaluated successfully.

III. Electronic Properties

1. ENERGY GAP

Most II–VI semiconductors including DMSs have a direct, fundamental energy gap at the Γ point of the first Brillouin zone. As for its x-dependence, the readers may refer to literature (Becker, 1988). The energy gap E_g shows a remarkable blueshift under hydrostatic pressure at a rate of 0.04–0.10 eV/GPa, depending on the substances. The shift is more or less quadratic because of a sublinear dependence of the bulk modulus on pressure P, so E_g can be expressed as

$$E_g(P) = E_g(0) + a_1 P + a_2 P^2. \quad (4)$$

Experimental values of the coefficients a_1 and a_2 in $A^{II}MnB^{VI}$ compounds are collected in Table III. The positive value of a_1 arises from the general trend that the upward shift of the s-like conduction band is larger than the shift of the p-like valence band (Martinez, 1980). The Mn $3d^5$ levels lie several electron volts below the top of the valence band, forming p–d hybridization (Wei and Zunger, 1987). The p–d hybridization would be strengthened as the Mn–B bond length diminishes. We see from Table III that the coefficient a_1 decreases significantly as x increases in all substances. Namely, the Mn substitution reduces the pressure-induced shift of the energy gap. Ambrazevicius *et al.* (1984) have suggested that the p–d hybridization functions to accelerate the upward shift of the valence band. This property may play an important role in $A^{II}B^{VI}/A^{II}MnB^{VI}$ superlattices under high pressure, as discussed later in Section IV.

TABLE III
ENERGY GAP PARAMETERS a_1 AND a_2

Substances		a_1 (10^{-2} eV/GPa)	a_2 (10^{-4} eV/GPa^2)	dE_g/dx^l (eV)
ZnMnSe	$x = 0$	7.17 ± 0.18,[a] 6.7[b]	-15 ± 1,[a] -14[b]	
	$0.1 \leq x \leq 0.25$[c]	6.3–6.7	$-(17–27)$	~0.58
	$x = 0.3$[d]	5.4	-20	
ZnMnTe	$x = 0$[e]	10.4 ± 0.5	-28 ± 5	0.5–0.8
	$x = 0.3$[e]	8.2 ± 0.4	-80 ± 8	
CdMnS	$x = 0$[f]	4.2 ± 0.3	—	
	$x = 0.05$[g]	3.7 ± 0.1	-29 ± 1	~0.9
	$x = 0.15$[g]	2.7 ± 0.1	-8 ± 1	
CdMnSe	$0 \leq x \leq 0.01$[h,i]	5.9 ± 0.1	-41 ± 6	
	$x = 0.1$[i]	5.4 ± 0.1	-38 ± 4	1.1–1.4
	$x = 0.25$[i]	4.4 ± 0.3	-11 ± 13	
CdMnTe	$0 \leq x \leq 0.25$[j]	7–8	$-(30–40)$	
	$x = 0.4$[k]	7.9	—	1.3–1.6
	$x = 0.7$[k]	4.6	—	

[a] Ves *et al.* (1985).
[b] Tuchman *et al.* (1992).
[c] Ves *et al.* (1986b); Strössner *et al.* (1987).
[d] Jiang *et al.* (1989b).
[e] Ves *et al.* (1986a); Strössner *et al.* (1987).
[f] Béliveau and Carlone (1989).
[g] Jiang *et al.* (1990).
[h] Mei and Lemos (1984); Lemos *et al.* (1986).
[i] Matsuda (1996).
[j] Mei and Lemos (1984); Lemos *et al.* (1986); Prakash *et al.* (1990); Jiang *et al.* (1992).
[k] Ambrazevicius *et al.* (1984).
[l] Becker (1988).

2. EXCHANGE INTERACTIONS

In wide-gap DMSs the band edge electrons have a strong Kondo-like exchange interaction with localized d electrons of magnetic ions. The spin splitting of a hole produced by an external magnetic field is more than an order of magnitude larger than the intrinsic Zeeman splitting. The strong exchange interaction arises from kinetic spin interactions, in addition to the ordinary potential exchange, between a hole of the p-like valence band and the d electrons of magnetic ions. This kinetic exchange interaction also gives rise to strong superexchange interactions between magnetic ions. In narrow-gap semiconductors, the Bloembergen–Rowland interaction, which is associated with virtual electron transitions across the energy gap, cannot be ignored. The Ruderman–Kittel–Kasuya–Yoshida (RKKY) interaction,

6 ELECTRONIC PROPERTIES OF DILUTED MAGNETIC SEMICONDUCTORS

in turn, plays a major role in spin interactions in degenerate semiconductors and metals.

Historically, interest in the pressure effects on exchange interactions was stimulated in early 1980s by electric and magnetic studies of the HgMn(Co)Se(Te) system, followed by the PbSnMnTe and $(CdZnMn)_3As_2$ systems. In the early 1990s, attention was extended to the cryobaric study of magneto-optical properties of CdMn(Co)Se and subsequently to quantum-well structures of CdTe/CdMnTe.

a. $Hg_{1-x}Mn_xSe$, $Hg_{1-x}Mn_xTe$

HgSe and HgTe have an inverted band structure with a bandgap E_g of about -0.3 eV as shown in Fig. 5. The Γ_8 conduction-band minimum and the Γ_8 valence-band maximum are degenerate at the Γ point. For increasing x, E_g varies continuously and goes to zero for some value x_0 of about 0.07. When this gapless state occurs, three bands are in contact. For $x > x_0$, E_g becomes positive, as it is in usual semiconductors having a zincblende crystal structure. For $x < x_0$, the absolute value of E_g is also reduced by hydrostatic pressure. The rate dE_g/dP is about 0.08 eV/GPa in HgMnSe and 0.10–0.15 eV/GPa in HgMnTe. Consequently, at some pressure P_i, which

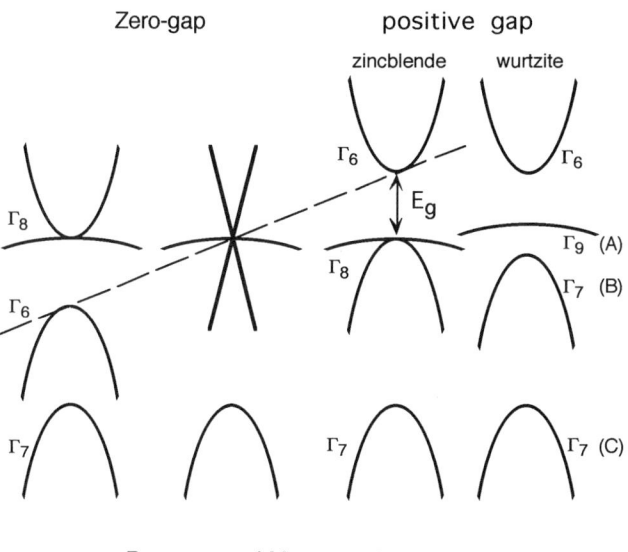

FIG. 5. Rearrangement of energy spectrum of $Hg^{II}MnB^{VI}$ compounds by Mn substitution and pressure.

would depend on x, the Γ_6 valence band and the Γ_8 conduction band are inverted, resulting in a positive E_g.

Brandt et al. (1983) have studied the spin-glass transition in p-type HgMnTe of x ranging from 0.020 to 0.075. The spin-glass transition temperature T_{SG} is found to decrease monotonically from 0.6 to 0.2 K as x decreases from 0.075 to 0.020, interpolating to $T_{SG} = 0$ K at $x = 0$. When the Γ_6 and Γ_8 bands are inverted by applying a pressure of $P > P_i$, the spin-glass state disappears. Brandt et al. have interpreted this observation as the breakdown of the Mn spin indirect exchange interaction of Bloembergen–Rowland type which is mediated by virtual electronic transitions between the Γ_8 zero-gap bands.

To study the pressure dependence of the d–d exchange interaction in zero-gap semiconductors in the paramagnetic phase in more detail, Chudinov et al. (1993) have examined the Shubnikov–de Haas oscillations in n-type HgMnSe under hydrostatic pressure up to 1.6 GPa using samples of $0 \leq x \leq 0.23$, in which the carrier concentration n_e ranges from 2×10^{18} to 6×10^{18} cm^{-3}. The Fermi energy ε_F reaches 0.4 eV for $n_e = 5 \times 10^{18}$ cm^{-3}. This value of ε_F is two orders of magnitude larger than the values for the HgMnTe crystals examined by Brandt et al. (1983). Figure 6 shows examples of the experimental results. The Shubnikov–de Haas oscillations exhibit a beating node, whose position shifts toward lower magnetic fields with increasing pressure. Under a given pressure the beating-node magnetic field H_b decreases with increasing temperature T. The H_b versus T curves measured under several pressures are shown in Fig. 7.

Under low magnetic field, the spin splitting of an electron is greater than the Landau splitting of the conduction band because of a strong electron–Mn^{2+} direct exchange interaction. In the mean field approximation the effective g-factor for the spin splitting is given by

$$g = g_0 + \frac{N_0 \alpha x^* \langle S_z \rangle}{\mu_B H}, \tag{5}$$

where g_0 is the intrinsic g-factor of the conduction electron at the Fermi energy, N_0 is the density of cation sites, α is the constant representing the electron–Mn^{2+} exchange, x^* is the content of isolated Mn ions, $\langle S_z \rangle$ is the mean value of S_z of the Mn spin **S**, μ_B is the Bohr magneton, and H is the external magnetic field. $\langle S_z \rangle$ can be expressed in terms of a modified Brillouin function as

$$\langle S_z(H) \rangle = S \mathscr{B}_{5/2} \left\{ \frac{g_M \mu_B S H}{k(T + T_0)} \right\}, \tag{6}$$

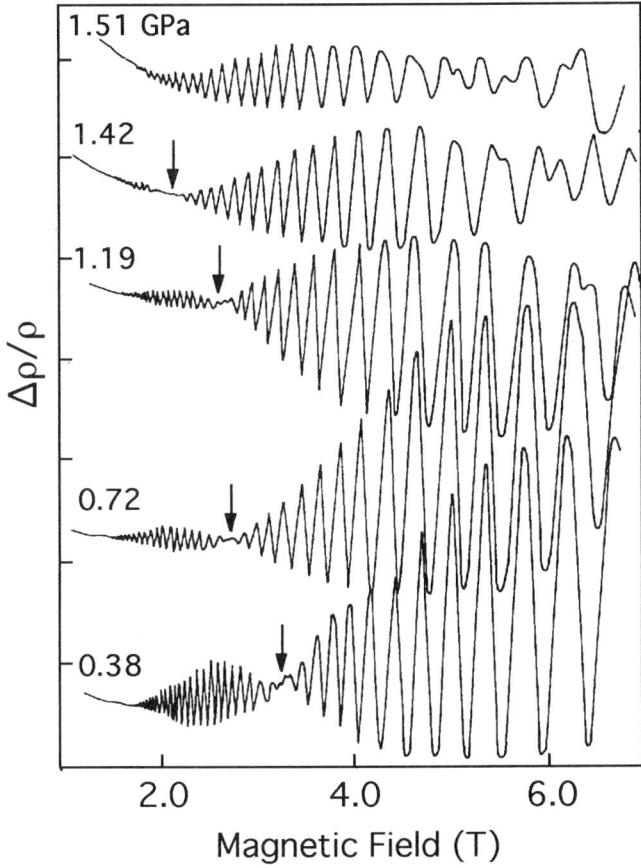

FIG. 6. Shubnikov–de Haas oscillations in HgMnSe of $x = 0.032$ at 2.1 K under various pressures. [Data taken from Chudinov et al. (1993).]

where g_M, k, and T_0 are the g-factor (≈ 2.0) of the ^6S state of Mn^{2+} $3d^5$ spins, the Boltzmann constant, and a temperature parameter characterizing the mean exchange fields due to spins of distant-neighbor Mn^{2+} ions (Barilero et al., 1987), respectively. As evident from Eqs. (5) and (6), the spin splitting tends to be saturated with increasing magnetic field. The observed beating node of the Shubnikov–de Haas oscillation occurs when the Landau and spin splittings cross. Thus, the beating node field H_b satisfies

$$\left(\frac{2m_0}{m_c} - g_0\right) \mu_B H_b = N_0 \alpha x^* \langle S_z(H_b) \rangle, \tag{7}$$

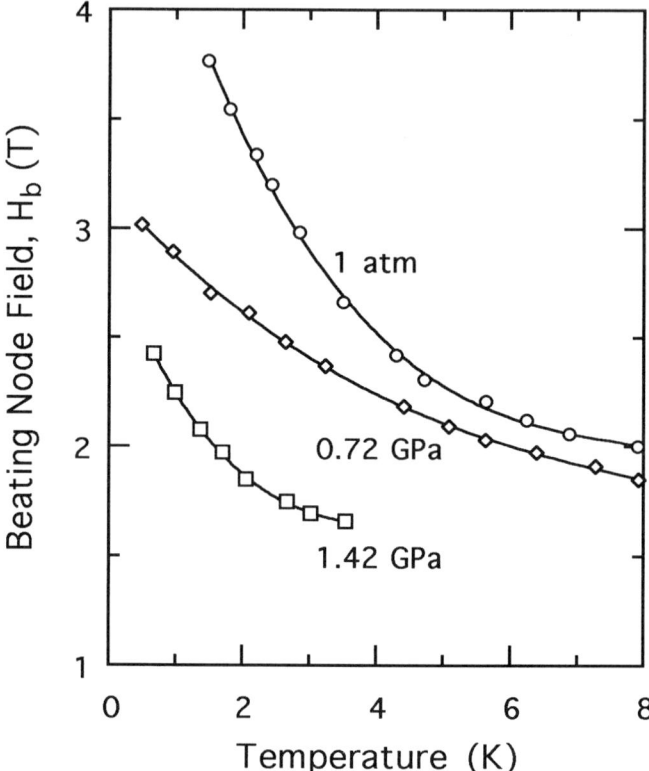

FIG. 7. Temperature dependence of the beating node field of Shubnikov–de Haas oscillation in HgMnSe of $x = 0.032$ under various pressures. [Data taken from Chudinov et al. (1993).]

where m_c and m_0 are the cyclotron effective mass of the conduction electron and bare electron mass, respectively. The quantities m_c and g_0 depend on pressure via the variations of E_g and the Fermi energy. Given the electron concentration, m_c and g_0 under a certain pressure can be calculated on the basis of Kane's $\mathbf{k} \cdot \mathbf{p}$ band model.

With Eqs. (6) and (7) the experimental data of $H_b(P, T)$ shown in Fig. 7 permit one to evaluate $N_0\alpha x^*$ and T_0 at various pressures. Figure 8 shows the results for a sample of $x = 0.032$, in which $\varepsilon_F = 0.4$ eV at 1 atm. We see that at low pressures T_0 is positive and thus the interaction between Mn^{2+} ions is suggested to be antiferromagnetic. As pressure increases, T_0 varies dramatically so as to change its sign, reflecting the switchover to a ferromagnetic regime, above about 0.3 GPa. Similar sign reversal of T_0 is also seen around $x = 0.04$ under ambient pressure as x increases from 0 to 0.07. Taking account of these findings, Chudinov et al. have argued for

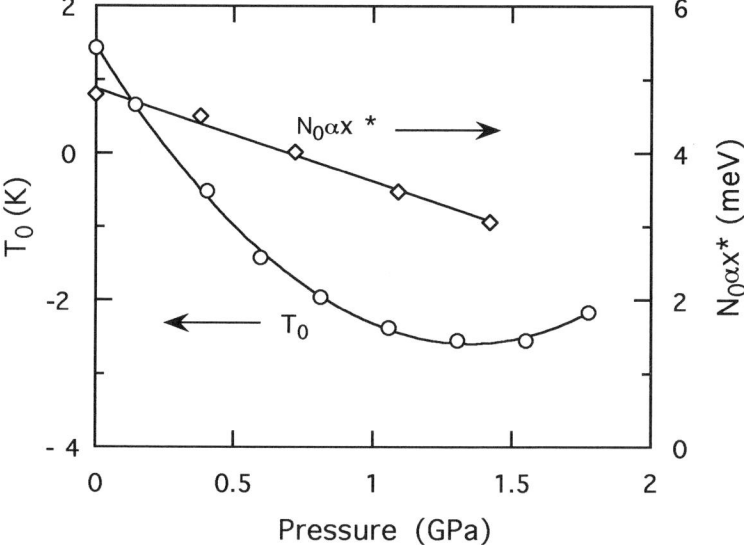

FIG. 8. Pressure dependence of $N_0 \alpha x^*$ and T_0 in HgMnSe of $x = 0.032$. [Data taken from Chudinov et al. (1993).]

the existence of RKKY interaction in addition to the Bloembergen–Rowland and/or kinetic superexchange mechanisms. It is also worth noting that the coefficient $N_0 \alpha x^*$ of the electron–Mn^{2+} exchange interaction decreases with increasing pressure, although no explanation is given.

b. $Pb_{1-x-y}Sn_yMn_xTe$

PbSnMnTe has a rock-salt structure under atmospheric pressure. The fundamental energy gap lies at the L point. The maximum of the next highest valence band is located at the Σ point and 0.21 eV in energy below the maximum of the L valence band, as shown in Fig. 9 (Story et al., 1989). The fundamental energy gap E_g is 0.275 eV. The effective masses of the L and Σ holes are about $0.2m_0$ and $1.3m_0$, respectively. Under hydrostatic pressure both E_g and the gap E_Σ at Σ point increase with a pressure coefficient of 0.08 and 0.06 eV/GPa, respectively.

In this system the concentration of holes of p-type samples can be controlled by annealing crystals in an appropriate atmosphere. Story et al. (1986) have found that ferromagnetic ordering occurs at a low temperature for the crystals containing holes more than 3×10^{20} cm^{-3}, and that the Curie temperature varies as the hole concentration increases. Figure 10 shows the result of magnetization measurement for a sample of $x = 0.03$.

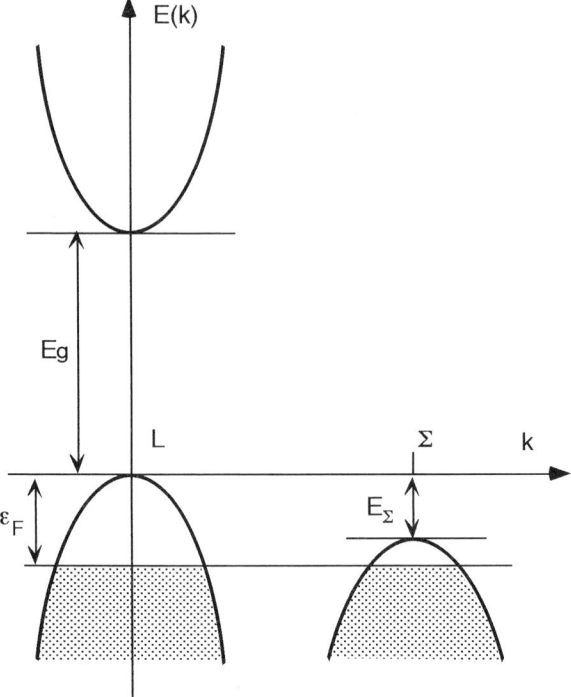

FIG. 9. Electronic structure of p-type $Pb_{1-x-y}Sn_yMn_xTe$.

The observed saturation magnetization of 3.1 emu/g confirms that the magnetization arises from Mn^{2+} ions. Figure 11 shows the dependence of the Curie temperature T_C and the temperature θ as determined from the Curie–Weiss law, $\chi \sim 1/(T - \theta)$, of paramagnetic susceptibility χ for $T > T_C$ on the hole concentration n_p. The fact that T_C and θ take a maximum around $n_p = 1 \times 10^{21}$ cm^{-3} has led Story et al. to ascribe the ferromagnetic ordering to the RKKY mechanism.

For the RKKY mechanism the exchange constant J of the coupling energy $-2J\,\mathbf{S}_1\mathbf{S}_2$ between spins $\mathbf{S}_1$ and $\mathbf{S}_2$ of magnetic ions, which are separated by a distance R, is given by

$$J = \frac{9\pi n^2 J_{s-d}^2}{2\varepsilon_F} \left\{ \frac{-2k_F R \cos(2k_F R) + \sin(2k_F R)}{(2k_F R)^4} \right\}, \tag{8}$$

where n is the concentration of the relevant carriers, J_{s-d} is the s–d exchange constant, and k_F is the Fermi momentum. As evident from Eq. (8), the exchange constant J is an oscillatory function of the product $k_F R$.

6 ELECTRONIC PROPERTIES OF DILUTED MAGNETIC SEMICONDUCTORS 531

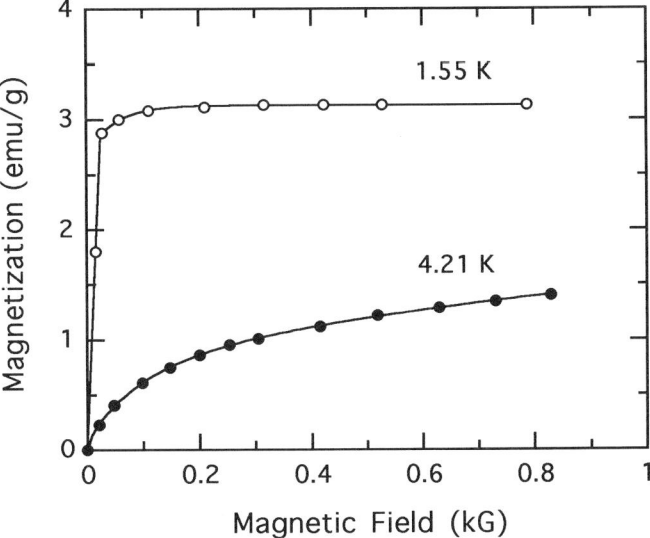

FIG. 10. Magnetization for a sample of PbSnMnTe. [Data taken from Story *et al.* (1986).]

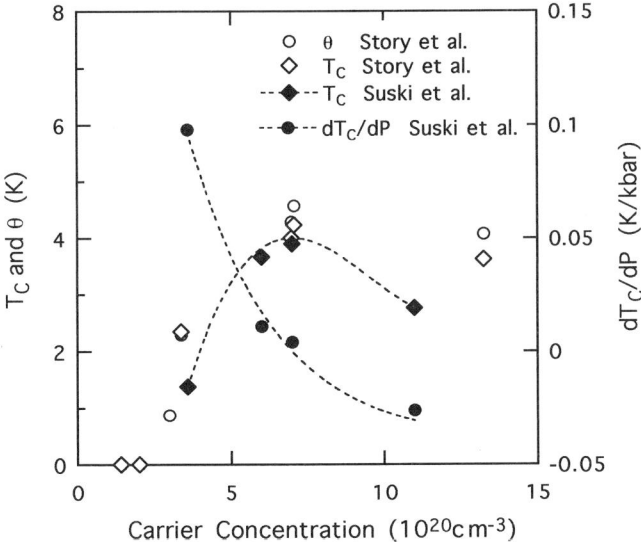

FIG. 11. Carrier concentration dependence of θ, T_C, and dT_C/dP in PbSnMnTe.

Clear experimental evidence that the Γ holes are closely related to the ferromagnetic ordering has been obtained from measurement of the pressure effect on T_C by Suski et al. (1987a, b). In the case of the PbSnMnTe system a pressure-induced increase in E_Σ causes some extent of holes in the Σ valence band to transfer to the Γ valence band, and thus causes an increase (decrease) in the concentration of $\Gamma(\Sigma)$ holes. In this situation the change in T_C due to the interband redistribution of holes is written as

$$\frac{dT_C}{dP} = \left(\frac{\partial T_C}{\partial n_{p\Gamma}}\right)\left(\frac{\partial n_{p\Gamma}}{\partial P}\right) + \left(\frac{\partial T_C}{\partial n_{p\Sigma}}\right)\left(\frac{\partial n_{p\Sigma}}{\partial P}\right), \tag{9}$$

where $n_{p\Gamma}$ and $n_{p\Sigma}$ denote the hole concentrations in the Γ and Σ valence bands, respectively. As shown in Fig. 11, the observed pressure coefficient of T_C varies as a function of n_p just like a derivative curve of T_C with respect to n_p. Note that the sign of $\partial n_p/\partial P$ responsible for the change in T_C is positive. The increase in $n_{p\Gamma}$ would enlarge k_F of Γ holes, supporting the notion that the RKKY mechanism is an important ingredient in the ferromagnetic ordering in PbSnMnTe.

c. $(Cd_{1-x-y}Zn_yMn_x)_3As_2$

Materials belonging to $(Cd_{1-x-y}Zn_yMn_x)_3As_2$ family have a complicated, tetragonal crystal structure. One of the parents, Cd_3As_2, has an inverted band structure with an energy gap of $E_g = -0.1$ eV, whereas Zn_3As_2 has a simple energy gap of $E_g = 1.1$ eV. For small Zn and Mn contents the mixed crystal is a narrow-gap semiconductor having a linear dependence of the energy gap on composition. The energy gap is conceived to increase with pressure at a rate of about 60 meV/ GPa (Cisowski et al., 1989). Pressure and magnetic field dependencies of the cyclotron effective mass and g-factor of conduction electrons show that the electronic structure around the fundamental energy gap can be described well in terms of Kane's $\mathbf{k \cdot p}$ band model (Fig. 5) (Laiho et al., 1996).

Substitution of Mn for Cd or Zn gives rise to a pronounced paramagnetism typical of DMS, as shown in Fig. 12 (Lubczynski et al., 1989). The saturation magnetization per manganese ion is about 4 μ_B (Denissen et al., 1987), which is a little smaller than the value ~ 5 μ_B expected for a divalent Mn^{2+} ion in the usual ^6S state. The data shown in Fig. 12 can be reproduced well in terms of the modified Brillouin function given by Eq. (6) in the magnetic field range examined. Seemingly, the saturation magnetization $x*S$ is suppressed or the temperature parameter T_0, which is positive, is enlarged by pressure, in contrast to the aforementioned case of HgMnSe. In addition, the beating node of the Shubnikov–de Haas oscillation shifts

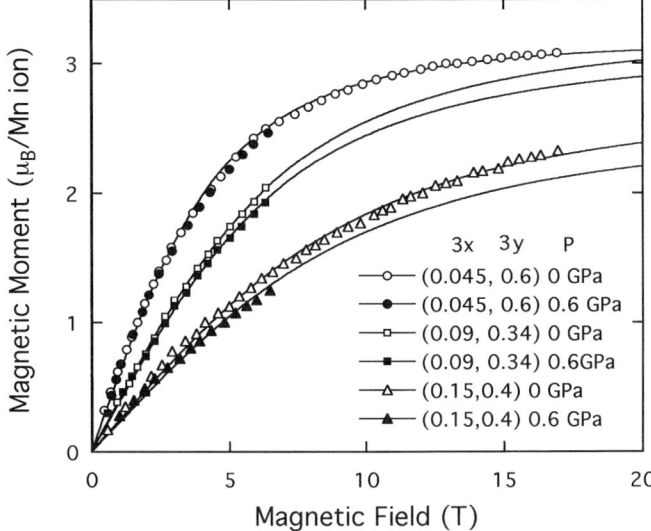

FIG. 12. Magnetization in $(Cd_{1-x-y}Zn_yMn_x)_3As_2$ under ambient pressure and 0.6 GPa at 4.2 K. Solid lines are fits of the modified Brillouin function to the experimental data. [Data taken from Lubczynski et al. (1989).]

toward lower magnetic field under hydrostatic pressure (Cisowski et al., 1989; Lubczynski et al., 1989). These findings suggest that there exists an antiferromagnetic coupling between Mn^{2+} ions and that the coupling is strengthened by pressure.

According to an analysis of magnetization by using a pair-approximation method generalized so as to be adaptable to the complicated crystal structure of this system (Bednarski et al., 1995), the superexchange and Bloembergen–Rowland exchange contribute additively to the antiferromagnetic coupling between spins of Mn^{2+} ions. The analysis has shown that in the sample of $(x, y) = (0.05, 0.14)$, for instance, the observed pressure-induced change in the magnetization curve can be explained in terms of the increase in the nearest-neighbor exchange constant J_{NN} from $J_{NN} = J_{NN}^{SE} + J_{NN}^{BR} = -(19 + 5.5) = -24.5$ K at 1 atm to $J_{NN} = J_{NN}^{SE} + J_{NN}^{BR} = -(20.4 + 6) = -26.4$ K at 0.6 GPa, where J_{NN}^{SE} and J_{NN}^{BR} are the interaction constants of the superexchange and the Bloembergen–Rowland exchange, respectively.

d. $Cd_{1-x}Mn_xSe$, $Cd_{1-x}Co_xSe$

Since CdMn(Co)Se compounds are wide-gap semiconductors, pressure effects on the exchange interactions in these materials have been studied by cryobaric experiments on magnetophotoluminescence properties. Because of the wurtzite structure of the crystal, the magnetophotoluminescence spectrum has a strong uniaxial anisotropy with respect to the direction of magnetic field against the c axis of the crystal (Matsuda et al., 1993; Matsuda and Kuroda, 1996). If x is so small that $x^* \approx x$, the energy of the lower magnetic sublevel of the A exciton, which consists of an electron of the Γ_6 conduction band and a hole of the Γ_9 valence band (Fig. 5), for $H \parallel c$ can be expressed as

$$E_A = E_0 - \frac{1}{2} N_0(\alpha - \beta) x^* \langle S_z \rangle - \frac{1}{2} |g_e - g_h| \mu_B H + \sigma H^2, \qquad (10)$$

where E_0 is the energy in the absence of magnetic field, β is the exchange constant of the interaction between a hole and Mn^{2+} or Co^{2+} ions, $g_{e(h)}$ is the intrinsic g-factor of an electron (hole), and σ is the coefficient of the diamagnetic shift of the exciton.

If x is elevated to the order of 0.05, a significant amount of magnetic ions are situated on nearest-neighbor cation sites to form pairs, of which the ground state is spin singlet because of the antiferromagnetic superexchange coupling. These pairs manifest themselves as a weak, stepwise development of magnetization under high magnetic fields and low temperatures. If the exchange energy is written as $-2J_{NN}\mathbf{S}_1\mathbf{S}_2 (J_{NN} < 0)$, the steps occur at (Larson et al., 1986; Forner et al., 1989)

$$H_l = -\frac{2lJ_{NN}}{g_M \mu_B} + H_d, \qquad l = 1, 2, \ldots, 2S, \qquad (11)$$

where H_d is a small correction ($\approx 1\,T$) due to distant-neighbor interactions. This stepwise magnetization contributes additionally to E_A by (Aggarwal et al., 1985)

$$E_{A\text{step}} = -\frac{1}{2} N_0(\alpha - \beta) x_p \sum_{l=1}^{2S} \left[1 + \exp\left\{ \frac{g_M \mu_B (H_l - H)}{k(T + T^*)} \right\} \right]^{-1}, \qquad (12)$$

where x_p is half of the content of transition-metal ions forming pairs and T^* is an effective temperature to represent the broadening of the steps.

Figure 13 shows the magnetic-field-induced shift of E_A in CdMnSe of $x = 0.05$ observed under several pressures (Kuroda and Matsuda, 1996a, b, c). The results can be reproduced fairly well by Eq. (10) combined with Eqs. (6), (11), and (12), where the calculation is made by choosing

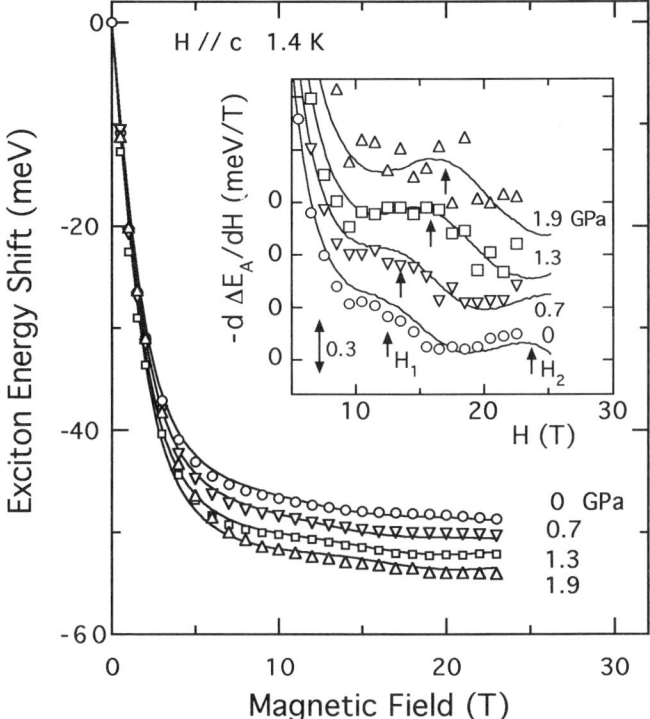

FIG. 13. Zeeman shift of the exciton energy in CdMnSe of $x = 0.05$ under various pressures at 1.4 K. The inset shows the derivative of the energy shift with respect to magnetic field. Solid lines are theoretical curves. Arrows show the positions of magnetization steps due to paired Mn^{2+} ions.

$N_0(\alpha - \beta)$ and J_{NN} as adjustable parameters; the other quantities, that is, T_0, T^*, x^*, x_p, g_M, $|g_e - g_h|$, and σ, are all fixed to the respective values at 1 atm. It turns out from these results that both $N_0(\alpha - \beta)$ and J_{NN} are enlarged significantly by pressure, as shown in Fig. 14. Almost the same behavior of $N_0(\alpha - \beta)$ is observed in CdMnSe of $x = 0.01$ and CdCoSe of $x = 0.012$.

Figure 15 depicts the single-particle density of states of electronic bands including the Hubbard states of Mn d electrons. The on-site Coulomb energy U shown there is the effective energy needed to take a d electron from a Mn^{2+} ion and put it on another Mn^{2+} ion (Fujimori et al., 1993), which can be represented as

$$d^5 + d^5 \rightarrow d^4 + d^6,$$

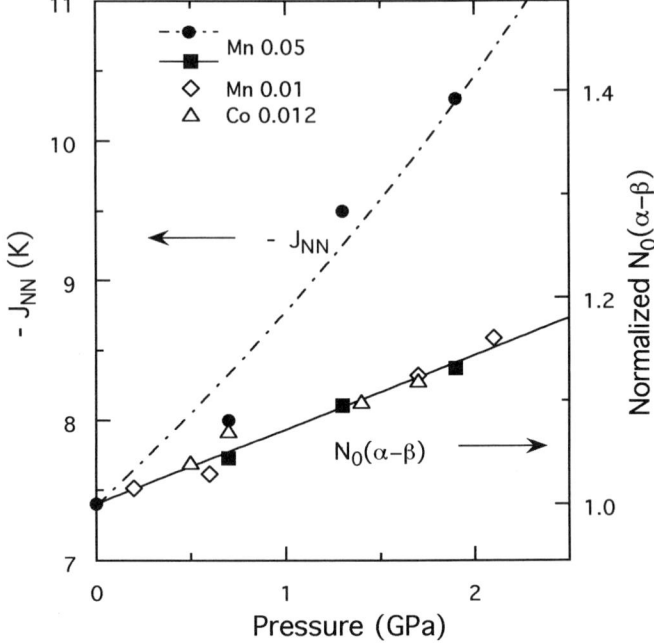

FIG. 14. Pressure dependence of $N_0(\alpha - \beta)$ in CdMnSe of $x = 0.01$ and $x = 0.05$ and CdCoSe of $x = 0.012$ and J_{NN} in CdMnSe of $x = 0.05$. The data of $N_0(\alpha - \beta)$ in CdMnSe and CdCoSe are normalized by the ambient value 1.37 and 2.52 eV, respectively. The solid line is the least-squares fit of a straight line to the experimental data of $N_0(\alpha - \beta)$ in CdMnSe of $x = 0.05$. The dash–dotted line is the curve of J_{NN} calculated from Eq. (14) with $V_{pd}(P = 0) = 0.79$ eV assuming U, Δ, and Λ decrease linearly with pressure. [After Kuroda and Matsuda (1996b, c).]

whereas Δ is the energy needed to transfer an electron from a ligand p orbital forming the valence band to a Mn^{2+} ion, leaving a hole in the valence band (Zaanen and Sawatzky, 1987), that is,

$$d^5 \rightarrow d^6 + \text{hole}.$$

Another charge-transfer energy $\Lambda(= U - \Delta)$ is defined as the energy to promote a d electron of a Mn^{2+} ion to the ligand p orbital to annihilate a hole:

$$d^5 + \text{hole} \rightarrow d^4.$$

The values of U, Δ, and Λ after Bhattacharjee (1995) for several DMS at ambient pressure are listed in Table IV.

6 ELECTRONIC PROPERTIES OF DILUTED MAGNETIC SEMICONDUCTORS 537

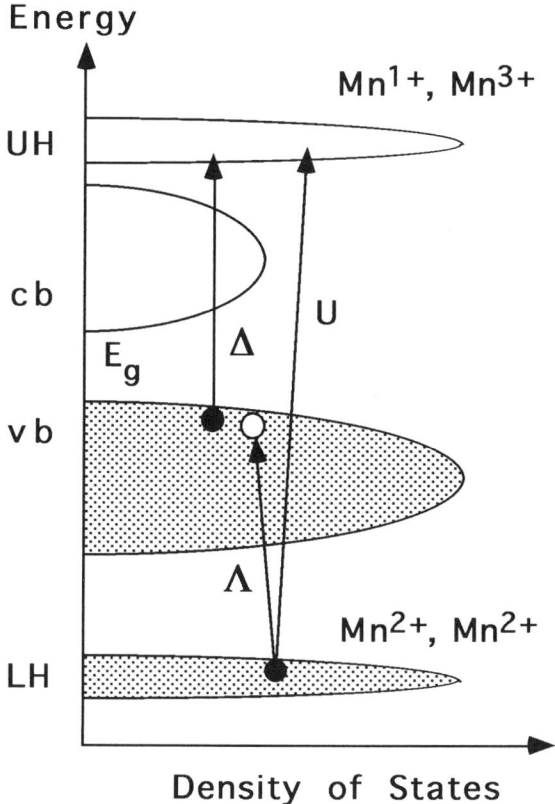

FIG. 15. Schematic representation of the single-particle density of states in CdMnSe. UH, cb, vb, and LH denote the upper Hubbard state, conduction band, valence band, and lower Hubbard state, respectively.

TABLE IV

ELECTRON CORRELATION ENERGIES IN SEVERAL DMSs AFTER BHATTACHARJEE (1995)

Substance	U (eV)	Δ (eV)	Λ (eV)
CdMnSe	7.6	4.2	3.4
CdCoSe	5.9	2.4	3.5
CdFeSe	6.8	3.1	3.7

The d electrons which are hybridized with the Se p valence electrons are those occupying Mn $d\epsilon$ orbitals forming the Hubbard states (Wei and Zunger, 1987). The interatomic p–d hopping of electrons via the hybridized orbitals dominates the exciton–Mn^{2+} exchange interaction (Larson et al., 1988; Bhattacharjee, 1992; Hamdani et al., 1992). On the basis of the three-level model shown in Fig. 15, $N_0(\alpha - \beta)$ is given to a good approximation by

$$N_0(\alpha - \beta) = \frac{16}{S} \frac{V_{pd}^2 U}{\Delta \Lambda}, \tag{13}$$

where V_{pd} is the transfer integral. Within the same model the d–d superexchange constant is given by (Zaanen and Sawatzky, 1987; Larson et al., 1985, 1988)

$$J_{NN} = J_{NN}^{SE} = -\frac{1}{2S^2} \frac{V_{pd}^4(U + \Delta)}{\Delta^3 U}. \tag{14}$$

The transfer integral V_{pd} varies with the bond length as $b^{-7/2}$ (Harrison, 1980). However, the change in V_{pd} expected from the linear compressibility 0.62×10^{-2} GPa^{-1} in CdSe (Martinez, 1980) yields only half of the observed changes in $N_0(\alpha - \beta)$ and J_{NN}. The remaining part arises from the pressure dependence of U, Δ, and Λ, as shown in Fig. 16. The correlation energies appear to diminish with increasing pressure, that is, $dU/UdP = (-2.5 \pm 0.5) \times 10^{-2}$ GPa^{-1}, $d\Delta/\Delta dP = (-3.0 \pm 1.0) \times 10^{-2}$ GPa^{-1}, and $d\Lambda/\Lambda dP = (-1.8 \pm 1.2) \times 10^{-2}$ GPa^{-1}, suggesting that the screening of the Coulomb interactions is enhanced as the degree of the p–d hybridization increases. Then we obtain the Grüneisen parameter of U to be

$$\gamma_U \equiv -\frac{bdU}{3Udb} = -1.3. \tag{15}$$

If a small splitting into R_1 and R_2 is neglected, the energy, say E_R, of the R lines in ruby is given by $3(3B_R + C_R)$, where B_R and C_R are the Racah parameters describing *intraion* Coulomb and exchange interactions in a given configuration of d electrons of a magnetic ion. The R lines in ruby are associated with the local structure $[CrO_6]^{9-}$ in Al_2O_3. The familiar redshift under pressure, that is, $\gamma_R \equiv -bdE_R/3E_R\,db = -0.125$, results from weakening of the d–d interactions in the $3d^3$ configuration of a Cr^{3+} ion due to enlargement of covalency with oxygen p orbitals (Ohnishi and Sugano, 1982). The large value of γ_U compared with γ_R of ruby illustrates the strength of the Mn d–Se p hybridization in CdMnSe. The values of γ

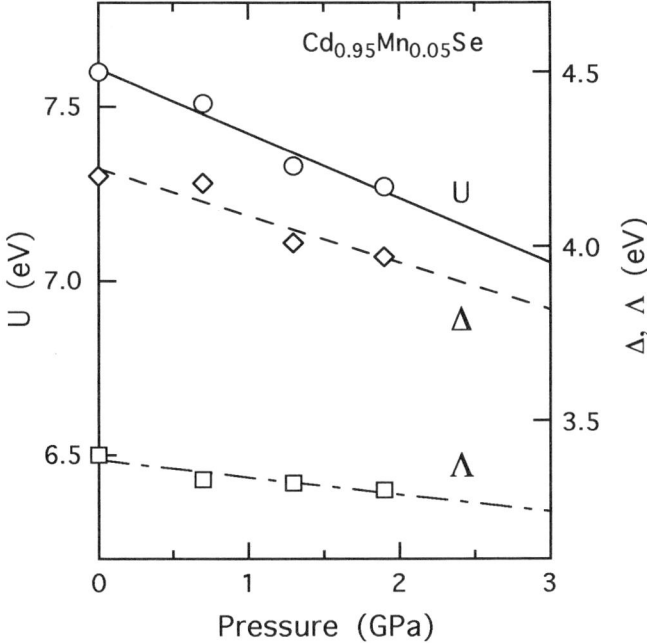

FIG. 16. Pressure dependence of U, Δ, and Λ in CdMnSe of $x = 0.05$. The solid, dashed, and dash–dotted lines are the least-squares fits of straight lines to the experimental data of U, Δ, and Λ, respectively. [After Kuroda and Matsuda (1996b, c).]

of the Racah parameters of Mn^{2+} and Co^{2+} d electrons in DMS will be discussed in the next section.

The correlation energies Δ and U decrease as if they become zero around 30–40 GPa to convert the material to a p- or d-metal, although the crystal actually transforms to a NaCl structure at about 2.5 GPa. It is known that NiI_2 is an antiferromagnetic charge-transfer semiconductor (Fig. 1) with the Néel temperature T_N of 75 K at ambient pressure. Chen et al. (1993) have found that the optical energy gap, being 1.25 eV at ambient pressure, closes at around 20 GPa and the electric conduction indeed becomes metallic above 20 GPa.

For $x > 0.1$, Mn^{2+} and Co^{2+} ions mostly form pairs or clusters. The magnetization of a larger cluster would spread over a wider region of magnetic field. Because of the antiferromagnetic character of Mn–Mn interactions and their strong sensitivity to pressure, the mean magnetization observed at a given magnetic field will diminish rapidly with pressure. In a description in terms of the modified Brillouin function, this behavior appears as a decrease in the saturation magnetization x^*S and/or an in-

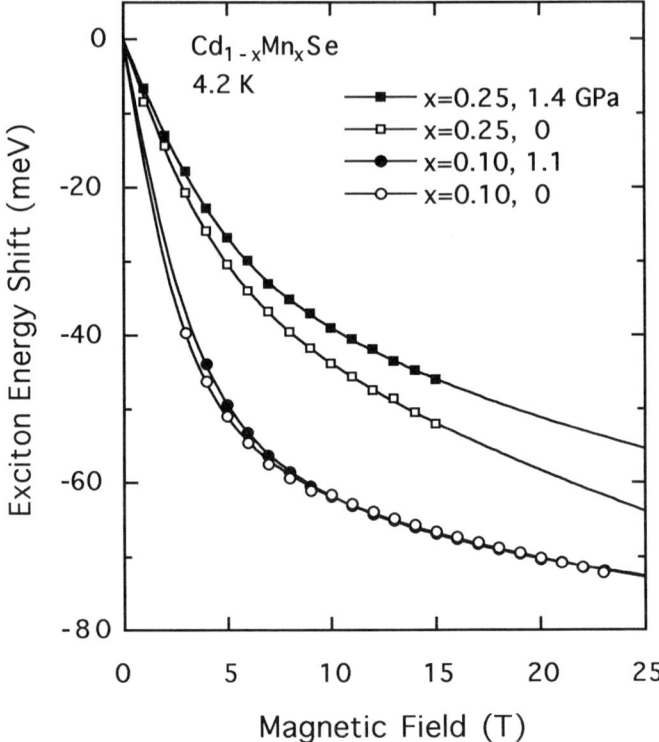

FIG. 17. Exciton energy shift under magnetic field at 4.2 K in CdMnSe of $x = 0.10$ under ambient pressure and 1.1 GPa, and in CdMnSe of $x = 0.25$ under ambient pressure and 1.4 GPa. [Data taken from Matsuda et al. (1993, 1994).]

crease in T_0, as has been observed in $(CdZnMn)_3As_2$ by Lubczynski et al. (1989) (Fig. 12). For $x = 0.1$ this effect almost cancels out the effect of increase in $N_0(\alpha - \beta)$, and hence the exciton energy shift undergoes only a slight change due to pressure, as shown in Fig. 17. Furthermore, for $x = 0.25$ the decrease of magnetization seems to overcome the increase of $N_0(\alpha - \beta)$ (Matsuda and Kuroda, 1996; Kuroda and Matsuda, 1996c).

3. INTRAION d–d OPTICAL TRANSITIONS

Pressure tunes the crystal field for the localized d electrons by changing the Racah parameters B_R and C_R as well as the point-charge potential

TABLE V

PRESSURE DEPENDENCE OF Mn^{2+} INTRAIONIC OPTICAL TRANSITION ENERGIES (eV)[a]

Substance	$^6A_1 \to {}^4T_1$	$^6A_1 \to {}^4T_2$	$^6A_1 \to {}^4A_1, {}^4E$	References
MnO	2.01–0.013P	2.50–0.011P	2.94–0.0014P	Kobayashi et al. (1995)
MnS	2.08–0.030P	2.45–0.028P	2.74–0.015P	Kobayashi et al. (1995)
ZnMnS	2.35 ($P=0$)	2.53 ($P=0$)	2.69 ($P=0$)	Langer and Ibuki (1965)
	2.13–0.030P^b	—	—	Anastassiadou et al. (1988)
ZnMnSe	2.38 ($P=0$)	2.57 ($P=0$)	2.68 ($P=0$)	Morales (1985)
	—	—	2.67–0.02P	Ves et al. (1986b)
	—	2.50–0.017P	2.68–0.015P	Jiang et al. (1989b)
	2.14–0.036P^b	—	—	Arora and Sakuntala (1995)
ZnMnTe	2.3 ($P=0$)	2.4 ($P=0$)	2.6 ($P=0$)	Morales et al. (1984)
	2.3–0.049P	—	2.6–0.043P	Ves et al. (1986a)
CdMnS	2.43 ($P=0$)	2.58 ($P=0$)	2.72 ($P=0$)	Ikeda et al. (1968)
	—	2.58–0.026P	2.71–0.020P	Jiang et al. (1990)
CdMnTe	2.2 ($P=0$)	2.5 ($P=0$)	—	Moriwaki et al. (1982)
	2.1–0.075P	—	—	Müller et al. (1983)
	2.18–0.051P	—	—	Shan et al. (1985)
	2.0–0.081P^b	—	—	Müller et al. (1983)
	2.0–0.060P^b	—	—	Jiang et al. (1992)

[a] P denotes pressure (GPa).
[b] Photoluminescence.

$10Dq$. Table V collects the experimental data on the pressure dependence of the lowest three optical transition energies of Mn^{2+} ions in various substances. Unless otherwise noted, the energies given in Table V are the values obtained from optical absorption experiments. In the visible and infrared regions no optical absorptions can be expected to arise from electronic transitions between the d^5 terms of the Mn^{2+} ions and the valence or conduction band of the host material, because as evident from Fig. 15 it costs an additional energy Δ or Λ, which is as large as ~ 4 eV, to change the number of the d electrons of a Mn^{2+} ion.

In the Sugano–Tanabe diagram, the ground state of a d^5 configuration is $^6A_1$ and the first and second excited levels are $^4T_1$ and $^4T_2$, respectively, while the third and fourth excited levels $^4A_1$ and ^4E are almost degenerate and thus unresolved experimentally. We see from Table V that the photon energies of the $^6A_1 \to {}^4T_1$, $^6A_1 \to {}^4T_2$, and $^6A_1 \to {}^4A_1, {}^4E$ transitions are about 2.2, 2.5, and 2.7 eV, respectively, being almost independent of the host materials. Furthermore these energies decrease noticeably with increasing pressure, despite the fact that $10Dq$ increases as b^{-5} with decreasing the bond length b. This fact implies that B_R and C_R are reduced by pressure.

The photon energy, say $E_{d-d,3}$, of the $^6A_1 \to {}^4A_1$, ^4E transitions directly reflects the changes in B_R and C_R, because the $^4A_1$ and ^4E states are obtained by simply reversing the spin of an electron of five d electrons of the $^6A_1$ state, in which spins are all aligned in the same direction and thus $E_{d-d,3}$ is independent of $10Dq$. Under a crystal field with a T_d symmetry $E_{d-d,3}$ is given by

$$E_{d-d,3} = 10B_R + 5C_R. \qquad (16)$$

The experimental data of $dE_{d-d,3}/dP$ yield $\gamma = -bdE_{d-d,3}/3E_{d-d,3}db = -0.43$ in ZnMnSe and -0.67 in ZnMnTe. In a similar way Bak *et al.* (1996) have obtained the pressure coefficient of the Racah parameter of Co^{2+} $3d^7$ configuration in ZnCoSe. Their result gives $\gamma = -0.57$. These γ values are comparable to $\gamma_U = -1.3$ of the on-site Coulomb energy U in CdMnSe given by Eq. (15). It is evident that the pressure-induced increase in the p–d hybridization also markedly reduces the intraion Coulomb and exchange interactions of d electrons.

Except zero-phonon transitions such as R_1 and R_2 lines of ruby, photoluminescence due to d–d transitions in solids generally manifests itself as a broad band and shows a large Stokes shift, because a strong coupling of the excited d electron with phonons induces a relaxation of the environmental lattice to produce vibronic states. In ZnMnS, ZnMnSe, and CdMnTe, as seen in Table V, a broad and yellow photoluminescence band appears at a photon energy about 0.2 eV below the $^6A_1 \to {}^4T_1$ absorption band at ambient pressure. Since application of pressure causes a red shift characteristic of intraion d–d transitions, the yellow photoluminescence band is attributable to the $^4T_1 \to {}^6A_1$ transition. An alternative assignment, that is, the *interion* recombination of an electron in a Mn^{2+} $3d$ level with a hole in the valence band, may be precluded from an energetic consideration taking account of the electron correlation of Mn $3d$ electrons.

In ZnMnSe the properties of the yellow photoluminescence are unaffected by the intermediate phase transition mentioned in Section II.3, which occurs in the substances of $x > 0.3$ under pressure (Arora *et al.*, 1988). This fact might be an indication of the strong localization of d orbitals. It has emerged from experimental studies of Nakahara and his co-workers (Nakahara *et al.*, 1993; Watanabe *et al.*, 1992; Gregus *et al.*, 1993) on the temperature and pressure dependencies of spectroscopic properties of the photoluminescence in CdMnTe that if a Mn^{2+} ion is excited up to the $^4T_1$ state, the ion can be displaced by 1.6 Å from the central position of the $MnTe_4$ tetrahedron like the DX and/or EL2 center in GaAs.

IV. Superstructures

1. ZnSe/Zn$_{1-x}$Mn$_x$Se

In contrast to bulk ZnMnSe, the ZnMnSe layers in ZnSe/ZnMnSe heterostructures maintain the zincblende structure for $x > 0.3$. The band offset of this system is of type I at ambient pressure. The ZnSe and ZnMnSe layers function as the wells and barriers, respectively, as shown in Fig. 18.

Tuchman et al. (1993) have measured photoluminescence of the superlattices of $x = 0.23$, 0.33, and 0.51 under hydrostatic pressure. Although the ZnSe and ZnMnSe layers are in tension and in compression, respectively, because of the difference in the lattice constant between the two layers, the exciton photoluminescence from ZnSe wells and the yellow photoluminescence due to the $^4T_1 \rightarrow {}^6A_1$ d–d transition from ZnMnSe barriers show almost the same pressure-induced shift as those in bulk ZnSe and ZnMnSe crystals, respectively. As mentioned in Section II, the pressure coefficient of the energy gap E_g of DMS is smaller than that of host materials. In the case of ZnMnSe the difference in the pressure coefficient a_1 between $x = 0$ and $x = 0.23$ is about 10 meV/GPa (Table III), so that the difference in E_g, which is 165 meV at ambient pressure, is to be reversed at about 16 GPa. In view of the band alignment shown in Fig. 18, if the valence band of the ZnMnSe layers goes up faster than the ZnSe layers because of p–d hybridization, as mentioned in Section III.1, the crossover of the valence band would occur at a significantly lower pressure. Indeed, the exciton photoluminescence from ZnSe layers begins abruptly to diminish around

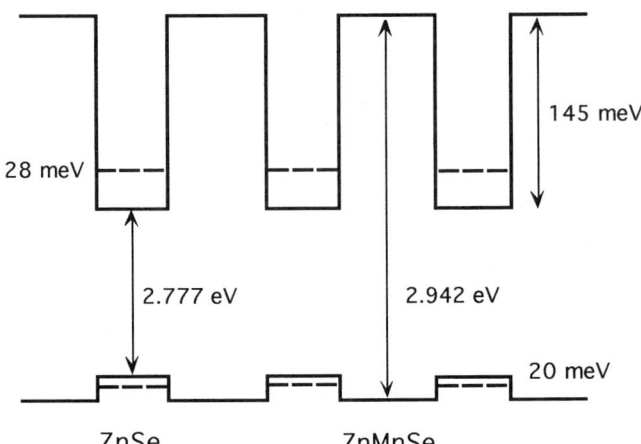

FIG. 18. Alignment of the conduction and light-hole valence bands for the ZnSe/Zn$_{0.77}$Mn$_{0.23}$Se superstructure at ambient pressure. [After Tuchman et al. (1993).]

5 GPa and is reduced by several orders of magnitude above 6 GPa. In contrast, the yellow band remains intense until the ZnMnSe layers undergo the structural phase transition.

The yellow photoluminescence band disappears once the structural phase transition occurs at 9.0, 8.0, and 7.0 GPa in the superlattices of $x = 0.23$, 0.33, and 0.52, respectively. The transition pressure P_t in the former two superlattices agrees well with the value 9.6 and 7.8 GPa, respectively, expected from the data (Fig. 4) for bulk ZnMnSe. However, since the ZnMnSe layers of $x = 0.52$ are superpressed, the observed value of P_t is significantly greater than the value 4.7 GPa expected from extrapolation of the bulk data in Fig. 4. It is widely known that in a semiconductor heterostructure the layer with lower P_t in the bulk phase can be superpressed.

2. $CdTe/Cd_{1-x}Mn_xTe$

This is another family of superstructures having type-I band offset. Perlin *et al.* (1995) have studied the pressure dependence of the exciton photoluminescence from CdTe layers and the yellow photoluminescence from CdMnTe layers for samples with a multiple quantum well structure, in which the width of the barriers of either $Cd_{0.5}Mn_{0.5}Te$ or $Cd_{0.32}Mn_{0.68}Te$ is fixed to be 50 nm. They report the pressure coefficient of the exciton energy to be $a_1 = 60 \pm 10$ meV/GPa and that of the yellow band to be -80 ± 10 meV/GPa. Meyer *et al.* (1995) and Yokoi *et al.* (1996) have made measurements on the exciton photoluminescence for single quantum well structures of $CdTe/Cd_{0.91}Mn_{0.09}Te$ and $CdTe/Cd_{0.76}Mn_{0.24}Te$, respectively. They evaluated the quadratic coefficient a_2, as well as a_1, of the exciton energy in the wells and barriers.

The latter two groups have measured the pressure dependence of exciton photoluminescence also under magnetic fields up to 6 and 30 T, respectively, at liquid helium temperatures. It is noteworthy that the degeneracy of the energies of heavy and light holes at the Γ point is lifted in quantum wells. This situation has a close resemblance to the case of bulk crystals of wurtzite CdMn(Co)Se (Matsuda and Kuroda, 1996) in the sense that the Zeeman shift of the energy of excitons in CdTe wells shows a uniaxial anisotropy with respect to the direction of magnetic field relative to the growth direction of the wells. The magneto-photoluminescence spectrum demonstrates that excitons in the wells, as well as those in barriers, undergo a large exchange-induced Zeeman shift since the wave function of excitons in the wells penetrates into barriers, and moreover, a significant number of Mn^{2+} ions are diffused into wells.

From their data taken in the instrumental geometry of the $H \perp$ growth direction, Meyer et al. (1995, 1996) have derived $dN_0(\alpha - \beta)/dP = 75 \pm 15$ meV/GPa for excitons in the 1.8-nm-wide CdTe well, and 46 ± 13 meV/GPa for excitons in the $Cd_{0.91}Mn_{0.09}Te$ barrier. The data of Yokoi et al. show that in the $Cd_{0.76}Mn_{0.24}Te$ barriers $dN_0(\alpha - \beta)/dP$ is even negative, just as it is in bulk $Cd_{0.75}Mn_{0.25}Se$ (Fig. 17), despite the fact that $dN_0(\alpha - \beta)/dP \approx 100$ meV/GPa in the 3.8-nm-wide CdTe well. Meyer et al. have noticed that regardless of excitons used as the probe the quantity T_0, which is a temperature parameter characterizing distant-neighbor Mn^{2+}–Mn^{2+} interactions [Eq. (6)], is about 3.5 K at ambient pressure but is enlarged to about 7.0 K at 3 GPa. This enlargement is 3–4 times greater than the enlargement of V_{pd}^4 postulated by the compressibility of the lattice.

Theories for the covalent spin interaction (Larson et al., 1988; Shen et al., 1995; Rusin, 1996) predict that the strength of nth-neighbor exchange interaction J_n has a spatial dependence of the form $J_n = J_{NN} f(R_n/d_0)$, where R_n is the separation of magnetic ions and d_0 is the lattice constant of the fcc (zincblende) or hcp (wurtzite) sublattice of cations. Although $f(R_n/d_0)$ is a rapidly decreasing function of R_n/d_0 (Twardowski et al., 1987), it is independent of bond length. According to a treatment of the effect of the distant-neighbor interactions by the molecular field approximation (Barilero et al., 1987), T_0 is related linearly to J_2 and J_3. Therefore, in view of the pressure dependence of J_{NN} in bulk CdMnSe, which has been described in Section III, changes in U, Δ, and Λ could partly contribute to the enlargement of T_0 observed. However, neither for bulk crystals nor for superlattices of DMS is systematic experimental information available on the pressure dependence of J_n of $n \geq 2$.

V. Miscellaneous

1. RAMAN SCATTERING

Raman scattering has been studied in ZnMnS (Anastassiadou et al., 1988), ZnMnSe (Arora et al., 1987, Arora and Sakuntala, 1995) and CdMnTe (Arora et al., 1987) under high pressure. In ZnMnS the TO and LO phonons show a one-mode behavior, whereas in CdMnTe and ZnMnSe they show a two-mode and intermediate behavior, respectively. In those studies the mode Grüneisen parameters and the pressure dependence of the transverse effective charges of the host and guest lattices are discussed.

Since the pressure-induced phase transition is of first order, crystals often break into pieces upon a pressure cycling across the transition point.

Kobayashi et al. (1996a) have examined the Raman spectrum due to the LO phonon in ZnMnS for the samples retrieved after compression up to 16 GPa, which is 2 GPa above the transition pressure. The Raman peak of the retrieved crystal is found to be slightly redshifted compared to the peak of the original sample. At the same time the spectrum is broadened significantly. From an analysis of the spectrum the retrieved sample is concluded to consist of nanocrystals with a diameter of 8 nm. Kobayashi et al. (1996b) have also studied pressure-made nanocrystals in CdS, of which the diameter ranges between 6.7 and 10 nm.

2. GALVANOMAGNETIC EFFECTS IN $Hg_{1-x}Fe_xSe$

In HgFeSe the solubility of iron reaches about 15%. For small values of x this system is a zero-gap semiconductor. Another remarkable aspect of the electronic structure of this system is the fact that the energy for the charge-transfer process of $Fe^{2+} \rightarrow Fe^{3+}$ + electron, which is very similar to the aforementioned Mn^{2+} + hole $\rightarrow Mn^{3+}$ process in CdMnSe, is small and negative, being -0.23 eV (Mycielski et al., 1986). Therefore, d electrons of the lower Hubbard state formed by Fe^{2+} ions fall into the Γ_8 conduction band, changing the valence of irons into Fe^{3+}, until the Fermi energy reaches 0.23 eV. As a consequence the electrons show a metallic conduction (Pool et al., 1987). In addition, if the iron density exceeds 5×10^{18} cm^{-3}, the Fermi level is pinned at 0.23 eV and thus the excess irons remain as Fe^{2+}. Interestingly, in this mixed-valence regime the conduction electrons have an anomalously high mobility at low temperatures. Furthermore, the mobility is enhanced by application of hydrostatic pressure (Skierbiszewski et al., 1989, 1990; Neifeld et al., 1996). At the same time the Dingle temperature of the Shubnikov–de Haas oscillation is lowered significantly. Skierbiszewski et al. have discussed their results in relation to the possibility of formation of a periodic lattice of Fe^{3+} ions which has been proposed by Mycielski (1986). Neifeld et al. have pointed out the importance of a reduction of the impurity-scattering cross-section of electrons which results from the pressure-induced reduction of the effective mass of electrons at the Fermi level.

Acknowledgments

Part of the contents of this article is based on author's own work, which has been carried out in collaboration with Dr. Y. Matsuda using the facilities of The High Field Laboratory for Superconducting Materials, Tohoku University. The author is indebted much to Dr. G. Kido, Dr. I. Mogi, Prof. J. R. Anderson (in Maryland), and Prof. Y. Nishina for their cordial assistance and fruitful discussions in the course of that work.

REFERENCES

Aggarwal, R. L., Jasperson, S. N., Becla, P., and Galazka, R. R. (1985). *Phys. Rev. B* **32,** 5132.
Ambrazevicius, G., Babonas, G., Marcinkevicius, S., Prochukhan, V. D., and Rud, Yu.V. (1984). *Solid State Commun.* **49,** 651.
Anastassiadou, A., Liarokapis, E., Stoyanov, S., Anastaskis, E., and Giriat, W. (1988). *Solid State Commun.* **67,** 633.
Arora, A. K., and Sakuntala, T. (1995). *Phys. Rev. B* **52,** 11052.
Arora, A. K., Bartholomew, D. U., Peterson, D. L., and Ramdas, A. K. (1987). *Phys. Rev. B* **35,** 7966.
Arora, A. K., Suh, E. K., Debska, U., and Ramdas, A. K. (1988). *Phys. Rev. B* **37,** 2927.
Bak, J., Mak, C. L., Sooryakumar, R., Venkateswaran, U. D., and Jonker, B. T. (1996). *Phys. Rev. B* **54,** 5545.
Balzarotti, A., Czyzyk, M., Kisiel, A., Motta, N., Podgorny, M., and Zimnal-Starnawska, M. (1984). *Phys. Rev. B* **30,** 2295.
Barilero, G., Rigaux, C., Hau, N. H., Picoche, J. C., and Giriat, W. (1987). *Solid State Commun.* **62,** 345.
Bartholomew, D. V., Furdyna, J. K., and Ramdas, A. K. (1986). *Phys. Rev. B* **34,** 6943.
Becker, W. M. (1988). *In* "Semiconductors and Semimetals," Vol. 25, "Diluted Magnetic Semiconductors" (Furdyna, J. K. and Kossut, J., eds.), Chapter 2. Academic Press, New York.
Bednarski, H., Cisowski, J., Lubczynski, W., Vioron, J., and Portal, J. C. (1995). *Acta Phys. Polonica A* **87,** 205.
Beliveau, A., and Carlone, C. (1989). *Semicond. Sci. Technol.* **4,** 277.
Bhattacharjee, A. K. (1992). *Phys. Rev. B* **46,** 5266.
Bhattacharjee, A. K. (1995). *Ann. Phys. Colloq. C2* **20,** 129.
Brandt, N. B., Moshchalkov, V. V., Orlov, A. O., Skrbek, L., Tsidil'kovskii, I. M., and Chudinov, S. M. (1983). *Sov. Phys. JETP* **57,** 614.
Chao, M. H., and Sladek, R. J. (1981). *J. Phys. Colloq. C* **6,** 667.
Chen, A. L., Yu, P. Y., and Tayler, R. D. (1993). *Phys. Rev. Lett.* **71,** 4011.
Chudinov, S. M., Rodichev, D. Tu., Mancini, G., and Stizza, S. (1993). *Phys. Stat. Sol: (b)* **175,** 213.
Cisowski, J., Lubczynski, W., Thuillier, J. C., and Portal, J. C. (1989). *Acta Phys. Polonica A* **75,** 301.
Demarest, H. H. (1972). *Phys. Earthplanet. Inter.* **6,** 146.
Denissen, C. J. M., Dankun, S., Kopinga, K., de Jonge, W. J. M., Nishihara, H., Sakakibara, T., and Goto, T. (1987). *Phys. Rev. B* **36,** 5316.
Ford, P. J., Millet, A. J., Saunders, G. A., Yogurtçu, Y. K., Furdyna, J. K., and Jaczynski, M. J. (1982). *J. Phys. C: Solid State Phys.* **15,** 657.
Forner, S., Shapira, Y., Heiman, D., Becla, P., Kershaw, R., Dwight, K., and Wold, A. (1989). *Phys. Rev. B* **39,** 11793.
Fujimori, A., Bocquet, A. E., Saitoh, T., and Mizokawa, T. (1993). *J. Electron Spectrosc. Relat. Phenom.* **62,** 141.

Furdyna, J. K. (1988). *J. Appl. Phys.* **64,** R29.
Gregus, J., Watanabe, J., and Nakahara, J. (1993). *Jpn. J. Appl. Phys.* **32,** Suppl. 32-3, 412.
Hamdani, F., Lascaray, J. P., Coquillat, D., Bhattacharjee, A. K., Nawrocki, M., and Golacki, Z. (1992). *Phys. Rev. B* **45,** 13298.
Harrison, W. A. (1980). *In* "Electronic Structure and the Properties of Solids," p. 451. Freeman, San Francisco.
Ikeda, M., Itoh, K., and Sato, H. (1968). *J. Phys. Soc. Jpn.* **25,** 455.
Jain, M., and Robins, J. L. (1991). *In* "Diluted Magnetic Semiconductors" (Jain, M., ed.), p. 1. World Scientific, Singapore.
Jiang, S., Shen, S. C., Li, Q. G., Zhu, H. R., Ju, G. L., and Giriat, W. (1989a). *Phys. Rev. B* **40,** 8017; (1989b). *Solid State Commun.* **70,** 1.
Jiang, S., Shen, S. C., and Giriat, W. (1990). *Phys. Stat. Sol. (b)* **161,** K71.
Jiang, S., Shen, S. C., and Li, Q. G. (1992). *J. Appl. Phys.* **72,** 1070.
Kobayashi, M., Nakai, T., Mochizuki, S., and Takayama, N. (1995). *J. Phys. Chem. Solids* **56,** 341.
Kobayashi, M., Iwata, H., Hanzawa, H., Yoshiue, T., and Endo, S. (1996a). *Phys. Stat. Solidi (b)* **198,** 515.
Kobayashi, M., Iwata, H., Horiguchi, T., and Endo, S. (1996b). *Phys. Stat. Solidi (b)* **198,** 521.
Kuroda, N., and Matsuda, Y. (1996a). *Sci. Rep. RITU A* **42,** 263; (1996b). *Phys. Rev. Lett.* **77,** 1111; (1996c). *Phys. Stat. Solidi (b)* **198,** 61.
Laiho, R., Lisunov, K. G., Shubnikov, M. L., Stamov, V. N., and Zakhvalinskii, V. S. (1996). *Phys. Stat. Solidi (b)* **198,** 135.
Langer, D., and Ibuki, S. (1965). *Phys. Rev.* **138,** A809.
Larson, B. E., Hass, K. C., Ehrenreich, H., and Carlsson, A. E. (1985). *Solid State Commun.* **56,** 347; (1988). *Phys. Rev. B* **37,** 4137.
Larson, B. E., Hass, K. C., and Aggarwal, R. L. (1986). *Phys. Rev. B* **33,** 1789.
Lawniczak-Jablonska, K., and Golacki, Z. (1994). *Acta Phys. Pol. A* **86,** 727.
Lemos, V., Moro, J. R., de Souza, Q. A. G., and Motisuke, P. (1986). *Solid State Commun.* **60,** 853.
Lubczynski, W., Voiron, J., Picoche, J. C., Portal, J. C., Cisowski, J., Thuillier, J. C., and Zdanowicz, W. (1989). *Semicond. Sci. Technol.* **4,** 223.
Maheswaranathan, P., and Sladek, R. J. (1990). *Phys. Rev. B* **41,** 12076.
Maheswaranathan, P., Sladek, R. J., and Debska, U. (1985). *Phys. Rev. B* **31,** 5212.
Martinez, G. (1980). *In* "Handbook on Semiconductors," Vol. 2 (Balkanski, M., ed.), p. 181. North-Holland, Amsterdam.
Matsuda, Y. (1996). Ph.D. Thesis, Tohoku University.
Matsuda, Y., and Kuroda, N. (1996). *Phys. Rev. B* **53,** 4471.
Matsuda, Y., Kuroda, N., Mogi, I., Kido, G., Nishina, Y., Nakagawa, Y., Anderson, J. R., and Giriat, W. (1993). *Jpn. J. Appl. Phys.* **32,** Suppl. 32-1, 270.
Matsuda, Y., Kuroda, N., Mogi, I., Kido, G., and Nishina, Y. (1994). *Physica B* **201,** 411.
Mayanovic, R. A., Sladek, R. J., and Debska, U. (1988). *Phys. Rev. B* **38,** 1311.
Mei, J. R., and Lemos, V. (1984). *Solid State Commun.* **52,** 785.
Meyer, R., Dahl, M., Schaack, G., Waag, A., and Boehler, R. (1995). *Solid State Commun.* **96,** 271; (1996). *J. Crystal Growth* **159,** 997.
Miller, A. J., Saunders, G. A., Yogurtçu, Y. K., and Abey, A. E. (1981). *Phil. Mag. A* **43,** 1447.
Montalvo, R. A., and Langer, D. W. (1970). *J. Appl. Phys.* **41,** 4101.
Morales Toro, J. E. (1985). Ph.D. Thesis, Purdue University. See also Becker (1988).
Morales Toro, J. E., Becker, W. M., Wang, B. I., Debska, U., and Richardson, J. W. (1984). *Solid State Commun.* **52,** 41.

Moriwaki, M. M., Becker, W. M., Gebhardt, W., and Galazka, R. R. (1982). *Phys. Rev. B* **26**, 3165.
Müller, E., Gebhardt, W., and Rehwald, W. (1983). *J. Phys. C: Solid State Phys.* **16**, L1141.
Mycielski, A. (1986). *Solid State Commun.* **60**, 165.
Mycielski, A., Dzwonkowski, P., Kowalski, B., Orlowski, B. A., Dobrowolska, M., Arciszewska, M., Dobrowolsi, W., and Baranowski, J. M. (1986). *J. Phys. C.: Solid State Phys.* **19**, 3605.
Nakahara, J., Watanabe, J., and Nouch, T. (1993). *Jpn. J. Appl. Phys.* **32**, Suppl. 32-1, 242.
Neifeld, E. A., Demchuk, K. M., Harus, G. I., Boubnova, A. E., Domanskaya, L. I., Shtrapenin, G. L., and Paranchich, S. Yu. (1996). *Phys. Stat. Sol. (b)* **198**, 143.
Nelmes, R. J., McMahon, M. I., Wright, N. G., and Allan, D. R. (1995). *J. Phys. Chem. Solids* **56**, 545.
Ohnishi, S., and Sugano, S. (1982). *Jpn. J. Appl. Phys.* **21**, L309.
Ohta, Y., Tohyama, T., and Maekawa, S. (1991). *Phys. Rev. Lett.* **66**, 1228.
Onodera, K., Masumoto, T., and Kimura, M. (1994). *Electronics Lett.* **30**, 1954.
Paul, W. (1961). *J. Appl. Phys.* **32**, 2082.
Perlin, P., Shilo, S., Sosin, T., Tyagur, Y., Suski, T., Trzeciakowski, W., Karczewski, G., Wojtowcz, T., Janik, E., Zakrzewski, A., Kutrowski, M., and Kossut, J. (1995). *J. Phys. Chem. Solids* **56**, 415.
Phillips, J. C. (1973). *In* "Bonds and Bands in Semiconductors" p. 200. Academic Press, New York; (1982). *Phys. Rev. B* **4**, 2310.
Pool, F. S., Kossut, J., Debska, U., and Reifenberger, R. (1987). *Phys. Rev. B* **35**, 3900.
Prakash, M., Chandrasekhar, M., Chandrasekhar, H. R., Miotkowski, I., and Ramdas, A. K. (1990). *Phys. Rev. B* **42**, 3586.
Qadri, S. B., Skelton, E. F., Webb, A. W., Carpentar, Jr., E. R., Schaefer, M. W., and Furdyna, J. (1987). *Phys. Rev. B* **35**, 6868.
Qadri, S. B., Skelton, E. F., Webb, A. W., Colombo, L., and Furdyna, J. K. (1989). *Phys. Rev. B* **40**, 2432.
Rodic, D., Spasojevic, V., Bajorek, A., and Onnerud, P. (1996). *J. Mag. Mag. Mater.* **152**, 159.
Rusin, T. M. (1996). *Phys. Rev. B* **53**, 12577.
Shan, R. K., Shen, S. C., and Zhu, H. R. (1985). *Solid State Commun.* **55**, 475.
Shen, Q., Luo, H., and Furdyna, J. K. (1995). *Phys. Rev. Lett.* **75**, 2590.
Singh, R. K. (1991). *In* "Diluted Magnetic Semiconductors" (Jain M., ed.), p. 193. World Scientific, Singapore.
Singh, R. K., and Arora, A. (1993a). *Phase Transitions* **46**, 57; (1993b). *Jpn. J. Appl. Phys.* **32**, Suppl. 32-3, 378.
Skierbiszewski, C., Suski, T., Litwin-Staszewska, E., Dobrowolski, W., Dybko, K., and Mycielski, A. (1989). *Semicond. Sci. Technol.* **4**, 293.
Skierbiszewski, C., Suski, T., Dobrowolski, W., and Kossut, J. (1990). *J. Crystal. Growth* **101**, 869.
Sladek, R. J. (1986). *Phys. Rev. B* **33**, 5899.
Story, T., Galazka, R. R., Frankel, R. B., and Wolff, P. A. (1986). *Phys. Rev. Lett.* **56**, 777.
Story, T., Karczewski, G., and Swierkowski, L. (1989). *Acta Phys. Polonica A* **75**, 277.
Strössner, K., Ves, S., Dieterich, W., Gebhardt, W., and Cardona, M. (1985). *Solid State Commun.* **56**, 563.
Strössner, K., Ves, S., Hönle, W., Gebhardt, W., and Cardona, M. (1987). *In* "Proc. 18th Int. Conf. Physics of Semiconductors" (Engström, O., ed.), p.1717. World Scientific, Singapore.
Suski, T., Igalson, J., and Story, T. (1987a). *J. Mag. Mag. Mater.* **66**, 325; (1987b). *In* "Proc. 18th Int. Conf. Physics of Semiconductors" (Engström, O., ed.), p. 1747. World Scientific, Singapore.

Taniguchi, M., Ley, L., Johnson, R. L., Ghijsen, J., and Cardona, M. (1986). *Phys. Rev. B* **33,** 1206.
Terauchi, H., Yoneda, Y., Kasatani, H., Sakaue, K., Koshiba, T., Murakami, S., Kuroiwa, Y., Noda, Y., Sugai, S., Nakashima, S., and Maeda, H. (1993). *Jpn. J. Appl. Phys.* **32,** Suppl. 32-2, 728.
Tuchman, J. A., Kim, S., Sui, J., and Herman, I. P. (1992). *Phys. Rev. B* **46,** 13371.
Tuchman, J. A., Sui, J., Kim, S., Herman, I. P. (1993). *J. Appl. Phys.* **73,** 7730.
Twardowski, A., Swagten, H. J. M., de Jonge, W. J. M., and Demianiuk, M. (1987). *Phys. Rev. B* **36,** 7013.
Van Vechten, J. A. (1969). *Phys. Rev.* **182,** 891.
Ves, S., Strössner, K., Christensen, N. E., Kim, C. K., and Cardona, M. (1985). *Solid State Commun.* **56,** 479.
Ves, S., Strössner, K., Gebhardt, W., and Cardona, M. (1986a). *Phys. Rev. B* **33,** 4077; (1986b). *Solid State Commun.* **57,** 335.
Watanabe, J., Arai, H., Nouchi, T., and Nakahara, J. (1992). *J. Phys. Soc. Japan* **61,** 2227.
Wei, S. H., and Zunger, A. (1987). *Phys. Rev. B* **35,** 2340.
Yogurtçu, Y. K., Miller, A. J., and Saunders, G. A. (1981). *J. Phys. Chem. Solids* **42,** 49.
Yokoi, H., Kakudate, Y., Schmiedel, T., Tozer, S., Jones, E. D., Takeyama, S., Wojtowicz, T., Karczewski, G., and Kossut, J. (1996). *In* "Proc. 23rd Int. Conf. Phys. Semicond." (Scheffler, M., and Zimmermann, R., eds.), p. 2039. World Scientific, Singapore.
Zaanen, J., and Sawatzky, G. A. (1987). *Can. J. Phys.* **65,** 1262.
Zaanen, J., Sawatzky, G. A., and Allen, J. W. (1985). *Phys. Rev. Lett.* **55,** 418.

Index

A

Ab initio calculations, 33, 92, 105, 110, 114, 116
Absolute deformation potentials (ADPs), 86
Absorption
  due to RL2 defect, 430–433
  interband, 397–400
Absorption near fundamental gap, 285–297
  See also Exciton absorption
Ad hoc LDA gap corrections, 66–69
AlAs (aluminum arsenide), structural transition of, 186
AlGaAs
  band offsets, 25–26
  donor states, 490–495
  influence of X-states on phenomena in, 26–27
AlN (aluminum nitride)
  structural transition of, 183–184
  transitions in and from Wurtzite structures, 107–113
AlP (aluminum phosphide), structural transition of, 184–185, 390
AlSb (aluminum antimonide), structural transition of, 187–188
A_1 (ab) states, 23
  co-existence with EMT and DX, 23–24
Amorphous semiconductors, 35–36
Angle-dispersive diffraction (ADX), use of, 147, 175–176
As_{Ga} antisite, 435–436
Atomic orbitals, linear combinations of, 260–264
Atomic-spheres approximation (ASA) and frozen-potential method, 69–79

B

Baldereschi point, 87
Band offsets, 25–26

Band structure
  empirical rule and changes in, 5–8
  parameters from binding energies, 316–321
Band structures, effect of pressure on bulk
  See also Electronic band structure under pressure
  dielectric theory of the chemical bond, 17
  pseudopotential calculations, 15–17
  volume deformation potential, 15
Barth–Hedin parameterization, 56–57
BAs, 183
BC8 structure, 171–172
  Ge, 182–183
  Si-III, 179–180
BeO, B3 → B1 pressure-induced transitions, 100–104
Bethe–Salpeter equation, 373
Bloch function, 267, 299, 330
Bloembergen–Rowland interaction, 524
BN, 183
Born–Oppenheimer approximation, 51–53, 130
Boron-V compounds, 183
Borrmann–Fulde calculations, 60, 61, 62
BP, 183
BSb, 183
B3 → B1 pressure-induced transitions, 99–104
B3 → Imma, 105–106
β-tin structure, 155–157
  See also Diatomic β-tin structure
  AlSb, 187
  CdTe, 223
  GaP, 190
  GaSb, 199
  Ge, 181
  HgSe, 229
  HgTe, 231–232
  InAs, 204

552 INDEX

InP, 202, 203
InSb, 205, 206, 210
Si, 178, 179
ZnTe, 216

C

C, pressure-induced transformations, 113–121
Capacitance transient techniques
　See also DX centers, capacitance techniques to study high pressure in
　at constant temperature, 463–465
　photocapacitance transient measurements, 466–467
　role of, 462–463
　scanning temperature, 465–466
$Cd_{1-x}Co_xSe$, 533–540
$Cd_{1-x}Mn_xSe$, 533–540
$(Cd_{1-x-y}Zn_yMn_x)_3As_2$, 532–533
CdS (cadmium sulfide), structural transition of, 218–220
CdSe (cadmium selenide), structural transition of, 220–222
CdTe (cadmium telluride), structural transition of, 222–224
$CdTe/Cd_{1-x}Mn_xTe$, 544–545
Charge-density excitations (CDEs), 353, 357–358
　asymmetric DQWs, 378–379
　inelastic light scattering and, 360, 362
　symmetric DQWs, 382, 385
　2-D electron gas and, 373–375
Cinnabar structure, 163–164
　CdSe, 221
　CdTe, 222–223
　GaAs, 196–197
　HgO, 225
　HgS, 226–227
　HgSe, 228–229
　HgTe, 230–231
　ZnS, 212
　ZnSe, 214
　ZnTe, 216–217
Clausius–Mossotti equation, 278
Cmcm structure, 164–168
　AlSb, 188
　CdS, 220
　CdSe, 221–222
　CdTe, 223–224
　GaAs, 193–195

GaP, 190–191
HgSe, 229–230
HgTe, 232
InAs, 204
InP, 203
InSb, 207–208
ZnS, 213
ZnSe, 214–215
ZnTe, 216–217
Covalent bond, dielectric theory of, 264–267
Covalent energy, 260, 262
Crossover pressure, 67, 68
CsCl structure, 169
　CdSe, 221
$C222_1$ structure, 168–169
　HgSe, 229
　HgTe, 231
Cubic compound semiconductors, shear deformation potentials in, 80–86

D

Debye–Scherrer powder rings, 148
Deep-level transient spectroscopy (DLTS), 20, 428–429, 461
　spectrum, 465–466
Deformation potentials
　hydrostatic, 86–97
　role of, 79–80
　shear, in cubic compound semiconductors, 80–86
Density-functional theory (DFT), 53–57
　gap problem, 61
Depletion approximation, 462
dhcp (double hexagonal close packed), Ge, 182
Diamond anvil cells (DACs)
　introducing wires into, 459–461
　performing capacitance measurements inside, 461–462
　use of, 18, 29, 148, 251–253
Diamond structure, 152–154
　Ge, 181
　Si, 177
Diatomic β-tin structure, 157
　GaP, 191
Dielectric function, volume dependence of low-frequency, 277–279
Dielectric midgap energy (DME) model, 87–88

Dielectric theory
  of chemical bond, 17, 249–250
  of covalent bond, 264–267
Diluted magnetic semiconductors (DMSs)
  $Cd_{1-x}Co_xSe$, 533–540
  $Cd_{1-x}Mn_xSe$, 533–540
  $(Cd_{1-x-y}Zn_yMn_x)_3As_2$, 532–533
  $CdTe/Cd_{1-x}Mn_xTe$, 544–545
  elastic properties, 517–518
  electronic properties, 523–542
  energy gap, 523–524
  exchange interactions, 524–540
  galvanomagnetic effects in $Hg_{1-x}Fe_xSe$, 546
  $Hg_{1-x}Fe_xSe$, 546
  $Hg_{1-x}Mn_xSe$, 525–529
  $Hg_{1-x}Mn_xTe$, 525–529
  intraion d-d optical transitions, 540–542
  $Pb_{1-x-y}Sn_yMn_xTe$, 529–532
  Raman scattering, 545–546
  role of, 513–515
  structural phase transitions, 518–523
  structural properties, 515–523
  superstructures, 543–545
  $ZnSe/Zn_{1-x}Mn_xSe$, 543–544
Donor-acceptor pair (DAP) transitions, 323–324
Double quantum wells (DQWs)
  asymmetric sample, 377–379
  electron-electron interactions in double-layer 2-D electron gases, 377–385
  symmetric sample, 379–385
Drude model, 353, 363
d-states, 122–130
DX centers, 20–22
  co-existence with A_1 (ab) and EMT states, 23–24
  discovery of, 467
  heterostructures and, 28–29
  high-pressure freezeout (HPFO), 497–499, 501
  ionization of, by phonons, 498
  ionization of, by photons, 498
  states, 22–23
  studies of spatial correlations for, 495–502
  temperature-induced capture of electrons in, 498
DX centers, capacitance techniques to study high pressure in
  background research, 457–459
  experimental studies, 467–481
  introducing wires into a DAC, 459–461
  models of, 474–481
  performing capacitance measurements inside a DAC, 461–462
  pressure-induced deep center (PIDC), 469–471
  substitutional donors, 469–473
Dyson equation, 355

E

Effective mass theory (EMT), 18, 21
  co-existence of effective mass state with A_1 (ab) and DX states, 23–24
  states, 22
Elastic properties, of diluted magnetic semiconductors (DMSs), 517–518
Electric field gradient (EFG), 128, 129
Electromagnetic waves in a medium, 394–397
Electron gas, 2D (2DEG)
  relaxation time and quantum transport of, 505–508
  remote impurity charges using, 502–504
Electron gases under pressure, optical properties of, 351
  electron-electron interactions in double-layer 2-D, 377–385
  elementary excitations of, 352–358
  elementary excitations of 2-D, 370–377
  inelastic light scattering, 358–363
  longitudinal optical (LO) phonon interaction with plasmons, 363–370
Electron-hole pairs, 352
Electronic band structure under pressure, 259
  $k \cdot p$ method, 267–269
  linear combinations of atomic orbitals and tight-binding method, 260–264
  Penn gap and dielectric theory of covalent bond, 264–267
  quantum well structures, 271–273
  uniaxial stress effects, 269–271
Electronic Hamiltonian, 51
Electronic structure calculations
  ad hoc LDA gap corrections, 66–69
  atomic-spheres approximation and frozen-potential method, 69–79
  basic theory, 50–68

B3 → B1 pressure-induced transitions, 99–104
B3 → Imma, 105–106
deformation potentials, 79–97
density-functional theory, 53–57
d-states, 122–130
excitation energies, 60–64
gap errors and deformation potentials, 64–66
gradient and generalized gradient approximations, 58–59
group IV elements and pressure-induced transformations, 113–121
local-density approximation, 55
local spin-density approximation, 55
phonon frequencies, pressure dependence of, 130–134
pressure-induced changes, 98–130
self-interaction-corrected potential, 57–58
transitions in and from Wurtzite structures, 106–113
zeroth Born-Oppenheimer approximation, 51–53
Electron paramagnetic resonance (EPR) spectroscopy
discovery of As_{Ga} antisite, 435–436
piezospectroscopic investigations of no-phonon lines of EL2, 436–440
Electron-photon interaction Hamiltonian, 358
Electron-spin resonance, 24
Electroreflectance spectroscopy, 343
EL2, 20–22
pressure experiments on, 24
EL2 defect in GaAs
basic properties of, 428–436
discovery of As_{Ga} antisite, 435–436
frequency of, 428–430
metastable properties, 433–435
metastable state, determining symmetry of, 447–453
mestastable state, nature of, 445–453
mestastable state, theoretical model of, 445–447
optical absorption, 430–433
piezospectroscopic investigations of no-phonon lines of, 436–440
role of, 427–428
studies of, under hydrostatic pressure, 440–445

Empirical pseudopotential method (EPM), 16
Empirical Rule, 4, 458
applications of, 11–15
band structure changes and, 5–8
band structures established with, 11–15
optical measurements, 7–8
pseudopotential calculations, 16
transport measurements, 5–7
Energy-dispersive diffraction (EDX), use of, 146–147, 175
Energy gap, diluted magnetic semiconductors (DMSs) and, 523–524
Envelope function approximation (EFA), 272, 273
Equations of state (EOS), role of, 49–50
Exchange interactions, diluted magnetic semiconductors (DMSs) and, 524–540
Excitation energies, 60–64
Exciton absorption, 297
band-structure parameters from binding energies, 316–321
coefficient for direct transitions, 303–305
description of, 303–307
energy spectrum of excitons, 298–303
finite linewidth and lineshape, 305–307
line broadening, 312–316
pressure effects on excitons, 307–316
strength and binding energy, 309–312
Excitonic shift, 357
Excitons
in bulk, 298–301
envelope function, 299, 300, 301
2-D, 301–303

F

Faraday rotation, 10
Far-infrared absorption, 18
Far-infrared magneto-optical techniques, 18
fcc, Si, 179
Fermi Golden Rule, 398
Fluctuation-dissipation theorem, 360
Fourier-transform spectrometer, 281
Franz–Keldysh effect, 343, 344
Free-exciton recombination, 324
Free-to-bound transitions, 323
Frozen-phonon approach and, Born-Oppenheimer approximation, 51–53, 130

Frozen-potential method, atomic-spheres approximation and, 69–79

G

GaAs (gallium arsenide)
  See also EL2 defect in GaAs
  ad hoc LDA gap corrections, 67–69
  band offsets, 25–26
  co-existence of EMT, DX, and A_1 (ab) states, 23–24
  deep impurity levels and conduction band extrema, 19–20
  donor states, 490–495
  influence of X-states on phenomena in, 26–27
  shear deformation potentials in cubic compound semiconductors, 85–86
  structural transition of, 191–198, 389–390
GaAs-II, 192, 193–195, 196
GaAs-III, 192–193, 195
GaAs-IV, 193
GaN (gallium nitride)
  gap errors and deformation potentials, 65–66
  GW corrections, 63
  structural transition of, 189
  transitions in and from Wurtzite structures, 107–113
GaP (gallium phosphide)
  band structure of, 11–13
  structural transition of, 190–181
GaSb (gallium antimonide), structural transition of, 198–200
Ge (germanium)
  optical measurements, 7–8
  pressure-induced transformations, 113–121
  structural transition of, 181–183
  transfer of hot electrons, 13–14
  transport measurements, 6–7
Ge-I, 181
Ge-II, 181
Ge-III, 182
Generalized gradient approximations (GGAs), 58–59
Gpa structure
  AlAs, 186
  AlN, 184
  AlP, 184–185
  AlSb, 187–188
  CdS, 218–220

CdSe, 221–222
CdTe, 222–224
GaAs, 191–197
GaN, 189
GaP, 190–191
GaSb, 199–200
Ge, 181
HgO, 225–226
HgS, 227
HgSe, 228–230
HgTe, 230–232
InAs, 203–205
InN, 201
InP, 202–203
InSb, 205, 207, 208, 210
Si, 178
ZnO, 211
ZnS, 212–213
ZnSe, 213–215
ZnTe, 215–217
Gradient approximations, 58–59
Green's-function techniques, 62
Group II semiconductors
  See also under name of
  research on, 146–150
Group III semiconductors
  See also under name of
  research on, 146–150
Group IV semiconductors
  See also under name of
  pressure-induced transformations and, 113–121
  research on, 146–150
Group V semiconductors
  See also under name of
  research on, 146–150
Group 6 elements, 34–35
Group VI semiconductors
  See also under name of
  research on, 146–150
Grüneisen parameters, 32, 131, 132, 134, 366
Gunn effect, 14–15
GW approximation, excitation energies, 60–64

H

Hamiltonian, orbital-strain (Pikus–Bir), 270, 272
Harrison's scalings, 65, 88, 262
Hartree–Fock calculations, 57, 60, 112

Hartree interactions, 353, 357–358
hcp (hexagonal close packed), Si, 179
Heterostructures
  band offsets, 25–26
  DX centers and, 28–29
  influence of X-states on phenomena in, 26–27
  research on, 25
  resonant tunneling, 28
$Hg_{1-x}Fe_xSe$, 546
$Hg_{1-x}Mn_xSe$, 525–529
$Hg_{1-x}Mn_xTe$, 525–529
HgO (mercury oxide), structural transition of, 225–226, 389
HgS (mercury sulfide), structural transition of, 236–227
HgSe (mercury selenide)
  band structure, 13
  structural transition of, 228–230
HgSe:Fe donor charges, 487–490
HgTe (mercury telluride)
  band structure, 13
  structural transition of, 230–232
HgTe-CdTe, band structure, 13
HgX, structural phase transitions, 31–32
High pressure
  See also DX centers, capacitance techniques to study high pressure in
  amorphous semiconductors, 35–36
  applications of empirical rule, 11–15
  band structure changes and the empirical rule, 5–8
  current studies on zincblende family, 18–25
  diamond anvil cells, benefits of, 3
  effects on bulk band structures, 15–17
  energy changes of deep impurity levels, 9–10
  energy changes of shallow hydrogenic impurity levels, 9
  experiments on bulk samples of zincblende family to 1970, 4–15
  group 6 elements, 34–35
  inequivalent extrema with different pressure coefficients, 10
  lead chalcogenides, 33–34
  nanocrystals and porous silicon, 36–37
  optical measurements, 7–8
  phases, 385–392
  scattering in a single band, 10–11
  semiconducting nitrides, 37
  structural phase transitions, 29–33
  transport measurements, 5–7
High-pressure freezeout (HPFO), 497–499, 501
Hydrostatic deformation potentials, 86–97
Hydrostatic pressure
  See also Optical properties of semiconductors under pressure
  studies of EL2 defect under, 440–445

I

Image-plate detector, use of, 147, 175, 176
Imma structure, 105–106, 159–160
  GaSb, 199–200
  Ge, 182
  InSb, 210
  Si, 178
Immm structure, 161
  InSb, 207
Imm2 structure, 160–161
  GaAs, 192–193
  GaSb, 200
  InSb, 206
Impurity-lattice coupling, 19
Impurity levels
  See also Spatial correlations of impurity charges
  deep, and conduction band extrema, 19–20
  energy changes of deep, 9–10
  energy changes of shallow hydrogenic, 9
InAs (indium arsenide), structural transition of, 203–205
Inelastic light scattering, 358–363
Inelastic neutron scattering, 32–33
InN (indium nitride)
  structural transition of, 200–201
  transitions in and from Wurtzite structures, 107–113
InP (indium phosphide)
  B3 → B1 pressure-induced transitions, 100–104
  structural transition of, 202–203
InSb (indium antimonide), structural transition id, 30, 205–210
Integer quantum Hall effect (IQHE), 505, 507–508
Intrasubband excitations, 380–381
Intrinsic conductivity, 6
Ionization of DX centers
  by phonons, 498
  by photons, 498

J

Jahn–Teller-like distortion, 126
Joint density-of-states, 399
Jones zone, 265

K

Kohn–Sham eigenvalue spectrum, 54–55, 57–58, 61
Kondo-like exchange interaction, 524
$k \cdot p$ method, 267–269
Kramers–Kronig transformation, 274, 275, 405–407
Kronig–Penney equations, 272–273

L

Lagrange multipliers, 54, 55
Lead chalcogenides, 33–34
Lindhard expression, 354–355, 364
Lindhard–Mermin function, 355, 364
Linear combination of atomic orbitals (LCAO) method, 112
Linear muffin-tin orbital (LMTO) method, 69, 77
   d-states, 122–130
   shear deformation potentials in cubic compound semiconductors, 80–86
Linear optical response of solids
   analysis of high-pressure reflectance spectra, 407–409
   dispersion relations and sum rules, 405–407
   electromagnetic waves in a medium, 394–397
   interband absorption, 397–400
   Lorentz–Drude oscillators, 402–405
   reflectance and transmittance, 400–402
LLR model, 21, 22
Local density (LD), 18, 22
Local-density approximation (LDA), 55
   ad hoc gap corrections, 66–69
   atomic-spheres approximation and frozen-potential method, 69–79
   d-states, 122–130
   gap errors and deformation potentials, 61, 64–66
   self-interaction correction and potentials, 57–58
Local-field equation, 278
Local spin-density approximation (LSDA), 55

self-interaction correction and potentials, 57–58
Longitudinal optical (LO) phonon interaction with plasmons, 363–370
Lonsdaleite, 154–155
Lorentz–Drude oscillators, 402–405
Lorentz–Lorentz formula, 278

M

Mixed valence regime (MVR), 487–488
Molecular beam epitaxy (MBE), 429
Monoclinic structure, Si, 179
Mössbauer spectroscopy, 128
Multiple quantum wells (MQWs)
   photoluminescence studies of, 346–351
   phototransmittance spectroscopy and, 343, 344–345
Murnaghan's equation of state, 281, 366

N

NaCl structure, 157
   AlN, 184
   AlSb, 187
   CdS, 218–220
   CdSe, 221
   CdTe, 222–223
   GaN, 189
   HgO, 225, 226
   HgS, 227
   HgSe, 229–230
   HgTe, 231
   InAs, 204
   InN, 201
   InP, 202–203
   InSb, 208
   ZnO, 211
   ZnS, 212–213
   ZnSe, 214
   ZnTe, 216, 217
Nanocrystals, 36–37
NiAs structure, 157, 159
   AlAs, 186
   AlP, 185
   AlSb, 188
Nitrides, semiconducting, 37

O

Optically detected electron nuclear double resonance, 437
Optical measurements, 7–8

Optical properties of semiconductors under pressure
  absorption near fundamental gap, 285–297
  diamond anvil cells (DACs), use of, 251–253
  electron gases and, 351–385
  electronic band structure under pressure, 259–273
  exciton absorption, 297–321
  high-pressure phases, 385–392
  optical spectroscopy design, 256–259
  photoluminescence studies, 321–342
  photomodulated reflectance, 342–351
  pressure measurement, 254–256
  pressure medium, 253–254
  refractive index dispersion, 273–285
  research on, 248–251
Optimized effective potentials (OEPs), 58, 65
Orbital-strain Hamiltonian, 270, 272

P

Pair-bubble Feynman diagram, 353
Pauli exclusion principle, 269, 294, 353, 475
Paul's Empirical Rule. *See* Empirical Rule
Pb, pressure-induced transformations, 113–121
$Pb_{1-x-y}Sn_yMn_xTe$, 529–532
Penn gap, 87, 88, 264–267
Perdew–Zunger parameterization, 56–57, 123
Persistent photoconductivity (PPC), 467–468, 493, 498
Pettifor's pressure relations, 75–76
Phillips–van Vechten theory/ionicity, 89, 90, 100, 264, 279, 284, 293
Phonon frequencies, pressure dependence of, 130–134
Phonon replica emission line, 323
Phonons, ionization of DX centers by, 498
Photocapacitance transient measurements, 466–467
Photoluminescence excitation (PLE), 326–327, 342–343
Photoluminescence spectra under pressure, 25
Photoluminescence studies, pressure and effects on bulk materials, 327–333
  low-dimensional structures and, 333–342
  optical emission in semiconductors, 321–327
Photomodulated reflectance (PR), 258
  optical response, 342–345
  quantum wells and epilayers, 346–351
  results for bulk materials, 345–346
Photons, ionization of DX centers by, 498
Phototransmittance spectroscopy, 343
Piezospectroscopic investigations of no-phonon lines of EL2, 436–440
Plane waves, 62
Plasmons, 353
  longitudinal optical (LO) phonon interaction with, 363–370
Pmmm structure, CdS, 219
Pmm2 structure
  CdTe, 223
  GaAs, 192–193
  GaSb, 199
Pm2m structure, GaAs, 193–195
Pnma structure, HgO, 225
Polar energy, 260–261, 262
Porous silicon, 36–37
Preferred orientation (PO), 148–149
Pseudopotential calculations, 15–17
Pseudopotential Hamiltonian, 16

Q

Quantum Hall effect, 370
  integer, 505, 507–508
Quantum wells
  See also Double quantum wells (DQWs); Multiple quantum wells (MQWs)
  DX centers and, 28
  electronic band structure under pressure and, 271–273
  influence of X-states on phenomena in, 26–27
Quasiparticle band structures, 62, 64

R

Raman scattering, 33, 36–37, 38, 545–546
Random phase approximation (RPA), 353–355
Refractive index dispersion, 273
  experimental results, 282–285
  method of optical interferences, 279–282
  optical dispersion and refractive index, 274–277

volume dependence of low-frequency
dielectric function, 277–279
R8 structure, 174
  Si, 179–180
Relaxation time, 2DEG and, 505–508
Resonant light scattering, 360–362
Resonant Raman spectroscopy, 257
Resonant tunneling, 28
Rietveld (full profile refinement), 147, 148
Ruby luminescence method, 254
Ruderman–Kittel–Kasuya–Yoshida
  (RKKY) interaction, 524–525

S

S
  structural phase transitions, 31
  work on, 34–35
Scattering in a single band, 10–11
Schrödinger equation, 53, 267, 272, 299,
  300, 301, 330
Scissors operator, 66–67
SC16 structure, 172–174
  GaAs, 197–198
  GaP, 191
Se
  structural phase transitions, 31
  work on, 34–35
Self-interaction-corrected (SIC) potential,
  57–58
Sellmeyer formula, 278
Shear deformation potentials in cubic
  compound semiconductors, 80–86
Shubnikov–de Haas oscillation, 506
Shubnikov–de Haas rotation, 10
Si (silicon)
  empirical rule and, 11
  optical measurements, 7–8
  porous, 36–37
  pressure-induced transformations,
    113–121
  pseudopotential calculations, 16
  structure of, 177–180
  transport measurements, 6–7
Si-II, 178
Si-III (BC8), 179–180
Si-IV, 180
Si-V, 178, 179
Si-VI, 179
Si-VII, 179
Si-VIII, 180
Si-IX, 180

Simple hexagonal (SH) structure, 161–162
  GaSb, 199
  Ge, 181, 182
  Si, 178
  ZnSe, 214
Single-particle excitations (SPEs), 352, 355
  asymmetric DQWs, 378
  inelastic light scattering and, 360,
    362–363
  symmetric DQWs, 381, 382, 383–385
  2-D electron gas and, 373–374
SLR model, 21
Sn, pressure-induced transformations,
  113–121
Spatial correlations of impurity charges
  AlGaAs donor states, 490–495
  CdTe:In donor states, 500
  DX center studies, 495–502
  experimental techniques, 487
  GaAs donor states, 490–495
  HgSe:Fe donor charges, 487–490
  relaxation time and quantum transport of
    2D electron gas, 505–508
  remote impurity charges, using 2D
    electron gas, 502–504
  research on, 485–486
Spin-density excitations (SDEs), 353, 355,
  357–358
  asymmetric DQWs, 378–379
  inelastic light scattering and, 362
  symmetric DQWs, 381, 382
  2-D electron gas and, 373, 375
Spin-density fluctuations, 361, 362
Spin–orbit interactions, 262, 263, 269, 362
SRS Daresbury, 148, 176
Structural phase transitions
  See also under name of
  of diluted magnetic semiconductors
    (DMSs), 518–523
  high-pressure phases, 385–392
  inelastic neutron scattering, 32–33
  other systems, 31–32
  research on, 29, 146–150
  S, Se, and Te, 31
  standard description/naming of, 151
  zincblende, 29–31
ST12 structure, 169–171
  Ge, 182
Sum rules, 277, 405–407
Superlattices, resonant tunneling and, 28

T

Te
  structural phase transitions, 31
  work on, 34–35
Thomas–Fermi screening, 64, 65
Tight-binding (TB) theory, 18, 21, 260–264
Transport measurements, 5–7
Tunnel diode effect, 13

U

Uniaxial stress effects, 269–271

V

Valence-band maximum (VBM), 80, 81, 122
Valence band offset (VBO), 26
Van Hove–type critical points, 399
Vegard's law, 516

W

Wigner crystal, 371, 376, 377
Wurtzite structure, 154–155
  AlN, 183–184
  CdS, 218
  CdSe, 220–221
  GaN, 189
  InN, 201
  transitions in and from, 106–113
  ZnO, 211
  ZnS, 212

X

X-states on phenomena in heterostructures, influence of, 26–27

Z

Zero-order discontinuities, 78–79
Zero-phonon line (ZPL), 323
  piezospectroscopic investigations of no-phonon lines of EL2, 436–440
Zeroth Born–Oppenheimer approximation, 51–53, 130
Zincblende (ZB)
  A_1 (ab) states, 23
  AlAs, 186
  AlP, 184
  AlSb, 187
  applications of empirical rule, 11–15
  CdS, 218
  CdSe, 220
  CdTe, 222
  co-existence of EMT, DX, and A_1 (ab) states, 23–24
  current studies on, 18–25
  deep impurity levels and conduction band extrema, 19–20
  DX and EL2, 20–22, 22–24
  effective mass theory (EMT), 18, 21, 22
  energy changes of deep impurity levels, 9–10
  energy changes of shallow hydrogenic impurity levels, 9
  experiments on bulk samples of, 4–15
  GaAs, 191
  GaP, 190
  GaSb, 198
  HgS, 226–227
  HgSe, 228
  HgTe, 230
  InAs, 203
  inequivalent extrema with different pressure coefficients, 10
  InP, 202
  InSb, 205
  optical measurements, 7–8
  scattering in a single band, 10–11
  shear deformation potentials in, 80–86
  structural phase transitions and, 29–31, 152–154
  transport measurements, 5–7
  ZnS, 212
  ZnSe, 213
  ZnTe, 215
Zinc compounds, d-states, 122–130
ZnO (zinc oxide)
  B3 → B1 pressure-induced transitions, 100–104
  structural transition of, 211
ZnS (zinc sulfide), structural transition of, 212–213
ZnSe (zinc selenide), structural transition of, 213–215, 389
ZnSe/Zn$_{1-x}$Mn$_x$Se, 543–544
ZnTe (zinc telluride), structural transition of, 215–217, 386–389

Contents of Volumes in This Series

Volume 1 **Physics of III–V Compounds**

C. Hilsum, Some Key Features of III–V Compounds
Franco Bassani, Methods of Band Calculations Applicable to III–V Compounds
E. O. Kane, The k-p Method
V. L. Bonch-Bruevich, Effect of Heavy Doping on the Semiconductor Band Structure
Donald Long, Energy Band Structures of Mixed Crystals of III–V Compounds
Laura M. Roth and Petros N. Argyres, Magnetic Quantum Effects
S. M. Puri and T. H. Geballe, Thermomagnetic Effects in the Quantum Region
W. M. Becker, Band Characteristics near Principal Minima from Magnetoresistance
E. H. Putley, Freeze-Out Effects, Hot Electron Effects, and Submillimeter Photoconductivity in InSb
H. Weiss, Magnetoresistance
Betsy Ancker-Johnson, Plasma in Semiconductors and Semimetals

Volume 2 **Physics of III–V Compounds**

M. G. Holland, Thermal Conductivity
S. I. Novkova, Thermal Expansion
U. Piesbergen, Heat Capacity and Debye Temperatures
G. Giesecke, Lattice Constants
J. R. Drabble, Elastic Properties
A. U. Mac Rae and G. W. Gobeli, Low Energy Electron Diffraction Studies
Robert Lee Mieher, Nuclear Magnetic Resonance
Bernard Goldstein, Electron Paramagnetic Resonance
T. S. Moss, Photoconduction in III–V Compounds
E. Antoncik and J. Tauc, Quantum Efficiency of the Internal Photoelectric Effect in InSb
G. W. Gobeli and I. G. Allen, Photoelectric Threshold and Work Function
P. S. Pershan, Nonlinear Optics in III–V Compounds
M. Gershenzon, Radiative Recombination in the III–V Compounds
Frank Stern, Stimulated Emission in Semiconductors

Volume 3 Optical of Properties III–V Compounds

Marvin Hass, Lattice Reflection
William G. Spitzer, Multiphonon Lattice Absorption
D. L. Stierwalt and R. F. Potter, Emittance Studies
H. R. Philipp and H. Ehrenveich, Ultraviolet Optical Properties
Manuel Cardona, Optical Absorption above the Fundamental Edge
Earnest J. Johnson, Absorption near the Fundamental Edge
John O. Dimmock, Introduction to the Theory of Exciton States in Semiconductors
B. Lax and J. G. Mavroides, Interband Magnetooptical Effects
H. Y. Fan, Effects of Free Carries on Optical Properties
Edward D. Palik and George B. Wright, Free-Carrier Magnetooptical Effects
Richard H. Bube, Photoelectronic Analysis
B. O. Seraphin and H. E. Bennett, Optical Constants

Volume 4 Physics of III–V Compounds

N. A. Goryunova, A. S. Borschevskii, and D. N. Tretiakov, Hardness
N. N. Sirota, Heats of Formation and Temperatures and Heats of Fusion of Compounds $A^{III}B^V$
Don L. Kendall, Diffusion
A. G. Chynoweth, Charge Multiplication Phenomena
Robert W. Keyes, The Effects of Hydrostatic Pressure on the Properties of III–V Semiconductors
L. W. Aukerman, Radiation Effects
N. A. Goryunova, F. P. Kesamanly, and D. N. Nasledov, Phenomena in Solid Solutions
R. T. Bate, Electrical Properties of Nonuniform Crystals

Volume 5 Infrared Detectors

Henry Levinstein, Characterization of Infrared Detectors
Paul W. Kruse, Indium Antimonide Photoconductive and Photoelectromagnetic Detectors
M. B. Prince, Narrowband Self-Filtering Detectors
Ivars Melngalis and T. C. Harman, Single-Crystal Lead-Tin Chalcogenides
Donald Long and Joseph L. Schmidt, Mercury-Cadmium Telluride and Closely Related Alloys
E. H. Putley, The Pyroelectric Detector
Norman B. Stevens, Radiation Thermopiles
R. J. Keyes and T. M. Quist, Low Level Coherent and Incoherent Detection in the Infrared
M. C. Teich, Coherent Detection in the Infrared
F. R. Arams, E. W. Sard, B. J. Peyton, and F. P. Pace, Infrared Heterodyne Detection with Gigahertz IF Response
H. S. Sommers, Jr., Macrowave-Based Photoconductive Detector
Robert Sehr and Rainer Zuleeg, Imaging and Display

Volume 6 Injection Phenomena

Murray A. Lampert and Ronald B. Schilling, Current Injection in Solids: The Regional Approximation Method
Richard Williams, Injection by Internal Photoemission
Allen M. Barnett, Current Filament Formation

R. Baron and J. W. Mayer, Double Injection in Semiconductors
W. Ruppel, The Photoconductor-Metal Contact

Volume 7 Application and Devices
Part A

John A. Copeland and Stephen Knight, Applications Utilizing Bulk Negative Resistance
F. A. Padovani, The Voltage-Current Characteristics of Metal-Semiconductor Contacts
P. L. Hower, W. W. Hooper, B. R. Cairns, R. D. Fairman, and D. A. Tremere, The GaAs Field-Effect Transistor
Marvin H. White, MOS Transistors
G. R. Antell, Gallium Arsenide Transistors
T. L. Tansley, Heterojunction Properties

Part B

T. Misawa, IMPATT Diodes
H. C. Okean, Tunnel Diodes
Robert B. Campbell and Hung-Chi Chang, Silicon Junction Carbide Devices
R. E. Enstrom, H. Kressel, and L. Krassner, High-Temperature Power Rectifiers of $GaAs_{1-x}P_x$

Volume 8 Transport and Optical Phenomena

Richard J. Stirn, Band Structure and Galvanomagnetic Effects in III–V Compounds with Indirect Band Gaps
Roland W. Ure, Jr., Thermoelectric Effects in III–V Compounds
Herbert Piller, Faraday Rotation
H. Barry Bebb and E. W. Williams, Photoluminescence I: Theory
E. W. Williams and H. Barry Bebb, Photoluminescence II: Gallium Arsenide

Volume 9 Modulation Techniques

B. O. Seraphin, Electroreflectance
R. L. Aggarwal, Modulated Interband Magnetooptics
Daniel F. Blossey and Paul Handler, Electroabsorption
Bruno Batz, Thermal and Wavelength Modulation Spectroscopy
Ivar Balslev, Piezopptical Effects
D. E. Aspnes and N. Bottka, Electric-Field Effects on the Dielectric Function of Semiconductors and Insulators

Volume 10 Transport Phenomena

R. L. Rhode, Low-Field Electron Transport
J. D. Wiley, Mobility of Holes in III–V Compounds
C. M. Wolfe and G. E. Stillman, Apparent Mobility Enhancement in Inhomogeneous Crystals
Robert L. Petersen, The Magnetophonon Effect

Volume 11 Solar Cells

Harold J. Hovel, Introduction; Carrier Collection, Spectral Response, and Photocurrent; Solar Cell Electrical Characteristics; Efficiency; Thickness; Other Solar Cell Devices; Radiation Effects; Temperature and Intensity; Solar Cell Technology

Volume 12 Infrared Detectors (II)

W. L. Eiseman, J. D. Merriam, and R. F. Potter, Operational Characteristics of Infrared Photodetectors
Peter R. Bratt, Impurity Germanium and Silicon Infrared Detectors
E. H. Putley, InSb Submillimeter Photoconductive Detectors
G. E. Stillman, C. M. Wolfe, and J. O. Dimmock, Far-Infrared Photoconductivity in High Purity GaAs
G. E. Stillman and C. M. Wolfe, Avalanche Photodiodes
P. L. Richards, The Josephson Junction as a Detector of Microwave and Far-Infrared Radiation
E. H. Putley, The Pyroelectric Detector–An Update

Volume 13 Cadmium Telluride

Kenneth Zanio, Materials Preparations; Physics; Defects; Applications

Volume 14 Lasers, Junctions, Transport

N. Holonyak, Jr. and M. H. Lee, Photopumped III–V Semiconductor Lasers
Henry Kressel and Jerome K. Butler, Heterojunction Laser Diodes
A Van der Ziel, Space-Charge-Limited Solid-State Diodes
Peter J. Price, Monte Carlo Calculation of Electron Transport in Solids

Volume 15 Contacts, Junctions, Emitters

B. L. Sharma, Ohmic Contacts to III–V Compounds Semiconductors
Allen Nussbaum, The Theory of Semiconducting Junctions
John S. Escher, NEA Semiconductor Photoemitters

Volume 16 Defects, (HgCd)Se, (HgCd)Te

Henry Kressel, The Effect of Crystal Defects on Optoelectronic Devices
C. R. Whitsett, J. G. Broerman, and C. J. Summers, Crystal Growth and Properties of $Hg_{1-x}Cd_x$Se alloys
M. H. Weiler, Magnetooptical Properties of $Hg_{1-x}Cd_x$Te Alloys
Paul W. Kruse and John G. Ready, Nonlinear Optical Effects in $Hg_{1-x}Cd_x$Te

Volume 17 CW Processing of Silicon and Other Semiconductors

James F. Gibbons, Beam Processing of Silicon
Arto Lietoila, Richard B. Gold, James F. Gibbons, and Lee A. Christel, Temperature Distributions and Solid Phase Reaction Rates Produced by Scanning CW Beams

Arto Leitoila and James F. Gibbons, Applications of CW Beam Processing to Ion Implanted Crystalline Silicon
N. M. Johnson, Electronic Defects in CW Transient Thermal Processed Silicon
K. F. Lee, T. J. Stultz, and James F. Gibbons, Beam Recrystallized Polycrystalline Silicon: Properties, Applications, and Techniques
T. Shibata, A. Wakita, T. W. Sigmon, and James F. Gibbons, Metal-Silicon Reactions and Silicide
Yves I. Nissim and James F. Gibbons, CW Beam Processing of Gallium Arsenide

Volume 18 Mercury Cadmium Telluride

Paul W. Kruse, The Emergence of $(Hg_{1-x}Cd_x)Te$ as a Modern Infrared Sensitive Material
H. E. Hirsch, S. C. Liang, and A. G. White, Preparation of High-Purity Cadmium, Mercury, and Tellurium
W. F. H. Micklethwaite, The Crystal Growth of Cadmium Mercury Telluride
Paul E. Petersen, Auger Recombination in Mercury Cadmium Telluride
R. M. Broudy and V. J. Mazurczyck, (HgCd)Te Photoconductive Detectors
M. B. Reine, A. K. Soad, and T. J. Tredwell, Photovoltaic Infrared Detectors
M. A. Kinch, Metal-Insulator-Semiconductor Infrared Detectors

Volume 19 Deep Levels, GaAs, Alloys, Photochemistry

G. F. Neumark and K. Kosai, Deep Levels in Wide Band-Gap III–V Semiconductors
David C. Look, The Electrical and Photoelectronic Properties of Semi-Insulating GaAs
R. F. Brebrick, Ching-Hua Su, and Pok-Kai Liao, Associated Solution Model for Ga-In-Sb and Hg-Cd-Te
Yu.Ya. Gurevich and Yu. V. Pleskon, Photoelectrochemistry of Semiconductors

Volume 20 Semi-Insulating GaAs

R. N. Thomas, H. M. Hobgood, G. W. Eldridge, D. L. Barrett, T. T. Braggins, L. B. Ta, and S. K. Wang, High-Purity LEC Growth and Direct Implantation of GaAs for Monolithic Microwave Circuits
C. A. Stolte, Ion Implantation and Materials for GaAs Integrated Circuits
C. G. Kirkpatrick, R. T. Chen, D. E. Holmes, P. M. Asbeck, K. R. Elliott, R. D. Fairman, and J. R. Oliver, LEC GaAs for Integrated Circuit Applications
J. S. Blakemore and S. Rahimi, Models for Mid-Gap Centers in Gallium Arsenide

Volume 21 Hydrogenated Amorphous Silicon
Part A

Jacques I. Pankove, Introduction
Masataka Hirose, Glow Discharge; Chemical Vapor Deposition
Yoshiyuki Uchida, di Glow Discharge
T. D. Moustakas, Sputtering
Isao Yamada, Ionized-Cluster Beam Deposition
Bruce A. Scott, Homogeneous Chemical Vapor Deposition

Frank J. Kampas, Chemical Reactions in Plasma Deposition
Paul A. Longeway, Plasma Kinetics
Herbert A. Weakliem, Diagnostics of Silane Glow Discharges Using Probes and Mass Spectroscopy
Lester Gluttman, Relation between the Atomic and the Electronic Structures
A. Chenevas-Paule, Experiment Determination of Structure
S. Minomura, Pressure Effects on the Local Atomic Structure
David Adler, Defects and Density of Localized States

Part B

Jacques I. Pankove, Introduction
G. D. Cody, The Optical Absorption Edge of a-Si: H
Nabil M. Amer and Warren B. Jackson, Optical Properties of Defect States in a-Si: H
P. J. Zanzucchi, The Vibrational Spectra of a-Si: H
Yoshihiro Hamakawa, Electroreflectance and Electroabsorption
Jeffrey S. Lannin, Raman Scattering of Amorphous Si, Ge, and Their Alloys
R. A. Street, Luminescence in a-Si: H
Richard S. Crandall, Photoconductivity
J. Tauc, Time-Resolved Spectroscopy of Electronic Relaxation Processes
P. E. Vanier, IR-Induced Quenching and Enhancement of Photoconductivity and Photoluminescence
H. Schade, Irradiation-Induced Metastable Effects
L. Ley, Photoelectron Emission Studies

Part C

Jacques I. Pankove, Introduction
J. David Cohen, Density of States from Junction Measurements in Hydrogenated Amorphous Silicon
P. C. Taylor, Magnetic Resonance Measurements in a-Si: H
K. Morigaki, Optically Detected Magnetic Resonance
J. Dresner, Carrier Mobility in a-Si: H
T. Tiedje, Information about band-Tail States from Time-of-Flight Experiments
Arnold R. Moore, Diffusion Length in Undoped a-Si: H
W. Beyer and J. Overhof, Doping Effects in a-Si: H
H. Fritzche, Electronic Properties of Surfaces in a-Si: H
C. R. Wronski, The Staebler-Wronski Effect
R. J. Nemanich, Schottky Barriers on a-Si: H
B. Abeles and T. Tiedje, Amorphous Semiconductor Superlattices

Part D

Jacques I. Pankove, Introduction
D. E. Carlson, Solar Cells
G. A. Swartz, Closed-Form Solution of I–V Characteristic for a-Si: H Solar Cells
Isamu Shimizu, Electrophotography
Sachio Ishioka, Image Pickup Tubes

P. G. LeComber and W. E. Spear, The Development of the a-Si: H Field-Effect Transistor and Its Possible Applications
D. G. Ast, a-Si: H FET-Addressed LCD Panel
S. Kaneko, Solid-State Image Sensor
Masakiyo Matsumura, Charge-Coupled Devices
M. A. Bosch, Optical Recording
A. D'Amico and G. Fortunato, Ambient Sensors
Hiroshi Kukimoto, Amorphous Light-Emitting Devices
Robert J. Phelan, Jr., Fast Detectors and Modulators
Jacques I. Pankove, Hybrid Structures
P. G. LeComber, A. E. Owen, W. E. Spear, J. Hajto, and W. K. Choi, Electronic Switching in Amorphous Silicon Junction Devices

Volume 22 Lightwave Communications Technology
Part A

Kazuo Nakajima, The Liquid-Phase Epitaxial Growth of IngaAsp
W. T. Tsang, Molecular Beam Epitaxy for III–V Compound Semiconductors
G. B. Stringfellow, Organometallic Vapor-Phase Epitaxial Growth of III–V Semiconductors
G. Beuchet, Halide and Chloride Transport Vapor-Phase Deposition of InGaAsP and GaAs
Manijeh Razeghi, Low-Pressure Metallo-Organic Chemical Vapor Deposition of $Ga_x In_{t-x} As P_{t-y}$ Alloys
P. M. Petroff, Defects in III–V Compound Semiconductors

Part B

J. P. van der Ziel, Mode Locking of Semiconductor Lasers
Kam Y. Lau and Ammon Yariv, High-Frequency Current Modulation of Semiconductor Injection Lasers
Charles H. Henry, Special Properties of Semiconductor Lasers
Yasuharu Suematsu, Katsumi Kishino, Shigehisa Arai, and Fumio Koyama. Dynamic Single-Mode Semiconductor Lasers with a Distributed Reflector
W. T. Tsang, The Cleaved-Coupled-Cavity (C^3) Laser

Part C

R. J. Nelson and N. K. Dutta, Review of InGaAsP InP Laser Structures and Comparison of Their Performance
N. Chinone and M. Nakamura, Mode-Stabilized Semiconductor Lasers for 0.7–0.8- and 1.1–1.6- μm Regions
Yoshiji Horikoshi, Semiconductor Lasers with Wavelengths Exceeding 2 μm
B. A. Dean and M. Dixon, The Functional Reliability of Semiconductor Lasers as Optical Transmitters
R. H. Saul, T. P. Lee, and C. A. Burus, Light-Emitting Device Design
C. L. Zipfel, Light-Emitting Diode-Reliability
Tien Pei Lee and Tingye Li, LED-Based Multimode Lightwave Systems
Kinichiro Ogawa, Semiconductor Noise-Mode Partition Noise

Part D

Federico Capasso, The Physics of Avalanche Photodiodes
T. P. Pearsall and M. A. Pollack, Compound Semiconductor Photodiodes
Takao Kaneda, Silicon and Germanium Avalanche Photodiodes
S. R. Forrest, Sensitivity of Avalanche Photodetector Receivers for High-Bit-Rate Long-Wavelength Optical Communication Systems
J. C. Campbell, Phototransistors for Lightwave Communications

Part E

Shyh Wang, Principles and Characteristics of Integrable Active and Passive Optical Devices
Shlomo Margalit and Amnon Yariv, Integrated Electronic and Photonic Devices
Takaoki Mukai, Yoshihisa Yamamoto, and Tatsuya Kimura, Optical Amplification by Semiconductor Lasers

Volume 23 Pulsed Laser Processing of Semiconductors

R. F. Wood, C. W. White, and R. T. Young, Laser Processing of Semiconductors: An Overview
C. W. White, Segregation, Solute Trapping, and Supersaturated Alloys
G. E. Jellison, Jr., Optical and Electrical Properties of Pulsed Laser-Annealed Silicon
R. F. Wood and G. E. Jellison, Jr., Melting Model of Pulsed Laser Processing
R. F. Wood and F. W. Young, Jr., Nonequilibrium Solidification Following Pulsed Laser Melting
D. H. Lowndes and G. E. Jellison, Jr., Time-Resolved Measurement During Pulsed Laser Irradiation of Silicon
D. M. Zebner, Surface Studies of Pulsed Laser Irradiated Semiconductors
D. H. Lowndes, Pulsed Beam Processing of Gallium Arsenide
R. B. James, Pulsed CO_2 Laser Annealing of Semiconductors
R. T. Young and R. F. Wood, Applications of Pulsed Laser Processing

Volume 24 Applications of Multiquantum Wells, Selective Doping, and Superlattices

C. Weisbuch, Fundamental Properties of III–V Semiconductor Two-Dimensional Quantized Structures: The Basis for Optical and Electronic Device Applications
H. Morkoc and H. Unlu, Factors Affecting the Performance of (Al, Ga)As/GaAs and (Al, Ga)As/InGaAs Modulation-Doped Field-Effect Transistors: Microwave and Digital Applications
N. T. Linh, Two-Dimensional Electron Gas FETs: Microwave Applications
M. Abe et al., Ultra-High-Speed HEMT Integrated Circuits
D. S. Chemla, D. A. B. Miller, and P. W. Smith, Nonlinear Optical Properties of Multiple Quantum Well Structures for Optical Signal Processing
F. Capasso, Graded-Gap and Superlattice Devices by Band-Gap Engineering
W. T. Tsang, Quantum Confinement Heterostructure Semiconductor Lasers
G. C. Osbourn et al., Principles and Applications of Semiconductor Strained-Layer Superlattices

Volume 25 Diluted Magnetic Semiconductors

W. Giriat and J. K. Furdyna, Crystal Structure, Composition, and Materials Preparation of Diluted Magnetic Semiconductors
W. M. Becker, Band Structure and Optical Properties of Wide-Gap $A_{1-x}Mn_xB^{IV}$ Alloys at Zero Magnetic Field
Saul Oseroff and Pieter H. Keesom, Magnetic Properties: Macroscopic Studies
Giebultowicz and T. M. Holden, Neutron Scattering Studies of the Magnetic Structure and Dynamics of Diluted Magnetic Semiconductors
J. Kossut, Band Structure and Quantum Transport Phenomena in Narrow-Gap Diluted Magnetic Semiconductors
C. Riquaux, Magnetooptical Properties of Large-Gap Diluted Magnetic Semiconductors
J. A. Gaj, Magnetooptical Properties of Large-Gap Diluted Magnetic Semiconductors
J. Mycielski, Shallow Acceptors in Diluted Magnetic Semiconductors: Splitting, Boil-off, Giant Negative Magnetoresistance
A. K. Ramadas and R. Rodriquez, Raman Scattering in Diluted Magnetic Semiconductors
P. A. Wolff, Theory of Bound Magnetic Polarons in Semimagnetic Semiconductors

Volume 26 III–V Compound Semiconductors and Semiconductor Properties of Superionic Materials

Zou Yuanxi, III–V Compounds
H. V. Winston, A. T. Hunter, H. Kimura, and R. E. Lee, InAs-Alloyed GaAs Substrates for Direct Implantation
P. K. Bhattachary and S. Dhar, Deep Levels in III–V Compound Semiconductors Grown by MBE
Yu. Yu. Gurevich and A. K. Ivanov-Shits, Semiconductor Properties of Supersonic Materials

Volume 27 High Conducting Quasi-One-Dimensional Organic Crystals

E. M. Conwell, Introduction to Highly Conducting Quasi-One-Dimensional Organic Crystals
I. A. Howard, A Reference Guide to the Conducting Quasi-One-Dimensional Organic Molecular Crystals
J. P. Pouquet, Structural Instabilities
E. M. Conwell, Transport Properties
C. S. Jacobsen, Optical Properties
J. C. Scott, Magnetic Properties
L. Zuppiroli, Irradiation Effects: Perfect Crystals and Real Crystals

Volume 28 Measurement of High-Speed Signals in Solid State Devices

J. Frey and D. Ioannou, Materials and Devices for High-Speed and Optoelectronic Applications
H. Schumacher and E. Strid, Electronic Wafer Probing Techniques
D. H. Auston, Picosecond Photoconductivity: High-Speed Measurements of Devices and Materials
J. A. Valdmanis, Electro-Optic Measurement Techniques for Picosecond Materials, Devices, and Integrated Circuits.

J. M. Wiesenfeld and R. K. Jain, Direct Optical Probing of Integrated Circuits and High-Speed Devices
G. Plows, Electron-Beam Probing
A. M. Weiner and R. B. Marcus, Photoemissive Probing

Volume 29 Very High Speed Integrated Circuits: Gallium Arsenide LSI

M. Kuzuhara and T. Nazaki, Active Layer Formation by Ion Implantation
H. Hasimoto, Focused Ion Beam Implantation Technology
T. Nozaki and A. Higashisaka, Device Fabrication Process Technology
M. Ino and T. Takada, GaAs LSI Circuit Design
M. Hirayama, M. Ohmori, and K. Yamasaki, GaAs LSI Fabrication and Performance

Volume 30 Very High Speed Integrated Circuits: Heterostructure

H. Watanabe, T. Mizutani, and A. Usui, Fundamentals of Epitaxial Growth and Atomic Layer Epitaxy
S. Hiyamizu, Characteristics of Two-Dimensional Electron Gas in III–V Compound Heterostructures Grown by MBE
T. Nakanisi, Metalorganic Vapor Phase Epitaxy for High-Quality Active Layers
T. Nimura, High Electron Mobility Transistor and LSI Applications
T. Sugeta and T. Ishibashi, Hetero-Bipolar Transistor and LSI Application
H. Matsueda, T. Tanaka, and M. Nakamura, Optoelectronic Integrated Circuits

Volume 31 Indium Phosphide: Crystal Growth and Characterization

J. P. Farges, Growth of Discoloration-free InP
M. J. McCollum and G. E. Stillman, High Purity InP Grown by Hydride Vapor Phase Epitaxy
T. Inada and T. Fukuda, Direct Synthesis and Growth of Indium Phosphide by the Liquid Phosphorous Encapsulated Czochralski Method
O. Oda, K. Katagiri, K. Shinohara, S. Katsura, Y. Takahashi, K. Kainosho, K. Kohiro, and R. Hirano, InP Crystal Growth, Substrate Preparation and Evaluation
K. Tada, M. Tatsumi, M. Morioka, T. Araki, and T. Kawase, InP Substrates: Production and Quality Control
M. Razeghi, LP-MOCVD Growth, Characterization, and Application of InP Material
T. A. Kennedy and P. J. Lin-Chung, Stoichiometric Defects in InP

Volume 32 Strained-Layer Superlattices: Physics

T. P. Pearsall, Strained-Layer Superlattices
Fred H. Pollack, Effects of Homogeneous Strain on the Electronic and Vibrational Levels in Semiconductors
J. Y. Marzin, J. M. Gerárd, P. Voisin, and J. A. Brum, Optical Studies of Strained III–V Heterolayers
R. People and S. A. Jackson, Structurally Induced States from Strain and Confinement
M. Jaros, Microscopic Phenomena in Ordered Superlattices

Volume 33 Strained-Layer Superlattices: Materials Science and Technology

R. Hull and J. C. Bean, Principles and Concepts of Strained-Layer Epitaxy
William J. Schaff, Paul J. Tasker, Marc C. Foisy, and Lester F. Eastman, Device Applications of Strained-Layer Epitaxy
S. T. Picraux, B. L. Doyle, and J. Y. Tsao, Structure and Characterization of Strained-Layer Superlattices
E. Kasper and F. Schaffer, Group IV Compounds
Dale L. Martin, Molecular Beam Epitaxy of IV–VI Compounds Heterojunction
Robert L. Gunshor, Leslie A. Kolodziejski, Arto V. Nurmikko, and Nobuo Otsuka, Molecular Beam Epitaxy of II–VI Semiconductor Microstructures

Volume 34 Hydrogen in Semiconductors

J. I. Pankove and N. M. Johnson, Introduction to Hydrogen in Semiconductors
C. H. Seager, Hydrogenation Methods
J. I. Pankove, Hydrogenation of Defects in Crystalline Silicon
J. W. Corbett, P. Deák, U. V. Desnica, and S. J. Pearton, Hydrogen Passivation of Damage Centers in Semiconductors
S. J. Pearton, Neutralization of Deep Levels in Silicon
J. I. Pankove, Neutralization of Shallow Acceptors in Silicon
N. M. Johnson, Neutralization of Donor Dopants and Formation of Hydrogen-Induced Defects in n-Type Silicon
M. Stavola and S. J. Pearton, Vibrational Spectroscopy of Hydrogen-Related Defects in Silicon
A. D. Marwick, Hydrogen in Semiconductors: Ion Beam Techniques
C. Herring and N. M. Johnson, Hydrogen Migration and Solubility in Silicon
E. E. Haller, Hydrogen-Related Phenomena in Crystalline Germanium
J. Kakalios, Hydrogen Diffusion in Amorphous Silicon
J. Chevalier, B. Clerjaud, and B. Pajot, Neutralization of Defects and Dopants in III–V Semiconductors
G. G. DeLeo and W. B. Fowler, Computational Studies of Hydrogen-Containing Complexes in Semiconductors
R. F. Kiefl and T. L. Estle, Muonium in Semiconductors
C. G. Van de Walle, Theory of Isolated Interstitial Hydrogen and Muonium in Crystalline Semiconductors

Volume 35 Nanostructured Systems

Mark Reed, Introduction
H. van Houten, C. W. J. Beenakker, and B. J. van Wees, Quantum Point Contacts
G. Timp, When Does a Wire Become an Electron Waveguide?
M. Büttiker, The Quantum Hall Effects in Open Conductors
W. Hansen, J. P. Kotthaus, and U. Merkt, Electrons in Laterally Periodic Nanostructures

Volume 36 The Spectroscopy of Semiconductors

D. Heiman, Spectroscopy of Semiconductors at Low Temperatures and High Magnetic Fields
Arto V. Nurmikko, Transient Spectroscopy by Ultrashort Laser Pulse Techniques

A. K. Ramdas and S. Rodriguez, Piezospectroscopy of Semiconductors
Orest J. Glembocki and Benjamin V. Shanabrook, Photoreflectance Spectroscopy of Microstructures
David G. Seiler, Christopher L. Littler, and Margaret H. Wiler, One- and Two-Photon Magneto-Optical Spectroscopy of InSb and $Hg_{1-x}Cd_xTe$

Volume 37 The Mechanical Properties of Semiconductors

A.-B. Chen, Arden Sher and W. T. Yost, Elastic Constants and Related Properties of Semiconductor Compounds and Their Alloys
David R. Clarke, Fracture of Silicon and Other Semiconductors
Hans Siethoff, The Plasticity of Elemental and Compound Semiconductors
Sivaraman Guruswamy, Katherine T. Faber and John P. Hirth, Mechanical Behavior of Compound Semiconductors
Subhanh Mahajan, Deformation Behavior of Compound Semiconductors
John P. Hirth, Injection of Dislocations into Strained Multilayer Structures
Don Kendall, Charles B. Fleddermann, and Kevin J. Malloy, Critical Technologies for the Micromachining of Silicon
Ikuo Matsuba and Kinji Mokuya, Processing and Semiconductor Thermoelastic Behavior

Volume 38 Imperfections in III/V Materials

Udo Scherz and Matthias Scheffler, Density-Functional Theory of sp-Bonded Defects in III/V Semiconductors
Maria Kaminska and Eicke R. Weber, El2 Defect in GaAs David C. Look, Defects Relevant for Compensation in Semi-Insulating GaAs
R. C. Newman, Local Vibrational Mode Spectroscopy of Defects in III/V Compounds
Andrzej M. Hennel, Transition Metals in III/V Compounds
Kevin J. Malloy and Ken Khachaturyan, DX and Related Defects in Semiconductors
V. Swaminathan and Andrew S. Jordan, Dislocations in III/V Compounds
Krzysztof W. Nauka, Deep Level Defects in the Epitaxial III/V Materials

Volume 39 Minority Carriers in III–V Semiconductors: Physics and Applications

Niloy K. Dutta, Radiative Transitions in GaAs and Other III–V Compounds
Richard K. Ahrenkiel, Minority-Carrier Lifetime in III–V Semiconductors
Tomofumi Furuta, High Field Minority Electron Transport in p-GaAs
Mark S. Lundstrom, Minority-Carrier Transport in III–V Semiconductors
Richard A. Abram, Effects of Heavy Doping and High Excitation on the Band Structure of GaAs
David Yevick and Witold Bardyszewski, An Introduction to Non-Equilibrium Many-Body Analyses of Optical Processes in III–V Semiconductors

Volume 40 Epitaxial Microstructures

E. F. Schubert, Delta-Doping of Semiconductors: Electronic, Optical, and Structural Properties of Materials and Devices

A. Gossard, M. Sundaram, and P. Hopkins, Wide Graded Potential Wells
P. Petroff, Direct Growth of Nanometer-Size Quantum Wire Superlattices
E. Kapon, Lateral Patterning of Quantum Well Heterostructures by Growth of Nonplanar Substrates
H. Temkin, D. Gershoni, and M. Panish, Optical Properties of Ga$1-_x$In$_x$As/InP Quantum Wells

Volume 41 High Speed Heterostructure Devices

F. Capasso, F. Beltram, S. Sen, A. Pahlevi, and A. Y. Cho, Quantum Electron Devices: Physics and Applications
P. Solomon, D. J. Frank, S. L. Wright, and F. Canora, GaAs-Gate Semiconductor–Insulator–Semiconductor FET
M. H. Hashemi and U. K. Mishra, Unipolar InP-Based Transistors
R. Kiehl, Complementary Heterostructure FET Integrated Circuits
T. Ishibashi, GaAs-Based and InP-Based Heterostructure Bipolar Transistors
H. C. Liu and T. C. L. G. Sollner, High-Frequency-Tunneling Devices
H. Ohnishi, T. More, M. Takatsu, K. Imamura, and N. Yokoyama, Resonant-Tunneling Hot-Electron Transistors and Circuits

Volume 42 Oxygen in Silicon

F. Shimura, Introduction to Oxygen in Silicon
W. Lin, The Incorporation of Oxygen into Silicon Crystals
T. J. Schaffner and D. K. Schroder, Characterization Techniques for Oxygen in Silicon
W. M. Bullis, Oxygen Concentration Measurement
S. M. Hu, Intrinsic Point Defects in Silicon
B. Pajot, Some Atomic Configurations of Oxygen
J. Michel and L. C. Kimerling, Electical Properties of Oxygen in Silicon
R. C. Newman and R. Jones, Diffusion of Oxygen in Silicon
T. Y. Tan and W. J. Taylor, Mechanisms of Oxygen Precipitation: Some Quantitative Aspects
M. Schrems, Simulation of Oxygen Precipitation
K. Simino and I. Yonenaga, Oxygen Effect on Mechanical Properties
W. Bergholz, Grown-in and Process-Induced Effects
F. Shimura, Intrinsic/Internal Gettering
H. Tsuya, Oxygen Effect on Electronic Device Performance

Volume 43 Semiconductors for Room Temperature Nuclear Detector Applications

R. B. James and T. E. Schlesinger, Introduction and Overview
L. S. Darken and C. E. Cox, High-Purity Germanium Detectors
A. Burger, D. Nason, L. Van den Berg, and M. Schieber, Growth of Mercuric Iodide
X. J. Bao, T. E. Schlesinger, and R. B. James, Electrical Properties of Mercuric Iodide
X. J. Bao, R. B. James, and T. E. Schlesinger, Optical Properties of Red Mercuric Iodide
M. Hage-Ali and P. Siffert, Growth Methods of CdTe Nuclear Detector Materials
M. Hage-Ali and P Siffert, Characterization of CdTe Nuclear Detector Materials
M. Hage-Ali and P. Siffert, CdTe Nuclear Detectors and Applications
R. B. James, T. E. Schlesinger, J. Lund, and M. Schieber, Cd$_{1-x}$Zn$_x$Te Spectrometers for Gamma and X-Ray Applications

D. S. McGregor, J. E. Kammeraad, Gallium Arsenide Radiation Detectors and Spectrometers
J. C. Lund, F. Olschner, and A. Burger, Lead Iodide
M. R. Squillante, and K. S. Shah, Other Materials: Status and Prospects
V. M. Gerrish, Characterization and Quantification of Detector Performance
J. S. Iwanczyk and B. E. Patt, Electronics for X-ray and Gamma Ray Spectrometers
M. Schieber, R. B. James, and T. E. Schlesinger, Summary and Remaining Issues for Room Temperature Radiation Spectrometers

Volume 44 II–IV Blue/Green Light Emitters: Device Physics and Epitaxial Growth

J. Han and R. L. Gunshor, MBE Growth and Electrical Properties of Wide Bandgap ZnSe-based II–VI Semiconductors
Shizuo Fujita and Shigeo Fujita, Growth and Characterization of ZnSe-based II–VI Semiconductors by MOVPE
Easen Ho and Leslie A. Kolodziejski, Gaseous Source UHV Epitaxy Technologies for Wide Bandgap II–VI Semiconductors
Chris G. Van de Walle, Doping of Wide-Band-Gap II–VI Compounds–Theory
Roberto Cingolani, Optical Properties of Excitons in ZnSe-Based Quantum Well Heterostructures
A. Ishibashi and A. V. Nurmikko, II–VI Diode Lasers: A Current View of Device Performance and Issues
Supratik Guha and John Petruzello, Defects and Degradation in Wide-Gap II–VI-based Structures and Light Emitting Devices

Volume 45 Effect of Disorder and Defects in Ion-Implanted Semiconductors: Electrical and Physiochemical Characterization

Heiner Ryssel, Ion Implantation into Semiconductors: Historical Perspectives
You-Nian Wang and Teng-Cai Ma, Electronic Stopping Power for Energetic Ions in Solids
Sachiko T. Nakagawa, Solid Effect on the Electronic Stopping of Crystalline Target and Application to Range Estimation
G. Müller, S. Kalbitzer and G. N. Greaves, Ion Beams in Amorphous Semiconductor Research
Jumana Boussey-Said, Sheet and Spreading Resistance Analysis of Ion Implanted and Annealed Semiconductors
M. L. Polignano and G. Queirolo, Studies of the Stripping Hall Effect in Ion-Implanted Silicon
J. Stoemenos, Transmission Electron Microscopy Analyses
Roberta Nipoti and Marco Servidori, Rutherford Backscattering Studies of Ion Implanted Semi-conductors
P. Zaumseil, X-ray Diffraction Techniques

Volume 46 Effect of Disorder and Defects in Ion-Implanted Semiconductors: Optical and Photothermal Characterization

M. Fried, T. Lohner and J. Gyulai, Ellipsometric Analysis
Antonios Seas and Constantinos Christofides, Transmission and Reflection Spectroscopy on Ion Implanted Semiconductors

Andreas Othonos and Constantinos Christofides, Photoluminescence and Raman Scattering of Ion Implanted Semiconductors. Influence of Annealing

Constantinos Christofides, Photomodulated Thermoreflectance Investigation of Implanted Wafers. Annealing Kinetics of Defects

U. Zammit, Photothermal Deflection Spectroscopy Characterization of Ion-Implanted and Annealed Silicon Films

Andreas Mandelis, Arief Budiman and Miguel Vargas, Photothermal Deep-Level Transient Spectroscopy of Impurities and Defects in Semiconductors

R. Kalish and S. Charbonneau, Ion Implantation into Quantum-Well Structures

Alexandre M. Myasnikov and Nikolay N. Gerasimenko, Ion Implantation and Thermal Annealing of III-V Compound Semiconducting Systems: Some Problems of III-V Narrow Gap Semiconductors

Volume 47 Uncooled Infrared Imaging Arrays and Systems

R. G. Buser and M. P. Tompsett, Historical Overview
P. W. Kruse, Principles of Uncooled Infrared Focal Plane Arrays
R. A. Wood, Monolithic Silicon Microbolometer Arrays
C. M. Hanson, Hybrid Pyroelectric-Ferroelectric Bolometer Arrays
D. L. Polla and J. R. Choi, Monolithic Pyroelectric Bolometer Arrays
N. Teranishi, Thermoelectric Uncooled Infrared Focal Plane Arrays
M. F. Tompsett, Pyroelectric Vidicon
T. W. Kenny, Tunneling Infrared Sensors
J. R. Vig, R. L. Filler and Y. Kim, Application of Quartz Microresonators to Uncooled Infrared Imaging Arrays
P. W. Kruse, Application of Uncooled Monolithic Thermoelectric Linear Arrays to Imaging Radiometers

Volume 48 High Brightness Light Emitting Diodes

G. B. Stringfellow, Materials Issues in High-Brightness Light-Emitting Diodes
M. G. Craford, Overview of Device issues in High-Brightness Light-Emitting Diodes
F. M. Steranka, AlGaAs Red Light Emitting Diodes
C. H. Chen, S. A. Stockman, M. J. Peanasky, and C. P. Kuo, OMVPE Growth of AlGaInP for High Efficiency Visible Light-Emitting Diodes
F. A. Kish and R. M. Fletcher, AlGaInP Light-Emitting Diodes
M. W. Hodapp, Applications for High Brightness Light-Emitting Diodes
I. Akasaki and H. Amano, Organometallic Vapor Epitaxy of GaN for High Brightness Blue Light Emitting Diodes
S. Nakamura, Group III-V Nitride Based Ultraviolet-Blue-Green-Yellow Light-Emitting Diodes and Laser Diodes

Volume 49 Light Emission in Silicon: from Physics to Devices

David J. Lockwood, Light Emission in Silicon
Gerhard Abstreiter, Band Gaps and Light Emission in Si/SiGe Atomic Layer Structures
Thomas G. Brown and Dennis G. Hall, Radiative Isoelectronic Impurities in Silicon and Silicon-Germanium Alloys and Superlattices
J. Michel, L. V. C. Assali, M. T. Morse, and L. C. Kimerling, Erbium in Silicon

Yoshihiko Kanemitsu, Silicon and Germanium Nanoparticles
Philippe M. Fauchet, Porous Silicon: Photoluminescence and Electroluminescent Devices
C. Delerue, G. Allan, and M. Lannoo, Theory of Radiative and Nonradiative Processes in Silicon Nanocrystallites
Louis Brus, Silicon Polymers and Nanocrystals

Volume 50 Gallium Nitride (GaN)

J. I. Pankove and T. D. Moustakas, Introduction
S. P. DenBaars and S. Keller, Metalorganic Chemical Vapor Deposition (MOCVD) of Group III Nitrides
W. A. Bryden and T. J. Kistenmacher, Growth of Group III–A Nitrides by Reactive Sputtering
N. Newman, Thermochemistry of III–N Semiconductors
S. J. Pearton and R. J. Shul, Etching of III Nitrides
S. M. Bedair, Indium-based Nitride Compounds
A. Trampert, O. Brandt, and K. H. Ploog, Crystal Structure of Group III Nitrides
H. Morkoc, F. Hamdani, and A. Salvador, Electronic and Optical Properties of III–V Nitride based Quantum Wells and Superlattices
K. Doverspike and J. I. Pankove, Doping in the III-Nitrides
T. Suski and P. Perlin, High Pressure Studies of Defects and Impurities in Gallium Nitride
B. Monemar, Optical Properties of GaN
W. R. L. Lambrecht, Band Structure of the Group III Nitrides
N. E. Christensen and P. Perlin, Phonons and Phase Transitions in GaN
S. Nakamura, Applications of LEDs and LDs
I. Akasaki and H. Amano, Lasers
J. A. Cooper, Jr., Nonvolatile Random Access Memories in Wide Bandgap Semiconductors

Volume 51A Identification of Defects in Semiconductors

George D. Watkins, EPR and ENDOR Studies of Defects in Semiconductors
J.-M. Spaeth, Magneto-Optical and Electrical Detection of Paramagnetic Resonance in Semiconductors
T. A. Kennedy and E. R. Glaser, Magnetic Resonance of Epitaxial Layers Detected by Photoluminescence
K. H. Chow, B. Hitti, and R. F. Kiefl, 'kSR on Muonium in Semiconductors and Its Relation to Hydrogen
Kimmo Saarinen, Pekka Hautojärvi, and Catherine Corbel, Positron Annihilation Spectroscopy of Defects in Semiconductors
R. Jones and P. R. Briddon, The *Ab Initio* Cluster Method and the Dynamics of Defects in Semiconductors

Volume 51B Identification of Defects in Semiconductors

Gordon Davies, Optical Measurements of Point Defects
P. M. Mooney, Defect Identification Using Capacitance Spectroscopy
Michael Stavola, Vibrational Spectroscopy of Light Element Impurities in Semiconductors
P. Schwander, W. D. Rau, C. Kisielowski, M. Gribelyuk, and A. Ourmazd, Defect Processes in Semiconductors Studied at the Atomic Level by Transmission Electron Microscopy
Nikos D. Jager and Eicke R. Weber, Scanning Tunneling Microscopy of Defects in Semiconductors

Volume 52 SiC Materials and Devices

K. Järrendahl and R. F. Davis, Materials Properties and Characterization of SiC
V. A. Dmitriev and M. G. Spencer, SiC Fabrication Technology: Growth and Doping
V. Saxena and A. J. Steckl, Building Blocks for SiC Devices: Ohmic Contacts, Schottky Contacts, and p–n Junctions
M. S. Shur, SiC Transistors
C. D. Brandt, R. C. Clarke, R. R. Siergiej, J. B. Casady, A. W. Morse, S. Sriram, and A. K. Agarwal, SiC for Applications in High-Power Electronics
R. J. Trew, SiC Microwave Devices
J. Edmond, H. Kong, G. Negley, M. Leonard, K. Doverspike, W. Weeks, A. Suvorov, D. Waltz, and C. Carter, Jr., SiC-Based UV Photodiodes and Light-Emitting Diodes
H. Morkoç, Beyond Silicon Carbide! III–V Nitride-Based Heterostructures and Devices

Volume 53 Cumulative Index

ISBN 0-12-752162-3